AF344478

Dynamics and Thermodynamics with Nuclear Degrees of Freedom

Società Italiana
di Fisica

 Springer

Dynamics and Thermodynamics with Nuclear Degrees of Freedom

edited by

Ph. Chomaz, F. Gulminelli,
W. Trautmann and S.J. Yennello

Dr. Philippe Chomaz
GANIL, CEA-DSM and CNRS-IN2P3
BP 55027, 14076 Caen cedex 5, France
chomaz@ganil.fr

Prof. Francesca Gulminelli
LPC, CNRS-IN2P3, ENSICAEN and Université
6 Bd du Maréchal Juin
14050 Caen cedex, France
gulminelli@lpccaen.in2p3.fr

Prof. Dr. Wolfgang Trautmann
GSI, Abteilung KP3
Planckstr. 1
64291 Darmstadt, Germany
w.trautmann@gsi.de

Prof. Sherry J. Yennello
Cyclotron Institute
Texas A&M University
3366 TAMUS, College Station
TX 77843-3366, USA
yennello@comp.tamu.edu

The articles in this book originally appeared in the journal
The European Physical Journal A − Hadrons and Nuclei
Volume 30, Number 1
ISSN 1434-6001
© SIF and Springer-Verlag Berlin Heidelberg 2006

ISBN-10 3-540-46494-8 Springer Berlin Heidelberg New York
ISBN-13 978-3-540-46494-5 Springer Berlin Heidelberg New York

Library of Congress Control Number: 2006933427

Springer is a part of Springer Science+Business Media
springer.com

© SIF and Springer-Verlag Berlin Heidelberg 2006
Printed in Italy

Typesetting and Cover design: SIF Production Office, Bologna, Italy
Printing and Binding: Tipografia Compositori, Bologna, Italy

Printed on acid-free paper   SPIN: 11684190 − 5 4 3 2 1 0

Eur. Phys. J. A **30**, III (2006)
DOI 10.1140/epja/i2006-10138-1
Published online: 30 October 2006
© Società Italiana di Fisica / Springer-Verlag 2006

THE EUROPEAN
PHYSICAL JOURNAL A

Preface

The journey leading to this topical issue began in the summer of 2003 at a small café in Montreal. The goal was to bring together many active members of the many-body reaction community to seek a consensus on the accomplishments in our field; the many things we have come to know and also, the things we do not yet know. As the reader will discover, consensus may sometimes mean agreeing on where we still disagree. The authors of this topical issue have tried to bring together diverse perspectives and present highlights of what we have learned and of some of the open questions.

The first stage was an online discussion of accomplishments and open questions. The most important topics emerged and discussion leaders were identified. After this we met in Catania, Sicily, in January, 2004, where we had a workshop with no invited talks. The workshop intensified the discussions and fostered debate. A set of working groups were formed. We then met at Smith College, Massachusetts, in June, 2004, for a day and a half of invigorating discussions. Each of the working groups reported its progress, and the community began to assess the status of our undertaking. The idea to collect the projected review papers in a common publication emerged at that time. We left with more homework and reconvened at Texas A&M University, Texas, in February, 2005, for our third workshop. It was there that the hard work of the first year and a half started to merge into the form that we present here.

This topical issue is the result of this extraordinary journey. We do not see it as an aim achieved, but rather as a milestone by which we can judge our progress as the journey continues toward an even greater knowledge of the dynamics and thermodynamics of heavy-ion collisions.

None of this would have been possible without the generous intellectual contributions of many individuals and the dedicated efforts of the local organizers of the workshops. We would also like to acknowledge the financial support of LNS-INFN Catania, Physics and Astronomy Department of the Catania University, Smith College, and Texas A&M University. To all of them we express our sincere thanks.

Thomas Walcher

Editor-in-Chief
EPJ A

Philippe Chomaz
Francesca Gulminelli
Wolfgang Trautmann
Sherry Yennello

The Editors

Organization

Overview of workshops

Organizers

Workshops

- LNS, INFN, Catania, Italy: M. Colonna, M. Di Toro, P. Sapienza, G. Ruggieri and G. Agnello
- Smith College, Northhampton, Massachusetts, USA: M.Z. Pfabé
- Texas A&M, College Station, Texas, USA: J.B. Natowitz and S.J. Yennello

Working groups

- Comparison of statistical models: M.B. Tsang and C. Bertulani
- Comparison of transport models: J. Aichelin
- Working group on fragment observables: F. Gulminelli

Related publications in the web

The website `http://cyclotron.tamu.edu/wci3/` contains further information about the initiative which has lead to the production of this topical issue and is updated by B. Hyman.

Our initiative has stimulated the production of a number of contributions and further review papers on different specific issues concerning nuclear dynamics and thermodynamics. Here is a list of these contributed papers:

- K.A. Bugaev, *Exactly solvable models: the road towards a rigorous treatment of phase transitions in small systems*, arXiv:nucl-th/0511031, contributed paper to the *Third World Consensus Initiative Workshop, WCI-III, Texas A&M University, College Station, Texas, USA, February 12-16, 2005*, `http://cyclotron.tamu.edu/wci3/contributed_papers.html`;
- D.H.E. Gross, *Equilibrium statistics applied to "small" systems: phase transitions in nuclei in particular*, contributed paper to the *Third World Consensus Initiative Workshop, WCI-III, Texas A&M University, College Station, Texas, USA, February 12-16, 2005*, `http://cyclotron.tamu.edu/wci3/contributed_papers.html`;
- V.A. Karnaukhov, *Nuclear multifragmentation and phase transitions in hot nuclei*, in Phys. Part. Nucl. **37**, 165 (2006) ISSN 1063-7796;
- L.G. Moretto, K.A. Bugaev, J.B. Elliott, L. Phair, *The quark gluon plasma: a perfect thermostat and a perfect particle reservoir*, contributed paper to the *Third World Consensus Initiative Workshop, WCI-III, Texas A&M University, College Station, Texas, USA, February 12-16, 2005*, `http://cyclotron.tamu.edu/wci3/contributed_papers.html`;
- M. Papa, *Recent results on pre-equilibrium GDR emission*, contributed paper to the *Third World Consensus Initiative Workshop, WCI-III, Texas A&M University, College Station, Texas, USA, February 12-16, 2005*, `http://cyclotron.tamu.edu/wci3/contributed_papers.html`;
- V.E. Viola, *Light-ion-induced multifragmentation: The ISiS project*, Phys. Rep. **434**, 1 (2006) arXiv:nucl-ex/0604012.

Cumulative list of participants

* Steering Committee members

Attilio Agodi
Dipartimento di Fisica
Università di Catania

Titti Agodi
INFN-LNS Catania

Jörg Aichelin
Subatech, Nantes

Francesca Amorini
INFN-LNS Catania

Marcello Baldo
INFN Sezione di Catania

Virgil Baran
NIPNE, Bucharest

Mihir Baran Chatterjee
INFN-LNS Catania

Sandro Barlini
LNL, Legnaro

Wolfgang Bauer
MSU, East Lansing

Vincenzo Bellini
Dipartimento di Fisica
Università di Catania

Aldo Bonasera
INFN-LNS Catania

Eric Bonnet
IPN, Orsay

Bernard Borderie
IPN, Orsay

Robert Botet
Laboratoire de Physique des Solides
Orsay

Alexander Botvina
Institute for Nuclear Research, RAS
Moscow

Rémi Bougault
LPC, Caen

Mauro Bruno
Università di Bologna

Kyrill Bugaev
LBL, Berkeley

Xavier Campi
LPTMS, Orsay

Giuseppe Cardella
INFN Sezione di Catania

Francesco Catara
Dipartimento di Fisica
Università di Catania

Salvatore Cavallaro
Dipartimento di Fisica
Università di Catania

Abdou Chbihi
GANIL, Caen

Philippe Chomaz*
GANIL, Caen

Marco Cinausero
LNL, Legnaro

Gianluca Colò
Dipartimento di Fisica
Università di Milano

Maria Colonna
INFN-LNS Catania

Rosa Coniglione
INFN-LNS Catania

Michela D'Agostino
Università di Bologna

Pawel Danielewicz*
MSU, East Lansing

Enrico De Filippo
INFN Sezione di Catania

Alberto Dellafiore
INFN Sezione di Firenze

Marzio De Napoli
INFN-LNS and Dipartimento di Fisica
Università di Catania

Romualdo deSouza
Indiana University/IUCF

Massimo Di Toro*
INFN-LNS Catania

Claudio Dorso
Universidad de Buenos Aires

James Elliott
LBL, Berkeley

Graziella Ferini
Dipartimento di Fisica
Università di Catania

William Friedman
University of Wisconsin, Madison

Christian Fuchs
University of Tübingen

Takuya Furuta
Department of Physics
Tohoku University

Theo Gaitanos
INFN-LNS Catania

Emmanuelle Galichet
IPN, Orsay

Elena Geraci
INFN and Università di Bologna

Roberta Ghetti
Lund University

Fabiana Gramegna
LNL, Legnaro

Dieter Gross
Hahn-Meitner Institut, Berlin

Benoit Guiot
INFN Sezione di Bologna

Francesca Gulminelli*
LPC and UCBN, Caen

Kris Hagel
Cyclotron Institute TAMU
College Station

Vladimir Henzl
GSI, Darmstadt

Daniela Henzlova
GSI, Darmstadt

Charles Horowitz
Indiana University

Sylvie Hudan
Indiana University

Jennifer Iglio
Cyclotron Institute TAMU
College Station

Marian Jandel
Cyclotron Institute TAMU
College Station

Vladimir Karnaukhov
JINR, Dubna

August Keksis
Cyclotron Institute TAMU
College Station

Aleksandra Kelić
GSI, Darmstadt

Sachie Kimura
INFN-LNS Catania

Vladimir Kolomietz
Institute for Nuclear Research
Kiev

Seweryn Kowalski
Cyclotron Institute TAMU
College Station

Arnaud Le Fèvre
GSI, Darmstadt

Gaetano Lanzanò
INFN Sezione di Catania

Nicolas Le Neindre
IPN, Orsay

Bao-An Li
Arkansas State University

Umberto Lombardo
Dipartimento di Fisica
Università di Catania

Olivier Lopez
LPC, Caen

Jerzy Łukasik
GSI, Darmstadt

William Lynch*
MSU, East Lansing

Yu-Gang Ma
Shanghai Institute of Applied Physics

Zbigniew Majka
Jagiellonian University
Kracow

Pierfrancesco Mastinu
LNL, Legnaro

Francesco Matera
Dipartimento di Fisica
Università di Firenze

Thomas Materna
Cyclotron Institute TAMU
College Station

Carl Metelko
Indiana University

Igor Mishustin
RRC, Kurchatov Institute
Moscow

Klaus Morawetz
Technische Universität Chemnitz

Luciano Moretto
LBL, Berkeley

Ranjini Murthy
Cyclotron Institute TAMU
College Station

Paolo Napolitani
GANIL, Caen

Joseph Natowitz*
Cyclotron Institute TAMU
College Station

Orazio Nicotra
Dipartimento di Fisica
Università di Messina

Alessandro Olmi
INFN Sezione di Firenze

Akira Ono
Department of Physics
Tohoku University

Angelo Pagano
INFN Sezione di Catania

Massimo Papa
INFN Sezione di Catania

Malgorzata Z. Pfabé
Smith College, Northampton

Larry Phair
LBL, Berkeley

Silvia Piantelli
INFN Sezione di Firenze

Paolo Piattelli
INFN-LNS Catania

Sara Pirrone
INFN Sezione di Catania

Eric Plagnol
APC Collège de France, Paris

Marek Ploszajczak
GANIL, Caen

Lijun Qin
Cyclotron Institute TAMU
College Station

Giovanni Raciti
Dipartimento di Fisica
Università di Catania

Alexandru H. Raduta
National Institute for Physics
Bucharest

Jørgen Randrup
LBL, Berkeley

Andrea Rapisarda
Dipartimento di Fisica
Università di Catania

Elisa Rapisarda
INFN-LNS and Dipartimento di Fisica
Università di Catania

Marie-France Rivet
IPN, Orsay

Vincent Regnard
GANIL, Caen

Willibrord Reisdorf
GSI, Darmstadt

Joseph Rizzo
INFN-LNS Catania

Giuseppina Rizzo
Dipartimento di Fisica
Università di Catania

René Roy
Laval University

Domenico Santonocito
INFN-LNS Catania

Piera Sapienza
INFN-LNS Catania

Rolf Scharenberg*
Purdue University

Karl-Heinz Schmidt
GSI, Darmstadt

Hans-Josef Schulze
INFN Sezione di Catania

Dinesh Shetty
Cyclotron Institute TAMU
College Station

Lijun Shi
McGill University, Montreal

Shalom Shlomo
Cyclotron Institute TAMU
College Station

Tappas Sil
Cyclotron Institute TAMU
College Station

Lee Sobotka*
Washington University

Sarah Soisson
Cyclotron Institute TAMU
College Station

George Souliotis
Cyclotron Institute TAMU
College Station

Loredana Spezzi
INFN-LNS and Dipartimento di Fisica
Università di Catania

Brijesh Srivastava
Purdue University

Brian Stein
Cyclotron Institute TAMU
College Station

Eric Suraud
LPT-IRSAMC, Toulouse

Bernard Tamain*
LPC and ENSICAEN, Caen

Salvatrice Terranova
Dipartimento di Fisica
Università di Catania

Jan Toke
University of Rochester

Wolfgang Trautmann*
GSI, Darmstadt

Betty Tsang
MSU, East Lansing

Giuseppe Verde
INFN Sezione di Catania

Vic Viola*
Indiana University

Roy Wada
Cyclotron Institute TAMU
College Station

Jean Pierre Wieleczko
GANIL, Caen

Hermann Wolter
LMU, München

Sherry Yennello*
Cyclotron Institute TAMU
College Station

The European Physical Journal A

Volume 30 · Number 1 · October · 2006

Eur. Phys. J. A **30**, 1–3 (2006)
DOI 10.1140/epja/i2006-10128-3

Challenges in nuclear dynamics and thermodynamics

F. Gulminelli[1], W. Trautmann[2,a], S.J. Yennello[3], and Ph. Chomaz[4]

[1] LPC (IN2P3-CNRS/Ensicaen and University of Caen), F-14050 Caen, France
[2] Gesellschaft für Schwerionenforschung mbH, D-64291 Darmstadt, Germany
[3] Cyclotron Institute, Texas A&M University, College Station, TX 77843, USA
[4] GANIL (DSM-CEA/IN2P3-CNRS), F-14076 Caen, France

Received: 18 September 2006 /
Published online: 30 October 2006 – © Società Italiana di Fisica / Springer-Verlag 2006

Abstract. The purpose and contents of this topical issue, *Dynamics and Thermodynamics with Nuclear Degrees of Freedom*, which grew out of a series of workshops in the years 2004 and 2005, are introduced. The central topics are the nuclear density functional, nuclear multi-fragmentation, and nuclear phase transitions.

PACS. 24.10.-i Nuclear reaction models and methods – 25.70.Pq Multifragment emission and correlations – 25.70.-z Low and intermediate energy heavy-ion reactions

1 Unveiling the nuclear density functional

Since the pioneering inclusive experiments of the Purdue group [1] and the first exclusive emulsion data [2] in the early eighties, it has been clear that the revolutionary discovery of Otto Hahn and Fritz Strassman in 1938 had been surpassed by new experimental evidence [3]. Not only is it possible to create smaller parts than the original nucleus by collisions of nuclear particles with atomic nuclei —what we know today as the phenomenon of nuclear fission— but also to break the nucleus into many different pieces. This intriguing phenomenon that we call nuclear multi-fragmentation has yet not revieled all its secrets and continues to fascinate the nuclear-physics community. This fascination, and our wish to share it with a wider community, motivates this topical issue.

Since the very early days of heavy-ion experiments, it has been understood that collisions with heavy ions could create nuclear material with density and excitation energy radically different from that of the ordinary matter surrounding us. Such reactions would then be the only terrestrial window that could provide a transient glimpse of the hot nuclear matter that is so abundantly present in the universe, in the last evolution steps of all dying massive stars.

These collisions would therefore provide unique constraints to pin down the energy-density functional of nuclear matter in a large domain of energy, density, and isotopic composition, as well as the underlying effective interaction that binds atomic nuclei.

To get quantitative information on the nuclear density functional, the so-called nuclear equation of state, the different functionals produced by theory have to be implemented in transport equations that predict the dynamics of the collisions, and robust observables have to be constructed that are selectively sensitive to the different parts of the effective interaction. This requirement has boosted the development of quantum transport models, as well as of sophisticated large-coverage, high-resolution detectors, able to reconstruct the one-body global observables accessible to the theory. Such observables, probing different times and consequently different densities, are mainly collective flows, energetic particle production, and diluted matter emitted at mid-rapidity in peripheral collisions. The different aspects of this fascinating adventure, and our present understanding of the nuclear equation of state, are reviewed in the monographs by C. Fuchs and H.H. Wolter, A. Andronic *et al.*, A. Bonasera *et al.*, and M. Di Toro *et al.* The theoretical methods to address this problem, namely mean-field theory and its extensions, are not specific to nuclear physics. Rather, they are also applicable to the study of many different fermionic systems, as helium droplets and metal clusters. The common methods and challenges in the field of small fermionic systems are reviewed in the monograph by J. Navarro *et al.*

If our understanding of the basic features of the nuclear density functional has greatly improved with these studies, many questions have not been answered yet and new problems have arisen. The interplay in the reaction

ᵃ e-mail: `w.trautmann@gsi.de`

dynamics between the density functional and transport properties leads to huge error bars in the quantitative estimation of fundamental quantities like nuclear incompressibility. Better constraints can be achieved if different independent observables are examined comparatively.

In particular, the study of isoscalar collective modes, reviewed by S. Shlomo *et al.*, provides such complementary information, which in recent years has allowed us to estimate the nuclear incompressibility around saturation within a few tens of MeV.

Both the initial high-density state of the collision and the later low-density finite-temperature stage leading to the formation of many-body correlations, are, in principle, extremely interesting to probe. The problem is that both stages are transient in time, and only the final outcome of the collision can be measured. As a result, any information on the nuclear equation of state necessarily is obtained through the comparison with a nuclear model. At the moment our most sophisticated transport theories solve the time-dependent problem without off-shell effects, and include correlations beyond the mean field at the classical level if at all. Moreover, the different approximation schemes developed in the different codes do not produce entirely compatible results. Improvement in the predictive power of transport models is certainly one of the greatest theoretical challenges facing the field.

The irreducible time dependence of the reaction problem can be addressed not only through improved theoretical models predicting asymptotic observables, but also through more sophisticated analyses of data reconstructing the reaction information backward in time. In particular, correlation functions provide an extremely powerful technique to, at least partially, disentangle the space and time information from particle and fragment yields. This, in turn, leads to the challenge of developing new third-generation nuclear detectors with high granularity and high resolution. The numerous fascinating applications of correlation functions and imaging techniques are reviewed in the monograph by G. Verde *et al.*

The availability of radioactive ion beams from various facilities either already constructed or planned for the near future (RIKEN, SPIRALII, RIA, FAIR, EURISOL, MSU, TAMU) has given a strong boost to the field. Not only does the comparison of nuclear dynamics of similar systems with different isotopic content provide new selective observables to probe our nuclear models, but also the possibility emerges to probe quantitatively the dependence of the density functional on the relative neutron/proton content —the so-called symmetry energy— through heavy-ion collisions. This quantity is of fundamental importance in a number of astrophysical situations, such as supernovae explosion dynamics and neutron star structure. These new applications generate great enthusiasm in the community, and the field is rapidly evolving. Although definitive answers will only be obtained with the advent of exotic beams in the energy regime of some tens of MeV per nucleon, the information already collected with stable beams is reviewed in monographs by M. Di Toro *et al.* and by M. Colonna and M.B. Tsang.

2 Experimental and theoretical challenges of nuclear multi-fragmentation

The field of multi-fragmentation deals first and foremost with the phenomenon of multiple fragment production. Such a phenomenon shows a clear character of universality: it has been observed with heavy-ion collisions using beams of a few tens of MeV per nucleon and in the relativistic target or projectile fragmentation regime as well as with light-particle, pion- or antiproton-induced reactions. The global characteristics of fragment production as revealed by the different experiments are reviewed in the contribution by B. Tamain.

Understanding the multi-fragmentation phenomenon raises the theoretical challenge of modeling the development of instabilities and correlations in finite quantum many-body systems. This interdisciplinary problem calls for important future developments that also will have implications to nuclear structure close to the driplines and to cluster physics. A partial review of this vast subject is presented in the monographs by V. Baran *et al.* and by A. Ono and J. Randrup.

The complexity of nuclear systems and the apparent universality of fragment observables encourages statistical treatments: together with the progress in transport theories, heavy-ion experiments have triggered an enormous theoretical effort in the development of sophisticated statistical descriptions of fragment production, reviewed in the article by A.S. Botvina and I.N. Mishustin. The predictive power of such models is tested through a detailed comparison of different codes associated with the models compiled by M.B. Tsang *et al.*

The construction of complex collective observables, as well as the exclusive analysis of isotopically resolved fragment yields, requires very sophisticated apparatuses capable of detecting all nuclear systems in the mass table from neutrons to uranium isotopes, and combining high granularity, geometrical coverage, and detection efficiency, with low thresholds for detection and identification. The present state of the art of nuclear detection and the challenges for third-generation detection systems are presented in the monograph by R.T. de Souza *et al.*

3 Phase transitions in finite, transient, non-extensive systems

Since the early days of multi-fragmentation, the breaking of a nucleus into many pieces has been tentatively associated with a phase transition. The intuitive association of fragmentation with a disordered phase is supported by different arguments. On the theoretical side, realistic mean-field calculations of the nuclear phase diagram consistently predict that nuclear matter should present a liquid-gas phase transition at sub-saturation densities and temperatures below 10–15 MeV. On the experimental side, evidence has been accumulated for fragment formation actually occurring under similar conditions of density and temperature. The rapid opening of the multi-fragmentation channel around a specific energy recalls the onset of a phase transition, and the observed scaling properties of fragment abundancies suggest a critical phenomenon.

The experimental measurement of the fragmentation phase transition would offer a unique possibility for settling the whole finite temperature phase diagram of nuclear matter. The extraordinary importance and ambition of this program can be appreciated if we consider that in the last 50 years of nuclear physics, only the saturation point of the nuclear matter phase diagram has been measured with good precision. The phase structure of exotic nuclear matter is also of particular interest for the static and transport properties of a number of compact objects in the universe, such as neutron star crusts and supernovae cores, as reviewed in the contribution by C. Horowitz.

To achieve this goal, sophisticated tools have been developed to measure the excitation energy and temperature of the transient multi-fragmenting stage of the collision. These tools are presented and discussed in the reviews by V.E. Viola *et al.* and by A. Kelić *et al.* A number of indications of such a phase transition have been accumulated, and are critically analyzed in the monographs by D. Santonocito and Y. Blumenfeld, by Y.G. Ma, by B. Borderie and P. Désesquelles, by F. Gulminelli and M. D'Agostino, and by O. Lopez and M.F. Rivet. Even if a global coherence of the different signals has been attained, and some quantitative estimates of the transition region start to be available, the order and the nature of the phase transition are still subject to uncertainties and debate.

The understanding of the phase transition is a particularly challenging problem since it is predicted theoretically that such a transition should be accompanied by numerous scalings and hyper-scalings as well as by thermodynamic anomalies like negative specific heat and bimodalities. Such anomalies are generic features of first-order phase transitions in non-extensive systems, and as such have been reported in other fields, such as cluster physics and self-gravitating systems. These interdisci-

plinary connections, which make heavy-ion thermodynamics a unique laboratory of statistical mechanics of finite systems, are explored in the monographs by D.H.E. Gross and by Ph. Chomaz and F. Gulminelli. Fragmentation is not the only phase transition expected in nuclear matter: at much higher energy density a transition towards a quark-gluon plasma has been predicted and is presently being explored at the high-energy heavy-ion facilities at CERN and Brookhaven National Laboratory. Some of the possible connections between the two phase transitions are addressed in the article by I.N. Mishustin.

4 Final remarks

Before concluding this introduction, a few remarks are necessary. The present topical issue gives a report of the momentary status of the field of dynamics and thermodynamics with nuclear degrees of freedom, as seen by the collection of authors. But it is not meant as a definitive and unique description of all of the relevant physics issues. The different articles are also different in character, reflecting both the preferences of the authors and the uneven state of the development of the various topics. In particular, the articles appearing in the last chapter making connections to neighboring fields have to be taken with a caveat. The authors have tried to make them as objective as possible in a developing field subject to discussions and controversies.

References

1. J.A. Gaidos *et al.*, Phys. Rev. Lett. **42**, 82 (1979); J.E. Finn *et al.*, Phys. Rev. Lett. **49**, 1321 (1982).
2. B. Jakobsson *et al.*, Z. Phys. A **307**, 293 (1982); C.J. Waddington, P.S. Freier, Phys. Rev. C **31**, 888 (1985).
3. For an early review, see J. Hüfner, Phys. Rep. **125**, 129 (1985).

Eur. Phys. J. A **30**, 5–21 (2006)
DOI 10.1140/epja/i2005-10313-x

THE EUROPEAN
PHYSICAL JOURNAL A

Modelization of the EOS

C. Fuchs[1,a] and H.H. Wolter[2]

[1] Institut für Theoretische Physik der Universität Tübingen, D-72076 Tübingen, Germany
[2] Sektion Physik der Universität München, D-85748 Garching, Germany

Received: 12 December 2005 /
Published online: 19 September 2006 – © Società Italiana di Fisica / Springer-Verlag 2006

Abstract. This paper summarizes theoretical predictions for the density and isospin dependence of the nuclear mean field and the corresponding nuclear equation of state. We compare predictions from microscopic and phenomenological approaches. An application to heavy-ion reactions requires to incorporate these forces into the framework of dynamical transport models. Constraints on the nuclear equation of state derived from finite nuclei and from heavy-ion reactions are discussed.

PACS. 21.65.+f Nuclear matter – 21.60.-n Nuclear structure models and methods – 25.75.-q Relativistic heavy-ion collisions – 24.10.Cn Many-body theory

1 Introduction

Heavy-ion reactions provide the only possibility to reach nuclear-matter densities beyond saturation density $\rho_0 \simeq 0.16\,\mathrm{fm}^{-3}$. Transport calculations indicate that in the low and intermediate energy range $E_{\mathrm{lab}} \sim 0.1$–$1\,\mathrm{AGeV}$ nuclear densities between 2–$3\rho_0$ are accessible while the highest baryon densities ($\sim 8\rho_0$) will probably be reached in the energy range of the future GSI facility FAIR between 20–$30\,\mathrm{AGeV}$. At even higher incident energies transparency sets in and the matter becomes less baryon rich due to the dominance of meson production. The isospin dependence of the nuclear forces which is at present only little constrained by data will be explored by the forthcoming radioactive beam facilities at FAIR/GSI [1], SPIRAL2/GANIL and RIA [2]. Since the knowledge of the nuclear equation of state (EOS) at supra-normal densities and extreme isospin is essential for our understanding of the nuclear forces as well as for astrophysical purposes, the determination of the EOS was already one of the primary goals when first relativistic heavy-ion beams started to operate in the beginning of the 80s [3]. In the following, we will briefly discuss the knowledge on the nuclear EOS from a theoretical point of view, then turn to the realization within transport models, and finally give a short review on possible observables from heavy-ion reactions to constrain the EOS.

2 Models for the nuclear EOS

Models which make predictions on the nuclear EOS can roughly be divided into three classes:

1. *Phenomenological density functionals*: These are models based on effective density-dependent interactions such as Gogny [4,5] or Skyrme forces [6,7] or relativistic mean-field (RMF) models [8]. The number of parameters which are fine tuned to the nuclear chart is usually larger than six and less than 15. This type of models allows the most precise description of finite nuclei properties.

2. *Effective-field theory approaches*: Models where the effective interaction is determined within the spirit of effective-field theory (EFT) became recently more and more popular. Such approaches lead to a more systematic expansion of the EOS in powers of density, respectively, the Fermi momentum k_F. They can be based on density functional theory [9,10] or, *e.g.*, on chiral perturbation theory [11–13]. The advantage of EFT is the small number of free parameters and a correspondingly higher predictive power. However, when high-precision fits to finite nuclei are intended this is presently only possible by the price of fine tuning through additional parameters. Then EFT functionals are based on approximately the same number of model parameters as phenomenological density functionals.

3. *Ab initio approaches*: Based on high-precision free-space nucleon-nucleon interactions, the nuclear many-body problem is treated microscopically. Predictions for the nuclear EOS are parameter free. Examples are variational calculations [14,15], Brueckner-Hartree-Fock (BHF) [16–19] or relativistic Dirac-Brueckner-Hartree-Fock (DBHF) [20–26] calculations and Green's functions Monte Carlo approaches [27–29].

Phenomenological models as well as EFT contain parameters which have to be fixed by nuclear properties around or below saturation density which makes the extrapolation to

a e-mail: christian.fuchs@uni-tuebingen.de

supra-normal densities somewhat questionable. However, in the EFT case such an extrapolation is safer due to a systematic density expansion. One has nevertheless, to keep in mind that EFT approaches are based on low-density expansions. Many-body calculations, on the other hand, have to rely on the summation of relevant diagram classes and are still too involved for systematic applications to finite nuclei.

2.1 Mean-field theory

Among non-relativistic density functionals, Skyrme functionals are the ones most frequently used. The Skyrme interaction contains an attractive local two-body part and a repulsive density dependent two-body interaction which can be motivated by local three-body forces. We will not consider surface terms which involve gradients as well as spin-orbit contributions since they vanish in infinite nuclear matter. For a detailed discussion of Skyrme functionals and their relation to relativistic mean-field (RMF) theory see, *e.g.*, [7]. The EOS of symmetric nuclear matter, *i.e.* the binding energy per particle has the simple form

$$E/A = \frac{3k_F^2}{10M} + \frac{\alpha}{2}\rho + \frac{\beta}{1+\gamma}\rho^\gamma , \qquad (1)$$

where the first term in (1) represents the kinetic energy of a non-relativistic Fermi gas and the remaining part the potential energy. To examine the structure of relativistic mean-field models it is instructive to consider the simplest version of a relativistic model, *i.e.* the $\sigma\omega$ model of quantum hadron dynamics (QHD-I) [30]. In QHD-I the nucleon-nucleon interaction is mediated by the exchange of two effective boson fields which are attributed to a scalar σ- and a vector ω-meson. The energy density in infinite cold and isospin-saturated nuclear matter is in mean-field approximation given by

$$\epsilon = \frac{3}{4}E_F\varrho + \frac{1}{4}m_D^*\varrho_S + \frac{1}{2}\left\{ \Gamma_V \varrho^2 + \Gamma_S \varrho_S^2 \right\}, \qquad (2)$$

where the Fermi energy is given by $E_F = \sqrt{k_F^2 + m_D^{*2}}$. We will denote m_D^* explicitly as Dirac mass in the following in order to distinguish it from its non-relativistic counterpart. The effective mass absorbs the scalar part of the mean field $m_D^* = M - \Gamma_S\varrho_S$. In the limit $m_D^* \longrightarrow M$ the first two terms in (2) provide the energy (kinetic plus rest mass) of a non-interacting relativistic Fermi gas.

A genuine feature of all relativistic models is the fact that one has to distinguish between the vector density $\varrho = 2k_F^3/3\pi^2$ and a scalar density ϱ_S. The vector density is the time-like component of a 4-vector current j_μ, whose spatial components vanish in the nuclear matter rest frame, while ϱ_S is a Lorentz scalar. The scalar density shows a saturation behavior with increasing vector density which is essential for the relativistic saturation mechanism. This becomes clear when the binding energy $E/A = \epsilon/\varrho - M$ is expanded in powers of the Fermi mo-

mentum k_F:

$$E/A = \left[\frac{3k_F^2}{10M} - \frac{3k_F^4}{56M^3} + \cdots\right] + \frac{1}{2}\left[\Gamma_V - \Gamma_S\right]\varrho$$
$$+ \Gamma_S\frac{\varrho}{M}\left[\frac{3k_F^2}{10M} - \frac{36k_F^2}{175M^3} + \cdots\right] + \mathcal{O}\left((\Gamma_S\varrho/M)^2\right) .$$
$$(3)$$

The first term in (3) contains the kinetic energy of a non-relativistic Fermi gas followed by relativistic corrections and the remaining terms are the contributions from the mean field. In QHD-I the scalar and vector field strengths are given by the coupling constants for the corresponding mesons $\Gamma_S = g_\sigma^2/m_\sigma^2$ and $\Gamma_V = g_\omega^2/m_\omega^2$ divided by the meson masses. The two parameters $\Gamma_{S,V}$ are now fitted to the saturation point of nuclear matter $E/A \simeq -16\,\mathrm{MeV}$, $\varrho_0 \simeq 0.16\,\mathrm{fm}^{-3}$ which follows from the volume part of the Weizsäcker mass formula. The saturation mechanism requires that both coupling constants are large. This leads automatically to the cancellation of two large fields, namely an attractive scalar field $\Sigma_S = -\Gamma_S\varrho_S$ and a repulsive vector field $\Sigma_V = \Gamma_V\varrho$. As a typical feature of relativistic dynamics, the single-particle potential $U = m_D^*/E^*\Sigma_S - \Sigma_V$ ($E^* = \sqrt{\mathbf{k}^2 + m_D^{*2}}$), which is of the order of $-50\,\mathrm{MeV}$, results from the cancellation of scalar and vector fields, each of the order of several hundred MeV.

However, with only two parameters QHD-I provides a relatively poor description of the saturation point with a too large saturation density and a very stiff EOS ($K = 540\,\mathrm{MeV}$). To improve on this, higher-order corrections in density have to be taken into account which can be done in several ways. In the spirit of the original Walecka model non-linear meson self-interaction terms have been introduced into the QHD Lagrangian [8,31]. An alternative are relativistic point-coupling models where the explicit meson exchange picture is abandoned. A Lagrangian of nucleon and boson fields with point couplings can be constructed in the spirit of EFT and expanded in powers of density [9,10]. Finite-range effects from meson propagators are replaced by density gradients [9,10]. A third possibility is the density-dependent hadron field theory DDRH [32,33]. In DDRH the scalar and vector coupling constants are replaced by density-dependent vertex functions $\Gamma_{S,V}(k_F)$. The density dependence of these renormalized vertices can either be taken from Brueckner calculations, thus parameterizing many-body correlations [32, 33], or be determined phenomenologically [34,35]. In all cases additional parameters are introduced which allow a description of finite nuclei with a precision comparable to the best fits from Skyrme functionals. Phenomenological density functionals provide high-quality fits to the known areas of the nuclear chart. Binding energies and rms radii are reproduced with an average relative error of about ~ 1–5%. However, when the various models are extrapolated to the unknown regions of extreme isospin or to super-heavies, predictions start to deviate substantially. This demonstrates the limited predictive power of these functionals.

2.2 Effective-field theory

When concepts of effective-field theory are applied to nuclear-physics problems one has to rely on a separation of scales. EFT is based on a perturbative expansion of the nucleon-nucleon (NN) interaction or the nuclear mean field within power-counting schemes. The short-range part of the NN interaction requires a non-perturbative treatment, *e.g.*, within the Brueckner ladder summation. The philosophy behind EFT is to separate short-range correlations from the long- and intermediate-range part of the NN interaction. This assumption is motivated by the fact that the scale of the short-range correlations, *i.e.* the hard core, is set by the ρ and ω vector meson masses which lie well above the Fermi momentum and the pion mass which sets the scale of the long-range forces. The density functional theory (DFT) formulation of the relativistic nuclear many-body problem [9,10] is thereby analogous to the Kohn-Sham approach in DFT. An energy functional of scalar and vector densities is constructed which by minimization gives rise to variational equations that determine the ground-state densities. Doing so, one tries to approximate the *exact* functional using an expansion in classical meson fields and their derivatives, based on the observation that the ratios of these quantities to the nucleon mass are small, at least up to moderate density. The exact energy functional which one tries to derive explicitly when using many-body techniques such as Brueckner or variational approaches contains exchange correlations and all other many-body and relativistic effects. The DFT interpretation implies that the model parameters fitted to nuclei implicitly contain effects of both short-distance physics and many-body corrections.

Recently also concepts of chiral perturbation theory (ChPT) have been applied to the nuclear many-body problem [12,13]. Doing so, the long- and intermediate-range interactions are treated explicitly within chiral pion-nucleon dynamics. This allows an expansion of the energy density functional in powers of m_π/M or in k_F/M. Like in DFT, short-range correlations are not resolved explicitly but handled by counter-terms (dimensional regularization) [11] or through a cut-off regularization [13]. Figure 1 shows the corresponding EOS obtained from chiral one- and two-pion exchange between nucleons. In order to account for the most striking feature of relativistic dynamics, expressed by the existence of the large scalar and vector fields, in refs. [12,13] iso-scalar condensate background nucleon self-energies derived from QCD sum rules have been added to the chiral fluctuations. To lowest order in density the QCD condensates give rise to a scalar self-energy $\Sigma_S = -\sigma_N M/(m_\pi^2 f_\pi^2)\varrho_S$ and a vector self-energy $\Sigma_V = 4(m_u + m_d)M/(m_\pi^2 f_\pi^2)\varrho$. It is remarkable that the total self-energies, *i.e.* condensates plus chiral fluctuations, are very close to those obtained from DBHF calculations [12,23]. The resulting EOS is also shown in fig. 1 in addition to that obtained after fine tuning to finite nuclei. Although the original EOS (case 1) is rather soft, the inclusion of the condensates and the adjustment to finite nuclei results in an EOS which is finally stiff.

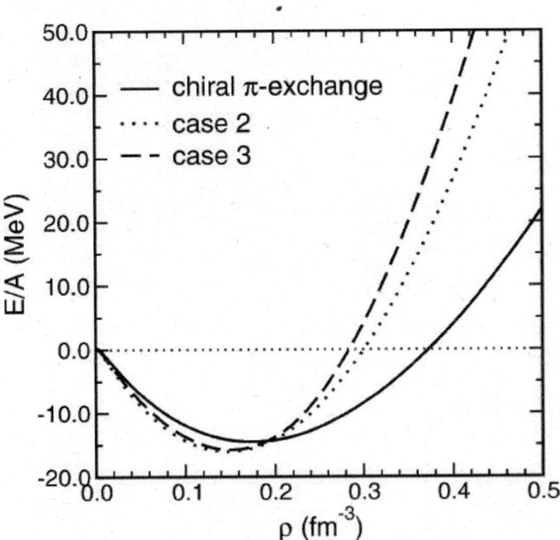

Fig. 1. EOS for symmetric nuclear matter obtained from chiral one- and two-pion exchange (case 1, solid line), by adding background fields from QCD sum rules (case 2, dotted line), and finally after fine tuning to finite nuclei properties (case 3, dashed line). The figure is taken from [12].

2.3 Ab initio calculations

In *ab initio* calculations based on many-body techniques one derives the energy functional from first principles, *i.e.* treating short-range and many-body correlations explicitly. A typical example for a successful many-body approach is Brueckner theory [16]. In the relativistic Brueckner approach the nucleon inside the medium is dressed by the self-energy Σ. The in-medium T-matrix which is obtained from the relativistic Bethe-Salpeter (BS) equation plays the role of an effective two-body interaction which contains all short-range and many-body correlations of the ladder approximation. Solving the BS equation the Pauli principle is respected and intermediate scattering states are projected out of the Fermi sea. The summation of the T-matrix over the occupied states inside the Fermi sea yields finally the self-energy in Hartree-Fock approximation. This coupled set of equations states a self-consistency problem which has to be solved by iteration.

In contrast to relativistic DBHF calculations which came up in the late 80s, non-relativistic BHF theory has already almost half a century's history. The first numerical calculations for nuclear matter were carried out by Brueckner and Gammel in 1958 [16]. Despite strong efforts invested in the development of improved solution techniques for the Bethe-Goldstone (BG) equation, the non-relativistic counterpart of the BS equation, it turned out that, although such calculations were able to describe the nuclear saturation mechanism qualitatively, they failed quantitatively. The results of a systematic study for a large number of NN interactions were found to be always located on a so-called *Coester-line* in the E/A-ρ plane which does not coincide with the empirical region of saturation. In particular, modern one-boson exchange (OBE) potentials lead to strong over-binding and too large saturation

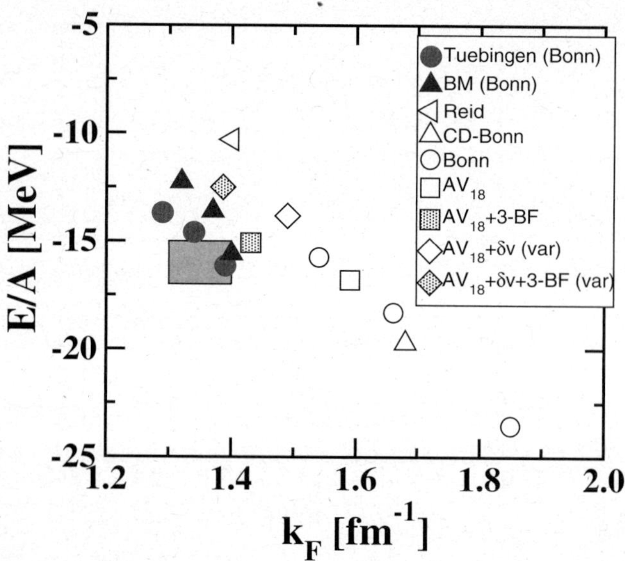

Fig. 2. Nuclear-matter saturation points from relativistic (full symbols) and non-relativistic (open symbols) Brueckner-Hartree-Fock calculations based on different nucleon-nucleon forces. The diamonds show results from variational calculations. Shaded symbols denote calculations which include 3-body forces. The shaded area is the empirical region of saturation.

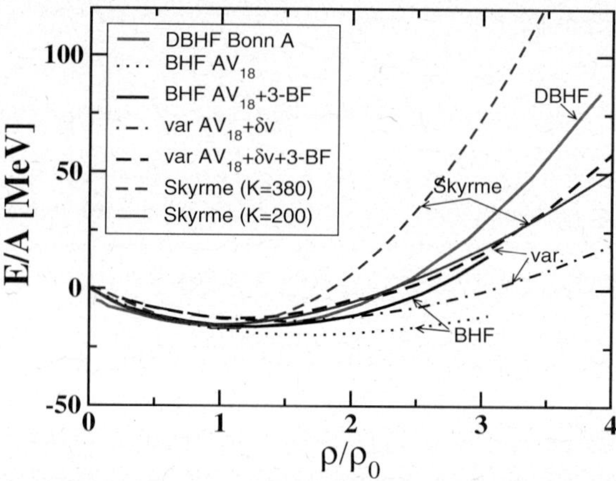

Fig. 3. Predictions for the EOS of symmetric nuclear matter from microscopic *ab initio* calculations, *i.e.* relativistic DBHF [23], non-relativistic BHF [18] and variational [15] calculations. For comparison also soft and hard Skyrme forces are shown.

densities whereas relativistic calculations do a much better job.

Figure 2 compares the saturation points of nuclear matter obtained by relativistic Dirac-Brueckner-Hartree-Fock (DBHF) calculations using the Bonn potentials [36] as bare NN interactions to non-relativistic Brueckner-Hartree-Fock calculations for various NN interactions. The DBHF results are taken from ref. [21] (BM) and more recent calculations based on improved techniques are from [23] (Tübingen). Several reasons have been discussed in the literature in order to explain the success of the relativistic treatment. The saturation mechanisms in relativistic and non-relativistic theories are quite different. In relativistic MFT the vector field grows linearly with density while the scalar field saturates at large densities. The magnitude and the density dependence of the scalar and vector DBHF self-energy is similar to MFT, *i.e.* the single-particle potential is the result of the cancellation of two large scalar and vector fields, each several hundred MeV in magnitude (see, *e.g.*, the effective mass in fig. 7). In BHF, on the other hand, the saturation mechanism takes place exclusively on the scale of the binding energy, *i.e.* a few tens of MeV. It cannot be understood by the absence of a tensor force. In particular, the second order 1-π exchange potential (OPEP) is large and attractive at high densities and its interplay with Pauli blocking leads finally to saturation. Relativistically, the tensor force is quenched by a factor $(m_D^*/M)^2$ and less important for the saturation mechanism [37].

Three-body forces (3-BFs) have been extensively studied within non-relativistic BHF [18] and variational calculations [15]. The contributions from 3-BFs are in total re-

pulsive which makes the EOS harder and non-relativistic calculations come close to their relativistic counterparts. The same effect is observed in variational calculations [15] shown in fig. 3. The variational results contain boost corrections (δv) which account for relativistic kinematics and lead to additional repulsion [15]. Both, BHF [18] and the variational calculations from [15] are based on the latest AV_{18} version of the Argonne potential. In both cases phenomenological 3-body forces are used, the Tucson-Melbourne 3-BF in [18] and the Urbana IX 3-BF[1] in [15]. It is often argued that in non-relativistic treatments 3-BFs play in some sense an equivalent role as the dressing of the two-body interaction by in-medium spinors in Dirac phenomenology. Both mechanisms lead indeed to an effective density-dependent two-body interaction V which is, however, of different origin. One class of 3-BFs involves virtual excitations of nucleon-antinucleon pairs. Such Z-graphs are in net repulsive and can be considered as a renormalization of the meson vertices and propagators. A second class of 3-BFs is related to the inclusion of explicit resonance degrees of freedom. The most important resonance is the $\Delta(1232)$ isobar which provides at low and intermediate energies large part of the intermediate-range attraction. Intermediate Δ states appear in elastic NN scattering only in combination with at least two-isovector-meson exchange ($\pi\pi, \pi\rho, \ldots$). Such box diagrams can satisfactorily be absorbed into an effective σ-exchange [36]. The maintenance of explicit Δ degrees of freedom (DoFs) gives rise to additional saturation, shifting the saturation point away from the empirical region [20]. However, as pointed out, *e.g.*, in ref. [27], the inclusion of non-nucleonic DoFs has to be performed with caution: freezing out resonance DoFs generates automatically a class of three-body forces which contains nucleon-resonance ex-

---

[1] Using boost corrections the repulsive contributions of the UIX interaction are reduced by about 40% compared to the original ones in [15].

citations. There exist strong cancellation effects between the repulsion due to box diagrams and contributions from 3-BFs. Non-nucleonic DoFs and many-body forces should therefore be treated on the same footing. Such a treatment may be possible with the next generation of nucleon-nucleon forces based on chiral perturbation theory [38,39] which allows a systematic generation of three-body forces. Next-to-leading order (NLO), all 3-BFs cancel while non-vanishing contributions appear at NNLO.

Figure 3 compares the equations of state from the different approaches: DBHF from ref. [23] based the Bonn-A interaction[2] [36], BHF [18] and variational calculations [15]. The latter ones are based on the Argonne AV_{18} potential and include 3-body forces. All the approaches use modern high-precision NN interactions and represent state-of-the-art calculations. Two phenomenological Skyrme functionals which correspond to the limiting cases of a soft ($K = 200\,\mathrm{MeV}$) and a hard ($K = 380\,\mathrm{MeV}$) EOS are shown as well. In contrast to the Skyrme interaction (1) where the high-density behavior is fixed by the compression modulus, in microscopic approaches the compression modulus is only loosely connected to the curvature at saturation density. DBHF Bonn-A has, *e.g.*, a compressibility of $K = 230\,\mathrm{MeV}$. Below $3\rho_0$, both are not too far from the soft Skyrme EOS. The same is true for BHF including 3-body forces.

When many-body calculations are performed, one has to keep in mind that elastic NN scattering data constrain the interaction only up to about $400\,\mathrm{MeV}$, which corresponds to the pion threshold. NN potentials differ essentially in the treatment of the short-range part. A model-independent representation of the NN interaction can be obtained in EFT approaches where the unresolved short-distance physics is replaced by simple contact terms. In the framework of chiral EFT the NN interaction has been computed up to N^3LO [39,40]. An alternative approach which leads to similar results is based on renormalization group (RG) methods [41]. In the $V_{\mathrm{low}\,k}$ approach a low-momentum potential is derived from a given realistic NN potential by integrating out the high-momentum modes using RG methods. At a cutoff $\Lambda \sim 2\,\mathrm{fm}^{-1}$ all the different NN potential models were found to collapse to a model-independent effective interaction $V_{\mathrm{low}\,k}$. When applied to the nuclear many-body problem low-momentum interactions do not require a full resummation of the Brueckner ladder diagrams but can already be treated within second-order perturbation theory [42]. However, without repulsive three-body-forces, isospin-saturated nuclear matter was found to collapse. Including 3-BFs first promising results have been obtained with $V_{\mathrm{low}\,k}$ [42], however, nuclear saturation is not yet described quantitativley. Moreover, one has to keep in mind that, due to the high-momentum cut-offs, EFT is essentially only suitable at moderate densities.

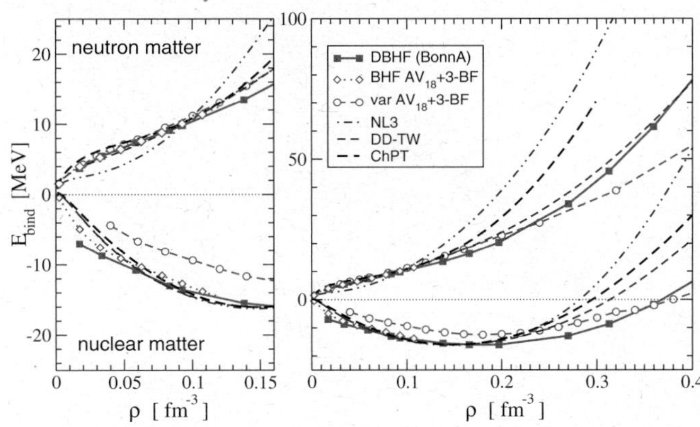

Fig. 4. EOS in nuclear matter and neutron matter. BHF/DBHF and variational calculations are compared to the phenomenological density functionals NL3 and DD-TW and ChPT+corr. The left panel zooms the low-density range.

3 EOS in symmetric and asymmetric nuclear matter

Figure 4 compares now the predictions for nuclear and neutron matter from microscopic many-body calculations —DBHF [26] and the "best" variational calculation with 3-BFs and boost corrections [15]— to phenomenological approaches and to EFT. As typical examples for relativistic functionals we take NL3 [43] as one of the best RMF fits to the nuclear chart and a phenomenological density dependent RMF functional DD-TW from [34]. ChPT+corr. is based on chiral pion-nucleon dynamics including condensate fields and fine tuning to finite nuclei (case 3 in fig. 1). As expected, the phenomenological functionals agree well at and below saturation density where they are constrained by finite nuclei but start to deviate substantially at supra-normal densities. In neutron matter the situation is even worse since the isospin dependence of the phenomenological functionals is less constrained. The predictive power of such density functionals at supra-normal densities is restricted. *Ab initio* calculations predict a soft EOS throughout the density range relevant for heavy-ion reactions at intermediate and low energies, *i.e.* up to about three times ρ_0. There seems to be no way to obtain an EOS as stiff as the hard Skyrme force shown in fig. 3 or NL3. Since the NN scattering lenght is large, neutron matter at subnuclear densities is less model dependent. The microscopic calculations (BHF/DBHF, variational) agree well and results are consistent with "exact" Quantum Monte Carlo calculations [29].

In isospin asymmetric matter the binding energy is a functional of the proton and neutron densities, characterized by the asymmetry parameter $\beta = Y_n - Y_p$ which is the difference of the neutron and proton fraction $Y_i = \rho_i/\rho_\ast, i = n, p$. The isospin dependence of the energy functional can be expanded in terms of β which leads to a parabolic dependence on the asymmetry parameter

$$E(\rho, \beta) = E(\rho) + E_{\mathrm{sym}}(\rho)\beta^2 + \mathcal{O}(\beta^4) + \cdots,$$

$$E_{\mathrm{sym}}(\rho) = \frac{1}{2}\frac{\partial^2 E(\rho, \beta)}{\partial \beta^2}\Big|_{\beta=0} = a_4 + \frac{p_0}{\rho_0^2}(\rho - \rho_0) + \cdots \quad (4)$$

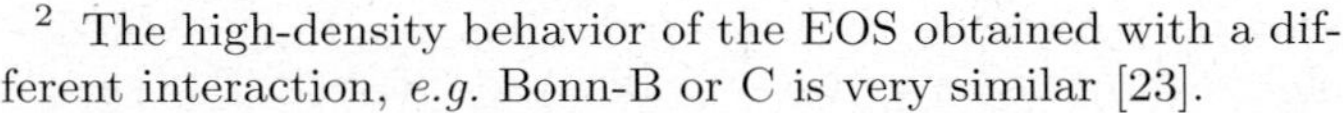

[2] The high-density behavior of the EOS obtained with a different interaction, *e.g.* Bonn-B or C is very similar [23].

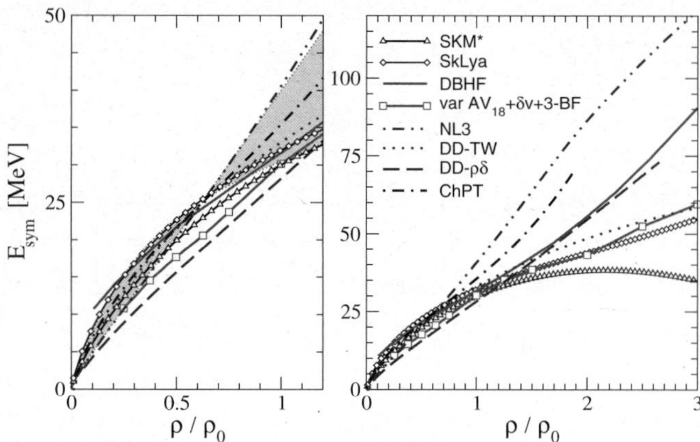

Fig. 5. Symmetry energy as a function of density as predicted by different models. The left panel shows the low-density region, while the right panel displays the high-density range.

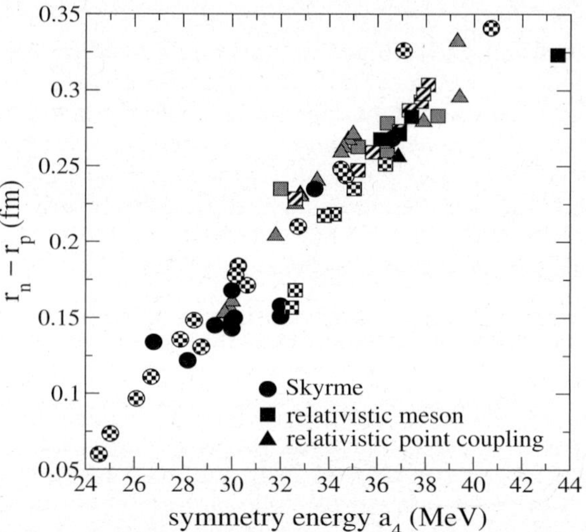

Fig. 6. Skin thickness in ^{208}Pb *versus* the linear symmetry energy parameter a_4 for various models. The figure is taken from [10].

Figure 5 compares the symmetry energy predicted by the DBHF and variational calculations to that of the empirical density functionals already shown in fig. 4. In addition, the relativistic DD-$\rho\delta$ RMF functional [44] is included. Two Skyrme functionals, SkM* and the more recent Skyrme-Lyon force SkLya, represent non-relativistic models. The left panel zooms the low-density region, while the right panel shows the high-density behavior of E_sym. It is remarkable that most empirical models coincide around $\rho \simeq 0.6\rho_0$, where $E_\mathrm{sym} \simeq 24\,\mathrm{MeV}$. This demonstrates that constraints from finite nuclei are active for an average density slightly above half-saturation density. However, the extrapolations to supra-normal densities diverge dramatically. This is crucial since the high-density behavior of E_sym is essential for the structure and the stability of neutron stars (see also the contribution VI.3 by Horowitz, this topical issue [45]). The microscopic models show a density dependence which can still be considered as *asy-stiff*. DBHF [26] is thereby stiffer than the variational results of [15]. The density dependence is generally more complex than in RMF theory, in particular at high densities where E_sym shows a non-linear and more pronounced increase. Figure 5 clearly demonstrates the necessity to constrain the symmetry energy at supra-normal densities with the help of heavy-ion reactions. The hatched area in fig. 5 displays the range of E_sym which has been obtained by constructing a density-dependent RMF functional varying thereby the linear asymmetry parameter a_4 from 30 to 38 MeV [35]. In ref. [35] it was concluded that charge radii, in particular the skin thickness $r_n - r_p$ in heavy nuclei, constrain the allowed range of a_4 to 32–36 MeV for relativistic functionals.

Figure 6 displays the correlation between the skin thickness in ^{208}Pb and a_4 obtained within various models. The skin thickness depends, however, not only on the symmetry energy but there exists a close correlation between a_4 and the compression modulus K [35]. This correlation is of importance when these quantities are extracted from finite nuclei (see the discussion by Shlomo *et al.*, contribution II.2, this topical issue [46]).

3.1 Effective nucleon masses

The introduction of an effective mass is a common concept to characterize the quasi-particle properties of a particle inside a strongly interacting medium. In nuclear physics, different definitions of the effective nucleon mass exist which are often compared and sometimes even mixed up: the non-relativistic effective mass m^*_{NR} and the relativistic Dirac mass m^*_D. These two definitions are based on different physical concepts. The non-relativistic mass parameterizes the momentum dependence of the single-particle potential. The relativistic Dirac mass is defined through the scalar part of the nucleon self-energy in the Dirac field equation which is absorbed into the effective mass $m^*_D = M + \Sigma_S(k, k_F)$. The Dirac mass is a smooth function of the momentum. In contrast, the non-relativistic effective mass —as a model-independent result— shows a narrow enhancement near the Fermi surface due to an enhanced level density [47]. For a recent review on this subject and experimental constraints on m^*_{NR}, see [48].

While the Dirac mass is a genuine relativistic quantity the effective mass m^*_{NR} is determined by the single-particle energy

$$m^*_{NR} = k[\mathrm{d}E/\mathrm{d}k]^{-1} = \left[\frac{1}{M} + \frac{1}{k}\frac{\mathrm{d}}{\mathrm{d}k}U\right]^{-1}; \qquad (5)$$

m^*_{NR} is a measure of the non-locality of the single-particle potential U (real part) which can be due to non-localities in space, resulting in a momentum dependence, or in time, resulting in an energy dependence. In order to clearly separate both effects, one has to distinguish further between the so-called k-mass and the E-mass [17]. The spatial non-localities of U are mainly generated by exchange Fock terms and the resulting k-mass is a smooth function of the momentum. Non-localities in time are generated by Brueckner ladder correlations due to the scattering to intermediate states which are off-shell. These are

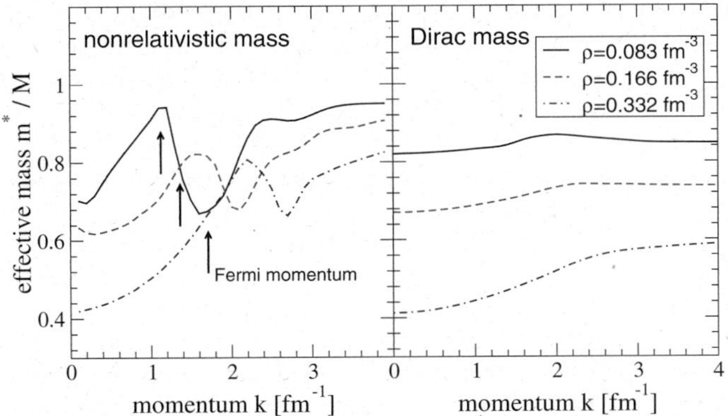

Fig. 7. The effective mass in isospin symmetric nuclear matter as a function of the momentum k at different densities determined from relativistic Brueckner calculations.

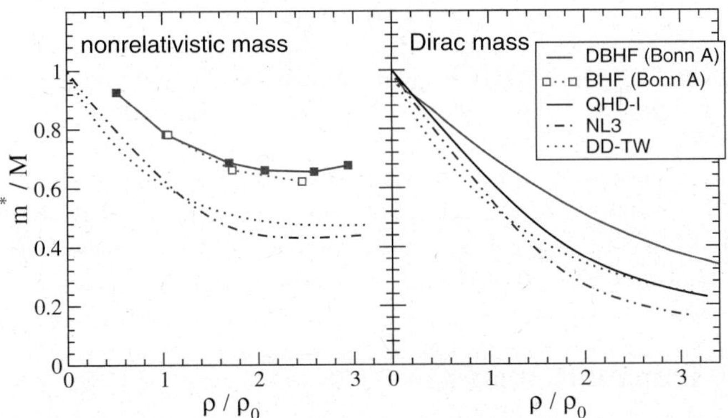

Fig. 8. Non-relativistic and Dirac effective mass in isospin symmetric nuclear matter as a function of the density for various models.

mainly short-range correlations which generate a strong momentum dependence with a characteristic enhancement of the E-mass slightly above the Fermi surface [47,17,49]. The effective mass defined by eq. (5) contains both, non-localities in space and time and is given by the product of k-mass and E-mass [17]. In fig. 7 the non-relativistic effective mass and the Dirac mass, both determined from DBHF calculations [50], are shown as a function of momentum k at different Fermi momenta of $k_F = 1.07, 1.35, 1.7\,\mathrm{fm}^{-1}$. m_{NR}^* shows the typical peak structure as a function of momentum around k_F which is also seen in BHF calculations [49]. The peak reflects the increase of the level density due to the vanishing imaginary part of the optical potential at k_F which is also seen, e.g., in shell model calculations [47,17]. One has, however, to account for correlations beyond mean field or Hartree-Fock in order to reproduce this behavior. Figure 8 compares the density dependence of the two effective masses determined at k_F. Both masses decrease with increasing density, the Dirac mass continously, while m_{NR}^* starts to rise again at higher densities. Phenomenological density functionals (QHD-I, NL3, DD-TW) yield systematically smaller values of m_{NR}^* than the microscopic approaches. This reflects the lack of

non-local contributions from short-range and many-body correlations in the mean-field approaches.

3.1.1 Proton-neutron mass splitting

A topic heavily discussed at present is the proton-neutron mass splitting in isospin-asymmetric nuclear matter. This question is of importance for the forthcoming new generation of radioactive beam facilities which are devoted to the investigation of the isospin dependence of the nuclear forces at its extremes. However, presently the predictions for the isospin dependences differ substantially. BHF calculations [18,49] predict a proton-neutron mass splitting of $m_{NR,n}^* > m_{NR,p}^*$. This stands in contrast to relativistic mean-field (RMF) theory. When only a vector isovector ρ-meson is included Dirac phenomenology predicts equal masses $m_{D,n}^* = m_{D,p}^*$ while the inclusion of the scalar isovector δ-meson, i.e. $\rho + \delta$, leads to $m_{D,n}^* < m_{D,p}^*$ [44]. When the effective mass is derived from RMF theory, it shows the same behavior as the corresponding Dirac mass, namely $m_{NR,n}^* < m_{NR,p}^*$ [44]. Conventional Skyrme forces, e.g. SkM*, lead to $m_{NR,n}^* < m_{NR,p}^*$ [51] while the more recent Skyrme-Lyon interactions (SkLya) predict the same mass splitting as RMF theory. The predictions from relativistic DBHF calculations are still controversial in the literature. They depend strongly on approximation schemes and techniques used to determine the Lorentz and the isovector structure of the nucleon self-energy. In the approach originally proposed by Brockmann and Machleidt [21] one extracts the scalar and vector self-energy components directly from the single-particle potential. Thus, by a fit to the single-particle potential mean values for the self-energy components are obtained where the explicit momentum dependence has already been averaged out. In symmetric nuclear matter this method is relatively reliable but the extrapolation to asymmetric matter is ambiguous [24]. Calculations based on this method predict a mass splitting of $m_{D,n}^* > m_{D,p}^*$ [52]. On the other hand, the components of the self-energies can directly be determined from the projection onto Lorentz invariant amplitudes [20,22–24,26, 53]. Projection techniques are involved but more accurate and yield the same mass splitting as found in RMF theory when the δ-meson is included, i.e. $m_{D,n}^* < m_{D,p}^*$ [22, 24,26]. Recently, also the non-relativistic effective mass has been determined with the DBHF approach and here a reversed proton-neutron mass splitting was found, i.e. $m_{NR,n}^* > m_{NR,p}^*$ [50]. Thus DBHF is in agreement with the results from non-relativistic BHF calculations.

Experimentally accessible is the p-n mass splitting, or the magnitude of the corresponding isovector effective mass m_V^*, $\left(\frac{\beta}{m_V^*} = \frac{\beta+1}{m_{NR}^*} - \frac{1}{m_{NR,n}^*}\right)$ through the electric dipole photoabsorption cross-section, i.e. through an enhancement of the Thomas-Reiche-Kuhn sum rule by the factor m/m_V^*. However, values derived from GDR measurements range presently from $m_V^*/m = 0.7$–1.05 [48,54,55]. The forthcoming radioactive beam facilites will certainly improve on this not yet satisfying situation.

3.2 Optical potentials

The second important quantity related to the momentum dependence of the mean field is the optical nucleon-nucleus potential. At subnormal densities the optical potential U_{opt} is constrained by proton-nucleus scattering data [56] and at supra-normal densities constraints can be derived from heavy-ion reactions [57–59]. In a relativistic framework the optical Schroedinger-equivalent nucleon potential (real part) is defined as

$$U_{\mathrm{opt}} = -\Sigma_S + \frac{E}{M}\Sigma_V + \frac{\Sigma_S^2 - \Sigma_V^2}{2M}. \qquad (6)$$

One should thereby note that in the literature sometimes also an optical potential, given by the difference of the single-particle energies in medium and free space $U = E - \sqrt{M^2 + \mathbf{k}^2}$ is used [57] which should be not mixed up with (6). In a relativistic framework momentum-independent fields $\Sigma_{S,V}$ (as, e.g., in RMF theory) lead always to a linear energy dependence of U_{opt}. As seen from fig. 9, DBHF reproduces the empirical optical potential [56] extracted from proton-nucleus scattering for nuclear matter at ρ_0 reasonably well up to a laboratory energy of about 0.6–0.8 GeV. However, the saturating behavior at large momenta cannot be reproduced by this calculations because of missing inelasticities, i.e. the excitation of isobar resonances above the pion threshold. When such continuum excitations are accounted for, optical model caculations are able to describe nucleon-nucleus scattering data also at higher energies [60]. In heavy-ion reactions at incident energies above 1 AGeV such a saturating behavior is required in order to reproduce transverse flow observables [59]. One has then to rely on phenomenological approaches where the strength of the vector potential is artificially suppressed, e.g. by the introduction of addi-

tional form factors [59] or by energy-dependent terms in the QHD Lagrangian [61] (D^{3}C model in fig. 9).

The isospin dependence, expressed by the isovector optical potential $U_{\mathrm{iso}} = (U_{\mathrm{opt},n} - U_{\mathrm{opt},p})/(2\beta)$ is much less constrained by data. The knowledge of this quantity is, however, of high importance for the forthcoming radioactive beam experiments. The right panel of fig. 9 compares the predictions from DBHF [26] and BHF [62] to the phenomenological Gogny and Skyrme (SkM* and SkLya) forces and a relativistic $T - \rho$ approximation [64] based on empirical NN scattering amplitudes [65]. At large momenta, DBHF agrees with the tree-level results of [64]. While the dependence of U_{iso} on the asymmetry parameter β is found to be rather weak [26,62], the predicted energy and density dependences are quite different, in particular between the microscopic and the phenomenological approaches. The energy dependence of U_{iso} is very little constrained by data. The old analysis of optical potentials of scattering on charge asymmetric targets by Lane [66] is consistent with a decreasing potential as predicted by DBHF/BHF, while more recent analyses based on Dirac phenomenology [67] come to the opposite conclusions. RMF models show a linearly increasing energy dependence of U_{iso} (i.e., quadratic in k) like SkLya, however generally with a smaller slope (see discussion in [44]). To clarify this question certainly more experimental efforts are necessary.

4 Transport models

The difficulty to extract information on the EOS from heavy-ion reactions lies in the fact that the colliding system is over a large time span of the reaction out of global and even local equilibrium. At intermediate energies the relaxation time needed to equilibrate coincides more or less with the high-density phase of the reaction. Hence, non-equilibrium effects are present all over the compression phase where one essentially intends to study the EOS at supra-normal densities. Experimental evidences for incomplete equilibration even in central collisions have been found by isospin tracing of projectile and target nuclei [68] and by different variances of longitudinal and transverse rapidity distributions [69]. To account for the temporal space-time evolution of the reactions requires dynamical approaches which are based on kinetic transport theory. In the following we briefly discuss the various approaches which are mainly used in order to describe the reaction dynamics at low and intermediate energies.

4.1 Boltzmann-type kinetic equations

The theoretical basis for the description of the collision dynamics at energies ranging from the Fermi regime up to 1–2 AGeV is the hadronic non-equilibrium quantum transport field theory [70]. The starting point of non-equilibrium QFT is the Schwinger-Keldysh formalism for many-body Green's functions in non-equilibrium configurations. The one-body Green's function is defined as the

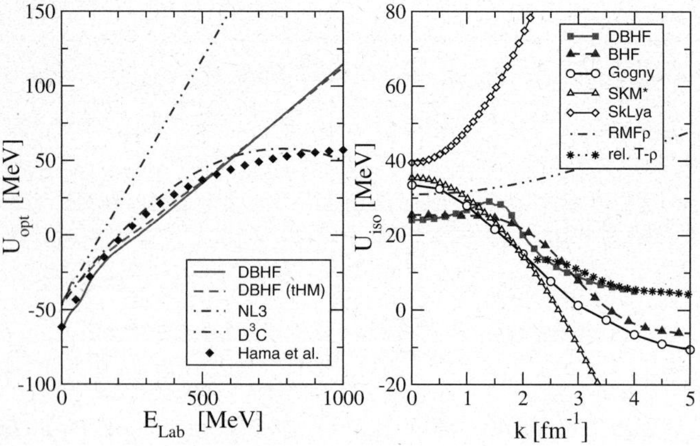

Fig. 9. Nucleon optical potential in nuclear matter at ρ_0. On the left side, DBHF calculations for symmetric nuclear matter from [20] and [23] are compared to the phenomenological models NL3 and D^{3}C [61] and to the p-A scattering analysis of [56]. The right panel compares the iso-vector optical potential from DBHF [26] and BHF [62] to phenomenological RMF [63], Gogny and Skyrme forces and to a relativistic $T - \rho$ approximation [64].

expectation value of the time-ordered product of fermionic field operators $G(1, 1') = (-i)\langle \mathcal{T}_{sk}(\Psi(1)\overline{\Psi}(1'))\rangle$, where $\mathcal{T}_{sk}$ defines the temporal sequence of the field operators. In non-equilibrium, time reversal invariance is violated and thus the application of $\mathcal{T}_{sk}$ leads to four possible combinations [70]:

$$G^c = -i\langle T^c[\Psi(1)\overline{\Psi}(1')]\rangle, \quad G^a = -i\langle T^a[\Psi(1)\overline{\Psi}(1')]\rangle,$$
$$G^> = -i\langle \Psi(1)\overline{\Psi}(1')\rangle, \qquad G^< = i\langle \overline{\Psi}(1')\Psi(1)\rangle, \qquad (7)$$

where T^c (T^a) is the causal (anti-causal) time-ordering operator. The physical quantity of interest is the correlation function $G^<$ since it corresponds in the equal time limit to the density $\lim_{t_{1'}\to t_1} G^<(1, 1') = (+i)\rho(\mathbf{x}_1, \mathbf{x}_{1'}, t)$. However, the four Green's functions are related through equations of motion (Kadanoff-Baym equations) for the correlation $G^{<,>}$ and the retarded and advanced $G^{\pm}$ functions (the retarded and advanced Green's functions are defined via $G^{+,-} = G^c - G^{<,>} = G^{>,<} - G^a$). From the Kadanoff-Baym equations one obtains a kinetic equation for the correlation function $G^<$:

$$DG^< - G^< D^* - \left(\mathrm{Re}\,\Sigma^+ G^< - G^< \mathrm{Re}\,\Sigma^+\right) - \left(\Sigma^< \mathrm{Re}\,G^+\right.$$
$$\left. - \mathrm{Re}\,G^+ \Sigma^<\right) = \frac{1}{2}\left(\Sigma^> G^< + G^< \Sigma^> - \Sigma^< G^> - G^> \Sigma^<\right). (8)$$

Here $D = -i\partial_{x_1}/2M$ is the Schrödinger operator or, in a relativistic framework, the Dirac operator ($D = i\gamma_\mu\partial_{x_1} - M$) and $\Sigma^{<,>,\pm}$ are the self-energies. The introduction of retarded and advanced functions allows to interpret the real part of the retarded self-energy as a mean field while the imaginary part describes the absorption or finite life times of quasi-particles (dressed nucleons) [70]. The self-energy Σ contains all higher-order correlations and couples the one-body kinetic equation (8) to the corresponding equations for the two- and 3-body densities and so forth. This requires to truncate the Dyson-Schwinger hierarchy which is usually done at the two-body level and leads to the ladder approximation for the T-matrix, i.e. the Bethe-Salpeter equation.

The formal structure of the kinetic equation (8) is complex and one should solve (8) together with the corresponding kinetic equations for $G^{\pm}$ which describe the *spectral properties* of the phase space distribution. Simultaneously, the self-energies should be derived for arbitrary non-equilibrium situations [70]. A solution of the full self-consistency problem has not yet been achieved. In practice, one applies further approximations. The most important ones are the *gradient expansion* (a semi-classical approximation to first order in $\hbar$) and the *quasi-particle approximation* which sets the particles on mass shell. The result is a Boltzmann-type transport equation, which is known as the Boltzmann-Uheling-Uhlenbeck (BUU) transport equation [71]. In its relativistic form the (R)BUU equation reads

$$\left[\left(m_D^* \partial_x^\mu m_D^* - k^{*\nu}\partial_x^\mu k_\nu^*\right)\partial_\mu^k - \left(m_D^* \partial_k^\mu m_D^* - k^{*\nu}\partial_k^\mu k_\nu^*\right)\partial_\mu^x\right] f$$
$$= \frac{1}{2(2\pi)^8}\int \frac{\mathrm{d}^3 k_2}{E_{k_2}^*}\frac{\mathrm{d}^3 k_3}{E_{k_3}^*}\frac{\mathrm{d}^3 k_4}{E_{k_4}^*}\, W\, \delta^4\left(k + k_2 - k_3 - k_4\right)$$
$$\times\left[f_3 f_4 (1-f)(1-f_2) - f f_2 (1-f_3)(1-f_4)\right], \qquad (9)$$

which describes the phase space evolution of the 1-particle distribution $f(\mathbf{x}, \mathbf{k}, t)$ under the influence of the mean field (which enters via the real part of the self-energy, i.e. via $m_D^* = M - \Sigma_S$ and $k^{*\mu} = k^\mu - \Sigma^\mu$) and binary collisions determined by the transition amplitude $W = m_D^{*4}|T(kk_2|k_3k_4)|^2$. Final-state Pauli blocking is accounted for by the blocking factors $(1 - f_i)$ in (9) with $f_i = f(\mathbf{x}, \mathbf{k}_i, t)$. The physical parameters entering into the kinetic equation are the mean field, i.e. the nuclear EOS, and elementary cross-sections for 2-particle scattering processes. Thus, one can test the high-density behavior of the nuclear EOS in heavy-ion collisions and the in-medium modifications of cross-sections, which also influence the stopping properties of the colliding system. Above the pion threshold where inelastic processes start to play an important role, eq. (9) becomes a *coupled-channel* problem for nucleonic, nucleon resonance and mesonic degrees of freedom. The collision integral, i.e. the right-hand side of eq. (9) has to be extended for the corresponding inelastic and absorptive processes and the new degrees of freedom must be propagated in their mean fields. In practice, the transport equation is solved within the *test particle method* which describes the phase space distribution f as an incoherent sum of point-like quasi-particles [71] or static Gaussians [72] which propagate on classical trajectories. Relativistic formulations of the two methods were developed in refs. [73] and [74].

4.2 Quantum molecular dynamics (QMD)

An alternative approach to the kinetic BUU equation is quantum molecular dynamics (QMD) [75–77]. QMD is a N-body approach which simulates heavy-ion reactions on an event-by-event basis taking fluctuations and correlations into account. The QMD equations are formally derived from the assumption that the N-body wave function Φ can be represented as the direct product of single coherent states $\Phi = \prod_i \phi_i$ which are described by Gaussian wave packets. Anti-symmetrization is *not* taken into account. A Wigner transformation yields the corresponding phase space representation of Φ. The equations of motion of the many-body system are obtained by the variational principle starting from the action $S = \int \mathcal{L}[\Phi, \Phi^*]$ (with the Lagrangian functional $\mathcal{L} = \langle\Phi|i\hbar\frac{\mathrm{d}}{\mathrm{d}t} - H|\Phi\rangle$). The Hamiltonian H contains a kinetic contribution and mutual two-body interactions V_{ij}. The variational principle leads finally to classical equations of motion for the generalized coordinates $\mathbf{q}_i$ and $\mathbf{k}_i$ of the Gaussian wave packets

$$\dot{\mathbf{q}}_i = \frac{\mathbf{k}_i}{m} + \nabla_{\mathbf{k}_i}\sum_{j\neq i}\langle V_{ij}\rangle = \nabla_{\mathbf{k}_i}\langle H\rangle,$$

$$\dot{\mathbf{k}}_i = -\nabla_{\mathbf{q}_i}\sum_{j\neq i}\langle V_{ij}\rangle = \nabla_{\mathbf{q}_i}\langle H\rangle.$$

The two-body interaction V_{ij} can, e.g., be taken from BHF calculations [77] or from local Skyrme forces which are usually supplemented by an empirical momentum dependence in order to account for the energy dependence of the optical nucleon-nucleus potential [75]. Binary collisions are treated in the same way as in BUU models. Furthermore, there exist relativistic extensions, *i.e.* RQMD and the UrQMD model which has been developed to simulate heavy-ion collisions at ultra-relativistic energies [76,78].

4.3 Antisymmetrized molecular dynamics (AMD/FMD)

An extension of QMD, in particular designed for low energies, are the antisymmetrized molecular dynamics (AMD) [79] and fermionic molecular dynamics (FMD) approaches [80]. In contrast to conventional QMD, the interacting system is represented by an *antisymmetrized* many-body wave function consisting of single-particle states which are localized in phase space. The equations of motion for the parameters characterizing the many-body state (*e.g.*, position, momentum, width and spin of the particles) are derived from a quantum variational principle. The models are designed to describe ground-state properties of nuclei as well as heavy-ion reactions at low energies (see also the contribution by Ono *et al.* [81]).

4.4 Off-shell transport

Essential for the validity of the classical equations of motion is the quasi-particle approximation (QPA) which assumes that the spectral strength of a hadron is concentrated around its quasi-particle pole. Particle widths can, however, dramatically change in a dense hadronic environment. To first order in density, the in-medium width of a hadron in nuclear matter can be estimated by the collision width $\Gamma^{\mathrm{tot}} = \Gamma^{\mathrm{vac}} + \Gamma^{\mathrm{coll}}$, $\Gamma^{\mathrm{coll}} = \gamma v \sigma \rho_B$ with v the hadron velocity relative to the surrounding matter and σ the total hadron-nucleon cross-section. A consistent treatment of the off-shell dynamics, *i.e.* a solution of the quantum evolution equations for the correlation functions $G^{<,>}$ has up to now only been performed for toy models and simplified geometries [82,83] or in first-order gradient approximation leading to an extended quasi-particle picture [84]. Comparing the non-local extension of BUU with standard simulations a visible effect of non-local correlations is seen and a better agreement with measured charge density distributions [85] or particle spectra [86] due to the virial corrections has been found. To develop a consistent lattice quantum transport for non-uniform systems and realistic interactions will be one of the future challenges in theoretical heavy-ion physics.

On the other hand, substantial progress has been made in recent years to map part of the off-shell dynamics on a modified test particle formalism [87,88]. This allows to apply off-shell dynamics, although in a simplified form, to the complex space-time evolution of a heavy-ion reaction. The present knowledge of off-shell matrix elements is, however, rather limited and theoretical investigations are scarce [89]. The off-shell T-matrix has been used in order to calculate the duration and non-locality of a nucleon-nucleon collision [90]. The question to what degree a depletion of the Fermi surface due to particle-hole excitations

and the high-momentum tails of the nuclear spectral functions will affect subthreshold particle production is not so obvious to answer. The high-momentum tails correspond to deeply bound states which are off-shell and to treat such states in a standard transport approach like on-shell quasi-particles would violate energy-momentum conservation. Energy-momentum conservation can be achieved consistently by the non-local kinetic theory [91] taking into account first-order off-shell effects. The contribution of the nuclear short-range correlations to subthreshold K^+ production in $p + A$ reactions have, *e.g.*, been estimated in [92]. The removal energy for a high-momentum state compensates the naively expected energy gain and the short-range correlations do therefore not significantly contribute to subthreshold particle production [92]. The situation changes, however, when the medium is heated up and high-momentum particles become on-shell or when the spectral distributions of the produced hadrons themselves are broadened.

5 Constraints from heavy-ion collisions

5.1 Flow and stopping

One of the most important observables to constrain the nuclear forces and the underlying EOS at supra-normal densities is the collective nucleon flow [93]. It can be characterized in terms of anisotropies of the azimuthal emission pattern. Expressed in terms of a Fourier series

$$\frac{\mathrm{d}N}{\mathrm{d}\phi} \propto 1 + 2v_1 \cos(\phi) + 2v_2 \cos(2\phi) + \ldots \qquad (10)$$

this allows a transparent interpretation of the coefficients v_1 and v_2. The dipole term v_1 arises from a collective sideward deflection of the particles in the reaction plane and characterizes the transverse flow in the reaction plane. The second harmonics describes the emission pattern perpendicular to the reaction plane. For negative v_2 one has a preferential out-of-plane emission. The phenomenon of an out-of-plane enhancement of particle emission at midrapidity is called *squeeze-out*.

The transverse flow v_1 has been found to be sensitive to the EOS and, in particular in peripheral reactions, to the momentum dependence of the mean field [57,58]. The elliptic flow v_2, in contrast, is very sensitive to the maximal compression reached in the early phase of a heavy-ion reaction. The crossover from preferential out-off-plane flow ($v_2 < 0$) to preferential in-plane flow ($v_2 > 0$) around 4–6 AGeV has also led to speculations about a phase transition in this energy region which goes along with a softening of the EOS [94].

The present situation between theory and experiment is illustrated in fig. 10 (from [95]). The BUU studies from Danielewicz *et al.* and the Giessen group (Larionov *et al.*) investigated the EOS dependence while Persram *et al.* find a sensitivity of v_2 to the medium dependence of the NN cross-sections. Finally, non-equilibrium effects have been investigated at the level of the effective interaction

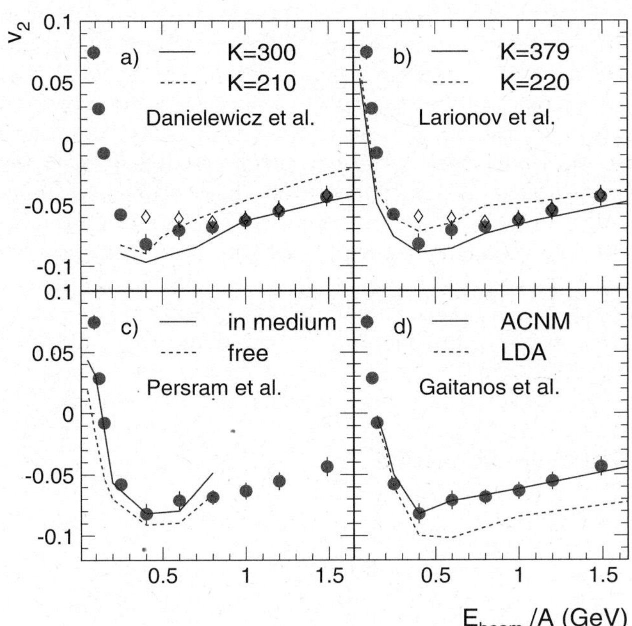

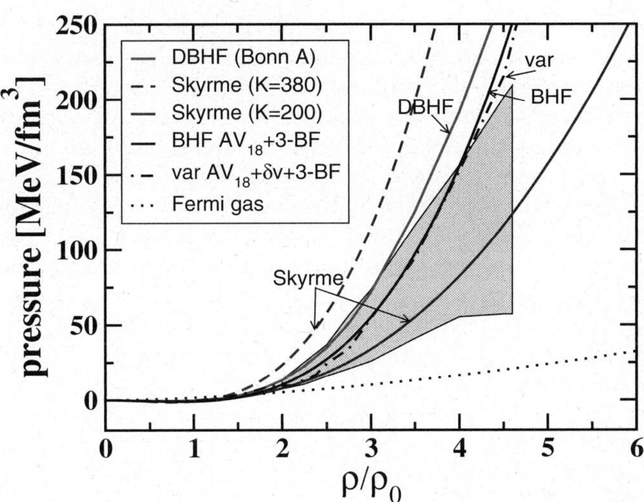

Fig. 11. Constraints on the nuclear EOS from heavy-ion flow data. The shaded area shows the pressure density which is compatible with heavy-ion flow data according the analysis on [97]. The equations of state from the models shown in fig. 3 are displayed.

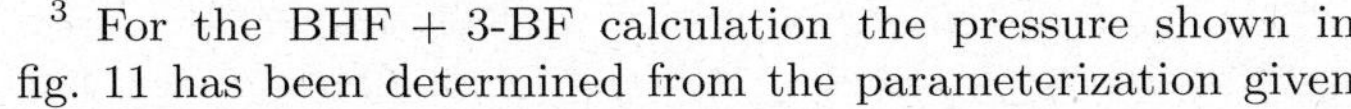

Fig. 10. Elliptic flow excitation function at SIS energies. Various theoretical studies using different EOSs (a,b), or different cross-sections (c) or DBHF mean fields in the LDA approach and including further non-equilibrium effects (ACNM) (d) are compared to FOPI data (symbols). The figure is taken from [95].

in [58,96]. It has been found that the local phase space anisotropies of the pre-equilibrium stages of the reactions reduce the repulsion of the mean field and soften the corresponding EOS which allows a good description of the v_2 data using microscopic DBHF mean fields (Gaitanos *et al.*). However, fig. 10 also demonstrates that v_2 is generated by the interplay of the mean field and binary collisions which makes it difficult to extract exclusive information on the EOS from the data. Here certainly furthergoing studies are required.

The following figure (fig. 11) is based on the studies of Danielewicz *et al.* [97]. It summarizes the status obtained within this model in terms of a band that represents the constraints from collective flow data. It is obtained from a compilation of analyses of sideward and elliptic anisotropies, studied at energies ranging from low SIS ($E_{\mathrm{lab}} \simeq 0.2$–$2\,\mathrm{AGeV}$) up to top AGS energies ($E_{\mathrm{lab}} \simeq 2$–$11\,\mathrm{AGeV}$). The conclusion of this study was that both, super-soft equations of state ($K = 167\,\mathrm{MeV}$) as well as hard EOSs ($K > 300\,\mathrm{MeV}$), are ruled out by the data. At SIS energies, existing flow data are consistent with a soft EOS [98,57] as, *e.g.*, the soft Skyrme EOS. In the models used by Danielewicz *et al.* [57,97], sideward flow favors a rather soft EOS with $K = 210\,\mathrm{MeV}$ while the development of the elliptic flow requires slightly higher pressures. The BHF and variational calculations including 3-body forces[3] fit well into the constrained area up to $4\rho_0$. At higher den-

sities the microscopic EOSs, also DBHF, tend to be too repulsive. However, conclusions from flow data are generally complicated by the interplay of the compressional part of the nuclear EOS and the momentum dependence of the nuclear forces. A detailed comparison to v_1 and v_2 data from FOPI [99,100] and KaoS [101] for v_1 and v_2 below $1\,\mathrm{AGeV}$ favors again a relatively soft EOS with a momentum dependence close to that obtained from microscopic DBHF calculations [57,58,102]. In fig. 11, the microscopic DBHF EOS ($K = 230\,\mathrm{MeV}$) lies at the upper edge of the boundary but is still consistent with it in the density range tested at SIS energies, *i.e.* up to at most $3\rho_0$. This fact is further consistent with the findings of Gaitanos *et al.* [58,102] where a good description of v_1 and v_2 data at energies between 0.2 and $0.8\,\mathrm{AGeV}$ has been found in RBUU calculations based on DBHF mean fields. As pointed out in [58,96,102] it is thereby essential to account for non-equilibrium effects and the momentum dependence of the forces which softens the EOS compared to the equilibrium case (shown in fig. 11).

As can be seen from fig. 10, not only the nuclear EOS, but also the cross-sections for elementary 2-particle scattering influences the collective dynamics, in particular, the degree of stopping and hence the maximum compression achieved in the fireball region. A challenge in this context is to reach a quantitative understanding of the recently observed strong correlations between maximum side flow v_1 and maximum stopping in the two excitation functions [69] (see also the contribution by A. Andronic *et al.* [103]). Most collective flow analyses performed so far were based on free cross-sections which works astonishingly well from a practical point of view. However, within a consistent picture one should treat the in-medium ef-

---

[3] For the BHF + 3-BF calculation the pressure shown in fig. 11 has been determined from the parameterization given in [19] which is based on the Urbana IX 3-BF different to that used in [18].

fects in both, the real (nuclear EOS) and the imaginary part (cross-sections) of the interaction, on the same footing. Many-body calculations (BHF/DBHF) predict an essential reduction of the elastic NN cross-section with increasing baryon density [77,104,89] and in [105] a similar reduction was proposed for the inelastic channels in order to describe pion multiplicities in the 1–2 AGeV region. One can, therefore, expect observable signals in heavy-ion collisions. In fact, recent QMD studies of stopping and transparency observables have shown that the data can be reproduced when the free cross-section is reduced by a factor of 0.5 [106]. These findings are supported by transport calculations using microscopic in-medium cross-sections [107,108]. Therefore, for a reliable extraction of the high-density nuclear EOS one should account for in-medium effects not only in the potential but also in the cross-sections.

5.1.1 Isospin dependence of the EOS

Another important aspect of heavy-ion collisions is the investigation of the density dependence of the EOS for asymmetric matter. There exist abundent studies on this sector, either non-relativistically or relativistically.

The momentum dependence of the isovector potential, fig. 9, which is also closely related to the proton/neutron mass splitting of both, the non-relativistic m^*_{NR} and the Dirac m^*_D effective mass, is one of the key questions which can be addressed by nuclear reactions induced by neutron-rich nuclei at RIA energies. Transverse and elliptic flow patterns as well p/n rapidity dsitributions have been suggested as possible observables to investigate the momentum dependence and the p/n mass splitting [109–111].

Promising observables to pin down the density dependence of the symmetry energy are the iso-scaling behavior of fragment yields and the isospin diffusion in asymmetric colliding systems. In both cases recent NSCL-MSU data in combination with transport calculations are consistent with a value of $E_{\rm sym} \approx 31$ at ρ_0 and rule out extremely "stiff" and "soft" density dependences of the symmetry energy [112,113] (see also the contribution IV.1 by M. Di Toro et al. [114]). The same value has been extracted [115] from low-energy elastic and (p,n) charge exchange reactions on isobaric analog states, i.e. $p(^6\mathrm{He},{}^6\mathrm{Li}^*)n$ measured at the HMI. Such a behavior is also consistent with the predictions from many-body theory [15,26]. Also the p/n ratio at mid-rapidity has been found to be sensitive to the high-density behavior of the nuclear symmetry energy [116].

In relativistic approaches, large attractive scalar and repulsive vector fields are required by Dirac phenomenology in order to describe simultaneously the central potential and the strong spin-orbit force in finite nuclei [8,9, 12]. The situation is, however, less clear in the iso-vector sector. There exist different possibilities to reproduce the same value of the a_4 coefficient (4): a) by only an iso-vector vector ρ field like in most RMF models (NL3 etc.), or b) by accounting for an additional iso-vector scalar δ field. Due to competing effects between attractive (scalar δ) and repulsive (vector ρ) fields, both alternatives can be fitted to the same empirical a_4 parameter. However, the inclusion of δ field leads to an essentially different high-density behavior of the symmetry energy [117]. The scalar δ field is suppressed at high densities, whereas the vector field is proportional to the baryon density which makes the symmetry energy stiffer at supra-normal densities. Recent transport studies have shown that these subtle relativistic effects can be observed in the intermediate-energy range by means of collective isospin flow, particle ratios and imbalance ratios of different particle species (protons, neutrons, pions and kaons) [117,111,44]. However, due to the lack of precise experimental data, no definitive conclusions could be drawn so far.

5.2 Particle production

5.2.1 Pions

With the start of the first relativistic heavy-ion programs the hope was that particle production would provide a direct experimental access to the nuclear EOS [118]. At twice saturation density which is reached in the participant zone of the reactions without additional compression, the difference between the soft and hard EOS shown in fig. 3 is about 13 MeV in binding energy. If the matter is compressed up to $3\rho_0$ the difference is already ~ 55 MeV. It was expected that the compressional energy should be released into the creation of new particles, primarily pions, when the matter expands [118]. However, pions have large absorption cross-sections and they turned out not to be suitable messengers of the compression phase. They undergo several absorption cycles through nucleon resonances ($N\pi \leftrightarrow \Delta$) and freeze out at final stages of the reaction and at low densities. Hence pions lose most of their knowledge on the compression phase and are not really sensitive probes for the stiffness of the EOS [119]. However, they carry information on the isotopic composition of the matter which is to some extent conserved until freeze-out. The final π^-/π^+ ratio was found to be sensitive to the initial n/p composition of the matter which, on the other hand, is influenced by the isospin dependence of the nuclear forces [120,121]. In [63] a reduction of the π^-/π^+ ratio was found when the δ-meson was included in the RMF approach. The effects are, however, moderate, i.e. at the 10–20% level, and most pronounced at extreme phase space regions, e.g. at the high-energy tails of p_t spectra [63,121,122]. Systematic measurements as, e.g., from the FOPI Collaboration may help to constrain the isospin dependence by pionic observables.

5.2.2 Kaons

After pions turned out to fail as suitable messengers, K^+-mesons were suggested as promising tools to probe the nuclear EOS, almost 20 years ago [124]. The cheapest way to produce a K^+-meson is the reactions $NN \longrightarrow N\Lambda K^+$ which has a threshold of $E_{\rm lab} = 1.58$ GeV kinetic energy for the incident nucleon. When the incident energy per nucleon in a heavy-ion reactions is below this values one speaks about *subthreshold* kaon production. Subthreshold kaon production is in particular interesting since it ensures that the kaons originate from the high-density phase of the reaction. The missing energy has to be provided either by the Fermi motion of the nucleons or by energy accumulating multi-step reactions. Both processes exclude significant distortions from surface effects if one goes sufficiently far below threshold. In combination with the long mean free path subthreshold K^+ production is an ideal tool to probe compressed nuclear matter in relativistic heavy-ion reactions.

Already in the first theoretical investigations by transport models it was noticed that the K^+ yield reacts rather sensitive to the EOS [125–127]. Both, in non-relativistic QMD calculations based on soft/hard Skyrme forces [125, 126, 128] and in RBUU [127, 129] with soft/hard versions of the (non-linear) $\sigma\omega$ model the K^+ yield was found to be about a factor 2–3 larger when a soft EOS is applied compared to a hard EOS. At that time, the available data favored a soft equation of state [126, 127, 129]. However, at that stage the theoretical calculations were still burdened with large uncertainties. First of all, it was noticed [125, 126] that the influence of the repulsive momentum-dependent part of the nuclear interaction leads to a strong suppression of the kaon abundances which made a quantitative description of the available data more difficult. Moreover, at that time the pion-induced reaction channels $\pi B \longrightarrow Y K^+$ have not yet been taken into account. These additional channels which contribute up to 30–50% to the total yield enabled to explain the measured yields with realistic momentum-dependent interactions [128, 130]. A breakthrough was achieved when the COSY-11 Collaboration measured the $pp \longrightarrow p K^+ \Lambda$ reactions at threshold [131] which constrains the strangeness production cross-sections $NN \longrightarrow NK^+Y$. Within the last decade the KaoS Collaboration has performed systematic measurements of the K^+ production far below threshold [119, 132, 133]. Based on the new data situation, the question if valuable information on the nuclear EOS can be extracted has been revisited and it has been shown that subthreshold K^+ production provides indeed a suitable and reliable tool for this purpose [134–136].

Figure 12 compares measured K^+ multiplicities as a function of the number of participating nucleons, $A_{\rm part}$, in Au+Au, Ni+Ni, C+Au and C+C reactions at 1 AGeV to QMD calculation using a soft/hard momentum-dependent Skyrme force [123]. This figure demonstrates thereby the interplay between $A_{\rm part}$, system size and EOS. A significant dependence of the kaon multiplicities on the nuclear EOS requires a large amount of collectivity which is easiest reached in central reactions of heavy-mass systems. Consequently, the EOS dependence is most pronounced in central Au+Au reactions. Also in Ni+Ni effects are still sizable while the small C+C system is completely insensitive to the nuclear EOS even in most central reactions. The data available for Au+Au and Ni+Ni support the soft EOS. Of particular interest is, in this context, the

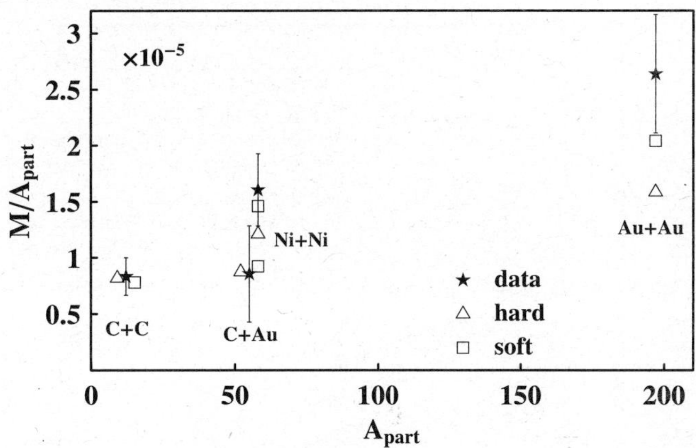

Fig. 12. K^+ multiplicities in inclusive C+C, Ni+Ni, Au+Au and C+Au reactions at 1 AGeV. QMD calculations using a hard/soft nuclear EOS are compared to KaoS data [123]. The figure is taken from [123].

asymmetric C+Au system: although in central C+Au reactions the number of participants is comparable to that in Ni+Ni, the K^+ yield does not depend on the EOS. This indicates again that a sensitivity to the EOS is not only a question of $A_{\rm part}$ but also of the compression which can be reached by the colliding system.

The next step is to consider now the energy dependence of the EOS effect. It is expected to be most pronounced far below threshold because there the highest degree of collectivity, reflected in multi-step collisions, is necessary to overcome the production thresholds. The effects become even more evident when the ratio R of the kaon multiplicities obtained in Au+Au over C+C reactions (normalized to the corresponding mass numbers) is considered [135, 133]. Such a ratio has, moreover, the advantage that possible uncertainties which might still exist in the theoretical calculations should cancel out to a large extent. This ratio is shown in fig. 13. Both, soft and hard EOS, show an increase of R with decreasing energy. However, this increase is much less pronounced when the stiff EOS is employed. The strong increase of R can be directly related to a higher compressibility of nuclear matter. The comparison with the experimental data from KaoS [133], where the increase of R is even more pronounced, strongly favors a soft equation of state. These findings were confirmed by independent IQMD transport calculations of the Nantes group [136]. Both, QMD and IQMD included also a repulsive kaon-nucleon potential as predicted by chiral perturbation theory [134]. The shaded area in the figure can be taken as the existing range of uncertainty in the theoretical model description of the considered observable. To estimate the stability of the conclusions, the IQMD calculations have been repeated with an alternative set of $N\Delta; \Delta\Delta \mapsto NYK^+$ cross-sections[4] which are almost one order of magnitude smaller than those used originally, but the EOS dependence remained stable [137].

---

[4] Cross-sections which involve Δ resonances in the initial or final states are not constrained by measurements.

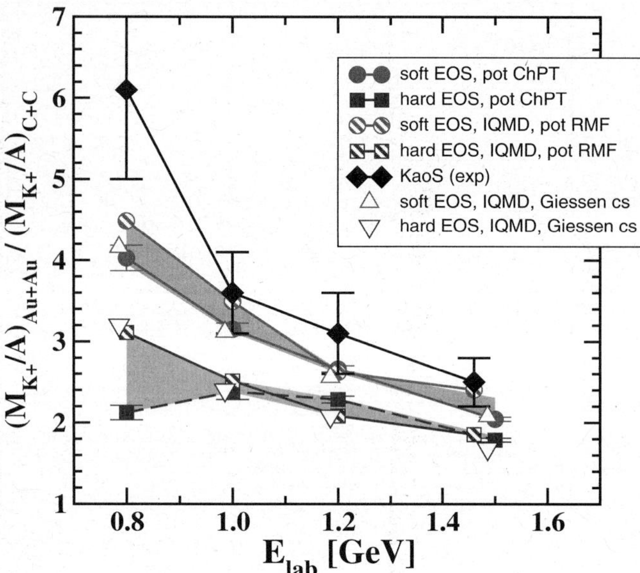

Fig. 13. Excitation function of the ratio R of K^+ multiplicities obtained in inclusive Au+Au over C+C reactions. QMD [135] and IQMD calculations [136] are compared to the KaoS data [133]. The shaded area indicates the range of uncertainty in the theoretical models. In addition, IQMD results based on an alternative set of elementary K^+ production cross-sections are shown.

6 Constraints from neutron stars

Measurements of "extreme" values, like large masses or radii, huge luminosities etc. as provided by compact stars offer good opportunities to gain deeper insight into the physics of matter under extreme conditions. There has been substantial progress in recent time from the astrophysical side.

The most spectacular observation was probably the recent measurements on PSR J0751+1807, a millisecond pulsar in a binary system with a helium white dwarf secondary, which implies a pulsar mass of $2.1 \pm 0.2(^{+0.4}_{-0.5})\,M_\odot$ with 1σ (2σ) confidence [138]. Therefore, a reliable EOS has to describe neutron star (NS) masses of at least $1.9\,M_\odot$ (1σ) in a strong, or $1.6\,M_\odot$ (2σ) in a weak interpretation. This condition limits the softness of EOS in NS matter. One might therefore be worried about an apparent contradiction between the constraints derived from neutron stars and those from heavy-ion reactions. While heavy-ion reactions favor a soft EOS, PSR J0751+1807 requires a stiff EOS. The corresponding constraints are, however, complementary rather than contradictory. Intermediate energy heavy-ion reactions, *e.g.* subthreshold kaon production, constrains the EOS at densities up to 2–$3\rho_0$ while the maximum NS mass is more sensitive to the high-density behavior of the EOS. Combining the two constraints implies that the EOS should be *soft at moderate densities and stiff at high densities*. Such a behavior is predicted by microscopic many-body calculations (see fig. 3). DBHF, BHF or variational calculations, typically, lead to maximum NS masses between 2.1–$2.3\,M_\odot$ and are therefore in accordance with PSR J0751+1807 [139].

There exist several other constraints on the nuclear EOS which can be derived from observations of compact stars, see, *e.g.*, [139–141]. Among these, the most promising one is the Direct Urca (DU) process which is essentially driven by the proton fraction inside the NS [142]. DU processes, *e.g.* the neutron β-decay $n \to p + e^- + \bar{\nu}_e$, are very efficient regarding their neutrino production, even in superfluid NM [143,144], and cool NSs too fast to be in accordance with data from thermally observable NSs. Therefore, one can suppose that no DU processes should occur below the upper mass limit for "typical" NSs, *i.e.* $M_{DU} \geq 1.5\,M_\odot$ $(1.35\,M_\odot$ in a weak interpretation). These limits come from a population synthesis of young, nearby NSs [145] and masses of NS binaries [138].

7 Summary and outlook

The quest for the nuclear equation of state is one of the longstanding problems in physics which has a history of more than 50 years in nuclear structure. Since about 30 years, one tries to attack this question with heavy-ion reactions. The exploration of the limits of stability, *i.e.* the regimes of extreme isospin asymmetry, is a relatively new field with rapidly growing importance in view of the forthcoming generation of radioactive beam facilities.

The status of theoretical models which make predictions for the EOS can roughly be summarized as follows: phenomenological density functionals such as the Skyrme, Gogny or relativistic mean-field models provide high precision fits to the nuclear chart but extrapolations to supranormal densities or to the limits of stability are highly uncertain. A more controlled way is provided by effective-field theory approaches which became quite popular in recent time. Effective chiral field theory allows, *e.g.*, a systematic generation of two- and many-body nuclear forces. However, these approaches are low-momentum expansions and, when applied to the nuclear many-body problem, low-density expansions. *Ab initio* calculations for the nuclear many-body problem such as variational or Brueckner calculations have reached a high degree of sophistication and can serve as guidelines for the extrapolation to the regimes of high-density and/or large isospin asymmetry. Possible future developments are to base such calculations on modern EFT potentials and to achieve a more consistent treatment of two- and three-body forces.

If one intends to constrain these models by nuclear reactions one has to account for the reaction dynamics by semi-classical transport models of a Boltzmann or molecular-dynamics type. Suitable observables which have been found to be sensitive to the nuclear EOS are directed and elliptic collective flow patterns and particle production, in particular kaon production, at higher energies. Heavy-ion data suggest that the EOS of symmetric nuclear matter shows a soft behavior in the density regime between one to about three times nuclear saturation density, which is consistent with the predictions from many-body calculations. Conclusions on the EOS are, however, complicated by the interplay between the density and the momentum dependence of the nuclear mean field. Data which constrain the isospin dependence of the mean field are still scarce. Promising observables are isospin diffusion, iso-scaling of intermediate mass fragments and particle ratios (π^+/π^- and eventually K^+/K^0). Here, the situation will certainly improve when the forthcoming radioactive beam facilities will be operating. This will also allow to measure the optical isospin potential in $p+A$ and $A+A$ reactions and to obtain more information on the symmetry energy and the proton/neutron mass splitting in asymmetric matter. From the theoretical side it will be unavoidable to invest significant efforts towards the development of quantum transport models with consistent off-shell dynamics.

We would like to thank K. Morawetz, T. Gaitanos and M. Di Toro for fruitful discussions.

References

1. GSI Conceptual Design Report, http://www.gsi.de/GSI-Future.
2. RIA homepage, http://www.orau.org/ria.
3. H.A. Gustafsson *et al.*, Phys. Rev. Lett. **52**, 1590 (1984).
4. J. Decharge, D. Gogny, Phys. Rev. C **21**, 1568 (1980).
5. M. Kleban, B. Nerlo-Pomorska, J.F. Berger, J. Decharge, M. Girod, S. Hilaire, Phys. Rev. C **65**, 024309 (2002).
6. B. Cochet, K. Bennaceur, J. Meyer, P. Bonche, T. Duguet, Int. J. Mod. Phys. E **13**, 187 (2004).
7. P.-G. Reinhard, M. Bender, Lect. Notes Phys. **641**, 249 (2004).
8. P. Ring, Prog. Part. Nucl. Phys. **73**, 193 (1996); Lect. Notes Phys. **641**, 175 (2004).
9. B.D. Serot, J.D. Walecka, Int. J. Mod. Phys. E **6**, 515 (1997).
10. R.J. Furnstahl, Lect. Notes Phys. **641**, 1 (2004).
11. M. Lutz, B. Friman, Ch. Appel, Phys. Lett. B **474**, 7 (2000).
12. P. Finelli, N. Kaiser, D. Vretenar, W. Weise, Eur. Phys. J. A **17**, 573 (2003); Nucl. Phys. A **735**, 449 (2004).
13. D. Vretenar, W. Weise, Lect. Notes Phys. **641**, 65 (2004).
14. V.R. Pandharipande, R.B. Wiringa, Rev. Mod. Phys. **51**, 821 (1979).
15. A. Akmal, V.R. Pandharipande, D.G. Ravenhall, Phys. Rev. C **58**, 1804 (1998).
16. K.A. Brueckner, J.L. Gammel, Phys. Rev. **107**, 1023 (1958).
17. M. Jaminon, C. Mahaux, Phys. Rev. C **40**, 354 (1989).
18. W. Zuo, A. Lejeune, U. Lombardo, J.F. Mathiot, Nucl. Phys. A **706**, 418 (2002).
19. X.R. Zhou, G.F. Burgio, U. Lombardo, H.-J. Schulze, W. Zuo, Phys. Rev. C **69**, 018801 (2004).
20. B. ter Haar, R. Malfliet, Phys. Rep. **149**, 207 (1987).
21. R. Brockmann, R. Machleidt, Phys. Rev. C **42**, 1965 (1990).
22. F. de Jong, H. Lenske, Phys. Rev. C **58**, 890 (1998).
23. T. Gross-Boelting, C. Fuchs, A. Faessler, Nucl. Phys. A **648**, 105 (1999).
24. E. Schiller, H. Müther, Eur. Phys. J. A **11**, 15 (2001).
25. C. Fuchs, Lect. Notes Phys. **641**, 119 (2004).
26. E. van Dalen, C. Fuchs, A. Faessler, Nucl. Phys. A **744**, 227 (2004); Phys. Rev. C **72**, 065803 (2005).
27. H. Müther, A. Polls, Prog. Part. Nucl. Phys. **45**, 243 (2000).
28. W.H. Dickhoff, C. Barbieri, Prog. Part. Nucl. Phys. **52**, 377 (2004).
29. J. Carlson, J. Morales, V.R. Pandharipande, D.G. Ravenhall, Phys. Rev. C **68**, 025802 (2003).
30. B.D. Serot, J.D. Walecka, Adv. Nucl. Phys. **16**, 1 (1986).
31. J. Boguta, Phys. Lett. B **109**, 251 (1982).
32. C. Fuchs, H. Lenske, H.H. Wolter, Phys. Rev. C **52**, 3043 (1995); H. Lenske, C. Fuchs, Phys. Lett. B **345**, 355 (1995).
33. F. Hofmann, C.M. Keil, H. Lenske, Phys. Rev. C **64**, 034314 (2001); C.M. Keil, F. Hofmann, H. Lenske, Phys. Rev. C **61**, 064309 (2000).
34. S. Typel, H.H. Wolter, Nucl. Phys. A **656**, 331 (1999).
35. T. Nikšić, D. Vretenar, P. Ring, Phys. Rev. C **66**, 064302 (2002).
36. R. Machleidt, K. Holinde, Ch. Elster, Phys. Rep. **149**, 1 (1987).
37. M.K. Banerjee, J.A. Tjon, Phys. Rev. C **58**, 2120 (1998); Nucl. Phys. A **708**, 303 (2002).
38. U. van Klock, Phys. Rev. C **49**, 2932 (1994).
39. D.R. Entem, R. Machleidt, Phys. Rev. C **66**, 014002 (2002); **68**, 041001 (2003).
40. E. Epelbaum, W. Glöckle, U.-G. Meissner, Nucl. Phys. A **747**, 362 (2005).
41. S.K. Bogner, T.T.S. Kuo, A. Schwenk, Phys. Rep. **386**, 1 (2003).
42. S.K. Bogner, A. Schwenk, R.J. Furnstahl, A. Nogga, Nucl. Phys. A **763**, 59 (2005).
43. G.A. Lalazissis, J. König, P. Ring, Phys. Rev. C **55**, 540 (1997).
44. V. Baran, M. Colonna, V. Greco, M. Di Toro, Phys. Rep. **410**, 335 (2005).
45. C.J. Horowitz, contribution VI.3, this topical issue.
46. S. Shlomo, V.M. Kolomietz, G. Colò, contribution II.2, this topical issue.
47. C. Mahaux, P.F. Bortignon, R.A. Broglia, C.H. Dasso, Phys. Rep. **120**, 1 (1985).
48. D. Lunney, J.M. Pearson, C. Thibault, Rev. Mod. Phys. **75**, 1021 (2003).
49. T. Frick, Kh. Gad, H. Müther, P. Czerski, Phys. Rev. C **65**, 034321 (2002).
50. E. van Dalen, C. Fuchs, A. Faessler, Phys. Rev. Lett. **95**, 022302 (2005).
51. J.M. Pearson, S. Goriely, Phys. Rev. C **64**, 027301 (2001).
52. D. Alonso, F. Sammarruca, Phys. Rev. C **67**, 054301 (2003); F. Sammarruca, W. Barredo, P. Krastev, Phys. Rev. C **71**, 064306 (2005).
53. C.J. Horowitz, B.D. Serot, Nucl. Phys. A **464**, 613 (1987).
54. H. Krivine, J. Treiner, O. Boghias, Nucl. Phys. A **336**, 155 (1980).
55. S. Goriely, E. Khan, Nucl. Phys. A **706**, 217 (2002).
56. S. Hama *et al.*, Phys. Rev. C **41**, 2737 (1990); E.D. Cooper *et al.*, Phys. Rev. C **47**, 297 (1993).
57. P. Danielewicz, Nucl. Phys. A **673**, 275 (2000).
58. T. Gaitanos, C. Fuchs, H.H. Wolter, A. Faessler, Eur. Phys. J. A **12**, 421 (2001).
59. P.K. Sahu, W. Cassing, U. Mosel, A. Ohnishi, Nucl. Phys. A **672**, 376 (2000).
60. H.F. Arellano, H.V. von Geramb, Phys. Rev. C **66**, 024602 (2002).
61. S. Typel, Phys. Rev. C **71**, 064301 (2005).

62. W. Zuo, L.G. Cao, B.A. Li, U. Lombardo, C.W. Shen, Phys. Rev. C **72**, 014005 (2005).
63. T. Gaitanos, M. Di Toro, S. Typel, V. Baran, C. Fuchs, V. Greco, H.H. Wolter, Nucl. Phys. A **732**, 24 (2004).
64. L.-W. Chen, C.M. Ko, B.-A. Li, Phys. Rev. C **72**, 064606 (2005).
65. J.A. McNeil, J.R. Shepard, S.J. Wallace, Phys. Rev. C **27**, 2123 (1983).
66. A.M. Lane, Nucl. Phys. **35**, 676 (1962).
67. R. Kozack, D.G. Madland, Phys. Rev. C **39**, 1461 (1989); Nucl. Phys. A **509**, 664 (1990).
68. FOPI Collaboration (F. Rami *et al.*), Phys. Rev. Lett. **84**, 1120 (2000).
69. FOPI Collaboration (W. Reisdorf *et al.*), Phys. Rev. Lett. **92**, 232301 (2004).
70. W. Botermans, R. Malfliet, Phys. Rep. **198**, 115 (1990).
71. G.F. Bertsch, S. Das Gupta, Phys. Rep. **160**, 190 (1988).
72. C. Gregoire, B. Remaud, F. Sebille, L. Vincet, Y. Raffay, Nucl. Phys. A **465**, 317 (1987).
73. B. Blättel, V. Koch, U. Mosel, Rep. Prog. Phys. **56**, 1 (1993).
74. C. Fuchs, H.H. Wolter, Nucl. Phys. A **589**, 732 (1995).
75. J. Aichelin, Phys. Rep. **202**, 233 (1991); C. Hartnack *et al.*, Eur. Phys. J. A **1**, 151 (1998).
76. S.A. Bass *et al.*, Prog. Part. Nucl. Phys. **41**, 225 (1998).
77. J. Jaenicke, J. Aichelin, N. Ohtsuka, R. Linden, A. Faessler, Nucl. Phys. A **536**, 201 (1992).
78. H. Sorge, H. Stöcker, W. Greiner, Ann. Phys. **192**, 266 (1989).
79. A. Ono, H. Horiuchi, T. Maruyama, A. Ohnishi, Prog. Theor. Phys. **87**, 1185 (1992).
80. H. Feldmeier, J. Schnack, Rev. Mod. Phys. **72**, 655 (2000).
81. A. Ono *et al.*, contribution III.3, this topical issue.
82. P. Danielewicz, Ann. Phys. **152**, 239; 305 (1984).
83. H.S. Köhler, Phys. Rev. C **51**, 3232 (1995).
84. K. Morawetz, P. Lipavský, V. Špička, Ann. Phys. **294**, 134 (2001).
85. K. Morawetz *et al.*, Phys. Rev. C **63**, 034619 (2001).
86. K. Morawetz *et al.*, Phys. Rev. Lett. **82**, 3767 (1999).
87. W. Cassing, S. Juchem, Nucl. Phys. A **665**, 377 (2000); **677**, 445 (2000).
88. J. Lehr, M. Effenberger, H. Lenske, S. Leupold, U. Mosel, Phys. Lett. B **483**, 324 (2000).
89. C. Fuchs, A. Faessler, M. El-Shabshiry, Phys. Rev. C **64**, 024003 (2001).
90. K. Morawetz, P. Lipavský, V. Špička, H.N. Kwong, Phys. Rev. C **59**, 3052 (1999).
91. P. Lipavský, V. Špička, K. Morawetz, Phys. Rev. E **59**, 1291 (1999).
92. M. Debowski *et al.*, Z. Phys. A **356**, 313 (1996).
93. N. Herrmann, J.P. Wessels, T. Wienold, Ann. Phys. **49**, 581 (1999).
94. P. Danielewicz *et al.*, Phys. Rev. Lett. **81**, 2438 (1998); C. Pinkenburg *et al.*, Phys. Rev. Lett. **83**, 1295 (1999).
95. FOPI Collaboration (A. Andronic *et al.*), Phys. Lett. B **612**, 173 (2005).
96. C. Fuchs, T. Gaitanos, Nucl. Phys. A **714**, 643 (2003).
97. P. Danielewicz, R. Lacey, W.G. Lynch, Science **298**, 1592 (2002).
98. A. Hombach, W. Cassing, S. Teis, U. Mosel, Eur. Phys. J. A **5**, 157 (1999).
99. FOPI Collaboration (A. Andronic *et al.*), Nucl. Phys. A **661**, 333c (1999); Phys. Rev. C **64**, 041604 (2001); **67**, 034907 (2003).
100. FOPI Collaboration (G. Stoicea *et al.*), Phys. Rev. Lett. **92**, 072303 (2004).
101. KaoS Collaboration (D. Brill *et al.*), Z. Phys. A **355**, 61 (1996).
102. T. Gaitanos, C. Fuchs, H.H. Wolter, Nucl. Phys. A **741**, 287 (2004); **650**, 97 (1999); C. Fuchs, T. Gaitanos, H.H. Wolter, Phys. Lett. B **381**, 23 (1996).
103. A. Andronic, J. Łukasik, W. Reisdorf, W. Trautmann, contribution II.3, this topical issue.
104. G.Q. Li, R. Machleidt, Phys. Rev. C **48**, 1702 (1993); **49**, 566 (1994).
105. A.B. Larionov, W. Cassing, S. Leupold, U. Mosel, Nucl. Phys. A **696**, 747 (2001).
106. FOPI Collaboration (B. Hong *et al.*), Phys. Rev. C **71**, 034902 (2005).
107. T. Gaitanos, C. Fuchs, H.H. Wolter, Phys. Lett. B **609**, 241 (2005).
108. P. Danielewicz, Acta Phys. Pol. B **33**, 45 (2002).
109. B.-A. Li, Nucl. Phys. A **708**, 365 (2002); Phys. Rev. C **69**, 064602 (2004).
110. B.-A. Li, C.B. Das, S. Das Gupta, Ch. Gale, Phys. Rev. C **69**, 011603 (2004).
111. J. Rizzo, M. Colonna, M. Di Toro, Nucl. Phys. A **732**, 202 (2004).
112. L.-W. Chen, C.M. Ko, B.-A. Li, Phys. Rev. Lett. **94**, 032701 (2005).
113. D.V. Shetty, S.J. Yennello, G.A. Souliotis, nucl-ex/0505011.
114. M. Di Toro, S.J. Yennello, Bao-An Li, contribution IV.1, this topical issue.
115. D.T. Khoa, H.S. Than, nucl-th/0502059; D.T. Khoa, W. von Oertzen, H.G. Bohlen, H.S. Than, nucl-th/0510048.
116. B.-A. Li, G.-C. Yong, W. Zuo, Phys. Rev. C **71**, 044604 (2005).
117. B. Liu, V. Greco, V. Baran, M. Colonna, M. Di Toro, Phys. Rev. C **65**, 045201 (2002).
118. R. Stock, Phys. Rep. **135**, 259 (1986).
119. P. Senger, H. Ströbele, J. Phys. G **25**, R59 (1999).
120. V.S. Uma Maheswari, C. Fuchs, A. Faessler, L. Sehn, D. Kosov, Z. Wang, Nucl. Phys. A **628**, 669 (1998).
121. B.-A. Li, G.-C. Yong, W. Zuo, Phys. Rev. C **71**, 014608 (2005).
122. Q. Li, Z. Li, S. Soff, M. Bleicher, H. Stöcker, Phys. Rev. C **72**, 034613 (2005).
123. KaoS Collaboration (A. Schmah *et al.*), Phys. Rev. C **71**, 064907 (2005).
124. J. Aichelin, C.M. Ko, Phys. Rev. Lett. **55**, 2661 (1985).
125. S.W. Huang, A. Faessler, G.Q. Li, R.K. Puri, E. Lehmann, D.T. Khoa, M.A. Matin, Phys. Lett. B **298**, 41 (1993).
126. C. Hartnack, J. Jaenicke, L. Sehn, H. Stöcker, J. Aichelin, Nucl. Phys. A **580**, 643 (1994).
127. G.Q. Li, C.M. Ko, Phys. Lett. B **349**, 405 (1995).
128. C. Fuchs, Z. Wang, L. Sehn, A. Faessler, V.S. Uma Maheswari, D.S. Kosov, Phys. Rev. C **56**, R606 (1997).
129. T. Maruyama, W. Cassing, U. Mosel, S. Teis, K. Weber, Nucl. Phys. A **573**, 653 (1994).
130. E.L. Bratkovskaya, W. Cassing, U. Mosel, Nucl. Phys. A **622**, 593 (1997).

131. COSY-11 Collaboration (J.T. Balewski *et al.*), Phys. Lett. B **338**, 859 (1996); **420**, 211 (1998).
132. KaoS Collaboration (D. Miskowiec *et al.*), Phys. Rev. Lett. **72**, 3650 (1994).
133. KaoS Collaboration (C. Sturm *et al.*), Phys. Rev. Lett. **86**, 39 (2001).
134. C. Fuchs, Prog. Part. Nucl. Phys. **56**, 1 (2006).
135. C. Fuchs, A. Faessler, E. Zabrodin, Y.M. Zheng, Phys. Rev. Lett. **86**, 1974 (2001); C. Fuchs, A. Faessler, S. El-Basaouny, E. Zabrodin, J. Phys. G **28**, 1615 (2002).
136. Ch. Hartnack, J. Aichelin, J. Phys. G **28**, 1649 (2002).
137. Ch. Hartnack, H. Oeschler, J. Aichelin, Phys. Rev. Lett. **96**, 012302 (2006).
138. D.J. Nice, E.M. Splaver, I.H. Stairs, O. Löhmer, A. Jessner, M. Kramer, J.M. Cordes, Astrophys. J. **634**, 1242 (2005).
139. T. Kähn *et al.*, nucl-th/0602038.
140. A.W. Steiner, M. Prakash, J.M. Lattimer, P.J. Ellis, Phys. Rep. **411**, 325 (2005).
141. B.-A. Li, A.W. Steiner, nucl-th/0511064.
142. J.M. Lattimer, C.J. Pethick, M. Prakash, P. Haensel, Phys. Rev. Lett. **66**, 2701 (1991).
143. D. Blaschke, H. Grigorian, D. Voskresensky, Astron. Astrophys. **424**, 979 (2004).
144. E.E. Kolomeitsev, D.N. Voskresensky, Nucl. Phys. A **759**, 373 (2005).
145. S. Popov, H. Grigorian, R. Turolla, D. Blaschke, Astron. Astrophys. **448**, 327 (2006).

Eur. Phys. J. A **30**, 23–30 (2006)

DOI 10.1140/epja/i2006-10100-3

THE EUROPEAN
PHYSICAL JOURNAL A

Deducing the nuclear-matter incompressibility coefficient from data on isoscalar compression modes

S. Shlomo[1], V.M. Kolomietz[2], and G. Colò[3,a]

[1] Cyclotron Institute, Texas A&M University, College Station, TX 77843, USA
[2] Institut for Nuclear Research, 03680 Kiev, Ukraine
[3] Dipartimento di Fisica, Università degli Studi, and INFN, via Celoria 16, 20133 Milano, Italy

Received: 15 February 2006 /
Published online: 28 September 2006 – © Società Italiana di Fisica / Springer-Verlag 2006

Abstract. Accurate assessment of the value of the incompressibility coefficient, K, of symmetric nuclear matter, which is directly related to the curvature of the equation of state (EOS), is needed to extend our knowledge of the EOS in the vicinity of the saturation point. We review the current status of K as determined from experimental data on isoscalar giant monopole and dipole resonances (compression modes) in nuclei, by employing the microscopic theory based on the random-phase approximation (RPA).

PACS. 21.65.+f Nuclear matter – 24.30.Cz Giant resonances – 21.60.Jz Hartree-Fock and random-phase approximation

1 Introduction

It is well known that the equation of state (EOS), $E/A = E(\rho)$, of symmetric nuclear matter (SNM) is a very important ingredient in the study of nuclear properties, heavy-ion collisions, neutron stars and supernovae. Experimentally, we have accurate data on the saturation point of the EOS, namely $(\rho_0, E(\rho_0))$. From electron and hadron scattering experiments on nuclei, one finds a constant central density of $\rho_0 = 0.16\,\mathrm{fm}^{-3}$, and from the extrapolation of empirical mass formula, we have $E(\rho_0) = -16\,\mathrm{MeV}$ for SNM. Since at saturation $\frac{\mathrm{d}E}{\mathrm{d}\rho}\big|_{\rho_0} = 0$, one has

$$E(\rho) = E(\rho_0) + \frac{1}{18}K\left(\frac{\rho - \rho_0}{\rho_0}\right)^2 + \ldots, \tag{1}$$

where

$$K = 9\rho_0^2 \frac{\mathrm{d}^2(E/A)}{\mathrm{d}\rho^2}\bigg|_{\rho_0} \tag{2}$$

is the SNM incompressibility coefficient. Therefore, a very accurate value of K is needed to extend our knowledge of the EOS in the vicinity of the saturation point.

There have been many attempts over the years to determine the value of K by considering properties of nuclei which are sensitive to a certain extent to K (see ref. [1]). In a macroscopic approach analysis of experimental data of a certain physical quantity, K appears in the expression for the physical quantity and the value of K is determined

by a direct fit to the data. In a microscopic approach, one considers various effective two-body interactions which are associated with different values of K but reproduce with comparable accuracies the experimental data of various properties of nuclei, such as binding energies and radii. One then determines the effective interaction which best fits the experimental data for a physical quantity which is sensitive to K. We mention, in particular, the attempts [1–5] of considering as physical quantities: nuclear masses, nuclear radii, nuclear scattering cross-sections, supernova collapses, masses of neutron stars, observables in heavy-ion collisions and the interaction parameters F_0 and F_1 in Landau's Fermi liquid theory for nuclear matter. Here we examine the most sensitive method [6,7] which is based on experimental data on the strength function distributions of the isoscalar giant monopole resonance (ISGMR), $T = 0, L = 0$, and the isoscalar giant dipole resonance (ISGDR), $T = 0, L = 1$, which are compression modes of nuclei, analyzed within the microscopic random-phase approximation (RPA) [8].

Over the last three decades, a significant amount of experimental work was carried out to identify strength distributions of the ISGMR and ISGDR in a wide range of nuclei [9–12]. The main experimental tool for studying isoscalar giant resonances is inelastic α-particle scattering. This is mainly because i) α-particles are selective as to exciting isoscalar modes, and ii) angular distributions of inelastically scattered α-particles at small angles are characteristic for some of the multipolar modes. Recent development in the area of experimental investigation of the

[a] e-mail: gianluca.colo@mi.infn.it

isoscalar giant resonances made it possible to measure the centroid energy (that is, the ratio of the energy-weighted and non–energy-weighted sum rules, m_1/m_0) E_0 of the ISGMR with an error $\delta E_0 \sim 0.1$–0.3 MeV [11,12]. Using the relation $(\delta K)/K = 2(\delta E_0)/E_0$ and, for example, the recent experimental value of $E_0 = 13.96 \pm 0.20$ MeV for the ISGMR in ^{208}Pb, one has an error of $\delta K = 6$–9 MeV for $K = 200$–300 MeV. This enhanced experimental precision calls for a critical accuracy check of the theoretical calculations. In fact, many available theoretical calculations, in which the monopole centroid is also determined only within about 0.2 MeV, due to various approximations, introduce a further contribution to δK which must be added quadratically to the experimental one, yielding a total error of 8–13 MeV (see [13]).

The extraction of K from experimental data on ISGMR is not straightforward. There have been several attempts [9] in the past to determine K simply by a least square fit to the ISGMR data of various sets of nuclei using a semi-empirical expansion in power of $A^{-1/3}$ of the nucleus incompressibility coefficient, K_A, obtained from E_0 using, for example, the scaling model assumption (we remind here that in the scaling model a simple shape of the ground-state density ρ_0 is assumed and its changes are associated to a single parameter λ, i.e., they are of the type $\rho_0 \to \rho_\lambda(\mathbf{r}) = \frac{1}{\lambda^3}\rho_0(\frac{\mathbf{r}}{\lambda})$). It was found [9] that the value deduced for K varied significantly, depending on the set of data of the ISGMR energies used in the fit. This is mainly due to the limited number of nuclei in which E_0 is known. We also point out that the scaling model assumption is not very reliable for medium and light nuclei.

If we have to resort to theory in order to extract K, we should start by discussing some principle remarks. The *static* incompressibility coefficient K of eq. (2) describes the propagation of the *first sound* excitations in nuclear matter having the sound velocity

$$c = c_1 = \sqrt{K/9m}. \tag{3}$$

However, the propagation of the first sound implies the regime of frequent inter-particle collisions [14] which is not realized in cold (and moderately heated) nuclei, where the compression modes are related to the zero sound (rare inter-particle collisions) regime. It is necessary to note that the sound velocity c and the eigenfrequency ω of the compression mode are, in principle, directly related to K for the first sound mode only. In general, the sound velocity c is a complicated function of both the incompressibility coefficient K and the dimensionless collisional parameter $\omega\tau$, where ω is the frequency of the mode and τ is its relaxation time. This complicated dependence is caused by the dynamic distortion of the Fermi surface (FSD) which accompanies the collective motion in a Fermi liquid. In cold nuclear matter, for the rare-collision regime $\omega\tau \to \infty$, one has, instead of eq. (3), the relation

$$c = c_0 = \sqrt{K'/9m}, \tag{4}$$

where K' is a strongly renormalized incompressibility coefficient which can be shown to obey [15]

$$K' \approx 3K. \tag{5}$$

Thus, within the theory of Fermi liquids, there is a significant difference between the static nuclear incompressibility coefficient, K, which is defined as the stiffness coefficient with respect to a change in the bulk density, and the dynamic one, K', associated with the zero sound velocity and the energy of the ISGMR or ISGDR. Nonetheless, the approximate relation (5) is consistent with the idea that the interaction which best fits the experimental data for ISGMR and ISGDR energies should also provide the correct value of K.

It can also be shown [15] that the consistent presence of the same FSD effects in the boundary condition strongly suppresses any increase of E_0 (the energy of lowest isoscalar giant monopole resonance) compared to the usual liquid-drop model where the FSD effects are not taken into account. We point out that the FSD effects are completely washed out from the dynamic incompressibility coefficient K' in the case of the scaling assumption. Note also that the effect of the FSD in the boundary condition is rather small for the overtone excitations. The dynamic and relaxation effects on the ISGMR and on the ISGDR are therefore significantly different. In contrast to the ISGMR, which is the lowest breathing mode, the ISGDR appears as the *overtone* to the lowest isoscalar dipole excitation, which corresponds to a spurious center-of-mass motion. Due to this fact, the energy of the ISGDR, E_1, varies with τ much more than the energy E_0 of the ISGMR.

If one wishes to make a link with microscopic effective interactions, the basic theory for the description of different giant resonance modes is self-consistent Hartree-Fock (HF) plus RPA [6,8]. The HF calculations using Skyrme-type interactions [16], which are density- and momentum-dependent zero-range interactions, have been very successful in reproducing experimental data on ground-state properties of nuclei. The parameters of the Skyrme interaction are varied so as to reproduce a selected set of experimental data of a wide range of nuclei on nuclear masses, charge and mass density distributions, etc. The nuclear response function is evaluated within RPA, which is a linear response theory suited for the description of small oscillations which can eventually accomodate a proper treatment of the particle continuum [8,17].

We emphasize that the values of E_0 and E_1 are correlated with the value of K which is associated with the effective nucleon-nucleon interaction adopted in the HF-RPA calculations, and thus can be used to extract an accurate value for K. This correlation has been explicitly shown, e.g., in refs. [18,19].

It is important to point out that the HF-RPA method solves the nuclear effective Hamiltonian in the space of one-particle–one-hole ($1ph$) excitations. Correlations, associated with excitations of $2ph$ and higher structures, are not accounted for explicitly. The effects of these correlations have been discussed in the literature, see for example the reviews in refs. [20–22]. The main effect is a collisional broadening of the strength distributions which can be accompanied by a certain shift of the resonance peak position. This shift grows with excitation energy and can be of the order of 1 MeV for the rather high-lying isovector modes (in the range above 20 MeV). However, in the case of the ISGMR the shift is quite small (of the order of few hundreds of keV [23], that is, comparable with the experimental uncertainty). This is not a numerical accident, rather a consequence of cancellations which arise when all diagrams corresponding to the coupling between $1ph$ and $2ph$ states are included (cf. [20] and references therein).

The first experimental identification of the ISGMR in ^{208}Pb at excitation energy of $E_0 = 13.7$ MeV [24] already triggered random-phase approximation (RPA) calculations using existing or modified effective interactions: those having $K = 210 \pm 30$ MeV gave results in agreement with experiment [25]. We point out, however, that i) in the early investigations, the experimental uncertainties for E_0 were relatively large, and only a limited class of effective interactions were explored; ii) many more recent calculations were not fully self-consistent [13,26]. Consequently, as we will see, we accept nowadays larger values for K.

The study of the isoscalar giant dipole resonance is very important since this compression mode provides an independent source of information on K. Early experimental investigation of the ISGDR in ^{208}Pb resulted in a value of $E_1 \sim 21$ MeV for the centroid energy [27,28]. It was first pointed out in ref. [29] that corresponding HF-RPA results for E_1, obtained with interactions adjusted to reproduce experimental values of E_0, are higher than the experimental value by more than 3 MeV and thus this discrepancy between theory and experiment raises some doubts concerning the unambiguous extraction of K from energies of compression modes. A similar result for E_1 in ^{208}Pb was obtained in more recent experiments [10,30]. Therefore, the value of K deduced from these early experimental data on ISGDR is significantly smaller than that deduced from ISGMR data.

Recent relativistic RPA (RRPA) calculations [31,32], with the inclusion of negative-energy states of the Dirac sea in the response function, yield a value of $K = 250$–270 MeV. This result has been obtained using different types of effective Lagrangians, including those having density-dependent coupling constants. Note that since an uncertainty of about 20% in the values of K is tantamount to an uncertainty of 10% in the value of E_0, the discrepancy in the value of K obtained from relativistic and non-relativistic models is quite significant in view of the accuracy of about 2% in the experimental data currently available on the ISGMR centroid energies. In refs. [19,33] it has been claimed that these significant differences are due to the model dependence of K. However, in the most recent works of refs. [13,34,35] this model dependence has been explained, as we shall discuss.

We should finally point out that it is quite common in theoretical work on giant resonances to calculate the strength function $S(E)$ for a certain simple scattering operator F, whereas in the analysis of experimental data of the excitation cross-section $\sigma(E)$ one carries out distorted-wave Born approximation (DWBA) calculations with a transition potential δU obtained from a collective model transition density ρ_{coll} using the folding model (FM) approximation. This may be a source of uncertainties, especially if most of the strength is not collective. Accordingly, it is important to examine the relation between $S(E)$ and the excitation cross-section $\sigma(E)$ of the ISGMR and the ISGDR, obtained by α-scattering, using the folding model DWBA method with ρ_t obtained from self-consistent HF-RPA.

In sect. 2 we review the basic elements of the microscopic HF-RPA theory for the strength function and the FM-DWBA method for the calculation of the excitation cross-sections of giant resonances by inelastic α-scattering. In sect. 3, we provide some results of the consequences of violations of self-consistency on the calculated strength function $S(E)$, the excitation cross-section $\sigma(E)$ and recent results of fully self-consistent HF-RPA calculations of the centroid energies (E_0 and E_1) for the ISGMR and ISGDR. We also present simple explanations for the discrepancies in the values deduced for K. Our conclusions are given in sect. 4.

2 Formalism

2.1 Self-consistent HF-RPA approach

In the microscopic and self-consistent HF-RPA approach, one starts by adopting a specific effective nucleon-nucleon interaction, V_{12}, and deriving the ground-state mean field. Then, the RPA equations are solved by using the particle-hole (p-h) interaction V_{ph} which is derived from the same mean field determined by V_{12} (in this sense, the calculation is self-consistent). Various numerical methods have been adopted in the literature to solve the RPA equations, see, for example, refs. [8,17,25,36,37]. In particular, in Green's function approach [8,17] one evaluates the RPA Green's function G, given by $G = G_0(1 + V_{ph}G_0)^{-1}$, where G_0 is the free p-h Green's function. Then, the strength function $S(E)$ and the transition density ρ_t, associated with the scattering operator $F = \sum_{i=1}^{A} f(\mathbf{r}_i)$, are obtained from

$$S(E) = \sum_n |\langle 0|F|n\rangle|^2 \, \delta(E - E_n) = \frac{1}{\pi} \mathrm{Im}\left[\mathrm{Tr}(fGf)\right], \quad (6)$$

$$\rho_t(\mathbf{r}, E) = \frac{\Delta E}{\sqrt{S(E)\Delta E}} \int f(\mathbf{r}') \left[\frac{1}{\pi} \mathrm{Im} G(\mathbf{r}', \mathbf{r}, E)\right] \mathrm{d}\mathbf{r}'. \quad (7)$$

Note that $\rho_t(\mathbf{r}, E)$, as defined in (7), is associated with the strength in the region of $E \pm \Delta E/2$. Green's function approach allows treating the continuum in a proper way. However, the RPA equations can also be solved on a discrete basis. Although the exact solution of RPA in the continuum may be crucial if one treats weakly bound nuclei or if one is interested in the particle decay of states which lie above the threshold, discrete RPA can nonetheless reproduce the main integral properties of giant resonances in stable nuclei.

There are also alternative methods to obtain these integral properties. For instance, the constrained energy

E_{-1} defined as $\sqrt{m_1/m_{-1}}$, where m_1 is the energy-weighted sum rule and m_{-1} is the inverse energy-weighted sum rule, can be calculated once m_1 is extracted from the double commutator $[F,[H,F]]$ while m_{-1} is obtained from constrained HF (CHF) calculations [38].

In fully self-consistent HF-RPA calculations, the spurious state (associated with the center-of-mass motion) $T = 0$, $L = 1$ must appear at zero excitation energy ($E = 0$), aside from small numerical inaccuracies, and no significant spurious state mixing (SSM) in the ISGDR must be expected. However, although not always stated in the literature, many actual implementations of HF-RPA (and relativistic RPA) are not fully self-consistent [26] (see, however, refs. [18,36,37,39–42]). Each approximation introduced in RPA may shift the centroid energies of giant resonances with respect to the exact value, and introduce a SSM in the ISGDR.

In refs. [26,43,44], in order to correct for the effects of the SSM on $S(E)$ and the transition density, the scattering operator $F = \sum_{i=1}^{A} f(\mathbf{r}_i)$ has been replaced by the projection operator

$$F_\eta = \sum_{i=1}^{A} f_\eta(\mathbf{r}_i) = \sum_{i=1}^{A} f(\mathbf{r}_i) - \eta f_1(\mathbf{r}_i), \qquad (8)$$

where $f(\mathbf{r}) = f(r)Y_{1M}(\Omega)$ and $f_1(\mathbf{r}) = rY_{1M}(\Omega)$. The value of η is obtained from the coherent spurious state transition density [45], $\rho_{ss}(\mathbf{r}) = \alpha_a \frac{\partial \rho_0}{\partial r} Y_{1M}(\Omega)$, where ρ_0 is the ground-state density of the nucleus. The result for $f(r) = r^3$ is $\eta = \frac{5}{3}\langle r^2 \rangle$ [46]. We point out that the IS-GDR transition density ρ_t is obtained [26] from eqs. (7) and (8) after subtracting the spurious state component ρ_{ss}. In ref. [47] it has been shown that the above procedure is equivalent to project out explicitly the spurious component from each excited state. Further discussions about the SSM can be found in refs. [48,49].

2.2 DWBA calculations of excitation cross-sections

The DWBA has been quite instrumental in providing a theoretical description of low-energy scattering reactions and is widely used in analyzing measured cross-sections of scattered probes. The folding model approach [50] to the evaluation of optical potentials appears to be quite successful and, at present, is extensively used in theoretical descriptions of α-particle scattering [51]. The main advantage of this approach is that it provides a direct link to the description of α-particle scattering reactions based on microscopic HF-RPA results.

The DWBA differential cross-section for the excitation of a giant resonance by inelastic α-scattering is

$$\frac{d\sigma^{DWBA}}{d\Omega} = \left(\frac{\mu}{2\pi\hbar^2}\right)^2 \frac{k_f}{k_i} |T_{fi}|^2, \qquad (9)$$

where μ is the reduced mass and k_i and k_f are the initial and final linear momenta of the α-nucleus relative motion,

respectively. The transition matrix element T_{fi} is given by

$$T_{fi} = \left\langle \chi_f^{(-)}\Psi_f | V | \chi_i^{(+)}\Psi_i \right\rangle, \qquad (10)$$

where V is the α-nucleon interaction, Ψ_i and Ψ_f are the initial and final states of the nucleus, and $\chi_i^{(+)}$ and $\chi_f^{(-)}$ are the corresponding distorted wave functions of the relative α-nucleus relative motion, respectively. To calculate T_{fi}, eq. (10), one can adopt the following approach which is usually employed by experimentalists. First, assuming that Ψ_i and Ψ_f are known, the integrals in (10) over the coordinates of the nucleons are carried out to obtain the transition potential $\delta U \sim \int \Psi_f^* V \Psi_i$. Second, the cross-section (9) is calculated using a certain DWBA code with δU and the optical potential $U(r)$ as input.

Within the FM approach, the optical potential $U(r)$ is given by

$$U(r) = \int d\mathbf{r}' V(|\mathbf{r} - \mathbf{r}'|, \rho_0(r'))\rho_0(r'), \qquad (11)$$

where $V(|\mathbf{r}-\mathbf{r}'|, \rho_0(r'))$ is the α-nucleon interaction, which is generally complex and density dependent, and $\rho_0(r')$ is the ground state HF density of a spherical target nucleus. To obtain the results given in the following, both the real and imaginary parts of the α-nucleon interaction were chosen to have Gaussian forms with density dependence [51], and parameters determined by a fit to the elastic scattering data. The radial form $\delta U_L(r, E)$ of the transition potential, for a state with the multipolarity L and excitation energy E, is obtained from:

$$\delta U(r, E) = \int d\mathbf{r}' \delta\rho_L(\mathbf{r}', E)\left[V(|\mathbf{r} - \mathbf{r}'|, \rho_0(r')) \right.$$
$$\left. + \rho_0(r')\frac{\partial V(|\mathbf{r} - \mathbf{r}'|, \rho_0(r'))}{\partial \rho_0(r')} \right], \qquad (12)$$

where $\delta\rho_L(\mathbf{r}', E)$ is the transition density for the considered state.

We point out that within the "microscopic" folding model approach to the α-nucleus scattering, both ρ_0 and ρ_L, which enter eqs. (11) and (12), are obtained from the self-consistent HF-RPA calculations (i.e., $\rho_L = \rho_t$, cf. eq. (7)). Within the "macroscopic" approach, one adopts collective transition densities, ρ_{coll}, which are assumed to have energy-independent radial shapes and are obtained from the ground-state density using a collective model. We stress that for a proper comparison between experimental and theoretical results for $S(E)$, one should adopt the "microscopic" folding model approach in the DWBA calculations of $\sigma(E)$.

3 Results and discussion

3.1 Consequences of the violation of self-consistency

Recently, the effects of common violations [26] of self-consistency in HF-RPA calculations of $S(E)$ and ρ_t of

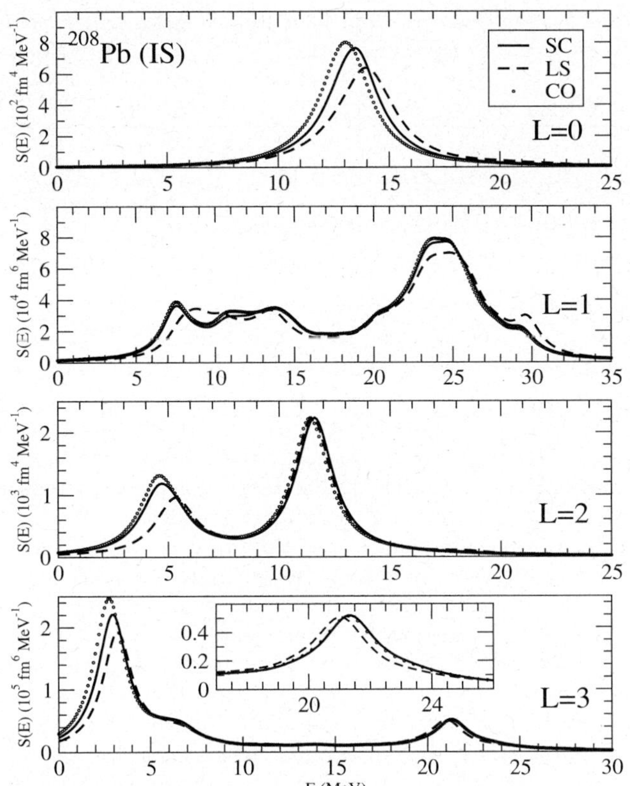

Fig. 1. Isoscalar strength functions of ^{208}Pb for $L = 0$–3 multipolarities are displayed. SC (full line) corresponds to the fully self-consistent calculation, whereas LS (dashed line) and CO (dotted line) represent the calculations without the residual spin-orbit and Coulomb interactions in the RPA, respectively. The interaction SGII [56] was used (taken from [54]).

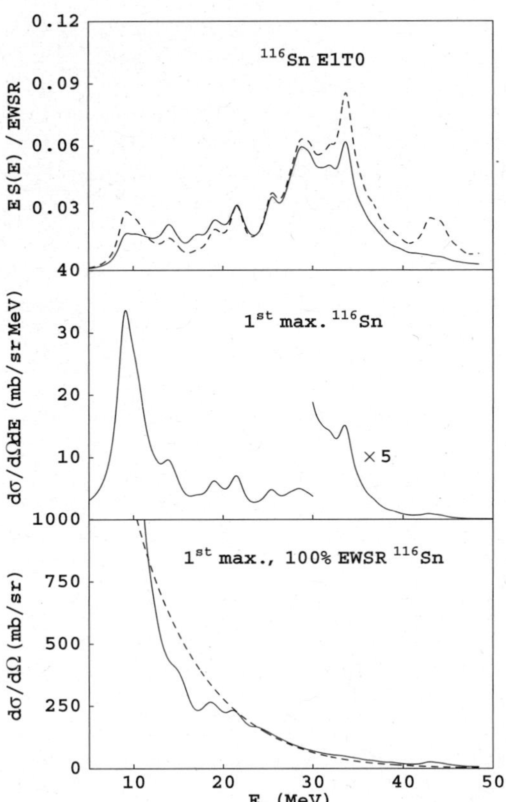

Fig. 2. Reconstruction of the ISGDR EWSR in ^{116}Sn from the inelastic α-particle cross-section. Middle panel: maximum double differential cross section obtained from ρ_t (RPA). Lower panel: maximum cross-section ($0°$) obtained with ρ_{coll} (dashed line) and ρ_t (solid line) normalized to 100% of the EWSR. Upper panel: the solid and dashed lines are the ratios of the middle panel curve with the solid and dashed lines of the lower panel, respectively (taken from ref. [26]).

various giant resonances were investigated in detail, see for example refs. [39–41,52–54]. To demonstrate the importance of carrying out fully self-consistent calculations, we present in fig. 1 recent results for $S(E)$ of isoscalar giant resonances in ^{208}Pb with multipolarities $L = 0$–3 using the fully self-consistent method described in refs. [36,55]. The interaction SGII [56] was used. It is seen (see also ref. [54]) from fig. 1 that the effects of violation of self-consistency due to the neglect of the particle-hole (p-h) spin-orbit or Coulomb interactions in the RPA calculations are most significant for the ISGMR. For the ISGMR in ^{208}Pb the shift in the centroid energy E_0 is about 0.8 MeV, which is 3 times larger than the experimental uncertainty. This is in agreement with fig. 1 of ref. [13], where a similar shift for E_{-1} has been obtained by means of CHF calculations.

We note that a shift of 0.8 MeV in E_0 correspond to a shift of about 25 MeV in K. In fact, this shift completely solves the issue of the previously advocated disagreement between values of K extracted from Skyrme and Gogny calculations. Fully self-consistent Skyrme calculations employing existing parametrizations do not point any more to the value of about 210 MeV quoted in the introduction, but to about 235 MeV in clear agreement with the Gogny-based extraction of K.

3.2 Nuclear compressibility from ISGMR and ISGDR

In contrast with the ISGMR, which presents a single peak, as a rule, in heavy nuclei, the dipole response displays a low-lying, fragmented part which lies below the giant resonance. This is a systematic feature of experimental and theoretical results in a number of isotopes. Different theoretical calculations [47,57] agree in indicating that the low-lying strength is not collective. In fact, while the centroids of the high-energy region, if calculated with interactions associated with different values of K, scale with these values, the centroids of the low-energy region do not. As far as the giant resonance centroid is concerned, discrete and continuum [58] RPA results are in good agreement with each other in ^{208}Pb. Coupling with $2ph$-type configurations is in this case relevant, as it shifts the centroid downwards by 1 MeV (leading to good agreement with experimental data) and produces a conspicous spreading width of about 6 MeV [59].

In refs. [26,60], numerical calculations were carried out for the $S(E)$, $\rho_t(r)$ within the HF-RPA theory and for $\sigma(E)$ as well, using the FM-DWBA method. The SL1 Skyrme interaction [61], which is associated with

Table 1. Fully self-consistent HF-RPA results [54] for the IS-GDR centroid energy (in MeV) in ^{90}Zr and ^{208}Pb, obtained using the interactions SGII [56] and SK255 [34], compared with the RRPA results obtained [19] with the NL3 interaction [62]. Also given are the values of K, and of the symmetry energy at saturation, J. The range of integration ω_1–ω_2 is given in the second column. The experimental data are from ref. [10] (a), ref. [11] (b), ref. [12] (c) and ref. [63] (d).

Nucleus	ω_1–ω_2	Experiment	NL3	SGII	SK255
^{90}Zr	18–50	25.7 ± 0.7^a	32.0	28.8	29.2
		26.7 ± 0.5^b			
		26.9 ± 0.7^d			
^{208}Pb	16–40	19.9 ± 0.8^a	26.0	24.1	24.5
		22.2 ± 0.5^c			
		22.7 ± 0.2^d			
K (MeV)			272	215	255
J (MeV)			37.4	26.8	37.4

Table 2. The same as table 1 for the ISGMR. Experimental data are taken from refs. [11, 12].

Nucleus	ω_1–ω_2	Experiment	NL3	SGII	SK255
^{90}Zr	0–60		18.7	17.9	18.9
	10–35	17.81 ± 0.30		17.9	18.9
^{208}Pb	0–60		14.2	13.6	14.3
	10–35	13.96 ± 0.20		13.6	14.4

$K = 230$ MeV, was employed. The density-dependent Gaussian α-nucleon interaction discussed in sect. 2.2 was used with parameters adjusted to reproduce the elastic cross-section, with ρ_0 taken from the HF calculations. In fig. 2, we present the results of this microscopic calculation of the fraction of the energy-weighted sum rule, and the excitation cross-section $\sigma(E)$ of the ISGDR in ^{116}Sn by 240 MeV α-particle scattering. It is seen from the upper panel that the use of the collective model transition densities ρ_{coll} in the whole energy range increases the EWSR by about 15%. However, the shift in the centroid energy is small (a few percents), similar in magnitude to the current experimental uncertainties. It was first pointed out in [26] that an important result of the calculation is that the maximum cross-section for the ISGDR decreases strongly at high energy and may drop below the experimental sensitivity for excitation energies above 30 MeV. This high excitation energy region contains about 20% of the EWSR. This missing strength leads to a reduction of about 3.0 MeV in the ISGDR energy which can significantly affect the comparison between theory and experiment.

In table 1, we give the results of fully self-consistent HF-RPA calculations for the ISGDR centroid energy (E_1) obtained (see ref. [54]) using the SGII [56] and SK255 [34] interactions and compare them with the RMF-based RPA results of ref. [57] for the NL3 interaction [62] and with the experimental data. The SGII result in ^{208}Pb compares well with 23.9 MeV obtained using discrete RPA in ref. [47] and with 23.4 MeV obtained using continuum RPA in ref. [48]. Note that the HF-RPA values for E_1 are larger than the corresponding experimental values of the early measurements of refs. [10, 27, 28, 30] by more than 3 MeV. The more recent results of refs. [11, 12, 63, 64], seem to better agree.

3.3 Nuclear compressibility in relativistic and non-relativistic models

To properly compare between the predictions of the relativistic and the non-relativistic models, parameter sets for Skyrme interactions were generated in ref. [34] by a least-square fitting procedure using exactly the same experimental data for the bulk properties of nuclei considered in ref. [62] for determining the NL3 parameterization of the effective Lagrangian used in the relativistic mean-field (RMF) models. The center-of-mass correction to the total binding energy, finite-size effects of the protons and Coulomb energy were calculated in a way similar to that employed in determining the NL3 parameter set in ref. [62]. Further, the values of the symmetry energy at saturation (J) and the charge rms radius of the ^{208}Pb nucleus were constrained to be very close to 37.4 MeV and 5.50 fm, respectively, as obtained with the NL3 interaction, and K was fixed in the vicinity of the NL3 value of $K = 271.76$ MeV. In particular, the Skyrme interactions SK272 and SK255, having $K = 272$ and 255 MeV, respectively, were generated in ref. [34]. It is seen from table 2 that the new Skyrme interaction SK255 yields for the ISGMR centroid energies (E_0) values which are close to the RRPA results obtained for the NL3 interaction, in good agreement with experimental data.

To better understand this result, a more systematic analysis has been made in ref. [35], in which a larger set of new Skyrme forces has been generated, built with the same protocol used for the Lyon forces [65] and spanning a wide range of values for K, for the symmetry energy at saturation and its density dependence. The main conclusions reached in that work are the following. The ISGMR energies, calculated by means of CHF, and consequently the extracted value of K, depend on a well-defined parameter (K_{sym}) which controls the slope of the symmetry energy curve as a function of density. The Skyrme forces having a density dependence characterized by an exponent $\alpha = 1/6$, like SLy4, predict K around 230–240 MeV. If this exponent is increased to values of the order of 1/3, and consequently the slope of the symmetry energy curve is made stiffer, one can produce forces which are compatible with K around 250–260 MeV. This result, obtained within the framework of a different protocol for fitting the Skyrme parameters, is nonetheless in full agreement with the result of [34]. The main results of ref. [35] are shown in fig. 3. It has to be noted that a further increase of α, and accordingly of K, would become difficult to obtain since the effective mass m^* would become too small.

One thus can make the clear and strong conclusion that the difference in the values of K obtained in the relativistic and non-relativistic models is not due to model dependence. It is mainly due to the different behavior of the symmetry energy within these models (cf. also [66]).

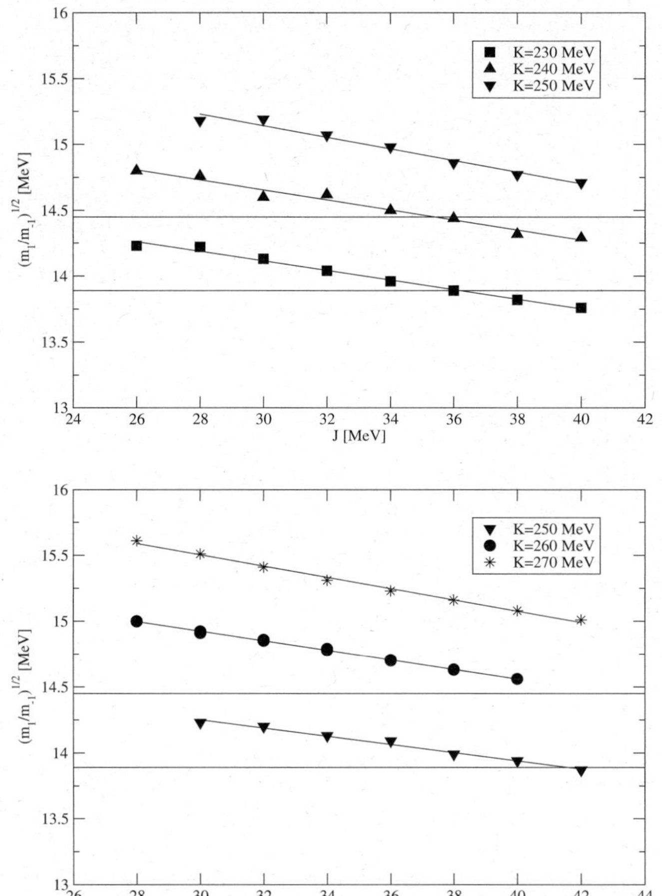

Fig. 3. Constrained ISGMR energies E_{-1} in ^{208}Pb obtained by using the Skyrme forces built in ref. [35], and having $\alpha = 1/6$ (upper panel) or $\alpha = 0.3563$ (lower panel). The two horizontal lines denote the experimental upper and lower bounds. See the text for a discussion (figure taken from ref. [35]).

4 Conclusions

Considering the status of determining the value of the nuclear-matter incompressibility coefficient, K, from data on the compression modes ISGMR and ISGDR of nuclei, we conclude that:

i) Recent improvement in the experimental techniques led to the identification of the ISGMR in light and medium nuclei and the observation of the ISGDR in nuclei. Currently, the centroid energy E_0 of the ISGMR can be deduced with very small experimental uncertainty of about 0.2 MeV, which corresponds to an uncertainty of about 7 MeV in the extracted value of K.

ii) Violations of self-consistency in HF-RPA calculations of the strength functions of giant resonances result in shifts in the calculated values of the centroid energies which may be larger in magnitude than the current experimental uncertainties. Thus, it is important to carry out fully self-consistent HF-RPA calculations in order to extract an accurate value of K from experimental data on the ISGMR and ISGDR. In fact, the prediction of K lying in the range 210–220 MeV were coming from not fully self-

consistent Skyrme calculations. Correcting for this drawback, Skyrme parametrizations of the SLy4 type predict values of K in the range 230–240 MeV.

iii) It is possible to build *bona fide* Skyrme forces so that the incompressibility is close to the relativistic value, namely 250–270 MeV.

iv) Therefore, from the ISGMR experimental data the conclusion can be drawn that $K = 240 \pm 20$ MeV. The uncertainty of about 20 MeV in the value of K is mainly due to the uncertainty in the value of the overall shape of the nuclear-matter symmetry energy curve, as a function of density.

v) The ISGDR data tend to point to lower values for K. However, there is consensus that the extraction of K is in this case more problematic for different reasons. In particular, the maximum cross-section for the ISGDR decreases very strongly at high excitation energy and may drop below the current experimental sensitivity for excitation energies above 30 and 26 MeV for ^{116}Sn and ^{208}Pb, respectively. More accurate experimental data, and analysis, on the ISGDR are very much needed.

This work was supported in part by the US National Science Foundation under Grant No. PHY-0355200 and the US Department of Energy under the Grant No. DOE-FG03-93ER40773. The authors would like to acknowledge the large benefit from either discussions or collaborations with B.K. Agrawal, V.K. Au, K. Bennaceur, P. Bonche, P.F. Bortignon, M. Centelles, H. Clark, S. Fracasso, U. Garg, I. Hamamoto, A. Kolomiets, Y.-W. Lui, J. Meyer, Nguyen Van Giai, J. Piekarewicz, M.R. Quaglia, P.-G. Reinhard, P. Ring, H. Sagawa, H. Sakaguchi, A.I. Sanzhur, T. Sil, M. Uchida, D. Vretenar and D. Youngblood.

References

1. N.K. Glendenning, Phys. Rev. C **37**, 2733 (1988).
2. W.D. Myers, W.J. Swiatecki, Phys. Rev. C **57**, 3020 (1998).
3. L. Satpathy, V.S.U. Maheswari, R.C. Nayak, Phys. Rep. **319**, 85 (1999).
4. W. Von Oertzen, H.G. Bohlen, D.T. Khoa, Nucl. Phys. A **722**, 202c (2003).
5. J.B. Natowitz, K. Hagel, Y. Ma, M. Murray, L. Qin, R. Wada, J. Wong, Phys. Rev. Lett. **89**, 21270 (2002).
6. A. Bohr, B.M. Mottleson, *Nuclear Structure II* (Benjamin, New York, 1975).
7. S. Stringari, Phys. Lett. B **108**, 232 (1982).
8. S. Shlomo, G.F. Bertsch, Nucl. Phys. A **243**, 507 (1975); K.F. Liu, Nguyen Van Giai, Phys. Lett. B **65**, 23 (1976).
9. S. Shlomo, D.H. Youngblood, Phys. Rev. C **47**, 529 (1993), and references therein; cf. also M. Pearson, Phys. Lett. B **271**, 12 (1991).
10. H.L. Clark, Y.W. Lui, D.H. Youngblood, Phys. Rev. C **63**, 031301 (2001) and references therein.
11. D.H. Youngblood, H.L. Clark, Y.W. Lui, Phys. Rev. C **69**, 034315 (2004).
12. D.H. Youngblood, H.L. Clark, Y.W. Lui, Phys. Rev. C **69**, 054312 (2004).
13. G. Colò, Nguyen Van Giai, Nucl. Phys. A **731**, 15c (2004).

14. A.A. Abrikosov, I.M. Khalatnikov, Rep. Prog. Phys. **22**, 329 (1959).
15. A. Kolomiets, V.M. Kolomietz, S. Shlomo, Phys. Rev. C **59**, 3139 (1999).
16. D. Vautherin, D.M. Brink, Phys. Rev. C **5**, 626 (1972); M. Beiner *et al.*, Nucl. Phys. A **238**, 29 (1975).
17. G.F. Bertsch, S.F. Tsai, Phys. Rep. **18**, 125 (1975).
18. J.P. Blaizot, J.F. Burger, J. Decharge, N. Girod, Nucl. Phys. A **591**, 435 (1995).
19. Nguyen Van Giai, P.F. Bortignon, G. Colò, Zhongyu Ma, M.R. Quaglia, Nucl. Phys. A **687**, 44c (2001).
20. G.F. Bertsch, P.F. Bortignon, R.A. Broglia, Rev. Mod. Phys. **55**, 287 (1983).
21. C. Mahaux, P.F. Bortignon, R.A. Broglia, C.H. Dasso, Phys. Rep. **120**, 1 (1985).
22. P.-G. Reinhard, C. Toepffer, Int. J. Mod. Phys. E **3**, 435 (1994).
23. G. Colò, P.F. Bortignon, N. Van Giai, A. Bracco, R.A. Broglia, Phys. Lett. B **276**, 279 (1992).
24. N. Marty *et al.*, Nucl. Phys. A **230**, 93 (1975); M.N. Harakeh *et al.*, Phys. Rev. Lett. **38**, 676 (1977); D.H. Youngblood *et al.*, Phys. Rev. Lett. **39**, 1188 (1977).
25. J.P. Blaizot, Phys. Rep. **64**, 171 (1980).
26. S. Shlomo, A.I. Sanzhur, Phys. Rev. C **65**, 044310 (2002); S. Shlomo, Pramana J. Phys. **57**, 557 (2001).
27. H.P. Morsch, M. Rogge, P. Turek, C. Mayer-Böricke, Phys. Rev. Lett. **45**, 337 (1980).
28. C. Djalali, N. Marty, M. Morlet, A. Willis, Nucl. Phys. A **380**, 42 (1982).
29. T.S. Dumitrescu, F.E. Serr, Phys. Rev. C **27**, 811 (1983).
30. B. Davis *et al.*, Phys. Rev. Lett. **79**, 609 (1997).
31. Zhong-yu Ma, Nguyen Van Giai, A. Wandelt, D. Vretenar, Nucl. Phys. A **686**, 173 (2001).
32. D. Vretenar, T. Nikšič, P. Ring, Phys. Rev. C **68**, 024310 (2003).
33. T. Nikšič, D. Vretenar, P. Ring, Phys. Rev. C **66**, 064302 (2002).
34. B.K. Agrawal, S. Shlomo, V.K. Au, Phys. Rev. C **68**, 031304(R) (2003); S. Shlomo, B.K. Agrawal, V.K. Au, Nucl. Phys. A **734**, 589 (2004).
35. G. Colò, Nguyen Van Giai, J. Meyer, K. Bennaceur, P. Bonche, Phys. Rev. C **70**, 024307 (2004).
36. P.-G. Reinhard, Ann. Phys. (Leipzig) **1**, 632 (1992).
37. T. Nakatsukasa, K. Yabana, Phys. Rev. C **71**, 024301 (2005).
38. O. Bohigas, A.M. Lane, J. Martorell, Phys. Rep. **51**, 267 (1979).
39. J. Terasaki, J. Engel, M. Bender, J. Dobaczewski, W. Nazarewicz, M. Stoitsov, Phys. Rev. C **71**, 034310 (2005).
40. B.K. Agrawal, S. Shlomo, A.I. Sanzhur, Phys. Rev. C **67**, 034314 (2003).
41. B.K. Agrawal, S. Shlomo, Phys. Rev. C **70**, 014308 (2004).
42. J. Piekarewicz, Phys. Rev. C **62**, 051304 (2000).
43. M.L. Gorelik, S. Shlomo, M.H. Urin, Phys. Rev. C **62**, 044301 (2000).
44. A. Kolomiets, O. Pochivalov, S. Shlomo, *Progress in Research, April 1, 1998 - March 31, 1999* (Cyclotron Institute, Texas A&M University, 1999) III-1.
45. G.F. Bertsch, Suppl. Prog. Theor. Phys. **74**, 115 (1983).
46. Nguyen Van Giai, H. Sagawa, Nucl. Phys. A **371**, 1 (1981).
47. G. Colò, Nguyen Van Giai, P.F. Bortignon, M.R. Quaglia, Phys. Lett. B **485**, 362 (2000).
48. I. Hamamoto, H. Sagawa, Phys. Rev. C **66**, 044315 (2002).
49. V.I. Abrosimov, A. Dellafiore, F. Matera, Nucl. Phys. A **697**, 748 (2002).
50. G.R. Satchler, *Direct Nuclear Reactions* (Oxford University Press, Oxford, 1983).
51. G.R. Satchler, D.T. Khoa, Phys. Rev. C **55**, 285 (1997).
52. S.A. Fayans, E.L. Trykov, D. Zawischa, Nucl. Phys. A **568**, 523 (1994).
53. S. Péru, J.F. Berger, P.F. Bortignon, Eur. Phys. J. A **26**, 25 (2005).
54. T. Sil, S. Shlomo, B.K. Agrawal, P.-G. Reinhard, Phys. Rev. C **73**, 034316 (2006).
55. P.-G. Reinhard, Nucl. Phys. A **646**, 305c (1999).
56. Nguyen Van Giai, H. Sagawa, Phys. Lett. B **106**, 379 (1981).
57. D. Vretenar, A. Wandelt, P. Ring, Phys. Lett. B **487**, 334 (2000).
58. I. Hamamoto, H. Sagawa, X.Z. Zhang, Phys. Rev. C **57**, R1064 (1998).
59. G. Colò, Nguyen Van Giai, P.F. Bortignon, M.R. Quaglia, *Proceedings of the RIKEN Symposium on Selected Topics in Nuclear Collective Excitations, RIKEN, Wako City 1999*, RIKEN Rev. **23**, 39 (1999).
60. A. Kolomiets, O. Pochivalov, S. Shlomo, Phys. Rev. C **61**, 034312 (2000).
61. K.-F. Liu, H.-D. Lou, Z.-Y. Ma, Q.-B. Shen, Nucl. Phys. A **534**, 1; 25 (1991).
62. G.A. Lalazissis, J. König, P. Ring, Phys. Rev. C **55**, 540 (1997).
63. M. Uchida *et al.*, Phys. Rev. C **69**, 051301 (2004).
64. M. Itoh *et al.*, Phys. Rev. C **68**, 064602 (2003).
65. E. Chabanat, P. Bonche, P. Haensel, J. Meyer, R. Schaeffer, Nucl. Phys. A **627**, 710 (1997); **635**, 231 (1998).
66. J. Piekarewicz, Phys. Rev. C **66**, 034305 (2002).

Eur. Phys. J. A **30**, 31–46 (2006)
DOI 10.1140/epja/i2006-10101-2

THE EUROPEAN
PHYSICAL JOURNAL A

Systematics of stopping and flow in Au+Au collisions

A. Andronic[1], J. Łukasik[1,2,a], W. Reisdorf[1], and W. Trautmann[1]

[1] GSI, D-64291 Darmstadt, Germany
[2] IFJ-PAN, Pl-31342 Kraków, Poland

Received: 27 April 2006 /
Published online: 6 October 2006 – © Società Italiana di Fisica / Springer-Verlag 2006

Abstract. Excitation functions of flow and stopping observables for the Au+Au system at energies from 40 to 1500 MeV per nucleon are presented. The systematics were obtained by merging the results of the INDRA and FOPI experiments, both performed at the GSI facility. The connection to the nuclear equation of state is discussed.

PACS. 25.70.-z Low and intermediate energy heavy-ion reactions – 25.75.Ld Collective flow – 25.70.Mn Projectile and target fragmentation

1 Introduction

The study of collective flow in nucleus-nucleus collisions has been an intense field of research for the past twenty years [1,2]. At beam energies below several GeV per nucleon, it is mainly motivated by the goal to extract the equation of state (EoS) of nuclear matter from the quantitative comparison of measurements with the results of microscopic transport model calculations [3–5]. Considerable progress has been made in this direction in recent years but the constraints on the EoS obtained so far remain rather broad [5,6].

The results of flow measurements performed before 1999 have been extensively reviewed in refs. [1,2]. In the meanwhile, a variety of new results has become available regarding the directed [7–27] and elliptic [22–33] flow. These recent experiments have expanded the study of flow over a broader range of incident energies. New results became available on collective motion of produced particles [12–15]. Several studies have focussed on balance (or transition) energies associated with sign changes of a flow parameter [20–22,28–31]. High-statistics measurements allowed to explore the transverse momentum dependence of flow [17–19,27,28].

Since flow is generated by pressure gradients, it is clear that its quantitative study reveals aspects of the EoS. However, by itself, flow is not sufficient to fix the EoS. We need to know, as a function of beam energy, what density was achieved in the collision. An optimal condition that matter be piled up to form a dense medium, is that the two colliding ions be stopped in the course of the collision, before the system starts to expand. Infor-

mation on the stopping can be obtained by studying the rapidity density distributions of the ejectiles in both the beam direction (the original direction) and the transverse direction. Recently [7], the ratio of the variances of the transverse to the longitudinal rapidities was proposed as an indicator of the degree of stopping and it was found to correlate with flow provided the incident energy E/A exceeded $150A$ MeV. While this flow-stopping correlation is only indirectly connected to a pressure-density correlation, it represents a potentially interesting constraint for microscopic simulations tending to extract the EoS from heavy-ion data.

The main purpose of this review is to present the excitation functions of flow (directed and elliptic) and of stopping in ^{197}Au + ^{197}Au collisions. This heavy, symmetric system has been studied with a variety of detectors in the intermediate energy domain throughout the last two decades:

Experiment	Reference	E/A (MeV)
PLASTIC-BALL	[34–37]	150–1050
MSU-ALADIN	[38–40]	100–400
LAND-FOPI	[41]	400
FOPI	[28,30,42]	90–1500
EOS	[43]	250–1150
MULTICS-MINIBALL	[44,45]	35
MSU-4π	[20]	25–60
INDRA-ALADIN	[22,46,47]	40–150
CHIMERA	[48]	15

The phase space coverage and the range of observables reported in these studies vary considerably. All these data sets could be and, in most cases, were indeed used for flow studies. However, except for the comparative study

a e-mail: j.lukasik@gsi.de

between the Plastic Ball and the EOS data on directed flow [43], and between the Plastic Ball, the FOPI and the INDRA data on elliptic flow [22,30], no detailed comparison has been made so far, in this energy domain, of the results obtained by different experimental groups with different detectors.

In this work we will concentrate on the results obtained with the 4π FOPI and INDRA detector systems in experiments performed at the heavy-ion synchrotron SIS at GSI Darmstadt [7,18,22,28]. The covered ranges of incident energies were $90A$ MeV to $1.5A$ GeV in the FOPI and $40A$ to $150A$ MeV in the INDRA experiments. By combining the results obtained with the two detectors, having well-adapted designs for the two different energy regimes, we were able to construct coherent systematics revealing a remarkable evolution of flow and stopping over a large range of incident energies.

The observed agreement in the overlap region will serve as a measure of the absolute accuracy of the experimental data. We will focus on two aspects in this context, the systematic errors associated with the unavoidable deficiencies of the experimental devices and on the systematic errors resulting from the analysis methods which are not necessarily independent of the former. Since the two detectors have different acceptances and the reaction mechanism evolves in the energy region covered by the two experiments, particular attention will be given to the problem of impact-parameter selection and to the corrections for the reaction plane dispersion, which need to be adapted accordingly. For the latter a new method has been devised and applied to the INDRA data.

2 The detectors

The INDRA detector is constructed as a set of 17 detection rings with azimuthal symmetry around the beam axis. The most forward ring ($2° \leq \theta_{lab} \leq 3°$) consists of 12 Si ($300\,\mu$m) – CsI(Tl) ($15$ cm long) telescopes. The angular range from $3°$ to $45°$ is covered by 8 rings of 192 telescopes in total, each with three detection layers: ionization chambers (5 cm of C_3F_8 at 50 mbar), Si detectors ($300\,\mu$m) and CsI(Tl) scintillators with lengths decreasing from 13.8 cm to 9 cm with increasing angle. The remaining 8 rings, covering the region $45° \leq \theta_{lab} \leq 176°$, have two detection layers: ionization chambers (5 cm of C_3F_8 at 30 mbar) and CsI(Tl) scintillators (7.6 to 5 cm). The total granularity is 336 detection cells covering 90% of the 4π solid angle.

In the forward region ($\theta_{lab} \leq 45°$), ions with $5 \leq Z \leq 80$ are identified using the $\Delta E - E$ method. Over the whole angular range, isotope identification is obtained for $1 \leq Z \leq 4$ using the technique of pulse-shape discrimination for the CsI(Tl) signals. A complete technical description of the detector and of its electronics can be found in [49], details of the calibrations performed for the GSI experiments are given in [47,50].

The FOPI detector [42,51] is comprised of two main components: the forward Plastic Wall and the Central Drift Chamber, covering regions of laboratory polar angles of $1.2° < \theta_{lab} < 30°$ and $34° < \theta_{lab} < 145°$, respectively. The Plastic Wall consists of 764 individual plastic scintillator units. Detected reaction products are identified according to their atomic number, up to $Z \simeq 12$, using the measured time-of-flight (ToF) and specific energy loss. Particles detected with the Central Drift Chamber ($Z \leq 3$) are identified according to their mass (A) by using the measured magnetic rigidity and specific energy loss. The 3-dimensional tracking profits from a high equivalent detector granularity. At beam energies of $400A$ MeV and above, the forward drift chamber Helitron can be employed for mass identification of light fragments ($Z \leq 2$) at angles $7° < \theta_{lab} < 29°$.

The FOPI detector has an effective granularity exceeding that of INDRA by about a factor of 4, a property matched to the increasing multiplicity of charged particles with rising beam energy[1]. Both, INDRA and FOPI detectors are essentially blind to neutral particles, such as neutrons, π^0 and γ's. The higher granularity is, however, not the only feature helping to cope with higher energies. As the energy of the emitted particles rises, a level is reached where the principle of stopping the particle in a sensitive detecting material in order to determine its energy is no longer adequate because the material depth needed leads to a high probability of nuclear reactions undermining the energy measurement. To avoid this difficulty, one switches to time-of-flight and magnetic rigidity (in addition to energy loss) measurements: the apparatus becomes larger and is no longer under vacuum. Hence the detection thresholds for the various ejectiles are raised. For the FOPI detectors this means that, *e.g.*, at $90A$ MeV fragments with $Z > 6$ cannot be detected at midrapidity anymore.

3 Impact parameter

In a binary collision of massive "objects", the transfer of energy, momentum, angular momentum, mass etc. between the two partners will be strongly affected by the impact parameter b. As a consequence, one expects to observe large event-to-event fluctuations due to impact parameter mixing. To be meaningful, a comparison of experimental observations among each other or with the predictions of theoretical simulations has to be performed for well defined and sufficiently narrow intervals of impact parameter. Generally, in microscopic physics and, in particular, in nuclear physics, the impact parameter is not directly measurable but has to be estimated from *global* observables g characterizing the registered events. Global observables are determined using all or a significant fraction of the detected particles.

The basic, so-called geometrical model assumption [52], underlying the association of an impact param-

---

[1] The 4π-integrated charged-particle multiplicities in central collisions increase from typically about 40 at $40A$ MeV to 95 at $150A$ MeV and exceed 200 (with one quarter of them being charged pions) at $1.5A$ GeV.

eter b with an observed value g is that g changes strictly monotonically with b allowing to postulate

$$\int_g^\infty \frac{\mathrm{d}\sigma(\bar g)}{\mathrm{d}\bar g}\,\mathrm{d}\bar g = \pi b^2(g) \quad \text{or} \quad \int_0^g \frac{\mathrm{d}\sigma(\bar g)}{\mathrm{d}\bar g}\,\mathrm{d}\bar g = \pi b^2(g), \quad (1)$$

where the left- (right-) hand equation holds for g decreasing (rising) with b. The distribution $\mathrm{d}\sigma(g)/\mathrm{d}g$ is determined experimentally in terms of differential cross-sections per unit of g in a minimum bias class of events, *i.e.* where a minimum number of conditions was required to trigger data taking.

At intermediate to low incident energies, especially for $E/A < 100\,\mathrm{MeV}$, the literature abounds with an impressive diversity in the choice of global observables that have been used in attempts to select either narrowly constrained impact parameters (keywords "highly exclusive" or "ultracentral") or events of special interest (keywords "fully equilibrated", "fully evaporated", "signals of phase coexistence"). The observables vary from very simple ones like proton, neutron or total charged-particle multiplicity to more specific ones as, *e.g.*, participant proton multiplicity (N_p) [53, 54], total (E_T) [55, 56] or light charged-particle transverse kinetic energy ($E_\perp^{12}$) [57], ratio of transverse-to-longitudinal kinetic energy ($Erat$) [58, 59], degree of isotropy of momenta (R) [60, 61], transverse momentum directivity (D) [62–65], longitudinal kinetic-energy fraction (E_e) [66, 67], linear momentum transfer [68], total kinetic-energy loss ($TKEL$) [69, 70], average parallel velocity (V_{av}) [71], midrapidity charge (Z_y) [72], total charge of $Z \geq 2$ products (Z_{bound}) [73, 74], longitudinal component of the quadrupole moment tensor (Q_{zz}) [75]. Even more complex observables are those obtained from sphericity [76, 77], from the kinetic-energy tensor [78–80] or momentum tensor [67, 81, 82], the thrust (T) [67, 83, 84], the deflection angle of the projectile (Θ_{defl}) [85], the flow angle (Θ_{flow}) [3, 86], the location in a "Wilczyński plot" [69, 86, 87], harmonic moments (H_2) [86, 88, 89], or combined global variables (ρ) [90]. The most sophisticated methods used for impact parameter selection are based on, *e.g.*, principal component analysis (PCA) [91–93] or on neural-network techniques (NN) [94–96].

There are also more technical event selection schemes involving the postulation of "complete" events by demanding that nearly the full system charge or the full total linear momentum is accounted for. These latter methods are specific for a given apparatus since these observables, strictly constrained by conservation laws, would not be impact parameter selective when using a perfect detection system. In this case, a comparison of different experimental data sets at a high level of precision is difficult and a comparison with theoretical approaches must use apparatus-specific filter software that reproduces the hardware cuts causing the observed selectivity. In the present study, aiming towards joining up the data of two rather different setups, we will try to avoid using such concepts. We will restrict ourselves to the use of "simple" global observables such as total charged-particle multiplicity M_c or transverse energy $E_\perp$ or its variants $E_\perp^{12}$ (limited

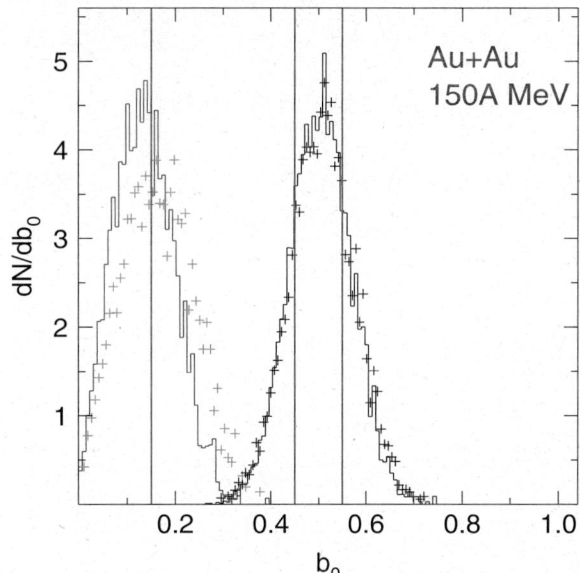

Fig. 1. Simulated reduced impact parameter distributions for Au+Au collisions at $150A\,\mathrm{MeV}$ using the global observables $Erat$ (histogram) or charged-particle multiplicity (crosses) for event selection. The two peaks correspond to nominal centralities $b_0 < 0.15$ and $0.45 < b_0 < 0.55$, respectively, as indicated by the vertical lines.

to $Z \leq 2$) and $Erat$ which, although it involves also the longitudinal kinetic energy, is highly correlated to $E_\perp$ due to energy conservation constraints.

The quality of the achieved selectivity in impact parameter is illustrated in fig. 1. It shows distributions of the scaled impact parameter $b_0 = b/b_{max}$ as obtained from the IQMD transport code [97] simulations for the reaction $^{197}\mathrm{Au} + ^{197}\mathrm{Au}$ at $150A\,\mathrm{MeV}$. We take $b_{max} = 1.15(A_P^{1/3} + A_T^{1/3})\,\mathrm{fm}$ and estimate b from the calculated differential cross-sections for the $Erat$ or multiplicity distributions, using the geometrical sharp-cut approximation. The figure gives an idea of the achievable impact parameter resolution, typically 1 to 2 fm for Au on Au, an unavoidable finite-size effect. The semi-central event class, at this energy, happens to be almost invariant against the choice of the selection method. For the central sample, about $130\,\mathrm{mb}$ here, the $Erat$ selection is somewhat more effective than the multiplicity selection, an observation [30] found to hold for all higher energies studied with FOPI. We also conclude that with this selection technique cross-section samples significantly smaller than $100\,\mathrm{mb}$ cannot be considered as representative of the chosen nominal b value.

In this simulation perfect 4π acceptance was assumed. In reality, limitations of the apparatus will further reduce the achievable selectivity. For the case of FOPI, extensive simulations suggested that the additional loss of performance is small, provided the incident energy per nucleon, E/A, is at least $150\,\mathrm{MeV}$ and the considered range of reduced impact parameter does not significantly exceed $b_0 = 0.5$.

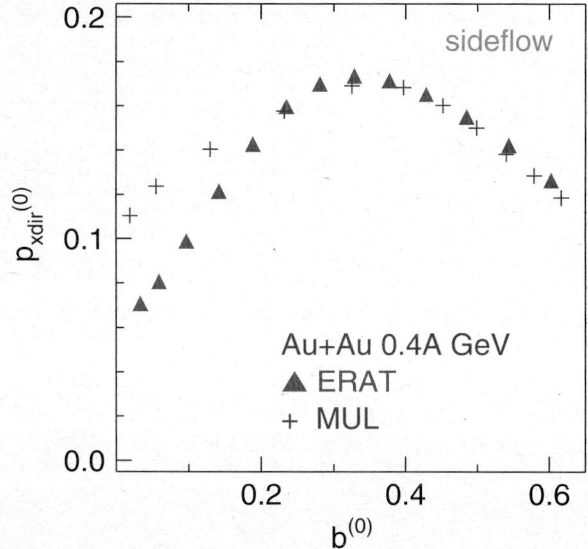

Fig. 2. Mean charge-integrated ($Z \leq 10$) scaled directed flow, $p^{(0)}_{xdir}$ measured with FOPI for Au+Au collisions at $400A$ MeV as a function of the scaled impact parameter b_0 as determined with $Erat$ (triangles) and with the multiplicity of charged particles (crosses), after [7].

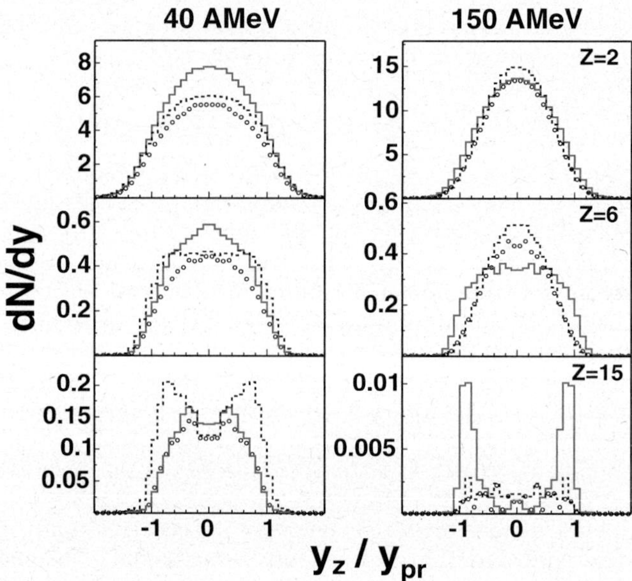

Fig. 3. Longitudinal scaled c.m. rapidity density distributions for $40A$ (left) and $150A$ (right) MeV Au+Au collisions at $b < 2$ fm from the INDRA experiment for selected charges as indicated. Solid histograms: multiplicity selection of the impact parameter, dashed histograms: $Erat$, circles: $E^{12}_\perp$.

At sufficiently high E/A, the measured directed flow can be used for a model-independent comparison of the relative performance of different selection methods. This is illustrated in fig. 2 with FOPI data for the reaction Au+Au at $400A$ MeV and for impact parameter selections using either $Erat$ or the multiplicity of charged particles in the geometrical sharp-cut approximation.

The scaled directed flow is $p^{(0)}_{xdir} \equiv p_{xdir}/u_{1cm}$ where $p_{xdir} = \sum sign(y) Z u_x / \sum Z$ (Z fragment charge, u_{1cm} spatial part of the center-of-mass projectile 4-velocity, $u_x \equiv \beta_x \gamma$ is the transverse projection of the fragment 4-velocity on the reaction plane [98]). The sum is taken over all measured charged particles with $Z < 10$, excluding pions, and y is the c.m. rapidity. For symmetry reasons, $p^{(0)}_{xdir}$ has to converge to zero as $b_0 \to 0$. The figure, therefore, indicates that i) the b resolution is not perfect in either case and ii) for the most central collisions the $Erat$ selection provides a more stringent impact parameter resolution than the multiplicity selection, as already expected on the basis of the simulations (fig. 1). The maximum value of $p^{(0)}_{xdir}$, on the other hand, and the b_0 interval where it is located are robust observables which do not significantly depend on the selection method. Based on these observations, when FOPI data is analyzed, in general one employs a mixed multiplicity-$Erat$ strategy for centrality selection.

Not all global observables behave monotonically with impact parameter, as evident for p_{xdir} from fig. 2. If they are used to select central collisions, an additional cut is required to suppress the high b_0 branch. A non-monotonic behaviour can also result from losses of heavy ejectiles close to zero degree or close to target rapidity. These losses tend to increase with decreasing E/A and (or) increasing

b_0. In the FOPI case we limit our analysis to $b_0 < 0.5$ and require that at least 50% of the total charge has been identified, a moderate, apparatus specific, constraint that does not significantly bias the topology of central collisions.

While for particle multiplicities the idea of a monotonic b correlation is intuitively expected, this is not self-evident for transverse energy. At sufficiently high energy ($\gtrsim 100A$ MeV), transverse energy is increasingly generated by the repeated action of many elementary collisions on the nucleonic level. Since the number of such collisions increases with increasing target-projectile overlap, high transverse energies are correlated with *low* impact parameters. If E/A is smaller than about 100 MeV mean-field effects involving the system as a whole dominate. One observes deflections of the projectile-like and target-like remnants to finite polar angles generating transverse energies that are associated with large impact parameters which carry large angular momenta. This complication can be avoided by using the sum $E^{12}_\perp$ of transverse momenta of light charged particles ($Z \leq 2$) which is more strongly related to the dissipated energy and does not involve properties of heavier fragments.

These complexities are illustrated in fig. 3 using INDRA data for Au+Au at $40A$ MeV and $150A$ MeV. Shown are charge-separated longitudinal rapidity distributions for central collisions, selected with three different observables, multiplicity, $Erat$ and $E^{12}_\perp$. To the extent that stronger yield accumulations near midrapidity indicate higher centrality, the multiplicity binning is more selective of central collisions than $Erat$ at $40A$ MeV while at $150A$ MeV the reverse is true. This appears more pronounced for the cases of larger fragments shown in the lower panels.

For the *Erat* and $E_\perp^{12}$ selections in fig. 3 the central-ity has been defined by all the relevant reaction prod-ucts except the one of interest. This method of excluding the "particle of interest" (POI) from the selection crite-ria allows to avoid autocorrelations between the studied observable and the one used for the estimation of central-ity. On the other hand, the exclusion of the POI makes the observable used for the impact parameter selection particle dependent, *i.e.* no longer globally event depen-dent. This may affect the partitions belonging to a given centrality bin since, depending on the particle, the event may, or may not fulfill the criteria for a given centrality class. It has serious consequences when the autocorrela-tion is strong, especially for low-energy collisions which are characterized by the presence of intermediate and heavy mass fragments carrying substantial amounts of momen-tum. Excluding, or missing, such a fragment unavoidably affects the measure of the impact parameter and increases its fluctuations. $E_\perp^{12}$ does not depend on the exclusion or detection of heavy fragments and thus is better suited for lower energies.

On the other hand, in the case of the INDRA detector, the multiplicity observable does not seem to be the opti-mal centrality selector at high energies (fig. 3, right bot-tom panel) where due to inefficiencies for light particles (multi-hits, punch throughs), this observable may admix less central events with higher multiplicities of fragments to the most central bin. Using $E_\perp^{12}$ as a centrality selec-tor avoids switching the selection method when studying excitation functions. As can be seen in the figure, $E_\perp^{12}$ performs similar to multiplicity at $40A$ MeV and similar to *Erat* at $150A$ MeV. Since molecular-dynamics simula-tions confirm this observation [99], we choose $E_\perp^{12}$ in the following as a centrality measure for the INDRA data, unless indicated otherwise.

4 Rapidity density and stopping

Rapidity distributions in longitudinal (y_z) and in an arbi-trarily fixed transverse direction (y_x) as obtained with the FOPI and INDRA detectors for central Au+Au collisions at $150A$ MeV are shown in fig. 4.

To allow a closer comparison of the shapes the dis-tributions have been normalized to the unit area, indi-vidually for each fragment charge. The *Erat* observable constructed from all detected reaction products except the particle of interest was used as impact parameter se-lector. In the case of the FOPI data, the distributions have been reconstructed for the uncovered phase space and symmetrized with respect to the c.m. rapidity us-ing two-dimensional extrapolation methods [100] in the transverse momentum *vs.* rapidity plane. For $Z = 1, 2$ these corrections represent less than 10% of the total yield, for heavier fragments they amount up to 30%, leading to estimated uncertainties of 10% near midrapidity and of 5% for $|y|/y_{pr} > 0.5$. The INDRA distributions have been corrected for the 10% geometrical inefficiency [49] by multiplying the yields with a factor of 1.11. The po-sitions of the detected particles and fragments were uni-

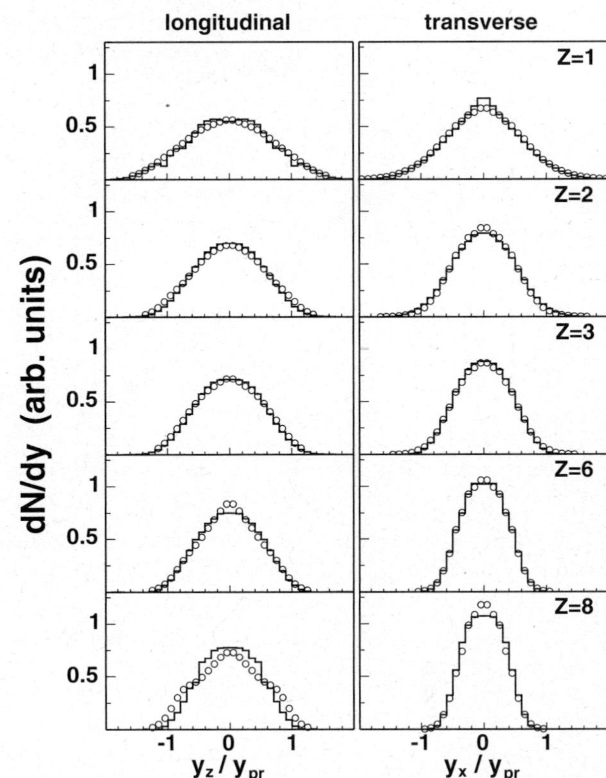

Fig. 4. Yield distributions as a function of the scaled longi-tudinal (left) and transverse (right) rapidity for several frag-ment species within $Z = 1$–8 for central Au+Au collisions at $150A$ MeV measured with FOPI (circles) and INDRA (his-tograms). The impact parameter $b_0 < 0.15$ is selected using *Erat* for both cases, the spectra are normalized to permit an easier comparison of their shapes.

formly randomized within the active area of the detection modules. For $Z = 1$ the backward c.m. distribution was used and reflected into the forward hemisphere which is affected by losses due to punch-through of energetic par-ticles. For heavier charges the forward part was used and symmetrized to profit from the higher granularity of the detector there and to avoid the higher thresholds affecting the yields at backward angles.

Taking into account the systematic errors (not shown in the figure), the agreement of the two independent mea-surements is very good. This feature is far from trivial: due to different acceptances, especially for heavier frag-ments, the composition of the global event selector cannot be made strictly identical for the two detectors. Since at this incident energy the difference between rapidity and velocity is small, one can say that in a naive thermal equilibrium model, ignoring flow and partial transparency effects, the two kinds of distributions, longitudinal and transverse, ought to be equal, with the common variances being a measure of the (kinetic) temperature. Clearly, this is not the case, the transverse widths are smaller than the longitudinal widths, even though the selection method, using maximal *Erat*, is definitely biased towards isolating the event sample (on the 130 mb level) with the largest

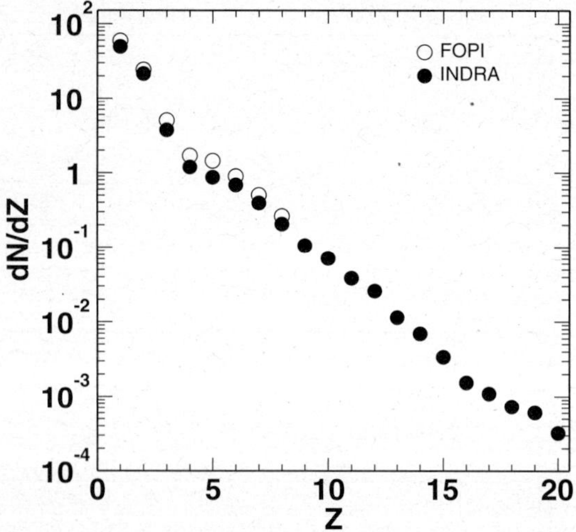

Fig. 5. Charge multiplicity distributions in central ($b_0 < 0.15$) collisions of Au+Au at $150A$ MeV. Open circles: FOPI data; closed circles: INDRA data.

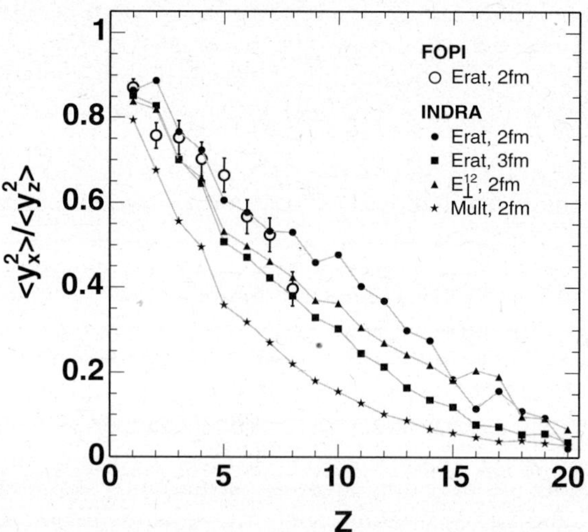

Fig. 6. Ratio of transverse to longitudinal variances for central Au+Au collisions at $150A$ MeV as measured with the FOPI and INDRA detectors (open and filled symbols, respectively). The variances were obtained for scaled c.m. rapidities in the range $-1 \leq y \leq 1$. The chosen selections of centrality are indicated in the legend.

ratio of transverse-to-longitudinal variances (although, as mentioned earlier, autocorrelations were removed).

The integration over rapidity yields absolute charged-particle distributions dN/dZ. The results for Au+Au at $150A$ MeV are shown in fig. 5. For FOPI, only $Z \leq 8$ yields are available at this energy while the INDRA data extend over almost 6 orders of magnitude up to $Z = 20$. The observed yields of heavy fragments are small, however, only about 2-3% of the available charge is clusterized in fragments with $Z > 8$, as expected at this energy where the c.m. collision energy amounts to four times the nuclear binding energy. With these 3% added, the FOPI data account for 97% of the total system charge which is consistent with the 4π-reconstruction method. The INDRA yields are systematically lower than FOPI by between 10% and 30%. The lower $Z = 1$ yield is mainly responsible for the detection of only 80% of the total system charge with INDRA but similar differences are also observed for larger Z. They are most likely caused by reaction losses and edge effects in the detectors which reduce the effective solid-angle coverage if Z identification is required. The light-particle yields may also be affected by the higher multi-hit probabilities at this incident energy at the upper end of the INDRA regime. Extrapolating these observations over the full range of incident energies studied in this work, one may expect that reaction losses and the multi-hit probability are considerably reduced at lower incident energies for INDRA while the missing yields at large Z in the FOPI case will be negligible at higher energies for the mainly central and mid-central collisions that are of interest here.

The ratio of the variances of the transverse and longitudinal rapidity distributions has recently been proposed as a measure of the degree of stopping reached in nuclear collisions [7]. The ratios obtained for central Au+Au collisions at $150A$ MeV, after integration over the range of scaled c.m. rapidity $-1 \leq y \leq 1$, are shown in fig. 6

as a function of Z. The open circles represent the FOPI data with error bars which include the systematic uncertainty of the reconstruction procedure. The INDRA data are shown for $Z \leq 20$ and for four different impact parameter selections as indicated in the figure.

The largest ratios from 0.8 to 0.9 are observed for light charged particles ($Z \leq 2$). With increasing fragment Z, the ratios decrease continuously to values of < 0.1 near $Z = 20$. In the common range of fragment Z and for the same impact parameter selection ($Erat, b \leq 2$ fm), the ratios measured with FOPI and INDRA are in good agreement. The selection with $Erat$ and $b \leq 3$ fm yields slightly smaller ratios as expected which, however, are similar to those obtained with $E_\perp^{12}$. Large transverse momenta of light charged particles and of fragments are apparently correlated. Autocorrelations are not present here because the particle of interest is removed from the impact parameter selector (see previous section). The smallest ratios of variances are obtained for selections according to multiplicity.

The trends as observed as a function of Z suggest that the heavier fragments, even in rather central collisions, experience less stopping than lighter ones and keep a strong memory of the entrance channel motion. Their transverse momenta seem to be, nevertheless, generated in collisions involving nucleons or light clusters as evident from the correlation with $E_\perp^{12}$. The momenta of struck nucleons absorbed in a cluster or the recoil momenta of nucleons knocked out from a cluster both contribute to their final momenta. Their relative weight will be smaller in larger fragments, consistent with the observed Z-dependence. Overall, these observations are clearly in contradiction to the assumption of global equilibrium including the kinetic degrees of freedom. Qualitatively, they agree with the pre-

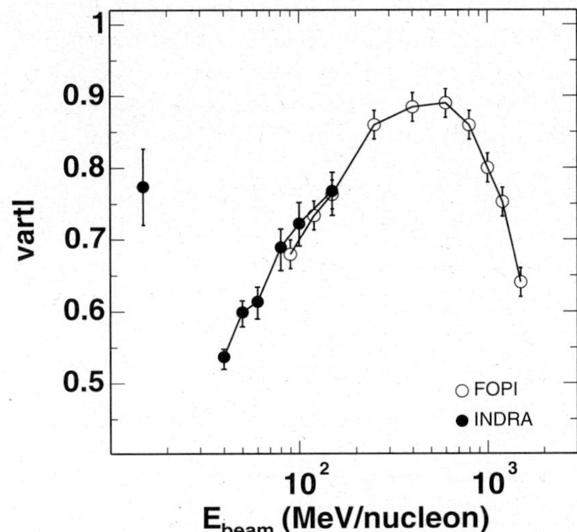

Fig. 7. Excitation function of the degree of stopping, *vartl*, in central Au+Au collisions ($b \leq 2\,$fm) obtained from the FOPI (open circles) and INDRA (dots) measurements. The result at $15A\,$MeV corresponds to a less central selection ($b \leq 5\,$fm).

dictions of quantum molecular-dynamics calculations for fragment production in this energy range [101].

A global observable to describe stopping, *vartl*, has been introduced in ref. [7]. It is defined as the ratio of the transverse over longitudinal variances of the summed and Z-weighted rapidity distributions. The excitation function of this observable is presented in fig. 7 for central Au+Au collisions ($b \leq 2\,$fm) and for the full energy range covered in the FOPI and INDRA experiments.

The FOPI results have been obtained with the *Erat* selection. The data have recently been reanalyzed by taking additional small corrections due to energy losses in structural parts of the detector into account. At the lowest three energies, this has led to an increase of *vartl* by up to about 10% compared to the data published in [7], while for the other energies the results are unaffected. The measured range of fragments extends up to $Z = 6, 8, 8$ for $E = 90A, 120A, 150A\,$MeV, respectively. The contribution of heavier fragments to the *vartl* observable has been estimated by extrapolating their weights and variance ratios to higher Z. At $90A\,$MeV this correction amounts to about 8%. The errors given in the figure are systematic and mainly reflect the uncertainty of the reconstruction procedure.

For the INDRA central event samples, the light-particle transverse energy $E_{\perp}^{12}$ has been used to select $b \leq 2\,$fm for $40A$ to $150A\,$MeV and $b \leq 5\,$fm for the data sample at $15A\,$MeV with low statistics, originally only taken for calibration purposes. For the charge-weighted average, fragments up to $Z = 60$ have been included. Except for the result at $15A\,$MeV, the error bars correspond to the variation of *vartl* for centrality selections within $b \leq 1.5\,$fm (upper end of the error bar) and $b \leq 2.5\,$fm (lower end) and thus represent the systematic uncertainty associated with the impact parameter determination. The statistical errors are below the percent level, except at

$15A\,$MeV, where they are the main contribution to the error shown in the figure.

The obtained excitation function of stopping is characterized by a broad plateau extending from about $200A$ to $800A\,$MeV with fairly rapid drops above and below. The highest value reached by *vartl* is about 0.9. With the INDRA data, the reduction of stopping at lower incident energies is followed down to $40A\,$MeV. In the overlap region, a very satisfactory agreement within errors is observed. The measurement at $15A\,$MeV suggests that stopping goes through a minimum at or below $40A\,$MeV.

It is clear that only a dynamical theory will be able to reproduce this excitation function. Using the relativistic Boltzmann-Uehling-Uhlenbeck (RBUU) transport model, an analysis of the combined FOPI stopping and flow (see later) data was recently presented [102]. The input to computer codes implementing transport theoretical models are the nuclear mean field U (or EoS) and the nucleon-nucleon (nn) cross-sections σ_{nn}. Although both are not independent in a consistent theory, it is useful to consider their effects separately. In general one finds that the cross-sectional part is dominant over the mean field part for a quantitative account of the observed incomplete stopping: note that if global equilibrium, or even local equilibrium (ideal hydrodynamics), were valid cross-sections would be irrelevant. Starting at the low-energy end one qualitatively expects, when raising the energy, that the increasingly repulsive mean field (due to increasing compression) and the drop in Pauli blocking of final and intermediate states in nn scattering (due to the increasing initial rapidity gap) conspire to raise rapidly the generation of transverse energy at the expense of the longitudinal energy. At the higher-energy end (say beyond $1A\,$GeV) again both aspects (mean-field and collisions) more or less may add up to make the drop faster. At $1.5A\,$GeV roughly one quarter of the nucleons are excited to a resonant state. The opening up of nucleonic degrees of freedom may lead to a softening of the EoS. On the other hand, the in-medium Dirac masses M_D^* are predicted to drop substantially in covariant theories [103–105], a fact that will seriously modify the phase-space and kinematical factors influencing the elementary cross-sections [106–108]. The calculations of ref. [102] suggest that these in-medium modifications of σ_{nn} are indeed necessary to reproduce the observed stopping.

Besides the "global" information shown in fig. 7 the "particle differential" information reveals additional information on the stopping mechanisms. Figure 6 shows that the partial transparency is predominantly experienced by the heavier fragments, which presumably have survived because their constituent nucleons have suffered a less violent average collision history. This feature is also observed at the high-energy end, although the "heavy-fragment" role is played there by mass $A = 2$–4 ejectiles [109]. Restricting the stopping observable to the lightest species at the various incident energies, one obtains higher *vartl* values and flatter excitation functions. The combined role of the mean field and of in-medium modified cross-sections will be picked up again in sect. 7 where the flow information will be added to the analysis.

5 Flow, reaction plane and corrections

Originally, the directed flow has been quantified by measuring the in-plane component of the transverse momentum [98] and the elliptic flow by parametrizing the azimuthal asymmetries using the Fourier expansion fits [110, 111]. More recently, it has been proposed [112] to express both, directed and elliptic flow in terms of the Fourier coefficients (v_1 and v_2, respectively) and also to investigate the higher flow components. The coefficients v_n are obtained by means of the Fourier decomposition [112–114] of the azimuthal distributions measured with respect to the true reaction plane:

$$\frac{\mathrm{d}N}{\mathrm{d}(\phi - \phi_R)} = \frac{N_0}{2\pi}\left(1 + 2\sum_{n \geq 1} v_n \cos n(\phi - \phi_R)\right) \quad (2)$$

with ϕ_R being the azimuth of the latter. In general, the coefficients $v_n \equiv \langle \cos n(\phi - \phi_R)\rangle$ may depend on the particle type, rapidity y and the transverse momentum p_T.

The standard methods of measuring flow can be split into those using explicitly the concept of the reaction plane [98, 112–114] and those based on the two-particle azimuthal correlations [115]. Still other methods have been proposed recently, satisfying the needs of high-energy experiments: the "cumulant" methods [116–118] using multiparticle correlations and the method based on the Lee-Yang theory of phase transitions [119, 120]. The latter is expected to perform well above about $100A$ MeV [119], while the three-particle variant of the "cumulant" method is claimed to be useful for extracting v_1 coefficients at energies near the balance energy and in the ultrarelativistic regime [118]. However, because the correlation methods require high event multiplicities and high-statistics data, and because the correlation between a particle and the flow vector is usually much stronger than that between two particles [121], the reaction plane methods are still more commonly used at intermediate energies. They have also been applied in the present case.

Since detectors do not allow to measure the angular momenta and spins of the reaction products, the orientation of the reaction plane can only be estimated using the momenta. The resulting azimuthal angle, ϕ_E, has a finite precision, and the measured coefficients v_n^{meas} are thus biased. They are related to the true ones through the following expression [113]:

$$v_n^{meas} \equiv \langle \cos n(\phi - \phi_E)\rangle = v_n \langle \cos n\Delta\phi\rangle, \quad (3)$$

where the average cosine of the azimuthal angle between the true and the estimated planes, $\langle \cos n\Delta\phi\rangle \equiv \langle \cos n(\phi_R - \phi_E)\rangle$, is the required correction (also referred to as "event plane resolution" or just "resolution") for a given harmonic. Note that, since the true values of flow are obtained by dividing by the average cosine, they are always larger than the measured ones.

The literature offers many different methods to estimate the reaction plane, like the flow-tensor method [78], the fission-fragment plane [122], the flow Q-vector method [98], the transverse momentum tensor [123] (also called "azimuthal correlation" [124]) method or others [125].

Among them, the Q-vector method has received special attention. Originally, the Q-vector has been defined as a weighted sum of the transverse momenta of the measured N reaction products [98]:

$$\mathbf{Q} = \sum_{i=1}^{N} \omega_i \mathbf{p}_i^{\perp} \quad (4)$$

with the weights ω chosen to be $+(-)1$ for reaction products in the forward (backward) c.m. hemisphere and with the possibility to exclude the midrapidity zone. The choice of the optimal weights is discussed in [72, 114, 117, 126, 127]. Definition (4) can be extended to Q-vectors built from higher harmonics [114], thus e.g. allowing to profit from strong elliptic flow, when applicable. Usually, in the flow studies, the POI is excluded from the sum in (4) to avoid autocorrelations. This does not concern the corrections, since the sub-events (see below) do not share particles.

The corrections for the reaction plane dispersion can be obtained using various methods [98, 112–114, 121, 123, 128–131]. What they all have in common, is the underlying assumption of the applicability of the central-limit theorem. In most of these methods the correction is searched for using the sub-event method [98], which consists in splitting randomly each event into two equal-multiplicity sub-events and getting the correction from the distribution of the relative azimuthal angle, $\Delta\Phi_{12}$, between their individual Q-vectors ("sub-Q-vectors"). This is done either by using the small-angle expansion [98] or by fitting with a theoretical distribution [113]. Instead of fitting the angular distributions one can alternatively fit the distributions of the magnitude of the total Q-vector itself [112, 114].

Assuming the Gaussian limit, ref. [113] gives an analytical formula for the distribution of $\Delta\Phi_{12}$ for the case that the distributions of sub-Q-vectors are independent and isotropic around their mean values. In refs. [112, 114] one can find the formulae relevant for the distributions of the magnitude of the Q-vector.

These methods proved their usefulness for correcting measured flow values at higher energies (see, e.g., [18, 28, 98, 132, 133]) which fulfill the high-multiplicity requirement. They are, however, not adequate for the intermediate-energy reactions, below about $100A$ MeV, where the particle multiplicities are lower and the events are characterized by a broad range of masses of the reaction products. Here, the applicability of the central limit theorem for devising the corrections is less obvious.

Figure 8 illustrates the difficulties one encounters at intermediate energies. It shows the experimental distributions of $\Delta\Phi_{12}$ as measured with the INDRA detector for the Au+Au reaction at $40A$ (top) and $150A$ (bottom) MeV and for two intermediate centrality bins. The lines represent the fits obtained with the method described

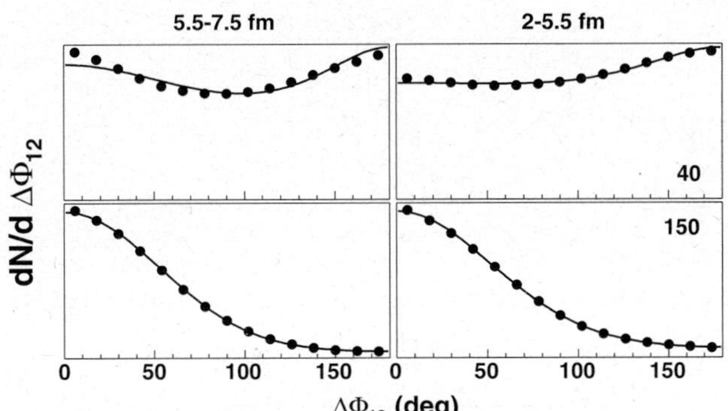

Fig. 8. Distributions of relative angle between reaction planes for two random sub-events (INDRA data, dots) and fits (lines) using the integrated product of bivariate normal distributions (eq. (5)) for centralities $b \simeq 5.5$–7.5 fm (left) and $b \simeq 2$–5.5 fm (right) and incident energies $40A$ MeV (top panels) and $150A$ MeV (bottom panels).

briefly below. The standard method [113] can, in principle, be used to derive the corrections for energies down to about $80A$ MeV, however it fails to describe distributions like those at $40A$ MeV with double maxima or maxima at backward angles, which reflect the presence and importance of the in-plane enhancement and of the correlation between sub-events.

Since at low and intermediate energies the sub-events are expected to be strongly correlated [129] and the distributions of the Q-vector no longer necessarily isotropic [113], we have extended the method of Olli-trault [113] by explicitly taking into account these two effects in the theoretical distribution of the sub-Q-vectors. The new method relies on the assumption of the Gaussian distribution of the flow sub-Q-vectors. This assumption has been verified to hold even at $40A$ MeV, except for very peripheral collisions, by performing tests with the CHIMERA-QMD model in which angular momentum is strictly conserved [134].

The form of the joint probability distribution of the random sub-Q-vectors has been searched for following the method outlined in appendix A of [121], by imposing the constraint of momentum conservation on the N-particle transverse momentum distribution and using the saddle-point approximation.

The resulting distribution has the form of a product of two bivariate Gaussians:

$$\frac{\mathrm{d}^4 N}{\mathrm{d}\mathbf{Q}_1 \mathrm{d}\mathbf{Q}_2} = \frac{1}{\pi^2 \sigma_{sx}^2 \sigma_{sy}^2 (1-\rho^2)}$$

$$\cdot \exp\left[-\frac{(Q_{1x}-\bar{Q}_s)^2 + (Q_{2x}-\bar{Q}_s)^2 - 2\rho(Q_{1x}-\bar{Q}_s)(Q_{2x}-\bar{Q}_s)}{\sigma_{sx}^2(1-\rho^2)} \right.$$

$$\left. -\frac{Q_{1y}^2 + Q_{2y}^2 - 2\rho\, Q_{1y}Q_{2y}}{\sigma_{sy}^2(1-\rho^2)} \right], \quad (5)$$

where we followed the convention of [113] of including the $\sqrt{2}$ in σ; the subscripts 1, 2 refer individually and s generally to sub-events; the subscripts x and y refer to the

in- and out-of-plane direction, respectively. This distribution differs from those proposed in [113, 121, 129] in that it combines all three effects that influence the reaction plane dispersion at intermediate energies, namely the directed flow (through the mean in-plane component $\bar{Q}_s$ or the resolution parameter $\chi_s \equiv \bar{Q}_s/\sigma_{sx}$ [113]), the elliptic flow (through the ratio $\alpha \equiv \sigma_{sx}/\sigma_{sy}$) and the correlation between the sub-events [129] (through the correlation coefficient $\rho \in [-1,1]$). It reduces to the one of [113] for $\alpha = 1$ and $\rho = 0$, and to the one of [121] for $\alpha = 1$. In deriving eq. (5) it was assumed that the in- and out-of-plane correlation coefficients are equal.

Making the division into sub-events random ensures that the distributions of the sub-Q-vectors are equivalent, in particular they have the same mean values and variances. Since the total-Q-vector is the sum of the sub-Q-vectors, $\mathbf{Q} = \mathbf{Q}_1 + \mathbf{Q}_2$, one finds the following relation between the resolution parameter obtained from the distribution of the Q-vector, χ, and that obtained from the distribution of sub-Q-vectors, χ_s:

$$\chi = \chi_s \sqrt{2/(1+\rho)}. \quad (6)$$

Relation (6) shows how the correlation between sub-events influences the reaction plane resolution. In particular, it indicates that the resolution improves in case the sub-events are anti-correlated ($\rho < 0$), which is predicted to be the case below about $150A$ MeV except for very peripheral collisions [134].

As in [113], the joint probability distribution (5) is used after integrating it over the magnitudes of the sub-Q-vectors and one angle, leaving the $\Delta\Phi_{12}$ as the only independent variable. Unlike in [113], the resulting distribution cannot be presented in an analytical form. It depends on 3 parameters (χ_s, α, ρ) which can be obtained from fits to the experimental or model data. The quality of the obtained fits is very good, even in the non-standard cases encountered at low energies (fig. 8).

The corrections for the n-th harmonic v_n, depending now on χ and α, can be calculated (also numerically) as the mean values of the $\cos n\Delta\phi$ obtained over the total-Q-vector distribution, in a similar way as in [113]. Figure 9 shows how the elongation of the Gaussian (α), resulting from elliptic flow, modifies the corrections for the first two harmonics. It demonstrates that the in-plane emissions ($\alpha > 1$) enhance slightly the resolution for v_1 and considerably for v_2, even in the absence of the directed flow. On the other hand, squeeze-out ($\alpha < 1$) deteriorates the resolution. The figure, in particular, shows that the correction can change the sign of elliptic flow in the case of small directed flow and squeeze-out.

Since the correlation between the sub-events increases at the lower energies, the knowledge of the correlation coefficient ρ becomes crucial. Estimated values, obtained from model calculations, can be useful as constraints for ρ in the fitting procedure. For example, the CHIMERA-QMD calculations predict ρ to be around -0.43 for $40A$ MeV and $2 < b < 8$ fm and about -0.2 at $150A$ MeV. Alternatively, mean values of some rotational invariants which are derived from the measured data can be used to

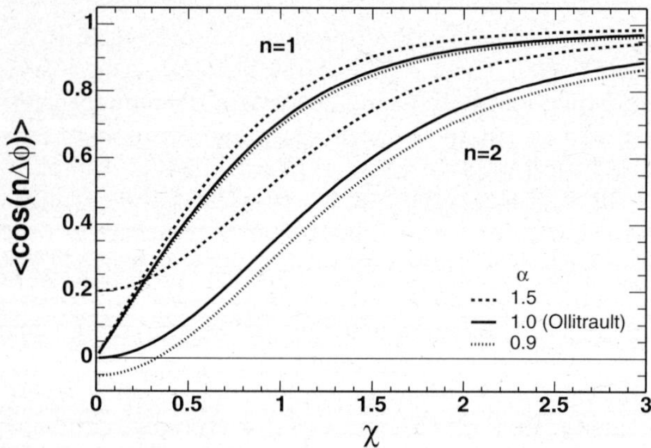

Fig. 9. Corrections for the first (3 upper curves) and the second (3 lower curves) harmonic as a function of the resolution parameter, χ, for different aspect ratios, $0.9 \leq \alpha \leq 1.5$, covering approximately the range of its variation.

reduce the number of fit parameters and to constrain the fitting routine to search for a conditional minimum [135].

Instead of fitting the azimuthal distributions one can express the probability distribution (5) in terms of the components of one of the sub-Q-vectors in the reference frame of the other, or in terms of the absolute values of the sum and of the difference of the sub-Q-vectors. The corresponding 2-dimensional experimental distributions can then be fit using such formulae. The method of fitting the distributions of components of the sub-Q-vector has been found sensitive enough to perform well without additional constraints.

The corrections obtained using various methods are presented in fig. 10. They are close to one, independent of the method, for the range of higher incident energies ($E > 100A$ MeV) where the directed flow is large and the reaction plane well defined by the high-multiplicity distribution of detected particles. At around $50A$ MeV, they go through a minimum and depend strongly on the chosen method. The FOPI flow results, as published in refs. [18, 28], have been corrected using the standard method, excluding the midrapidity region of ± 0.3 of the scaled c.m. rapidity from the Q-vector to improve the resolution. The corrections used here for the INDRA data are obtained with the new method in two ways, by fitting the azimuthal distributions and by fitting the distributions of components of the sub-Q-vectors. The mean values are given in fig. 10 (full circles) with error bars representing the systematic uncertainty as given by the difference of these results. At $15A$ MeV, the statistical errors dominate. Even at their minima, the corrections are not smaller than 0.6 and 0.5 for directed and elliptic flow, respectively, indicating that the measured flow values will increase, after applying the corrections, by no more than about a factor of two.

For a comparison of the different methods and of their applicability, also the corrections according to the standard method of Ollitrault [113] have been determined. This corresponds to fixing the parameters $\alpha = 1$ and $\rho = 0$

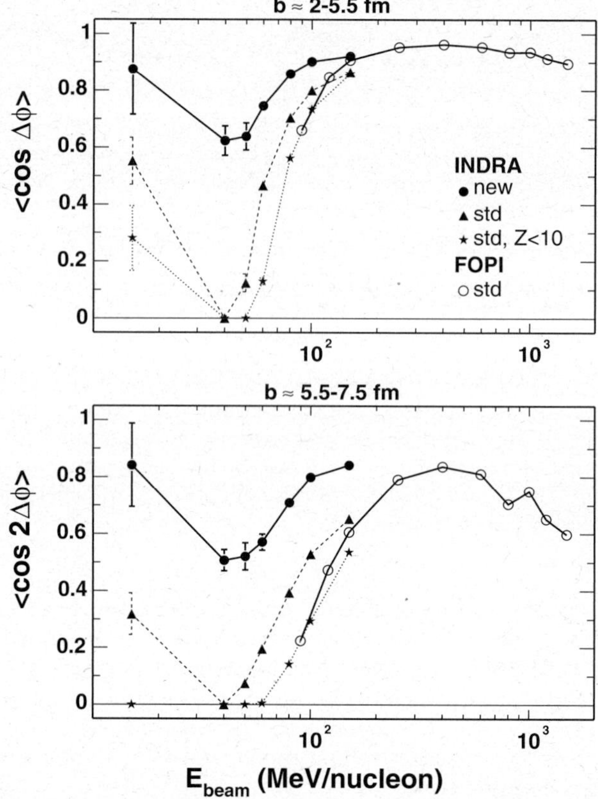

Fig. 10. Corrections for the first harmonic and impact parameter $b = 2$–5.5 fm (top) and for the second harmonic and $b = 5.5$–7.5 fm (bottom) used in the flow analysis of the Au+Au reaction. The corrections for the INDRA data (full symbols) are shown as obtained with the new method (dots, see text), with the standard (std) method (ref. [113], triangles), and with the standard method and the restriction $Z < 10$ (stars). Open circles represent the results for the FOPI data obtained with the standard method.

in the new method. Near $50A$ MeV, the results are close to zero which would require nearly infinitely large corrections (triangles in fig. 10). The figure, furthermore, shows the same corrections according to the standard method as obtained for the FOPI case (circles). They are very similar and, in the overlap region, virtually identical to the result for INDRA if the limit $Z < 10$ of the FOPI acceptance is applied in the INDRA case (stars). This very close agreement is not unexpected because good agreement was already observed for the uncorrected flow data obtained with the two detection systems [22, 28, 33]. The standard method, nevertheless, fails below about $80A$ MeV. As mentioned above, the independent, isotropic Gaussian approximation is no longer confirmed by satisfactory fits of the experimental distributions.

Several additional observations and comments can be made. Comparing the results of the new and standard methods (filled circles and triangles) shows a dramatic improvement of the resolution obtained by taking the effects of the correlation between the sub-events and of the in-plane enhancement into account. It should be stressed, that both these effects are responsible for the finite correction for the directed flow at $40A$ MeV, near the expected balance energy [20]. Non-vanishing resolution, suggested by the new method, indicates that even here the reaction plane can be defined, and apparently, questions the occurrence of the "global" balance, which otherwise would manifest itself with the vanishing of $\langle \cos \Delta\phi \rangle$. However, the finiteness of the corrections around $40A$ MeV may also partially result from the incompleteness of the experimental data and from the mixing of events with different centralities, which may add up to mask the signal of the "global" balance. The fitting procedure yields relatively accurate results for the corrections for the first two harmonics in case of the complete results of the simulations (*e.g.*, 2–5% accuracy for $40A$ MeV and 0.2–0.4% for $150A$ MeV and $4 < b < 8$ fm [134]), and in particular, is able to reveal the signal of the "global" balance, but in the case of experimental data it will certainly return some effective corrections biased by the experimental uncertainties and inefficiencies. The effects of the latter may not necessarily drop out by applying eq. (3) to correct the measured observables, but may require additional corrections. At the higher energies, the results of the standard and the new methods approach each other but, in the overlap region of the FOPI and INDRA experiments, the differences are still significant, and need further investigation.

Comparing the less and the more complete data sets (stars and triangles, respectively) shows that the resolution improves with the completeness of the data. Triangles represent INDRA events with at least 50% of the system charge collected to which an additional single fragment carrying the missing momentum and charge was added. This artificial completion of events was found important for peripheral collisions where, due to the energy thresholds, the heavy target-like fragment is always lost. The distributions of the relative angle between sub-events become then narrowly peaked at small relative angles which improves the resolution of the reaction plane.

However, it is not only the reaction plane correction that relies on the completeness of the measured data. Also the measured v_n^{meas} parameters are affected by the non-isotropic loss of particles due to multi-hits (INDRA) or unresolved tracks in high-track-density regions (FOPI). A rough estimate of the correction [134] due to multi-hit losses for v_2 can be obtained, for segmented detectors like INDRA, by using the unfolded "true" in- and out-of-plane multiplicities and calculating the true and measured mean v_2 by integrating the azimuthal distribution (2) over the in- and out-of-plane quadrants. The "true" multiplicities can be estimated using the calculated (*e.g.*, in a way similar to that of ref. [136]) or simulated (using the detector filter and the model data) multiplicity response function specific for a given detector. An analogous procedure can be applied also for v_1; however, due to the lack of the forward-backward center-of-mass symmetry of the detector, the results may be less accurate. The flow parameters obtained from the INDRA data presented in the next section have been additionally corrected for the multi-hit

losses using the above procedure. For v_1, these additional corrections vary from about 7% at $40A$ MeV to about 33% at higher energies for $Z = 1$ and do not exceed 15% for $Z = 2$. For v_2 and $Z = 1$ they increase from about 18% at $40A$ MeV to about 36% at $100A$ MeV and about 70% at $150A$ MeV, for the centrality bins in question. Within this simple procedure, the corrections depend essentially on the average of the in- and out-of-plane multiplicities and only weakly on their difference, that is why the corrections basically increase with the increasing multiplicity (thus with the centrality and incident energy). This explains the large correction factor at $150A$ MeV. Nevertheless, since v_2 is small at this energy, the absolute change of the measured value due to the correction is small compared to that at lower energies.

Generally, one may remark that, at energies below about $100A$ MeV, devising the corrections becomes a delicate task. The corrections are no longer those in the usual sense, say, of a few percent. Depending on the method, they may change the measured results by a large factor, mainly because of the smallness of directed flow around $40A$–$50A$ MeV. The accuracy relies in addition on the completeness of the data. Flow data free of reaction plane dispersions are, nevertheless, very desirable since they allow to compare the results obtained with different detectors. They are also of great interest from the theoretical point of view, by permitting the direct comparison with the model predictions. In problematic cases, however, detailed filtering of the model results and treating them with the experimental type of analysis may still be necessary, if not for the direct comparison on the level of uncorrected observables, then for the reliable estimate of the systematic uncertainties associated with the correction scheme.

The effects of momentum conservation, distortions due to removal of the particle of interest (expected to be important at low multiplicities (energies)) and possible corrections to the reaction plane resolution due to the detector inefficiencies (missing part of the Q-vector) remain a subject for future study.

6 Directed and elliptic flow

The rapidity dependence of the slope of the directed-flow $\partial v_1/\partial y$ at midrapidity for $Z = 1$ and 2 particles, integrated over transverse momentum, is shown in fig. 11. The INDRA data is combined with the FOPI data (published for $Z = 2$ in [18]), both measured for mid-central collisions with impact parameters of 2–5.5 fm and shown after correcting for the reaction plane dispersion. The FOPI data has been corrected using the method of [113] while the INDRA data has been corrected using the method outlined in sect. 5. In both data sets the reaction plane has been reconstructed using the Q-vector method with the weights $\omega = sign(y_{cm})$, excluding the POI. In case of the FOPI data the midrapidity region of ± 0.3 of the scaled rapidity has been excluded from the Q-vector to improve the resolution. The INDRA data has been corrected for the effects of momentum conservation [137]. In both cases linear fits have been performed in the range of ± 0.4 of the

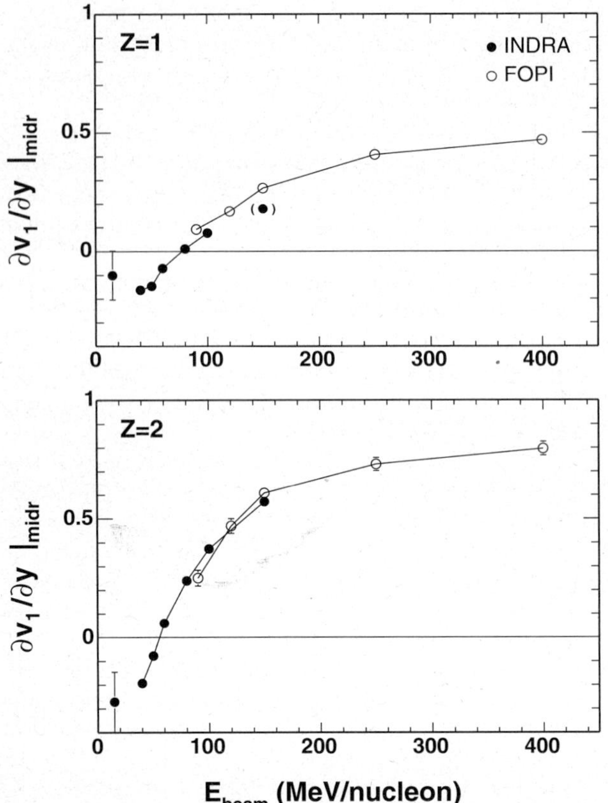

Fig. 11. Slopes of directed flow $\partial v_1/\partial y$ for $Z = 1$ (top) and $Z = 2$ (bottom) particles integrated over p_T for mid-central collisions (2–5.5 fm). The open and filled symbols represent the FOPI [18] and the INDRA data, respectively. The uncertainty at $15A$ MeV is mainly statistical. The INDRA point, in brackets, at $150A$ MeV in the top panel is biased due to experimental inefficiencies for $Z = 1$ at this energy.

scaled c.m. rapidity, except for the $15A$ MeV data where the range of ± 0.55 was used.

The excitation function of the slope of the directed flow at midrapidity for $Z = 1$ changes sign around $80A$ MeV. The apparent minimum around $40A$ MeV is mostly suggested by the $15A$ MeV data point and should be confirmed by other measurements. The FOPI data has been additionally corrected for the effects of unresolved tracks in the in-plane high-track-density region. This correction influences also the slope of the v_1 rapidity distribution, increasing it by up to 15% for $Z = 1$ and up to 5% for $Z = 2$. The INDRA results have been corrected for the effect of multi-hit losses (see sect. 5). The still apparent discrepancy between the INDRA and FOPI results at $150A$ MeV can be partially attributed to the losses of $Z = 1$ particles due to punch-through effects in the INDRA detector at high energies. Up to 10% of the difference may also come from the different methods used for correcting the reaction plane dispersions (see fig. 10).

For $Z = 2$, the slope of v_1 is seen to rise monotonically with energy over the full range of 15 to 400 MeV per nucleon which is covered by the two experiments. Here, the agreement in the overlap region is slightly better re-

flecting the better efficiency of the INDRA detector for $Z = 2$ particles. The trends observed for the uncorrected data [22] for v_1 are preserved. Unlike in ref. [20], the excitation function does not show a clean signature of a minimum (see ref. [22] for discussion). It changes sign between 50 and 60 MeV per nucleon, in agreement with the extrapolated values of the balance energy, E_{bal}, obtained from the higher-energy measurements [43, 138, 139].

Negative flow is observed not only for $Z = 1, 2$ (fig. 11) but with even larger slopes also for other light fragments. This intriguing phenomenon has already been reported for the lighter systems ^{40}Ar + ^{58}Ni, ^{58}Ni + ^{58}Ni, and ^{129}Xe + ^{nat}Sn, provided the "1-plane-per-particle" method was used for estimation of the reaction plane [21]. For these systems, a balance energy has been determined by associating it with the minima of the approximately parabolic excitation functions of the flow parameter which, in the cases of ^{40}Ar + ^{58}Ni and ^{58}Ni + ^{58}Ni, appeared at negative flow values. Negative flow values of light reaction products can indeed be measured experimentally, provided the detector is able to measure "quasi-complete" events, including the heavy fragments. Then, the observed anti-flow of light products is measurable relative to the reaction plane fixed and oriented by the heavy remnants.

A possible scenario of the anti-flow has been proposed for the lighter systems in [21], and for the heavy systems, emphasizing the role of the strong Coulomb field, in [22]. Despite the appeal of a globally defined balance energy, it is worth noticing that directed flow apparently never vanishes completely. It was shown with BUU calculations that at the balance energy the flow cancellation results from a complex transverse momentum dependence and that the flow pattern is influenced by EoS and σ_{nn} [140]. The presently available differential data, measured by FOPI down to $90A$ MeV [18] suggest that the change of sign of v_1 is dependent, in addition to transverse momentum, also on particle type and rapidity.

The results on v_2 measured at midrapidity are summarized in fig. 12. Elliptic flow varies as a function of energy from a preferential in-plane, rotational-like [141–143], emission ($v_2 > 0$) to an out-of-plane, or "squeeze-out" [37] ($v_2 < 0$) pattern, with a transition energy of about $150A$ MeV. This transition energy is larger than that for the directed flow (see above and the discussion in ref. [139]) and was shown to depend on centrality, particle type and transverse momentum [28, 30]. For higher energies, the strength of the collective expansion overcomes the rotational-like motion, leading to an increase of out-of-plane emission. A maximum is reached at $400A$ MeV, followed by a decrease towards a transition to preferential in-plane emission [29, 144]. This behavior is the result of a complex interplay between fireball expansion and spectator shadowing [28], with the spectators acting as clocks of the expansion times. For instance, in the energy range $400A$–$1500A$ MeV, the passing time of the spectators decreases from 30 to 16 fm/c, implying that overall the expansion gets about two times faster in this energy range. This interpretation is supported by the observed

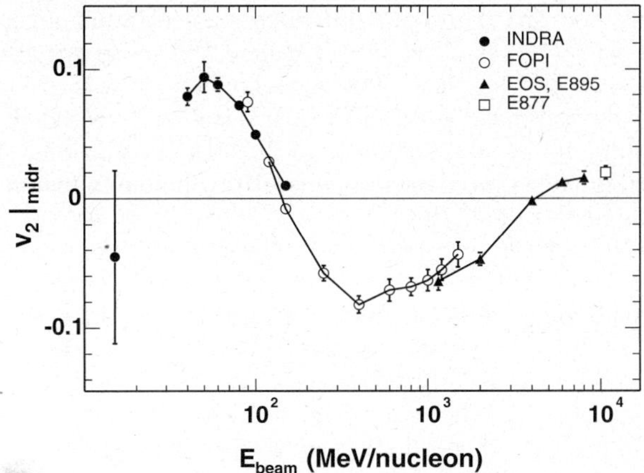

Fig. 12. Elliptic flow parameter v_2 at mid-rapidity for collisions at intermediate impact parameters (about 5.5–7.5 fm) as a function of incident energy, in the beam frame. The filled and open circles represent the INDRA and FOPI [28] data, respectively, for $Z = 1$ particles, the triangles represent the EOS and E895 [29] data for protons and the square represents the E877 data [144] for all charged particles.

scaling of elliptic flow as a function of transverse momentum scaled with beam momentum [28]. We note that the energy dependence of elliptic flow is similar to that of directed flow [1,2,7], with the extra feature of the transition to in-plane flow at $4A$ GeV [29]. This high-energy transition has received particular interest as it is expected to provide a sensitive probe of the EoS at high densities [145]. At SPS and RHIC energies, the in-plane elliptic flow is determined by the pressure gradient-driven expansion of the almond-shaped isolated fireball [146] and is currently under intensive experimental investigation [147–149].

The agreement between the corrected INDRA and FOPI data is good. The INDRA results have been corrected using the new method, including the correction for the multi-hit losses (see sect. 5). According to IQMD calculations, the reaction plane correction for the lowest FOPI energy of $90A$ MeV appears to be somewhat overestimated. On the other hand, this may partially compensate for the lack of corrections due to unresolved tracks which were not applied for v_2 in the FOPI case. Overall, the differences between the corrections is small enough, so that comparisons of uncorrected data sets are already meaningful. A good agreement was found to exist for the INDRA [22,33], FOPI [28] and Plastic Ball [37] data in the reference frame of the directed flow and without the correction for reaction plane resolution [28,33].

A remarkable feature of the v_2 observable is that it allows to show a continuous evolution over a region covering completely different reaction mechanisms, from those dominated by the mean field near the deep inelastic domain, and the multifragmentation in the Fermi energy domain towards the participant-spectator regime at relativistic energies.

7 Correlation between stopping and flow

Information on stopping and flow in heavy-ion collisions represents part of the input to theoretical efforts to deduce constraints on the EoS. Remembering that the EoS is a relation between pressure and density, it is intuitively understandable that these two heavy-ion observables are related to the EoS: flow is generated by pressure gradients established in compressed matter, while the achieved density is connected to the degree of stopping. Recently, it was observed [7] that a strong correlation exists between the stopping, measured in central collisions and the directed flow measured at impact parameters where it is maximal (see fig. 2). The relevant data are shown in fig. 13 in the upper left panel. Plotted against each other are two dimensionless global event observables characterizing stopping, *vartl*, and global scaled directed flow, $p_{xdir}^{(0)}$, both defined earlier.

The data points correspond to 21 system energies with varying system size (from Ca+Ca to Au+Au) and energy (from $150A$ to $1930A$ MeV). The straight correlation line represents a linear least-squares fit to the data and is repeated in the other panels. These other panels show the location along the correlation line of theoretical simulations using the IQMD code for Au+Au at $400A$, $1000A$ and $1500A$ MeV as indicated. The points are marked HM and SM, respectively, for a stiff (incompressibility $K = 380$ MeV) and a soft ($K = 200$ MeV)

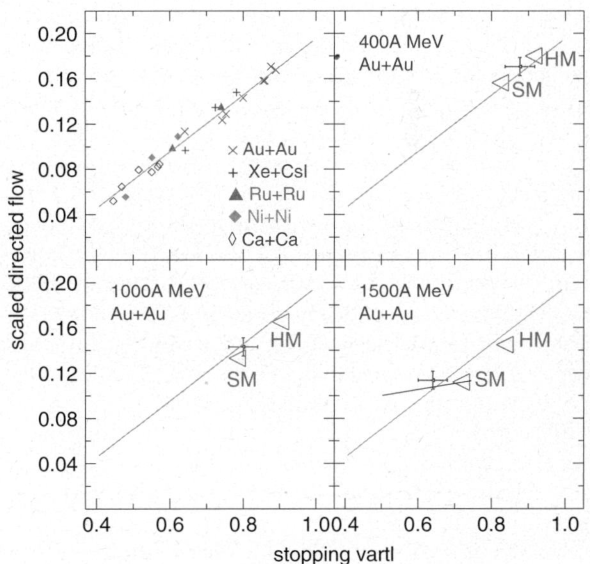

Fig. 13. Upper left panel: Correlation between the maximal directed sideflow and the degree of stopping, after [7]. The line is a linear least-squares fit to the data, which extend from $0.15A$ to $1.93A$ GeV. The correlation line is repeated in the other panels which show results of simulations for Au+Au at three incident energies using two different equations of state, SM and HM, together with the experimental points. The short segment passing through the SM point in the lower right panel shows an estimate of the trajectory using the SM EoS and modifying the in-medium cross-sections in a way that is compatible with [102].

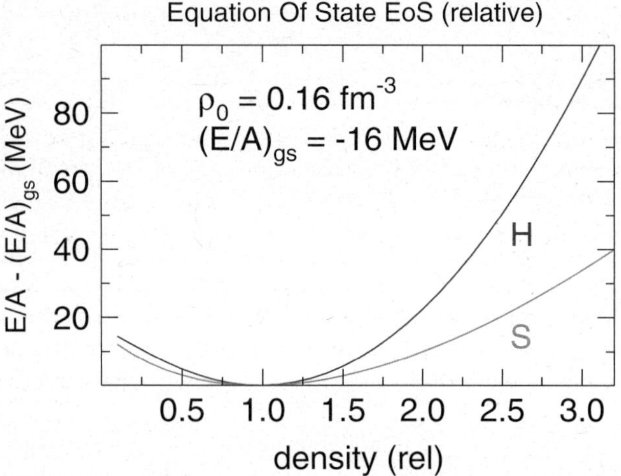

Fig. 14. Equations of state (relative to the ground state) used in the calculation.

EoS. The M in HM and SM stands for the momentum dependence of the nn interaction. IQMD incorporates a phenomenological Ansatz fitted to experimental data on the real part of the nucleon optical potential. The relevant experimental points are given together with their estimated systematic errors (these errors were omitted for clarity in the upper left panel, but are of comparable magnitude for all the data). The main purpose is to show the sensitivity of these combined observables to variations of the zero-temperature EoS as compared to the uncertainty of the data. The EoS, that are purely technical, are shown in fig. 14. The trajectory of the simulation when changing K seems to follow the correlation line, the distance between SM and HM is larger at $1000A$ MeV than at $400A$ MeV (*i.e.* the sensitivity is increased at the higher energy), but then does not seem to further increase at the highest energy, possibly due to the increase of transparency suggested by fig. 7. Data measured at energies below $150A$ MeV do not continue the same linear correlation, an interesting topic that deserves further studies.

In our exploratory IQMD simulations [97,150] we have not tried to be realistic with regard to in-medium modifications of the nucleon-nucleon cross-sections σ_{nn}, using instead the vacuum values standardly implemented in the code [97]. An estimate of the trajectory in the flow *versus* stopping plot when the EoS is kept constant, but the σ_{nn} are decreased is shown in the right-hand lower panel of fig. 13. For this estimate we rely on the more sophisticated calculations of ref. [102] which show that a switch to more realistic smaller σ_{nn} decreases the stopping by about 20% and the scaled sideflow by 6–7%, *i.e.* σ_{nn} acts more strongly, relatively speaking, on the stopping than on the scaled flow, as expected. The σ_{nn} modification trajectory crosses the correlation line because it has a different, flatter, slope than the EoS modification trajectory joining the SM to the HM point (which is not plotted) which happens to have a slope very similar to that of the experimental correlation line. Generally speaking, one can say that an underestimation of the apparent transparency will lead to an underestimation of the stiffness of the EoS. Never-

theless, the procedure just outlined suggests that an EoS closer to SM than HM would seem to be more appropriate to describe the data. The same conclusion was reached from the comparison of the FOPI data on directed flow, including its p_T-dependence, to IQMD calculations [19] and from the comparison of the out-of-plane expansion to BUU calculations [32].

Despite this encouraging result we would like to stress at this time that it would be premature to draw firm conclusions from one particular transport code and it is beyond the scope of this experimental contribution to the subject to conclusively settle the question of the EoS. Besides trying to predict correctly the global observables just shown, probably a good strategy to start with, the simulations must then proceed to reproduce the more differential data such as the variations of the stopping and of the various flow components with the particle type, as shown here in figs. 6 and 11, respectively. Another important physics quantity one would like to have under theoretical control, in order to be convincing on the conclusion side, is the created entropy. Although this is not a direct observable, the entropy at freeze-out is strongly constrained by the degree of clusterization (of which we showed an example in fig. 5) and the degree of pionization. An idea of the freeze-out volume can be obtained from two-particle correlations [151], or even multi-particle correlations [152,153]. All this is a rather challenging task. We refer to the work of Danielewicz, Lacey and Lynch [5] for a summary of the situation obtained a few years ago using a subset of the then available heavy-ion data reaching up to the AGS energies.

8 Summary and outlook

We have presented a systematics of directed and elliptic flow and of stopping for ^{197}Au $+$ ^{197}Au reactions in the intermediate range of energies from 40 to 1500 MeV per nucleon by merging the data from INDRA and FOPI experiments performed at the SIS synchrotron at GSI. The overlap region of the two data sets, 90 to 150 MeV per nucleon incident energy, has been used to confirm their accuracy on an absolute scale, and a very satisfactory agreement has been found.

Particular emphasis was given to the experimental reconstruction of the impact parameter and to the corrections required by the dispersion of the reconstructed azimuthal orientation of the reaction plane. The superiority of observables based on transverse energy, either the ratio $Erat$ of transverse to longitudinal energy or the transverse energy $E_\perp^{12}$ of light charged particles with $Z \leq 2$, over multiplicity for the selection of central collisions has been demonstrated. A new method, derived by extending the Gaussian approximation of the sub-Q-vector distribution to the non-isotropic case and by including the effect of correlation between the sub-events, has been presented and applied to the data at the lower incident energies at which the multiplicities are still moderate and the range of emitted fragment Z is still large, even in the most central collisions. The differences between the standard and the new corrections of derived flow parameters are significant up to incident energies as high as 150 MeV per nucleon.

The deduced excitation functions of the v_1 and v_2 observables describing directed and elliptic flow exhibit several changes of sign which reflect qualitative changes of the underlying dynamics as a function of the bombarding energy. The transition from mean-field–dominated attractive sidewards flow to repulsive dynamics is observed for $Z = 1$ and $Z = 2$ particles at 80 MeV and 60 MeV per nucleon, respectively, in mid-central collisions. The transition from predominantly in-plane to out-of-plane emissions occurs at 150 MeV per nucleon for $Z = 1$ particles. The second change of sign at several GeV per nucleon marks the transition to the ultrarelativistic regime. These transition points are quite well established and not very sensitive to the chosen correction method. The present study shows that also the maxima reached by the flow parameters are reliable within the typically 5% systematic uncertainties due to the corrections and the impact parameter selection. Within this margin they may be used to test transport model predictions and their sensitivity to the chosen parameterization of the nuclear EoS.

It has, furthermore, been shown that the significance of the comparison can be enhanced by including the experimentally observed stopping as represented by the ratio of the variances of the integrated transverse and longitudinal rapidity distributions. This observable can best be determined for central collisions at which the directional flow vanishes for symmetry reasons whereas the compression in the collision zone presumably reaches its maximum. The common origin of the observed stopping and flow is evident from the strict correlation of the two observables, including finite-size effects. However, their individual sensitivity to the magnitude of the nucleon-nucleon cross-sections and to the flow parameters is different and can be used to resolve ambiguities between these two main ingredients of the models. The sensitivity to parameters of the equation of state is shown to increase with bombarding energy over the present energy range, and a soft EoS is clearly favored by the data.

Further constraints for the determination of the parameters of the equation of state can be obtained by including the detailed dependences of flow on the fragment Z, the impact parameter and the accepted ranges of transverse momentum and rapidity into the comparison with theory. These data, for the present reactions, are either available already or in preparation. This will have to be accompanied by theoretical studies of the still existing systematic differences between specific code realizations. The importance or necessity of full antisymmetrization at low energies or of a covariant treatment at high bombarding energies and the role of nucleonic excitations will have to be assessed.

On the experimental side, a gap of missing flow data for the Au+Au system exists at energies below 40 MeV per nucleon where interesting information on transport coefficients as, *e.g.*, shear *versus* bulk viscosity or thermal conductivity may be obtained. The origin of the observed negative flow should be confirmed and clarified. At higher energies, new information, possibly also on the symmetry part of the equation of state, can be expected from new experiments involving isotopically pure projectiles and targets and detector systems permitting mass identification at midrapidity.

The authors would like to thank Y. Leifels for the implementation of the IQMD code at GSI, the FOPI and INDRA-ALADIN Collaborations for the permission to include partially unpublished data in this comparative study, and J.-Y. Ollitrault for stimulating discussions on flow evaluation and corrections.

References

1. W. Reisdorf, H.G. Ritter, Annu. Rev. Nucl. Part. Sci. **47**, 663 (1997).
2. N. Herrmann *et al.*, Annu. Rev. Nucl. Part. Sci. **49**, 581 (1999).
3. H. Stöcker, W. Greiner, Phys. Rep. **137**, 277 (1986).
4. P. Danielewicz, Nucl. Phys. A **673**, 375 (2000).
5. P. Danielewicz *et al.*, Science **298**, 1592 (2002).
6. C. Fuchs, H.H. Wolter, *Modelization of the EOS*, this topical issue.
7. W. Reisdorf *et al.*, Phys. Rev. Lett. **92**, 232301 (2004).
8. R.C. Lemmon *et al.*, Phys. Lett. B **446**, 197 (1999).
9. M.M. Htun *et al.*, Phys. Rev. C **59**, 336 (1999).
10. E.P. Prendergast *et al.*, Phys. Rev. C **61**, 024611 (2000).
11. D.J. Magestro *et al.*, Phys. Rev. C **62**, 041603(R) (2000).
12. P. Crochet *et al.*, Phys. Lett. B **486**, 6 (2000).
13. P. Crochet *et al.*, J. Phys. G **27**, 267 (2001).
14. A. Devismes *et al.*, J. Phys. G **28**, 1591 (2002).
15. F. Uhlig *et al.*, Phys. Rev. Lett. **95**, 012301 (2005).
16. G.D. Westfall, Nucl. Phys. A **681**, 343c (2001).
17. H. Liu *et al.*, Phys. Rev. Lett. **84**, 5488 (2000).
18. A. Andronic *et al.*, Phys. Rev. C **64**, 041604 (2001).
19. A. Andronic *et al.*, Phys. Rev. C **67**, 034907 (2003).
20. D.J. Magestro *et al.*, Phys. Rev. C **61**, 021602 (2000).
21. D. Cussol *et al.*, Phys. Rev. C **65**, 044604 (2002).
22. J. Łukasik *et al.*, Phys. Lett. B **608**, 223 (2005).
23. N.N. Abd-Allah, J. Phys. Soc. Jpn. **69**, 1068 (2000).
24. L. Chkhaidze *et al.*, Phys. Lett. B **479**, 21 (2001).
25. L.J. Simić, J. Milošević, J. Phys. G **27**, 183 (2001).
26. N. Bastid *et al.*, Nucl. Phys. A **742**, 29 (2004).
27. N. Bastid *et al.*, Phys. Rev. C **72**, 011901 (2005).
28. A. Andronic *et al.*, Phys. Lett. B **612**, 173 (2005).
29. C. Pinkenburg *et al.*, Phys. Rev. Lett. **83**, 1295 (1999).
30. A. Andronic *et al.*, Nucl. Phys. A **679**, 765 (2001).
31. P. Chung *et al.*, Phys. Rev. C **66**, 021901(R) (2002).
32. G. Stoicea *et al.*, Phys. Rev. Lett. **92**, 072303 (2004).
33. J. Łukasik *et al.*, Prog. Part. Nucl. Phys. **53**, 77 (2004).
34. K.G.R. Doss *et al.*, Phys. Rev. Lett. **57**, 302 (1986).
35. K.G.R. Doss *et al.*, Phys. Rev. Lett. **59**, 2720 (1987).
36. H.H. Gutbrod *et al.*, Rep. Prog. Phys. **52**, 1267 (1989).
37. H.H. Gutbrod *et al.*, Phys. Rev. C **42**, 640 (1990).
38. M.B. Tsang *et al.*, Phys. Rev. Lett. **71**, 1502 (1993).
39. W.C. Hsi, Phys. Rev. Lett. **73**, 3367 (1994).
40. M.B. Tsang *et al.*, Phys. Rev. C **53**, 1959 (1996).
41. Y. Leifels *et al.*, Phys. Rev. Lett. **71**, 963 (1993).
42. J. Ritman *et al.*, Nucl. Phys. B (Proc. Suppl.) **44**, 708 (1995).
43. M.D. Partlan *et al.*, Phys. Rev. Lett. **75**, 2100 (1995).

44. M. D'Agostino *et al.*, Phys. Rev. Lett. **75**, 4373 (1995).
45. M. D'Agostino *et al.*, Phy. Lett. B **371**, 175 (1996).
46. A. Le Fèvre *et al.*, Nucl. Phys. A **735**, 219 (2004).
47. J. Łukasik *et al.*, Phys. Rev. C **66**, 064606 (2002).
48. A. Pagano, private communication; see also *Proceedings of the IWM2005 International Workshop on Multifragmentation and related topics, Catania, Italy, 2005*, edited by R. Bougault *et al.*, Conf. Proc., Vol. **91** (Italian Physical Society, Bologna, 2006).
49. J. Pouthas *et al.*, Nucl. Instrum. Methods Phys. Res. A **357**, 418 (1995).
50. A. Trzciński *et al.*, Nucl. Instrum. Methods Phys. Res. A **501**, 367 (2003).
51. A. Gobbi *et al.*, Nucl. Instrum. Methods A **324**, 156 (1993).
52. C. Cavata *et al.*, Phys. Rev. C **42**, 1760 (1990).
53. J. Gosset *et al.*, Phys. Rev. C **16**, 629 (1977).
54. K.G.R. Doss *et al.*, Phys. Rev. C **32**, 116 (1985).
55. L. Phair *et al.*, Nucl. Phys. A **548**, 489 (1992).
56. R. Pak *et al.*, Phys. Rev. C **53**, R1469 (1996).
57. J. Łukasik *et al.*, Phys. Rev. C **55**, 1906 (1997).
58. C. Kuhn *et al.*, Phys. Rev. C **48**, 1232 (1993).
59. W. Reisdorf *et al.*, Nucl. Phys. A **612**, 493 (1997).
60. H. Ströbele *et al.*, Phys. Rev. C **27**, 1349 (1983).
61. Y. Larochelle *et al.*, Phys. Rev. C **53**, 823 (1996).
62. P. Beckmann *et al.*, Mod. Phys. Lett A **2**, 163 (1987).
63. R. Bock *et al.*, Mod. Phys. Lett A **2**, 721 (1987).
64. J.P. Alard *et al.*, Phys. Rev. Lett. **69**, 889 (1992).
65. L. Phair *et al.*, Nucl. Phys. A **564**, 453 (1993).
66. G. Bertsch, A.A. Amsden, Phys. Rev. C **18**, 1293 (1978).
67. J. Cugnon, D. L'Hôte, Nucl. Phys. A **397**, 519 (1983).
68. T.C. Awes *et al.*, Phys. Rev. C **24**, 89 (1981).
69. R.J. Charity *et al.*, Z. Phys. A **341**, 53 (1991).
70. S. Piantelli *et al.*, Phys. Rev. Lett. **88**, 052701 (2002).
71. J. Pèter *et al.*, Nucl. Phys. A **519**, 611 (1990).
72. C.A. Ogilvie *et al.*, Phys. Rev. C **40**, 654 (1989).
73. J. Hubele *et al.*, Z. Phys. A **340**, 263 (1991).
74. A. Schüttauf *et al.*, Nucl. Phys. A **607**, 457 (1996).
75. W. Bauer, Phys. Rev. Lett. **61**, 2534 (1988).
76. G. Hanson *et al.*, Phys. Rev. Lett. **35**, 1609 (1975).
77. S. Brandt, H.D. Dahmen, Z. Phys. C **1**, 61 (1979).
78. M. Gyulassy *et al.*, Phys. Lett. B **110**, 185 (1982).
79. P. Danielewicz, M. Gyulassy, Phys. Lett. B **129**, 283 (1983).
80. G. Buchwald *et al.*, Phys. Rev. **28**, 2349 (1983).
81. S.L. Wu, G. Zobernig, Z. Phys. C **2**, 107 (1979).
82. J. Cugnon *et al.*, Phys. Lett. B **109**, 167 (1982).
83. E. Farhi, Phys. Rev. Lett. **39**, 1587 (1977).
84. J. Kapusta, D. Strottman, Phys. Lett. B **106**, 33 (1981).
85. G. Buchwald *et al.*, Phys. Rev. C **24**, 135 (1981).
86. J.D. Frankland *et al.*, Nucl. Phys. A **689**, 905 (2001).
87. J.F. Lecolley *et al.*, Phys. Lett. B **387**, 460 (1996).
88. G.C. Fox, S. Wolfram, Phys. Rev. Lett. **41**, 1581 (1978).
89. G.C. Fox, S. Wolfram, Phys. Lett. B **82**, 134 (1979).
90. P. Pawłowski *et al.*, Z. Phys. A **357**, 387 (1997).
91. P. Désesquelles, Ann. Phys. (Paris) **20**, 1 (1995).
92. P. Désesquelles *et al.*, Phys. Rev. C **62**, 024614 (2000).
93. E. Geraci *et al.*, Nucl. Phys. A **734**, 524 (2004).
94. S.A. Bass *et al.*, J. Phys. G **20**, L21 (1994).
95. C. David *et al.*, Phys. Rev. C **51**, 1453 (1995).
96. F. Haddad *et al.*, Phys. Rev. C **55**, 1371 (1997).
97. C. Hartnack *et al.*, Eur. Phys. J. A **1**, 151 (1998).
98. P. Danielewicz, G. Odyniec, Phys. Lett. B **157**, 146 (1985).
99. E. Plagnol *et al.*, Phys. Rev. C **61**, 014606 (2000).
100. W. Reisdorf *et al.*, Phys. Lett. B **595**, 118 (2004).
101. K. Zbiri *et al.*, preprint nucl-th/0607012.
102. T. Gaitanos *et al.*, Phys. Lett. B **609**, 241 (2005).
103. R. Brockmann *et al.*, Phys. Rev. C **42**, 1965 (1990).
104. E.N.E. van Dalen *et al.*, Phys. Rev. Lett. **95**, 022302 (2005).
105. A. Mishra *et al.*, Phys. Rev. C **69**, 024903 (2004).
106. R. Malfliet *et al.*, Prog. Part. Nucl. Phys. **21**, 207 (1988).
107. C. Fuchs *et al.*, Phys. Rev. C **64**, 024003 (2001).
108. A.B. Larionov, U. Mosel, Nucl. Phys. A **728**, 135 (2003).
109. FOPI Collaboration, in preparation.
110. H.H. Gutbrod *et al.*, Phys. Lett. B **216**, 267 (1989).
111. M. Demoulins *et al.*, Phys. Lett. B **241**, 476 (1990).
112. S. Voloshin, Y. Zhang, Z. Phys. C **70**, 665 (1996).
113. J.-Y. Ollitrault, preprint nucl-ex/9711003.
114. A.M. Poskanzer, S.A. Voloshin, Phys. Rev. C **58**, 1671 (1998).
115. S. Wang *et al.*, Phys. Rev. C **44**, 1091 (1991).
116. N. Borghini *et al.*, Phys. Rev. C **64**, 054901 (2001).
117. N. Borghini *et al.*, Phys. Rev. C **63**, 054906 (2001).
118. N. Borghini *et al.*, Phys. Rev. C **66**, 014905 (2002).
119. R.S. Bhalerao *et al.*, Nucl. Phys. A **727**, 373 (2003).
120. R.S. Bhalerao *et al.*, Phys. Lett. B **580**, 157 (2004).
121. N. Borghini *et al.*, Phys. Rev. C **66**, 014901 (2002).
122. M.B. Tsang *et al.*, Phys. Rev. Lett. **52**, 1967 (1984).
123. J.-Y. Ollitrault, Phys. Rev. D **48**, 1132 (1993).
124. W.K. Wilson *et al.*, Phys. Rev. C **45**, 738 (1992).
125. G. Fai *et al.*, Phys. Rev. C **36**, 597 (1987).
126. M.B. Tsang *et al.*, Phys. Rev. C **44**, 2065 (1991).
127. P. Danielewicz, Phys. Rev. C **51**, 716 (1995).
128. P. Danielewicz *et al.*, Phys. Rev. C **38**, 120 (1988).
129. J.-Y. Ollitrault, Nucl. Phys. A **590**, 561c (1995).
130. J.-Y. Ollitrault, Nucl. Phys. A **638**, 195c (1998).
131. S.A. Voloshin *et al.*, Phys. Rev. C **60**, 024901 (1999).
132. J. Barrette *et al.*, Phys. Rev. Lett. **73**, 2532 (1994).
133. M.M. Aggarwal *et al.*, Phys. Lett. B **403**, 390 (1997).
134. J. Łukasik *et al.*, in preparation.
135. J. Łukasik, W. Trautmann, in *Proceedings of the IWM2005 International Workshop on Multifragmentation and related topics, Catania, Italy, 2005*, edited by R. Bougault *et al.*, Conf. Proc., Vol. **91** (Italian Physical Society, Bologna, 2006) p. 387; preprint nucl-ex/0603028.
136. S.Y. van der Werf, Nucl. Instrum. Methods **153**, 221 (1978).
137. C.A. Ogilvie *et al.*, Phys. Rev. C **40**, 2592 (1989).
138. W.M. Zhang *et al.*, Phys. Rev. C **42**, 491 (1990).
139. P. Crochet *et al.*, Nucl. Phys. A **624**, 755 (1997).
140. B.-A. Li, A.T. Sustich, Phys. Rev. Lett. **82**, 5004 (1999).
141. M.B. Tsang *et al.*, Phys. Lett. B **148**, 265 (1984).
142. W.K. Wilson *et al.*, Phys. Rev. C **41**, R1881 (1990).
143. R.A. Lacey *et al.*, Phys. Rev. Lett. **70**, 1224 (1993).
144. P. Braun-Munzinger, J. Stachel, Nucl. Phys. A **638**, 3c (1998).
145. P. Danielewicz *et al.*, Phys. Rev. Lett. **81**, 2438 (1998).
146. J.-Y. Ollitrault, Phys. Rev. D **46**, 229 (1992).
147. S.A. Voloshin, Nucl. Phys. A **715**, 379 (2003).
148. F. Retière, J. Phys. G **30**, S827 (2004).
149. R.A. Lacey, Nucl. Phys. A **774**, 199 (2006); preprint nucl-ex/0510029.
150. The code IQMD was implemented at GSI by Y. Leifels (FOPI Collaboration), more results will be published elsewhere.
151. R. Kotte *et al.*, Eur. Phys. J. A **23**, 271 (2005).
152. S. Piantelli *et al.*, Phys. Lett. B **627**, 18 (2005).
153. G. Tabacaru *et al.*, Nucl. Phys. A **764**, 371 (2006).

Eur. Phys. J. A **30**, 47–64 (2006)
DOI 10.1140/epja/i2006-10107-8

THE EUROPEAN
PHYSICAL JOURNAL A

High-energy probes

A. Bonasera[a], R. Coniglione, and P. Sapienza

INFN - Laboratori Nazionali del Sud, Via S. Sofia 62, I-95123 Catania, Italy

Received: 13 February 2006 /
Published online: 12 October 2006 – © Società Italiana di Fisica / Springer-Verlag 2006

Abstract. We review some results on energetic particle production in heavy-ion collisions below roughly 100 A·MeV, both theoretically and experimentally. We discuss the possible mechanisms of particle production, as well as the possibility to gather information on the nuclear equation of state (EOS) from data. Results on subthreshold pions, energetic photons, nucleons and light charged particles ($Z \leq 2$) are discussed and contrasted to microscopic models. Important information about the first stages of the reaction are obtained by such probes. At present, we can conclude that we have at least a qualitative understanding of the processes involved when such particles are produced. However, a quantitative determination of relevant EOS parameters is still missing. The production mechanism close to the kinematical threshold (incoherent, cooperative or statistical) is not completely elucidated either. This calls for new data using more modern detector systems and comparison to more refined microscopic models.

PACS. 25.70.-z Low and intermediate energy heavy-ion reactions – 25.75.Dw Particle and resonance production

1 Introduction

One among the many purposes to collide heavy ions at beam energies below 100 A·MeV is the study of the nuclear-matter equation of state (EOS) at finite densities and temperatures. In fact, in such reactions the colliding nuclei are compressed and heated up. After some tens of fm/c, a maximum compression is reached and a compound system is formed, which then expands and, depending on the excitation energy reached, might break into many pieces (multi-fragmentation). In such a scenario there are many factors at play. In the compression stage the dynamics is ruled by the EOS of the system and by the viscosity. Thus data sensitive to the early stage such as energetic protons, neutrons and more complex fragments, as well as photons and pions, will give valuable information and put constraints on these fundamental ingredients of the nuclear interaction.

In the intermediate-energy regime, different powerful detection systems available (MEDEA [1], INDRA [2], NIMROD [3], CHIMERA [4], TAPS [5]), MINIBALL [6], MULTICS [7] (see the contribution by de Souza *et al.* in this topical issue) allow to study with great accuracy energetic and subthreshold particle emission. The comparison of experimental data (impact parameter dependence of particle multiplicity, removed excitation energy, angular distributions, slopes of energy spectra, . . .) with the prediction of transport models can put constraints on basic

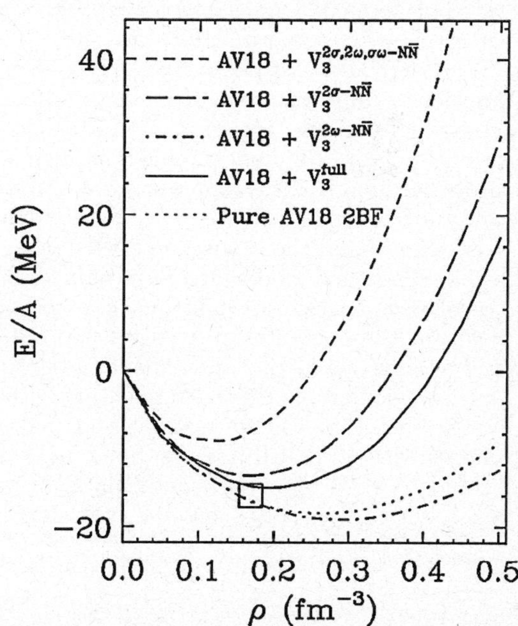

Fig. 1. BHF calculations of the nuclear-matter EOS including three-body forces. The binding energy per nucleon is reported as a function of density [8].

properties of hadronic matter as the in medium Nucleon-Nucleon (NN) cross-section, nuclear-matter compressibility, mean-field properties, or the relevance of two-body *versus* three-body forces. It has been shown, for exam-

---

[a] e-mail: `bonasera@lns.infn.it`

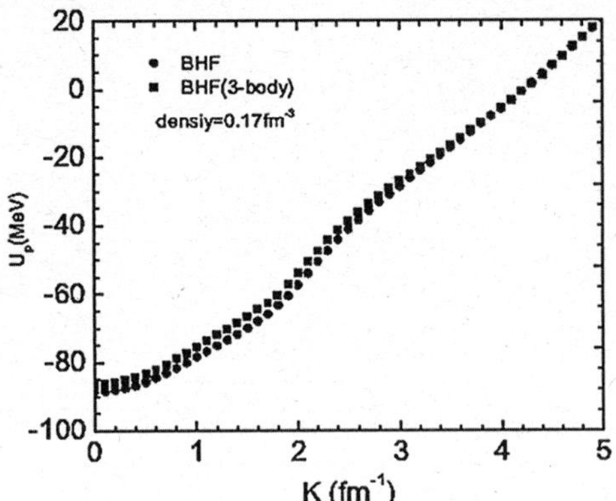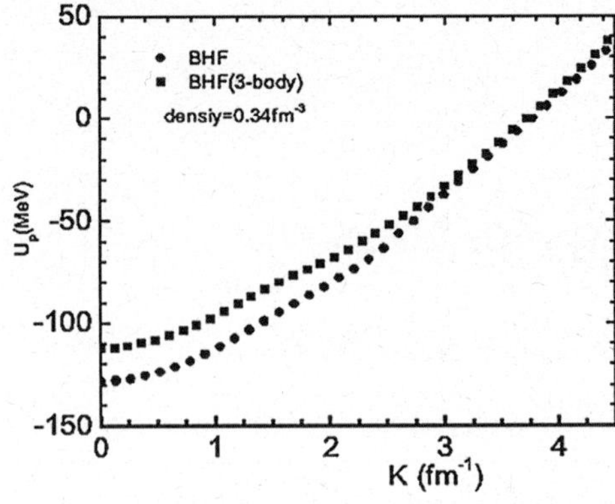

Fig. 2. BHF calculations. The proton potentials are reported as a function of the proton momenta for two different nuclear-matter densities.

ple, that the interplay between two- and three-body forces is very subtle and it turns out that, to fit the ground-state properties of nuclear matter in non-relativistic microscopic calculations, it is necessary to introduce a three-body force [8]. This is demonstrated in fig. 1 for microscopic Bruckner-Hartree-Fock (BHF) calculations [8] where the binding energy per nucleon is reported as a function of the density of nuclear matter at zero temperature. In the figure, the calculations with only two-body forces are given by the dotted line. The two-body force is parameterized to fit the NN data. We clearly see that the approach does not work, in fact it gives a ground-state density of about $0.3\,\mathrm{fm}^{-3}$ and a binding energy of about $-20\,\mathrm{MeV}$, while experimental data correspond to $0.15\,\mathrm{fm}^{-3}$ and $-16\,\mathrm{MeV}$, respectively (square symbol in fig. 1). In order to improve the agreement to data, a genuine three-body force was included in the calculations. The contribution from different channels is displayed in the figure and the final result is represented by the full line. The effect is indeed dramatic. The ground-state density is shifted to $0.19\,\mathrm{fm}^{-3}$ and the binding energy to the experimental value. The three-body force is obtained through a fit to the binding energies of light nuclei, t and ^{3}He essentially [9], but the one reported in fig. 1 is obtained by the meson-exchange model of Fujita-Miazawa [8]. In these nuclear-matter calculations there are no adjusted free parameters. The fact that calculations are not yet perfect implies that something is still missing. Some light on this problem could be shed by experimental data on nucleon production in heavy-ion collisions. We will show below that such data do not support the need for a strong three-body force.

The equation of state discussed so far is at zero temperature. A complete knowledge of the EOS requires, however, information at finite temperatures. Microscopic calculations performed at finite temperatures show, as expected, that the EOS of nuclear matter looks like a Van Der Waals (VDW) EOS. In fact, the NN force has an attractive tail and a repulsive hard core such as many classical systems. At variance with classical systems the ground state is not a solid but a Fermi liquid. However, other properties such as the liquid-to-gas phase transition

at finite temperatures and small densities, are of the VDW type [10]. An important feature that makes nuclei different from classical systems is the strong momentum dependence of the mean field. Microscopic BHF calculations give strong indications on how the momentum dependence force should look like in nuclei and nuclear matter. Typical results of non-relativistic BHF calculations are given in fig. 2 where the potential for protons is reported as a function of the nucleon momentum transfer for two different nuclear-matter densities [8]. Also in this case the difference between 2- and 3-body forces is large especially for low momenta and at high densities. Notice the difference between two- and three-body forces at the two different densities. These features might be revealed through a careful analysis near the Fermi energy for the first case and at higher incident energies in the second one.

To study the momentum dependence of the nuclear mean field we have already many findings coming from electron scattering [11]. However, in those experiments the mean field can be only tested at ground-state densities relevant for the results of fig. 2 left panel. At variance, in a heavy-ion collision, depending on the beam energy, higher densities can be explored. The momentum dependence of the force, as well as the compressibility of the EOS might be inferred, or at least strongly constrained, by subthreshold production of pions, gamma-rays and energetic particle emission. Concerning the relevance of NN collisions and in-medium effects on the NN cross-section, effects can be seen in proton experiments [12,13].

In this work, we will restrict ourselves to energies below roughly $100\,A\cdot\mathrm{MeV}$, where non nucleonic degrees of freedom (such as Δ excitation) are not so relevant yet. This energy region, we believe, carries important information on the EOS near the ground-state density and moderate temperature. In fig. 3 the maximum and average densities estimated by VUU calculations for central collisions as a function of the incident energies are reported for the ^{40}Ca + ^{40}Ca reaction [14]. The understanding of nuclear-matter properties at moderate densities is crucial if we want to understand the EOS at higher densities and temperatures where other degrees of freedom become relevant. In particular, in our contribution we will not discuss

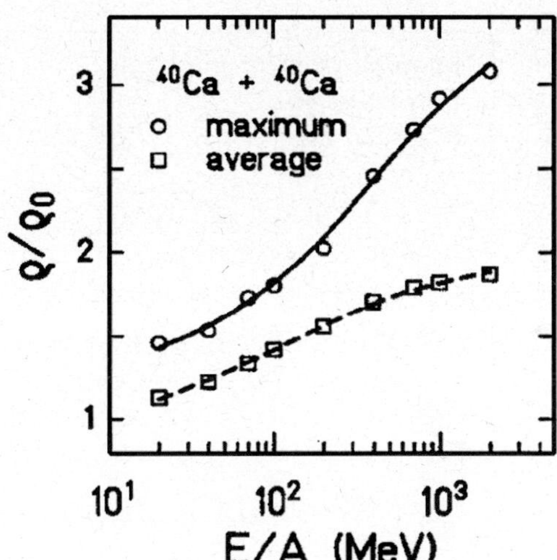

Fig. 3. Maximum and average densities as a function of the bombarding energy per nucleon as estimated by VUU calculations for central ^{40}Ca + ^{40}Ca reactions [14].

the kaon and η production because they require quite a different production mechanism with respect to the one relevant at the lower energies discussed here.

The understanding of the EOS in the region of our interest has to go step by step with the elaboration of microscopic models. While some features of the EOS can be obtained directly from the data, a more quantitative understanding of nuclear properties must be obtained from a detailed comparison between models and data. For instance, we expect that energetic photon and pion production must be very sensitive to the momentum-dependent part of the mean field. In fact, assuming incoherent NN collisions, the final momenta of the colliding nucleons after producing a pion or a photon are decreased. This results in a strong (repulsive) mean field which acts against the production of the new particle essentially because the colliding nucleons must provide an extra energy to overcome the repulsive field. Their final momenta are further reduced and they must not be Pauli blocked, otherwise the collision is not allowed. Unfortunately, we have not been able to find in the literature microscopic calculations of subthreshold particle production with momentum-dependent forces in our region of interest. Many calculations exist for nucleons or more complex particle spectra and as a general feature a reduction of incoherent NN collisions has been found with momentum-dependent forces. Comparison of yields and slopes of energy spectra of energetic protons with predictions of transport models that include a local and a momentum-dependent potential have been published and will be discussed later in this chapter. If one extends this result to subthreshold particle production, a difference of the calculated yield, compared to the results for momentum-independent forces, is expected. This aspect should be carefully (re)analyzed also for π and hard-photon production.

Most of the microscopic calculations, Boltzmann or molecular-dynamics–based [14–17] have two main ingre-

dients. One is the mean field which is parameterized to fit some general results such as electron scattering data, ground-state properties of nuclei etc. The other feature is a collision term which is composed of a probability inferred from NN data, and a Pauli blocking which forbids that particles undergoing an elastic or inelastic scattering, end up in an occupied state. These two ingredients of the models are usually uncoupled, while in principle they should come from the same microscopic interaction. Few attempts exist to date to calculate these ingredients microscopically from the same interaction and to implement them in a transport code (see the contribution by Fuchs and Wolter, *Modelization of the EOS*, this topical issue). In most calculations the phenomenological approach underlined above is used, and one tries to put constraints from a comparison to data. The problem is that most often data are sensitive to both ingredients and it is not easy to disentangle them. However, a systematic comparison of the models to the data should give some constraints on the mean field and the collision term which are included in the calculations.

It is important to stress that in this approach the determination of physical parameters completely relies on the comparison with a transport code, *i.e.* it is fully model dependent. It is, therefore, essential that the different dynamical models and their different numerical implementations within the same parameter set, give compatible results for the observables. The detailed comparison between different codes is an important part of the WCI initiative [18], and tends to show that ambiguities exist among numerical codes, which should be solved before any physical conclusion can be obtained.

Apart from constraining EOS parameters, the interest in studying particle production also relies in the understanding of the production mechanism itself. As we will show in great detail in the following, there are three types of observations:

- particles with an energy much greater than the beam energy in the center of mass. These particles are most likely emitted in cooperative processes that require the collaboration of many nucleons;
- particles with an energy around the beam energy in the center of mass. These particles are presumably created via incoherent processes at the beginning of the reaction when beam particles have still their initial energy;
- particles with an energy much lower than the beam energy in the center of mass. These particles test presumably the late stage of the interaction and may reflect the temperature of the (sub)system.

If this general classification is well established, the transition between the different mechanisms and their detailed modelization are not yet clear and depend on the particle type. For this reason, we have chosen to review particle production according to the particle type.

In the next sections we will discuss some relevant features of data and comparison to models. Indeed, probes of the different stages of the reaction are necessary to achieve a complete picture of the reaction dynamics and to gather information about the EOS of nuclear matter. Energetic

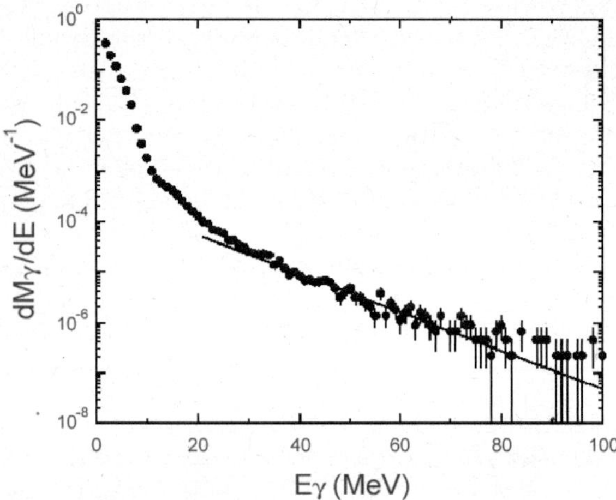

Fig. 4. Gamma multiplicity spectrum for incomplete-fusion reactions in ^{36}Ar + ^{98}Mo reactions at 37 A·MeV. The continuous line indicates the exponential fit of the hard component ($E_\gamma \geq 35$ MeV).

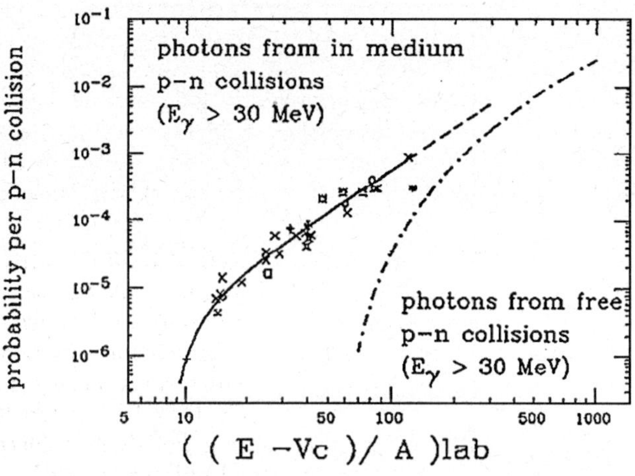

Fig. 5. Emission probability of photons with $E_\gamma \geq 30$ MeV for in-medium NN collisions as a function of the incident energy per nucleon. V_c is the Coulomb barrier [49].

particles like protons, neutrons, pions, and gamma-rays were originally proposed to characterize the initial stage of the reaction. It is clear that energetic photons can do that since their mean free path in nuclear matter is very long, thus once they are produced, they are not scattered again. This is similar for kaons. We have to cite an early experiment at Ganil at 94 A·MeV where kaons were detected [19]. Since then no further data at that energy regime has been discussed, but we believe we have nowadays very performing detectors which could study kaons produced in heavy-ion collisions at around 100 A·MeV. The potential interest of kaons will be clearer after discussing nucleons, pions and hard photons.

2 Hard-photon production

The spectrum of photons emitted in heavy-ion collisions at intermediate energies carries much information on the system evolution from the very early stage of the collision to the late phase at the end of the de-excitation process. Hard photons are particularly appealing probes, since they do not interact again with the surrounding nuclear matter after the production and, therefore, might also provide information on the nuclear-dynamics chronology at various stages of the reaction. A very good review of hard-photon production is given by ref. [14], where the possibility of exploiting energetic particles as probes of the first stages of the reaction is deeply investigated.

A typical spectrum of photons emitted in heavy-ion collisions at intermediate energies is reported in fig. 4 for the reaction ^{36}Ar + ^{98}Mo at 37 A·MeV [20]. As a first rough classification we can divide the spectrum in three main regions with increasing energies:

– in the energy range from some hundreds of keV to approximately 10 MeV the spectrum is dominated by

the statistical emission from excited nuclei occurring at the end of the de-excitation process;

– in the energy range between roughly 10 MeV and 20 MeV a bump due to the γ decay of the giant dipole resonance (GDR), which is a major isovector collective mode in nuclei, can be observed [21,22]. The γ decay of the GDR has been extensively investigated also in hot nuclei in order to gather information on the maximum temperature that a nucleus can hold [23]. This subject is covered in the contribution *Evolution of the Giant Dipole Resonance properties with excitation energy* by Santonocito and Blumenfeld in this topical issue;

– in the energy range beyond 30 MeV the spectrum is characterized by a large inverse slope parameter increasing with energy and by a yield that, for a given incident energy per nucleon, increases with the size of the colliding nuclei. These high-energy gamma-rays ($E_\gamma \geq 30$ MeV) are the so-called hard photons, and are the subject of this part of our review. Their production has been deeply investigated. The main results will be reported in the following with the current understanding of the hard-photon emission and open problems which still need further investigations.

The first experimental observation of an unexpected hard component in the photon spectrum emitted during nucleus-nucleus collisions was found in the ^{12}C + ^{12}C reaction at 84 A·MeV. This experiment aimed at the study of subthreshold neutral pions π^0's, which decay by the emission of two energetic photons, hard photons representing a background [24]. The analysis of these first data [25] revealed many interesting features from both the experimental and theoretical point of view. Indeed, in spite of the very low hard-photon cross-section, several (inclusive) experiments followed with various projectile-target combinations on a rather broad energy range between 10 A·MeV [26] and 124 A·MeV [27]. Indeed hard photons are expected to be a good probe up to 90 A·MeV, since at higher incident energies the contributions of the π^0 decay cannot be neglected (see fig. 5 and fig. 10).

Inclusive experiments yield information about hard-photon cross-sections, inverse-slope parameters, angular distributions and source velocities. From a theoretical point of view, the question about the origin, *i.e.* the production mechanisms, of hard-photon emission was also faced and the proposed solutions can be summarized as follows:

- nucleus-nucleus collective bremsstrahlung [28–30] where photons are emitted as a consequence of the coherent deceleration of the electric field of the two colliding nuclei. The photon yield strongly increases with increasing energy and the spectrum slope depends on the deceleration time;
- incoherent bremsstrahlung as a consequence of NN, in particular n-p, collisions occurring in the interaction region in the first stages of the reaction [31,32], as radiation due to proton deceleration. Several prescriptions have been used for the np-$np\gamma$ cross-section such as the semi-classical one [33], the neutral scalar meson exchange model, and others [34,35]. If these prescriptions are sometimes contradictory, the situation is presently largely clarified after the results of recent experiments which studied the proton-induced reactions at 190 MeV on a liquid-hydrogen target [36]. In this experiments high p-p bremsstrahlung data have been obtained. Moreover, studies on a deuterium target [37, 38] have allowed to investigate all the channels leading to gamma bremsstrahlung, including the coherent bremsstrahlung contribution;
- statistical emission either from a compound-like emission [39] or from a "fireball"-like system [40], where the spectrum slope should reflect the temperature of the emitting source;
- cooperative effects where several nucleons group together into virtual clusters which provide the extra energy for the hard-photon production [41].

The inclusive data systematics gave evidence of a source velocity close to half the beam velocities and an angular emission pattern consistent with an isotropic plus a dipole-like emission in the source reference frame [14]. These results are consistent with the n-p bremsstrahlung mechanism as a dominant process in the hard-photon emission in heavy-ion collisions at intermediate energy. The absolute yield is however much larger than expected from free n-p collisions, and this difference strongly increases with decreasing beam energy. For this reason, hard photons are considered as "subthreshold particles" using the same definition that applies for mesons produced at an incident energy per nucleon lower than the energy threshold for free NN collision. This feature has been successfully explained in terms of the Fermi boost provided by in-medium nucleon-nucleon collisions (see fig. 5).

Alternative approaches have also been proposed. In particular, some results have been rather well reproduced by a statistical approach, *i.e.* the experiments ^{92}Mo + ^{92}Mo at 19 A·MeV [39] and N + (C, Zn, Pb) at 20, 30 and 40 A·MeV [42]. However, it is important to mention that hard-photon angular distributions in asymmetric systems [43,44] are consistent with a source velocity close to half the beam velocity with the presence of a dipole component, rather than to the compound nucleus velocity.

The unique result indicating evidences of coherent bremsstrahlung [45] was not confirmed. Anyway, the expected yield for collective nucleus-nucleus bremsstrahlung [29] is much lower than the observed ones. Dynamical calculations indicate that at most 10% of the observed hard-photon ($E_\gamma \geq 50$ MeV) [31] cross-section is consistent with nucleus-nucleus bremsstrahlung, while the dominant part of the total yield is due to first chance n-p collisions. It is important to note, however, that inclusive data are also consistent with n-p bremsstrahlung in a nuclear fireball.

In order to disentangle the different hypotheses concerning the origin of the hard-photon emission and to get a deeper insight on the phenomenon, exclusive measurements were necessary. A first group of measurements faced the issue of the impact parameter dependence of the hard-photon production. The first experiments indicated that the hard-photon multiplicity increases with increasing reaction centrality and the slope slightly decreases with increasing impact parameter [42,46–48]. These experimental facts are, unfortunately, again well accounted for by different theoretical models. In the fireball model [40], the hard-photon multiplicity scales with the volume of the interaction zone, which increases with increasing centrality. In dynamical calculations like BUU [31], where hard photons ($E_\gamma \geq 50$ MeV) are mostly emitted as a consequence of first n-p collisions, a hard-photon multiplicity dependence on the overlap of target and projectile nuclei is found consistent with emission in the first stages of the reaction [31]. Moreover, in dynamical calculations, the decreasing of the slope with increasing impact parameter is interpreted as due to the fact that nucleons with softer momentum are mostly located at the nuclear surface.

The quality of exclusive data was strongly boosted by the high efficiency of two multidetector apparatuses for hard photons, MEDEA, by which hard photons and light charged particles can be detected simultaneously [1], and TAPS [5]. In particular, the dependence of hard-photon multiplicity M_γ on the impact parameter b was investigated quantitatively. In models based on n-p bremsstrahlung, the hard-photon multiplicity scales with the number of n-p collisions (N_{np}) and therefore with the size of the interaction zone (A_{part}):

$$M_\gamma(b) = P_\gamma \cdot N_{np}(b) \propto P_\gamma \cdot A_{part}(b).$$

Here P_γ is the probability of emitting a hard photon in a single n-p collision (p-p collisions are not considered since they provide a much smaller contribution $\leq 10\%$) at a given incident energy for a heavy-ion reaction. This probability is usually extracted from inclusive data (see [49] for systematics) within the approximation that P_γ in nuclei only depends on the incident energy per nucleon. Several experiments [50–52] were run and compared with the results of dynamical models based on a transport equation to simulate nucleus-nucleus collisions, whereby the

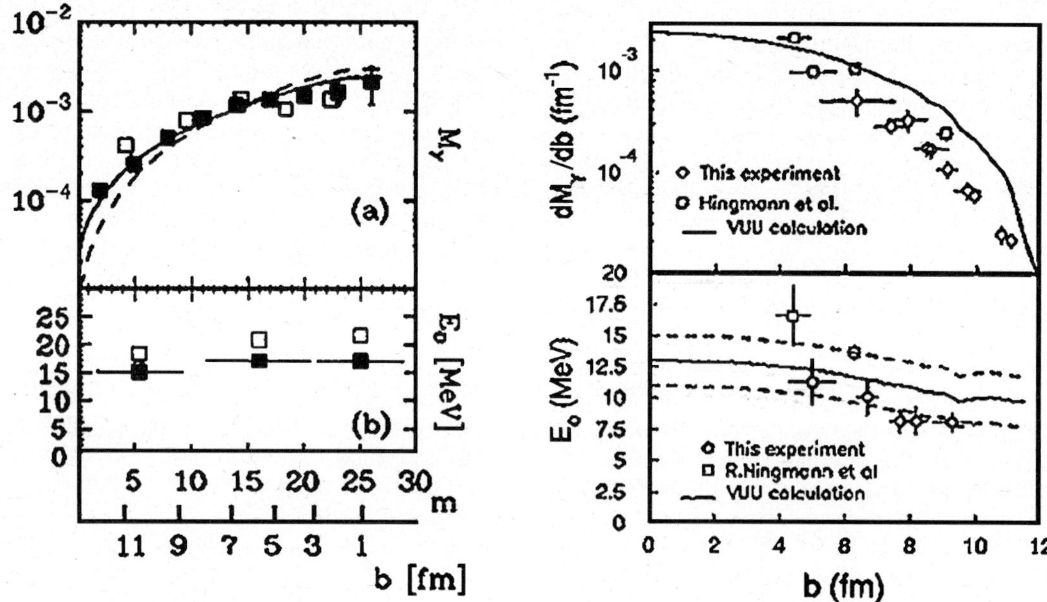

Fig. 6. Gamma-ray multiplicity for $E_\gamma \geq 40\,\mathrm{MeV}$ (left top panel) and for $E_\gamma \geq 25\,\mathrm{MeV}$ (right top panel) and inverse-slope parameters (bottom panels) as a function of the centrality for the reaction $\mathrm{Xe} + \mathrm{Au}$ at $44\,A\cdot\mathrm{MeV}$ (solid squares in the left panels, ref. [51]) and for the reaction $\mathrm{Ar} + \mathrm{Gd}$ at $44\,A\cdot\mathrm{MeV}$ (open symbols in the right panels, ref. [50]). In the left panels, for comparison, BNV calculations are reported (open symbols) while the continuous line is calculated assuming M_γ proportional to the surface of the overlap zone and the dashed line is calculated assuming M_γ proportional to the volume of the overlap zone. In the right panels VUU simulations are reported (full lines, dashed lines are the uncertainties).

hard-photon production is treated in a perturbative way. In fig. 6 the measured and calculated M_γ and inverse-slope parameters as a function of the estimated impact parameters are reported for two different reactions measured with two different apparatuses. In general, a rather good qualitative agreement between the measured and calculated hard-photon multiplicity for several reactions is observed. Moreover, the γ-ray multiplicity appears to scale with the surface of the overlap region in the reaction $^{129}\mathrm{Xe} + ^{197}\mathrm{Au}$ at $44\,A\cdot\mathrm{MeV}$ (see fig. 6 top left panel). For the same reaction a good agreement with BNV calculation was found [51]. These results provide a further support for the models based on n-p bremsstrahlung. Similar information was gathered from inclusive data varying the size of the colliding nuclei. Eventually, these experimental data provide a measurement of the spatial origin of the hard photons from the interaction zone. It is important to underline the fact that the hard-photon multiplicity scales linearly with the participant region, does not necessarily imply that a fireball is formed as an independent source, but, especially around the Fermi energy, rather provides a snapshot of the interaction zone in the early stage of the reaction. For this reason the hard-photon multiplicity can also be used as a quantitative measure of the impact parameter in heavy-ion collisions.

This large set of data, although consistent with BUU calculations, cannot rule out a fireball scenario especially at the higher incident energies. For this reason, a deeper insight into hard-photon production was attempted by looking at hard-photon–particle correlations and hard two-photon correlations. In calculations in which

hard photons mainly arise from first chance n-p collisions, if the photon and proton energies are high enough, an anticorrelation is expected due to the kinematical limit imposed on the total proton and gamma-ray energy in the bremsstrahlung process. In the reaction $^{14}\mathrm{N} + \mathrm{Zn}$ at $40\,A\cdot\mathrm{MeV}$ the gamma-ray proton coincident ratio was found to be independent of proton energy for $E_\gamma \geq 20\,\mathrm{MeV}$, thus suggesting that high-energy photon production, for $E_\gamma \geq 20\,\mathrm{MeV}$, may arise in part from n-p bremsstrahlumg in a later stage of the collision [53]. On the other hand, high-statistics data on the reaction $^{40}\mathrm{Ar} + ^{51}\mathrm{V}$ at $44\,A\cdot\mathrm{MeV}$ [54] were also analyzed. These data show that, while hard photons with $E_\gamma \geq 25\,\mathrm{MeV}$ exhibit a slight anticorrelation constant with increasing energy, the very energetic photons, namely with $E_\gamma \geq 70\,\mathrm{MeV}$, exhibit a much stronger anticorrelation increasing with increasing proton energy as expected in a first chance n-p bremsstrahlung scenario. This result, which confirms the expectations of dynamical models (see [14]), provide an experimental evidence of the hypothesis that very energetic photons are mostly produced in the early stage of the reaction. This signature qualifies high-energy photons (as well as protons) as probes of the momentum and energy distributions of the nucleons in the early stages of heavy-ion collisions. Hard two-photon correlations were investigated in different systems, and the results reported in refs. [55,56] also support the idea that hard photons originate from an early stage of the reaction.

Besides the understanding of the production mechanism, the study of the hard-photon emission allows to use them as probes of the nuclear matter. In particular, due to the nature of the electromagnetic radiation, they can carry unperturbed information on the nuclear matter at the moment of their production and are not affected by subsequent stages of the reaction. Information on nuclear dynamics, on the contribution of the mean field and two-body collisions in dissipative heavy-ion reactions, and a time scale for multifragmentation have been deduced by detailed investigation of the hard-photon emission in heavy-ion collisions at intermediate energy and the main results are reviewed in the following.

In collisions around the Fermi energy, the nuclear dynamics is governed by the interplay between one- and two-body dissipation, namely between the mean-field and NN collisions. At energies around and above the Fermi energy (about 35 MeV) the role of NN collisions increases with increasing incident energy due to a reduced contribution of the Pauli blocking, which inhibits NN collisions at low energy. Two different experiments addressed the problem of one- and two-body dissipation mechanisms at intermediate energy via the study of hard-photon emission in peripheral and central reaction, respectively. Hard photons were measured in coincidence with projectile-like fragments in the reaction ^{36}Ar $+$ ^{159}Tb at 44 A·MeV in the peripheral events [57]. The hard-photon multiplicity, which scales with the number of n-p collisions, as discussed above, and therefore represent a measure of two-body dissipation, was measured as a function of the mass of the primary projectile-like mass and was found to increase linearly with the transferred mass, showing the importance of two-body collisions. Moreover, the multiplicity value depends on the direction of transfer. Indeed more collisions are needed to transfer mass from the heavier target than viceversa; this effect is understood in terms of the action of the mean field that favors nucleon transfer from the lighter to the heavier partner of the collision. A complementary study for central collisions leading to incomplete-fusion residues was reported in ref. [58]. In this case, hard photons ($E_\gamma \geq 35$ MeV) were measured in coincidence with heavy residues emitted in the reactions ^{36}Ar $+$ ^{90}Zr at 27 A·MeV and ^{36}Ar $+$ ^{98}Mo at 37 A·MeV. In this incident energy regime, central and semi-central collisions lead to incomplete fusion and this process is usually described by the Viola systematics that shows a decrease of the momentum transfer as a function of the incident energy [59]. In the two ^{36}Ar-induced reactions cited above, the ratio between the residue velocity and the center-of-mass velocity (v_r/v_{cm}) was measured as a function of the reduced impact parameter (b/b_{max}) given by the hard-photon multiplicity. A strong correlation is found between v_r/v_{cm} and b/b_{max}. Most remarkable, data coincide for both incident energies. This indicates that the fraction of linear momentum transfer for events giving rise to a residue depends only on the impact parameter and not on the bombarding energy. This demonstrates the role of two-body collisions in the transfer process leading to the production of highly excited nuclei. In conclusion, the hard-photon detection has played an important role in elucidating the interplay between mean-field and NN collisions around the Fermi energy both in peripheral and central collision, leading re-

spectively to projectile-like fragments (PLFs) and heavy residues.

High-energy photons do not only probe the first stage of the collision. A contribution from a second, less energetic, photon source emitted in a later phase of the reaction was put in evidence in several reactions at various incident beam energies. The feature of this so-called "thermalized" hard-photon component and its consistency with statistical and dynamical model calculations are discussed in the following.

Hard photons ($E_\gamma \geq 30$ MeV) were measured in the reactions induced by an ^{36}Ar beam at 95 A·MeV on ^{197}Au and ^{12}C. It is important to note that in this case the incident energy per nucleon is well above the free NN threshold for the production of photons with $E_\gamma \geq 30$ MeV. It was shown [54] that while very energetic photons originate from the first phase of the reaction, the bulk of photons ($E_\gamma \geq 30$ MeV) is produced over a longer time span and could probe if the phase that leads to the thermalization of the fireball is formed in the reaction [60]. From an experimental point of view, the very good statistics hard-photon measurements with the MEDEA [1] and TAPS [5] detectors demonstrate that a good description of hard-photon spectra is obtained with the superposition of two components with different slopes and yields[1]. Moreover, hard two-photon interferometry measurements in the reactions ^{86}Kr $+$ ^{nat}Ni at 60 A·MeV and ^{181}Ta $+$ ^{197}Au at 39.5 A·MeV are well described within the hypothesis of two different sources [61]. The softer component has been associated with photon emission in a later stage of the reaction. In BUU calculations, besides the dominant hard-photon contribution produced by n-p bremsstrahlung during the compression phase at the early stage of the reaction ("direct" photons), "thermal" hard photons are also emitted in a later stage from less energetic n-p collisions inside a thermalized source during the resilience of the system after the expansion phase. Experimentally, inclusive and exclusive photon spectra consistent with a "thermal" and a "direct" component were measured in the reactions ^{86}Kr $+$ ^{nat}Ni at 60 A·MeV, ^{181}Ta $+$ ^{197}Au at 39.5 A·MeV, ^{208}Pb $+$ ^{197}Au at 29.5 A·MeV [62] and in the reactions ^{36}Ar $+$ ^{197}Au, ^{107}Ag, ^{58}Ni and ^{12}C at 60 A·MeV [63] and in the reactions ^{58}Ni $+$ ^{27}Al, ^{58}Ni and ^{197}Au at 30 A·MeV [64] and ^{58}Ni $+$ ^{197}Au at 45 A·MeV [65–67].

An appealing aspect of thermal photons is that their emission signals that a big piece of nuclear matter still exists at the end of the dynamical evolution of the collision, and their slopes can be related to the temperature of such a system. This can be exploited to get information on other processes using thermal photons as a probe. For instance, the nuclear caloric curve has been investigated using thermal photons as a new "thermometer" for hot nuclear matter [68]. Moreover, thermal photons have been used as a "clock" to deduce the time scale of intermediate mass fragment (IMF) emission. Studying the

---

[1] However, we would like to notice that the change in slope of the photon yield could be also affected by the $1/E_\gamma$ factor which enters the elementary np-$np\gamma$ bremsstrahlung probability [15].

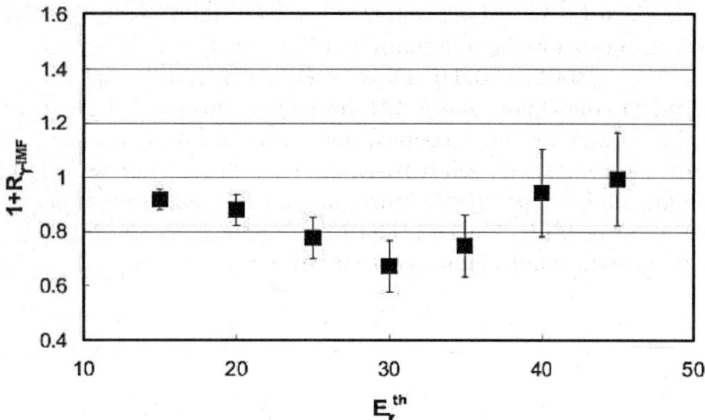

Fig. 7. Experimental photon-IMF correlation factor *versus* the threshold E_γ for IMFs in the velocity window near the center-of-mass velocity. Data are shown for the Ni + Au reaction at 45 A·MeV for central collisions.

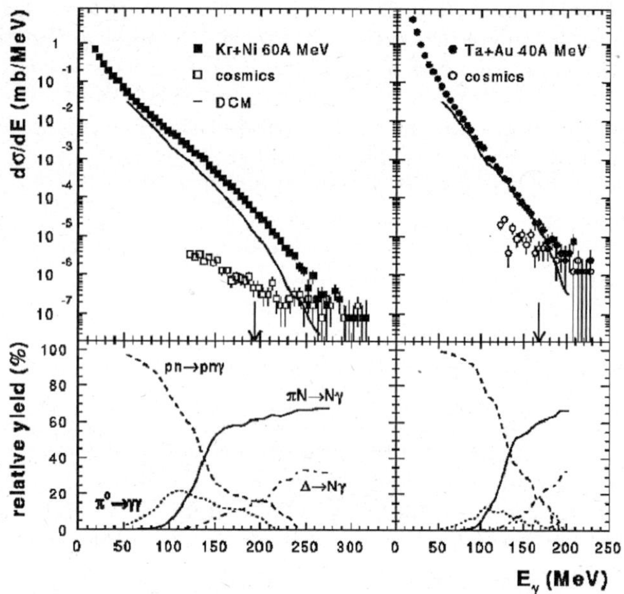

Fig. 8. Measured photon spectrum (full symbols) in the reaction ^{86}Kr + ^{nat}Ni at 60 A·MeV (left panel) and ^{181}Ta + ^{197}Au at 40 A·MeV (right panel) after subtraction of the cosmic-ray contribution (open symbols). The solid line represents calculations. In the lower part, the calculated spectrum is decomposed into fractions corresponding to the following mechanisms: $pn \rightarrow pn\gamma$, $\pi N \rightarrow N\gamma$, $\pi^0 \rightarrow \gamma\gamma$, $\Delta \rightarrow N\gamma$ [69].

thermal photon-IMF correlations, an anticorrelation signal with IMFs in the nucleus-nucleus center-of-mass velocity region has been observed in central Ni + Au collisions at 45 A·MeV (fig. 7) [66], while the same has not been seen in the data for the same system at 30 A·MeV [67]. This indicates that around 45 A·MeV a transition occurs from late to prompt IMF emission, where "prompt" has to be interpreted as faster than the emission time associated with thermal photons. Stochastic mean-field simulations performed for these two reactions are consistent with the data. At 30 A·MeV for most of the events the system, after the initial compression and expansion, recombines leading to the formation of an heavy excited system with $Z \approx 80$. On the other hand, at 45 A·MeV a dominant role of a prompt IMF formation is observed [65,66].

The last part of the γ-yield concerns the very high-energy part of the spectrum. Deep sub-threshold particles, with respect to the kinematical limit expected for NN collisions including the boost due to the Fermi motion, are observed on a broad range of incident energies addressing the question of which mechanism allows to concentrate such a relevant fraction of the total available energy in the production of a single energetic or massive "particle". Several hypotheses have been considered such as nucleon off-shell effects, three-body collisions, dynamical fluctuations or multi-step processes involving pion and Δ's.

High-statistics data exhibit the presence of hard photons with energy well above the kinematical limit for NN collisions. In the reactions ^{86}Kr + ^{nat}Ni at 60 A·MeV and ^{181}Ta + ^{197}Au at 40 A·MeV [69], hard-photon spectra, with energy extending up to 5 times the beam energy per nucleon, were measured. The data were compared with a cascade model which takes into account several channels (see fig. 8) including the radiative channel $\pi + N \rightarrow N + \gamma$. The calculations are in good agreement with the data for the reaction ^{181}Ta + ^{197}Au at 40 A·MeV, while undershoot the data of the reaction ^{86}Kr + ^{nat}Ni at 60 A·MeV, both concerning the highest-energy component of the photon spectrum and slope and yield of the π^0 energy spectrum. In summary, theoretical calculations, in spite of the many

hypotheses proposed, are not satisfactory in explaining deep subthreshold data and a more detailed comparison between data and models is desirable. We will come back to the problem of deep subthreshold particle production in the next section, devoted to pions.

3 Subthreshold pion production

Pions (and nucleons), at variance with photons, after being produced, can interact with nuclear matter again and be scattered and/or reabsorbed. These multiple interactions of pions with the surrounding matter explain the early success of statistical models [70–72]. On the other hand, rescatterings lead to ambiguities in transport approaches where pion production is calculated perturbatively, similarly to hard photons. This method is in principle not applicable because pion dynamics should be followed microscopically. What is generally done in the literature is to correct the results with an absorption factor expressed in terms of a pion mean free path, and in turn obtain a value for the mean free path for each experimental condition [14–16,73,74]. For this procedure to be fully consistent, the other parameters entering the calculation should be fixed from independent observables, for instance, photon production in the same reaction. Unfortunately, this is rarely reported in the literature we are aware of.

Keeping this problem in mind, in this section we would like to review some of the experimental data on pion pro-

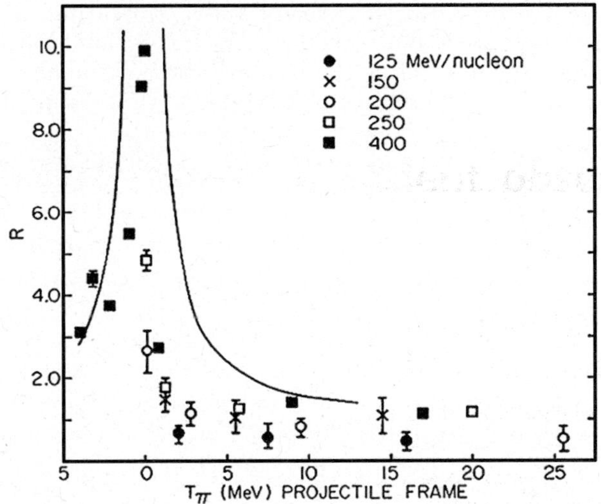

Fig. 9. Ratio R of π^- to π^+ cross-section in Ne + NaF at different incident energies as a function of the pion energy in the projectile frame. The solid curve shows the Coulomb energy as described in the text [75].

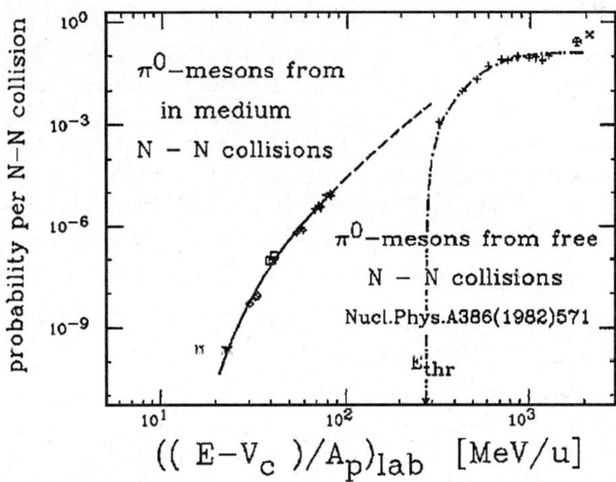

Fig. 10. Emission probability of neutral pions for in-medium NN collisions as a function of the incident energy per nucleon. V_c is the Coulomb barrier [14].

duction at energies ranging from the absolute kinematical threshold to about 100 A·MeV. An abundant literature on higher energies exists, which is however outside the scope of this work.

First data on pion production at subthreshold energies (corresponding to a beam energy per particle in the laboratory below twice the pion mass) were obtained by Benenson *et al.* at LBL above 100 A·MeV [75] and subsequently at much lower beam energies, ^{16}O + Al, Ni at 25 A·MeV [76]. The latter result was somehow surprising, at that time, because one would estimate a higher threshold for pion production around 50 A·MeV by coupling the relative to the Fermi motion in first-chance NN collisions. In the work of Benenson *et al.* the ratio of π^+/π^-, reported in fig. 9, was measured and successfully explained in terms of a statistical model which invoked the ratios of the absorption cross-sections and a Coulomb shift (see the full line in fig. 9) [72].

A collection of available data was analyzed in terms of probability of elementary NN collisions folded with the number of possible collisions in a nucleus-nucleus interaction. This gave a scaling approximation similar to the one reported for photon production, which is displayed in fig. 10 [46,49]. This scaling shows that at least the gross features could be understood in terms of single NN collisions. Below, we will show that this mechanism is however insufficient to explain data very close to the kinematical threshold.

To enter more in detail into the microscopic calculations, we briefly recall how pion production is simulated in kinetic models. This is very similar to photon production discussed above, *i.e.* for each elementary NN collision the production probability is calculated perturbatively. This means that for each elementary collision a pion of a given charge (charge conservation enforced) and energy is produced according to an emission probability extracted from a parametrization of the free $NN \to NN\pi$ cross-section. This cross-section is dominated by the excitation of a Δ-resonance as a doorway state. The underlying hypothesis is the neglect of the finite Δ lifetime. The emission angles are randomly chosen and the final momenta of the nucleons are calculated to conserve total energy and momentum in the collision. The obtained probability is multiplied by the Pauli-blocking factors $(1 - f) \cdot (1 - f)$ for each chosen emission angles of the pion (and averaged over the different emission angles). This approach has been used by many authors [14–16], but one different idea was proposed by W. Bauer in ref. [73]. There it was assumed that pions may be produced through a Δ-resonance whose decay would be hindered by the blocking effect of the Fermi sea. The crucial point for this scenario is to get some information on the mean field seen by Δ's. In fact, if Δ's see a strong repulsive mean field, their formation will not be favored and the process discussed above will be unphysical. Gathering some information about the properties of the mean field seen by the Δ at moderate densities and temperatures could be possible through a careful experimental and theoretical analysis of pion and photon production in the nucleus-nucleus collisions. This would call for a second campaign of coincidence experiments with more performing detectors to study the Δ propagation in the medium supported by a deep theoretical analysis with more recent and refined models that are now available. Since in the world there are many laboratories able to deliver beams of high quality at the energies of interest and there are very good and performing detectors that with small modifications could be suitable to study the type of physics discussed here, there is no need for large financial efforts and to leave the field not fully explored would be a real pity. But in order to disentangle the main features we have understood so far and the questions still not answered, we will go in more detail into the results that have been obtained.

In terms of microscopic models, the first ingredient to be considered is the ground-state momentum distribution. In the first codes this distribution was given within a degenerate Thomas-Fermi approximation, *i.e.* for each local density a Fermi momentum was calculated and the particles (usually test particles in a Boltzmann transport equation) randomly distributed as a step function with the corresponding local Fermi momentum. This approach lacks self-consistency and the high-momentum tail of the distribution is neglected [14–16]. More recent approaches such as Fermionic molecular dynamics (FMD) [77], antisymmetrized molecular dynamics (AMD) [78], constrained molecular dynamics (CoMD) [79] and more recent Vlasov approaches [80,81] calculate in a self-consistent way the ground state used for the time evolution of the collisions. However, detailed calculations for pion production within the framework of those more refined models are not available to our knowledge.

In the old calculations a large dependence of the pion production on the Fermi momenta was observed. We notice in passing that if the nuclear ground state is not obtained self-consistently, one can change the Fermi energy by modifying slightly the surface term or the momentum-dependent part of the potential (or some parameters of the Gaussian or delta test particles) keeping the nuclear binding energy unmodified. Changing the Fermi energy will change the pion distributions by orders of magnitude [15]. The use of a realistic correlated ground-state momentum distribution fitted from electron scattering experiments [14] naturally allows to increase the maximal Fermi boost, and consequently decrease the threshold energy for pions and all other subthreshold particles. It is, however, worth mentioning that, from a numerical point of view, the particle production rate is not entirely trustable in this approach. Indeed, due to the semiclassical nature of the simulation, test particles initialized in the high-energy tails are not Pauli blocked and may lead to a spurious production of particles already in the ground state. Because of this ambiguity, the question whether energetic particles are produced in incoherent NN encounters or not is not completely solved. Recent data on proton production as a function of the number of participant nucleons [13] clearly demonstrate that cooperative processes have to be invoked when detecting protons whose energies are close to the NN kinematical limit. Similar exclusive data at the energies of interest here and for pion production are unfortunately not available to our knowledge. This kind of data would provide more precise information on the type of mechanism responsible for pion and more general particle production near the kinematical threshold. Furthermore, they will give a more stringent test to the more refined models available nowadays.

As discussed in the introduction, one of the main motivations in studying energetic particle production is the possible sensitivity to the characteristics of the nuclear EOS properties and in-medium cross-sections. Information on these issues coming from pion production is discussed below. The nucleon-nucleon cross-section appears in transport models in the collision term. Most calculations include a two-body collision term which takes into account, in a semiclassical way, the effect of Pauli block-

ing. However, when the density and temperature of the system increase Pauli blocking relaxes and the dilute-gas approximation which is the basis of the Boltzmann collision term is no longer valid. Attempts have been made to include three-body collisions [82–84] to calculate not only particle production and collective effects [83] but also more complex particle production [84].

It is important to notice that the 3-body collision term is not unique and its expression depends on the adopted approximation scheme [82–84]. In [82,84] a 3-body collision can happen if the particles did not undergo a 2-body collision while in [83] the probabilities for two- and three-body collisions are calculated independently. This latter assumption leads to a decrease of the particle mean free path while the previous one does not necessarily. The two approaches give different values for physical observables such as collective flow under fixed conditions for the other parameters, *i.e.* same elementary NN cross-section and similar mean fields. To discriminate between different treatments, codes should be confronted with analytical solutions accessible in model cases [83], and independent observables sensitive to the nucleon mean free path such as nuclear stopping, should be systematically compared to experimental data. In any case, when three-body collisions are included energetic particles are produced with higher probability as compared to the two-body case and with higher energy [15].

Another important physical ingredient is the nuclear mean field. Many calculations have shown a modest sensitivity to the compressibility of the EOS for pion and photon production. This has been explored especially for momentum-independent interactions. Of course, it is well known that the mean field is momentum dependent (fig. 2) thus models should take into account this feature also for particle production. When the momenta of two (or three) colliding nucleons change because of the scattering, the field changes as well because of its momentum dependence. What one does in practical calculations is to modify the momenta of the particles in such a way that the total energy is conserved. If this is not possible the collision is rejected. This generally results in a reduced number of NN collisions and possibly in a global transparency effect as compared to calculations with momentum-independent forces. The role of the momentum-dependent force should be further investigated when a particle, a pion or a photon is produced. In such a case the final momenta of the nucleons are further reduced because some energy and momentum is carried away from the produced particles. Thus, on top of the Pauli-blocking effect one should consider the effect of the momentum-dependent mean field which being usually repulsive will result in a need for more energy to produce a particle and in turn to a reduction of its formation probability. No microscopic models with momentum-dependent forces that calculate subthreshold particle production are available to our knowledge. Some exist at higher energies where the calculations are non-perturbative [85] and a sensitivity to the EOS is demonstrated. Even in the perturbative regime, calcula-

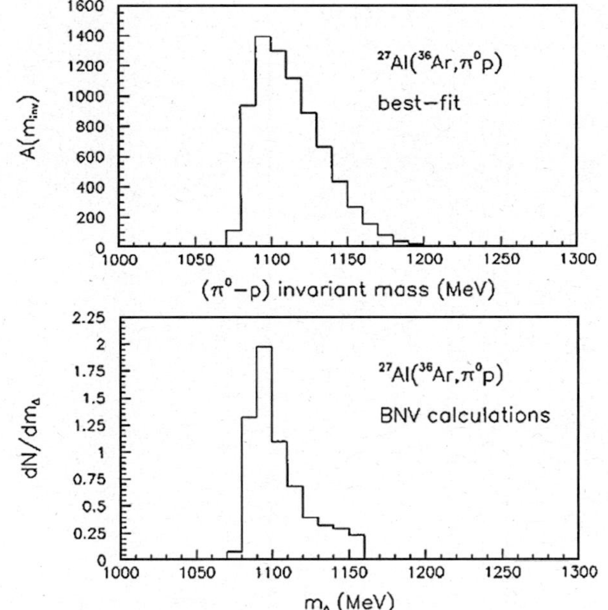

Fig. 11. Comparison between the "indirect" channel (π^0-p) invariant-mass distribution $A(m_{inv})$ (upper panel), extracted from a best-fit procedure, and Δ mass distribution (lower panel) predicted by the BNV theoretical calculation for the same system at the same bombarding energy [86].

tions should be feasible nowadays with the more performing computers.

For the particular case of pion production, if the process occurs through Δ formation and if the Δ cannot decay because of the Pauli principle until it is in vacuum (or in a low-density and high-temperature region), one could try to study the properties of the mean field seen by the Δ. Some experimental data on Δ production in nucleus-nucleus collisions at energy below 100 A·MeV have been discussed in [86,87] and investigated in BNV calculations (see fig. 11). Data show that the Δ width is reduced from its free value to about 25–50 MeV. Naively this would indicate that the Δ lifetime in matter is increased to about 10 fm/c which could be a sufficient time to say that the delta does not decay before the nuclei disassemble. On the other hand, the width of the resonance is roughly reproduced in the BNV calculation as due to the folding of the delta width in vacuum and energy conservation (or Pauli blocking) that blocks the higher-momentum part of the pion distribution. The model has the features discussed above, *i.e.* momentum-independent mean field, Fermi-gas approximation for the initial distribution and two-body collisions only. A more refined approach and more data for other systems at different energies would be needed to study the dynamics of Δ's in moderately excited nuclear matter.

4 Energetic light-particle emission

Together with pions and hard photons, energetic nucleons are a powerful probe to get information on the initially compressed nuclear phase and on the following dynamics, because their emission drives the system towards an expanded and more thermalized stage or a fragmentation stage. Moreover, the knowledge of the energetic particle multiplicities and of the energy dissipated in the first reaction phase is of particular interest for the understanding of the role of the isospin degree of freedom in nuclear reactions and in the EOS for asymmetric nuclear matter.

Energy spectra of light particles (p, d, t, He) have been measured for a large variety of reactions in a wide range of incident energies with apparatuses covering the whole angular range. In fig. 12 the experimental proton energy spectra (dots) for ^{64}Zn + ^{92}Mo collected at different incident energy and detection polar angles measured with the NIMROD apparatus [3] are reported. A frequently used technique to study light-particle emission mechanisms is a simultaneous fit (in energy and in angle) of these spectra assuming isotropic emission from sources with a Maxwellian spectrum in their center of mass. Such analysis performed on energy spectra, collected in inclusive [88–93] data and in data sorted as a function of centrality, [94–97] shows that the procedure is able to give a qualitative characterization of the light-particle emission process.

The source velocities (v_s), the inverse-slope parameters (T), the multiplicities (M) and the Coulomb emission barriers (E_c) are the fit parameters. A good reproduction of the experimental data, in the whole angular range, is possible only if three sources are taken into account: a projectile-like source (PLF) ($v_s \approx v_{beam}$) that dominates at forward angles, a target-like source (TLF) ($v_s \approx 0$) localized at low energies, and an intermediate-velocity source (IS) ($v_s \approx v_{beam/2}$) that dominates at high energies and at larger polar angles. The relative yields of the sources depend on the system asymmetry, on the reaction centrality and incident energy [95,96]. The values of the inverse-slope parameters (T) are of the order of 4–6 MeV for TLF and PLF while, for the IS source, T is much higher, depending on the incident energy. The presence of these three sources is clearly evidenced also in the Lorentz-invariant differential cross-section plots for light particles [95,98].

Light particles emitted from TLF and PLF sources are interpreted as particles evaporated from equilibrated systems with a statistically predicted Maxwellian spectrum. Such interpretation is strengthened by the analysis at lower bombarding energy and/or excitation energies. The exponential slope T reflects the "apparent temperature" of the emitting systems averaged over the whole de-excitation cascade. Applying corrections using statistical models, it is possible to estimate the initial temperatures, that are in agreement with the values estimated with other methods [99].

Protons and neutrons emitted from a source with half beam velocity (IS), which accounts for the most energetic part of the spectra at around 90° in the nucleon-nucleon reference frame, are interpreted as emitted in a non-equilibrated phase of the reaction as a consequence of NN collisions. In ref. [100] mid-velocity emission is already found at 25 A·MeV while the onset of hard-photon

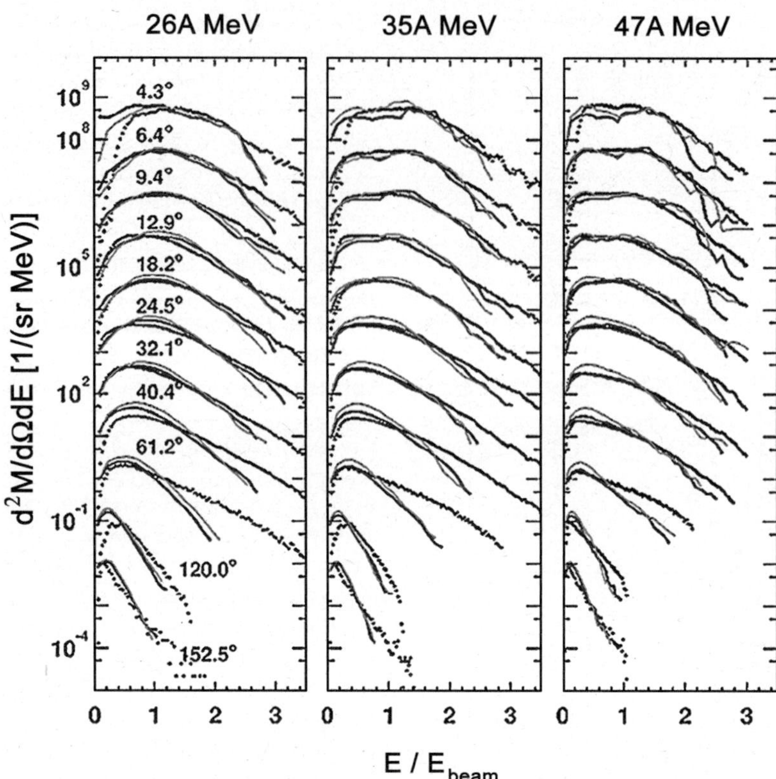

Fig. 12. Proton energy spectra for violent collisions in ^{64}Zr $+$ ^{92}Mo reactions at different incident energies (indicated on top) and different detection polar angles (indicated in the left column) are reported. Experimental data are shown by dots. Thick and thin solid lines refer to AMD-V simulation results for soft and stiff EOS with different prescriptions for the in-medium cross-section [3].

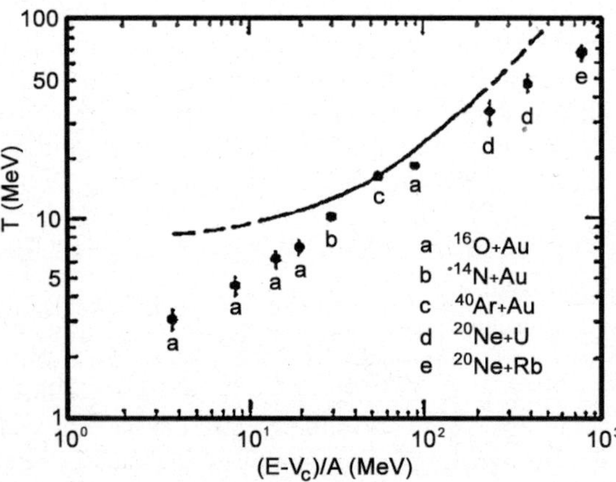

Fig. 13. Experimental slope parameters of the IS in proton spectra for various systems as a function of the bombarding energy above the Coulomb barrier V_c. The curve represents the estimate for quasi-free NN scattering [101].

emission, discussed in the previous paragraph, is found at 10 A·MeV [26].

Originally, the Maxwellian parameterization was introduced following statistical arguments (Boltzmann gas). The momentum distribution components are assumed to be independent Gaussians with a mean square value $\sigma^2 =$ $m \cdot T$, where m is the nucleon mass, and T a temperature parameter. In the case of IS, the parameter T rather represents the random composition of the beam velocity with the Fermi momenta of the nucleons inside the nucleus [101]. The systematics of the extracted slope parameters T for IS for proton spectra is reported in fig. 13 and explained in this framework. Deviations from this picture, especially at low incident energy, have been explained as due to Pauli-blocking effects [101].

By analyzing the neutron energy spectra we expect to find the same characteristics as for protons (the only difference is the lack of Coulomb repulsion). From the experimental point of view neutron detection, especially at high energies, is quite difficult and poor experimental data with respect to charged particles are available. In the ^{35}Cl $+$ ^{nat}Ta reaction at 43 A·MeV [102] neutron spectra have been measured up to 50 MeV from 60° to 150° polar angles as a function of the PLF excitation energies and pre-equilibrium neutrons from the IS source have been evidenced. In this work from the Maxwellian fit a IS velocity lower than half the beam velocity for all the PLF excitation energies has been extracted and reproduced by BNV calculations. This trend has been explained as due to the attractive mean field from the target that the emitted neutrons of moderately high energy still feel. In the scenario where pre-equilibrium neutrons are emitted as a consequence of NN collisions, if the emitted neutron undergoes more than one collision, a velocity lower than the NN velocity is expected. Recently, although neutron spectra were measured only up to 25 MeV in the ^{36}Ar $+$ ^{58}Ni reaction at 50 A·MeV from 60° to 150° [103], they were analyzed as a function of the centrality and the IS velocity was extracted. The IS velocity is found closer and closer to the NN center-of-mass velocity with increasing centrality. The analysis of proton spectra, presented in the same work but in the ^{58}Ni $+$ ^{58}Ni reaction at 52 A·MeV, does not show the same trend; the IS velocity is near to the NN velocity for all the centrality bins. This trend can be explained if neutrons are emitted both from NN collisions in the interaction zone in the first stage of the reaction (predominant at central collisions) and/or by a delayed emission that occurs after the neck rupture and enhanced near the heaviest partner of the reaction thus explaining the lower IS velocity at peripheral and semi-central collisions for neutrons. This delayed emission near the quasi-target is enhanced for neutrons with respect to protons due to the lack of the Coulomb repulsion. To confirm this interpretation simultaneous measurements of protons and neutrons for the same reactions are necessary [104].

An analysis in terms of Maxwellian source emission applied to light-cluster (d, t, He) energy spectra shows a similar scenario. The emission of the most energetic particles from the IS has been explained with a coalescence model where the emission of light clusters is related to the momentum space densities of nucleons in the collision [88,92,93,105]. The coalescence radius, P_0, is the single free parameter of the model once proton and neutron energy spectra are known. P_0 is the radius of a sphere in momentum space where coalescence occurs. Recently, the coalescence model has been coupled with dynamical models describing the collision [98,106,107] to explain the energy spectra of complex light particles. The percentage of particles emitted promptly at intermediate velocity (pre-equilibrium particles) decreases with increasing mass of the cluster [98]. In ref. [106] a self-consistent coalescence model analysis has been used to determine the size of the system as a function of time and to follow the evolution of density and temperature during the reaction. Recently, the emission of light cluster $(Z \geq 3)$ at mid-velocities has been interpreted as emitted from a neck-like structure, formed dynamically during the reaction, joining the quasi-projectile and the quasi-target [108,109]. The interest in these studies is focused on understanding the nature of "neck" formation (see the contribution by Di Toro *et al.*, *Neck dynamics*, this topical issue). One of the most relevant features of IMF emission at mid-velocities is a neutron enrichment with respect to IMF emission from the projectile. Four reactions ^{124,136}Xe $+$ ^{112,124}Sn [110] were studied at 55 A·MeV supporting the idea that IMFs are emitted from a multiple neck rupture from a material that is "surface-like" thus enhancing the N/Z ratio. These results were confirmed by the chemical analysis of the mid-velocity component measured [100] in peripheral and semi-central collisions induced by Xe and Sn at energies between 25 and 50 A·MeV. The most neutron-rich isotopes are favored at lower energies and in peripheral

collisions, where the emission is globally more neutron rich than evaporative processes. Similar results have been found in [111] where more neutron-rich He isotopes are found in mid-peripheral emission from the neck zone with respect to He isotopes emitted from PLF. Exclusive measurements of neutron and proton emission characteristics from intermediate-velocity sources, measured in the same reactions with different N/Z ratio, and comparison with dynamical calculations can add information on the mechanism leading to the neutron enrichment of the neck region.

In the energy regime considered in this review this well-established scenario, in terms of three emitting sources, is a way to mimic the emission of particles dynamically originating during the whole reaction time. In particular, pre-equilibrium particles are not necessarily emitted from a source well located in time. The comparison of experimental data with dynamical model predictions [12] and more complex analyses, as particle-particle correlations (see the contribution by Verde *et al.* in this topical issue and refs. [112,113]) allow to infer a space-time characterization of the emission mechanisms. However, from the experimental point of view, the emitting source parameterization is able to give an estimate of the number of nucleons and of the energy removed at each step in the reaction [95] and to "isolate", by selecting detection angles and energies, energetic particles emitted from the IS (pre-equilibrium particles) for more complex event-based analyses [13,94].

In order to estimate the amount of pre-equilibrium emission, besides the integration of the Maxwellian fitted curves, alternative methods have been applied for the ^{58}Ni on ^{58}Ni reaction at 32 A·MeV [114] in complete events detected with the INDRA apparatus. In this work all the methods applied lead to the same estimate of the mass and the energy removed in the pre-equilibrium stage. In [115] a balance of mass, momentum and energy has been performed for several central reactions with different mass asymmetries and energies (from 17 to 115 A·MeV). These works prove that a large fraction of the initial mass and available energy is removed in the pre-equilibrium stage confirming that the estimate of pre-equilibrium emission is of crucial importance for the study of "hot" systems formed in central heavy-ion collisions. In particular in [116] the isospin content of the pre-equilibrium nucleon emission at high transverse momentum is suggested as a probe to explore the momentum dependence of the symmetry term potential in asymmetric nuclear matter.

The space characterization of pre-equilibrium emission can also be inferred by studying the impact parameter dependence of pre-equilibrium particles [94,117,118]. Energetic protons at large polar angles, measured as a function of the impact parameter for reactions with different mass asymmetry at 44 A·MeV [94], show that pre-equilibrium proton multiplicities increase with the size of the overlapping region and, from system to system, with the number of protons in the collision zone. The trend as a function of impact parameter b can be understood if we assume that proton yields scale with the overlap surface of the colliding system thus indicating that pre-equilibrium protons are emitted mainly from first NN collisions as already reported for hard-photon emission. This surface dependence had already been predicted for high-energy gamma emission by BUU calculations [31]. In ref. [119] from light particles (p, d, t, ^{3}He and ^{4}He) measured in Ar + Ni collisions from 52 to 95 A·MeV, the amount of matter and energy associated with the IS are estimated. The results indicate that the total mass is directly correlated to the impact parameter and it does not depend on the incident energy, while the energy carried by light particles at intermediate velocity is not strongly dependent on the impact parameter but it depends on the incident energy.

From the analysis of proton angular distributions in reactions with different mass asymmetry at 44 A·MeV [12], a reminiscence of the elementary NN cross-section in the observed anisotropy in central collisions for quasi-symmetric systems confirms the hypothesis that pre-equilibrium protons are emitted as a consequence of first NN collisions. These results, compared with BNV predictions, are consistent with a scenario in which particles are emitted in the first phase of the reaction mainly from the first NN collisions in the interaction zone. Due to the short proton mean free path a strong screening effect is evident, which distorts the expected angular-distribution trends in peripheral reactions and heavy systems. A clear signature of this scenario is provided by γ-p correlation results [54], as described in sect. 2. These results suggest that protons produced in light symmetric systems should be good probes to gather information on the in-medium NN cross-section. Indeed light-cluster formation has been calculated in microscopic transport approaches including nucleon-nucleon cross-section in-medium effects, which depend on the density and energies deposited in the system [120]. This study shows that the number and the spectra of light charged particles change in a significant manner [121,122].

High-efficiency apparatuses are capable of measuring differential energy spectra that span several orders of magnitude. Particles with energy per nucleon up to 3–4 times the incident energy [3,12,13,123] have thus been measured. With the hypothesis that energetic protons are emitted as a consequence of first-chance NN collisions, and that the momentum distribution can be approximated as a degenerate Fermi gas, a kinematical limit in proton energies is expected. The observation in the energy spectra of protons far exceeding this limit is a puzzle not yet resolved. The mechanism able to concentrate in few nucleons so much energy is not yet known, but its knowledge can be of crucial importance to shed light on all sub-threshold particle emission in heavy-ion reaction at intermediate energy. Mechanisms as cooperative effects [15], high-momentum tail of the nucleon Fermi distribution [124], fluctuations in the momentum space [125] or properties of the potential have been proposed. Attempts to reproduce the extremely high-energy tail in proton spectra with dynamical models [3,13,123,125] have been done. In these works, different prescriptions for the effective mean-field potential and for in-medium properties of the two-body collision cross-section have been used.

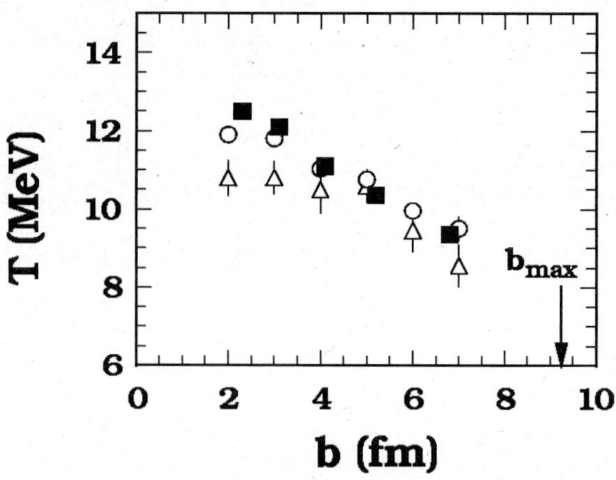

Fig. 14. Inverse-slope parameters extracted from proton spectra as a function of the impact parameter for the ^{58}Ni + ^{58}Ni reaction at 30 A·MeV. Full squares represent the experimental data, open triangles the BNV calculations with a local mean field and open circles the BNV calculations with a momentum-dependent potential [126].

The most promising mechanism responsible for the production of extremely energetic protons seems to be a cooperative mechanism by which more nucleons act together to produce a high-energy nucleon. A similar production of high-energy gamma-rays exceeding the kinematical limit for n-p collisions has been observed in ref. [69]. Experimental inclusive proton spectra measured in Ar + Ta collisions at 95 A·MeV at large polar angles [123] were compared with QMD [125] and BNV [123] calculations that include, besides the usual local mean-field potential and two-body collisions, three-body collisions which succeeded in the explanation of sub-threshold pion production (see section on pions). The comparison with the experimental data shows good agreement in the reproduction of the high-energy slope.

Protons up to twice the NN kinematical limit (5 times the beam energy per nucleon) have been measured as a function of the centrality in the ^{58}Ni + ^{58}Ni reaction at 30 A·MeV. The slope and yield of energetic proton spectra collected at different impact parameters have been extracted and the comparison with results of BNV calculations with a local Skyrme potential and Gale-Bertsch-Das Gupta momentum-dependent potential is reported in fig. 14 [126]. The slopes are well reproduced by a momentum-dependent potential, while the local potential fails especially at central collisions. Since momentum-dependent effects could be expected also from a stiff potential, the dependence on the compressibility term has also been investigated for the same reaction [127]. The proton high-energy spectra predicted by BNV calculations with a hard and a stiff compressibility for a local Skyrme interaction have been compared to the experimental spectra, and the results indicate that the proton spectra are not sensitive to the compressibility term while they are sensitive, both in yield and slope, to a momentum-dependent potential.

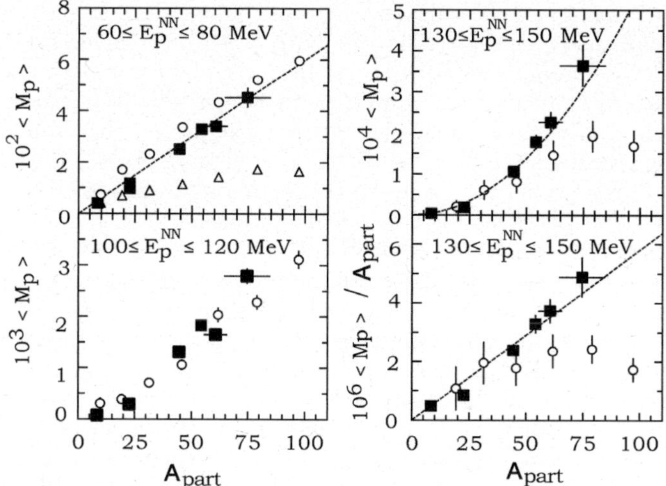

Fig. 15. Proton multiplicities as a function of the number of the participant nucleons (A_{part}) for different proton energies in the reaction ^{58}Ni + ^{58}Ni at 30 A·MeV. Full squares represent the experimental data, open triangles the BNV calculations with a local potential and open circles the BNV calculations with a momentum-dependent potential. In the bottom right panel the points for $130 \leq E_p \leq 150$ are reported, divided by A_{part}. The BNV calculations are scaled by a factor 0.6 in order to take into account the yield reduction due to complex particle emission not predicted by the calculations.

A more detailed analysis for the same reaction has been performed in [13]. High-energy protons were detected in coincidence with heavy fragments in order to select classes of events with different centralities. For each class of events high-energy gamma-rays were measured in coincidence and a quantitative measure of the size of the interaction zone (A_{part}) was determined. In fig. 15 high-energy proton multiplicities as a function of A_{part} (*i.e.* impact parameter) for different energy values are reported. A linear dependence on A_{part} (dashed line in the top left panel) is observed for protons between 60 and 80 MeV, thus confirming that the high-energy proton multiplicity increases linearly with increasing A_{part}. With increasing proton energy a stronger deviation from linearity is observed. The dashed line in the top right panel of fig. 15 shows a quadratic dependence. A similar behavior is observed for π^0 and η at much higher incident energies [128] and interpreted as due to multi-step processes. The experimental data were compared with microscopic BNV calculations where two different potentials were included: a local Skyrme interaction (open triangles in fig. 15) and a Gale-Bertsch-Das Gupta momentum-dependent interaction (open circles). Momentum-dependent BNV calculations reproduce well the data up to 120 MeV (left panels in fig. 15) but fail in reproducing the extremely high-energy proton multiplicity. BNV with local potential undershoot the data already at low energies and central collisions. These results call for the introduction in the transport models of ingredients that are beyond the one- and two-body effects as cooperative effects (three- or higher-order collisions) [15]. As quoted in the introduction, the fact that

three-body effects might be important has been advocated also in microscopic Bruckner HF calculations [8] to reproduce the experimental ground-state energy and density of nuclear matter. The information that can be extracted from energetic proton data is that, even though cooperative effects are important, their relevance compared to nucleons produced from two-body collisions is negligible. In BHF microscopic calculations instead the relevance of the three-body force is large, in fact the ground-state density in the calculations decreases almost of a factor two when three-body forces are included [8,129]. We would expect that when increasing the excitation energy of the system three-body forces would become even more important as compared to cold nuclear matter because of some relaxation of the Pauli blocking.

This aspect calls for a further and more detailed experimental and theoretical analysis of these reactions at various beam energies to try to pin down the relevance of two-body *versus* three-body forces, which could be relevant for microscopic calculations of the nuclear-matter EOS and the binding energies of light nuclei.

A detailed comparison between experimental results from several reaction at different incident energies (^{64}Zn + ^{58}Ni, ^{92}Mo and ^{197}Au at 26, 35, 47 A·MeV) with dynamical model predictions has been presented in ref. [3]. The general aim of this work is to get information on the reaction mechanisms by comparing the model results with a wide set of experimental data. Direct experimental observables as velocity and energy spectra, multiplicity and charge distribution for light particles (p, d, t, α) and IMF ($Z \geq 3$) have been compared with a modified antisymmetrized molecular dynamics (AMD-V) that takes into account different prescriptions for the in-medium NN cross-section and a Gogny effective interaction with a momentum-dependent mean field and two different compressibility values. In fig. 12 the experimental proton energy spectra (dots) and AMD-V calculations with two different compressibility values and in-medium NN cross-section description are reported. Results corresponding to a compressibility for infinite nuclear matter $K = 228$ MeV and an empirical in-medium NN cross-section prescription, where no distinction is made between n-n and n-p cross-sections (soft EOS+NN_{emp}) are represented as thick lines while results corresponding to $K = 360$ MeV and an in-medium NN cross section with different-cross section prescription for n-n and n-p cross-sections (stiff EOS+NNLM) are represented as thin lines. Both calculations with different compressibilities are in qualitative agreement with the bulk of data, but none of them is able to reproduce the high-proton energy tails (see fig. 12) especially at around 50° polar angle, which corresponds to emission near 90° in the center-of-mass system, where the high-energy IS component can be clearly evidenced.

5 Conclusions and future perspectives

In this work we have reviewed the production of hard photons, subthreshold pions and energetic nucleons in nucleus-nucleus collisions at intermediate energies

(10 A·MeV $\leq E \leq$ 100 A·MeV). A first remark concerns the fact that the bulk of data are qualitatively rather well reproduced by dynamical calculations like BUU. The nuclear dynamics is described in terms of a mean field and two-body collisions, thus confirming the dominant role of NN collisions, boosted by Fermi motion of the colliding nuclei. This result opened the possibility of carrying on a more detailed comparison between experimental data and calculations in order to put constraints on the in-medium NN cross-section, as well as on the momentum dependence of the nuclear mean field, and on the equation of state of nuclear matter.

As a second general result, the general agreement between high-energy particle data and dynamical transport models calculations qualifies these particles as probes for the early non-equilibrated stage of the reaction. This allows to extract information regarding the pre-equilibrium phase and the following evolution towards equilibration. Some interesting results from a detailed comparison between data and calculations have been obtained:

- a clear evidence of momentum-dependent interactions has been gathered;
- particle production is very sensitive to the NN cross-section, however more work should be done to draw conclusive results on σ_{NN} modifications in the nuclear medium taking also into account the effects on the reaction dynamics.

A word of caution is however necessary. All these conclusions crucially rely on the comparison of data with transport calculations, and one should make clear that they do not depend on the numerical implementation of the model. One main suggestion could be to compare the results of several dynamical calculations with many different experimental data, for example for both energetic nucleon, pion and hard-photon production and, on the other hand, to verify that the same prescriptions allow also a good description of other features of the nucleus-nucleus dynamics both in peripheral (neck, PLF fragmentation, etc.) and central collisions (fusion, multifragmentation, etc.). This procedure should allow to constrain the parameters of interest for the EOS. Moreover, the fact that a probe is sensitive only to some parameters of the EOS and not to others is important since it allows to disentangle the contribution of various parameters. This is the case of hard photons and energetic nucleon spectra that are not particularly sensitive to the stiffness of the EOS.

It is also important to mention that, for energetic proton production, a strong improvement in the agreement between data and calculations is achieved, for central collisions, only including a momentum dependence of the nuclear mean-field interaction. To our knowledge, momentum-dependent calculations have not been carried out for the hard-photon production and, since the effect of Pauli in the final state is much stronger in this case, this comparison is expected to provide additional valuable information.

Hard photons, which are unperturbed probes due to the fact that they once produced do not interact anymore with the surrounding nuclear matter, provide a clean chronology of the various stages of the reaction. In particular, energetic hard photons ($E_\gamma \geq 50$ MeV) gave access to the momentum Fermi distributions of the colliding nuclei in the early non-equilibrated stages of the reaction, while "thermal" hard photons provide a clock for multifragmentation.

The observation of deep-subthreshold or extremely energetic "particles" addressed the question of which mechanisms could allow to concentrate a relevant fraction of the available energy in the production of a single energetic or massive "particle". For hard photons, pions and energetic protons as described in this report, as far as we know, there is a lack of theoretical models to compare with existing deep-subthreshold data. Cooperative effects, where more nucleons or cluster of nucleons participate in the collisions, seem very promising and more theoretical effort should be devoted to this issue. One extreme case which we would like to stress is pionic fusion where all the beam energy is transformed into the pion mass and a compound nucleus is formed very close to its ground state [130].

In the near future the investigation of the isospin degree of freedom will be boosted by the new facilities providing exotic beams and its impact on the EOS of asymmetric nuclear matter will be the next challenge for heavy-ion nuclear physics. In this field, n, p, pion and hard-photon detection is expected to provide very important pieces of information, especially due to the fact that these probes are sensitive to the first stage of the reaction where the largest asymmetry in isospin and densities can be reached. In asymmetric matter a splitting of neutron and proton effective masses is expected, but the sign of the splitting is quite controversial giving opposite results for various Skyrme forces. The investigation of pre-equilibrium particles, for which the high-momentum components have a crucial role, could provide sensitive probes. In particular, the neutron proton ratio of fast nucleon emission as a function of centrality and the slopes and yields of hard-photon spectra, which can provide complementary pieces of information with respect to the nucleon emission thanks to the fact that they are not affected by final-state interaction, should be carried out [131].

Concerning pion emission, several theoretical works have been published, investigating the sensitivity of the π^+/π^- ratio to the isospin degree of freedom at incident energies of about 400 A·MeV, and its dependence on the neutron and proton chemical potentials and on the symmetry energy has been put in evidence [132,133]. However, it is important to underline that these works deal with equilibrated dense nuclear matter, while the possibility of investigating the π^+/π^- ratio at incident nucleon energy below the NN pion energy threshold, could give access to the early non-equilibrated stage of the reaction.

References

1. E. Migneco *et al.*, Nucl. Instrum. Methods. Phys. Res. A **314**, 31 (1992).
2. J. Pouthas *et al.*, Nucl. Instrum. Methods. Phys. Res. A **357**, 418 (1995).

3. R. Wada *et al.*, Phys. Rev. C **69**, 044610 (2004).
4. A. Pagano *et al.*, Nucl. Phys. A **681**, 331 (2001).
5. G. Martinez *et al.*, Nucl. Instrum. Methods. Phys. Res. A **391**, 435 (1997).
6. R.T. De Souza *et al.*, Nucl. Instrum. Methods. Phys. Res. A **295**, 109 (1990).
7. I. Iori *et al.*, Nucl. Instrum. Methods. Phys. Res. A **325**, 458 (1993).
8. W. Zuo *et al.*, Eur. Phys. J. A **14**, 469 (2002); Zuo W. *et al.*, Nucl. Phys. A **706**, 418 (2002).
9. A. Kievski, *Proceedings of the 10th Conference on Problems in Theoretical Nuclear Physics, Cortona 6-9 October 2004* (World Scientific, 2005).
10. A. Bonasera, M. Bruno, C.O. Dorso, P.F. Mastinu, Riv. Nuovo Cimento **23**, 1 (2000).
11. B. Povh, K. Rith, C. Scholz, F. Zetsche, *Particles and Nuclei ? An Introduction to the Physical Concepts* (Springer, 1995).
12. R. Coniglione *et al.*, Phys. Lett. B **471**, 339 (2000).
13. P. Sapienza *et al.*, Phys. Rev. Lett. **87**, 072701 (2001).
14. W. Cassing *et al.*, Phys. Rep. **188**, 363 (1990).
15. A. Bonasera, F. Gulminelli, J. Molitoris, Phys. Rep. **243**, 1 (1994).
16. G.F. Bertsch, S. Das Gupta, Phys. Rep. **160**, 189 (1988).
17. J. Aichelin, Phys. Rep. **202**, 233 (1991).
18. J. Aichelin *et al.*, *Comparison between the different transport codes*, in preparation.
19. J. Julien *et al.*, Phys. Lett. B **264**, 269 (1991).
20. P. Piattelli *et al.*, Nucl. Phys. A **649**, 181c (1996).
21. K.A. Snover, Annu. Rev. Nucl. Part. Sci. **36**, 545 (1986).
22. J.J. Gardhøje, Annu. Rev. Nucl. Part. Sci. **42**, 483 (1992).
23. J.H. Le Faou *et al.*, Phys. Rev. Lett. **72**, 3321 (1994).
24. H. Noll *et al.*, Phys. Rev. Lett. **52**, 1284 (1984).
25. E. Grosse, Nucl. Phys. A **447**, 611c (1985); E. Grosse *et al.*, Europhys. Lett. **2**, 9 (1986).
26. N. Gan *et al.*, Phys. Rev. C **49**, 298 (1994).
27. J. Clayton *et al.*, Phys. Rev. C **40**, 1207 (1989).
28. D. Vasak *et al.*, Nucl. Phys. A **428** (1984).
29. Che Ming Ko, G. Bertsch, J. Aichelin, Phys. Rev. C **31**, R2324 (1985).
30. K. Nakayama, G.F. Bertsch, Phys. Rev. C **36**, 1848 (1987).
31. W. Bauer *et al.*, Phys. Rev. C **34**, 2127 (1986); T.S. Biro *et al.*, Nucl. Phys. A **475**, 579 (1987).
32. B.A. Remington, M. Blann, G.F. Bertsch, Phys. Rev. Lett. **57**, 2909 (1986); B.A. Remington, M. Blann, Phys. Rev. C **36**, 1387 (1987).
33. J.D. Jackson, *Classical Electrodynamics* (J. Wiley and Sons).
34. W. Cassing *et al.*, Phys. Lett. B **181**, 217 (1986).
35. K. Nakayama, Phys. Rev. C **42**, 1009 (1989).
36. H. Huisman *et al.*, Phys. Rev. Lett. **83**, 4017 (1999).
37. M. Volkerts *et al.*, Phys. Rev. Lett. **90**, 062301-1 (2003).
38. M. Volkerts *et al.*, Phys. Rev. Lett. **92**, 202301-1 (2004).
39. N. Herrmann *et al.*, Phys. Rev. Lett. **60**, 1630 (1988).
40. H. Nifenecker, J.P. Bondorf, Nucl. Phys. A **442**, 478 (1985); H. Nifenecker, J.A. Pinston, Annu. Rev. Nucl. Part. Sci. **40**, 113 (1990).
41. R. Shyam, J. Knoll, Nucl. Phys. A **448**, 322 (1986).
42. J. Stevenson *et al.*, Phys. Rev. Lett. **57**, 555 (1986).
43. G. Breitbach *et al.*, Phys. Rev. C **40**, 2893 (1989).
44. C.L. Tam *et al.*, Phys. Rev. C **38**, 2526 (1988).
45. N. Alamanos *et al.*, Phys. Lett. B **173**, 392 (1986).
46. M. Kwato Njock *et al.*, Nucl. Phys. A **489**, 368 (1988).
47. R. Hingmann *et al.*, Phys. Rev. Lett. **58**, 759 (1987).
48. T. Reposeur *et al.*, Phys. Lett. B **276**, 418 (1992).
49. V. Metag, Nucl. Phys. A **488**, 483c (1988).
50. S. Riess *et al.*, Phys. Rev. Lett. **69**, 1504 (1992).
51. E. Migneco *et al.*, Phys. Lett. B **298**, 46 (1993).
52. G. Martinez *et al.*, Phys. Lett. B **334**, 23 (1994).
53. A.R. Lampis *et al.*, Phys. Rev. C **38**, 1961 (1988).
54. P. Sapienza *et al.*, Phys. Rev. Lett. **73**, 1769 (1994).
55. A. Badalá *et al.*, Phys. Rev. Lett. **74**, 4779 (1995).
56. H.W. Barz *et al.*, Phys. Rev. C **53**, R553 (1996).
57. J.H.G. Van Pol *et al.*, Phys. Rev. Lett. **76**, 1425 (1996).
58. P. Piattelli *et al.*, Phys. Lett. B **442**, 48 (1998).
59. V.E. Viola *et al.*, Phys. Rev. C **26**, 178 (1982).
60. A. Schubert *et al.*, Phys. Rev. Lett. **72**, 1608 (1994).
61. F.M. Marqués *et al.*, Phys. Lett. B **349**, 30 (1995).
62. G. Martinez *et al.*, Phys. Lett. B **349**, 23 (1995).
63. D.G. D'Enterria *et al.*, Phys. Rev. Lett. **87**, 022701 (2001).
64. R. Alba *et al.*, Nucl. Phys. A **654**, 761c (1999).
65. R. Alba *et al.*, Nucl. Phys. A **681**, 339c (2001).
66. R. Alba *et al.*, Nucl. Phys. A **749**, 98c (2005).
67. R. Alba *et al.*, arXiv:nucl-ex/0507028, 22-July-2005.
68. D.G. D'Enterria *et al.*, Phys. Lett. B **538**, 27 (2002).
69. K.K. Gudima *et al.*, Phys. Rev. Lett. **76**, 2412 (1996).
70. J. Aichelin, G. Bertsch, Phys. Lett. B **138**, 350 (1984).
71. R. Shyam, J. Knoll, Nucl. Phys. A **426**, 606 (1984).
72. A. Bonasera, G.F. Bertsch, Phys. Lett. B **195**, 521 (1987).
73. W. Bauer, Phys. Rev. C **40**, 715 (1989).
74. A. Badalá *et al.*, Phys. Rev. C **48**, 2350 (1993).
75. W. Benenson *et al.*, Phys. Rev. Lett. **43**, 683 (1979).
76. G.R. Young *et al.*, Phys. Rev. C **33**, 742 (1986).
77. H. Feldmeier, J. Schnack, Rev. Mod. Phys. **72**, 655 (2000).
78. A. Ono *et al.*, Phys. Rev. Lett. **68**, 2898 (1992).
79. M. Papa, T. Maruyama, A. Bonasera, Phys. Rev. C **64**, 024612 (2001).
80. D. Lacroix D. Ph., Chomaz, Nucl. Phys. A **636**, 85 (1998).
81. A. Bonasera, arXiv:nucl-th/0110068.
82. T. Kodama *et al.*, Phys. Rev. C **29**, 2146 (1984).
83. A. Bonasera, F. Gulminelli, Phys. Lett. B **275**, 24 (1992).
84. P. Danielewicz, Q. Pan, Phys. Rev. C **46**, 2002 (1992).
85. J. Aichelin, C.M. Ko, Phys. Rev. Lett. **55**, 2661 (1985)
86. A. Badalá *et al.*, Phys. Rev. C **54**, 2138 (1996).
87. A. Badalá *et al.*, Phys. Rev. C **57**, 166 (1998).
88. T.C. Awes *et al.*, Phys. Lett. B **103**, 417 (1981); T.C. Awes *et al.*, Phys. Rev. C **24**, 89 (1981).
89. T.C. Awes *et al.*, Phys. Rev. C **25**, 2361 (1982).
90. G.D. Westfall *et al.*, Phys. Lett. B **116**, 118 (1982).
91. G.D. Westfall *et al.*, Phys. Rev. C **29**, 861 (1984).
92. B.V. Jacak *et al.*, Phys. Rev. C **35**, 1751 (1987).
93. T. Fukuda *et al.*, Nucl. Phys. A **425**, 548 (1984).
94. R. Alba *et al.*, Phys. Lett. B **322**, 38 (1994).
95. D. Santonocito *et al.*, Phys. Rev. C **66**, 044619 (2002).
96. R. Wada *et al.*, Phys. Rev. C **39**, 497 (1989).
97. B.E. Hasselquist, Phys. Rev. C **32**, 145 (1985).
98. P. Pawlowski *et al.*, Eur. Phys. J. A **9**, 371 (2000) and references therein.
99. J.B. Natowitz *et al.*, Phys. Rev. C **65**, 034618 (2002).
100. E. Plagnol *et al.*, Phys. Rev. C **39** (2000).
101. H. Fuchs, K. Möhring, Rep. Prog. Phys. **57**, 231 (1994).
102. Y. Larochelle *et al.*, Phys. Rev. C **59**, R565 (1999).
103. D. Thieriault *et al.*, Phys. Rev. C **71**, 014610 (2005).

104. R. Ghetti *et al.*, Nucl. Phys. A **674**, 277 (2000).
105. B.V. Jacak *et al.*, Phys. Rev. C **31**, 704 (1985).
106. K. Hagel *et al.*, Phys. Rev. C **62**, 034607 (2000).
107. V. Avdeichikov *et al.*, Nucl. Phys. A **736**, 22 (2004).
108. P.M. Milazzo *et al.*, Nucl. Phys. A **756**, 39 (2005).
109. E. De Filippo *et al.*, Phys. Rev. C **71**, 044602 (2005) and reference therein.
110. J.F. Dempsey *et al.*, Phys. Rev. C **54**, 1710 (1996).
111. P.M. Milazzo *et al.*, Phys. Lett. B **509**, 204 (2001).
112. R. Ghetti *et al.*, Phys. Rev. Lett. **87**, 102701 (2001).
113. G. Verde *et al.*, Phys. Rev. C **65**, 054609 (2002).
114. P. Lautesse *et al.*, Phys. Rev. C **71**, 034602 (2005).
115. Rulin Sun *et al.*, Phys. Rev. Lett. **84**, 43 (2000).
116. J. Rizzo *et al.*, Phys. Rev. C **72**, 064609 (2005).
117. J. Peter *et al.*, Phys. Lett. B **237**, 187 (1990).
118. D. Prindle *et al.*, Phys. Rev. C **48**, 291 (1993).
119. T. Lefort *et al.*, Nucl. Phys. A **662**, 397 (2000).
120. C. Kuhrts *et al.*, Phys. Rev. C **63**, 034605 (2001).
121. E.A. Remler, Ann. Phys. **119**, 326 (1979).
122. E.A. Remler, Phys. Rev. C **25**, 2974 (1982).
123. M. Germain *et al.*, Nucl. Phys. A **620**, 81 (1997).
124. I. Bobeldijk *et al.*, Phys. Lett. B **353**, 32 (1995).
125. M. Germain *et al.*, Phys. Lett. B **437**, 19 (1998).
126. P. Sapienza *et al.*, Nucl. Phys. A **734**, 601 (2004).
127. P. Sapienza *et al.*, LNS Activity Report 2001, p. 46 www.lns.infn.it.
128. A.R. Wolf *et al.*, Phys. Rev. Lett. **80**, 5281 (1998).
129. A. Bonasera, *Proceedings of the 10th Conference on Problems in Theoretical Nuclear Physics, Cortona 6-9 October 2004* (World Scientific, 2005) p. 171.
130. D. Horn *et al.*, Phys. Rev. Lett. **77**, 2408 (1996).
131. The Physics Objectives SPIRAL 2 Project (2005) see www.ganil.fr.
132. Bao An Li, Phys. Rev. C **67**, 017601 (2003).
133. T. Gaitanos *et al.*, Phys. Lett. B **595**, 209 (2004); G. Ferrini *et al.*, nucl-th/0504032.

Eur. Phys. J. A **30**, 65–70 (2006)

DOI 10.1140/epja/i2006-10106-9

THE EUROPEAN
PHYSICAL JOURNAL A

Neck dynamics

M. Di Toro[1,a], A. Olmi[2], and R. Roy[3]

[1] Laboratori Nazionali del Sud INFN, Department of Physics and Astronomy, University of Catania, Via S. Sofia 62, I-95123, Italy

[2] INFN - Sezione di Firenze, Via G. Sansone 1, I-50019 Sesto Fiorentino, Italy

[3] Université Laval, Département de physique, de génie physique et d'optique, Cité universitaire, Québec, Qc, Canada, G1K 7P4

Received: 28 February 2006 /
Published online: 6 October 2006 – © Società Italiana di Fisica / Springer-Verlag 2006

Abstract. Intermediate-energy heavy-ion reactions produce a mid-rapidity region or neck, mostly in the semiperipheral collisions. Brief theory and experiment surveys are presented. General properties of the mid-rapidity zone are reviewed and discussed in the framework of reaction dynamics. Hierarchy effect, neutron enrichment, isospin diffusion are all new neck phenomena which are surveyed. The main neck observables are also examined, mainly in the context of the symmetry term of the nuclear equation of state.

PACS. 25.70.-z Low and intermediate energy heavy-ion reactions – 25.70.Lm Strongly damped collisions – 25.70.Mn Projectile and target fragmentation

1 Neck fragmentation in semiperipheral collisions at Fermi energies

1.1 Theory survey

The possibility of observation of new effects, beyond the deep-inelastic binary picture, in fragment formation for semicentral collisions with increasing energy was advanced on the basis of the reaction dynamics studied with transport models [1–3]. The presence of a time matching between the instability growth in the dilute overlap zone and the expansion-separation time scale suggested the observation of mean-field instabilities first at the level of anomalous widths in the mass/charge/... distributions of Projectile-Like or Target-Like (PLF/TLF) residues in binary events, then through a direct formation of fragments in the *neck region* [4,5]. It is clear that in the transport simulations stochastic terms should be consistently built in the kinetic equations in order to have a correct description of instability effects. Stochastic Mean-Field approaches have been introduced, reproducing the presently available data and having a large predictive power [6–9].

In conclusion, at the Fermi energies, we expect an interplay between binary and *neck fragmentation* events, where Intermediate Mass Fragments (IMF, in the range $3 \leq Z \leq 10$) are directly formed in the overlapping region, roughly at mid-rapidity in semicentral reactions. The competition between the two mechanisms is expected to be

rather sensitive to the nuclear equation of state, in particular to its compressibility that will influence the interaction time as well as the density oscillation in the *neck region*. In the case of charge asymmetric colliding systems the poorly known stiffness of the symmetry term will also largely influence the reaction dynamics. An observable sensitive to the stiffness of the symmetry term can be just the relative yield of incomplete fusion *vs.* deep-inelastic in neck fragmentation events [10]. Moreover, for neutron-rich systems, in the asy-soft case we expect more interaction time available for charge equilibration. This means that even the binary events will show a sensitivity through a larger isospin diffusion. At variance, in the asy-stiff case the two final fragments will keep more memory of the initial conditions.

Systematic transport studies of isospin effects in the neck dynamics have been performed so far for collisions of Sn-Sn isotopes at 50 AMeV [6,7], Sn-Ni isotopes at 35 AMeV [8,9] and finally Fe-Fe and Ni-Ni, mass 58, at 30 and 47 AMeV [11].

1.2 Experimental survey

It is now quite well established that a large part of the reaction cross-section for dissipative collisions at Fermi energies goes through the *Neck Fragmentation* channel, with IMFs directly produced in the interacting zone in semiperipheral collisions on very short time scales. Before a clear *Neck Fragmentation* was proposed, out-of-equilibrium emission at mid-rapidity was observed for

---

a e-mail: ditoro@lns.infn.it

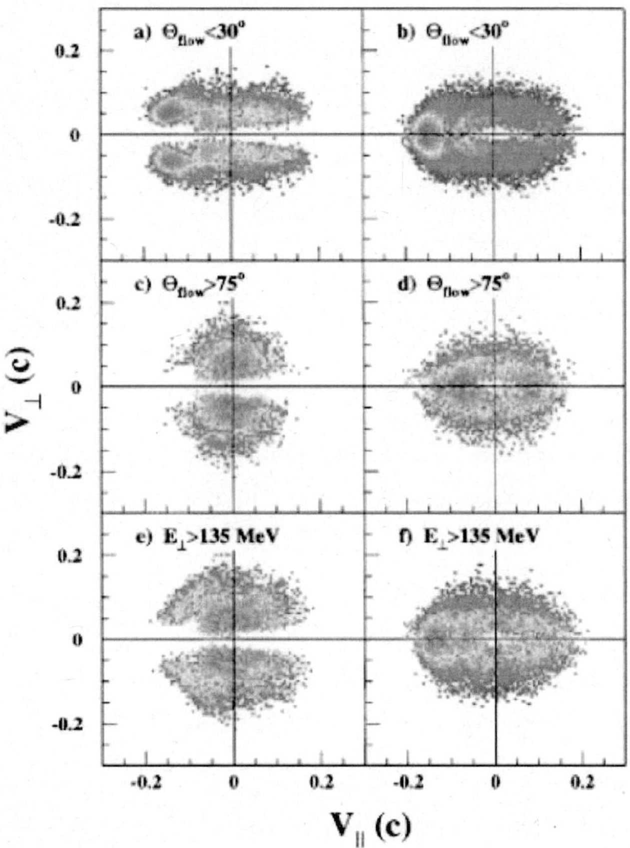

Fig. 1. Galilean-invariant perpendicular *vs.* parallel velocity in the c.m. frame for $Z = 3$ fragments. Parallel velocities are along the beam axis ((a),(c),(e)) and the main axis of the momentum tensor ((b),(d),(f)). Cuts on $\Theta_{flow} < 30°$ ((a),(b)), $\Theta_{flow} > 75°$; ((c),(d)), and on $E_\perp > 135\,\text{MeV}$ which is the top 10% of the $E_\perp$ distribution: ((e),(f)) are made. The count yield is in a logarithmic scale. (From [24].)

IMFs and light charged particles [12–15]. From ternary fission, a non-statistical angular emission pattern was also observed for the IMFs [16–18]. We can expect different isospin effects for this new fragment formation mechanism since clusters are formed still in a dilute asymmetric matter but always in contact with the regions of the projectile-like and target-like remnants almost at normal densities.

A first evidence of this new dissipative mechanism was suggested at quite low energies, around 19 AMeV, in semi-central ^{100}Mo + ^{100}Mo, ^{120}Sn + ^{120}Sn reactions [19,20]. A transition from binary, deep-inelastic, to ternary events was observed, with a dynamically formed fragment that influences the fission-like decay of the primary projectile-like PLF and target-like TLF partners. From the in-plane fragment angular distribution a decrease of scission-to-scission lifetimes with the mass asymmetry of the PLF or TLF "fission-fragments" down to $200\,\text{fm}/c$ has been deduced. Similar conclusions were reached in [21,22]. Consistent with the dynamical scenario was the anisotropic azimuthal distribution of IMFs. In fact the IMF *alignment*

with respect to the (PLF*) velocity direction has been one clear property of the "neck fragments" first noticed by Montoya *et al.* [23] for ^{129}Xe + ^{63,65}Cu at 50 AMeV. As an example in the same context, fig. 1, from [24], displays an important neck component for $Z = 3$ particles in a much lighter system, ^{35}Cl + ^{12}C at 43 AMeV. The flow angle and the total transverse energy are the observable used for impact parameter selection.

In peripheral collisions around 30 AMeV, the IMF production cross-section presents a maximum at mid-rapidity, but their experimental emission pattern cannot be reproduced without a sizable contribution of fragments emitted on a rather short time scale ($< 300\,\text{fm}/c$) and almost at rest in the PLF or TLF reference frame [25]. Light charged particles also show a short time scale at mid-rapidity [26, 27]. With increasing bombarding energy, the mid-rapidity region of peripheral collisions becomes progressively depleted, while the IMFs are increasingly concentrated on Coulomb-like rings around the projectile and target rapidities [28]. However, their distribution on these rings is anisotropic, with a strong preference for emissions toward mid-rapidity. This behavior is particularly evident in the ^{197}Au + ^{197}Au data of ref. [29].

The velocity of the projectile remnants and the distribution of mid-rapidity particles indicate that the mid-rapidity or neck emission mechanism represents an important effect in the excitation energy deposition [30]. Indeed, a recent comparison of the emissions from mid-rapidity with the evaporative emissions from the excited PLF shows that this mechanism has an important role in the overall balance of the reaction (both in terms of emitted mass or charge and energy) and that an important part of the dissipated energy is localized at mid-rapidity [31]. This suggests that a rather large energy density is stored in the contact region of the colliding nuclei and may also explain the well-established feature of an enhanced emission of mid-rapidity IMFs.

A rise and fall of the neck mechanism for mid-rapidity fragments with the centrality, with a maximum for intermediate impact parameters $b \simeq \frac{1}{2}b_{max}$, as observed in [32, 33], suggests the special physical conditions required. The size of the participant zone is of course important but it also appears that a good time matching between the time scales of the reaction and the neck instabilities is also needed, as suggested in refs. [1,4]. In fact a simultaneous presence, in non-central collisions, of different IMF production mechanisms at mid-rapidity was inferred in several experiments [34–44].

An accurate analysis of charge, parallel velocity, and angular distributions has been extended to high fragment multiplicities by Colin *et al.* [43]. They have noticed a "hierarchy effect": the ranking in charge induces on average a ranking in the velocity component along the beam, v_{par}, and in the angular distribution. This means that the heaviest IMF formed in the mid-rapidity region is the fastest and the most forward-peaked, consistent with the formation and breakup of a neck structure or a strongly deformed quasiprojectile. A very precise and stimulating study of the time scales in neck fragmentation can be carried out using the new 4π detectors with improved performances on mass resolution and thresholds for fragment measurements. Such kind of data are now appearing from the Chimera Collaboration [44].

We can immediately expect an important isospin dependence of the neck dynamics, from the presence of large density gradients and from the possibility of selecting various time scales for the fragment formation. The first evidences of isospin effects in neck fragmentation were suggested by Dempsey *et al.* [45] from semiperipheral collisions of the systems ^{124,136}Xe $+$ ^{112,124}Sn at 55 AMeV, where correlations between the average number of IMFs, N_{IMF}, and neutron and charged-particle multiplicities were measured. The variation of the relative yields of ^{6}He/^{3,4}He, ^{6}He/Li with v_{par} for several Z_{PLF} gates shows that the fragments produced in the mid-rapidity region are more neutron rich than are the fragments emitted by the PLF. Enhanced *triton* production at mid-rapidity was considered in ref. [33], and more recently in [46], as an indication of a neutron neck enrichment.

Milazzo *et al.* [47–49] analyzed the IMF parallel velocity distribution for ^{58}Ni $+$ ^{58}Ni semiperipheral collisions at 30 AMeV. The two-bump structure for IMFs with $5 \leq Z \leq 12$, located around the center-of-mass velocity and close to the projectile (PLF*) source, respectively, was explained assuming the simultaneous presence of two production mechanisms: the statistical disassembly of an equilibrated PLF* and the dynamical fragmentation of the participant region. The separation of the two contributions allows for several interesting conclusions. The average elemental event multiplicity $N(Z)$ exhibits a different trend for the two processes: in particular, the fragments with $5 \leq Z \leq 11$ are more copiously produced at the mid-rapidity region. This experiment has a particular importance since isospin effects were clearly observed, *in spite of the very low initial asymmetry*. The measured isotopic content of the fragments is clearly different in the two mechanisms. The experimental heavy-isotope/light-isotope yield ratios, ^{14}C/^{12}C, ^{12}B/^{10}B, ^{10}Be/^{7}Be, ^{8}Li/^{6}Li, show a systematic decreasing trend as a function of parallel velocity from c.m. to PLF values.

All these results indicate a neutron enrichment of the neck region, *even when initially the system N/Z is close to unity*. The same reactions have recently been studied at the Cyclotron Institute of Texas A&M at various beam energies with measurements of the correlations of fragment charge/mass *vs.* dynamical observables (emission angles and velocities) [50,51].

Plagnol *et al.* [52] have examined, for the system Xe $+$ Sn between 25 and 50 AMeV, the competition between mid-rapidity dynamical emission and equilibrium evaporation as well as its evolution with incident energy. The onset of the neck emission takes place around 25 AMeV and rises with the energy while the evaporative part remains quite invariant for a selected centrality. Neck matter is found to be more charge asymmetric: more neutron-rich isotopes are favored at mid-rapidity in comparison to evaporation. Evidence of a neck-like structure and its neutron enrichment has been seen even in collisions

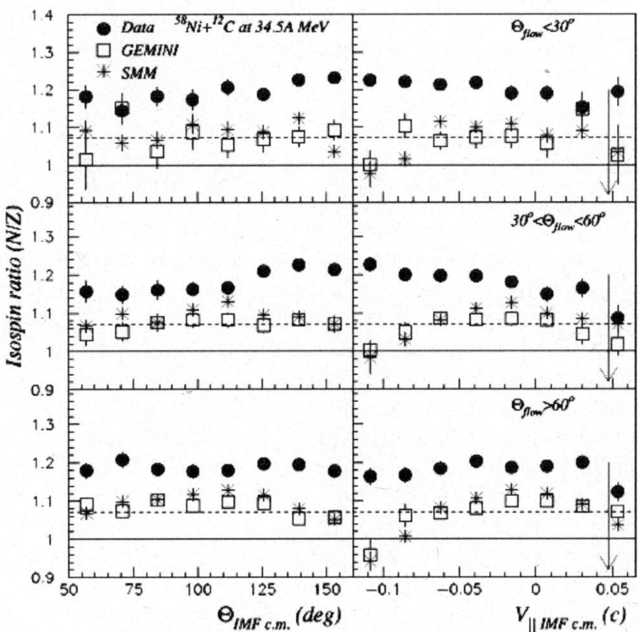

Fig. 2. Average isospin ratios (N/Z) for well-identified IMFs with $Z = 3$ or 4 as a function of the emission angle in the center-of-mass reference frame (left) and the center-of-mass particle velocity parallel to the beam axis (right) for the reaction ^{58}Ni $+$ ^{12}C at 34.5 AMeV (full dots). Cuts are made on $\Theta_{flow} < 30°$ (top), $30° < \Theta_{flow} < 60°$ (middle), and $\Theta_{flow} > 60°$ (bottom). Open boxes represent filtered GEMINI simulations and stars are results from filtered SMM simulations. Error bars are the statistical errors for a given angle or velocity bin. When no error bar is present, the error is smaller than the size of the symbol. The dotted lines show the isospin ratio for ^{58}Ni (1.07) and the full line for ^{12}C (1.00). The arrow shows the velocity of the ^{58}Ni projectile in the center-of-mass frame for the ^{58}Ni $+$ ^{12}C reaction. (From [53].)

with a rather light symmetric target, ^{58}Ni $+$ ^{12}C, ^{24}Mg, at 34.5 AMeV [53]. The average N/Z ratio for isotopes with $Z = 3, 4$ exhibits a clear increase from the PLF to the mid-rapidity zone. Figure 2 from [53] shows part of those measurements. Also, a combined analysis of the ^{58}Ni $+$ ^{58}Ni and ^{36}Ar $+$ ^{58}Ni systems around 50 AMeV has evidenced an asymmetric migration of neutrons and protons between the quasiprojectile and the mid-rapidity region [54]. Time sequence emission is another way to probe the neck rupture processes and to characterize the fragments [55].

That indicates the need of new, possibly more exclusive, data. The reasons for a preponderance of neutron-rich isotopes emitted from the neck region are a matter of debate. Possible explanations being, apart the density dependence of symmetry energy [56], also a fast light cluster production, especially of α-particles, which promptly leads to an amplification of neutron excess in the participant matter [57]. For completeness, we have to mention that different analyses even give conflicting results on the neutron enrichment of the clusters produced at mid-rapidity [58,59]. This shows that the reaction dynamics is in general very complicated, and there could even be different isospin effects in competition.

2 Isospin diffusion

The isospin equilibration appears of large interest also for more peripheral collisions, where we have shorter interaction times, less overlap and a competition between binary and neck fragmentation processes. The specific feature at Fermi energies is that the interaction times are close to the specific time scales for isospin transport allowing a more detailed investigation of isospin diffusion and equilibration in reactions between nuclei with different N/Z asymmetries. The low-density neck formation and the pre-equilibrium emission are adding essential differences with respect to what is happening in the lower-energy regime. Tsang *et al.* [60] have probed the isospin diffusion mechanism for the systems ^{124}Sn $+$ ^{112}Sn at $E = 50$ AMeV in a peripheral impact parameter range $b/b_{max} > 0.8$, observing the isoscaling features of the light isotopes $Z = 3$–8 emitted around the projectile rapidity. An incomplete equilibration has been deduced. The value of the isoscaling parameter $\alpha = 0.42 \pm 0.02$ for ^{124}Sn $+$ ^{112}Sn differs substantially from $\alpha = 0.16 \pm 0.02$ for ^{112}Sn $+$ ^{124}Sn. The isospin imbalance ratio [61], defined as

$$R_i(x) = \frac{2x - x^{124+124} - x^{112+112}}{x^{124+124} - x^{112+112}} \tag{1}$$

($i = P, T$ refers to the projectile/target rapidity measurement, and x is an isospin-dependent observable, here the isoscaling α parameter) was estimated to be around $R_P(\alpha) = 0.5$ (*vs.* $R_P(\alpha) = 0.0$ in full equilibrium). This quantity can be sensitive to the density dependence of symmetry energy term since the isospin transfer takes place through the lower-density neck region.

3 Neck observables

– Properties of neck fragments, mid-rapidity IMF produced in semicentral collisions: correlations between N/Z, *alignment* and size.
  The alignment between PLF-IMF and PLF-TLF directions represents a very convincing evidence of the dynamical origin of the mid-rapidity fragments produced on short time scales [8]. The form of the Φ_{plane} distributions (centroid and width) can give a direct information on the fragmentation mechanism [19,20, 62]. Recent calculations confirm a general feature, predicted for that rupture mechanism: the light fragments are emitted first, as displayed in fig. 3.
– Time scale measurements.
  The estimation of time scales for fragment formation from velocity correlations appears to be a very exciting possibility [25,44,63]. With a good event-by-event detection of the projectile(target) residues we can measure the violations of the Viola systematics for the TLF-IMF and PLF-IMF systems which tell us how

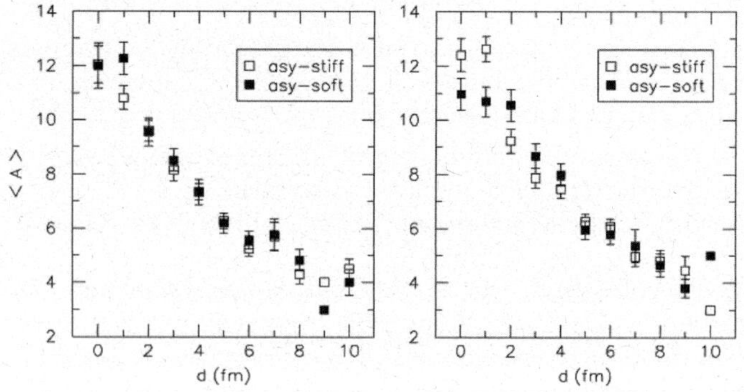

Fig. 3. Average IMF mass as a function of the distance from the PLF-TLF axis at the freeze-out time for 47 AMeV, for collisions at a reduced impact parameter of 0.5. Left panel: Fe + Fe. Right panel: Ni + Ni. Empty squares: asy-soft symmetry term. Full squares: asy-stiff. (From [11].)

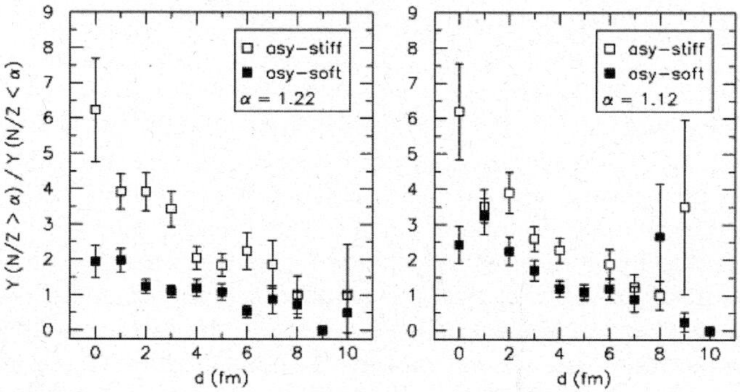

Fig. 4. Ratio of the IMF yields, with N/Z larger and smaller than the value α obtained just after pre-equilibrium emission, as a function of the distance from the PLF-TLF axis at the freeze-out time for 47 AMeV, for collisions at a reduced impact parameter of 0.5. Left panel: Fe + Fe. Right panel: Ni + Ni. Empty squares: asy-stiff symmetry term. Full squares: asy-soft. (From [11].)

much the IMFs are uncorrelated to the *spectator* remnants [64].
With appropriate cuts in the velocity correlation plots we can follow the properties of clusters produced from sources with a "controlled" different degree of equilibration. We can figure out a continuous transition from fast-produced fragments via neck instabilities to clusters formed in a dynamical fission of the projectile(target) residues up to the evaporated ones (statistical fission). Along this line it would be even possible to disentangle the effects of volume and shape instabilities. The isospin dynamics will look different in the various scenarios and rather dependent on the symmetry term of the EOS.
– Isospin dynamics.
Isospin effects on the reaction dynamics and *Isospin Migration*: an interesting neutron enrichment of the overlap ("neck") region is expected, due to the neutron migration from higher (spectator) to lower (neck)

density regions. This effect is also nicely connected to the slope of the symmetry energy. Neutron and/or light isobar measurements in different rapidity regions appear important. Moreover, in moving from mid- to "spectator" rapidities, an increasing hierarchy in the mass and N/Z of the fragments is expected [11]. Some experimental evidences are in ref. [43]. An interesting related observable is the corresponding angular correlation due to the driving force of the projectile(target)-like partners [11].

– Isospin diffusion.

With measurements of charge equilibration in the "spectator" region in semicentral collisions, we can get the *Imbalance Ratios* for different isospin properties. It is a test of the interplay between concentration and density gradients in the isospin dynamics [9,65,66]. For the reasons noted before we expect to see a clear difference in the isospin diffusion between binary (deep-inelastic like) and neck fragmentation events. Moreover, in the mid-rapidity emission, there is a clear neutron enrichment predicted [11] for neutron-rich and neutron-poor systems (see fig. 4, from [11]).

– "Pre-equilibrium" emissions.

As already noted, the isospin content of the fast particle emission can largely influence the subsequent reaction dynamics, in particular, the isospin transport properties (charge equilibration, isospin diffusion). We can reach the paradox of a detection of isospin dynamics effects in charge symmetric systems.

Finally, the simultaneous measurements of properties of fast nucleon emissions and of the neck dynamics can even shed light on the very controversial problem of the isospin momentum dependence [9,66].

We stress the richness of the phenomenology and nice opportunities of getting several cross-checks from completely different experiments. Apart from the interest of this new dissipative mechanism and the amazing possibility of studying properties of fragments produced on an almost continuous range of time scales, we remark the expected dependence on the isovector part of the nuclear EOS. From transport simulations we presently get some indications of "asy-stiff" behaviors, *i.e.* increasing repulsive density dependence of the symmetry term, but not more fundamental details. Moreover, all the available data are obtained with stable beams, *i.e.* within low asymmetries.

References

1. A. Bonasera, G.F. Bertsch, E.N. El-Sayed, Phys. Lett. B **141**, 9 (1984).
2. M. Colonna, N. Colonna, A. Bonasera, M. Di Toro, Nucl. Phys. A **541**, 295 (1992).
3. L.G. Sobotka, Phys. Rev. C **50**, 1272R (1994).
4. M. Colonna, M. Di Toro, A. Guarnera, Nucl. Phys. A **589**, 160 (1995).
5. M. Di Toro *et al.*, Progr. Part. Nucl. Phys. **42**, 125 (1999).
6. M. Di Toro *et al.*, Nucl. Phys. A **681**, 426c (2001).
7. V. Baran, M. Colonna, V. Greco, M. Di Toro, M. Zielinska-Pfabé, H.H. Wolter, Nucl. Phys. A **703**, 603 (2002).
8. V. Baran, M. Colonna, M. Di Toro, Nucl. Phys. A **730**, 329 (2004).
9. V. Baran, M. Colonna, V. Greco, M. Di Toro, Phys. Rep. **410**, 335 (2005).
10. M. Colonna, M. Di Toro, G. Fabbri, S. Maccarone, Phys. Rev. C **57**, 1410 (1998).
11. R. Lionti, V. Baran, M. Colonna, M. Di Toro, Phys. Lett. B **625**, 33 (2005).
12. L. Stuttgé *et al.*, Nucl. Phys. A **539**, 511 (1992).
13. R. Wada *et al.*, Nucl. Phys. A **548**, 471 (1992).
14. D.E. Fields *et al.*, Phys. Rev. Lett. **69**, 3713 (1992).
15. J.E. Sauvestre *et al.*, Phys. Lett. B **335**, 300 (1994).
16. J. Boger *et al.*, Phys. Rev. C **41**, 801 (1990).
17. S.L. Chen *et al.*, Phys. Rev. C **54**, R2114 (1996).
18. R. Yanez *et al.*, Phys. Rev. Lett. **82**, 3585 (1999).
19. G. Casini *et al.*, Phys. Rev. Lett. **71**, 2567 (1993).
20. A.A. Stefanini *et al.*, Z. Phys. A **351**, 167 (1995).
21. J. Tõke *et al.*, Phys. Rev. Lett. **75**, 2920 (1995).
22. J.F. Lecolley *et al.*, Phys. Lett. B **354**, 202 (1995).
23. C.P. Montoya *et al.*, Phys. Rev. Lett. **73**, 3070 (1994).
24. L. Beaulieu *et al.*, Phys. Rev. Lett. **77**, 462 (1996).
25. S. Piantelli *et al.*, Phys. Rev. Lett. **88**, 052701 (2002).
26. P. Pawlowski *et al.*, Eur. Phys. J. A **9**, 371 (2000).
27. D. Doré *et al.*, Phys. Rev. C **63**, 034612 (2001).
28. W.G. Lynch, Nucl. Phys. A **583**, 471c (1995).
29. J. Łukasik *et al.*, Phys. Lett. B **566**, 76 (2003).
30. Y. Larochelle *et al.*, Phys. Rev. C **59**, R565 (1999).
31. A. Mangiarotti *et al.*, Phys. Rev. Lett. **93**, 232701 (2004).
32. J. Péter *et al.*, Nucl. Phys. A **593**, 95 (1995).
33. J. Łukasik *et al.*, Phys. Rev. C **55**, 1906 (1997).
34. J. Tõke *et al.*, Nucl. Phys. A **583**, 519c (1995).
35. J. Tõke *et al.*, Phys. Rev. Lett. **77**, 3514 (1996).
36. Y. Larochelle *et al.*, Phys. Rev. C **55**, 1869 (1997).
37. P. Pawlowski *et al.*, Phys. Rev. C **57**, 1771 (1998).
38. T. Lefort *et al.*, Nucl. Phys. A **662**, 397 (2000); D. Doré *et al.*, Phys. Lett. B **491**, 15 (2000).
39. F. Bocage *et al.*, Nucl. Phys. A **676**, 391 (2000).
40. B. Grabez, Phys. Rev. C **64**, 057601 (2001).
41. L. Gingras *et al.*, Phys. Rev. C **65**, 061604 (2002).
42. B. Davin *et al.*, Phys. Rev. C **65**, 064614 (2002).
43. J. Colin *et al.*, Phys. Rev. C **67**, 064603 (2003).
44. A. Pagano *et al.*, Nucl. Phys. A **734**, 504c (2004).
45. J.F. Dempsey *et al.*, Phys. Rev. C **54**, 1710 (1996).
46. G. Poggi, Nucl. Phys. A **685**, 296c (2001).
47. P.M. Milazzo *et al.*, Phys. Lett. B **509**, 204 (2001).
48. P.M. Milazzo *et al.*, Nucl. Phys. A **703**, 466 (2002).
49. P.M. Milazzo *et al.*, Nucl. Phys. A **756**, 39 (2005).
50. D.V. Shetty *et al.*, Phys. Rev. C **68**, 021602(R) (2003).
51. D.V. Shetty *et al.*, Phys. Rev. C **70**, 011601(R) (2004).
52. E. Plagnol *et al.*, Phys. Rev. C **61**, 014606 (2000).
53. Y. Larochelle *et al.*, Phys. Rev. C **62**, 051602(R) (2000).
54. D. Thériault *et al.*, Phys. Rev. C **71**, 014610 (2005).
55. Chimera Collaboration (E. De Filippo *et al.*), Phys. Rev. C **71**, 044602 (2005).
56. S. Hudan *et al.*, Phys. Rev. C **71**, 054604 (2005).
57. L.G. Sobotka *et al.*, Phys. Rev. C **55**, 2109 (1997).
58. L.G. Sobotka *et al.*, Phys. Rev. C **62**, 031603(R) (2000).
59. H. Xu *et al.*, Phys. Rev. C **65**, 061602(R) (2002).
60. M.B. Tsang *et al.*, Phys. Rev. Lett. **92**, 062701 (2004).
61. F. Rami *et al.*, Phys. Rev. Lett. **84**, 1120 (2000).
62. Chimera Collaboration (E. De Filippo, A. Pagano, E. Piasecki *et al.*), Phys. Rev. C **71**, 064604 (2005).

63. Chimera Collaboration (J. Wilczyński *et al.*), Int. J. Mod. Phys. E **14**, 353 (2005); Chimera Collaboration (E. De Filippo *et al.*), Phys. Rev. C **71**, 044602 (2005).

64. It has been proposed to call the Viola-violation-correlation plot as *Wilczyński-2 Plot*. Indeed this correlation, very important to rule out a statistical fission scenario for fragments produced at mid-rapidity, nicely emerged during hot discussions of one of us (M.D.T.) with J. Wilczyński at the LNS-INFN, Catania. In fact this correlation represents also a *chronometer* of the fragment formation mechanism. In this sense it is the nice Fermi energy complement of the famous *Wilczyński Plot* which gives the time scales in Deep-Inelastic Collisions.

65. V. Baran, M. Colonna, M. Di Toro, M. Zielinska-Pfabé, H.H. Wolter, Phys. Rev. C **72**, 064620 (2005).

66. L.-W. Chen, C.M. Ko, B.-A. Li, Phys. Rev. Lett. **94**, 032701 (2005).

Eur. Phys. J. A **30**, 71–79 (2006)
DOI 10.1140/epja/i2006-10108-7

THE EUROPEAN
PHYSICAL JOURNAL A

Systematics of fragment observables

B. Tamain[a]

LPC Caen (IN2P3-CNRS/Ensicaen et Université), F-14050 Caen Cédex, France

Received: 25 April 2006 /
Published online: 19 October 2006 – © Società Italiana di Fisica / Springer-Verlag 2006

Abstract. Multifragmentation is observed in many reaction types: light-ion–induced reactions at large incident energies (in the GeV region), central heavy-ion collisions from 30 to 100 MeV/u, and peripheral heavy-ion collisions between 30 and 1000 MeV/u or above. When nucleus-nucleus collisions are considered, another entrance channel parameter is the corresponding mass asymmetry. The first question which is addressed in this contribution is: do we observe similar reactions in each case? Multifragmentation may be related to a phase transition of nuclear matter. Some others features indicate that dynamical features are dominant. It is *a priori* possible that the underlying mechanisms are different in proton- and nucleus-induced reactions, in central and in peripheral collisions, at limited and at large bombarding energies. In order to see to what extent they can reflect similar behaviour, it is useful to compare the results of various reactions. The observables can be the fragment multiplicity, the mass distributions or the kinematical properties. In this contribution, we are looking for such general features. We will limit the discussion to the observations themselves, rather than the interpretation, which is the subject of numerous entries in this volume. The experimental results indicate that multifragmentation exhibits at the same time universal and entrance-channel–dependent properties.

PACS. 24.10.Pa Thermal and statistical models – 25.70.Pq Multifragment emission and correlations – 68.35.Rh Phase transitions and critical phenomena

1 The necessity and the difficulty of the sorting

A first difficulty in comparing nucleus-nucleus collision data lies in the fact that they can differ significantly according to the impact parameter. Now, the impact parameter cannot be directly measured: it can be only estimated from other more direct observables. Depending on the experiment, various sorting parameters have been used: neutron or light charged particle (LCP) or total charged particle multiplicity [1–4], or LCP (or total) transverse energy [5], or flow angle [6], or specific quantities like $Erat$ (ratio between the total perpendicular and parallel kinetic energy) [7] or Z_{bound} (the total charge bound in fragments) [8,9]. One may also use more sophisticated methods as the principal component analysis method [10] or calorimetry [11,12] (see also the contribution V.3 in this topical issue [13]).

The sorting aims either at following the evolution of the mechanism when the violence of the collision is increased (from peripheral to central collisions for nucleus-nucleus collisions), or at selecting something which is generally labelled "a source". An example is the selection of central collisions in nucleus-nucleus collisions. In the pre-

vious sentences, we have two concepts: "the collision violence" and "the source".

The violence is linked with the proportion of the initial aligned energy (the kinetic energy of the beam) that is shared among other degrees of freedom. It may be linked with a thermal energy if the available phase space is fully explored for the ensemble of selected events. A "source" is a piece of nuclear matter that is localized in momentum space. It is not necessarily equilibrated.

An important question is the quality of the sorting: to what extent is the selection efficient? The sorting cannot be precise for several reasons: finite-size effects; detection inefficiency (dead areas and thresholds); fluctuations in the energy sharing in multi-source processes (for instance, in binary processes). One may have an idea of this precision by looking at the correlation between various sorting variables. Data have been obtained for instance at MSU [1] in which particle multiplicities and transverse energies have been correlated. Another example has been obtained by the INDRA-ALADIN Collaboration [14]: in this case, binary symmetrical collisions have been studied and transverse energy correlations have been obtained between the projectile-like (PLS) and the target-like (TLS) sources. The correspondences are not better than about 20 percent. This means that for a selected value of a selected sorting variable, the variation range of a second sorting

---

[a] e-mail: tamain@in2p3.fr

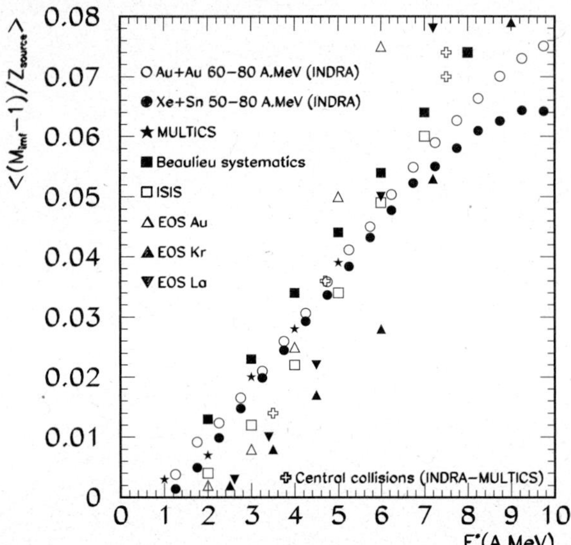

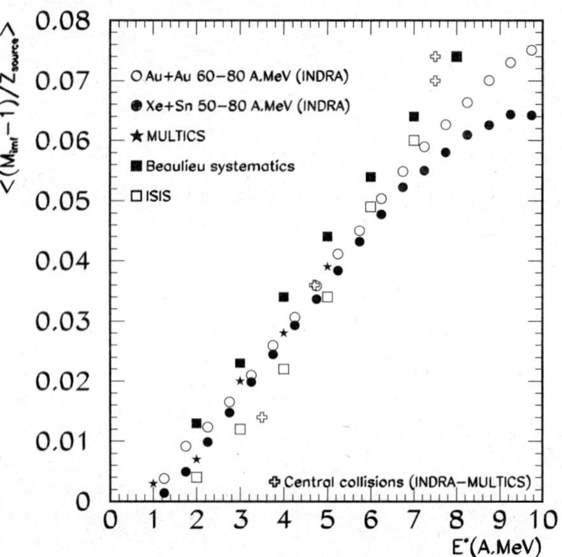

Fig. 1. Correlation between $(M_{IMF} - 1)/Z_{source}$ (ordinate) and the dissipated energy (abscissa). The excitation (dissipated) energy has been corrected for pre-equilibrium and for expansion (if any: it is especially the case for the EOS data [4]). In that sense, the dissipated energy is mainly thermal. Several systems have been "summarized" in a single data set when the results were very close: it is the case for the Laval and ALADIN data [11] and for the INDRA@GSI data [15]. Other data are extracted from refs. [2,10,16,17]. This figure has been prepared with the help of R. Bougault.

Fig. 2. This figure is similar to the previous one but the EOS data have been removed. It turns out that the coherence between various reactions is very good in spite of the fact that one has included in the figure peripheral and central collisions, light- and heavy-nucleus–induced rections.

variable covers about 20 percent of its mean value: sorting is efficient but not very precise. In any case, the detection has to be as complete as possible. It is possible to study the continuous evolution of the sorting variable keeping in mind that some mixing cannot be avoided. It is difficult to isolate a definite class of events without encountering one or another drawback: either a mixing with other event classes; or a cut in the available phase space for the selected event class. This difficulty is very well understood in simulations. Sorting from a mixing of various variables (principal component analysis) can slightly improve the quality of the selection [10,18,19].

2 Fragment observables

The raw multifragmentation observables are multiplicities, mass or charge distributions, isotopic distributions, kinetic energy and angular distributions. They can have various meanings depending on the collision nature: nucleus-nucleus collision *versus* light-projectile (p, d, α, $\bar{p}$, π) induced reactions; peripheral *versus* central collisions.

Various observables can also magnify different collision features. This can be illustrated from what is well known at low bombarding energies, below 10 MeV/u. In this case, deep inelastic reactions are dominant and it is well known that, depending on the observable, one is focussing on various aspects of the collision: fragment angular and kinetic energy distributions (Wilczynski plots)

reflect the dynamics of the process: complete damping and isotropy is not observed for most of the events. On the other hand, mass transfer is described with Fokker-Planck equations for which some degrees of freedom (the fast ones) are thermally treated (heat bath) whereas some others (mass transfer) are slowly evolving and do not reach necessarily equilibrium.

3 Fragment production: a hierarchy

At low bombarding energy, it is well known that the decay of an excited nucleus ends with residue production. This decay product has a specific role among all the disentegration products. This feature is clear at low energy. A recent compilation [20] and the results of figs. 1 and 2 indicate that this specific role of the largest fragment is often evidenced. It is the reason why, in the next sections, one will distinguish the largest fragment from the others.

4 Fragment production: multiplicities

We label "fragments" as the detected products with an atomic number Z of at least 3, which are generally named intermediate mass fragments (IMF). The lighter products (Z smaller than 3) are labelled light charged particles (LCP). The fragment multiplicity is M_{IMF}.

To what extent is M_{IMF} correlated with energy dissipation? At low bombarding energies, it is established that $M_{IMF} - 1$ is close to zero since no IMFs are emitted other than the residue: only LPCs remove the deposited energy. The situation is more complicated for larger energy deposition for which the pre-equilibrium energy contribution is significant and not uniquely defined. It has

Table 1. This table is a non-exhaustive compilation of many experiments in which the IMF multiplicities have been measured as a function of the excitation (dissipated) energy. The systems involved are indicated in the first column. The projectiles can be light (pions) or heavy (up to gold nuclei); the selected collisions can be central (one single source), or peripheral (projectile-like source). The references are also indicated in the first column. The second column indicates the method that has been used to determine the excitation energy given in the fourth column. Two excitation energies have been selected: around 4 MeV and aroud 8 MeV/u, corresponding to close to and above the multifragmentation threshold. The fragment multiplicities (except for the heaviest fragment) normalized to the source size are given in the last column.

System	Method	Z_{source}	E^*/u (MeV)	$(M_{IMF} - 1)/Z_{source}$
π + Au 8 GeV/c; [11]	cal	67	4	0.022
Cl + Au 43 MeV/u; peripheral; [21]	cal	17	4	0.035
Ge + Ti 35 MeV/u; peripheral; [21]	cal	32	4	0.035
Nb + Mg 30 MeV/u; central; [16]	cal	45	3.4–3.8	0.014
Au + Au 35 MeV/u; peripheral; [2]	cal	≈ 75	4	0.030
Au + Au 600 MeV/u; peripheral; [9]	cal	≈ 75	4	0.035
System	Method	Z_{source}	E^*/u (MeV)	$(M_{IMF} - 1)/Z_{source}$
π + Au 8 GeV/c; [11]	cal	59	7–8	0.068
Cl + Au 43 MeV/u; peripheral; [21]	cal	17	8	0.071
Ge + Ti 35 MeV/u; peripheral; [21]	cal	32	8	0.071
Ni + Au 90 MeV/u; central; [10]	cal/SMM	86	7.5	0.070
Xe + Sn 50 MeV/u; central; [10]	cal/SMM	85	7–8	0.074
Xe + Sn 80 MeV/u; peripheral; [15]	cal	48	8	0.077
Au + Au 80 MeV/u; peripheral; [15]	cal	70	7	0.069
Au + Au 600 MeV/u; peripheral; [9]	cal	55	8	0.073
Au + C 1000 MeV/u; semi-peripheral; [4]	cal	53–40	7.5	0.10
La + C 1000 MeV/u; semi-peripheral; [4]	cal	40–34	7.5	0.077
Kr + C 1000 MeV/u; semi-peripheral; [4]	cal	26–23	7.5	0.07

to be subtracted. After this subtraction, excitation energy is usually measured by calorimetry. It can be also obtained from the comparison with a model (for instance, SMM [22]) in which equilibrium is assumed. When the bombarding energy is large, some compression effect may also be present and the corresponding expansion energy can be taken away. All these procedures can be disputed. Nevertheless, we have compared many data obtained in various ways to try to evidence some general behaviours. In table 1, such a compilation is shown for two values of the "measured" excitation energy: 4 MeV/u and 8 MeV/u. The list is not exhaustive. Since the expansion energy has been subtracted, the word "thermal" energy could be more appropriate but its use can be considered as too precise. For this reason, we will use the word "dissipated" for which the consensus may be better obtained. Very different reaction types are considered in table 1: pion-induced reactions, central or peripheral heavy-ion reactions, intermediate (35 MeV/u) or large (1000 MeV/u) bombarding energies. The method used to estimate the excitation energy can be calorimetry or comparison with SMM (indicated in the second column). The source size Z_{source} is also estimated in various ways [23]. Nevertheless, it appears that the ratio $(M_{IMF} - 1)/Z_{source}$ seems to be about the same for a defined excitation energy. This result is a first indication that multiframent production could be correlated with the dissipated the energy.

This tendency is confirmed from figs. 1 and 2 which show the correlation between $(M_{IMF} - 1)/Z_{source}$ and the measured excitation (dissipated) energy. All the systems considered in table 1 have been used. In order to clarify the figure, several systems are sometimes "summarized" by a single result. This is the case for the INDRA@GSI data or for the Laval + ALADIN data [21]. The general tendency is again the same for any system whatever the entrance channel is: light or heavy projectile; low or large incident energy; central or peripheral collisions (see also ref. [24]). The coherence is especially good for high excitation (dissipation) energy and in fig. 2 in which the EOS data have not been included. The fact that the EOS data do not fit so well with others can be understood since in this case, the non-thermal contributions which are subtracted are huge and difficult to estimate with a good precision.

The results plotted in figs. 1 and 2 indicate a continuous increase of the ordinate. One knows also that at larger dissipations, the fragment multiplicities decrease: *i.e.* the rise and fall of multifragment emission [4,8] for which a universal behaviour is also established (see figs. 3 and 4). Altogether, there is a continuous evolution from low-energy collisions with a large released residue to complete vaporization with only LCPs. The specific role of the largest fragment is evident at low exitation and disappears when complete vaporization sets in; in between, the meaning of the fragment hierarchy is still open to debate and is interpreted either as a dynamical effect reflecting the

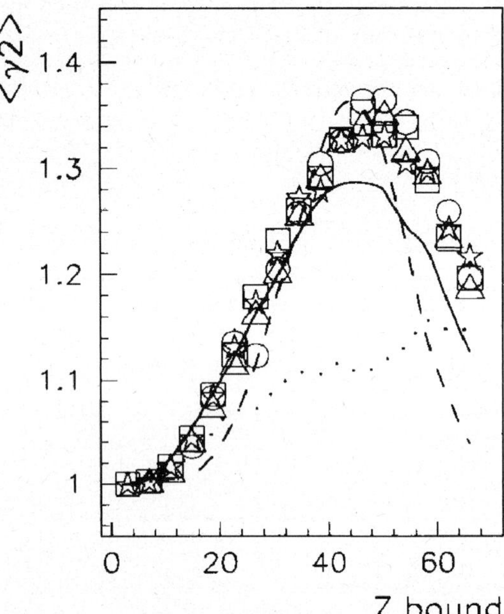

Fig. 3. Rise and fall of the fragment multiplicity as a function of the total detected bound charge which is expected to be related to the dissipated energy. Different symbols correspond to PLS sources produced in Au collisions on different targets ranging from C to Pb. Extracted from ref. [8].

collision geometry, or in terms of liquid-gas coexistence. Further experimental and calculation results are needed in order to progress on this point.

Fragment multiplicities are in any case correlated with the energy dissipated in the collision. This property has sometimes been described in terms of reducibility [5,25] in the sense that the probability for emitting several fragments can be reduced to the probability for emitting a single fragment and to the corresponding energy cost. Such a result is quite coherent with the above discussion of figs. 1 and 2.

Thus it seems that multifragment production is to a large extent defined by the energy dissipated during the collision. Of course, the correlation obtained from the data cannot be perfect for two reasons. First of all, it is impossible to measure properly the "dissipated" energy because it is not possible to separate clearly in the data the relative contributions of pre-equilibrium, compression or thermal parts. A second feature is that many aspects of the collisions reflect an important role of the dynamics which is observed in mid-rapidity and in forward-backward emissions. These contributions are to a large extent responsible for the deviations observed between the data at low dissipations in figs. 1 and 2. They are discussed in the next section.

5 Pre-equilibrium emissions

5.1 General observations

Pre-equilibrium emissions correspond to particles or fragments that are not randomly emitted from identified

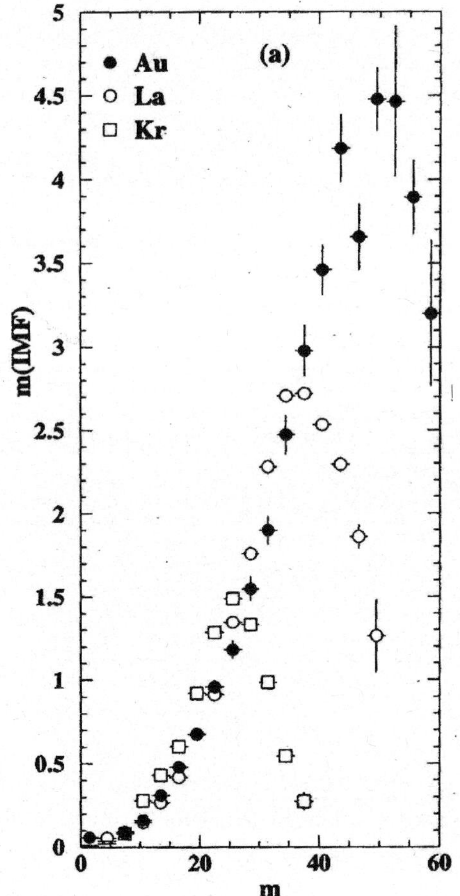

Fig. 4. Rise and fall of the fragment multiplicity as a function of the total particle multiplicity which is correlated with the energy dissipated in the PLS released in several nucleus-nucleus collisions for various systems. Extracted from ref. [4]

sources (no isotropic emission in the plane perpendicular to the angular momentum). Besides the key quantities such as energy and angular momentum, they have kept some memory of the entrance channel, *i.e.* of the beam direction and/or velocity. From the time scale point of view, pre-equilibrium particles are emitted early. Their center-of-mass kinetic energies are generally larger than expected after full equilibrium, reflecting the fact that the incident beam energy has not been shared among all the available degrees of freedom. The energy relaxation step brings energy in various degrees of freedom: the stored energy can be thermal if the whole available phase space has been occupied. The energy can also partially be stored as compression energy of nuclear matter, thus leading to an additional expansion contribution. A fraction of the available energy can also be stored as deformation energy of the hot source. The distinction between pre-equilibrium, expansion and thermal contributions is not trivial since the mean thermal decay time becomes very short for large excitations.

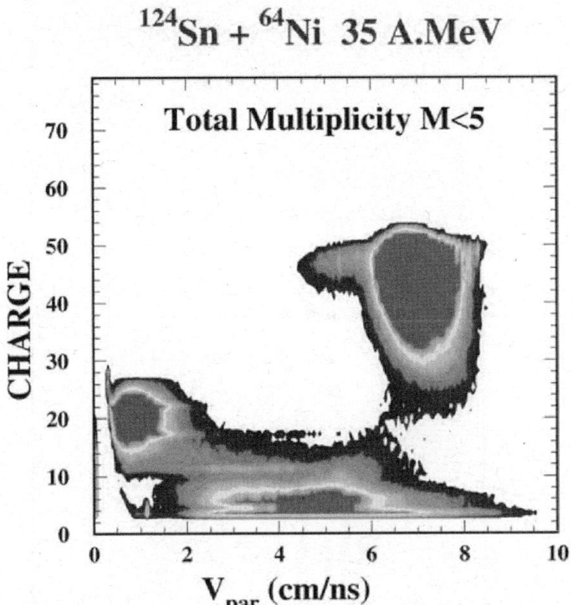

Fig. 5. Correlation between the charges and the velocities of products emitted in semi-peripheral nucleus-nucleus collisions. Mid-rapidity is evidenced for light IMFs. Similar results are published in ref. [26].

5.2 Angular distributions as signatures of pre-equilibrium

Two kinds of pre-equilibrium emissions are recognized in the data. In particle-nucleus or in central nucleus-nucleus collisions, pre-equilibrium LCP angular distributions are mostly forward or backward peaked relative to the beam direction.

In semi-peripheral reactions, mid-rapidity neck emission occurs both for LCP and IMF. Pre-equilibrium LCPs result mainly from direct nucleon-nucleon collisions in the overlap zone (see sect. 5.3). Concerning fragments, a general observation is that the largest decay fragment from a projectile-like source (PLS) is mostly faster than the lighter IMFs that are detected forward in the c.m. frame [26–28]: these lighter IMFs are accumulated close to the backward part of the Coulomb ring associated to the PLS whatever the bombarding energy is [29]. If the incident energy is limited (40 MeV/u or below), this backward part of the Coulomb ring is close to the c.m. velocity (mid-rapidity). The data of fig. 5 correspond to this situation. Neck emission is clearly an entrance dynamical effect that leads to ambiguities in the measurement of the dissipated energy in a projectile-like source. It affects the projectile-like source velocity if it is reconstructed from the detected fragments. It affects also the excitation energy calculated from calorimetry. This ambiguity is larger when neck contribution is a sizeable fraction of the whole total yield. This is especially true for limited excitations and for symmetric heavy-ion collisions [30]. This can explain partially the relative dispersion of data in figs. 1 and 2 at limited dissipations.

Depending on the observable, one may focus more or less on dynamical features. Neck emission is used in this context. On the contrary, one may subtract identified pre-equilibrium particles to try to isolate sources and try to get their excitation energies. Finally, one may select events for which the pre-equilibrium energy is small and can be neglected [2,23]. This procedure is never perfect especially for symmetric collisions in the entrance channel. Nevertheless, it is possible to isolate events for which most of the available energy has been shared among many degrees of freedom. The deviations from full exploration of the available phase space can be to some extent "summarized" in collective variables such as deformation or expansion, which can be associated to Lagrange parameters [31].

The fact that the results of figs. 1 and 2 are coherent indicate that extracting dissipated energies from the data is a meaningful procedure. Similarly, we will see in sect. 6 that the released IMF observables indicate that the process reflects to a large extent the available phase space.

5.3 Kinetic energies as signatures of pre-equilibrium

Another indication of pre-equilibrium can be found in the measured kinetic energies of the emitted LCP and IMF. In semi-peripheral collisions, entrance channel effects are clearly evidenced [32,33]. For instance, in ref. [32], it is shown that the transverse LCP energy at mid-rapidity does not depend on the violence of the collision at variance with the energy of LCP emitted from the PLS (fig. 6). LCPs emitted at mid-rapidity reflect the incident energy per nucleon and the Fermi motion of the projectile and target nucleons whereas LCPs emitted at velocity closer to the PLS one reflect the dissipated energy. Depending

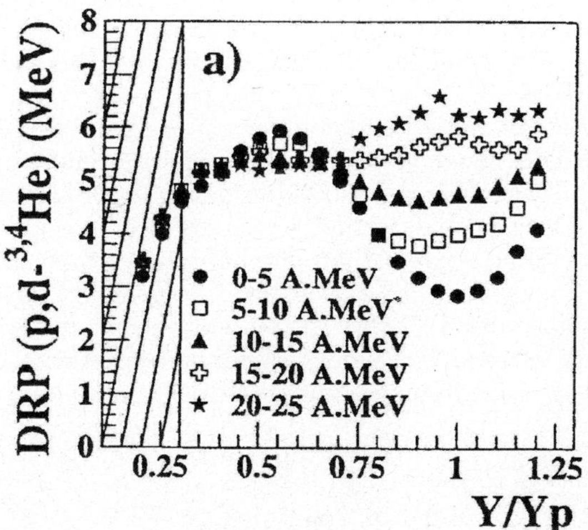

Fig. 6. Abscissa: rapidity of selected LCPs in beam rapidity units; ordinate: double ratio parameter (p, d, He thermometer) corresponding to the abscissa rapidity. Various curves correspond to various energy dissipations (the excitation energy per nucleon has been measured by calorimetry in assuming a binary reaction: see ref. [32] for details). The dissipated energy has no influence on the results obtained at mid-rapidity.

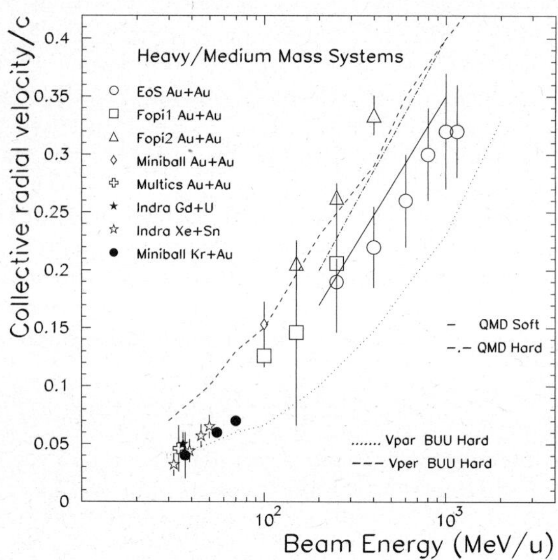

Fig. 7. Non-exhaustive compilation for the collective radial velocity as a function of the beam energy for medium- and heavy-mass systems in central collisions. Lines correspond to the predictions of transport models; for BUU calculations, the collective motion is found to be anisotropic so that both v_{par} and v_{perp} are shown. From ref. [34] and references therein.

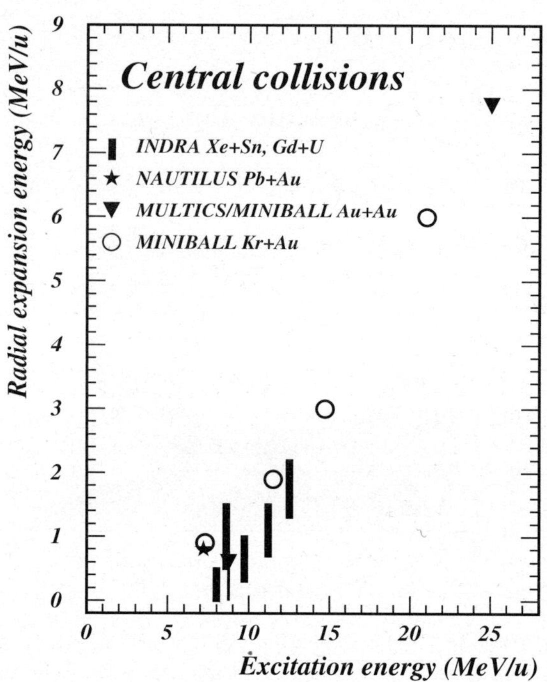

Fig. 8. Systematics of the collective expansion energy as a function of the available center-of-mass energy per nucleon in central collisions [35].

on the location in the velocity plane, we observe entrance channel or dissipation effects.

In central collisions, the LCP kinetic energy spectra exhibit non-Maxwellian shapes especially along the beam direction. Many data have been interpreted in unfolding the measured spectra in order to separate two components: pre-equilibrium on the one side and an equilibrated part on the other side. Their relative contributions depend strongly on the emission angle which is a help to succeed in the unfolding. For the equilibrated part, the mean c.m. kinetic energy $\langle \epsilon \rangle$ depends on the mass of emitted LCP or IMF. This result indicates that a non-thermal component is present. It is generally attributed to an expansion energy reflecting nuclear-matter compression properties. Figure 7 is a non-exhaustive compilation showing that expansion energy (or radial velocity) is small for incident energies lower than 30 MeV/u [34]. Conversely, fig. 8 indicates that expansion is significant for measured excitation (deposited) energies exceeding 5 to 6 MeV/u [35].

6 Charge or mass distributions

We have already noticed in sect. 3 that the heaviest fragment emitted from a selected source plays a significant role among all the outgoing fragments. This is the case at excitation energies below the multifragmentation threshold since, in this case, the largest fragment is an evaporation residue. When multifragmentation occurs, the largest

fragment has no longer this specific role and its mass becomes much lower. It is observed in the experiments that this change from evaporation-like events to multifragmentation is rather abrupt when the dissipation is increased. In some cases, the coexistence of evaporation-like and multifragmentation events has been observed for comparable dissipations: it is the bimodality signal that is a possible signature of a phase transition of the system (see O. Lopez and M.F. Rivet, this topical issue). It is only stressed here that bimodality can be a first indication of a statistical behaviour for defining the masses or charges of the products released in nucleus-nucleus collisions.

More generally speaking, many data indicate that the overall charge and mass distributions can be described by statistical models, *i.e.* in models in which the main ingredient is the available phase space. This is true for the total charge distribution [2,23] and for the distributions associated with the largest or the second and third largest fragment [18,17]. This is true for limited excitations [2] for which few fragments are released up to very large ones leading to vaporization [36]. In this last case, only LCP are detected but their relative abundances are also understood in a statistical approach [37]. An interesting compilation is shown in fig. 9. It concerns several experiments with quite different entrance channels and for which the measured mass distributions seem to reflect mainly the deposited energy in MeV/u. Similarly, it has been shown in ref. [2] that similar results are found in peripheral and central collisions, indicating that the dissipated energy seems again to be the main ingredient which defines the splitting of the system.

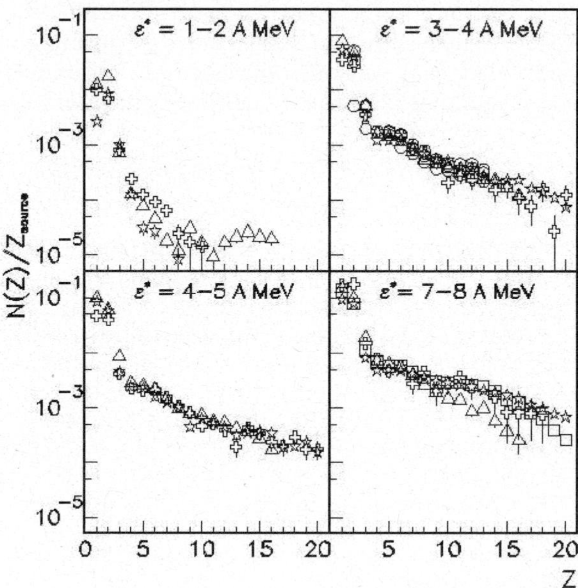

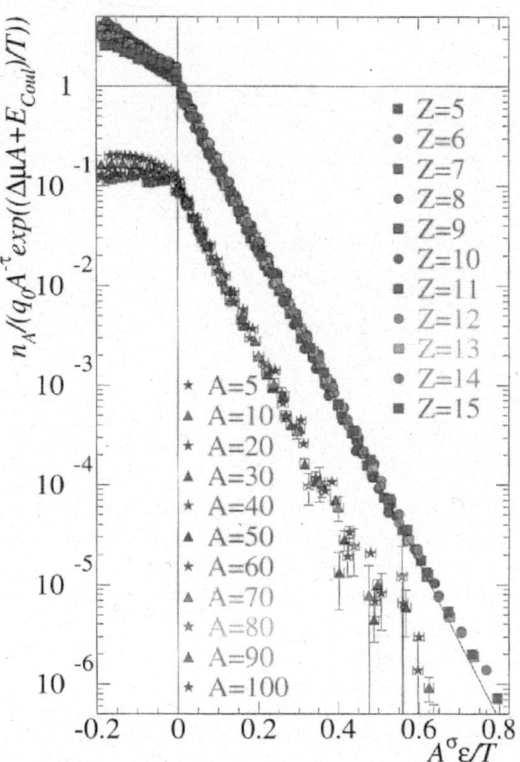

Fig. 9. Mass distribution for various excitation energy ranges obtained in several experiments involving very different entrance channels: stars: Au + Au central collisions from MULTICS data [2]; crosses: peripheral collisions with MULTICS [17]; hexagons: FASA data [38]; squares: 32 MeV/u Xe + Sn INDRA data [10]; triangles: ISIS data [11]. This figure has been prepared by M. D'Agostino [39, 40].

Fig. 10. Analysis of the ISIS data showing that the probability for emitting a given fragment can be fitted in the Fisher formalism in which the emission is mainly governed by statistical properties of nuclei [44].

Many results are reproduced by models like SMM [22] or MMMC [41] in which full statistical equilibrium is assumed. Of course, one may argue that such agreements can be obtained only by adjusting parameters such as the density at freeze-out. The total mass and the excitation energy of the initial source are also adjusted to reproduce the data, but their values are in agreement with calorimetric measurements when they are available. The excitation energies and source masses are smaller than the available energy and mass simply because of pre-equilibrium emission. More direct data are also available in which several systems are compared independently of a model. For instance, in ref. [42] it is shown that the systems Xe + Sn and Gd + U exhibit similar mass distributions at similar measured excitation energies in MeV/nucleon. Similarly, in ref. [10], central Xe + Sn and Ni + Au collisions (same fusion-like source mass at similar excitation energies) exhibit similar mass distributions. This dominance of phase space is also evidenced by the fact that the observed multifragmentation mass distributions can be reproduced simply in cutting at random a rope in a number of elements equal to the observed multiplicity [43]. The multifragmentation mass distribution would hence be constrained only by the mass conservation for a given fragment multiplicity.

One of the most spectacular results indicating a statistical behaviour is the reducibility property [25] which indicates that fragment production probabilities can be put together in Arrhenius plots and the very beautiful fits obtained in the so-called scaling analysis. Figure 10 is the most famous one but similar fits have been obtained with

other data [2, 45]. Even if such an analysis relies on several adjusted parameters (which are consistent with theoretical expectations) and in spite of the blurring effects of secondary decay, this property is a further evidence of statistical behaviour.

An isospin analysis of the released products is also in agreement with this statement [18, 46, 47]: isoscaling is the observation that the probability ratio for producing a defined isotope in two different reactions may be expressed as:

$$R_{1,2} = \exp(\alpha N + \beta Z), \qquad (1)$$

where N and Z are the neutron and proton numbers of the isotope. Even if the physics is not transparent for the values of the parameters α and β [28], the validity of eq. (1) indicates that statistical features are present everywhere. Figure 11 is an illustration showing that this description is valid over a wide range of incident energies and reactions. In this figure, $R_{1,2}$ has been multiplied by $\exp(\beta Z)$ in order to express the results only as a function of N. Scaling is observed for deep inelastic collisions, for evaporation and for multifragmentation as well. It is consistent with the fact that all these processes are governed by the available final states [48]. It does not seem that the sequential decay affects significantly the results [49]. Nevertheless, such observations do not mean that full equilibrium is achieved and some FOPI data [7] indicate that the full mixing be-

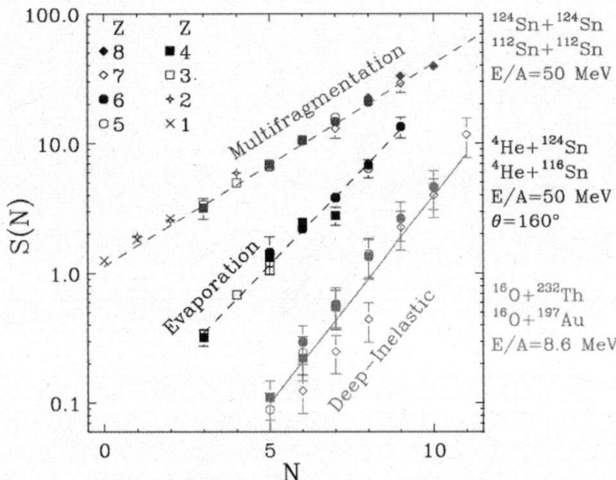

Fig. 11. Production ratio for various isotopes and several reaction pairs. The isoscaling as a function of the neutron number is observed for very different reaction types, from evaporation and deep inelastic collisions to multifragmentation. See text for details. Extracted from ref. [46].

tween the projectile and the target is not achieved even in the most central collisions. Hence, the available phase space is widely opened but is still constrained by some entrance channel memory.

7 Conclusion

From many data, fragment production exhibits both dynamical and statistical aspects. The multiplicity is mainly governed by the dissipated energy. It increases from a single residue (or two fission fragments) for limited excitations up to large values in the multifragmentation regime, the rise and fall leading to a vanishing multiplicity when the dissipated energy is sufficient to allow vaporization. In the multifragmentation case, some fragments can be released in dynamical processes such as neck emission observed in semi-peripheral collisions. In any case, fragments are accompagnied by multiple light particles, some of which show dynamical features.

The size distributions of the detected fragments are also mainly governed by the available phase space; the heaviest fragment has specific properties at least for limited excitations, below 3–5 MeV/u, *i.e.* below the threshold energy for which the multifragmentation channel sets in significantly. Above this threshold, the specificity of the heaviest fragment is weaker.

However, many kinematical properties of the fragments reflect dynamics in the sense that they have retained some memory of the entrance channel. This is clearly the case for their angular distributions and also for their kinetic energies which are not purely thermal for nucleus-nucleus collisions at bombarding energies exceeding 50 MeV/u even if central collisions are selected. This deviation from a thermal behaviour can sometimes be interpreted as a collective deformation [50] or compres-

sion effect initiated by the early compression phase in the collision. In such a case, a statistical description can be used provided that one introduces in the description a constraint summarizing the dynamical behaviour.

References

1. L. Phair *et al.*, Nucl. Phys. A **548**, 489 (1992).
2. MULTICS Collaboration (M. D'Agostino *et al.*), Nucl. Phys. A **724**, 455 (2003).
3. R. Wada *et al.*, Phys. Rev. C **69**, 044610 (2004).
4. EOS Collaboration (L. Hauger *et al.*), Phys. Rev. C **62**, 024616 (2000).
5. L. Moretto *et al.*, Phys. Rev. Lett. **74**, 1530 (1995).
6. INDRA Collaboration (J. Frankland *et al.*), Nucl. Phys. A **689**, 905 (2001).
7. FOPI Collaboration (F. Rami *et al.*), Phys. Rev. Lett. **84**, 1120 (2000).
8. ALADIN Collaboration (P. Kreutz *et al.*), Nucl. Phys. A **556**, 672 (1993).
9. ALADIN Collaboration (W. Trautmann *et al.*), in *Proceedings of the XXXIII International Winter Meeting on Nuclear Physics, Bormio, Italy, 1995*, edited by by I. Iori, Suppl. # 101 (Ricerca Scientifica ed Educazione Permanente, Milano, 1995) p. 372.
10. INDRA Collaboration (N. Bellaize *et al.*), Nucl. Phys. A **709**, 367 (2002).
11. L. Beaulieu *et al.*, Phys. Rev. C **64**, 064604 (2001).
12. Y.G. Ma *et al.*, Phys. Rev. C **69**, 031604 (2004).
13. V.E. Viola, R. Bougault, contribution V.3, this topical issue.
14. M. Pichon *et al.*, unpublished data.
15. INDRA-ALADIN Collaboration, unpublished results.
16. L. Manduci, PhD Thesis, Caen, 2005.
17. M. D'Agostino *et al.*, Nucl. Phys. A **650**, 329 (1999).
18. CHIMERA Collaboration (E. Geraci *et al.*), Nucl. Phys. A **734**, 524 (2004).
19. INDRA Collaboration (P. Lautesse *et al.*), Phys. Rev. C **71**, 034602 (2005).
20. R. Bougault *et al.*, to be published.
21. L. Beaulieu *et al.*, Phys. Rev. C **54**, R973 (1996).
22. J. Bondorf *et al.*, Phys. Rep. **257**, 133 (1995).
23. INDRA Collaboration (N. Marie *et al.*), Phys. Lett. B **391**, 15 (1997); Phys. Rev. C **58**, 256 (1998).
24. INDRA Collaboration (V. Metivier *et al.*), Nucl. Phys. A **672**, 357 (2000).
25. L. Moretto *et al.*, Phys. Rep. **287**, 249 (1997).
26. A. Pagano *et al.*, Nucl. Phys. A **734**, 504 (2004).
27. INDRA Collaboration (J. Colin *et al.*), Phys. Rev. C **67**, 064603 (2003).
28. S.R. Souza *et al.*, Phys. Rev. C **69**, 031607R (2004).
29. INDRA-ALADIN Collaboration (J. Łukasik *et al.*), Phys. Lett. B **566**, 76 (2003).
30. INDRA Collaboration (F. Bocage *et al.*), Nucl. Phys. A **676**, 391 (2000).
31. P. Chomaz *et al.*, Ann. Phys. **320**, 135 (2005).
32. T. Lefort *et al.*, Nucl. Phys. A **662**, 397 (2000).
33. INDRA-ALADIN Collaboration (J. Łukasik *et al.*), Phys. Rev. C **66**, 064606 (2002).
34. D. Durand, Nucl. Phys. A **654**, 273c (1999); D. Durand *et al.*, *Nuclear Dynamics in the Nucleonic Regime* (IOP Publishing Ltd, 2001).

35. M.F. Rivet *et al.*, *Proceedings of the XXVII International Workshop on Gross Properties of Nuclei and Nuclear Excitations, Hirschegg, Austria, 1999*, edited by H. Feldmeier *et al.* (GSI, Darmstadt, 1999) p. 293.
36. INDRA Collaboration (M.F. Rivet *et al.*), Phys. Lett. B **388**, 219 (1996).
37. INDRA Collaboration (B. Borderie *et al.*), Eur. Phys. J. A **6**, 197 (1999).
38. FASA Collaboration (V.A. Karnaukhov *et al.*), Nucl. Phys. A **734** 520 (2004).
39. M. D'Agostino *et al.*, Nucl. Phys. A **749**, 55 (2005).
40. M. D'Agostino, unpublished.
41. D.H.E. Gross, Rep. Prog. Phys. **53**, 605 (1990).
42. INDRA Collaboration (M.F. Rivet *et al.*), Phys. Lett. B **430**, 217 (1998); INDRA Collaboration (J. Frankland *et al.*), Nucl. Phys. A **689**, 940 (2001).
43. R. Dayras, unpublished calculations.
44. J.B. Elliott *et al.*, Phys. Rev. Lett. **88**, 042701 (2002).
45. INDRA Collaboration (N. Leneindre), submitted to Eur. Phys. J. A.
46. M.B. Tsang *et al.*, Phys. Rev. Lett. **86**, 5023 (2001).
47. G.A. Souliotis *et al.*, Phys. Rev. C **68**, 024605 (2003).
48. D.V. Shetty *et al.*, Phys. Rev. C **71**, 024602 (2005).
49. M.B. Tsang *et al.*, Phys. Rev. C **64**, 054615 (2001).
50. INDRA-ALADIN Collaboration (A. Lefèvre *et al.*), Nucl. Phys. A **735**, 219 (2004).

Eur. Phys. J. A **30**, 81–108 (2006)

DOI 10.1140/epja/i2006-10109-6

THE EUROPEAN
PHYSICAL JOURNAL A

Correlations and characterization of emitting sources

G. Verde[1,2,a], A. Chbihi[2], R. Ghetti[3], and J. Helgesson[3]

[1] INFN, Sezione di Catania, 64 Via Santa Sofia, I-95123 Catania, Italy
[2] GANIL, IN2P3-CNRS, BP 5027, F-14076 CAEN Cedex 5, France
[3] School of Technology and Society, Malmö University, SE-20506 Malmö, Sweden

Received: 24 May 2006 /
Published online: 16 October 2006 – © Società Italiana di Fisica / Springer-Verlag 2006

Abstract. Dynamical and thermal characterizations of excited nuclear systems produced during the collisions between two heavy ions at intermediate incident energies are presented by means of a review of experimental and theoretical work performed in the last two decades. Intensity interferometry, applied to both charged particles (light particles and intermediate mass fragments) and to uncharged radiation (gamma rays and neutrons) has provided relevant information about the space-time properties of nuclear reactions. The volume, lifetime, density and relative chronology of particle emission from decaying nuclear sources have been extensively explored and have provided valuable information about the dynamics of heavy-ion collisions. Similar correlation techniques applied to coincidences between light particles and complex fragments are also presented as a tool to determine the internal excitation energy of excited primary fragments as it appears in secondary-decay phenomena.

PACS. 24.10.-i Nuclear reaction models and methods – 25.70.-z Low and intermediate energy heavy-ion reactions – 25.70.Pq Multifragment emission and correlations – 25.75.Gz Particle correlations

1 Introduction

Heavy-ion collisions are the only terrestrial means to explore the properties of nuclear matter under extreme conditions. In order to extract such nuclear-matter properties, a clear understanding of the complex dynamics of heavy-ion collisions is required [1–10]. The detected particles are indeed produced by different emission mechanisms and at different stages whose experimental identification is challenging. Researchers have therefore intensively focused on obtaining a well-defined dynamical and thermal characterization of particle- and fragment-emitting sources.

Where and when are fragments produced? What are their thermal properties, *i.e.* excitation energy, internal temperature, or spin? At what density do nuclear multifragmentation phenomena occur? Can we learn something about their link to a liquid-gas phase transition in nuclear matter [11–15]?

In this paper we will present a review of those research activities that have been devoted to finding answers to these questions. We will first focus on the space-time characterization of particle-emitting sources, namely the estimation of their sizes, shapes, densities, lifetimes and emission chronology. This task has been extensively addressed with intensity interferometry studies by exploring light-particle–light-particle and IMF-IMF (Intermediate Mass Fragment) correlation functions. The last section

[a] e-mail: `verde@ct.infn.it`; `verde@ganil.fr`

will be devoted to the thermal characterization of emitting sources by means of light-particle–IMF correlation function techniques, providing information about fragment internal excitation energies and the relative proportion of the thermal component. We will finally conclude with some remarks and perspectives for future research in this field.

2 Intensity interferometry and light-particle emission

The space-time properties of heavy-ion collisions can be accessed by intensity interferometry [16–18]. This technique was originally introduced in astronomy by Hanbury-Brown and Twiss to measure astronomical distances, such as the radii of stars and galaxies [19,20]. The technique was later extended to subatomic physics by Goldhaber *et al.* who studied distributions of K mesons in proton-antiproton annihilation processes [21]. Due to their bosonic nature, two-pion correlation functions show an enhancement at zero relative momentum. The width of this enhancement provided information about the volume of the region emitting pions in the studied processes. Pion-pion interferometry plays still today a key role in the study of heavy-ion collisions at ultra-relativistic energies where it is an important observable to investigate the production of the Quark Gluon Plasma [22].

The use of intensity interferometry in heavy-ion collisions at intermediate energies is generally characterized by a more complicated scenario, as compared to the case of astronomical applications. During a nuclear reaction not only photons but several particle species can be emitted: neutrons, protons, complex particles or fragments. These emitted radiations can be either bosons or fermions, interacting with one another by means of repulsive Coulomb and attractive nuclear forces. Another important complication inherent to heavy-ion collisions is represented by the fact that the produced nuclear systems live for a very short time ranging between 10^{-22} and 10^{-15} seconds. This situation is very different from the case of astronomical objects, where the geometry of a static object is studied. Moreover, in nuclear reactions different particles can be produced at different times and by different sources. Therefore, only a full space-time characterization of all these multiple emitting sources can improve our understanding of heavy-ion collision dynamics.

2.1 Measuring two-particle correlation functions

Given two particles with momenta $\mathbf{p}_1$ and $\mathbf{p}_2$, total momentum $\mathbf{P} = \mathbf{p}_1 + \mathbf{p}_2$ and momentum of relative motion $\mathbf{q} = \mu(\mathbf{p}_1/m_1 - \mathbf{p}_2/m_2)$, the two-particle correlation function, $1+R(\mathbf{q},\mathbf{P})$ is defined experimentally by the following equation:

$$\sum Y_{12}(\mathbf{p}_1, \mathbf{p}_2) = C_{12} \cdot [1 + R(\mathbf{q}, \mathbf{P})] \cdot \sum Y_1(\mathbf{p}_1) \cdot Y_2(\mathbf{p}_2). \tag{1}$$

In this equation, $Y_{12}(\mathbf{p}_1, \mathbf{p}_2)$ is the two-particle coincidence yield while $Y_1(\mathbf{p}_1)$ and $Y_2(\mathbf{p}_2)$ are the single-particle yields. The normalization constant C_{12} is commonly determined by the requirement $R(\mathbf{q}) = 0$ at large relative momentum values, q. In order to obtain sufficient statistics, the sums in eq. (1) are performed over all detector and particle energy combinations satisfying a specific gating condition. Experimental studies have therefore focused on two types of observables: *directionally gated* and *angle-averaged* correlation functions. Directionally gated correlation functions are constructed by selecting particle pairs with specific conditions on the relative direction between the relative momentum, $\mathbf{q}$, and the total momentum, $\mathbf{P}$ [23,24]. For instance, correlation functions with the vector $\mathbf{q}$ either parallel or perpendicular to the vector $\mathbf{P}$ have been extensively constructed [25,26]. Studies with such directional gates generally require high statistics. Directional effects might indeed be small and difficult to analyze [25]. Alternatively, one can study angle-averaged correlation functions by integrating over the relative angle between the vectors $\mathbf{q}$ and $\mathbf{P}$. The resulting correlation function depends only on the magnitude of the relative momentum, q:

$$\sum Y_{12}(\mathbf{p}_1, \mathbf{p}_2) = C_{12} \cdot [1 + R(q)] \cdot \sum Y_1(\mathbf{p}_1) \cdot Y_2(\mathbf{p}_2). \tag{2}$$

Experimentally, the product of the single yields, $Y_1(\mathbf{p}_1) \cdot Y_2(\mathbf{p}_2)$ in eqs. (1) and (2), has often been approximated with the uncorrelated two-particle yields, $Y_{12}^{unco}(\mathbf{p}_1, \mathbf{p}_2)$,

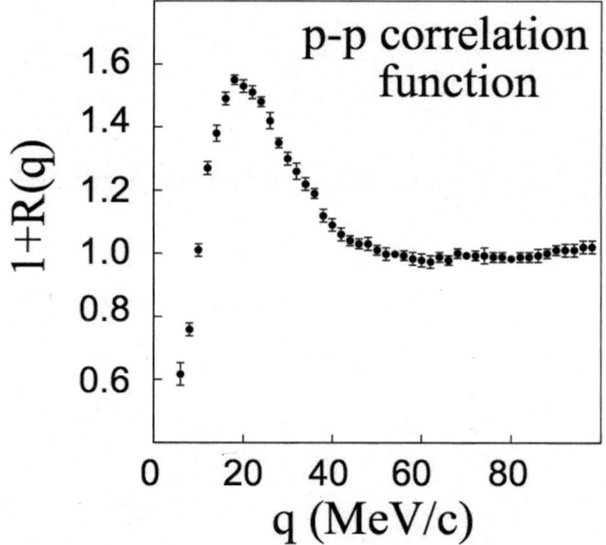

Fig. 1. Two-proton correlation function measured in ^{14}N $+$ ^{197}Au collisions at $E/A = 75$ MeV (from refs. [27,28]).

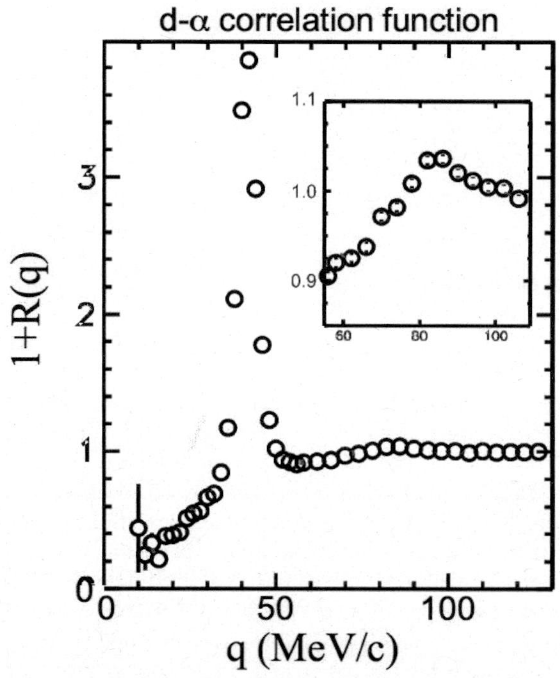

Fig. 2. Deuteron-alpha correlation function measured in Sn + Sn collisions at $E/A = 50$ MeV with the LASSA detector array [29].

constructed via the so-called event-mixing technique: particles 1 and 2 are taken from two different events and the correlation function is calculated as $1 + R(q) = C_{12} \cdot Y_{12}/Y_{12}^{unco}$.

Figures 1 and 2 show examples of angle-averaged proton-proton and deuteron-alpha correlation functions, respectively, represented as a function of the relative momentum q between the particle pairs [29–31]. If the two emitted particles were totally uncorrelated, the probability of detecting them in coincidence would be equal to the product of the probabilities of detecting the singles, *i.e.* $Y_{12}(\mathbf{p}_1, \mathbf{p}_2) \approx Y_1(\mathbf{p}_1) \cdot Y_2(\mathbf{p}_2)$, resulting in a flat correlation function, $R(q) = 0$, at all q-values. Experimentally, it is easily observed that this is not the case. As one detects pairs at small relative momentum, strong deviations from unity are observed. These deviations are due to quantum statistics and to the so-called final-state interactions (FSI) [23]. In the case of identical fermions (bosons), the relative wave function must respect anti-symmetrization (symmetrization) rules that induce measurable effects in the correlation function at small relative momenta [19–21,23]. Furthermore, the coincident particles can interact with their mutual Coulomb and nuclear interaction. The Coulomb repulsion is responsible for the anti-correlation at small q-values. The nuclear attractive force is responsible for the observed prominent peaks both in *p-p* and *d-α* correlation functions.

At large relative momenta, $q \approx \infty$, the correlation functions shown in figs. 1 and 2 appear as flat, $R(q) \approx 0$, indicating the absence of correlations between the coincident particles. However, the presence of collective motion can generate correlation effects even at large relative momenta where the correlation function may significantly deviate from the limit, $R(q) \approx 0$. If collective motion exists, the uncorrelated relative momentum distribution, constructed by mixing particles from different events, can contain additional collective components that do not exist in the coincidence spectrum. These additional contributions may affect the correlation function constructed from the ratio of the coincidence and the uncorrelated spectra [32,33].

In general, the shape of the correlation function is sensitive to the space-time properties of particle-emitting sources produced during the reaction [23]. In order to properly access information about these emitting sources, the use of detector arrays with a high angular and energy resolution is required. Especially, the angular resolution plays an important role in determining the exact location and the shape of the resonance peaks and in accessing the correlation function at very low relative momentum. In this respect, position-sensitive detectors and silicon strips have been quite successful thanks to their capability of providing relative angle measurements as small as 0.1°–0.3°.

3 Space-time properties from two-proton correlation functions

Intensity interferometry has extensively been used with protons, these particles being abundantly produced at all incident energies and easily detected with high resolution. Theoretically, the proton-proton correlation function is calculated by the so-called *Koonin-Pratt equation* (KP equation) [24]:

$$1 + R(\mathbf{q}) = 1 + \int d\mathbf{r}\, S(\mathbf{r}) \cdot K(\mathbf{r}, \mathbf{q}). \qquad (3)$$

The goal of intensity interferometry consists of solving eq. (3): from the measured correlation function on the

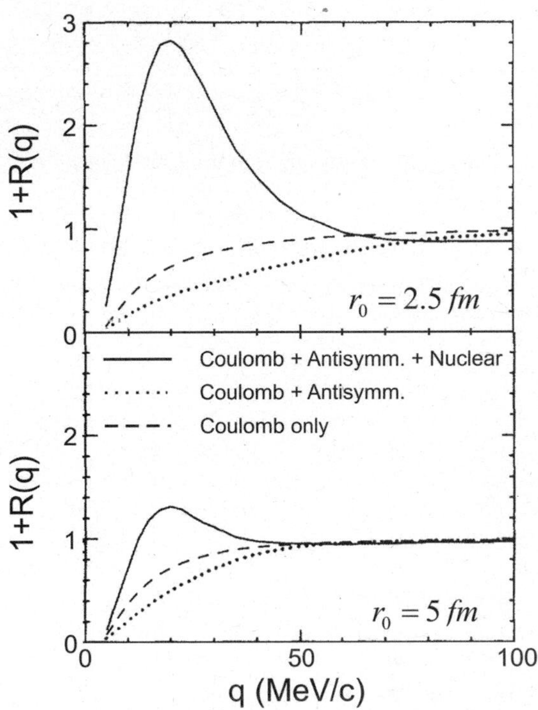

Fig. 3. Two-proton correlation functions calculated by means of eq. (3) assuming a Gaussian spherically symmetric source function, $S(\mathbf{r}) \propto \exp(-r^2/r_0^2)$, with source size $r_0 = 2.5$ (top panel) and 5 fm (bottom panel). The different lines correspond to calculations performed by considering the antisymmetrization and all the final-state interactions (solid line), only the Coulomb interaction (dashed line), and the Coulomb interaction and the anti-symmetrization of the wave function (dotted line).

l.h.s of eq. (3) one needs to extract the unknown *source function*, $S(\mathbf{r})$. This source function is defined as the probability of emitting two particles at relative distance $\mathbf{r}$, calculated at the time when the last of the two particles is emitted. The so-called *kernel function*, $K(\mathbf{r}, \mathbf{q})$, can be calculated as $K(\mathbf{r}, \mathbf{q}) = |\Psi_{\mathbf{q}}(\mathbf{r})| - 1$, where $\Psi_{\mathbf{q}}(\mathbf{r})$ is the proton-proton scattering wave function [19,20,24,30]. The kernel contains all the information about the antisymmetrization of the proton-proton wave function, due to their Fermionic nature, and the mutual Coulomb and nuclear final-state interactions (FSI).

Figure 3 shows correlation functions calculated with eq. (3) by using a Gaussian-shaped source function, $S(\mathbf{r}) \propto \exp(-r^2/r_0^2)$, with width parameter values of $r_0 = 2.5$ fm (small source, top panel) and $r_0 = 5$ fm (large source, bottom panel) [16]. Gaussian sources have extensively been used in the literature due to their simplicity. The dashed lines in fig. 3 correspond to the correlation functions obtained for protons interacting only with the mutual Coulomb force. This repulsive interaction between protons is responsible for the anti-correlation at small relative momentum, $q < 15\,\mathrm{MeV}/c$. The dotted line is obtained by adding the two-fermion anti-symmetrization in the two-proton wave function, inducing a further anti-correlation in the region $q = 15$–$60\,\mathrm{MeV}/c$, due to the

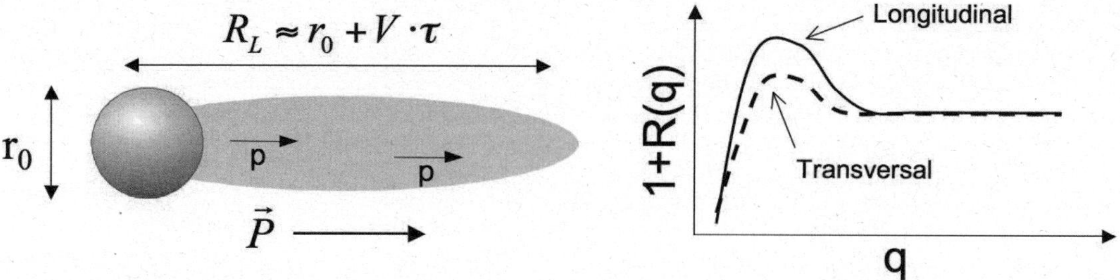

Fig. 4. Effects of finite emission lifetimes on two-proton source functions (left side) and directionally gated correlation functions (see text for details).

Pauli principle that prevents protons from occupying relative momentum states over an interval of $\Delta p_x = h/\Delta x$, where Δx is the spread of the proton spatial distribution in a certain direction x. Finally, when also the final-state mutual nuclear interaction is included, the p-p correlation functions displays a prominent peak at $20\,\mathrm{MeV}/c$. This peak is due to the s-partial wave of the proton-proton scattering problem and strongly depends on the volume of the emitting source. Indeed, when a large emitting source is used (bottom panel in fig. 3), the peak height reduces significantly.

The correlation functions displayed in fig. 3 have been determined assuming that the two protons are simultaneously emitted by a source with zero lifetime. However, proton emission is known to occur over finite timescales (ranging from few tens of fm/c in the case of pre-equilibrium emission to thousands of fm/c in secondary-decay processes). If the particle pairs are not emitted simultaneously, the source function is affected by a space-time ambiguity and will appear deformed, as it is schematically shown in fig. 4 (left panel) [23,25]. If r_0 is the actual geometrical source size, the source function appears elongated and with a larger size in the direction defined by the total momentum vector **P**. The elongation is approximately given by $V\tau$, with V and τ being, respectively, the average pair velocity and the emission lifetime. Due to this deformation of the source function, the Pauli suppression effect, described in fig. 3 (dotted lines), plays a key role in determining the shape of the correlation function. As it is schematically shown in the right panel of fig. 4, the transverse correlation function, constructed with the relative momentum, **q**, perpendicular to the total momentum, **P**, will undergo a larger Pauli suppression as compared to the correlation function constructed by selecting a relative momentum, **q**, parallel to the total momentum, **P**. This simple qualitative argument shows that directionally gated correlation functions allow one to disentangle the space and time information hidden in the source function, providing quantitative estimates of both source size and lifetime [25,26].

3.1 Volumes and lifetimes from directionally gated correlation functions

The sensitivity of directional correlation functions to finite lifetimes and emission volumes has stimulated an extensive research activity in the field of heavy-ion collisions.

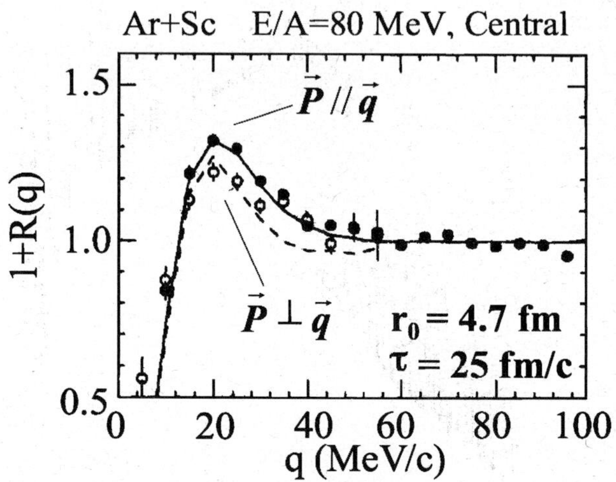

Fig. 5. Data points: directional two-proton correlation functions measured in Ar + Sc collisions at $E/A = 80\,\mathrm{MeV}$ (from ref. [25]). The solid and dashed lines correspond, respectively, to longitudinal and transversal calculated correlation functions.

Figure 5 shows two-proton correlation functions measured in central Ar + Sc collisions at $E/A = 80\,\mathrm{MeV}$ and for proton pairs with total momenta $P = 400$–$600\,\mathrm{MeV}/c$ [25]. The filled and open circles correspond, respectively, to longitudinal and transverse correlation functions. A stronger Pauli suppression in the transverse direction is observed, as compared to the longitudinal direction. These results are one of the first experimental evidences of the predicted lifetime effects in two-proton correlation functions. In the same reaction, protons at higher total momenta showed no directional dependence [31], consistent with negligible lifetime effects. Indeed, these fast protons are dominated by rapidly decaying pre-equilibrium sources. The directional correlations in fig. 5 were analyzed by using the Koonin-Pratt equation, eq. (2), assuming a source function composed by a Gaussian spatial distribution and an exponentially decaying time profile, $S(\mathbf{r},t) \propto \exp(-r^2/2r_0^2) \cdot \exp(-t/\tau)$. The simultaneous best fit of longitudinal and transverse correlations provided source radii and lifetimes of $r_0 = 4.5$–$4.8\,\mathrm{fm}$ and $\tau = 10$–$30\,\mathrm{fm}/c$, respectively. The obtained geometrical size, $r_0 = 4.5$–$4.8\,\mathrm{fm}$, is comparable to the size of the overlapping region in central Ar + Sc collisions. The extracted short lifetime was interpreted as consistent with proton sources dominated by the early pre-equilibrium stage of the reaction. These results were also predicted by BUU transport models that are expected to well describe the dynamical early stages of the reaction [31].

Directional correlation functions have also been used to characterize proton evaporation from long-lived nuclear systems at relatively low excitation energies [34, 35]. In the study of quasi-compound projectiles produced in ^{129}Xe + ^{27}Al reactions at $E/A = 31$ MeV [34], a lifetime of the order of $\tau \approx 1300$ fm/c and a source size of $r_0 \approx 2.2$ fm were obtained. However, this source size appeared small as compared to the size of the evaporating compound system. The difficulty in explaining the results was attributed by the authors to possible model dependences in the source parametrization. Similar difficulties have been encountered in the study of ^{32}S + ^{nat}Ag reactions at $E/A = 22.3$ MeV/u [30]. Directional two-proton correlation functions were used to explore both pre-equilibrium emission (fast sources) and proton evaporation (slow source) from the produced compound nuclei. In both cases no significant difference between transverse and longitudinal correlation functions was observed. While this observation can be expected in the case of pre-equilibrium emission, the absence of any directional effects for low energy protons evaporated from compound nuclei appeared controversial.

The space-time properties of heavy-ion collisions at relativistic incident energies were explored by the FOPI and the ALADiN Collaborations [26, 32, 33, 36–38] to access information about nuclear densities and to study the temporal evolution of excited nuclear systems. Figure 6 shows directional two-proton correlation functions measured in central ^{96}Zr(^{96}Ru) + ^{96}Zr(^{96}Ru) collisions at $E/A = 400$ MeV [26]. A very small Pauli suppression in the transverse correlation function (solid point) as compared to the longitudinal correlation function (open points) is observed. This suppression is consistent with a slightly deformed Gaussian source characterized by a negligible lifetime and a geometrical size of about $r_0 \approx 5.4$ fm [26]. The obtained negligible lifetime can be attributed to the presence of a strong collective motion and the consequent decrease of emission timescales in participant matter. Longer emission lifetimes can be expected in the decay of spectator matter produced in peripheral collisions [11]. The left panels in fig. 7 show transverse (closed symbols) and longitudinal (open symbols) two-proton correlation functions measured in the decay of target spectators produced in Au + Au collisions at $E/A = 1$ GeV [37]. The data from the top to the bottom are obtained by imposing different gates on the Z_{bound} observable, defined as the sum of the charges of all fragments with $Z \geq 2$ and emitted by projectile spectators [11]. Z_{bound} increases with increasing impact parameter and decreasing spectator excitation energy. These directional correlation functions provided lifetimes $\tau \leq 20$ fm/c and source radii $r_0 \approx 8$ fm, independently of Z_{bound}. From an estimate of the amount of nucleons in the decaying spectator it was possible to deduce densities ranging between $\rho/\rho_0 \approx 0.15$ and $\rho/\rho_0 \approx 0.4$. Even in the case of spectator decay, the

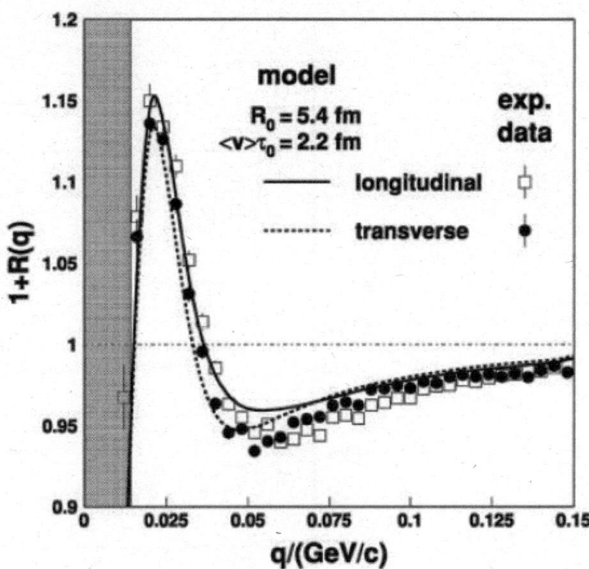

Fig. 6. Longitudinal (open dots and solid line) and transversal (close dots and dotted line) two-proton correlation functions measured in ^{96}Zr(^{96}Ru) + ^{96}Zr(^{96}Ru) central collisions at $E/A = 400$ MeV with the FOPI detector [26].

obtained lifetimes appear to be rather short. High proton identification thresholds [32,37] could result in correlation functions that are dominated by fast particles emerging from the short-lived pre-equilibrium emission stages of the reaction. On the other hand, other authors have investigated the intrinsic small sensitivity of directional correlation functions to those long lifetimes typical of evaporation processes and secondary decays. Can one probe the time structure of proton emissions when lifetimes of the order of $\tau > 10^3$–10^4 fm/c exist? In ref. [30] it was observed that, if the emission lifetime is very long, longitudinal and transverse correlation functions become undistinguishable and, as a result, a quantitative extraction of the lifetime itself becomes difficult.

3.2 Space-time source extents from angle-averaged two-proton correlations

Angle-averaged correlation functions are constructed experimentally by removing any conditions on the angle between the relative and the total momentum of the emitted proton pairs. From this point of view, these observables do not require as high statistics as in the case of directionally gated correlation functions. In order to extract physics information, one needs to use the *angle-averaged* Koonin-Pratt equation [17, 23, 27, 28, 39]:

$$R(q) = 4\pi \int \mathrm{d}r \cdot r^2\, S(r) \cdot K(r, q), \qquad (4)$$

where $K(r, q)$ is the angle-averaged kernel. The resulting spherically symmetric source function, $S(r)$, contains information about its spatial extent and its finite lifetimes folded together in the value of the relative distance, r.

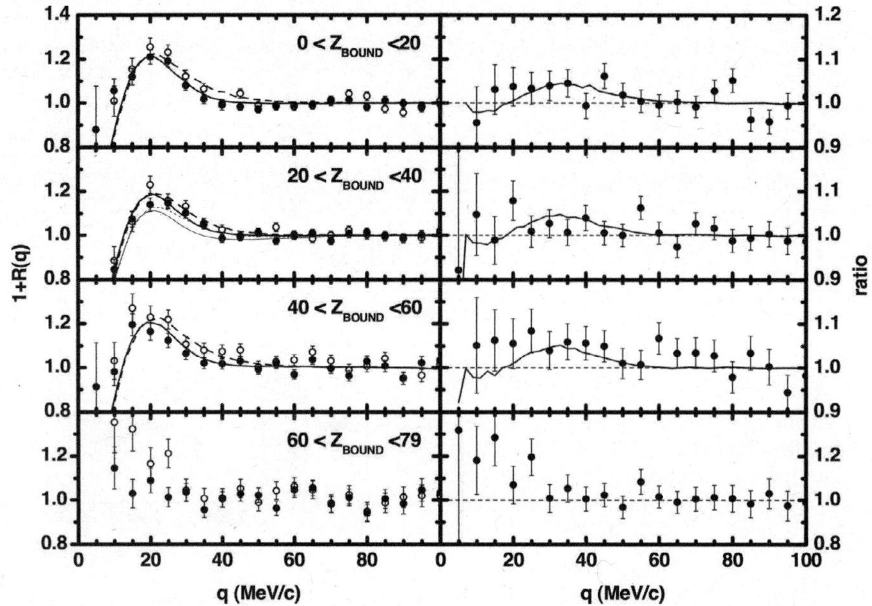

Fig. 7. Left panel: transverse (closed symbols) and longitudinal (open symbols) two-proton correlation functions measured by the ALADiN collaboration in the decay of target spectators formed in Au + Au collisions at $E/A = 1\,$GeV [37]. The data from the top to the bottom panel correspond to increasing impact parameter, selected by means of the Z_{bound} observable [11,37]. Right panel: ratio between longitudinal and transverse correlation functions.

Therefore, it is not easy to disentangle the space-time ambiguity contained in the emitting source. However, even under these limited conditions, angle-averaged correlation functions have provided space-time information about particle emission mechanisms.

In order to solve the angle-averaged Koonin-Pratt equation, eq. (4), and extract information about the emitting-source function, $S(r)$, different approaches have been proposed. In the following subsections we will present the main results obtained with *Gaussian source approaches* and *imaging techniques*.

3.2.1 Gaussian source approaches

In fig. 8 angle-averaged two-proton correlation functions measured in ^{14}N $+$ ^{197}Au reactions at $E/A = 75\,$MeV are shown for three different gates on the total momentum of detected proton pairs, $P = 270$–390, 450–780 and 840–1230 MeV$/c$ [27,28]. The dashed lines correspond to best fits of the experimental data with eq. (4), where the source function is assumed to be characterized by a Gaussian shape, $S_{\mathbf{P}}(\mathbf{r}) \propto \exp(-r^2/2r_0^2)$. Gaussian source sizes, $r_0 = 5.9$, 4.2 and 3.4 were obtained, respectively, for $P = 270$–390, 450–780 and 840–1230 MeV$/c$.

The total momentum dependence of two-proton correlation functions, shown in fig. 8, has been extensively explored in the literature. Figure 9 shows a collection of Gaussian source sizes measured in reactions induced by different projectiles (^{3}He, ^{14}N, ^{16}O and ^{40}Ar) impinging on Au and Ag targets [16,17,40]. Source sizes are represented as a function of the average velocity of the coincident proton pairs, $v_p = 1/2(p_1+p_2)/M_p$, normalized to the

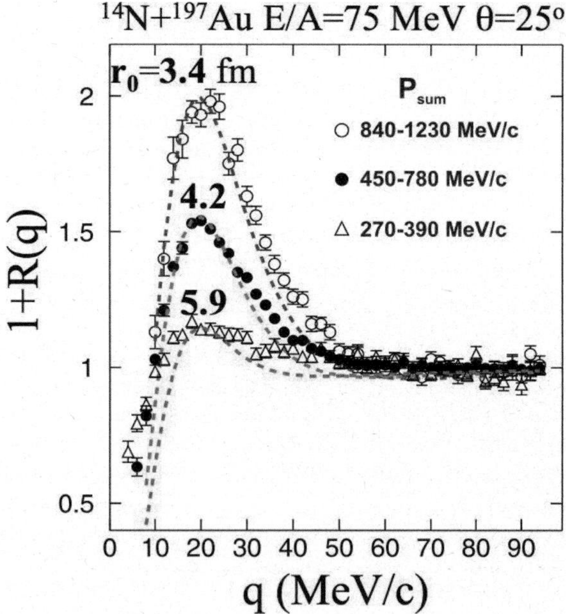

Fig. 8. Data points: two-proton angle-averaged correlation functions measured in ^{14}N $+$ ^{197}Au at $E/A = 75\,$MeV [27,28]. The different symbols correspond to different gates in the total momentum of proton pairs, as indicated on the figure. The dashed lines are calculated assuming Gaussian-shaped source functions and correspond to best fits of the height of the peaks at $20\,$MeV$/c$.

beam velocity, v_{beam}. The observed decrease of the size of the emitting source with increasing proton velocities has been interpreted as a consequence of the cooling dynamics

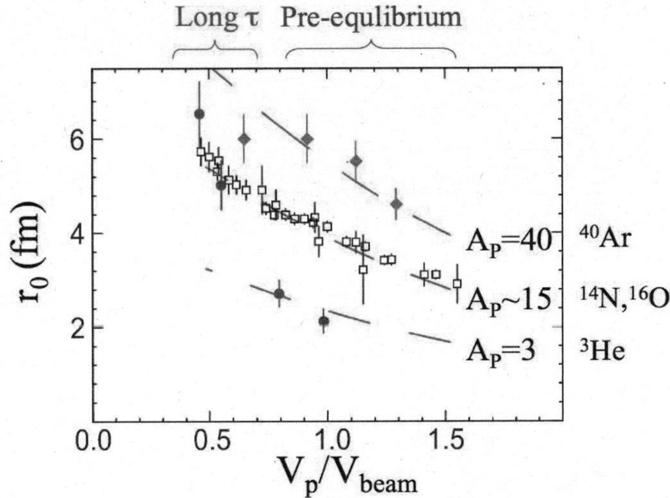

Fig. 9. Gaussian source sizes obtained from the study of reactions induced by ^{14}N and ^{16}O projectiles (open symbols), ^{3}He projectiles (closed circles) and ^{40}Ar projectiles (closed diamonds) on ^{197}Au targets (adapted from refs. [16,17]).

of the produced nuclear systems [16,17,28]. Energetic protons are associated with small sources as one would expect in the case of emissions at the early stages of the reaction. For less energetic protons, however, the extracted source radii increase as one would expect at the later stage when the system is cooled and expanded. The systematics of two-proton radii with the size of the interacting projectile and target nuclei is more involved. The data points corresponding to reactions induced by ^{14}N and ^{16}O projectiles overlap and are interpolated by a dashed line. The other two dashed lines are obtained by multiplying and dividing the ^{14}N/^{16}O line by a factor proportional to $(A_P/14)^{1/3}$, where A_P is the mass number of the projectile nuclei. It is clear that for high-energy protons, $v_p/v_{beam} > 0.5$, where fast pre-equilibrium emission plays an important role, the two-proton source radius scales approximately with the radius of the projectile. This scaling with the radius of projectile nuclei is not observed in the case of low-energy protons that are known to be predominantly emitted by evaporation and secondary-decay processes. The presence of such long-lifetime emissions makes two-proton correlation functions more difficult to be interpreted. This is confirmed by the difficulty in reproducing the shape of the experimental data for pairs at low total momenta (see fig. 8).

The Gaussian radii r_0 shown in fig. 9 must be considered as upper bounds for the actual source sizes. Indeed, since angle-averaged source function fold space and time extents together, a small size could correspond both to a small emitting volume or to short proton emission times. On the other hand, the use of gates on total momentum or, alternatively, on proton velocities is expected to reduce the space-time ambiguities contained in the angle-averaged correlation function. Indeed, these gates should help to isolate, to some extent, more localized emitting sources. Stimulated by these ideas, the CHIC Collaboration has used statistical event generators to simulate

total-momentum–gated correlation functions and estimate emission lifetime and spatial extents in Ni + Al reactions at $E/A = 45$ MeV [41]. The authors explored also the existence of a possible overlap of multiple sources with different lifetimes (corresponding to pre-equilibrium and evaporative emissions). Due to the selectivity of the total momentum gate, it was still possible to estimate both a spatial Gaussian source size, $R_G = 2.7 \pm 0.3$, and an emission lifetime, $\tau_p \approx 400 \pm 200$ fm/c.

Systematic studies of source sizes in central Ca + Ca and Au + Au collisions with varying incident energies have been recently performed by the FOPI Collaboration [36]. The size of the two-proton source is observed to decrease as the incident energy is increased from $E/A = 400$ MeV to $E/A = 1500$ MeV. In the case of Au + Au reactions, the source size decreases from 5.0 fm to 4.1 fm. A smaller source size change is observed in the case of Ca + Ca collisions. This result is consistent with participant matter which is compressed to higher and higher densities as one increases the incident energy. This indication is quite attractive and further work on the details of these two-proton correlation functions can provide useful tools to investigate the nuclear equation of state.

3.2.2 Imaging analyses

The importance of studying the detailed shape of two-proton correlation functions was recently addressed by introducing an imaging approach to intensity interferometry [42–45]. This imaging technique consists of extracting the source profile, $S(r)$, by a numerical inversion of the Koonin-Pratt equation, eq. (4). No *a priori* assumptions about the source shape are made, thus considerably reducing model dependences. Figure 10 shows an application of the imaging technique to two-proton correlation functions measured in ^{14}N + ^{197}Au collisions at $E/A = 75$ MeV [28, 42]. The thick lines on the left panel of the figure represent the imaged measured correlation functions. The imaging technique reproduces in details the entire shape of the correlation functions. On the right panel of fig. 10 the extracted imaging profiles are also shown. Contrary to the case of Gaussian source analyses shown in fig. 8, the source size does not decrease with increasing proton pair total momentum, remaining constant at about 3 fm. This result is very different from the systematics shown in fig. 9, where the source size was mostly extracted from the height of the peak at 20 MeV/c. This constancy of the source size with increasing total momentum of the protons shows that the height of the peak at 20 MeV/c does not provide unambiguous information about the size of the source [42]. These results have been explained within a simple scenario where protons are emitted by the overlap of a fast source, corresponding to pre-equilibrium emissions, and a slow source representing the last stages dominated by secondary decays of excited fragments [42,46]. In this schematic model, fast emissions provide a fraction f of the total proton yields, $Y_{fast} = f \cdot Y$, while the slow secondary decay sources provide the remainder, $Y_{slow} = (1 - f) \cdot Y$. In this limit, the profiles extracted

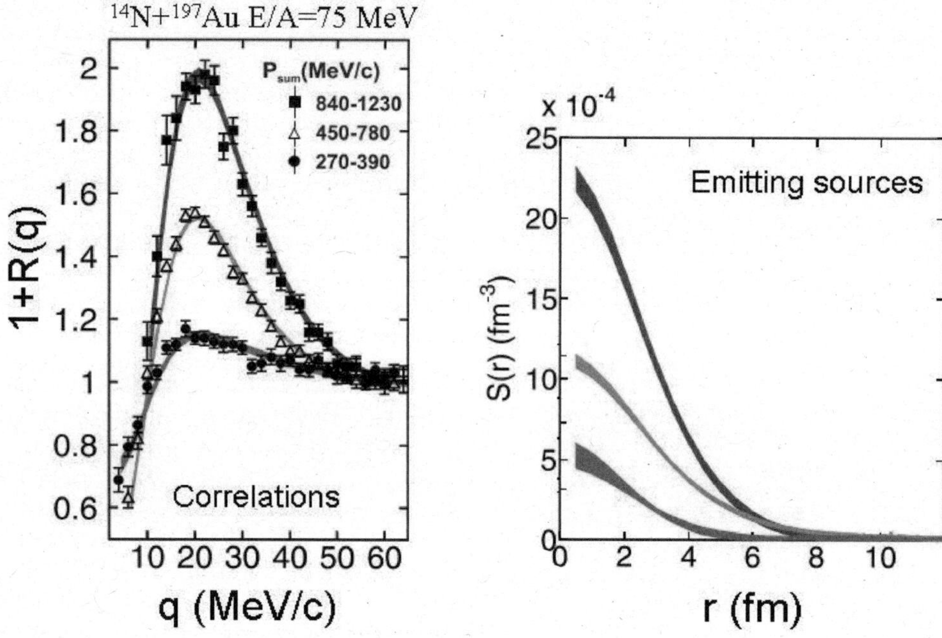

Fig. 10. Imaging analysis of data points already shown in fig. 8 and corresponding to two-proton correlation functions measured in ^{14}N $+$ ^{197}Au reactions at $E/A = 75$ MeV [28]. The left panel shows imaged correlation functions by thick bands. The right panel displays the extracted source function profiles [42].

from the imaging analysis of the experimental $R(q)$ at $q > 10$ MeV/c and shown on the right panel of fig. 10 are strongly dominated by the dynamical source. The size of this source is determined by the width or, even better, by the shape of the peak at 20 MeV/c. The detailed study of the shape provides also a measure of the relative contributions of the fast source, f, and of the slow source, $1 - f$. This fast/slow contributions represent new physics information contained in proton-proton correlation functions, opening the opportunity to constrain the importance of secondary decays. The detailed shape of the long-lived portion of the emitting source contributes to the source function $S(r)$ at large relative distance values, $r > 10$ fm, where the kernel $K(q, r)$ is dominated by the Coulomb interaction. These large distances influence the correlation function only at very small relative momenta, $q < 10$ MeV/c, where measurements are difficult because of the finite angular resolution typical of the used experimental setups [42]. Therefore, the details of these slow emitting sources $r > 10$ fm are hard to access. In fig. 10, the observed quasi-Gaussian shapes refer only to the fast source. This is demonstrated by the fact that their integral over relative distances r is less than unity. If the slowly emitting sources could be experimentally accessed by measuring very low relative momentum particle pairs, it would be possible to image even the tails of the source profiles at large relative distances. In that case, one would clearly observe how the emitting source can deviate significantly from the simplified Gaussian shapes assumed in most works on two-proton correlation functions. Such detailed studies on the shape of two-proton correlation func-

tions require high-resolution experiments that represent a challenge for the future.

With the introduction of imaging analyses, the systematics of proton-proton source sizes shown in fig. 9 is expected to change. On the other hand, such high-order shape analyses require high-resolution measurements that certainly represent a challenging perspective for the future.

4 Space-time characterization from uncharged radiations

Considerable experimental efforts have also been devoted to the measurement of correlation functions involving non-charged radiations (neutrons and photons).

Figure 11 shows two-neutron correlation functions measured in Ni + Al, Ni, Au reactions at $E/A = 45$ MeV [41,46,47]. These measurements are very complicated because of cross-talk problems: the interaction and registration of the same neutron in two different detectors can substantially affect neutron coincidence measurements [48,49].

The neutron-neutron correlation function is dominated by the Fermionic nature of the neutrons and by their mutual nuclear final-state interaction. Unlike the case of proton-proton correlation functions, no Coulomb anti-correlation at small relative momenta exists. The strong maximum at zero relative momentum is caused by the same resonance that brings about the maximum in the p-p correlation function and has been used to investigate

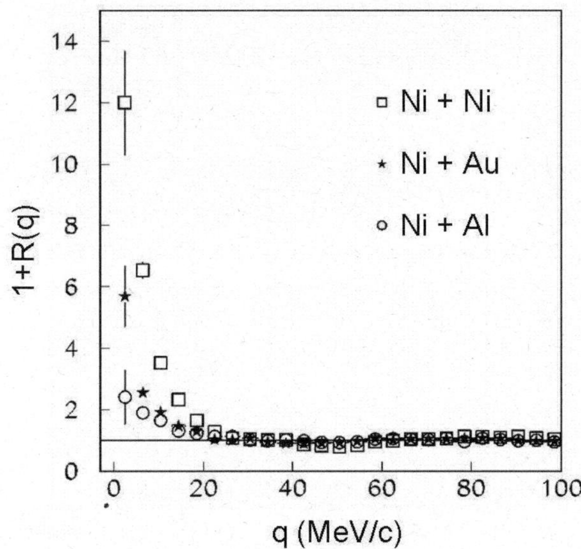

Fig. 11. Two-neutron correlation functions measured in Ni + Ni, Ni + Au and Ni + Al collisions at $E/A = 45\,\mathrm{MeV}$ by the CHIC Collaboration [47].

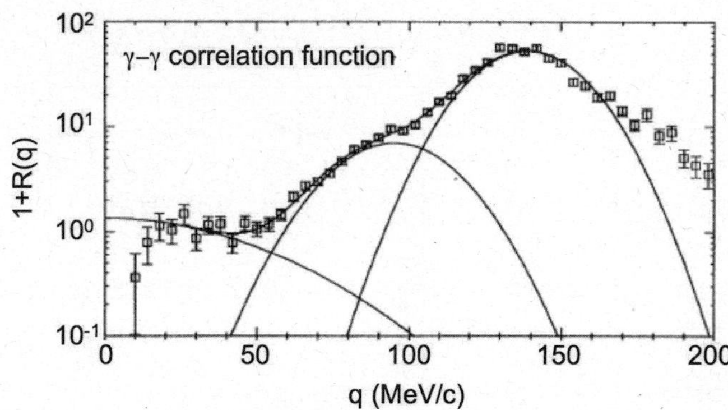

Fig. 12. Data points: two-photon correlation function measured in Ar + Al reactions at $E/A = 95\,\mathrm{MeV}$ with the MEDEA array [52]. The solid lines show calculations of the different effects contributing to the overall shape of the correlation function.

the timescale of the emission and to study the interplay between dynamical and statistical effects [41]. Similarly to what was observed in the case of protons, the height of this maximum increases with increasing two-neutron total momentum [41,46,47], indicating reduced timescales for the emission of more energetic particles emerging from the pre-equilibrium stage. The simultaneous study of p-p and n-n correlation functions in Ni + Al reactions at $E/A = 45\,\mathrm{MeV}$ by means of the Csorgo-Helgesson statistical model (ref. [41] and references therein) has provided comparable source radii of $R_G = 2.7 \pm 0.3\,\mathrm{fm}$ for both protons and neutrons. However, slightly larger timescales, with lifetimes of the order of $\tau_n \approx 600 \pm 200\,\mathrm{fm}/c$, appeared to be associated with neutron emission, if compared to protons emitted with lifetimes of about $\tau_p \approx 400 \pm 200\,\mathrm{fm}/c$ [41].

The enhancement at small relative momenta observed in the n-n correlation function in fig. 11 is due to the short-range nuclear final-state interaction. However, if neutrons are emitted from a long-lived decaying system, the increased average spatial separation between successive nucleons will reduce the significance of the final-state strong interaction. The correlation function is then expected to be dominated by quantum statistics effects. An attempt to observe such an effect was successfully performed by the authors of ref. [50] in the study of compound nuclei formed in $^{18}\mathrm{O} + ^{26}\mathrm{Mg}$ reactions at $E = 60$ and $71\,\mathrm{MeV}$. The n-n correlation function was constructed with the evaporated neutrons and a dip at small relative energies was observed. This anti-correlation is generated by the anti-symmetrization of the n-n wave function and represents the first experimental evidence of a pure Fermionic HBT effect. The n-n correlation length is longer than the nuclear interaction length and final-interaction effects are negligible.

Two-photon correlation functions have been measured with the TAPS and MEDEA arrays [51,52] to extract space-time information about γ-emitting sources. Figure 12 shows a two-photon correlation function measured in $^{36}\mathrm{Ar} + ^{27}\mathrm{Al}$ reactions at $E/A = 95\,\mathrm{MeV}$ [52]. The pure intensity interferometry effect is observed only at small relative momentum ($q < 45\,\mathrm{MeV}/c$). The peak at $140\,\mathrm{MeV}/c$ is due to the neutral pion decay into two photons while the bump at intermediate q-values is produced by badly detected pion decays [52]. Because of these complications, two-photon correlation functions were studied with sophisticated analysis techniques yielding Gaussian source sizes of the order of $r_0 \approx 2\,\mathrm{fm}$ and emission lifetimes of the order $\tau \approx 4\,\mathrm{fm}/c$. These results seem to indicate that γ-γ correlation functions probe mostly the dynamical stage of the reaction. Similar conclusions were also presented in ref. [53] where anti-correlations between energetic γ-rays and protons were studied with the MEDEA array in Ar + V collisions at $E/A = 44\,\mathrm{MeV}$. Energetic gamma rays can be considered very good probes of the spatial and temporal properties of the geometric overlap region developed in the very early stages of the collision process when nucleon-nucleon collisions are more important.

5 Space-time characterization from light complex particles

During heavy-ion collisions at intermediate energies a large variety of particles and fragments are produced. A complete space-time characterization of the reaction therefore requires multiple intensity interferometry studies extended to several nuclear species. Correlation functions between light charged particles other than protons have indeed been investigated by different authors (ref. [17] and references therein).

Figure 13 shows a d-α correlation function measured in the reaction $^{16}\mathrm{O} + ^{197}\mathrm{Au}$ at $E/A = 94\,\mathrm{MeV}$ [54]. The large minimum at small relative momentum is due to the mu-

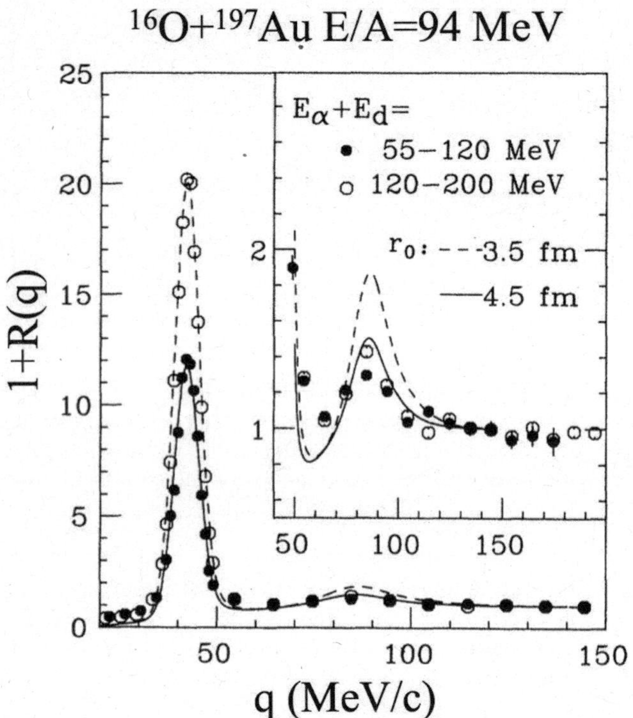

Fig. 13. Data points: deuteron-alpha correlation functions measured in ^{16}O + ^{197}Au reactions at $E/A = 94$ MeV [54] for different gates on the total kinetic energy of the particle pairs. The solid and dashed lines correspond to Gaussian analyses of the first peak after taking into account the finite angular resolution of the used experimental setup.

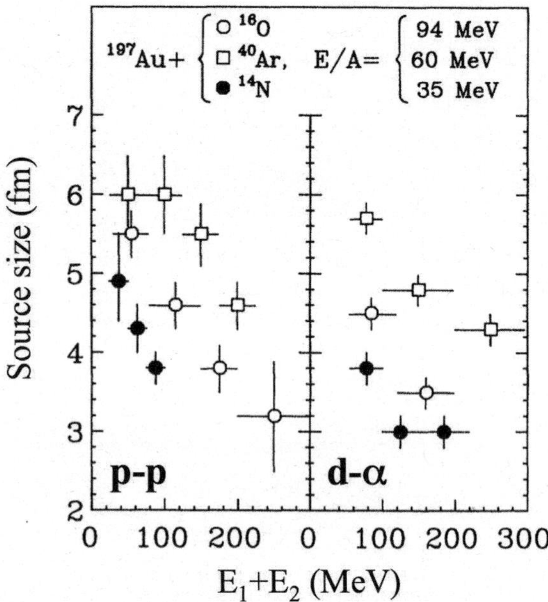

Fig. 14. Comparison between proton-proton (left side) and deuteron-alpha (right side) Gaussian source sizes measured in the same reaction systems and gated on the total kinetic energy of the particle pairs [17].

tual Coulomb repulsion and the peaks at relative momentum $q > 30$ MeV/c are due to the nuclear interaction. In particular, the sharp peak around 42 MeV/c corresponds to the first excited state of ^{6}Li at $E^* = 2.186$ MeV and the broad peak around 84 MeV/c stems mainly from the resonance at $E^* = 4.31$ MeV with small contributions from the resonance at $E^* = 5.65$ MeV. The solid and dashed lines in fig. 13 show d-α correlation functions calculated from eq. (4) with the kernel function, $K(q,r)$, constructed from the d-α relative scattering wave function [55]. The best fit of the integral of the first peak at 42 MeV/c provides Gaussian source radii $r_0 = 4.5$ and 3.5 fm for d-α pairs with total energies $E_\alpha + E_d = 55$–120 MeV and 120–200 MeV, respectively. As already observed in the case of two-proton correlation functions, the height of the peaks increases with increasing velocity of selected d-α pairs. A closer look at fig. 13 clearly shows that the whole shape of the d-α correlation function is not reproduced. Equation (4) does not fit simultanously the first peak at 42 MeV/c and the second peak at 84 MeV/c. The broad peak at 84 MeV/c is better described by using a source radius larger by about 1 fm as compared to the radius extracted from the best fit of the sharp peak at 42 MeV/c. These difficulties have not been resolved yet and require further research.

Attempts to perform systematic size measurements by using the first sharp peak at $q = 42$ MeV/c can be found in

the literature [17]. For instance, fig. 14 shows the comparison between source sizes extracted from p-p correlation functions (left side) and source sizes extracted from the best fit of the first peak at 42 MeV/c in d-α correlation functions (right side) measured in the same reaction systems, *i.e.* ^{16}O + ^{197}Au at $E/A = 94$ MeV, ^{40}Ar + ^{197}Au at $E/A = 60$ MeV and ^{14}N + ^{197}Au at $E/A = 35$ MeV (ref. [17] and references therein). Both p-p and d-α source radii decrease with increasing energy of the detected pairs. Furthermore, d-α radii appear to be smaller than p-p source radii. Even if this result seems to be intriguing and requires further research, the difficulties in reproducing the overall shape of the d-α correlation function described in fig. 13 indicate that these comparisons between d-α and p-p source radii must be taken very cautiously.

Examples of deuteron-deuteron correlation functions measured in Ar + Au at $E/A = 60$ MeV are shown in fig. 15 [56,57] for different gates in the total energy of the deuteron pairs (see different symbols). The anticorrelation at small relative momentum, due to the repulsive Coulomb interaction is the main visible feature of these observables. These correlation functions have been analyzed with final-state interaction models that include both the Coulomb and the nuclear interaction [56, 58–63]. Source radii from d-d correlation functions have been found to be generally larger than both p-p and d-α radii [17,56,59–61].

More recently d-d correlation functions from projectile-like sources produced in Xe + Sn reactions at $E/A = 50$ MeV [63] have been measured with the INDRA 4π array [64–66]. By means of a quantum model including both Coulomb and nuclear final-state interactions [67], deuteron emission times from the decay of the projectile-

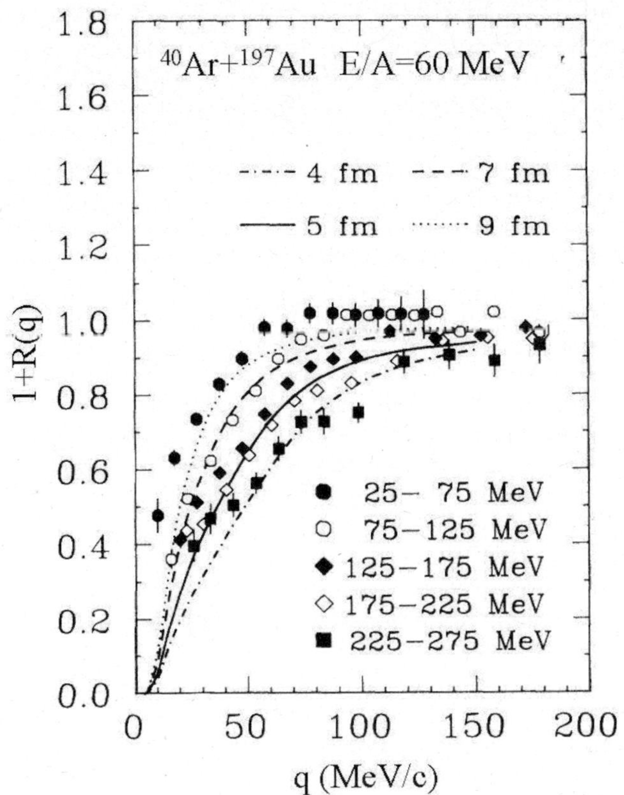

Fig. 15. Deuteron-deuteron correlation functions measured in ^{40}Ar $+$ ^{197}Au collisions at $E/A = 60$ MeV [56,57]. The different symbols correspond to different gates on the total kinetic energy of the detected deuteron pairs. The lines show calculated correlation functions constructed by considering both the Coulomb and the nuclear final-state interactions.

like fragment at different impact parameters were estimated. As the impact parameter decreases from peripheral to more central collisions, the emission times were observed to decrease from 200 fm/c up to 25 fm/c. This result was interpreted as an increasingly important contribution of out-of-equilibrium emission as one moves towards more and more central events where more excited projectile-like fragment are produced. A better angular resolution to measure the d-d correlation function at very low relative momentum [63] and an improved knowledge of the d-d final-state interaction [58] could improve our understanding of deuteron emission mechanisms.

Triton-triton correlation functions have been also studied by some authors with models including both Coulomb and nuclear final-state interactions or with simplified prescriptions using only the Coulomb repulsion between the coincident tritons [56,59,61,62]. The extracted source radii are comparable to those extracted from d-d correlation functions, even if uncertainties in the knowledge of the triton-triton nuclear interaction still exist [17].

In this section we have not discussed the possibility that the correlation function might be distorted by the Coulomb field of the residual system. Only the mutual two-body final-state interactions are indeed often used in the literature. This approximation is reasonable only in

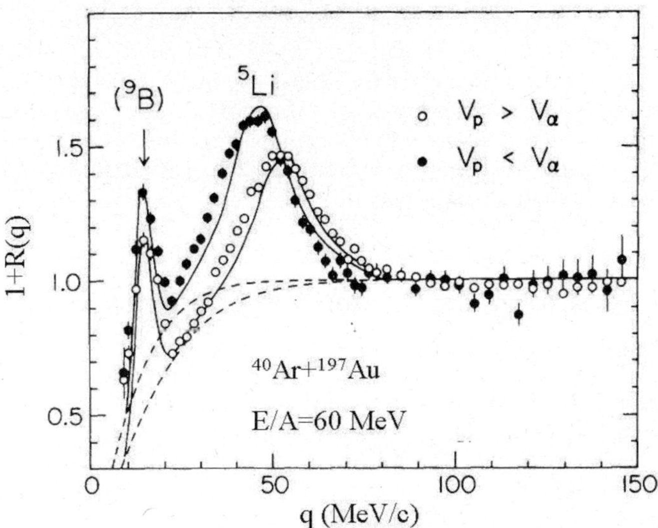

Fig. 16. Proton-alpha correlation functions measured in ^{40}Ar $+$ ^{197}Au collisions at $E/A = 60$ MeV [58,68]. The closed and open circles are obtained by choosing particle pairs in which the proton is, respectively, faster or slower than the alpha particle.

the case of correlations between particles with identical charge-to-mass ratios causing them to experience a similar acceleration in the Coulomb field of the residual system. Figure 16 shows proton-alpha correlation functions measured in ^{40}Ar $+$ ^{197}Au reactions at $E/A = 60$ MeV [58, 68]. The open (solid) points refer to particle pairs with the proton being faster (slower) than the alpha, $v_p > v_\alpha$ ($v_p < v_\alpha$). The correlation function displays two peaks at about 15 and 50 MeV/c. The first peak is due to the sequential decay of ^{9}B, $i.e.$ ^{9}B $\rightarrow p + {}^8$Be $\rightarrow p + (\alpha + \alpha)$ with detection of a proton and an α-particle only. The second peak at 50 MeV/c is related to the unbound ground state of ^{5}Li, which has a lifetime of about 130 fm/c.

While the location of the peak at 15 MeV/c is not altered by the gate on the proton and alpha-particle velocities, the second peak is observed at slightly higher relative momenta if pairs with $v_p > v_\alpha$ are selected. Since the charge-to-mass ratio of protons is greater than that of the alpha particles, the proton will experience a larger acceleration in the Coulomb field of the residual nuclear system. As a consequence, the relative velocity of the pair will be decreased if $v_p < v_\alpha$ and increased if $v_p > v_\alpha$, resulting in a shift of the peak position. After taking these distortions into account, p-α correlation functions have been analyzed in ^{14}N $+$ ^{197}Au at $E/A = 35$ MeV [56,58] and the extracted source radii appear to be comparable to the radii obtained from d-d and t-t correlations and larger than those extracted from d-α correlation functions.

The existing systematics [17] shows that two-particle source sizes change when different nuclear species are selected. The obtained differences need further investigations but they all point to the fact that different particles originate from different emitting stages and sources produced during the reaction. The complex scenario emerging from these studies calls for higher-resolution detectors and improved theoretical descriptions of final-state interaction models for two-body correlations in heavy-ion collisions. In the next sections we will discuss the work that has been done to try to understand time ordering in the emission of light particles. The study of emission chronology is indeed very important in order to improve our understanding of particle production mechanisms in heavy-ion collisions.

6 Particle emission time and chronology

Particles originating from one specific source are emitted during a certain time interval. Normally, the rate of emitted particles changes during this time interval, leading to a specific time distribution for the source. If a two-particle correlation function of identical particles could be constructed from particles emitted from a single source (*i.e.* if the particle source could be isolated), and if the spatial distribution were known, then the shape of the correlation function would yield information on the shape of the time distribution. Normally, neither the spatial nor the temporal distributions are known. In this case it is not straightforward to extract the spatial and time distributions from the shape of the correlation function. To interpret the experimental results, source models are often used. These source models contain some pre-assumption on the spatial and temporal distributions. The shape of these distributions can, to some extent, be varied by varying parameters of the model (such as radius and temporal width parameters). Such parameters then represent average emission point and average emission time, though it should be remembered that such average values are model dependent. In experimental data the situation is much more complex, since it is never possible to completely isolate one source from all the other sources present during the reaction. The contribution from several sources leads to a total time distribution with a complex shape, where the effective average emission time will depend on different factors.

What one is usually referring to with *emission chronology* is a difference in the average emission times between two particle types. Though, it is worth noticing that the emission times of the two particles may overlap to a large extent, and the difference in their respective average emission times can be small compared with the width of the emission time distributions. Furthermore, if different sources contribute, like in peripheral reactions, they contribute to a complex time distribution of the emitted particles. As an example, we can consider two particle types emitted from two sources. Different origins for different average emission times can be recognized: one origin could be a simple shift of two similar time distributions. Another origin could be that the width of the time distributions are different, while their shape is quite similar. Finally, it is also possible to think that, if more than one source is contributing, the relative weight of the sources is different for the two particle types, leading to different average emission times. In a real reaction, all of the above reasons contribute, with a different weight depending for example on applied gatesor selected angular ranges.

Even though the extraction of the emission times and sequences is quite involved, there is a wealth of experimental information that is available. A great advantage is to perform simultaneous measurements of both like- and unlike-particle correlation functions. By applying different gates, such as polar angle gates, total momentum gates, directional gates, or velocity gates for unlike particles, certain sources are enhanced relative to others. For instance, non-equilibrium emission can be enhanced by high and intermediate total momentum gates, while particles from evaporation and excited fragments can be suppressed by shape analyses disregarding the very low relative momentum region. Furthermore, single-particle information, such as energy spectra at different angles, should be used in the analysis. A systematic study can, therefore, to a large extent disentangle the space-time characteristics of the contributing sources, thereby putting strong constraints on theoretical models.

6.1 Emission chronology from like-particle correlations

The emission chronology between two particle types can, under certain conditions, be determined from like-particle correlation functions. If it is valid to assume that both particle types are emitted from the same spatial region, a fit with a calculated correlation function, based on some source function, to the experimental correlation function, will yield an average emission time for each particle type. By comparing these average emission times, an emission chronology can be inferred. For a review, see ref. [69] and references therein. The same method can also be used to determine differences in emission times for the same particle type emitted from different systems but with the same spatial region, for instance, different neutron emission times from systems similar in size and energy content, but with different isospin content (see next section).

The drawback of this method is that the results are sensitive to the assumption of emission from the same spatial region. Furthermore, the extracted average emission times are model dependent, since the average emission times depend on the pre-assumption of the shape of the (spatial and) temporal distributions assumed by the specific source model.

6.2 Emission chronology from unlike-particle correlations

Model independent information on the emission chronology of two particle types can instead be obtained from unlike-particle correlation functions. If there is a difference in the average emission times, it is possible, by suitable gates, to divide the particle pairs into two classes with different average distances when the two particles "start to interact". Since the strength of the final-state interaction depends on the distance between the particles, this will lead to a different strength of the correlation function for the two classes. By comparing these correlation functions, the emission chronology can then be inferred without any model assumptions.

Below we present the details of this technique. An important assumption for this method to be valid is that the particles are emitted independently. Furthermore, certain assumptions (which depends on the used gates) must be made on the spatial region from which the particles are emitted.

6.2.1 Velocity-gated correlation functions

A technique to probe the emission sequence and time delay of ejectiles in nuclear reactions was first suggested for charged-particle pairs, based on the idea that mutual Coulomb repulsion would be experienced by pairs of charged particles emitted with a short time delay. Comparison of the velocity difference spectra with trajectory calculations would thus give a measure of the average particle emission sequence [70,71,57].

The technique was extended to any kind of interacting, non-identical particle pairs in the theoretical study of ref. [72]. There it was demonstrated that the sensitivity of the correlation function to the asymmetry of the distribution of the relative space-time coordinates of the particle emission points can be used to determine the differences in the mean emission times by applying energy or velocity gates. This effect has been proposed for particle pairs such as pd and np [72], $p\pi$ [73], K^+K^- [74].

Velocity-gated correlation functions of non-identical particles is a very powerful tool to investigate emission sequences in nuclear collisions [75]. The basic idea is that, if there is an average time difference in the emission times of two particles types, there will also be a difference in the average distance for particle pairs selected with the condition $v_1 > v_2$ as compared to the pairs selected with the complementary condition $v_1 < v_2$. This is because the particle emitted first will, on average, travel a different distance in the two complementary classes (due to the different velocities) before the second particle is emitted. In particular, the interaction is enhanced for those pairs for which the average distance is smaller. This can be easily seen if one compares the correlation function $C_1(q)$, gated on pairs $v_1 > v_2$, with the correlation function $C_2(q)$, gated on pairs $v_1 < v_2$. If particle 2 is emitted earlier (later) than particle 1, than the condition $v_1 > v_2$ will sample smaller distances (larger correlations) than the complementary condition $v_1 < v_2$. Therefore, the ratio C_1/C_2 will show a peak (dip) in the region of relative momentum q where there is a correlation and a dip (peak) where there is an anti-correlation. Furthermore, the ratio C_1/C_2 will approach unity both for $q \to 0$ (since the velocity difference of the two emitted particles is negligible) and $q \to \infty$ (since modifications of the two-particle phase-space density arising from final-state interactions are negligible). The exact location of the peak and dip in the ratio depends on the source, and in particular on the origin of the difference in the average emission times.

6.2.2 Experimental results

At low energies, the analysis of two-particle correlations and velocity difference spectra has allowed to find results consistent with the statistical compound nucleus decay for the light products emitted from the reactions $^{16}O + ^{10}B$ ($E_{lab} = 62.5\,\mathrm{MeV}$), and $^{16}O + ^{12}C$ (64 MeV). The method could also be used for the determination of fission timescales [76].

At intermediate energies, the time sequence of p and d has been deduced for the $E/A = 50\,\mathrm{MeV}$ Xe + Sn reaction studied by the INDRA Collaboration. An average emission of deuterons $\approx 250\,\mathrm{fm}/c$ earlier than protons has been explained as the result of averaging over a long time sequence between pre-equilibrium and thermal emission for protons, whereas deuteron emission, resulting mainly from hard nucleon-nucleon collisions, is concentrated at a few tens of fm/c [63].

First experimental evidence of the emission chronology of neutrons and protons deduced from the np correlation function, was reported in ref. [75] for the $E/A = 45\,\mathrm{MeV}$ $^{58}Ni + ^{27}Al$ reaction measured by the CHIC Collaboration at LNS, Catania. The experimental results from differently gated correlation functions were in qualitative agreement with the Koonin-Pratt formalism [23,24]. It was claimed that for events selected for high parallel velocity and high total momentum which enhance projectile residue and/or intermediate velocity sources the proton is, on average, emitted earlier than the neutron.

In ref. [77] the emission time chronology of neutrons, protons, and deuterons was presented for the $E/A = 61\,\mathrm{MeV}$ $^{36}Ar + ^{27}Al$ reaction measured at KVI, Groningen. The experimental results showed that the angular and total-momentum dependences of the pp and np correlation functions support a dissipative binary reaction scenario, where early dynamical emission is followed by statistical evaporation. The reverse kinematics utilized in the experiment, and fairly high-energy thresholds, enhanced the early dynamical emission component in the backward measurement. The analysis of velocity-gated correlations of non-identical particle pairs yielded detailed information about the particle emission time sequence. The results from the np backward measurement are shown in fig. 1, right column (from ref. [77]). The dip in the C_n/C_p ratio indicates that neutrons are, on the average, emitted earlier than protons. For the forward measurement (fig. 17, left column), the shape of the correlation function exhibits a correlation at $q < 40\,\mathrm{MeV}/c$ and a small anti-correlation at $40 < q < 100\,\mathrm{MeV}/c$, and the pairs with $v_n > v_p$ contributing to C_n interact more strongly. This indicates that protons are, on the average, emitted earlier than neutrons, a result in agreement with that obtained for the $\theta_{lab} \approx 45°$ $E/A = 45\,\mathrm{MeV}$ $^{58}Ni + ^{27}Al$ reaction [75]. The complete sequence of average emission times, τ, extracted from the $E/A = 61\,\mathrm{MeV}$ $^{36}Ar + ^{27}Al$ reaction was the following: for the dynamical emission source, $\tau_n < \tau_d < \tau_p$; for the projectile residue emission, $\tau_d < \tau_p < \tau_n$. The interpretation of these results, presented in ref. [77], highlights the importance of the contribution from the different emission sources.

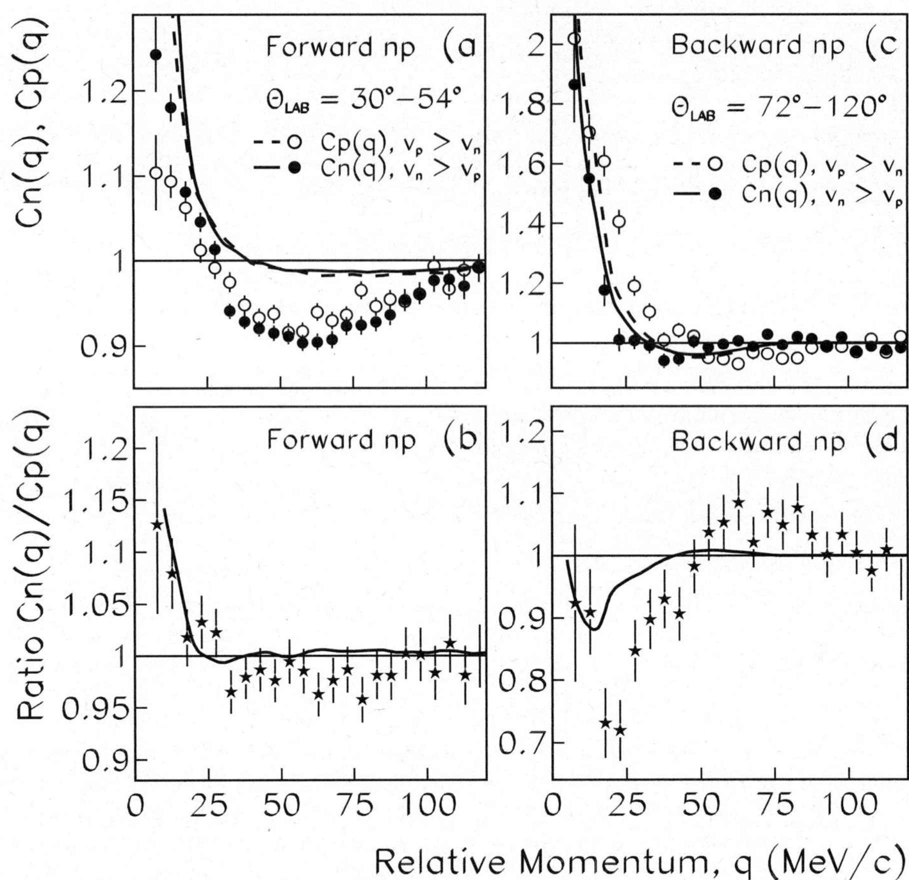

Fig. 17. From the $E/A = 61\,\mathrm{MeV}\ ^{36}\mathrm{Ar} + \mathrm{Al}$ reaction measured at KVI, velocity-gated np correlation functions (C_n, filled circles, and C_p, open circles) and their ratio (C_n/C_p). Left column: forward measurement. Right column: backward measurement. The lines are results from model calculations. From ref. [77].

The technique of ref. [72] has also been applied to pairs of non-identical light charged particles produced in central collisions of heavy ions in the $A = 100$ mass region at a beam energy of $400\,\mathrm{MeV/nucleon}$, measured with the FOPI detector system at GSI [78]. The difference between longitudinal correlation functions with the relative velocity parallel and anti-parallel to the center-of-mass velocity of the pair in the central source has allowed the extraction of apparent space-time differences of the emission of the charged particles. Comparing the correlations with results of a final-state interaction model delivered quantitative estimates of these asymmetries. Time delays as short as $1\,\mathrm{fm}/c$ or —alternatively— source radius differences of a few tenth fm were resolved. The strong collective expansion of the participant zone introduces not only an apparent reduction of the source radius but also a modification of the emission times. After correcting for both effects a complete sequence of the space-time emission of p, d, t, $^3\mathrm{He}$, α particles was extracted.

At even higher-energy regimes, the above method has been used to tackle the problem of the possible observation of strangelets in the frame of the distillation process following the creation of a quark gluon plasma. In this case, strange and anti-strange particles may not be pro-duced at the same time in a baryon-rich system under low bag constant scenarios. Such a prediction has been tested using K^+K^- correlations in ref. [79].

The knowledge of the emission times and chronology is very important in order to extract information on the nuclear interaction from complex nuclear collisions. It is, however, not straightforward to obtain emission times and chronology from a nuclear reaction, and different methods need to be combined. Using velocity gates is a relative-ley new and promising method to, model independently, extract the emission chronology. Examples of first experimental results using velocity gates have been presented, together with their interpretations. To exclude other explanations, the method needs to be combined with other probes and methods to draw the correct conclusions. This is a field where further developments can be made.

7 Isospin effects and perspectives for the asy-EOS

The isospin dependence of the nuclear equation of state is probably the most uncertain property of neutron-rich matter. This property is essential for the understanding

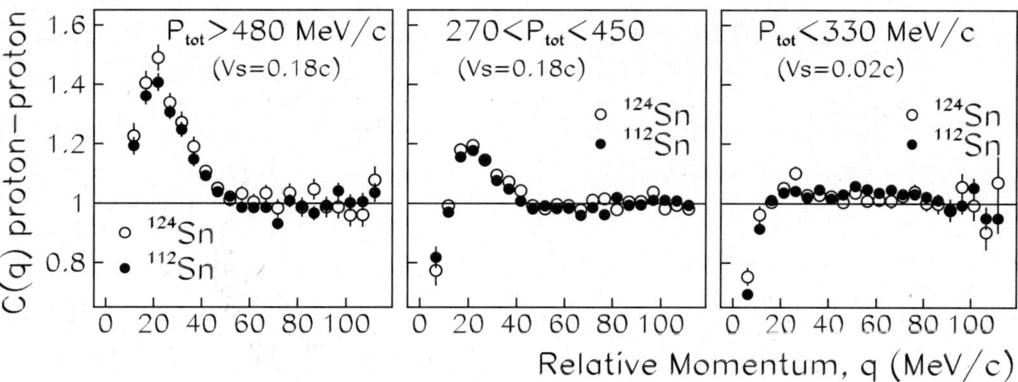

Fig. 18. For ^{36}Ar + ^{112}Sn (filled circles) and ^{36}Ar + ^{124}Sn (open circles) collisions, pp correlation functions. Left: pp, high P_{tot}, $30° \leq \theta \leq 42°$. Middle: pp, intermediate P_{tot}, $30° \leq \theta \leq 42°$. Right: pp, low P_{tot}, $54° \leq \theta \leq 114°$. From ref. [85].

of extremely asymmetric nuclei and nuclear matter as it may occur in the r-process of nucleosynthesis or in neutron stars. In order to study the isospin-dependent EOS, heavy-ion collisions with isotope-separated beam and/or target nuclei can be utilized. In these collisions, excited systems are created with varying degree of proton-neutron asymmetry and a noticeable isospin dependence of the decay mechanism is expected [80]. By selecting semi-peripheral collisions, the symmetry term of the EOS is expected to be probed at low densities, while central collisions are expected to be sensitive to the high-density dependence.

7.1 Theoretical predictions

Recently, the two-nucleon correlation function has been considered as a probe for the density dependence of the nuclear symmetry energy [81–83]. In these theoretical studies with an isospin-dependent IBUU transport model, it was shown that the density dependence of the symmetry term of the EOS affects the temporal and spatial structure of reaction dynamics by affecting the average emission times of neutrons and protons as well as their relative emission sequence. For central collisions, a stiff EOS causes high-momentum neutrons and protons to be emitted almost simultaneously, thereby leading to strong correlations. A soft EOS delays proton emission, which weakens the np correlation [81,82]. It was found that the symmetry energy effect becomes weaker with increasing impact parameter and incident energy. Also, the strength of the nucleon-nucleon correlation function is reduced in collisions of heavier reaction systems as a result of larger nucleon emission source [82]. It was further found that the momentum dependence of both isoscalar nuclear potential and the symmetry potential influences significantly the space-time properties of the nucleon emission source. Specifically, the momentum dependence of the nuclear potential reduces the sensitivity of two-nucleon correlation functions to the stiffness of the nuclear symmetry energy [83].

In spite of the large uncertainties in the symmetry potential, and in particular in the momentum-dependent part, the exploratory studies in [81–83] are very encouraging for using two-particle correlations to study the sym-

metry potential. It is also important to remember that the symmetry interaction does not only influence the dynamical emission of particles from the overlap and neck-like regions, but also the formation of the residues. Therefore also the particles emitted from the residues contain information on the symmetry energy. To make improvements in the future on the understanding of the symmetry potential, it is necessary to have models that consistently can describe both the dynamical emission of particles and the formation of residues and their subsequent pre-equilibrium and equilibrium emission of particles. By applying the experimental conditions and systematically comparing calculated and experimental energy spectra and gated correlation functions for like and unlike particles, it will be possible to obtain hard constraints on the symmetry potential.

7.2 Experimental results

First experimental results have been obtained on two-particle correlation functions from systems similar in size, but with different isospin content [84]. Small-angle two-particle correlation functions with neutrons and protons have been obtained from semi-peripheral $E/A = 61$ MeV Ar + ^{112,124}Sn collisions measured at the AGOR cyclotron of KVI.

The emission from the different sources was enhanced or suppressed by introducing angular cuts (intermediate velocity source emission is enhanced at forward, and target residue emission at backward angles) and cuts in the total momentum (P_{tot}) of the particle pair, calculated in the relevant emission source frame. Figure 18 (from ref. [85]) presents the pp correlation function for particle pairs selected within the three different gates.

1. Particles emitted by the intermediate velocity source at prompt dynamical emission stage, (*e.g.*, first-chance nucleon-nucleon collisions), are enhanced by selecting high-P_{tot} pairs in the intermediate velocity source frame. For the sample of pp pairs, the angular range $30° \leq \theta \leq 42°$ is used for this gate.

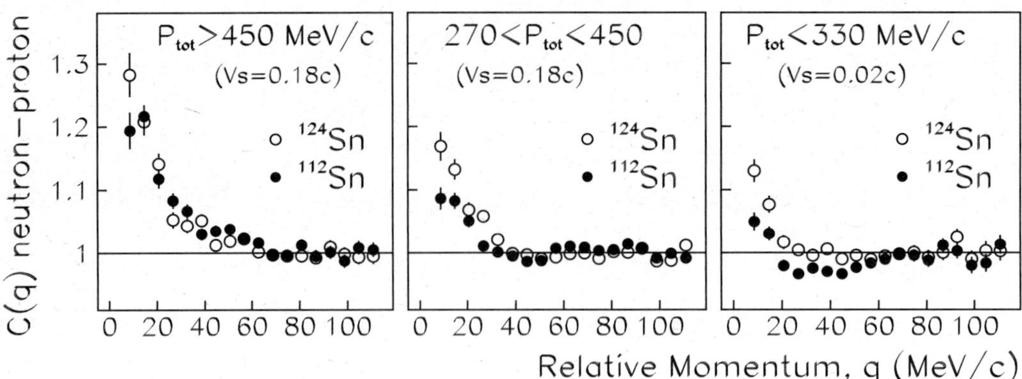

Fig. 19. For ^{36}Ar + ^{112}Sn (filled circles) and ^{36}Ar + ^{124}Sn (open circles) collisions, np correlation functions. Left: np, high P_{tot}, $30° \leq \theta \leq 120°$. Middle: np, intermediate P_{tot}, $30° \leq \theta \leq 120°$. Right: np, low P_{tot}, $54° \leq \theta \leq 120°$. From ref. [85].

2. Particles emitted by the intermediate velocity source at a later dynamical emission stage (*e.g.*, neck emission), are enhanced by selecting intermediate-P_{tot} pairs in the intermediate velocity source frame. Again, the angular range $30° \leq \theta \leq 42°$ is utilized for pp pairs.
3. Particles emitted by the target residue are enhanced by selecting low-P_{tot} pairs in the target residue frame. The angular range $54° \leq \theta \leq 120°$ is applied to both pp and np pairs. The neutron energy threshold is set to 8 MeV for this gate.

One can notice that the height of the peak at $q \approx 20\,\mathrm{MeV}/c$ is progressively reduced going from gate 1 to gate 3, indicating an increase in the particle emission time. Figure 19 presents the corresponding results for the np correlation function. By comparing the results for the two Sn targets, one can see an isospin effect, particularly in gate 1 for pp and in gates 2 and 3 for np, where the height of the correlation function is larger for the more neutron-rich target. This indicates a shorter average emission time for this system.

For the interpretation of the correlation data, it is important to note that the correlation function depends on the space-time extent of the emitting source. From the size of the source, a stronger correlation is expected for the smaller ^{36}Ar + ^{112}Sn system, an effect expected also because of the larger excitation energy per particle available for this system (yielding a shorter emission time). On the other hand, the change in neutron number implies a different symmetry energy which also affects the n (and p) emission times. Neutrons are expected to be emitted faster in the neutron-rich system, which would lead to an enhancement of the correlation strength for ^{36}Ar + ^{124}Sn. Thus, the net influence on the correlation function is not easily predictable, both due to the uncertainty in the symmetry energy and to the presence of more than one source of emission. The stronger np correlation observed for the larger Ar + ^{124}Sn system in the low total momentum gate may indicate that the more asymmetric system generates a more asymmetric and excited target residue that, consequently, decays on a faster timescale.

More insight into these results has been gained by performing an analysis of the particle emission time sequence in ref. [86]. In all studied angle and total-momentum gates, it was found that neutrons are, on average, emitted earlier than protons. Furthermore, the shorter np emission timescale for the Ar + ^{124}Sn system results from a faster emission of the neutrons, as compared to the Ar + ^{112}Sn system. This is particularly true for the particles emitted from the target residue, indicating that the residues in the two reactions were formed differently due to the symmetry interaction. Further experimental results of particle emission sequence involving deuterons indicate that for the Ar + ^{124}Sn system neutrons are emitted slightly earlier than deuterons, again a result pointing to a faster neutron emission for the more neutron-rich system. Deuterons, being formed mainly by coalescence, appear to have emission times that fall in-between that of neutrons and protons. No sizeable isospin effects in the emission sequence of deuterons and protons were found for the above-mentioned systems.

8 Accessing the space-time properties at freeze-out

IMF-IMF correlation functions are expected to provide the space-time properties of nuclear systems produced in nuclear reactions at the time when they approach the freeze-out stage. This can offer a unique opportunity to better understand the mechanisms of multifragmentation phenomena and their possible links to a liquid-gas phase transition in nuclear matter.

IMF-IMF correlation functions are usually constructed by combining together fragments with different charges and masses into a single correlation function. This is accomplished by sorting all IMF-IMF pairs with respect to the so-called reduced velocity, $v_{red} = v_{rel}/\sqrt{Z_1 + Z_2}$, where Z_1 and Z_2 are the charges of the two IMFs [87–94]. Typical measured correlation functions are shown in figs. 20-24 and 26.

Theoretically, final-state interaction models including only the Coulomb repulsion have often been used to analyze IMF-IMF correlation functions. Indeed, fragment spatial separations are on average expected to be larger than

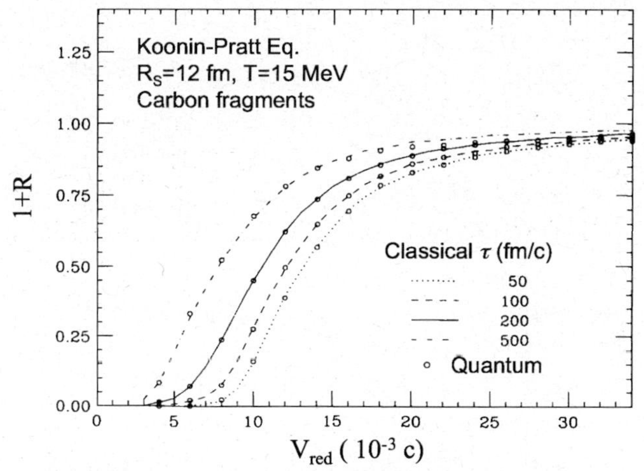

Fig. 20. IMF-IMF correlation functions calculated for different emitting-source lifetimes [87]. The lines and dots correspond, respectively, to classical and quantum calculations.

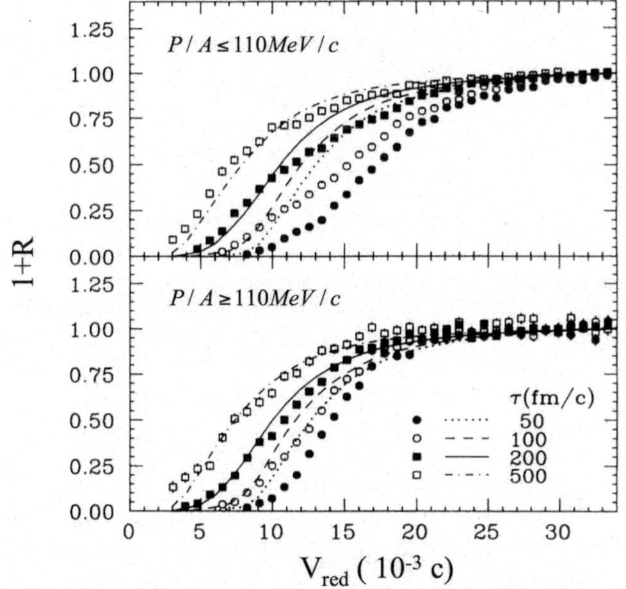

Fig. 21. Two-body (lines) and three-body (dots) trajectory calculations for IMF pairs with total momenta per nucleon, $P/A \leq 110\,\mathrm{MeV}/c$ (top panel) and $P/A \geq 110\,\mathrm{MeV}/c$ (bottom panel). The calculations are performed for different values of the IMF emission lifetimes.

the range of the nuclear force. This approximation needs certainly to be verified within the context of more quantitative models. However, the most prominent feature observed in IMF-IMF correlation functions is represented by the strong Coulomb anti-correlation at small relative velocities (see figs. 20-24, 26). Possible resonances corresponding to the decay of heavier fragments either occur at higher relative velocities not easily accessed experimentally or require higher energy and angular resolution in order to be clearly resolved. Such limitations limit the study of IMF-IMF correlation functions by means of simplified Coulomb interaction models. The IMF-IMF Kernel function in eq. (4) has been calculated either from

the relative Coulomb wave function or by using a classical Coulomb treatment leading to the simple expression, $K(r,q) = (1 - r/r_c)^{1/2} - 1$, with $r_c = 2\mu Z_1 Z_2 e^2/q^2$ and Z_1 and Z_2 being the charges of the two coincident fragments [87]. The emitting-source function is commonly assumed to have a spherical shape with radius R_S and it is assumed to be characterized by a temperature T. Fragments are then emitted according to an exponential time profile, $P(t) \propto \exp(-t/\tau)$. The radius, R_S, and the lifetime, τ, are deduced as free parameters from the best fit of the experimental correlation functions, thus providing the space-time characterization of IMF emission. Figure 20 shows correlation functions calculated with sources having emission lifetimes between 50 and 500 fm/c. It can be observed that the shape of the correlation function at small reduced velocities ($V_{red} < 30 \cdot 10^{-3}\, c$) is strongly affected by fragment emission times. The calculations in fig. 20 are performed by using both the classical (lines) and the quantum (dots) two-body kernel function in eq. (4), providing virtually identical results [87]. The width of the Coulomb anti-correlation at small reduced velocities is used to extract IMF emission times.

Alternative 3-body approaches to IMF-IMF correlation functions are based on trajectory calculations where two IMFs are emitted sequentially with some average time delay, and their motion is propagated to the detectors under the influence of both their mutual final-state Coulomb interaction and the repulsion induced by the Coulomb field of the residual system [88–90]. The correlation function is then deduced from eq. (2) applied to the simulated coincidence and single-particle yields [88,89,91]. Figure 21 is taken from ref. [87] and shows the comparison between two-body classical calculations (lines) and three-body Coulomb trajectories calculations (points) for a source having charge $Z_S = 93$, radius $R_S = 12\,\mathrm{fm}$, temperature $T = 15\,\mathrm{MeV}$ and emission lifetimes varying between $\tau = 50\,\mathrm{fm}/c$ and $\tau = 500\,\mathrm{fm}/c$. The upper and lower panels show the results for carbon pairs emitted with total momenta, $P/A \leq 110\,\mathrm{MeV}/c$ and $P/A \geq 110\,\mathrm{MeV}/c$, respectively. The use of a very large mass number for the source ($A_S = 10000$) ensures that recoil effects are not taken into account and only the effects of the residual Coulomb field are explored. For low-momentum gates, fragments are emitted with small initial velocities and distortions in the Coulomb field of the source are large. Indeed, significant differences between three- and two-body calculations exist up to $\tau = 200\,\mathrm{fm}/c$. For longer emission lifetimes, the two-body approach still remains valid. As it can be easily expected, in the case of fragment pairs at high total momentum (lower panel), fragments move so quickly that the effects induced by the Coulomb field of the residual system acting as a third body can be neglected. The two-body approach is found to be still reasonable if the emission times are longer ($\tau > 200\,\mathrm{fm}/c$). For shorter emission times, the disagreement is substantial and three-body correlations seem to be important, even if the effect is less pronounced than in the case of low-momentum fragments (upper panel in fig. 21). Three-body trajectories and even more realistic N-body Coulomb tra-

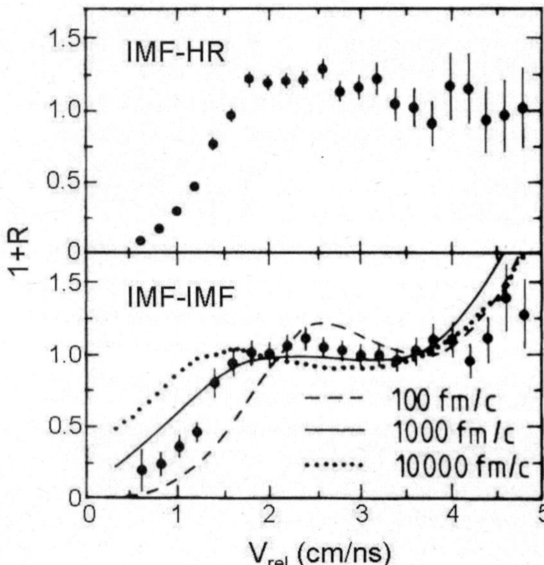

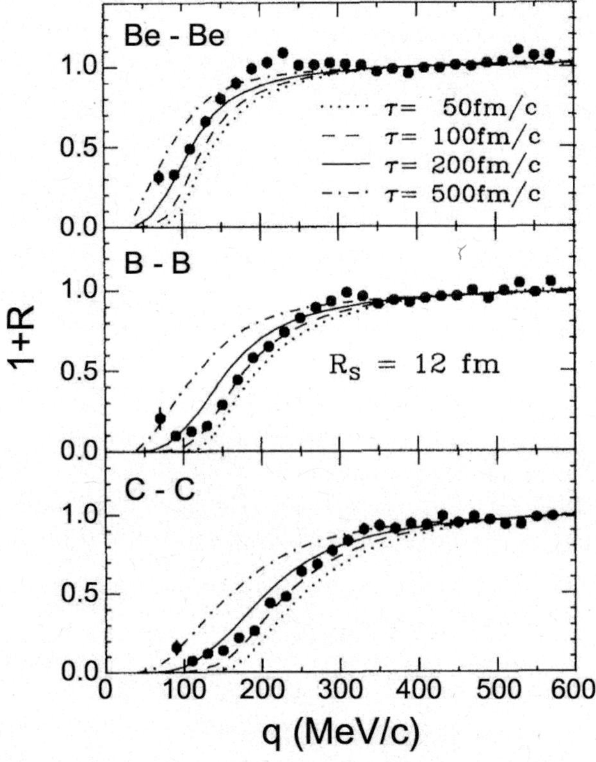

Fig. 22. Bottom panel: the data points show the IMF-IMF correlation function measured in ^{18}O $+$ ^{nat}Ag reactions at $E/A = 84\,$MeV [88] in events where at least two IMS fragments are detected. The lines correspond to calculations performed with three-body Coulomb trajectories assuming different IMF emission lifetimes. The top panel refers to events where only one IMF and one heavy residue (HR) are observed and shows their correlation function.

Fig. 23. Data points: beryllium-beryllium (top panel), boron-boron (middle panel) and carbon-carbon (bottom panel) correlation functions measured in Au $+$ Au central collisions at $E/A = 35\,$MeV [89]. Lines: calculations performed with eq. (4) using a classical approximation to the kernel function and assuming different emission lifetimes and a spherical source size of $12\,$fm.

jectory calculations have extensively been used in the literature [88, 92–94].

More recently, IMF-IMF correlation functions have also been compared to predictions of microscopic models where both the Coulomb and the nuclear interactions are taken into account [95, 96]. Microscopic models provide a more realistic approach and represent certainly a promising opportunity for the future in order to have a clear and unambiguous space-time characterization of complex fragment emission mechanisms.

9 Characterizing multifragmentation phenomena

Understanding whether multifragmentation results from a sequence of binary splittings or rather from a simultaneous break-up of an excited nuclear system has certainly represented one of the main questions raised in the last two decades. A sequential binary splitting would correspond to cluster emission from the surface of an excited source (similar to fission). This process is associated with long emission times of the order of 10^{-20}–10^{-21} s, necessary for shape deformation. In contrast, if multifragmentation corresponds to a simultaneous breakup of nuclear matter, the system is expected to fall apart over shorter times (10^{-22}–10^{-23} s), comparable to the timescales involved in the growth of density fluctuations in the spinodal instability region of the nuclear phase diagram. In the following part of this section we will present some of the main re-

sults that can be found in the literature on the study of fragment emission times in heavy-ion collisions.

One of the earliest studies of fragment-fragment correlation functions was presented in ref. [88] where the authors have studied ^{18}O $+$ ^{197}Au and ^{18}O $+$ ^{nat}Ag reactions at $E/A = 84\,$MeV. The IMF-IMF correlation function in events where two or more IMFs were observed (fig. 22, bottom panel), was compared to the correlation function IMF-HR (IMF-Heavy Residue, fig. 22 top panel) in events where only one IMF and one heavy fragment were observed. The lines in the bottom panel of fig. 22 show the results of three-body Coulomb trajectory calculations. IMF emission times of the order of $\tau = 1000\,$fm/c were obtained in both classes of events selected by the authors. These results suggest that the production of more than two IMFs in the studied reactions is characterized by a mechanism similar to the evaporation of one IMF by a heavy fragment. This result supports a sequential binary evaporation mechanism as responsible for IMF emission.

A similar analysis was performed also in ref. [90] where incomplete-fusion reactions with the production of two or three heavy ($Z \geq 10$) fragments in ^{22}Ne $+$ ^{197}Au collisions at $60\,$MeV/u were investigated. The analysis of relative velocities and angles in the center of mass of the coincident fragments was found to favour a sequential emission of

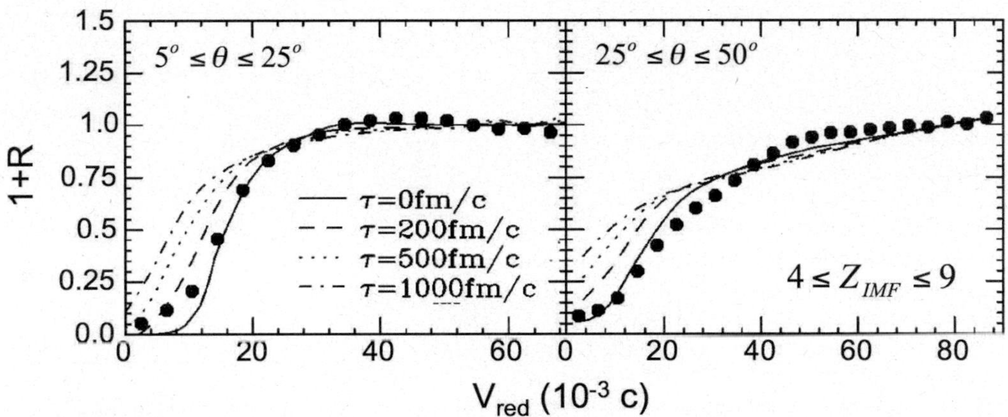

Fig. 24. Data points: IMF-IMF correlation functions measured in central collisions ^{86}Kr + ^{93}Nb at $E/A = 50$ MeV at polar angles $5° \leq \theta \leq 25°$ (left panel) and $25° \leq \theta \leq 50°$ [93]. N-Coulomb trajectory calculations are shown by lines and correspond to different emission lifetimes for the detected fragments.

fragments with short time steps between two consecutive binary decays [90].

The data points in fig. 23 show Be-Be, B-B and C-C correlation functions measured in central Ar + Au collisions at $E/A = 35$ MeV at polar angles $12° < \theta < 35°$ [89]. This angular region allowed the authors to select IMFs produced from incomplete-fusion mechanisms (ref. [89] and references therein). The curves in fig. 23 represent calculations performed with eq. (4) using a classical approximation for the two-fragment Coulomb kernel function [89]. IMFs were assumed to be emitted from a spherical source of radius $R_S = 12$ fm and with emission times distributed according to an exponential law, $P(t) \propto \exp(-t/\tau)$. The comparison to experimental data suggests that fragments are emitted with timescales of about $\tau \approx 100$–200 fm/c. Comparable emission times were obtained with three-body Coulomb trajectory calculations where the IMFs move towards the detectors under the influence of both their mutual final-state Coulomb interaction and the repulsion induced by the Coulomb field of the residual system [89]. The obtained emission times are shorter than the timescales characteristic of evaporation processes from compound nuclei. An analysis of total-momentum–gated IMF-IMF correlation functions in the same set of data showed that high-energy IMFs are produced with even shorter emission times, of the order of 50 fm/c, thus indicating that fragment emission in central collisions begins in early stages of the reaction and continues throughout the later equilibrium stages [89,90].

The reaction system ^{36}Ar + ^{197}Au has also been studied at incident energies, $E/A = 50$ MeV, using an N-body Coulomb approach [92]. Angle-averaged correlation functions contain space-time ambiguities that are difficult to resolve: the wide minimum of the correlation function at small reduced velocities cannot be associated with a unique combination of source size, R_S, and emission time, τ. A long-lifetime emission produces emitting sources elongated in the longitudinal direction defined by the total momentum vector, **P**. Similarly to the case of p-p correlation functions already discussed previously, IMF-IMF directional correlation functions were used to extract emission times, $\tau = 50$ fm/c, from the surface of a dilute source having a density of about $\rho/\rho_0 \approx 0.4$ [92].

Central collisions between nearly symmetric systems, ^{86}Kr + ^{93}Nb, at $E/A = 50$ MeV were studied in ref. [93] with the goal of disentangling instantaneous from sequential multifragmentation break-up scenarios. Figure 24 shows correlation functions constructed with IMF fragments having charges $4 \leq Z_{IMF} \leq 9$ detected at polar angles $5° \leq \theta_{lab} \leq 25°$ (left panel) and $25° \leq \theta_{lab} \leq 50°$ (right panel). N-body Coulomb trajectory calculations (lines in fig. 24) indicated that the data are consistent with very short emission times, $\tau < 100$ fm/c. The authors of ref. [93] have also observed that the study of kinetic-energy–gated IMF-IMF correlation functions is sensitive to the details of the fragment emission topology (surface/volume emission processes) in the initial state. However, no definitive quantitative results could be deduced because of the unknown sensitivity to other correlations and conservation laws in the initial state [93].

The perspective of extracting emission times from IMF-IMF correlation functions seems to be quite attractive in order to better understand even the evolution of emission mechanisms with the excitation energy deposited into the excited system undergoing the multifragment decay. The study of such evolution with the excitation energy has been performed both in central collisions at intermediate energies and in peripheral collisions at relativistic energies. In the next section we will present the results obtained from IMF emission time measurements in central collisions where the excitation energy is controlled by the incident beam energy. In the following section we will show similar emission time studies extended to the decay of excited target spectators after bombardment by relativistic-energy light probes. In this case, the decay of the system is mostly governed by the deposited excitation energy, without strong contributions from collective motion that represent a relevant phenomenon in the case of central collisions. These two sections will allow us to have an idea about how the emission times of IMFs are correlated with the degree of excitation of the decaying system.

9.1 Emission timescales in central collisions: sensitivity to the incident energy

Mean emission lifetimes for multifragment final states produced in Kr + Nb reactions at incident energies $E/A = 35$, 45, 55, 65 and 75 MeV/nucleon were studied in ref. [97]. The width of the Coulomb dip in the IMF-IMF correlation function at small reduced velocities was observed to increase as the bombarding energy increases from $E/A = 35$ to $E/A = 55$ MeV. This result is consistent with a reduction of IMF emission times as the violence of the collision is increased. The measured IMF-IMF correlation functions were compared to classical three-body Coulomb trajectory calculations performed with the code MENEKA [98] and the extracted emission times are represented in fig. 25 as a function of the incident energy. Long emission times, $\tau \approx 400$ fm/c, are observed at lower incident energies, $E/A = 35$ MeV. As the incident energy is increased, the emission of IMFs is observed to occur over shorter timescales, until a saturation at a value of $\tau \approx 100$ fm/c is observed at $E/A = 55$ MeV [98]. This result is consistent with an evolution of the fragment emission mechanism from long-lived sequential processes, at lower incident energies, to an almost simultaneous scenario, at higher incident energies. The extracted minimum lifetime, $\tau \approx 100$ fm/c, is comparable to the timescales of growing density fluctuations for nuclear matter in the low-density instability regions of the nuclear phase diagram.

The results shown in fig. 25 are based on the assumption that the fragments are emitted from a system with a single freeze-out condition well localized in time. This scenario might be too simplified. Figure 26 shows IMF-IMF correlation functions measured in Kr + Au reactions at $E/A = 55$ MeV for three different values of v_{min}, defined as the minimum velocity of the less energetic fragment of each pair. It is clearly seen that faster IMFs exhibit a wider Coulomb hole, indicating shorter emission timescales. Similar results were observed also at $E/A = 35$ and 70 MeV [99]. The dependence of the emission lifetime on the velocity of the emitted fragments can be considered as an indication that a single freeze-out condition for fragment emission might not exist. This indication was confirmed by the calculations performed with the Expanding Evaporating Source model [100] that predicted decreasing emission times for fragments with higher velocities [99]. Models based on a simultaneous multifragmentation scenario were not capable of reproducing such evolutionary emission pattern. A similar evolutionary scenario in multifragment emission is also supported by the extraction of different emission times for different nuclear species [101]. For instance, in the study of Kr + Au collisions at $E/A = 70$ MeV, emission times of about 50 fm/c and 100 fm/c were found, respectively, for carbon and beryllium isotopes [101].

The described results obtained in the study of central collisions have provided evidence for short emission times

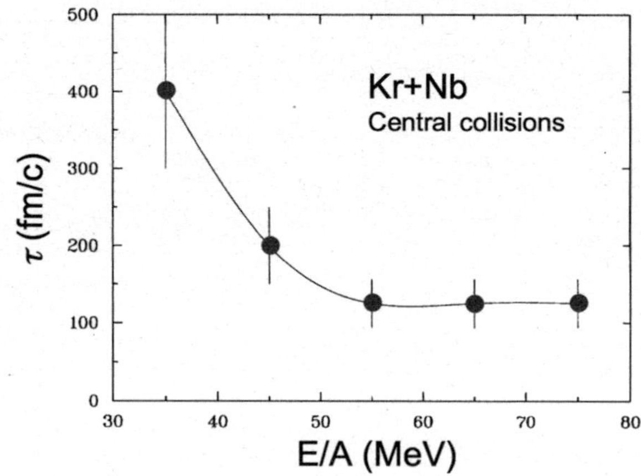

Fig. 25. IMF emission lifetimes extracted in the study of central Kr + Nb collisions at incident energies between $E/A = 35$ and 75 MeV [97].

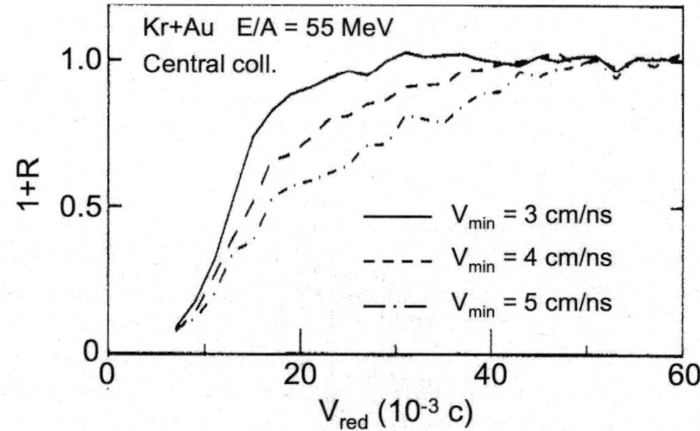

Fig. 26. IMF-IMF correlation functions measured in Kr + Au central collisions at $E/A = 55$ MeV and corresponding to different gates on the velocity of the slower of the two coincident fragments [99].

at higher energies. While this finding suggests a simultaneous scenario for multifragmentation, velocity-gated and Z-gated correlation studies suggest that evolutionary decay processes cannot be excluded [99–101]. The definition of a freeze-out stage might be considered more involved than expected, requiring further experimental and theoretical investigations. In this respect, it would be important to compare experimental correlation functions to predictions of microscopic models that describe the whole dynamical evolution of multifragmenting nuclear systems. Similar studies have recently been performed [96, 102] by comparing IMF-IMF correlation functions for central Xe + Sn and Gd + U collisions at $E/A = 32$ MeV, measured with the INDRA detector, to simulations performed with the BOB model (Brownian One-Body dynamics) based on the BNV (Boltzmann-Nordheim-Vlasov) approach [95, 103, 104]. The authors have studied higher-order correlations by selecting different decay channels based on fragment charges and specific event topologies [96, 102]. The conclusions of these studies seems to support freeze-out times of the order of 200–240 fm/c and fragment spatial

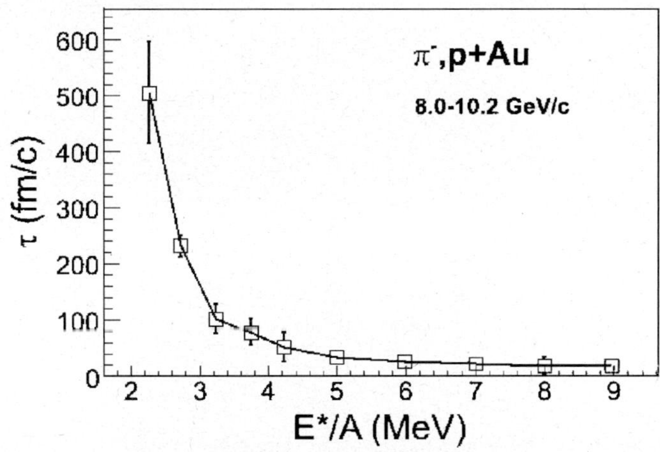

Fig. 27. IMF emission lifetimes as a function of the deposited excitation energy measured in target spectators decay induced by the bombardment π^- and p beams at incident momentum of 8.0 and 10.2 GeV/c [94].

distributions of the order of 3–4 V_0 in Xe + Sn and 8 V_0 in Gd + U reactions, with V_0 being the volume of the source at normal density, $V_0 = (4\pi/3)(1.2)^3 A_{tot}$ fm³.

The study of central collisions is generally complicated by the presence of collective motion. The explosive nature of such systems can decrease the effective IMF emission times. It is, therefore, not easy to correlate the IMF emission times shown in fig. 25 with the deposited excitation energy.

In the next subsection we will present measurements of IMF emission times in target spectators where estimates of the actual thermal deposited excitation energy were made. Such studies are less dependent on the dynamical effects induced by collective motion and can provide better links to the occurrence of a liquid-gas phase transition in nuclear matter driven by thermal effects.

9.2 Evolution of emission times with excitation energy in target spectators

Fragmentation phenomena of target-spectator fragmentation induced by light probes at relativistic energies are one of the best tools to investigate thermally driven phase transitions. In the study of π^-, p + Au at 8.0, 8.2, 9.2 and 10.2 GeV/c, IMF-IMF correlation functions from the decay of Au target spectators were studied [94]. The evolution of IMF emission times with the excitation energy per nucleon is represented in fig. 27. As the deposited excitation energy increases, emission lifetimes decrease from $\tau \approx 500$ fm/c at excitation energies E^*/A 2.5 MeV to a saturating value of about $\tau \approx 20$–50 fm/c for excitation energies above 5 MeV/nucleon. These results indicate a transition from a surface evaporation-like emission at low excitation energies towards a bulk simultaneous multifragmentation scenario above excitation energies of the order of $E^*/A = 5$ MeV [94]. Furthermore, the extracted emission times seem to be comparable with timescales of thermodynamical fluctuations leading to liquid-gas phase

transitions in nuclear matter. The decreasing emission times should be related to the increasing thermal excitation energy deposited into the system. Contrary to the case of central collisions presented in the previous subsection, only a small collective-motion component caused by the thermal expansion of the system should exist and the dynamics of the decay is mostly dominated by thermodynamical aspects.

10 Internal excitation energy of complex fragments

Light-particle–fragment correlation functions have been used to perform a thermal characterization of the emitting source by providing experimental information on the size and the internal excitation energy of the primary fragments [105–109].

Several statistical and dynamical models predict very different internal excitation energies of the primary fragments. The quantum molecular dynamic (QMD) model [4, 110–112] and the microcanonical metropolis Monte Carlo (MMMC) model [113] predict a rather low excitation energy for primary fragments. Anti-symmetrized molecular dynamics [114–116] (AMD), stochastic mean-field simulations [95,103,104,117], the statistical multifragmentation model [118,119] (SMM) and microcanonical multifragmentation models [120–122] predict moderately hot primary fragments. Moreover, the mechanism responsible for the formation of fragments differs from a model to another. In dynamical models, the formation of the fragments and their excitation energy depend not only on the collision dynamics but also on the procedure employed for performing the cluster recognition [123–125]. On the other hand, in the case of statistical models, the excitation energy of the fragments is an assumption of the model itself used to calculate the statistical weights of the partitions. Therefore, experimental estimates of the importance of secondary decays and of the size of the primary fragments can provide very significant constraints and crucial tests for the theoretical models. A direct measurement of these quantities would be also interesting for various aspects: the extraction of certain physical information, such as the rate of statistically to dynamically emitted particles or the caloric heat capacity [14,15], would be less dependent on model assumptions.

Previous studies have shown that it is possible to extract the intrinsic properties of the fragments independently of the mechanism of their formation. These studies are based on the measurement of relative-velocity correlation functions between fragments and light charged particles (LCP). The aim of this section is to give a review of the work dedicated to the determination of the sizes and excitation energies of primary fragments. An excitation function of these quantities will be given for Xe + Sn system for the incident energy range $E/A = 30$–50 MeV [105, 107–109]. We will present results on the study of central Kr + Nb collisions at $E/A = 45$ MeV [106] and of quasi-projectiles produced in peripheral Xe + Sn collisions at

$E/A = 100\,\mathrm{MeV}$ [126]. These studies were conducted by using either the Washington University Dwarf Ball/Wall array [127] or the 4π INDRA multidetector [64–66]. It is important to note that the same data can provide information on the space-time extent of the outgoing fragments, but in this section we will concentrate only in the measurement of their excitation energy.

10.1 Reconstruction of the primary fragments

At intermediate energies, there are at least two extreme mechanisms for the production of light particles, i) in nucleon-nucleon collisions and ii) by statistical evaporation from excited sources. The former occurs during the first stage of the collision when the dynamics play an important role. This process is called direct or preequilibrium emission. The second process is supposed to occur at later stages of the reaction as a secondary emission from excited primary fragments. Between these two stages, continuous emission may occur from other sources that are difficult to define. In multifragmentation events, it is possible to isolate the secondary statistical component if the fragments formed are not too excited, so that the timescale associated with their decay is much larger than the timescale of their production.

In this context the primary fragment excitation energy can be deduced from the multiplicity of its associated evaporated LCPs. Figure 28, taken from ref. [106], shows the projection of the relative velocity between the heaviest fragment and the protons detected in coincidence and in the case of Kr + Nb collisions at $E/A = 45\,\mathrm{MeV}$. The components $V_{\parallel}$ and $V_{\perp}$ are the projections of the relative velocity onto the axis representing the fragment direction in the center-of-mass frame and into a plane perpendicular to that axis. The top panel corresponds to raw coincidence data, the middle panel corresponds to the background distribution and the bottom panel displays the result of a subtraction of this background from the total LCP emission. This bottom panel clearly shows a ring surrounding the IMF location in velocity space ($V_{\parallel} = V_{\perp} = 0$). This feature may correspond to the Coulomb ring associated with the proton emission from the heaviest fragment. In order to estimate quantitatively the amount of evaporated protons from secondary-decay processes, one needs to estimate the contributions induced by other background effects that have to be properly subtracted. Few methods have been employed to perform such background estimation and subtraction [105–109] and they are all based on a scenario deduced from Boltzmann-Nordheim-Vlasov [128] calculations. Multifragmentation is described as a two-step process. The first step is the cooling of the initial fused system through a sequence of light-particle-emission processes. The second step is the fragmentation of the smaller remaining source where the remaining excitation energy is shared between a fixed number of primary fragments. These fragments then decay sequentially while moving apart under the influence of the Coulomb force. An initial radial velocity can be added to the Coulomb motion in order to mimic a possible expansion of the

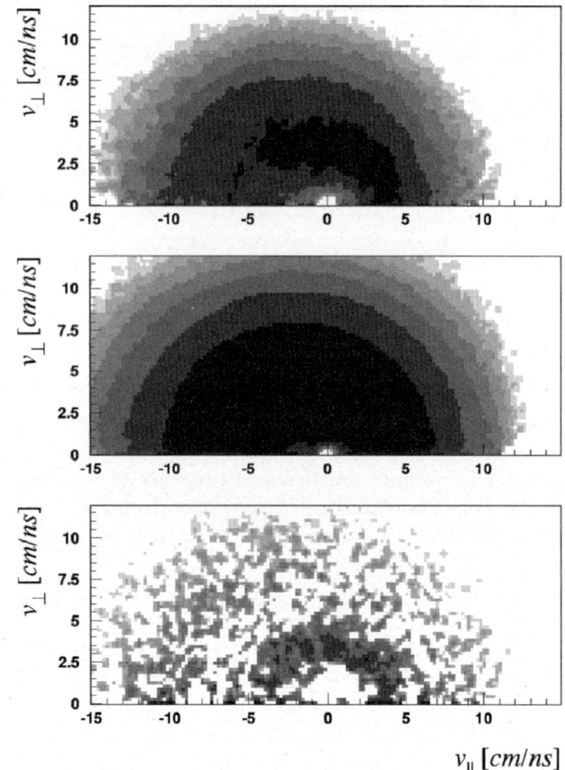

Fig. 28. Reaction system: Kr + Nb at $E/A = 45\,\mathrm{MeV}$. Parallel velocity-transverse velocity invariant diagram for protons in the center of mass of the heaviest fragment detected in coincidence in each event. Top panel: correlated events; middle panel: uncorrelated events; bottom panel: difference between correlated and uncorrelated events [106].

source. Based on such a scenario and on model predictions, two approaches have been suggested to estimate the background of the correlation function, differing in the way one compares calculations to experimental data. In one case, the shape of the background is deduced, normalized to the data and then subtracted [105, 107–109]. In the other case, the input parameters of the calculations are adjusted until the data points are reproduced (kind of back-tracing) [106]. After the background is subtracted, one can access the amount of evaporated particles from the primary fragments. In the following subsection we will describe how the technique is applied to experimental data and what physical information has been extracted.

10.2 Application to data and experimental results

Figure 29 shows the experimental relative velocity correlation function (a) and the difference function (b) of the phosphorus-alpha pairs measured in the central collisions Xe + Sn at $E/A = 32\,\mathrm{MeV}$. To build the correlation function shown in this figure, eq. (2) has been used by replacing the momenta $\mathbf{p}_1$ and $\mathbf{p}_2$ with the velocities $\mathbf{v}_1$ and $\mathbf{v}_2$ of a given fragment (here is a phosphorus) and the LCP (here an α-particle). The uncorrelated events have been

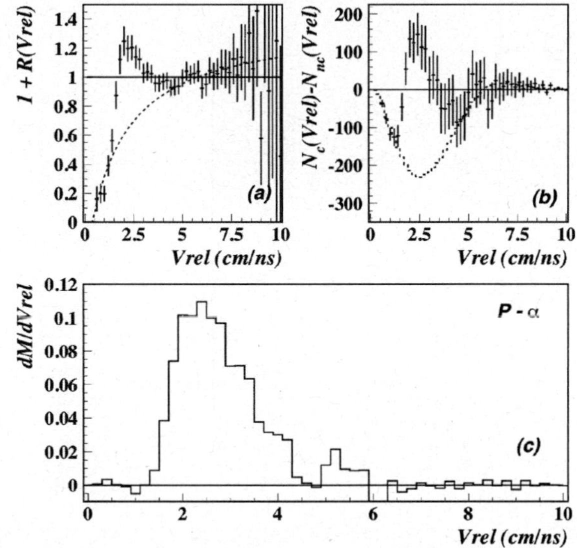

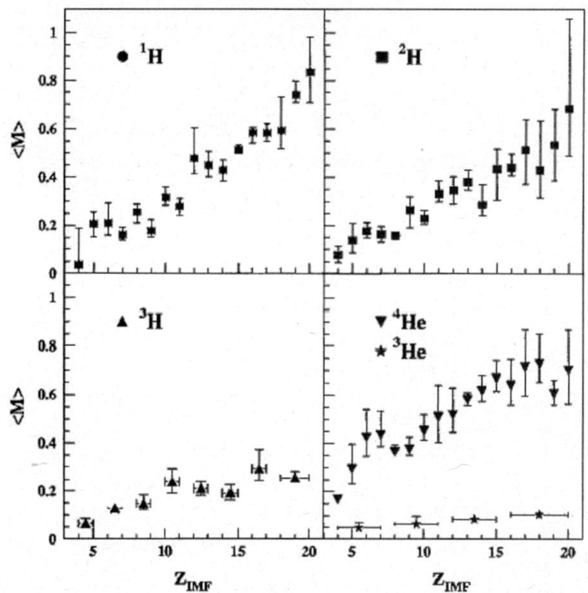

Fig. 29. (a) Phosphorus-alpha correlation measured in central collisions of $Xe + Sn$ at $E/A = 32$ MeV. (b) Difference function. (c) Velocity spectrum of α-particles in the center of mass of the phosphorus fragments, obtained by subtracting the background (dashed line in (b)) from the the difference function (data points in (b)), from ref. [109].

Fig. 30. Average secondary multiplicities per fragment of the evaporated hydrogen and helium isotopes as a function of the atomic number of the detected fragments for $Xe + Sn$ central collisions at $E/A = 50$ MeV [105].

constructed by the event-mixing technique: For a given fragment in an event i having n LCP, we take randomly n LCP from n different events, then the relative velocity of this fake event is calculated. The dotted lines represent the background [105–109]. This background should contain any other contribution, which is not evaporated from parents of phosphorus. The relative velocity distribution presented in fig. 29c, obtained by subtracting the background from the difference function, can be considered as the deduced contribution of α-particles evaporated by parents of phosphorus.

This procedure was applied to all fragment-LCP pairs produced in the central collisions $Xe + Sn$ at $E/A = 32$, 39, 45 and 50 MeV. Thus, for each detected fragment, the average multiplicities of the light particles evaporated by the primary fragment was determined. These multiplicities increase slightly with the size of the fragment. However they remain weak, not exceeding the value of 1.5 at all incident energies, implying that the excitation energy of the primary fragments is particularly moderate. An example of the extracted multiplicity is given in fig. 30 for central collision of $Xe + Sn$ at $E/A = 50$ MeV. The multiplicity of a given LCP does not change much with incident energy. From the spectra of the evaporated light charged particles, the average kinetic energy has been extracted.

Both observables (average multiplicity and kinetic energy) have been used in order to deduce the average charge of the primary fragments, $\langle Z_{pr} \rangle$, their average mass, $\langle A_{pr} \rangle$ and their excitation energy, E^*. The average charge of the primary fragment is calculated as the sum of the charge of the detected fragment, Z, and the charge of all light charged particles correlated to that fragment, $\langle Z_{LCP} \rangle$. This last quantity is defined by $\langle Z_{LCP} \rangle = \sum_i z_i \cdot \langle M_i \rangle$,

where z_i and $\langle M_i \rangle$ are the charge and the average multiplicity of the evaporated particles, $i = p, d, t, {}^3\text{He}, \alpha$. Since the isotopes of the detected fragments are not resolved and the neutrons are not detected, two extreme assumptions were necessary in order to determine the primary mass of the fragment. The first assumption consists of assuming that the detected fragment is produced in the valley of stability. The second assumption considers two cases: i) the primary fragments are produced in the valley of stability; ii) the primary fragments keep the same ratio N/Z as the initial system. The obtained charge of the primary fragments varies between 1 and 5 units of charge in addition to the charge of the detected fragment. The mass of the primary fragments depends also on the considered assumptions. The average multiplicity of the neutrons is deduced by the mass conservation, knowing the mass of the primary fragment, of the detected fragment and that of the secondary light particles. Multiplicities of about 7 are reached, but depend strongly on the assumption made on the mass of the primary fragment.

Finally, with the help of all these variables, it was possible to apply a calorimetry method to determine the excitation energy of the primary fragments E^*_{pr}. Figure 31 shows the average excitation energy per nucleon E^*_{pr}/A of the primary fragment as a function of its detected charge, and this for four incident energies and the two assumptions on the mass of the primary fragment. The horizontal line in fig. 31 represents the average value $\langle e^*_{pr} \rangle$ within the range of all studied primary fragments. Besides nuclei with low charges, all the experimental points lie on this straight horizontal line within error bars. In other words, whatever the incident energy and the mass assumption considered,

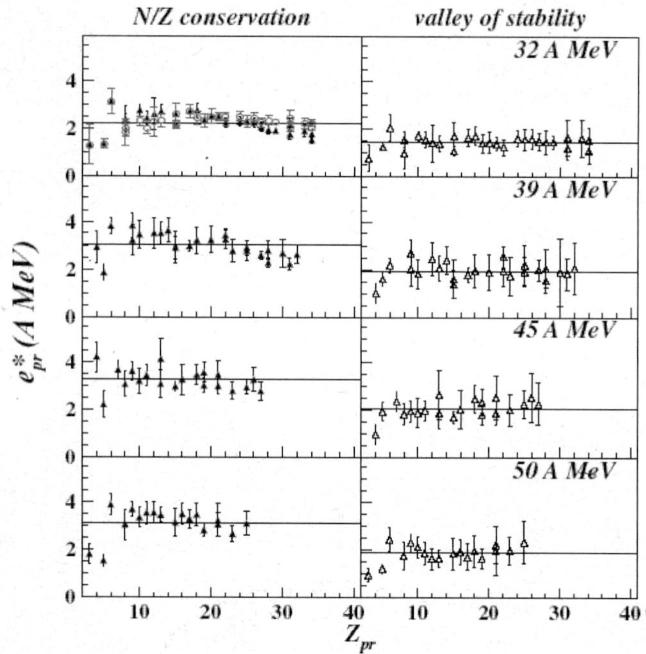

Fig. 31. Average excitation energy per nucleon of the primary fragments as a function of their charge for central Xe + Sn collisions at $E/A = 32$, 39, 45 and 50 MeV. The horizontal lines represent a fit to the average excitation energies with a Z-independent value for each bombarding energy. Left panels: the primary fragments have the same N/Z as the combined system. Right panels: the fragments are produced in the valley of stability. The masses of detected fragments are assumed to follow the valley of stability except for the data points represented by open circles where the EAL assumption is used [109].

the excitation energies per nucleon of the primary fragments are constant. Figure 32 shows the evolution of the average value with the bombardment energy. The vertical bars represent the standard deviations from the mean values. They are small and do not exceed 1 AMeV, consolidating the constancy of the value $\langle e^*_{pr} \rangle$. In the case of the N/Z conservation assumption, the excitation energy per nucleon increases from 2.2 AMeV at $E/A = 32$ MeV up to a saturation value of 3 AMeV at $E/A \geq 39$ MeV. In the case of the other assumption, where the primary fragments are produced on the valley of stability, the values saturate also but at lower energy. The constancy observed in fig. 32 of the excitation energy per nucleon for the various primary fragments can suggest that thermodynamical equilibrium was reached during the disintegration of the system. On the other hand, the saturation of $\langle e^*_{pr} \rangle$ can indicate that the fragments have reached their limiting excitation energy per nucleon (or their limiting temperature) [129,130].

How to choose between the two assumptions of mass on the primary fragments? This choice was dictated by calculations with a statistical code GEMINI [131]. This code is very well suited at low excitation energies not exceeding 3–4 AMeV. The procedure consists in using the deduced primary-fragment characteristics, Z_{pr}, A_{pr} and E^*_{pr} as input parameter, letting them decay and compar-

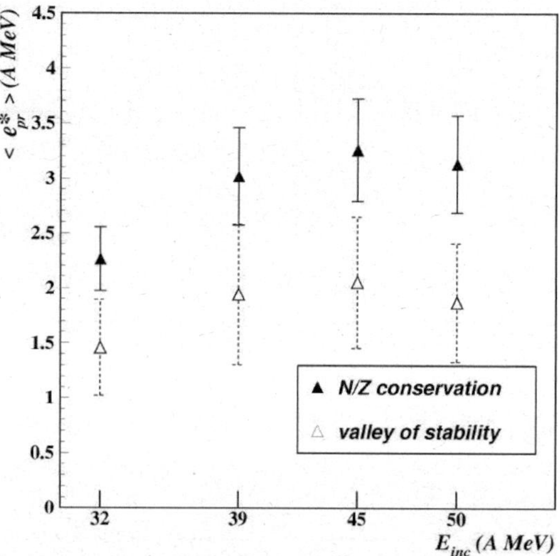

Fig. 32. Average excitation energy per nucleon of primary fragments produced in central collisions of the Xe + Sn system as a function of incident energy. The black and open symbols correspond to two hypotheses of the fragment masses. The vertical lines represent the standard deviation from the mean values [109].

ing the resulting multiplicities of the evaporated particles to the experimental values. It was shown in ref. [105] that the assumption of the N/Z conservation was reasonable.

The average multiplicities of secondary light charged particles allowed an intrinsic characterization of the primary fragments, but it can also be used to give valuable information on multifragmentation events. Indeed, the ratio of the secondary evaporated LCP multiplicities to the total detected LCP multiplicities gives the fraction of thermally produced particles. It has been shown in refs. [108, 109] that the maximum proportion of evaporated particles does not exceed, on average, 35% of the total number of produced light charged particles. For incident energies between $E/A = 32$ and 39 MeV, the proportion of evaporated particles increases, thus reflecting the increase of the excitation energy of the fragments observed in fig. 32. Beyond $E/A = 39$ MeV, this proportion decreases to reach 23% at $E/A = 50$ MeV while the excitation energy of the primary fragments remains constant. It should be noted that the extracted proportion of secondarily evaporated particles constitutes a lower limit, because it does not contain the contributions which can come from the disintegration of the unstable nuclei like the ^{8}Be, ^{5}Li etc. and the decay of the short-lived excited states [132]. Similar results have been extracted also from the study of Kr + Nb reactions at $E/A = 45$ MeV [106]. It has been shown that i) the excitation energy of the primary fragments does not exceed 2.5 AMeV, ii) about 80% of the detected LCPs do not originate from the secondary statistical decay of the primary fragments.

10.3 Comparison to models

The described techniques have been applied to the quasi-projectile (QP) nuclei formed in Xe + Sn collisions at

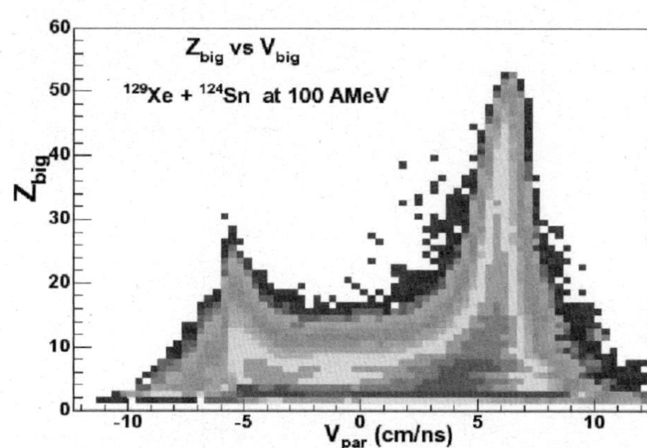

Fig. 33. Atomic number of the heaviest fragment in the event as a function of center-of-mass parallel velocity, for Xe + Sn at $E/A = 100\,\mathrm{MeV}$ reaction.

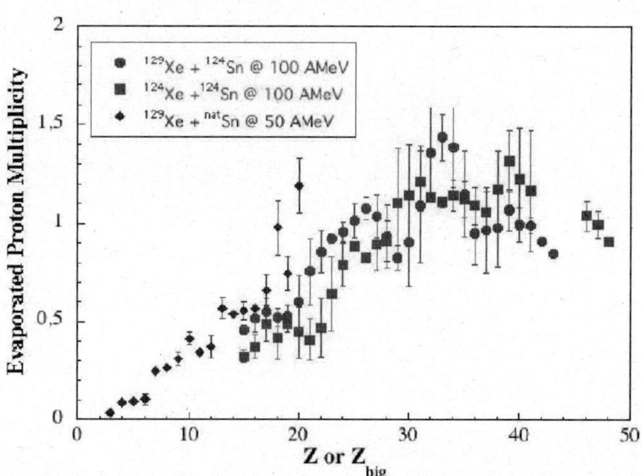

Fig. 34. Evaporated proton multiplicity as function of the atomic number of the emitter: quasi-projectile or fragments formed in central collisions. The systems presented here are indicated in the figure.

$E/A = 100\,\mathrm{MeV}$ in order to extract their thermal characteristics [126]. Figure 33 gives an overview of the collision. It shows the atomic number of the heaviest fragments in the event as a function of its velocity parallel to the beam direction. One can observe clearly two main components, one centered at the projectile velocity and the other one centered at the target velocity (the heavy quasi-target fragments are not detected since their velocity do not exceed the energy threshold of the used detectors). Between the two components one can also observe an emission of fragments at mid-rapidity and characterized by a low charge and a high cross-section.

This preliminary result is surprising since the excitation energies of the spectator at $E/A = 100\,\mathrm{MeV}$ and participant fragments at $E/A = 50\,\mathrm{MeV}$ are almost the same. Is it an indication of thermal energy saturation? Or are we dealing with the same fragment production mechanism? The question is open.

The correlation function method has been applied to extract the secondary evaporated light charged particles from each QP. Figure 34 shows results for two systems having different isospin: $^{129}\mathrm{Xe} + ^{124}\mathrm{Sn}$, $^{124}\mathrm{Xe} + ^{124}\mathrm{Sn}$. The figure represents the extracted proton multiplicity as a function of the charge of the QP emitter. The data points obtained above for central collisions and same system but at a lower beam energy $E/A = 50\,\mathrm{MeV}$ are superimposed on the same figure. All data points follow the same systematic: the evaporated proton multiplicity increases with the charge of the quasi-projectile or with the charge of the fragment (in the case of central collisions). The proton multiplicity values do not exceed 1.5 and are compatible with an excitation energy of 2–3 AMeV.

The experimental estimate of the secondary-decay component can be compared with the predictions of statistical multifragmentation models such as the SMM [118, 119] or the micro-canonical model of Raduta *et al.* [120–122] which explicitly consider fragment excitation. The comparison of the extracted quantities with the calculations can indeed constitute a crucial test of the basic assumptions of these models. In the MMMC [113] approach,

the excitation of primary fragments is implicitly included in the neutron production at freeze-out, and a direct comparison is not possible.

Statistical Multifragmentation Model (SMM) calculations have been performed in order to reproduce the thermal component. The version of SMM used in [108, 109] gives access to the freeze-out configuration, *i.e.*, to primary-fragment characteristics before secondary decay. Figure 35 shows the results of this calculation. The left panel in this figure represents the excitation energy in MeV of the primary fragments as a function of their atomic number (same data points as in fig. 31 with the only difference being the use of MeV/nucleon as units of excitation energies). The right panel represents the total charge contributions of secondary evaporated particles, Z_{LCP}. The data indicate that a maximum LCP evaporation is obtained for $E/A = 39\,\mathrm{MeV}$ and a decrease of thermal contribution takes place above this energy. Although the internal excitation energy of the primary fragments is well reproduced at each incident energy, the trend of the total charge of secondary LCPs, Z_{LCP}, with the beam energy is not reproduced. This discrepancy can be understood if we consider the increasingly important effects of the collision dynamics as the beam energy increases. Direct emissions of LCP increase with increasing incident energy, while the proportion of the thermal contribution decreases.

The obtained results have also been compared to dynamical calculations. In order to understand the mechanism responsible for the saturation of the excitation energy of the primary fragments observed in fig. 33, AMD calculations [114–116] for central collisions of Xe + Sn at $E/A = 50\,\mathrm{MeV}$ have been performed. A reasonably good agreement for the charge distribution, the charge of the heaviest fragment in the event and average kinetic energy of the fragments has been obtained [108, 126, 133]. The comparison of such calculations to the experimental

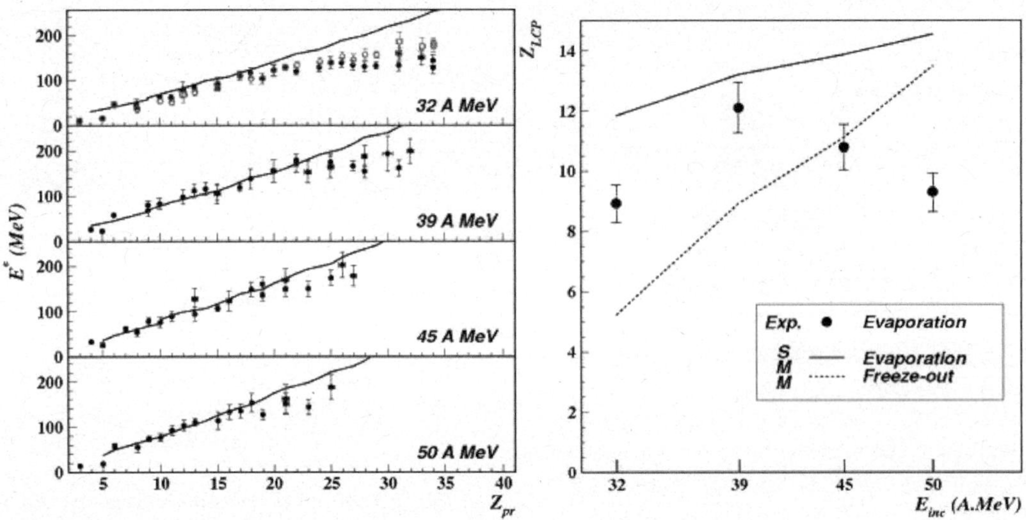

Fig. 35. Comparison between SMM calculations (curves) and data (symbols) are presented. Left panel: average excitation energy of the primary fragments as a function of their atomic number for central Xe + Sn collisions at $E/A = 32\text{–}50\,\text{MeV}$. Right panel: total charge contributions of secondary evaporated particles Z_{LCP}. The dotted histogram represents the calculated freeze-out contribution [109].

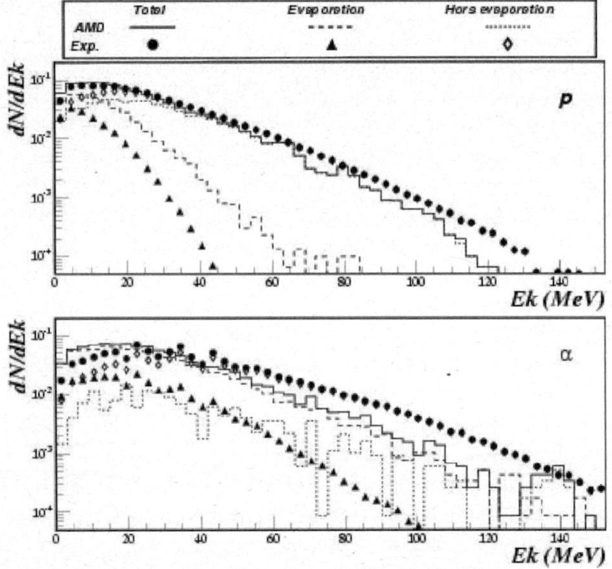

Fig. 36. Proton and alpha energy spectra produced in central collisions of Xe + Sn at $E/A = 50\,\text{MeV}$. The symbols are the data and curves represent AMD calculations. The dots (continuous histogram) represent the total energy spectra, the triangles (dashed histogram) represent the secondary evaporation contribution and the open symbols (dotted histogram) represent the direct early emission.

data is shown in fig. 36. It represents the energy spectra of proton and alpha particles produced as free particles in AMD (direct emission) and secondary particles evaporated from the calculated decay of the excited primary fragments. The two contributions are compared to experimentally deduced secondary particles with the method described above. AMD predicts reasonably well the total energy spectra of protons (78% of total emission), but it fails to reproduce the thermal contribution (22%).

11 Conclusions

The study of heavy-ion collisions at intermediate energies has provided important information about the mechanisms of multifragmentation and their links to a possible nuclear liquid-gas phase transition. Our capability of fully understanding these phenomena depends strongly on how well one can identify and characterize the thermal and dynamical properties of fragment and particle emitting sources. In this paper we have presented a review of the most significant results that different research groups have achieved in the last decades by using two-particle correlation techniques.

Intensity interferometry techniques have demonstrated that we have *space-time probes capable of measuring sizes as small as* 10^{-15} *m and time intervals as short as several* 10^{-23} *s*. This can be viewed as an important advance in the field of heavy-ion physics. These space-time characterization techniques have provided valuable information about the size/volume, the density of decaying nuclear systems, as well as quantitative estimates of emission lifetimes and particle chronology. The extracted results demonstrate that the emitting sources produced in heavy-ion collisions form a quite complex system: different particles are emitted at different times and by different mechanisms. In this respect, it seems clear that a complete space-time characterization of nuclear reactions requires a study of multiple correlations extended to all particle species and including neutrons. Proton-neutron chronology is also indicated as an important candidate to explore the density dependence of the symmetry energy which represents one of the most challenging perspectives for the future.

The recent introduction of imaging techniques has certainly changed our interpretation of correlation observables. The source sizes extracted with this technique can differ significantly from those extracted from other approaches. This difference suggests that a detailed shape analysis of correlation observables is necessary in order to obtain a correct determination of nuclear densities and lifetimes. In this chapter we have also shown how techniques were found to quantify the strength of secondary decays. Light-particle–fragment correlations have been used to extract information about the excitation energy of primary unstable fragments. The possibility of constraining secondary decays represents an important perspective: several observables are indeed affected by the unknown contributions to the experimental spectra from different emission mechanisms dominating different stages of the reaction.

Despite the large amount of information that the scientific community has been able to extract, certain aspects have not been fully explored and require further research. For instance, from an experimental point of view, higher isotopic resolution could provide an important opportunity to investigate isotopically resolved fragment-fragment correlation functions. This perspective is very important in view of clarifying the indications of evolutionary freeze-out conditions in heavy-ion collisions and their possible links to the equation of state of asymmetric nuclear matter, relevant to astrophysical environments like neutron stars and supernova explosions. Higher angular resolution and a large solid-angle coverage are important requirements in order to increase the quality of measured correlation observables and to explore their features while having a complete characterization of the collision event (impact parameter, reaction plane, exact determination of velocity vectors, etc.). Higher-resolution devices will be the key to the perspective of extending imaging techniques to several particle species, improving complex particle correlation analyses that provide information about the space-time properties at freeze-out and about the excitation energies of primary fragments. Other fields of nuclear-physics research will certainly profit from the existence of detector setups that are characterized by outstanding correlation capabilities. This is particularly the case for the nuclear structure groups working on spectroscopic properties of exotic nuclear systems explored with the future radioactive-ion-beam accelerator facilities.

The close interaction between the theoretical and experimental communities will certainly contribute significantly in improving our capability of characterizing emitting sources. The use of more powerful experimental setups and the implementation of full quantum multi-body approaches to correlation functions promise to provide unambiguous information about the thermal and dynamical scenarios characterizing multifragmention phenomena and their links to the equation of state of nuclear matter.

References

1. W.G. Lynch, Annu. Rev. Nucl. Part. Sci **37**, 493 (1987).
2. L.G. Moretto, G.J. Wozniak, Annu. Rev. Nucl. Part. Sci. **43**, 379 (1993).
3. H. Feldmeier *et al.* (Editors), *Proceedings of the XXVII International Workshop on Gross Properties of Nuclei and Nuclear Excitations, Hirschegg, Austria* (GSI, 1999) p. 200.
4. J. Aichelin, Phys. Rep. **202**, 233 (1991).
5. C.P. Montoya *et al.*, Phys. Rev. Lett. **73**, 3070 (1994).
6. Y. Larochelle *et al.*, Phys. Rev. C **59**, R565 (1999).
7. E. Plagnol *et al.*, Phys. Rev. C **61**, 014606 (1999).
8. J. Lukasik *et al.*, Phys. Lett. B **566**, 76 (2003).
9. T. Lefort *et al.*, Nucl. Phys. A **662**, 397 (2000).
10. Ph. Eudes, Z. Basrak, F. Sebille, Phys. Rev. C **56**, 2003 (1997).
11. J. Pochodzalla *et al.*, Phys. Rev. Lett. **75**, 1040 (1995).
12. A. Chbihi *et al.*, Eur. Phys. J. A **5**, 251 (1999).
13. F. Gulminelli, Ph. Chomaz, Phys. Rev. Lett. **82**, 1402 (1999).
14. Ph. Chomaz, V. Duflot, F. Gulminelli, Phys. Rev. Lett. **85**, 3587 (2000).
15. M. D'Agostino *et al.*, Nucl. Phys. A **699**, 795 (2002).
16. W. Bauer, C.K. Gelbke, S. Pratt, Annu. Rev. Nucl. Part. Sci. **42**, 77 (1992).
17. D.H. Boal, C.K. Gelbke, B.K. Jennings, Rev. Mod. Phys. **62**, 553 (1990).
18. U. Heinz, B.V. Jacak, Annu. Rev. Nucl. Part. Sci. **49**, 529 (1999).
19. R. Hanbury Brown, R.Q. Twiss, Philos. Mag. **45**, 663 (1954); Nature **177**, 27 (1956).
20. R. Hanbury-Brown, R.Q. Twiss, Nature **178**, 1046 (1956).
21. G. Goldhaber *et al.*, Phys. Rev. Lett. **3**, 181 (1959); Phys. Rev. **120**, 300 (1960).
22. M. Lisa, S. Pratt, R. Soltz, U. Wiedemann, Annu. Rev. Nucl. Part. Sci. **55**, 357 (2005).
23. S.E. Koonin, Phys. Lett. B **70**, 43 (1977).
24. S. Pratt, M.B. Tsang, Phys. Rev. C **36**, 2390 (1987).
25. M.A. Lisa *et al.*, Phys. Rev. Lett. **70**, 2545 (1993).
26. R. Kotte *et al.*, Eur. Phys. J. A **6**, 185 (1999).
27. W.G. Gong *et al.*, Phys. Rev. C **43**, 1804 (1991).
28. W.G. Gong *et al.*, Phys. Rev. Lett. **65**, 2114 (1990).
29. B. Davin *et al.*, Nucl. Instrum. Methods A **473**, 302 (2001).
30. T.C. Awes *et al.*, Phys. Rev. Lett. **61**, 2665 (1988).
31. D.O. Handzy *et al.*, Phys. Rev. C **50**, 858 (1994).
32. S. Fritz *et al.*, Phys. Lett. B **461**, 315 (1999).
33. R. Kotte *et al.*, Phys. Rev. C **51**, 2686 (1995).
34. M.A. Lisa *et al.*, Phys. Rev. C **49**, 2788 (1994).
35. P.A. DeYoung *et al.*, Phys. Rev. C **39**, 128 (1989).
36. R. Kotte *et al.*, Eur. Phys. J. A **23**, 271 (2005).
37. C. Schwarz *et al.*, Nucl. Phys. A **681**, 279 (2001).
38. V. Serfling *et al.*, Phys. Rev. Lett. **80**, 3928 (1998).
39. W.G. Gong, W. Bauer, C.K. Gelbke, S. Pratt, Phys. Rev. C **43**, 781 (1991).
40. F. Zhu *et al.*, Phys. Rev. C **44**, R582 (1991).
41. R. Ghetti *et al.*, Nucl. Phys. A **674**, 277 (2000).
42. G. Verde *et al.*, Phys. Rev. C **65**, 054609 (2002).
43. D.A. Brown, P. Danielewicz, Phys. Lett. B **398**, 252 (1997).

44. D.A. Brown, P. Danielewicz, Phys. Rev. C **57**, 2474 (1998).
45. D.A. Brown, P. Danielewicz, Phys. Lett. C **64**, 014902 (2001).
46. R. Ghetti *et al.*, Phys. Rev. C **64**, 017602 (2001).
47. R. Ghetti *et al.*, Phys. Rev. C **62**, 037603 (2000).
48. B. Jakobsson *et al.*, Phys. Rev. C **44**, R1238 (1991).
49. N. Colonna *et al.*, Nucl. Instrum. Methods A **421**, 542 (1999).
50. W. Dünnweber *et al.*, Phys. Rev. Lett. **65**, 297 (1990).
51. M. Marques *et al.*, Phys. Rev. Lett. **73**, 34 (1994).
52. A. Badala *et al.*, Phys. Rev. Lett. **74**, 4779 (1995).
53. P. Sapienza *et al.*, Phys. Rev. Lett. **73**, 1769 (1994).
54. Z. Chen *et al.*, Phys. Rev. C **36**, 2297 (1987).
55. D.H. Boal, J.C. Shillcock, Phys. Rev. C **33**, 549 (1986).
56. J. Pochodzalla *et al.*, Phys. Lett. B **174**, 36 (1986).
57. C.J. Gelderloos *et al.*, Phys. Rev. C **52**, R2834 (1995).
58. J. Pochodzalla *et al.*, Phys. Rev. C **35**, 1695 (1987).
59. C.B. Chitwood *et al.*, Phys. Rev. Lett. **54**, 302 (1985).
60. Z. Chen *et al.*, Phys. Lett. B **199**, 171 (1987).
61. D. Fox *et al.*, Phys. Rev. C **38**, 146 (1988).
62. D.A. Cebra *et al.*, Phys. Lett. B **227**, 336 (1989).
63. D. Gourio *et al.*, Eur. Phys. J. A **7**, 245 (2000).
64. INDRA Collaboration (J. Pouthas *et al.*), Nucl. Instrum. Methods A **357**, 418 (1995).
65. J.C. Steckmeyer *et al.*, Nucl. Instrum. Methods Phys. Res. A **361**, 472 (1995).
66. INDRA Collaboration (J. Pouthas *et al.*), Nucl. Instrum. Methods A **369**, 222 (1996).
67. R. Lednicky, V.L. Lyuboshitz, Sov. J. Nucl. Phys. **35**, 770 (1982).
68. J. Pochodzalla *et al.*, Phys. Lett. B **175**, 275 (1986).
69. D. Ardouin, Int. J. Mod. Phys. E **6**, 391 (1997).
70. C.J. Gelderloos, J.M. Alexander, Nucl. Instrum. Methods A **349**, 618 (1994).
71. C.J. Gelderloos *et al.*, Phys. Rev. Lett. **75**, 3082 (1995).
72. R. Lednicky, V.L. Lyuboshitz, B. Erazmus, D. Nouais, Phys. Lett. B **373**, 30 (1996).
73. S. Voloshin, R. Lednicky, S. Panitkin, N. Xu, Phys. Rev. Lett. **79**, 4766 (1997).
74. D. Ardouin *et al.*, Phys. Lett. B **446**, 191 (1999).
75. R. Ghetti *et al.*, Phys. Rev. Lett. **87**, 102701 (2001).
76. M.M. de Moura *et al.*, Nucl. Phys. A **696**, 64 (2001).
77. R. Ghetti *et al.*, Phys. Rev. Lett. **91**, 092701 (2003).
78. R. Kotte *et al.*, Eur. Phys. J. A **6**, 185 (1999).
79. S. Soff *et al.*, J. Phys. G **23**, 789 (1997).
80. M. Di Toro, S.J. Yennello, Bao-An Li, this topical issue.
81. L.W. Chen, V. Greco, C.M. Ko, Bao-An Li, Phys. Rev. Lett. **90**, 162701 (2003).
82. L.W. Chen, V. Greco, C.M. Ko, Bao-An Li, Phys. Rev. C **68**, 014605 (2003).
83. L.W. Chen, C.M. Ko, Bao-An Li, Phys. Rev. C **69**, 054606 (2004).
84. R. Ghetti *et al.*, Phys. Rev. C **69**, 03160 (2004).
85. R. Ghetti, J. Helgesson, Nucl. Phys. A **752**, 480c (2005).
86. R. Ghetti *et al.*, Phys. Rev. C **70**, 034601 (2004).
87. Y.D. Kim *et al.*, Phys. Rev. C **45**, 387 (1992).
88. R. Trockel *et al.*, Phys. Rev. Lett. **59**, 2844 (1987).
89. Y.D. Kim *et al.*, Phys. Rev. Lett. **67**, 14 (1991).
90. R. Bougault *et al.*, Phys. Lett. B **232**, 291 (1994).
91. D.R. Bowman *et al.*, Phys. Rev. C **52**, 818 (1995).
92. T. Glasmacher *et al.*, Phys. Rev. C **50**, 952 (1994).
93. R. Popescu *et al.*, Phys. Rev. C **58**, 270 (1998).
94. L. Beaulieu *et al.*, Phys. Rev. Lett. **84**, 5791 (2000).
95. Ph. Chomaz *et al.*, Phys. Rev. Lett. **73**, 3512 (1994).
96. M. Parlog *et al.*, Eur. Phys. J. A **25**, 223 (2005).
97. E. Bauge *et al.*, Phys. Rev. Lett. **70**, 3705 (1993).
98. E. Elmaani *et al.*, Nucl. Instrum. Methods A **313**, 401 (1992).
99. E. Cornell *et al.*, Phys. Rev. Lett. **75**, 1475 (1995).
100. W.A. Friedman, Phys. Rev. C **42**, 667 (1990).
101. E. Cornell *et al.*, Phys. Rev. Lett. **77**, 4508 (1995).
102. G. Tabacaru *et al.*, Nucl. Phys. A **764**, 371 (2006).
103. M. Colonna *et al.*, Phys. Rev. C **51**, 2671 (1995).
104. A. Guarnera *et al.*, Phys. Lett. B **403**, 191 (1997).
105. INDRA Collaboration (N. Marie *et al.*), Phys. Rev. C **58**, 256 (1998).
106. P. Staszel *et al.*, Phys. Rev. C **63**, 064610 (2001).
107. S. Hudan *et al.*, in *Proceedings of the XXXVIII International Winter Meeting on Nuclear Physics, Bormio, Italy, 2000*, edited by I. Iori, A. Moroni, Suppl. # 116 (Ricerca Scientifica ed Educazione Permanente, Milano, 2000) p. 443.
108. S. Hudan, PhD Thesis, Université de Caen (2001) GANIL T01 07.
109. INDRA Collaboration (S. Hudan *et al.*), Phys. Rev. C **67**, 064613 (2003).
110. O. Tirel, PhD Thesis, Université de Caen (1998) GANIL T98 02.
111. R. Nebauer, J. Aichelin, Nucl. Phys. A **650**, 65 (1999).
112. INDRA Collaboration (R. Nebauer *et al.*), Nucl. Phys. A **658**, 67 (1999).
113. D.H.E. Gross, Rep. Prog. Phys. **53**, 605 (1990).
114. A. Ono *et al.*, Phys. Rev. Lett. **68**, 2898 (1992).
115. A. Ono *et al.*, Prog. Theor. Phys. **87**, 1185 (1992).
116. A. Ono *et al.*, Phys. Rev. C **59**, 853 (1999).
117. Ph. Chomaz *et al.*, Phys. Lett. B **254**, 340 (1991).
118. A.S. Botvina *et al.*, Nucl. Phys. A **475**, 663 (1987).
119. J.P. Bondorf *et al.*, Phys. Rep. **257**, 133 (1995).
120. A.H. Raduta, A.R. Raduta, Phys. Rev. C **55**, 1344 (1997).
121. A.H. Raduta, A.R. Raduta, Phys. Rev. C **56**, 2059 (1997).
122. A.H. Raduta, A.R. Raduta, Phys. Rev. C **59**, 323 (1999).
123. D. Cussol, Phys. Rev. C **68**, 014602 (2003).
124. A. Strachan, C. Dorso, Phys. Rev. C **59**, 285 (1999).
125. C. Dorso, J. Randrup, Phys. Lett. B **301**, 328 (1993).
126. C. Escano *et al.*, *Proceedings of the International Workshop on Multifragmentation and related topics, IWM2003, GANIL, Caen, France, November 5-7, 2003*, p. 197.
127. D.W. Stracener *et al.*, Nucl. Instrum. Methods Phys. Res. A **294**, 485 (1990).
128. A. Bonasera *et al.*, Phys. Rep. **243**, 1 (1994).
129. S. Levit, P. Bonche, Nucl. Phys. A **437**, 426 (1985).
130. S.E. Koonin, J. Randrup, Nucl. Phys. A **474**, 173 (1987).
131. R.J. Charity *et al.*, Nucl. Phys. A **483**, 371 (1988).
132. T. Nayak *et al.*, Phys. Rev. C **45**, 132 (1992).
133. O. Ono, S. Hudan, A. Chbihi, J.D. Frankland, Phys. Rev. C **66**, 014603 (2002).

Eur. Phys. J. A **30**, 109–120 (2006)
DOI 10.1140/epja/i2006-10110-1

THE EUROPEAN
PHYSICAL JOURNAL A

Dynamical models for fragment formation

A. Ono[1,a] and J. Randrup[2]

[1] Department of Physics, Tohoku University, Sendai 980-8578, Japan
[2] Lawrence Berkeley National Laboratory, Berkeley, CA 94720, USA

Received: 21 June 2006 /
Published online: 17 October 2006 – © Società Italiana di Fisica / Springer-Verlag 2006

Abstract. The various dynamical models for fragment formation in nuclear collisions are discussed in order to bring out their relative advantages and shortcomings. After discussing the general requirements for dynamical models that aim to describe fragment formation, we consider the various mean-field models that incorporate fluctuations and then turn to models based on molecular dynamics.

PACS. 24.10.-i Nuclear reaction models and methods – 05.60.Gg Quantum transport – 24.60.Ky Fluctuation phenomena – 25.70.Pq Multifragment emission and correlations

1 Introduction

Nuclear collisions in the medium-energy regime (from several tens to several hundreds MeV/nucleon) typically yield several intermediate-mass fragments (IMFs) (see Tamain [1]). Thus, at relatively low energies dissipative binary reactions may create IMFs at midrapidity, while higher-energy central collisions create expanding systems that produce clusters copiously; and peripheral collisions produce excited projectile-like fragments that multifragment. The IMFs typically carry a major part ($\sim 50\%$) of the nucleons involved.

Since the fragments are formed in dynamical reactions where equilibrium is not guaranteed *a priori*, there is a need for developing microscopic dynamical descriptions for fragment formation. This poses a significant theoretical challenge because of the basic quantal nature of the many-body nuclear system. Although it is possible to derive such models by truncating a hierarchy of quantum many-body equations, it is difficult to ensure that the error would remain small throughout the rather long duration of fragmentation reactions. Therefore, most of the currently employed models have been developed by performing certain drastic simplifications while seeking to retain a quantitatively useful description for those aspects that are deemed to be of most interest. Consequently, models with different characteristics have been developed and applied to fragmentation reactions with reasonable successes in specific cases.

The purpose of this paper is to summarize the main requirements for models that aim to describe fragment formation and to elucidate their relative merits and short-

comings. First, sect. 2 discusses those model features that are of largest importance. Then, within that framework, we discuss in sect. 3 models that have been developed on the basis of mean-field theory, while sect. 4 covers those that involve molecular dynamics.

Since this paper focuses on fragment formation, we do not intend to evaluate the overall utility of individual models. Indeed, treatments that do not seek to describe fragment formation, such as the nuclear Boltzmann equation (see Fuchs and Wolter [2]) are not addressed here, even if they may have proven to be very important generally for the study of heavy-ion reactions. (An early guide to microscopic models for intermediate-energy nuclear collisions was given in ref. [3].) Neither does this paper cover the statistical models for fragment formation (see Botvina and Mishustin [4]) which provide us with a powerful tool for understanding fragmentation.

2 General requirements

We discuss here the general features that are required by any dynamical model aiming to describe nuclear fragmentation.

2.1 General framework for the time evolution

Ideally, any such model should be derivable from the underlying quantum many-body description by means of well-defined approximations. Most of the models for nuclear dynamics are based on the mean-field picture, exemplified by the time-dependent Hartree-Fock (TDHF)

---

a e-mail: ono@nucl.phys.tohoku.ac.jp

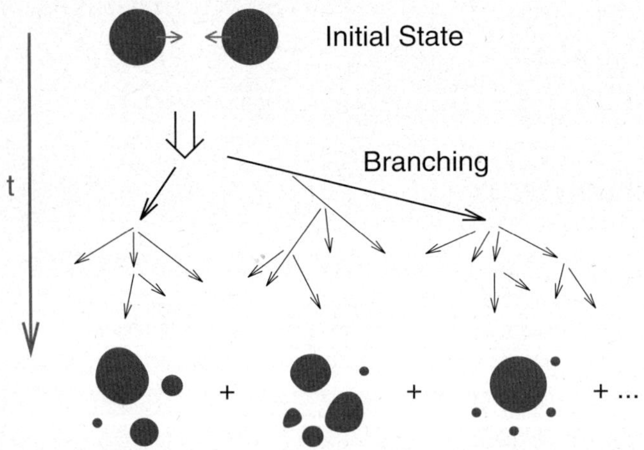

Fig. 1. A schematic picture of a fragmentation reaction in which a given initial channel may develop into many different fragmentation channels during the dynamical evolution.

treatment or its semiclassical analogue, the Vlasov equation. However, the direct two-nucleon collisions grow increasingly important when the collision energy reaches and surpasses the Fermi energy, as the Pauli exclusion principle becomes ever less effective in blocking the two-nucleon collisions. Therefore, the models should incorporate both single-particle motion in a mean field and Pauli-suppressed two-nucleon collisions. This is indeed the case for most of the models discussed here.

2.2 Dynamical bifurcations

The most important special challenge associated with fragmentation is the occurrence of dynamical bifurcations, the feature that a given initial configuration may lead to many different fragmentations, as illustrated in fig. 1. Since the number of final channels is huge, even if only the most important fragmentations are considered, a standard coupled-channel treatment would be practically impossible. On the other hand, this very feature makes it natural to employ concepts and methods from transport theory. Consequently, most models involve some stochastic agent as a simple way to produce spontaneous fluctuations and the associated trajectory branchings.

2.3 Basic quantum statistics

Any quantitatively useful model of nuclear systems must take account of the basic quantum-statistical feature that causes Pauli blocking and endows the nucleons with Fermi motion. Therefore, the initial nucleon momenta are usually sampled from a Fermi distribution, even if the specific model does not inherently contain such a feature. This ensures that the one-body phase-space density is reasonably consistent with the corresponding single-nucleon wave functions. Furthermore, the final states of the direct two-body collisions are usually suppressed by suitable Pauli-blocking factors, thus helping to prevent the nucleons to revert to the Maxwell-Boltzmann form characteristic of classical systems. Even though the various models tend to include these basic features, the specific manner in which this is actually done varies greatly from one model to another.

2.4 Macroscopic nuclear properties

The dynamical models should have stationary solutions that reproduce the most important macroscopic nuclear properties, such as density distributions and binding energies, whereas shell and pairing effects are not very important because the produced fragments are usually excited by several MeV/nucleon. This requirement can be met if the model yields proper values of the nuclear saturation density and the associated binding energy (including its isospin dependence), together with especially the nuclear surface tension. It is thus important that these key quantities be known for the various models.

2.5 Thermal nuclear properties

Because of the complexity of the fragmentation process, statistical features play a large role in determining the relative fragment yields. It is therefore quantitatively important that the nuclear level densities, as reflected in the specific heat, have realistic magnitudes. In particular, in a quantum system the excitation energy grows quadratically with temperature, $E^* = aT^2$, while the relation tends to be linear in a classical system. The desirability of this characteristic property poses a significant problem for the dynamical models and, as we shall discuss, most models are inadequate in this particular regard.

2.6 Interactions

It is preferable that the models contain only a minimal number of parameters. In fact, the mean-field Hamiltonian should in principle be known from static nuclear properties and thus not be subject to adjustment. It is in the context of fragment production especially important that the employed interaction yields a liquid-gas phase transition in uniform matter.

With regard to the residual two-body interaction, it is most often represented by means of differential scattering cross-section which may, in principle, be modified by the local density and temperature. Although such medium modifications might be calculable, they may also be taken as somewhat adjustable.

In any case, both the long-range interaction responsible for the mean field and the residual interaction causing the collisions should already have been fixed from applications that do not involve fragmentation. So, consequently, there should ideally be no new parameters associated with the treatment of fragmentation processes.

2.7 Particle emission from hot nuclei

The fragments produced after the violent stage of the reactions are still excited by typically several MeV/nucleon. Such fragments de-excite by light-particle emission over a time scale that is very long in comparison with that of the collision. Although it would be impractical to propagate the dynamical models for such long times, it would still be desirable that the models in principle describe the de-excitation processes. However, this is generally not the case, in large part because of the rather rough character of the dynamical models relative to the more refined treatments required for such emission processes. Indeed, the proper description of particle emission from hot nuclei usually requires a quantum-mechanical treatment. Therefore, when particle-stable fragments are needed, it is necessary to apply suitable de-excitation treatments to each of the (pre)fragments formed in the course of the collision.

2.8 Correlations and fragmentation mechanisms

It is desirable that the models can describe many-body correlations beyond those of the mean-field description. This is particularly important for a proper description of light fragments, such as alpha particles. One of the most important advantages of treatments based on molecular dynamics is that such correlations are included automatically (though not necessarily correctly, of course). It is important to recognize that one-body models also contain non-trivial correlation features when augmented by a stochastic agency that produces an entire ensemble of one-body systems from a single initial configuration. Thus, while the mean-field models may not be suitable for the description of very light fragments, they may be quite reasonable for fragments that lend themselves to a mean-field description, such as typical IMFs.

3 Mean-field models with fluctuations

Significant advances in our understanding of nuclear dynamics have been achieved within the mean-field framework. Just as the Hartree-Fock treatment provides a useful starting point for the discussion of static nuclear properties, its time-dependent version, TDHF, presents a good conceptual starting point for the treatment of nuclear dynamics. An early study of multifragmentation within the TDHF framework was made by Knoll and Strack [5], who considered the evolution of individual Slater determinants that had been sampled from a statistical ensemble representing a hot source.

The main shortcoming of pure mean-field treatments is the omission of the short-range residual interaction. An attempt to include this important physical ingredient is presented by the stochastic TDHF model [6] in which the many-body system continually jumps from one Slater determinant to another. Though conceptually appealing, this approach has not yet been developed into a practical

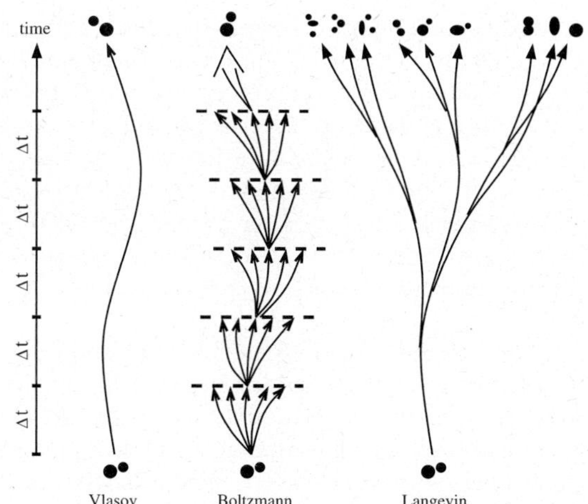

Fig. 2. Characterization of dynamical models. The various semiclassical treatments of microscopic nuclear dynamics can be characterized by the manner in which the single-particle phase-space density is being propagated from one time step to the next. In the Vlasov treatment, the particles experience only the self-consistent effective field, leading to a single dynamical history $f(\boldsymbol{r}, \boldsymbol{p}, t)$. At the Boltzmann level, the various possible outcomes of the residual collisions are being averaged at each step, leading then to a different but still single dynamical trajectory. Finally, the Boltzmann-Langevin model allows the various stochastic collision outcomes to develop independently, thus leading to a continual trajectory branching and a corresponding ensemble of histories.

tool (but it provides a basis for deriving the Boltzmann-Langevin treatment discussed below).

Indeed, the residual interaction is more readily included within the framework of semi-classical descriptions of the Nordheim type [7], often referred to as Boltzmann-Ühling-Uhlenbeck (BUU) or Vlasov-Ühling-Uhlenbeck models, in which the collisionless mean-field evolution is augmented by a Pauli-blocked Boltzmann collision term. There are various techniques for solving the nuclear Boltzmann equation. One common approach introduces a (usually large) number of pseudo-particles for each nucleon present, $\mathcal{N}$, with correspondingly reduced interacting cross-sections. This method makes it possible to achieve an arbitrarily fine coverage of phase space and, in principle, the resulting solution approaches the exact solution as $\mathcal{N}$ is increased. However, it requires a fairly cumbersome programming to prevent the computational task from increasing quadratically with $\mathcal{N}$. Therefore, it is often preferable to use the parallel-ensemble method, in which $\mathcal{N}$ individual A-body systems are treated in parallel in a common mean field that is obtained by averaging, at each time, over the $\mathcal{N}$ systems (see fig. 2). This treatment retains some correlation, so although it does not converge to the Boltzmann solution it may well present a more useful model.

In the standard Boltzmann treatment, only the average effect of the collisions between the particles is included, thus yielding a deterministic evolution of the one-particle phase-space density $f(\boldsymbol{r}, \boldsymbol{p})$ (see fig. 2). While this simplification may be suitable in many physical scenarios in which the macroscopic dynamics is stable (such as the early stages of a nuclear collision when the system is hot and compressed), it is inadequate for processes involving instabilities, bifurcations, or chaos. In particular, if the combined expansion and cooling brings the system into the spinodal zone of the phase diagram, it is essential to admit the occurrence of fluctuations and allow their subsequent self-consistent development.

Various attempts to overcome this problem have been made. On the more formal side, transport theory was invoked to treat the effect of many-body correlations as a stochastic process and thus derive a transport equation for the one-particle phase-space density $f(\boldsymbol{r}, \boldsymbol{p})$ [8–10]. In particular, in ref. [10] the general transport equation was reduced to coupled equations for the mean evolution of $f(\boldsymbol{r}, \boldsymbol{p})$ and its fluctuations around this average trajectory. The evolution is then determined by the transport coefficients, namely the *drift* coefficients $V[f](\boldsymbol{r}, \boldsymbol{p})$ that govern the average change of $f(\boldsymbol{r}, \boldsymbol{p})$ (and is given by the usual Boltzmann equation) and the *diffusion* coefficients $D[f](\boldsymbol{r}, \boldsymbol{p}; \boldsymbol{r}', \boldsymbol{p}')$ governing the correlation between changes at two different phase-space locations. These coefficients are given in terms of the differential cross-section and this fundamental relationship ensures that they satisfy the fluctuation-dissipation theorem. While this approach yields a more exhaustive description, it is applicable only when the dynamics is macroscopically stable so all dynamical histories remain fairly similar.

However, in many situations of actual interest in heavy-ion physics, such as multifragmentation processes, the dynamical trajectories branch into configurations that are qualitatively different, as illustrated in fig. 2. It is therefore necessary to devise methods that can admit and propagate arbitrary fluctuations. This need has led to the development of the nuclear Boltzmann-Langevin (BL) model which is briefly described below.

3.1 Boltzmann-Langevin model

The Boltzmann-Langevin equation of motion for $f(\boldsymbol{r}, \boldsymbol{p})$ can be written on a condensed form as [11,12]

$$\dot{f} \equiv \frac{\partial f}{\partial t} - \{h[f], f\} = C[f] \equiv \bar{C}[f] + \delta C[f]. \qquad (1)$$

Here the mean-field evolution of $f(\boldsymbol{r}, \boldsymbol{p})$ on the left is governed by the effective one-body Hamiltonian $h[f](\boldsymbol{r}, \boldsymbol{p})$, which depends self-consistently on $f(\boldsymbol{r}, \boldsymbol{p})$. The collision term $C[f]$ on the right represents the effect of the two-body collisions and is therefore stochastic in nature. As such, it can be separated into its average, $\bar{C}[f]$, which is the term retained in the standard Boltzmann equation, and its fluctuating part, $\delta C[f]$. The two parts can be expressed in terms of the elementary collision process $\boldsymbol{p}_1 \boldsymbol{p}_2 \to \boldsymbol{p}_1' \boldsymbol{p}_2'$, for which the expected number of occurrences within a small time interval, $\bar{\nu}$, is equal to the associated variance σ_ν^2, as in a standard random walk.

This fundamental relationship leads to the fluctuation-dissipation theorem.

After it had been demonstrated [11] that the fluctuating collision term $C[f]$ produces the correct quantum-statistical equilibrium fluctuations and correlations in a uniform gas, the transport theory was turned into a practical tool by the development of a numerical method for the direct simulation of the stochastic part $\delta C[f]$ [12]. With this model, the dynamical clusterization in the presence of instabilities was then addressed [13] and explicit numerical studies were made for two-dimensional matter in the phase-space region of spinodal instability. The fluctuating part of the collision term acts as a source of irregularities in the density which may then be amplified by the self-consistent mean field. The corresponding dispersion relation (the growth rate $\gamma_k = 1/t_k$ as a function of the wave number of the distortion) was extracted from the numerical simulations and shown to exhibit a maximum which identifies the characteristic length scale for the clusterization, as reflected in the Fourier transform of the spatial density. A more detailed treatment of the linear response in stochastic mean-field theories and the onset of instabilities was subsequently made [14].

It thus appears that the Boltzmann-Langevin model offers a suitable one-body framework for the study of unstable nuclear dynamics, such as fragmentation processes. Nevertheless, it appears that an accurate description of the agitation of unstable modes in nuclear matter generally requires the inclusion of memory time effects resulting from the basic quantal nature of the system [15].

Furthermore, the numerical treatment of the fluctuating collision term presents a formidable challenge and is not yet feasible in three dimensions. Therefore a number of approximate treatments have been developed. However, typically, these approaches introduce the fluctuations by fiat in a manner that is inconsistent with the general relaxation properties of the one-body density, as expressed through the fluctuation-dissipation theorem. We discuss those various approaches in the following.

3.2 Brownian one-body dynamics

A powerful approximate treatment of the BL model was obtained by approximating the effect of the fluctuating part of the collision term, δC, by that of a suitable stochastic one-body potential, $\delta U(\boldsymbol{r}, t)$,

$$\delta C[f] \to -\delta \boldsymbol{F}[f] \cdot \frac{\partial f}{\partial \boldsymbol{p}}, \qquad (2)$$

with the Brownian force $\delta \boldsymbol{F} \equiv \partial \delta U / \partial \boldsymbol{r}$ being tuned at each point in time and space to ensure that the dynamics of important collective modes emulates the results of the complete Boltzmann-Langevin model [16].

In the resulting Brownian one-body (BOB) model for nuclear dynamics, the stochastic force is adjusted to ensure the correct growth of the fastest-growing unstable spinodal mode, as obtained by making a local-density approximation. Since the local adjustment of the Brownian force can be made on the basis of simple analytical approximations [17,18], the BOB scheme can be implemented by making a relatively straightforward modification of a standard BUU code [16], thus providing a powerful tool for studies of fragment formation.

The BOB model was subsequently applied to dynamical scenarios where spinodal fragmentation occurs. One study considered the multifragmentation of an initially compressed gold nucleus [19] and found, in accordance with an earlier BUU-based study [20], that the system quickly expands into a hollow and unstable configuration, where the irregularities resulting from the stochastic force are then amplified by the self-consistent mean field, resulting in several intermediate-mass fragments, together with a large number of unbound nucleons.

It thus appears that the stochastic mean-field model framework is suitable for the treatment of nuclear fragmentation dynamics, provided that the self-consistent propagation of fluctuations has been suitably incorporated.

3.3 Other approximate Boltzmann-Langevin methods

An attempt to introduce spontaneous fluctuations in a practically realizable manner was made by Bauer *et al.* [21]. Their method can be implemented relatively easily into standard BUU codes that use the pseudo-particle method of solution and it consists essentially in forcing similar two-body collisions to occur for neighboring pseudo-particles so that effectively two entire nucleons are involved in each particular collision event. Employing an idealized two-dimensional nucleon gas as a test case, Chapelle *et al.* [22] examined this intuitively appealing method. They found that it is able to produce fluctuations of the correct general magnitude, provided that a suitable coarse graining of the phase space is performed, and that these display some of the correlation features expected from the basic characteristics of the two-body collision process. These features can be improved by suitable tuning of the phase-space metric (the concept of a distance in phase space is required for the selection of the "neighboring" pseudo-particles). However, for any tuning, the detailed momentum dependence of the variance in phase-space occupancy deviates significantly from what is dictated by quantum statistics. Therefore this simple prescription may be unsuitable for problems in which these properties are important.

On a more formal basis, Ayik and Gregoire [9] proposed an approximate method for numerical implementation of the Boltzmann-Langevin theory. The method reduces the Boltzmann-Langevin equation for the microscopic one-body phase-space density $f(\boldsymbol{r},\boldsymbol{p})$ to stochastic equations for a set of macroscopic variables, namely the local or global quadrupole moment of the momentum distribution. A random change of the quadrupole moment is then made at each time step and a suitable stretching of $f(\boldsymbol{r},\boldsymbol{p})$ is performed subsequently in order to reconstruct the entire phase-space density. This method was also examined by in ref. [22] and, although several variations

of the proposed scheme were examined, it was generally found that the results were far from satisfactory, since the resulting correlations associated with the fluctuating one-body density will tend to reflect the symmetries and other characteristics of the employed reconstruction procedure rather than those of the underlying physical fluctuations. Therefore this method appears unsuitable for calculating quantities that depend sensitively on the details of the momentum distribution.

For the purpose of addressing catastrophic phenomena in nuclear dynamics, such as multifragmentation, Colonna *et al.* [23] explored the possibility of simulating the stochastic part of the collision integral in the Boltzmann-Langevin model by the numerical noise $\sigma_k(0)$ associated with the finite number of pseudo-particles $\mathcal{N}$ employed in the ordinary BUU treatment. This idea is based on the observation that for large times, $t \gg t_k$, the fluctuation of density undulations of a given wave number k is given by $\sigma_k^2(t) = D_k t_k \mathrm{e}^{2t/t_\nu}$ in the Boltzmann-Langevin treatment, whereas it is $\sigma_k^2(t) = (D_k t_k/\mathcal{N} + \sigma_k(0))\mathrm{e}^{2t/t_\nu}$ in the BUU pseudo-particle treatment. Since $\sigma_k(0)$ also scales as $1/\mathcal{N}$, the matching of those two asymptotic fluctuations yields a relation determining the value of $\mathcal{N}$. For idealized two-dimensional matter, which presents a suitable test case, as it is here practical to simulate the Boltzmann-Langevin equation directly, they demonstrated that $\mathcal{N}$ can be adjusted so that the corresponding BUU calculation yields a good reproduction of the spontaneous clusterization occurring inside the spinodal region. This approximate method may therefore provide a relatively easy way to introduce meaningful fluctuations in simulations of unstable nuclear dynamics. This method was subsequently extended to 3D nuclear matter, allowing the direct extraction of the growth times t_k of the unstable modes and the associated diffusion coefficients D_k [24].

Guarnera *et al.* [25] studied the spinodal fragmentation of a hot and dilute nucleus by first expanding the system into a spinodally unstable configuration and then adding a stochastic density fluctuation that is carefully tuned to reflect the degree of fluctuation in the most unstable mode, as determined by the corresponding linear-response analysis of the unstable sphere. They found that the early clusterization appears to be dominated by unstable modes whose spatial structure is similar to the fastest growing spinodal modes in infinite matter at similar density and temperature. They followed the development of the instabilities until multifragmentation had occurred and then made an analysis of the resulting fragment size distribution. As expected from the fact that only a few modes dominate, the clusterization pattern has a large degree of regularity which in turn favors breakup into fragments of nearly equal size, with a corresponding paucity of small clusters.

Subsequently, Colonna *et al.* [26] introduced a method that roughly approximates the Boltzmann-Langevin model by adding a suitable noise to the collision term in the usual BUU treatment. The noise employed corresponds to the thermal fluctuation in the local phase-space occupancy, $\sigma_f^2(\boldsymbol{r},\boldsymbol{p}) = f(1-f)$, where $f(\boldsymbol{r},\boldsymbol{p})$ is the local

Fermi-Dirac equilibrium distribution. By performing such a local momentum redistribution at suitable intervals in the course of the evolution, the inherently stochastic nature of the two-body collision processes is mimicked. The method has the advantage that it is readily tractable and it applies equally well to both stable and unstable parts of the phase diagram. The method has been applied to multifragmentation in central collisions in the Fermi energy domain [27], showing spinodal decomposition in expanding systems.

A different approach was taken by Matera and Dellafiore [28] who applied white noise to a Vlasov system. The noise term was determined self-consistently by invoking the fluctuation-dissipation theorem and, within the linear approximation, the time evolution of the density fluctuations was found to be given by the same closed form as was found in ref. [14]. The authors showed that while a white-noise form of the stochastic field is in general *not* consistent with the fluctuation-dissipation theorem, it may provide a good approximation when the free response function is sufficiently peaked.

It is important to note that all of the methods described above employ an *ad hoc* procedure to generate fluctuations. Therefore the microscopic structure in phase space of the produced correlations is typically very different from the prediction of the Boltzmann-Langevin model. However, as stressed first in ref. [23], in situations where the dynamics is dominated by only a few modes (such as the fastest growing spinodal modes) it may suffice to require equivalence with the exact Boltzmann-Langevin approach for only those few degrees of freedom. As a consequence, some of the approaches [23–25] have carefully designed the fluctuation source so as to mimic the effects of the stochastic Boltzmann-Langevin term on the dynamics of the most unstable modes and the main dynamics of the spinodal decomposition can then be simulated.

Several studies aimed directly at cases of experimental interest [25, 29, 30] have found that central collisions of Xe and Sn should lead to spinodal fragmentation and display corresponding correlations in the resulting IMF sizes (see Borderie and Désesequelles [31]). This system has been investigated experimentally at INDRA [32] and a signal in quantitatively good agreement with the transport calculations was indeed observed. A comprehensive review of nuclear spinodal fragmentation was given in ref. [33].

3.4 Drawbacks of mean-field dynamics

The mean-field models treat the reduced single-particle phase-space density $f(\boldsymbol{r}, \boldsymbol{p})$ and they are therefore most suitable for the calculation of quantities that can be expressed as expectation values of one-body observables. But the extraction of more complicated observables (such as two-body correlations) is problematic. This inherent problem is particularly evident when fragmentation processes are considered. For example, any emerging "fragments" need not have integer particle numbers. Fortunately, this principal problem is usually unimportant in actual applications, especially when the observables of interest can be

expressed in terms of moments of the mass distribution (such as the mean IMF charge).

In this connection, it is important to recognize that although mean-field models treat only the one-body phase-space density, the stochastic versions generate entire ensembles of one-body densities. Therefore, insofar as the different fragmentations may each be satisfactorily described within the one-body framework, stochastic one-body models may in fact be well suited for multifragmentation processes.

Although quantum statistics is taken into account by the inclusion of the appropriate Fermi blocking or Bose enhancement factors in the collision term, the numerical treatments are generally classical in nature and, consequently, the occupation coefficients will eventually revert to their classical (Maxwell-Boltzmann) form. (This feature was discussed in refs. [34, 35] for Vlasov dynamics.) Fortunately, because the associated time scale is usually fairly long, this principal problem is of little practical import for applications to nuclear collisions. But it does make it somewhat tricky to use the models to study equilibration phenomena.

A common problem with existing semi-classical one-body microscopic models of nuclear dynamics is their failure to provide an accurate description of the thermal properties of ordinary nuclei at only moderate excitation. As a consequence of this and the basically classical nature of the equations of motion, the de-excitation of produced prefragments is not well described and to make contact with experiment it is necessary to switch from the dynamical model to an "afterburner" that treats the de-excitation of each individual prefragment. This problem is also commonly encountered with molecular dynamics.

4 Molecular dynamics

A more direct connection to the observable physical states is provided by the molecular-dynamics many-body models. These models have been developed to ever higher levels of refinement and we can here give only a rough overview with some illustrative examples.

4.1 Classical molecular dynamics

Generally, classical molecular dynamics (CMD) solves the classical equation of motion for the positions and momenta of A particles,

$$\frac{\mathrm{d}}{\mathrm{d}t}\boldsymbol{r}_i = \{\boldsymbol{r}_i, \mathcal{H}\}, \qquad \frac{\mathrm{d}}{\mathrm{d}t}\boldsymbol{p}_i = \{\boldsymbol{p}_i, \mathcal{H}\}, \qquad (3)$$

where the many-body Hamiltonian is of the form

$$\mathcal{H}\{\boldsymbol{r}_n, \boldsymbol{p}_n\} = \sum_{i=1}^{A} \frac{\boldsymbol{p}_i^2}{2m_i} + \sum_{i<j} V(|\boldsymbol{r}_i - \boldsymbol{r}_j|). \qquad (4)$$

The nucleon-nucleon potential $V(r)$ (which may depend on the particle species) generally consists of a short-range repulsive part and a long-range attractive part, so that the resulting matter equation of state (EOS) is of the Van der Waals type. The work by Lenk and Pandharipande [36, 37] provides a good illustration of this type of model.

The CMD equation of motion is entirely deterministic. Nevertheless, the collision dynamics has a chaotic character so that small differences in the initial states may lead to quite different final states. This feature automatically gives access to many fragmentation channels. Furthermore, while it is hard to justify CMD as a good approximation for the dynamics of the nuclear many-body quantum system, CMD does have the virtue of retaining all the orders of many-body correlations at the classical level.

Indeed, CMD simulations have provided useful insight into the general features of fragmenting finite systems, such as critical phenomena [38], phase evolution [39], the caloric curve [40] and isoscaling [41]. The character of the fragment emission has also been elucidated [42]. A particularly intriguing result was obtained by Dorso et $al.$ [43] who employed a criterion that considers the binding of each particle in its host cluster and found that the fragment size distribution may be extracted rather early, already when the system is still quite dense.

4.2 Quasi-classical molecular dynamics

One of the problems with classical molecular dynamics for nuclear systems is that the fermion nature of the nucleons cannot readily be incorporated. Indeed, in the ground state of the classical Hamiltonian $\mathcal{H}$ all particles have vanishing velocities. This basic feature makes it hard to emulate the most basic features of nuclear systems.

One partial remedy for this problem is the introduction of a so-called Pauli potential, a momentum-dependent repulsion that serves to emulate the exclusion principle, as first proposed by Wilets et $al.$ [44,45].

This approach was pursued in more detail by Dorso et $al.$ [46] with a Gaussian repulsion depending on the phase-space separation s_{ij}, with $s_{ij}^2 = r_{ij}^2/q_0^2 + p_{ij}^2/p_0^2$. They first demonstrated that such a repulsion leads to a reasonable emulation of the Fermi-Dirac momentum distribution in thermal equilibrium, over a broad energy range of interest [46]. Furthermore, when augmented by a Lennard-Jones potential, the model yields a reasonable reproduction of the nuclear equation of state and hence appears to be suitable for instructive simulations of nuclear collisions [47]. Indeed, a first application to an initially compressed and heated nucleus allowed the extraction of its thermodynamic phase evolution, showing that the spinodal region was entered, and the resulting fragmentation exhibited characteristic signs of filamentation [39].

4.3 Quantum molecular dynamics

It is possible to go beyond deterministic molecular dynamics by introducing a Pauli-blocked collision term in a manner similar to what is done in the nuclear Boltzmann

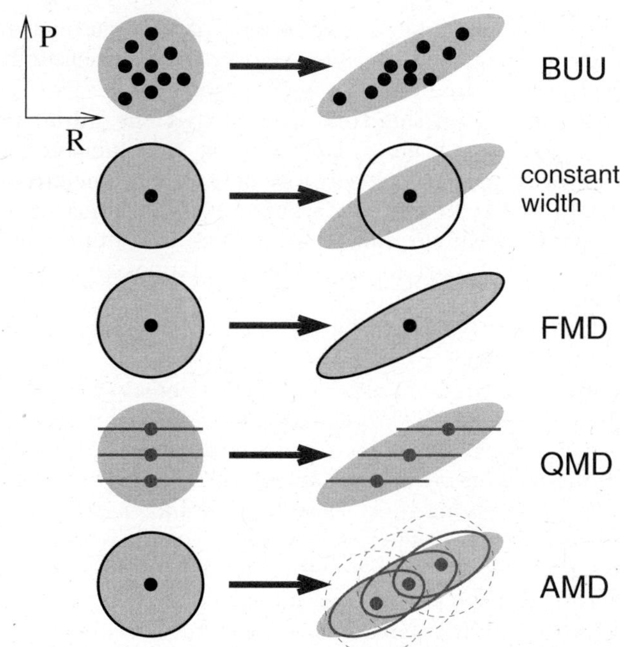

Fig. 3. Schematic depiction of the free time evolution of the phase-space distribution of a single nucleon as described by various models when the initial state is represented by a wave packet having both spatial and momentum widths. The exact evolution is indicated by the gray (yellow on-line) area.

(BUU-type) treatments. The resulting model is then technically identical to the parallel-ensemble treatment of the nuclear Boltzmann equation with $\mathcal{N} = 1$. But an important difference from the usual BUU treatment is that the fluctuations produced by the stochasticity automatically develop self-consistently for each individual collision event and thus allow the emergence of different fragmentation channels.

In addition, a Gaussian smearing is performed to obtain the spatial density of the nucleons at any point in time, which is intended to emulate the effect of individual wave packets. The resulting class of models is usually referred to as quantum molecular dynamics (QMD) [48–52]. The spatial smearing causes the force acting on each nucleon to be much smoother than the bare nucleon-nucleon force used in CMD. Furthermore, since the interaction used does not have a repulsive core the resulting force becomes rather similar to that of the mean-field description.

Even though the spatial smearing was introduced to emulate the effect of individual wave packets, the corresponding effect on the momentum distribution is ignored. Indeed, the momenta are treated as in CMD, with the kinetic energy of a nucleon taken as $\boldsymbol{p}_i^2/2m$ without any zero-point energy, while the momentum distribution is treated by a random sampling of $\boldsymbol{p}_i$. This treatment can be regarded as a practical method for including the effect of the momentum distribution on the time evolution: The nucleons will have different velocities in different events so they will be found at different positions in the final states.

In the simple example of free motion of a single nucleon, as illustrated in fig. 3, the average over the event ensemble will yield the correct free-time evolution.

A drawback of this treatment is that the initial nuclei, which are obtained by sampling the nucleon momenta $\boldsymbol{p}_i$ from a Fermi sphere, are not in their true ground states in which all nucleons would have vanishing velocities. While this undesirable feature is of little import during the violent part of the collision, it does play a significant role for the production of fragments, since their yield is largely governed by the statistical weights of the dynamical model.

QMD has been applied mainly to collisions at relatively high energy. In low-energy processes with long time scales it is difficult to keep the phase-space occupation below unity merely by means of the Pauli suppression in the two-body collisions since these become increasingly rare. In an attempt to remedy this problem, some approaches have employed a Pauli potential [53,54], but there is no quantum-mechanical foundation for such a force in the equation of motion. Furthermore, statistical particle emission from excited fragments cannot be reliably described by QMD. Therefore the dynamical QMD calculation has to be stopped at a certain time and the decay of the fragments should be calculated by a statistical decay code. The long-time simulation for equilibrated systems is beyond the limit of applicability of QMD and the result of such a simulation would not be consistent with quantum statistics.

QMD simulations of energetic nuclear collisions typically lead to copious production of fragments whose multiplicities are often comparable with the experimental data [55–59]. However, these fragments are extracted at the end of the violent stage of the collision and are generally excited by several MeV/nucleon. Therefore, the subsequent decay processes tend to significantly reduce the IMF yield [60], thus leaving a persistent discrepancy between the QMD results and the data.

Furthermore, multifragmentation of projectile-like fragments in peripheral collisions is underestimated more seriously in some QMD calculations [55,60]. However, using an algorithm based on the early cluster recognition method that invokes the single-particle binding energies in the candidate preclusters and is applicable even at high densities [43], Gossiaux *et al.* [61] were able to reproduce the observed multiplicity. This issue may be related to the fact that the effective interaction range is quite large in QMD because of the spatial smearing [61,62].

4.4 Constrained molecular dynamics

The Pauli principle requires that the phase-space density should not exceed one nucleon per phase-space volume $(2\pi\hbar)^3$ for each spin-isospin state in semiclassical descriptions. The exact treatment of the Pauli principle (see sects. 4.5 and 4.6) requires significant computational power for large systems. In order to overcome the computational difficulty, an approximate implementation of the Pauli principle has been proposed as constrained molecular dynamics (CoMD) [63,64]. In this approach, a stochastic process is added to the usual QMD in order to prevent the violation of the Pauli principle. The process is invoked when the phase-space density f_i around a nucleon i becomes greater than 1. The momenta of the nucleon i and other nucleon(s) are changed as in the two-nucleon scattering so that the Pauli principle $f_i \leq 1$ is finally satisfied after several trials. This is one of the ways to satisfy the Pauli principle, though it is not derived from first principles.

Due to the stochastic process for the Pauli principle, the condition $f_i \leq 1$ remains satisfied when a ground-state nucleus is propagated for a long time. The properties of hot nuclei may be better described by CoMD than QMD.

CoMD can reproduce the multifragmentation data at the incident energy of 35 MeV/nucleon [63,64]. The effect of the stochastic process for the Pauli principle is to reduce the Pauli-blocking factor for the two-nucleon collisions when the two nuclei overlap, which results in stronger stopping and expansion towards instability of multifragmentation. The charge distribution of intermediate-mass fragments are reasonably reproduced, except for the problems in the light-particle multiplicities.

4.5 Fermionic molecular dynamics

Fermionic molecular dynamics (FMD) [65–69] is a true quantum treatment that represents the many-body state as an antisymmetrized Slater determinant of wave packets having a Gaussian form,

$$\varphi_i(\boldsymbol{r}) \sim \exp\left[-\nu_i(\boldsymbol{r} - \boldsymbol{Z}_i)^2\right]. \tag{5}$$

The wave packet centroids $\{\boldsymbol{Z}_i\}$ and widths $\{\nu_i\}$ are complex dynamical variables whose equations of motion can be derived from the time-dependent variational principle. The nucleons are assumed to move in a mean field and their spin and isospin degrees of freedom may be included. Thus FMD is a constrained form of TDHF with nonorthogonal single-particle states for which the overlap matrix $\langle\varphi_i|\varphi_j\rangle$ should be properly considered. The derived equation of motion shows that $\{\boldsymbol{Z}_i\}$ and $\{\nu_i\}$ are not canonical variables.

Since the FMD wave function is a Slater determinant, the effective interactions developed for mean-field calculations are basically applicable to FMD. It is also possible to employ realistic nuclear forces by means of unitary correlation operators [70,71]. Furthermore, the FMD wave function provides a reasonable approximate description of ground-state nuclei [67], obtained by minimizing the energy of the constrained wave function. Properties such as binding energies and radii can be reproduced well with a reasonable effective interaction. Contrary to the molecular-dynamics models discussed above, the energy minimization yields a unique FMD ground state which is invariant under the FMD time evolution.

Even though FMD utilizes a quantum wave function, the dynamics is fully deterministic and the system remains a single Slater determinant at all times. This is inadequate for fragmentation processes, where many different configurations are reachable. This problem could be remedied by the introduction of direct two-body collisions, which would bring the description close to Stochastic TDHF (sect. 3). While this has not yet been done in the full FMD model, it has been successfully carried out in AMD (sect. 4.6) where the width parameters $\{\nu_i\}$ are kept fixed.

Since the many-body state is described by a Slater determinant, FMD incorporates the Pauli principle perfectly, of course. Furthermore, the dynamical growth of the imaginary part of the width parameter ν_i produces a correlation between positions and momenta in the course of time, thereby ensuring that the single-particle motion is described correctly for both free motion (see fig. 3) and for nucleons in a harmonic-oscillator potential.

The deterministic character of FMD has the drawback that it does not offer a natural description of dynamical bifurcations. An important example is nucleon emission which occurs with some probability, while the nucleon remains in the source with the complementary probability. A reasonable description would yield an increasing width of the emitted part of the wave packet, while the residual packet should remain rather compact, something that is clearly beyond the reach of a single Gaussian wave packet. Clearly, such branchings could be described by the introduction of suitable stochasticity in the dynamics (see sects. 4.6 and 4.7).

By enclosing the system in a large harmonic-oscillator potential well and coupling the system weakly to a virtual thermometer while examining its long-time behavior, it has been possible to study the thermodynamic properties of FMD [72]. The model was shown to exhibit a liquid-gas phase transition and the associated caloric curves were extracted. They are similar to those obtained experimentally, with a low-temperature liquid-like region, an intermediate plateau associated with the coexistence region, and a high-temperature gas-like region. However, the contact with experiment could be firmed up by deriving the temperature from observed quantities such as isotope ratios and kinetic energies of gaseous nucleons.

There have been several other works based on molecular dynamics with dynamical wave packet widths. It was found that the inclusion of a dynamical width improves the agreement with data in some cases, such as fusion cross-section above the Coulomb barrier [73]. Kiderlen *et al.* [74] studied the fragmentation of excited systems. In response to the initial pressure, the excited system begins to expand but clusters were *not* produced even though Gaussian wave packets with many-body correlations were employed. When the excited system expands, the widths of wave packets grow and then, in turn, the interaction between the packets weakens. The mean field for such a configuration is very shallow and smooth and there is then little chance for clusters to appear. This feature conflicts with the general expectation that clusters should appear in such situations. Similarly, studies of spinodal instability [75] showed that the zero sound is significantly affected

when the width grows large and this spreading of the nucleon wave packet then inhibits cluster formation.

4.6 Antisymmetrized molecular dynamics

Antisymmetrized molecular dynamics (AMD) [76–78] is similar to FMD in that the system is represented by a Slater determinant and that a part of the equation of motion is derived from the time-dependent variational principle. An important difference from FMD is that stochastic terms have been added to the equation of motion so that many configurations can appear through the reaction dynamics.

On the other hand, AMD usually treats the width parameters $\{\nu_i\}$ of the single-particle wave packets as a constant parameter common to all the nucleons. This simplification reduces the computational burden but limits the flexibility of the description, compared to the FMD description, as long as the stochastic extension terms are ignored. Nevertheless, the constant width parameter guarantees that there is no spurious coupling of the internal motion and the center-of-mass motion of a cluster or a nucleus. Furthermore, the presence of trajectory branching due to the stochasticity avoids the creation of spurious correlations in the wave function. For example, for the nucleon emission process, channels with and without nucleon emission will not mix in a single AMD wave function.

It is a very attractive feature of AMD that it provides, with a conventional effective interaction and a reasonable value of the width parameter, a quite good description of not only the basic properties of ground-state nuclei but also many detailed structure features, such as the excitation level spectra of light nuclei [79], with some extensions such as the parity and angular-momentum projections.

Recent versions of AMD [80–82] seek to take account of the dynamics of the wave packet width and shape by splitting the wave packet into components by means of a stochastic term that is calculated based on the single-particle motion in the mean field (see fig. 3). It assumes that the coherence of the single-particle wave function is lost and it branches into incoherent Gaussian wave packets at a certain time due to many-body effects. This quantum branching process makes possible the coexistence of the single-particle dynamics in the mean field and the fragment formation, which requires spatial localization and the emergence of many configurations. The resulting extended AMD may be regarded as a specific case of the stochastic mean-field equation (sect. 3) with the correlations of the fluctuation $\delta C[f]$ designed in such a way that a Gaussian wave packet appears.

The introduction of two-nucleon collisions is similar to QMD (sect. 4.3), with some differences described below. The antisymmetrization implies that the wave packet centroids $\{Z_i\}$ cannot be interpreted as the positions and momenta of nucleons. Rather, the physical coordinates are introduced as nonlinear functions of the centroids [77] and the two-nucleon collisions are performed by using these physical coordinates. There then appear Pauli-forbidden phase-space regions into which the physical coordinates will never enter, for any values of the centroid variables $\{Z_i\}$. These regions are regarded as Pauli-blocked and not allowed as final state of a collision. Another difference from QMD is the fact that the physical momentum in AMD is the momentum centroid of a Gaussian phase-space distribution, while the momentum variable in QMD usually represents the definite momentum of a nucleon.

The equilibrium properties of AMD have been studied by solving the time evolution of a many-nucleon system in a container for a long time to obtain a microcanonical ensemble. When the liquid phase, in the form of a nucleus, is embedded in a nucleon gas of temperature T, the characteristic quantum relation $E^*_{\text{liq}} \sim T^2$ was obtained [83] and the resulting caloric curves show that AMD is consistent with the liquid-gas phase transition [84–86].

The wave packet branching plays an essential role for obtaining such physically reasonable equilibrium properties in AMD. The importance of a stochastic term has also been demonstrated within the Quantal Langevin model discussed below (sect. 4.7). On the other hand, as mentioned above (sect. 4.5), in FMD the nucleon emission from a nucleus in the liquid-gas phase coexistence region is described by a deterministic motion of the nucleon wave packet with a variable width. The differences between these approaches have not yet been fully explored.

Although it has not been studied very carefully, the approximate reproduction of the quantum relation $E^*_{\text{liq}} \sim T^2$ suggests that statistical nucleon emission from an excited fragment may be qualitatively well described in AMD, while a quantitative description would require that the model gives the correct value for the nuclear level density parameter a. If this were indeed the case, then AMD should be able to describe the statistical decay of fragments produced in collisions if the time evolution could be calculated for a sufficiently long time. Fortunately, the final results do not depend very much on the time at which the dynamical calculation is connected to the statistical decay calculation.

When the wave packet branching is included by means of a stochastic term, the resulting state must be adjusted to ensure energy conservation. This is achieved by means of a dissipative term in the equation of motion. Although this dissipative term has been constructed carefully in order to obtain a reasonable time evolution, its form has not been derived from a basic principle.

AMD has been successfully applied to fragmentation reactions, such as central collisions in the energy region of several tens of MeV/nucleon for light and heavy systems [80,87]. The fragment isospin composition obtained in dynamical collisions is consistent with statistical predictions, such as the isoscaling relation and the dependence on the symmetry energy term of the effective force [88,89]. These results are consistent with the idea that the fragment isospin composition is determined when the density is low ($\rho \approx \frac{1}{2}\rho_0$), and reflects the symmetry energy of dilute nuclear matter.

The description of few-body correlations in AMD is probably rather crude in some situations. In particular, when the incident energy is high ($\gtrsim 50\,\text{MeV/nucleon}$),

the nucleon multiplicity is strongly overestimated, which is probably because of the too small probability of forming light clusters from highly excited matter. The correlation needed to form light clusters should probably be treated more quantum mechanically than the accidental merging of randomly distributed wave packets.

4.7 Quantal Langevin dynamics

A more formal development of trajectory branching in wave packet dynamics has led to the Quantal Langevin (QL) model [90,91]. The motivation for this work lies in the fact that the nuclear liquid-gas phase transition differs significantly from the usual liquid-gas phase transition in macroscopic matter primarily in the role played by quantum statistics. For usual macroscopic matter, the total energies are to a good approximation linear functions of the temperature in both the liquid and gas phases. Thus the effective number of degrees of freedom is essentially constant in each phase. In contrast to this familiar situation, the liquid phase of a nucleus exhibits an increase in the number of activated degrees of freedom as the temperature is raised. In particular, the excitation energy of a nucleus at low temperature increases like $E^* = aT^2$ (where the level density parameter is $a \approx A/(8\,\text{MeV})$), which is a typical quantal behavior, while the gas phase is characterized by the usual classical relation $E^*/A = \frac{3}{2}T$. The two curves intersect at $T \approx 12\,\text{MeV}$, which is much higher than the transition temperature suggested by experimental data. This indicates that the quantal statistical nature of the nuclear system plays an important role for the phase transition and, presumably, for the associated nuclear multifragmentation processes.

Part of the reason for the persistent shortcoming of wave packet dynamics for the description of multifragmentation (see sect. 4.3) may be found in the fact that the equation of motion for the wave packet centroids is not consistent with the quantal statistical nature, because quantum fluctuations inherent in the wave packets are neglected. The presence of quantum fluctuations is signaled by the fact that a given wave packet is a superposition of many energy eigenstates. Therefore the fluctuations should be taken into account in such a way that the different components are properly explored in the course of time.

This fundamental problem can be clearly brought out by making a cumulant expansion of the canonical weight of a given wave packet, at the temperature $T = 1/\beta$ [92],

$$\ln \mathcal{W}_\beta = \ln \langle \exp(-\beta \hat{H}) \rangle = \beta \mathcal{H} + \frac{1}{2}\beta^2 \sigma_H^2 + \mathcal{O}(\beta^3)\,. \quad (6)$$

Here $\mathcal{H} \equiv \langle \hat{H} \rangle$ is the usual expectation value of the energy in the given wave packet and it is evident that the weight $\mathcal{W}_\beta$ is affected by its energy spread σ_H. Truncation of the cumulant expansion at second order, corresponding to a Poisson energy distribution in each packet (as a Gaussian would have) leads to a much improved global description of the quantum-statistical properties of the many-body system.

This approach was extended to dynamical scenarios by the introduction of a Langevin force emulating the transitions between the wave packets [90,91]. The corresponding transport process in wave packet space can be described as a Langevin process and the general form of the associated transport coefficients was derived. The ensuing diffusive wave packet evolution exhibits appealing physical properties, including relaxation towards the appropriate microcanonical quantum-statistical equilibrium distribution in the course of the time evolution. Specific expressions for the transport coefficients were subsequently derived on the basis of Fermi's golden rule and it was verified that they satisfy the associated fluctuation-dissipation theorem.

This approach is not specific to nuclear dynamics but has general applicability. For example, it was used to study the effect of quantum fluctuations on the critical properties of noble gases [93]. In nuclear physics it has been applied to hyperfragment formation from Ξ^- absorption on ^{12}C where it was found that quantum fluctuations affect the outcome qualitatively [94,95] and to multifragmentation [96] which is of particular interest here and will be briefly summarized below.

The Langevin force enables the wave packet system to explore its entire energy spectral distribution, rather than being restricted to its average value. This leads to a much improved description of the quantum-statistical features. In particular, the resulting specific heat now exhibits the characteristic evolution from a quantum fluid towards a classical gas as a function of temperature [92], in contrast to the behavior emerging with the usual treatment. Since a change of a fragment's specific heat is associated with a change in its statistical weight, the effect is clearly relevant for the fragment production problem.

The key new features of the results obtained with the quantal Langevin model are the occurrence of larger fluctuations and an enhancement of stable configurations, such as bound fragments, as a result of the need to take account of the spectral distortion of the wave packets. The former feature arises from the fact that the wave packet parameter of each nucleon is populated according to the strength of the eigen components for the given energy expectation value, and therefore the wave packet parameter can have larger fluctuations than when the energy is fixed to the expectation value. On the other hand, in order to project out the appropriate energy component from the wave packet, it is necessary to take account of its internal distortion. The combination of these two basic features then enhances the average IMF multiplicity at the final stage, especially in central collisions, as was demonstrated for Au+Au at 100–400 MeV/nucleon [96]. While the larger fluctuations allow the system to explore more configurations and thus enhances the yield of primary fragments, the latter stabilizes the fragments, since the compensation for the quantum distortion effectively acts as a cooling mechanism.

These studies suggest that the underlying quantal nature of the nuclear many-body system may indeed play a significant role in fragmentation reactions.

5 Concluding remarks

The development of a suitable dynamical description of fragment formation in nuclear collisions is a daunting task that poses many interesting challenges and makes contact with other areas of modern many-body and mesoscopic physics. We have here given a brief overview of the most commonly employed models and sought to bring out their relative merits and shortcomings. Although much progress has been made over the past couple of decades, we are still far from having models that are formally well founded, practically applicable, and sufficiently realistic to be quantitatively useful. As our discussion has brought out, the description of nuclear fragmentation dynamics requires that proper account be taken of the basic quantal nature of the system. This requirement renders purely classical equations of motion inadequate and calls for the development of quantal transport theory. Further advances along this line are likely to be of broad physical interest.

References

1. B. Tamain, *Systematics of fragment observables*, this topical issue.
2. C. Fuchs, H.H. Wolter, *Modelization of the EOS*, this topical issue.
3. G.F. Bertsch, S. Das Gupta, Phys. Rep. **160**, 189 (1988).
4. A.S. Botvina, I.N. Mishustin, *Statistical description of nuclear break-up*, this topical issue.
5. J. Knoll, B. Strack, Phys. Lett. B **149**, 45 (1984).
6. P.G. Reinhard, E. Suraud, Ann. Phys. **216**, 98 (1992).
7. L.W. Nordheim, Proc. R. Soc. London, Ser. A **119**, 689 (1928).
8. S. Ayik, C. Grégoire, Phys. Lett. B **212**, 269 (1988).
9. S. Ayik, C. Grégoire, Nucl. Phys. A **513**, 187 (1990).
10. J. Randrup, B. Remaud, Nucl. Phys. A **514**, 339 (1990).
11. Ph. Chomaz, G.F. Burgio, J. Randrup, Phys. Lett. B **254**, 340 (1991).
12. G.F. Burgio, Ph. Chomaz, J. Randrup, Nucl. Phys. A **529**, 157 (1991).
13. G.F. Burgio, Ph. Chomaz, J. Randrup, Phys. Rev. Lett. **69**, 885 (1992).
14. M. Colonna, Ph. Chomaz, J. Randrup, Nucl. Phys. A **567**, 637 (1993).
15. S. Ayik, J. Randrup, Phys. Rev. C **50**, 2947 (1994).
16. Ph. Chomaz, M. Colonna, A. Guarnera, J. Randrup, Phys. Rev. Lett. **73**, 3512 (1994).
17. J. Randrup, S. Ayik, Nucl. Phys. A **572**, 489 (1994).
18. J. Randrup, Nucl. Phys. A **583**, 329 (1995).
19. A. Guarnera, Ph. Chomaz, M. Colonna, J. Randrup, Phys. Lett. B **403**, 191 (1997).
20. G. Batko, J. Randrup, Nucl. Phys. A **563**, 97 (1993).
21. W. Bauer, G.F. Bertsch, S. Das Gupta, Phys. Rev. Lett. **58**, 863 (1987).
22. F. Chapelle, G.F. Burgio, Ph. Chomaz, J. Randrup, Nucl. Phys. A **540**, 227 (1992).
23. M. Colonna, G.F. Burgio, Ph. Chomaz, M. Di Toro, J. Randrup, Phys. Rev. C **47**, 1395 (1993).
24. M. Colonna, Ph. Chomaz, Phys. Rev. C **49**, 637 (1994).
25. A. Guarnera, M. Colonna, Ph. Chomaz, Phys. Lett. B **373**, 267 (1996).

26. M. Colonna, M. Di Toro, A. Guarnera, S. Maccarone, M. Zielinska-Pfabé, H.H. Wolter, Nucl. Phys. A **642**, 449 (1998).
27. M. Colonna, G. Fabbri, M. Di Toro, F. Matera, H.H. Wolter, Nucl. Phys. A **742**, 337 (2004).
28. F. Matera, A. Dellafiore, Phys. Rev. C **62**, 0044611 (2000).
29. M.F. Rivet *et al.*, Phys. Lett. B **430**, 217 (1998).
30. J.D. Frankland *et al.*, Nucl. Phys. A **689**, 940 (2001).
31. B. Borderie, P. Désesequelles, *Many-fragment correlations and possible signature of spinodal fragmentation*, this topical issue.
32. INDRA Collaboration (B. Borderie *et al.*), Phys. Rev. Lett. **86**, 3252 (2001).
33. Ph. Chomaz, M. Colonna, J. Randrup, Phys. Rep. **389**, 263 (2004).
34. P.G. Reinhard, E. Suraud, Ann. Phys. **239**, 193; 216 (1995).
35. C. Jarzynski, G.F. Bertsch, Phys. Rev. C **53**, 1028 (1996).
36. R.J. Lenk, V.R. Pandharipande, Phys. Rev. C **34**, 177 (1986).
37. R.J. Lenk, V.R. Pandharipande, Phys. Rev. C **36**, 162 (1987).
38. C.O. Dorso, V.C. Latora, A. Bonasera, Phys. Rev. C **60**, 34606 (1999).
39. C. Dorso, J. Randrup, Phys. Lett. B **232**, 29 (1989).
40. A. Chernomoretz, C.O. Dorso, J.A. López, Phys. Rev. C **64**, 044605 (2001).
41. C.O. Dorso, C.R. Escudero, M. Ison, J.A. López, Phys. Rev. C **73**, 044601 (2006).
42. M.J. Ison, C.O. Dorso, Phys. Rev. C **71**, 064603 (2005).
43. C. Dorso, J. Randrup, Phys. Lett. B **301**, 328 (1993).
44. L. Wilets, E.M. Henley, M. Kraft, A.D. Mackellar, Nucl. Phys. A **282**, 341 (1977).
45. L. Wilets, Y. Yariv, R. Chestnut, Nucl. Phys. A **301**, 359 (1978).
46. C. Dorso, S. Duarte, J. Randrup, Phys. Lett. B **188**, 287 (1987).
47. C. Dorso, J. Randrup, Phys. Lett. B **215**, 611 (1988).
48. J. Aichelin, H. Stöcker, Phys. Lett. B **176**, 14 (1986).
49. J. Aichelin, Phys. Rep. **202**, 233 (1991).
50. G. Peilert, H. Stöcker, W. Greiner, A. Rosenhauer, A. Bohnet, J. Aichelin, Phys. Rev. C **39**, 1402 (1989).
51. T. Maruyama, A. Ohnishi, H. Horiuchi, Phys. Rev. C **42**, 386 (1990).
52. Ch. Hartnack, R.K. Puri, J. Aichelin, J. Konopka, S.A. Bass, H. Stöcker, W. Greiner, Eur. Phys. J. A **1**, 151 (1998).
53. G. Peilert, J. Konopka, H. Stöcker, W. Greiner, M. Blann, M.G. Mustafa, Phys. Rev. C **46**, 1457 (1992).
54. D.H. Boal, J.N. Glosli, Phys. Rev. C **38**, 1870 (1988).
55. M.B. Tsang *et al.*, Phys. Rev. Lett. **71**, 1502 (1993).
56. J.P. Bondorf, D. Idier, I.N. Mishustin, Phys. Lett. B **359**, 261 (1995).
57. H.W. Barz, J.P. Bondorf, D. Idier, I.N. Mishustin, Phys. Lett. B **382**, 343 (1996).
58. R. Nebauer *et al.*, Nucl. Phys. A **658**, 67 (1999).
59. K. Hagel *et al.*, Phys. Rev. C **50**, 2017 (1994).
60. Toshiki Maruyama, K. Niita, Tomoyuki Maruyama, A. Iwamoto, Prog. Theor. Phys. **98**, 87 (1997).
61. P.B. Gossiaux, R. Puri, Ch. Hartnack, J. Aichelin, Nucl. Phys. A **619**, 379 (1997).
62. Tomoyuki Maruyama, K. Niita, Prog. Theor. Phys. **97**, 579 (1997).
63. M. Papa, Toshiki Maruyama, A. Bonasera, Phys. Rev. C **64**, 024612 (2001).
64. M. Papa, G. Giuliani, A. Bonasera, J. Comput. Phys. **208**, 403 (2005).
65. H. Feldmeier, Nucl. Phys. A **515**, 147 (1990).
66. H. Feldmeier, J. Schnack, Nucl. Phys. A **583**, 347 (1995).
67. H. Feldmeier, K. Bieler, J. Schnack, Nucl. Phys. A **586**, 493 (1995).
68. H. Feldmeier, J. Schnack, Prog. Part. Nucl. Phys. **39**, 393 (1997).
69. H. Feldmeier, J. Schnack, Rev. Mod. Phys. **72**, 655 (2000).
70. H. Feldmeier, T. Neff, R. Roth, J. Schnack, Nucl. Phys. A **632**, 61 (1998).
71. T. Neff, H. Feldmeier, Nucl. Phys. A **713**, 311 (2003).
72. J. Schnack, H. Feldmeier, Phys. Lett. B **409**, 6 (1997).
73. Toshiki Maruyama, K. Niita, A. Iwamoto, Phys. Rev. C **53**, 297 (1996).
74. D. Kiderlen, P. Danielewicz, Nucl. Phys. A **620**, 346 (1997).
75. M. Colonna, Ph. Chomaz, Phys. Lett. B **436**, 1 (1998).
76. A. Ono, H. Horiuchi, Toshiki Maruyama, A. Ohnishi, Phys. Rev. Lett. **68**, 2898 (1992).
77. A. Ono, H. Horiuchi, Toshiki Maruyama, A. Ohnishi, Prog. Theor. Phys. **87**, 1185 (1992).
78. A. Ono, H. Horiuchi, Prog. Part. Nucl. Phys. **53**, 501 (2004).
79. Y. Kanada-En'yo, M. Kimura, H. Horiuchi, C. R. Phys. (Paris) **4**, 497 (2003).
80. A. Ono, H. Horiuchi, Phys. Rev. C **53**, 2958 (1996).
81. A. Ono, Phys. Rev. C **59**, 853 (1999).
82. A. Ono, S. Hudan, A. Chbihi, J.D. Frankland, Phys. Rev. C **66**, 014603 (2002).
83. A. Ono, H. Horiuchi, Phys. Rev. C **53**, 2341 (1996).
84. Y. Sugawa, H. Horiuchi, Phys. Rev. C **60**, 064607 (1999).
85. Y. Sugawa, H. Horiuchi, Prog. Theor. Phys. **105**, 131 (2001).
86. T. Furuta, A. Ono, Phys. Rev. C **74**, 014612 (2006).
87. R. Wada *et al.*, Phys. Rev. C **62**, 034601 (2000).
88. A. Ono, P. Danielewicz, W.A. Friedman, W.G. Lynch, M.B. Tsang, Phys. Rev. C **68**, 051601(R) (2003).
89. A. Ono, P. Danielewicz, W.A. Friedman, W.G. Lynch, M.B. Tsang, Phys. Rev. C **70**, 041604(R) (2004).
90. A. Ohnishi, J. Randrup, Phys. Rev. Lett. **75**, 596 (1995).
91. A. Ohnishi, J. Randrup, Ann. Phys. **253**, 279 (1997).
92. A. Ohnishi, J. Randrup, Nucl. Phys. A **565**, 474 (1994).
93. A. Ohnishi, J. Randrup, Phys. Rev. A **55**, R3315 (1997).
94. Y. Hirata, A. Ohnishi, Y. Nara, T. Harada, J. Randrup, Nucl. Phys. A **639**, 389 (1998).
95. Y. Hirata, Y. Nara, A. Ohnishi, T. Harada, J. Randrup, Prog. Theor. Phys. **102**, 89 (1999).
96. A. Ohnishi, J. Randrup, Phys. Lett. B **394**, 260 (1997).

Eur. Phys. J. A **30**, 121–128 (2006)
DOI 10.1140/epja/i2005-10316-7

THE EUROPEAN
PHYSICAL JOURNAL A

Statistical description of nuclear break-up

A.S. Botvina[1,a] and I.N. Mishustin[2,3]

[1] Institute for Nuclear Research, Russian Academy of Sciences, 117312 Moscow, Russia
[2] Frankfurt Institute for Advanced Studies, J.W. Goethe University, D-60438 Frankfurt am Main, Germany
[3] Kurchatov Institute, Russian Research Center, 123182 Moscow, Russia

Received: 27 October 2005 /
Published online: 17 October 2006 – © Società Italiana di Fisica / Springer-Verlag 2006

Abstract. We present an overview of concepts and results obtained with statistical models in the study of nuclear multifragmentation. Conceptual differences between statistical and dynamical approaches and the selection of experimental observables for identification of these processes are outlined. New and perspective developments, like inclusion of in-medium modifications of the properties of hot primary fragments, are discussed. We list important applications of statistical multifragmentation in other fields of research.

PACS. 25.70.Pq Multifragment emission and correlations – 24.60.-k Statistical theory and fluctuations – 21.65.+f Nuclear matter

1 Introduction

Statistical models have proven to be very successful in nuclear physics. They are used for the description of nuclear decay when an equilibrated source can be identified in the reaction. The most famous example of such a source is the "compound nucleus" introduced by Niels Bohr in 1936 [1]. It was clearly seen in low-energy nuclear reactions leading to excitation energies of a few tens of MeV. It is remarkable that this concept works also for nuclear reactions induced by particles and ions of intermediate and high energies, when nuclei break up into many fragments (multifragmentation). According to the statistical hypothesis, initial dynamical interactions between nucleons lead to a re-distribution of the available energy among many degrees of freedom, and the nuclear system evolves towards equilibrium. In the most general consideration the process may be subdivided into several stages: 1) a dynamical stage leading to formation of equilibrated nuclear system, 2) the disassembly of the system into individual primary fragments, 3) the de-excitation of hot primary fragments. Below we consider these stages step by step. In this paper we give an overview of main results obtained with statistical models in multifragmentation studies, and analyze the most important problems (see also reviews [2–4]). Several hundred papers concerning multifragmentation were published during the last two decades, and we apologize that in a short review we cannot mention all works related to this field.

[a] e-mail: a.botvina@gsi.de

2 Formation of a thermalized nuclear system

At present, a number of dynamical models is used for the description of nuclear reactions at intermediate energies. The intranuclear cascade model was the first one used for realistic calculations of ensembles of highly excited residual nuclei which undergo multifragmentation, see, *e.g.* [5]. Other more sophisticated models were also used for dynamical simulations of heavy-ion reactions, such as quantum molecular dynamics (QMD), Boltzmann (Vlasov)-Uehling-Uhlenbeck (BUU, VUU) and other similar models (see, *e.g.*, ref. [6]). All dynamical models agree that the character of the dynamical evolution changes after a few rescatterings of incident nucleons, when high-energy particles ("participants") leave the system. This can be seen from distributions of nucleon velocities and density profiles in remaining spectators [7–10]. However, the time needed for equilibration and transition to the statistical description is still under debate. This time is estimated around or less than $100 \, \mathrm{fm}/c$ for spectator matter, however, it slightly varies in different models. Apparently, this time should be shorter for the participant zone produced in heavy-ion collisions at energies above the Fermi energy, as a result of initial compression. Parameters of the predicted equilibrated sources, *i.e.* their excitation energies, mass numbers and charges vary significantly with this time. We believe that the best strategy is to use results of the dynamical simulations as a qualitative guide line, but extract parameters of thermalized sources from the analysis of experimental data. In this case, one can avoid uncertainties of dynamical models in describing thermalization processes.

3 Break-up of a thermalized system into hot primary fragments

3.1 Evolution from sequential decay to simultaneous break-up

After dynamical formation of a thermalized source, its further evolution depends crucially on the excitation energy and mass number. The standard compound nucleus picture is valid only at low excitation energies when sequential evaporation of light particles and fission are the dominant decay channels. Some modifications of the evaporation/fission approach were proposed in order to include emission of fragments heavier than α-particles, see e.g. [11–13]. However, the concept of the compound nucleus cannot be applied at high excitation energies, $E^* \gtrsim 3\,\mathrm{MeV/nucleon}$. The reason is that the time intervals between subsequent fragment emissions, estimated both within the evaporation models [14] and from experimental data [15], become very short, of the order of a few tens of fm/c. In this case there will be not enough time for the residual nucleus to reach equilibrium between subsequent emissions. Moreover, the produced fragments will be in the vicinity of each other and, therefore, should interact strongly. The rates of the particle emission calculated as for an isolated compound nucleus will not be reliable in this situation. On the other hand, the picture of a nearly simultaneous break-up in some freeze-out volume seems more justified in this case. Indeed, the time scales of less than $100\,\mathrm{fm}/c$ are extracted for multifragmentation reactions from experimental data [16,17]. Sophisticated dynamical calculations have also shown that a nearly simultaneous break-up into many fragments is the only possible way for the evolution of highly excited systems, e.g. [10,18]. Theoretical arguments in favor of a simultaneous break-up follows also from the Hartree-Fock and Thomas-Fermi calculations which predict that the compound nucleus will be unstable at high temperatures [19].

There exist several analyses of experimental data which reject the binary decay mechanism of fragment production via sequential evaporation from a compound nucleus, at high excitation energy. For example, this follows from the fact that the popular sequential GEMINI code cannot describe the multifragmentation data [20–22]. We believe that a formal reason for this failure is that the evaporation approaches always predict larger probabilities for emission of light particles (in particular, neutrons) than for intermediate mass fragments (IMFs). We mention also attempts to extend the compound nucleus picture by including its expansion within the harmonic-interaction Fermi gas (HIFGM) model [23], and within the expanding emitting-source (EES) model [24]. However, these models have the same theoretical problem with short emission times. Unfortunately, the EES model has never been compared with multifragmentation experiment in a comprehensive way since it is limited by considering emission of light IMFs with charges $Z \lesssim 10$ only.

As was shown already in early statistical model calculations, see e.g. [25], the entropy of the compound nucleus dominates over entropies of multifragmentation channels at low energies, but this trend reverses at high excitation energies. This means that the evaporation/fission based models can only be used at excitation energies below the multifragmentation threshold, $E_{th} = 2\text{–}4\,\mathrm{MeV/nucleon}$. At higher excitations, a simultaneous emission must be the preferable assumption. Close to the onset of multifragmentation the most probable decay channels contain one (compound-like), or two (fission-like) fragments, and a few small fragments. With increasing excitation energy the break-up into several IMFs becomes more probable, and at very high excitation energies the decay channels with nucleons and lightest fragments (vaporization) dominate. Such evolution of nuclear-decay mechanisms is predicted by all statistical models.

3.2 Statistical models of multifragmentation

The main concepts of the statistical approach to nuclear multifragmentation have been formulated in the 1980s by Randrup et al. [26], Gross et al. (MMMC) [27], and Bondorf et al. (SMM) [25,28,12]. This approach is based on the assumption that the relative probabilities of different break-up channels are determined by their statistical weights, which include contributions of phase space (spatial and momentum) factors and level density of internal excitations of fragments. Different versions of the model differ in details of the description of individual fragments, Coulomb interaction and choice of statistical ensembles (grand-canonical, canonical, or microcanonical). Usually, all these details do not affect significantly qualitative features of the statistical break-up. For example, the differences in ensembles can hardly be seen in fragment distributions at high excitation energies [3], unless the observables are selected in a very special way. As was later demonstrated in experiments of many groups: ALADIN [29], EOS [30], ISIS [31], Miniball-Multics [32], INDRA [33], FASA [34], NIMROD [35] and others, equilibrated sources are indeed formed in nuclear reactions, and statistical models are very successful in describing the fragment production from them. This proves that the multifragmentation process to a large extent is controlled by the available phase space including internal excitations of fragments. Furthermore, systematic studies of such highly excited systems have brought important information about a liquid-gas phase transition in finite nuclear systems [36,37].

The success of the first statistical models has stimulated the appearance of their new versions in the next decades. The models MMM [38] and ISMM [39] are based on the same principles and use the same methods with small modifications. In the SIMON code [40], fragments evaporated from the compound nucleus are placed in a common volume in order to simulate a simultaneous break-up. There were also developments of the original models: SMM [41], and MMMC [42], bringing some improvements seen as necessary from the analysis of experimental data. An interesting mathematical development has been made in ref. [43], where a canonical version of the

SMM with simple partition weights was exactly analytically resolved by using recursive relations for the partition sum. Most models use the Boltzmann statistics, since the number of particular fragments in the freeze-out volume is typically of order 1. Calculations of ref. [44] have demonstrated that the quantum statistical effects do not play a role for all species, but only for nucleons at excitation energies and entropies characterizing multifragmentation. The same conclusion has been made in ref. [45] by direct comparisons of SMM with a quantum statistical model (QSM) [46].

As a rule, all above-mentioned statistical models give very similar results concerning the description of mean characteristics of multifragmentation. For example, the description of ALADIN experiments requires ensembles of emitting sources which in SMM, MMMC and MMM models differ within 10% of their masses and excitation energies [29,47,48]. Such an uncertainty is of the same magnitude as the precision of most experimental data. One can see some differences between the models only in more sophisticated observables. For example, the isotope properties of produced fragments, especially the isoscaling observables, may allow for better discrimination between different approaches, as well as between parameters within a specific model [49,50].

3.3 Fragment formation and freeze-out volume

In a simplified consideration, all simultaneously produced fragments are placed within a fixed freeze-out volume. It is assumed that nuclear interactions between the fragments cease at this point and that, at later times, fragments propagate independently in the mutual Coulomb field. In fact, there is a deep physical idea behind this simple picture. During the fragment formation the nucleons move in a common mean field and experience stochastic collisions. When collisions practically cease, the relatively cold group of nucleons gets trapped by the local mean field and forms fragments [51]. It is assumed that there exists a certain point in the space-time evolution which is crucial for the final fate of the system. This is a so-called "saddle" point, and the freeze-out volume provides a space for the "saddle" point configurations. According to the statistical approach, the probabilities of the fragment partitions are determined by their statistical weights at the "saddle" point. Actually, the nuclear interactions between fragments may not cease completely after the "saddle" point, however, they do not change the fragment partitions which have been decided at this point. Only when the system reaches the "scission" point the contact between the fragments is finally disrupted. This picture may be justified by the analogy with nuclear fission, where the existence of "saddle" and "scission" points is commonly accepted.

In most statistical models one assumes that "saddle" and "scission" points coincide and the statistical weight is characterized by a single freeze-out volume. On the other hand, one should distinguish the full geometrical volume and a so-called "free" volume, which is available for the fragment translational motion in coordinate space. Due to the final size of fragments and their mutual interaction this free volume is smaller than the physical freeze-out volume, at least, by the proper volume of all produced fragments. This "excluded volume" can be included in statistical models with different prescriptions, which, however, must respect the conservation laws [52]. In the SMM there are two distinct parameters which control the free volume and the freeze-out volume. In some respects these two different volumes are introduced similar to "saddle" and "scission" points discussed above. In principal, the different volumes should be extracted from the analysis of experimental data [30,53]. Since the entropy associated with the translational motion is typically much smaller than the entropy associated with the internal excitation of fragments, uncertainties in the determination of the free volume do not affect significantly the model predictions, especially in the case of break-up into few fragments.

There are several schematic views of how the fragments are positioned in coordinate space. The most popular picture assumes expansion of uniform nuclear matter to the freeze-out volume, accompanied by its "cracking" and fragment formation. However, this picture is more appropriate for the processes with a large excitation energy and flow, and corresponds to the transition of the nucleon "gas" to the "liquid" drops by cooling during the expansion. This picture cannot be applied at energies close to the multifragmentation threshold since they are not sufficient for the essential uniform expansion of the nucleus. At $E^* \gtrsim E_{th}$ the picture of a simultaneous "fission" into several fragments seems more appropriate. One should bear in mind that for the statistical description it is not important how the system has evolved toward the "saddle" point. The only assumption in this case is that the phase space and level density factors dominate over the transition matrix elements. This explains why different models are rather consistent with each other irrespective of the way how the fragment positioning is made.

The average density which corresponds to the freeze-out volume is usually taken in the range between 1/3 and 1/10 of the normal nuclear density $\rho_0 \approx 0.15\,\mathrm{fm}^{-3}$. In the case of thermal multifragmentation the freeze-out density can be reliably estimated from experimental data on fragment velocities since, to 80–90%, they are determined by the Coulomb acceleration after the break-up. The experimental analyses of the kinetic energies, angle and velocity correlations of the fragments indeed point to values of $(0.1$–$0.4)\,\rho_0$ [54,55].

3.4 Fragments in the statistical approach

Another important concept refers to "primary fragments", *i.e.*, the fragments which are produced in the freeze-out volume. The properties of these fragments essentially determine the statistical weights of the partitions. The simplest approximation is to use the masses (or binding energies) of the nuclei from the nuclear data tables referring to cold isolated nuclei, for example, as it is done in MMMC, or in ISMM. In order to calculate the contribution of internal fragment excitations to the statistical weight one should introduce additional assumptions concerning their level densities. For example, the MMMC prescription is 1) to limit the internal excitation of fragments by particle stable levels only (this leads to relatively cold fragments), and 2) to include in the statistical weight the contribution of secondary neutrons, which are assumed to be evaporated instantaneously from primary fragments in the freeze-out volume [2]. Randrup *et al.* [26] and MMM [38] use a Fermi-gas type approximation with a cut-off at high temperatures. In the ISMM this is done via level density expressions motivated by empirical information for isolated nuclei [39]. However, as clear from the previous discussion, the approximations used for isolated nuclei may not be true in the freeze-out volume since the fragments can still interact and, therefore, have modified properties. For example, as was noted long ago, the neutron content of primary fragments can be changed due to a reduced Coulomb interaction in the hot environment of nucleons and other fragments [12,25,41].

In order to include possible in-medium effects, the SMM has adopted a liquid-drop description of individual fragments ($A > 4$) extended for the case of finite temperatures and densities [3]. Smaller clusters are considered as elementary particles. At low excitation energies this description corresponds to known properties of cold nuclei, but it is generalized for the consideration of highly excited nuclei in the medium. The parameters of the liquid-drop description change as a result of interactions between the fragments leading, in particular, to modifications of bulk, surface and Coulomb terms. These parameters can be evaluated from the analysis of experimental data. As shown in ref. [50], information on possible changes in the symmetry energy of hot fragments can be extracted from the isoscaling data. This method is presently under discussion in the community. There are already experimental evidences that the simmetry energy of hot fragments in multifragmentation is significantly reduced as compared with cold nuclei [56–58]. Forthcoming experiments should clarify this conclusion.

We emphasize that in-medium modifications of fragment properties is a natural way to include the interaction between fragments within the statistical approach. Recently, an attempt has been made [59] to consider the evolution of the fragments after freeze-out within the framework of a dynamical model with explicit inclusion of nuclear interactions. As reported, this interaction results in a fusion (recombination) of primary fragments, and thus modifies the fragment partitions. However, the dynamical fragment formation after the "statistical freeze-out" leads to violation of fundamental assumptions of the statistical approach, such as the ergodicity and detailed balance principles. Generally, an application of a time-dependent approach (dynamics) to a statistical ensemble would be a controversial operation since, according to the ergodicity principle, the time average over microscopic configurations must be equivalent to the ensemble average. The dynamical consideration may be justified only for the long-range

Coulomb forces influencing fragments' motion after their formation. Therefore, results of ref. [59] are misleading and cannot be considered as an improvement of the statistical approach.

3.5 Influence of flow on fragment formation

As was established experimentally, an "ideal" picture of thermal multifragmentation begins to fail at excitation energies of about 5–6 MeV/nucleon [60]. At higher excitations a part of the energy goes into a collective kinetic energy of the produced fragments, without thermalization. This energy is defined as the flow energy, and its share depends on the kind of reaction. For example, at thermal excitation energy of $E^* \approx 6$ MeV/nucleon, the additional flow energy is around 0.2 MeV/nucleon in hadron-induced reactions, and it is around 1.0 MeV/nucleon in central heavy-ion collisions around the Fermi energy. Since a dynamical flow itself can break matter into pieces, it is necessary to understand limits of the statistical description in the case of a strong flow.

This problem was addressed in a number of works within dynamical and lattice-gas models [10,61–63]. Their conclusion is that a flow does not change statistical model predictions, if its energy is essentially smaller than the thermal energy. This justifies a receipt often used in statistical models, when the flow energy is included by increasing the velocities of fragments in the freeze-out volume according to the flow velocity profile [3]. This is in agreement with many experimental analyses. However, statistical models work surprisingly well even when the flow energy is comparable with the thermal energy, or even higher [42,64]. This observation requires additional study.

3.6 Nuclear liquid-gas phase transition within statistical models

Many statistical models have demonstrated that multifragmentation is a kind of a phase transition in highly excited nuclear systems. In the SMM a link to the liquid-gas phase transition is especially strong. In particular, the surface energy of hot primary fragments is parametrized in such a way that it vanishes at a certain critical temperature. The SMM has predicted distinctive features of this phase transition in finite nuclei, such as the plateau-like anomaly in the caloric curve [3,28], which have been later observed in experiments [36,65]. Many other manifestations of the phase transition, such as large fluctuations and bimodality [29,37,66], critical behavior and even values of critical exponents [37,67], have been investigated within this model. The experimental data are usually in agreement with the predictions.

Nevertheless, the properties of this phase transition are not yet fully understood. The critical behavior observed in experimental data can also be explained within a percolation model [68], or a Fisher's droplet model [69], which correspond to a second-order phase transition in the vicinity of the critical point. We must note, however, that the finiteness of the systems under investigation plays a crucial role. To connect this anomalous behavior with a real phase transition one should study it in a thermodynamical limit. Within the SMM this was done in ref. [70], where multifragmentation of an equilibrated system was identified as a first-order phase transition. The mixed phase in this case consists of an infinite liquid condensate and gas of nuclear fragments of all masses. In a finite system this mixed phase corresponds to U-shaped fragment distributions with a heaviest fragment representing the liquid phase. Thus one can connect multifragmentation of finite nuclei with the fragmentation of a very big system. This is important for the application of statistical models in astrophysical environments (neutron stars, supernova explosions), where nuclear statistical equilibrium can also be expected [71].

3.7 Relation between statistical and dynamical descriptions

One of the problems, which is highly debated now, is if dynamical models alone can describe (at least qualitatively) the same evolutionary scenario leading to equilibration and multifragmentation as assumed by statistical models. In other words, is it possible to use only a "universal" dynamical description, instead of subdividing the process into dynamical and statistical stages? Some dynamical approaches try to reach this goal starting from "first principles" like Fermionic molecular dynamics (FMD) [72], or antisymmetrized molecular dynamics (AMD) [73]. Other approaches, like QMD [6,9], NMD [10], or BNV [74] use classical equations including two-body collisions and some elements of stochasticity. In all cases dynamical simulations are more complicated and time-consuming as compared with statistical models. This is why full calculations, e.g. with FMD and AMD models, can only be done for relatively light systems. By using simplified receipts, like a coalescence for final fragment definition in AMD, one may reduce the computing time, but it still remains rather long. This prevents using these codes in practical calculations of nuclear fragmentation in an extended complex medium, which are required, e.g., in medicine, space research, and other fields. A natural solution of this problem is to develop hybrid approaches which combine dynamical models for describing the nonequilibrium early stages of the reaction with statistical models for describing the fragmentation of equilibrated sources. In this respect, these two approaches are complementary.

One should bear in mind that the statistical and dynamical approaches are derived from different physical principles. The time-dependent dynamical approaches are based on Hamiltonian dynamics (the principle of minimal action), whereas the statistical models employ the principle of uniform population of the phase space. Actually, these two principles are not easily reducible to each other, and they represent complementary methods for describing the physical reality. There are numerous studies of the phase space population with dynamical models (see, e.g. [75]), which, however, have never shown the uniform population in the limit of long times. Therefore, a decision of using statistical or dynamical approaches for the description of nuclear multifragmentation should be made after careful examination of the degree of equilibration expected in particular cases, and it can only be justified by the comparison with experiment.

There is still a large difference in details between "statistical" and "dynamical" descriptions of individual fragments as finite quantum systems. As a rule, the realistic description of clustering is difficult to achieve in dynamical models dealing with individual nucleons, but it is easily done in statistical models, considering nuclear fragments as independent degrees of freedom. In the case of equilibrated sources predictions of statistical models are usually in better agreement with experimental data. This especially concerns isospin observables. Experimental data demonstrate a characteristic trend: increasing neutron richness of intermediate mass fragments in collisions of nuclei with increasing centrality, i.e. with increasing excitation energy, both in the "neck region" [76] and in the equilibrated sources [77]. This trend can easily be explained in the framework of the statistical model [41]. Until now dynamical models are not very successful in describing isoscaling observables (e.g., the slope coefficients) [78], while they are naturally explained within statistical approaches [49,50].

4 De-excitation and propagation of hot primary fragments

After their production in the freeze-out volume, primary fragments will propagate in the mutual Coulomb field and undergo de-excitation. It is usually assumed that the long-range Coulomb force, which has participated only partly in the fragment formation, is fully responsible for the post–freeze-out acceleration of the fragments. All statistical models solve classical Newton equations, taking into account the initial positions of fragments inside the freeze-out volume and their thermal velocities. At this stage the collective flow of fragments can also be taken into consideration.

The hot fragments will lose excitation in the course of their propagation to the detectors. There are different secondary de-excitation codes used in multifragmentation studies. The standard fission-evaporation and Fermi-break-up codes described in [12] were used in SMM [3] and MMM [48]. Another procedure, which includes GEMINI for de-excitation of big fragments, was adopted in the ISMM [39]. In the MMMC [2] a schematic model was used which takes into account only early emission of secondary neutrons. Apparently, this oversimplification is responsible for deviations of the MMMC predictions from other models in description of correlations between neutrons and charged particles [79]. It should be emphasized that most de-excitation models are based on properties of cold isolated nuclei, known from experiments at low energy. At present there is a need in more advanced models, which take into account possible in-medium modifications of primary fragments in the freeze-out volume, e.g., changing their symmetry and surface energy. An example of such a model is presented in ref. [66].

The de-excitation process depends strongly on the nuclear content of the primary fragments. For example, in the SMM at E^* slightly above E_{th}, almost all nucleons are contained in fragments, and the fraction of free nucleons is negligible. This shows an analogy with the fission process. As a result the neutron content of primary fragments is nearly the same as in the initial source. The outcome of de-excitation depends on the actual code used. Generally, in realistic statistical models most neutrons come from the secondary de-excitation stage, for example, more than 90% in the SMM. If one takes into account a reduction of the symmetry energy of primary fragments, and includes its restoration in the course of de-excitation, the neutron richness of cold final fragments will be larger than predicted by standard codes [66].

5 Conclusions

We believe that statistical models suit very well for the description of such a complicated many-body process as nuclear multifragmentation. If a thermalized source can be recognized in a nuclear reaction, the main features of multiple fragment production can be well described within the statistical approach. The success of statistical models in describing a broad range of experimental data gives us confidence that this approach will be used and further developed in the future. We especially stress two main achievements of statistical models in the theory of nuclear reactions: first, a clear understanding has been reached that sequential decay via compound nucleus must give way to a nearly simultaneous break-up of nuclei at high excitation energies; and, second, the character of this change can be interpreted as a liquid-gas–type phase transition in finite nuclear systems.

The results obtained in the nuclear multifragmentation studies can be applied in several other fields. First, the mathematical methods of the statistical multifragmentation can be used for developing thermodynamics of finite systems [80,81]. These studies were stimulated by the recent observation of extremely large fluctuations of the energy of produced fragments, which can be interpreted as the negative heat capacity [82]. At this point one can see links with cluster physics and condensed-matter physics [80]. These methods might also be useful for investigating possible phase transitions from hadronic matter to quark-gluon plasma in relativistic heavy-ion collisions.

Another conclusion is related to the fact that the multifragmentation channels take as much as 10–15% of the total cross-section in high-energy hadron-nucleus reactions, and about twice more in high-energy nucleus-nucleus collisions. Moreover, multifragmentation reactions are responsible for the production of some specific isotopes. The importance of multifragmentation reactions is now widely recognized and, in recent years, the attention given to them has risen in several domains of research. Indeed, practical calculations of fragment production and

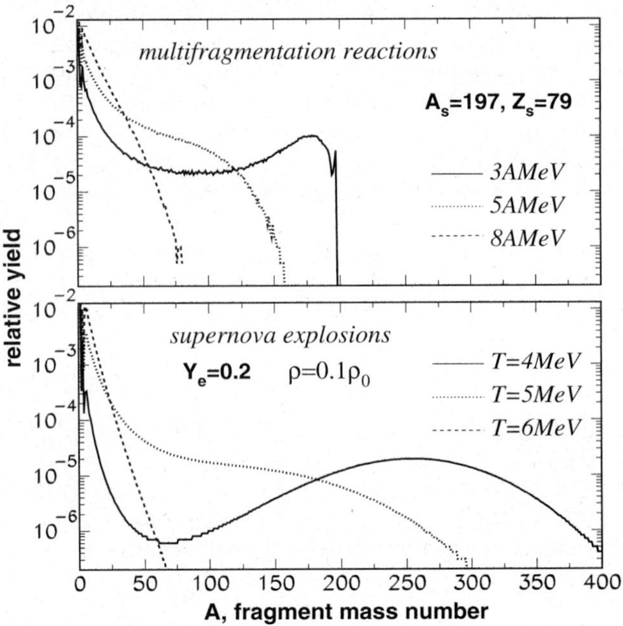

Fig. 1. Mass distributions of hot fragments calculated with SMM: top panel —for multifragmentation of Au nuclear sources at excitation energies of 3, 5, and 8 MeV per nucleon; and, bottom panel —for stellar matter at the baryon density ρ and electron fraction Y_e, typical for supernova explosions, at temperatures 4, 5, and 6 MeV.

transport in complex medium are needed for: nuclear-waste transmutation (environment protection), electro-nuclear breeding (new methods of energy production), proton and ion therapy (medical applications), radiation protection of space missions (space research). Until recently, only evaporation and fission codes have been used for describing the nuclear de-excitation. We believe that the state of the art today requires the inclusion of multi-fragmentation reactions in these calculations. The SMM is especially suitable for this purpose because of its multifunctional code structure: Besides the multifragmentation channels it includes also compound nucleus decays via evaporation and fission, and takes into account competition between all channels. Encouraging attempts to construct hybrid models, combining dynamical and statistical approaches, were undertaken in refs. [5,7,83]. The hybrid models are quite successful in describing data, including correlation observables between dynamical and statistical stages [3,84,85]. Several multi-purpose codes, like GEANT4 [86], have been developed to describe transport of hadrons and ions in an extended medium. The SMM was included in this code as important part responsible for fragment production.

It is important that nuclear multifragmentation reactions allow for the experimental determination of in-medium modifications of hot nuclei/fragments in hot and dense environments. This opens the unique possibility for investigating the phase diagram of nuclear matter at temperatures $T \approx 3$–8 MeV and densities around $\rho \approx 0.1$–$0.3\rho_0$, which are expected in the freeze-out volume. These studies are complementary to the previous studies of isolated nuclei existing in the matter with terrestrial densities, and at low temperatures, $T < 1$–$2\,\mathrm{MeV}$. The experimental information on properties of hot nuclei in dense surroundings is crucial for the construction of a reliable equation of state of stellar matter and for the modeling of the nuclear composition in supernovae [71].

As illustration of this connection we show in fig. 1 examples of fragment mass distributions produced in multifragmentation reactions (see [37,41]), and in astrophysical conditions associated with supernova type-II explosions. The calculations of nuclear compositions of stellar matter at subnuclear densities were carried out with the SMM generalized for astrophysical conditions [71,87]. One can see that the evolution of the mass distribution with excitation energy is qualitatively the same for both, the nuclear multifragmentation reactions and the supernova explosions. However, in the supernova environments much heavier and neutron-rich nuclei can be produced because of screening of their charge by surrounding electrons. This shows that studying the multifragmentation reactions in the laboratory is important for understanding how heavy elements were synthesized in the Universe.

We thank S. Das Gupta for careful reading this paper and positive comments. We also thank the participants of the WCI workshops for stimulating discussions and the organizers for hospitality and support. This work was supported in part by the grant RFFR 05-02-04013 (Russia).

References

1. N. Bohr, Nature **137**, 344 (1936).
2. D.H.E. Gross, Rep. Prog. Phys. **53**, 605 (1990).
3. J.P. Bondorf, A.S. Botvina, A.S. Iljinov, I.N. Mishustin, K. Sneppen, Phys. Rep. **257**, 133 (1995).
4. S. Das Gupta, A.Z. Mekjian, B. Tsang, Adv. Nucl. Phys. **26**, 89 (2001).
5. A.S. Botvina, A.S. Iljinov, I.N. Mishustin, Nucl. Phys. A **507**, 649 (1990).
6. H. Stoecker, W. Greiner, Phys. Rep. **137**, 277 (1986); J. Aichelin et al., Phys. Rev. C **37**, 2451 (1988); W. Bauer et al., Annu. Rev. Nucl. Part. Sci. **42**, 77 (1992); B.A. Li, Phys. Rev. C **47**, 693 (1993); J. Konopka et al., Prog. Part. Nucl. Phys. **30**, 301 (1993); M. Colonna et al., Nucl. Phys. A **589**, 160 (1995); C. Fuchs, H.H. Wolter, Nucl. Phys. A **589**, 732 (1995).
7. H.W. Barz et al., Nucl. Phys. A **561**, 466 (1993).
8. J.P. Bondorf, A.S. Botvina, I.N. Mishustin, S.R. Souza, Phys. Rev. Lett. **73**, 628 (1994).
9. A.S. Botvina, A.B. Larionov, I.N. Mishustin, Phys. At. Nucl. **58**, 1703 (1995).
10. J.P. Bondorf, O. Friedrichsen, D. Idier, I.N. Mishustin, Nucl. Phys. A **624**, 706 (1997).
11. L.G. Moretto, Nucl. Phys. A **247**, 211 (1975).
12. A.S. Botvina et al., Nucl. Phys. A **475**, 663 (1987).
13. R.J. Charity et al., Nucl. Phys. A **483**, 371 (1988).
14. R.J. Charity, Phys. Rev. C **61**, 054614 (2000).
15. M. Jandel et al., J. Phys. G **31**, 29 (2005).
16. L. Beaulieu et al., Phys. Rev. Lett. **84**, 5971 (2000).
17. V.A. Karnaukhov et al., Phys. At. Nucl. **66**, 1242 (2003).
18. S. Das Gupta et al., Phys. Rev. C **35**, 556 (1987); B. Strack, Phys. Rev. C **35**, 691 (1987); D.H. Boal, J.N. Gloshi, Phys. Rev. C **37**, 91 (1988).
19. P. Bonche, S. Levit, D. Vautherin, Nucl. Phys. A **436**, 265 (1985); E. Suraud, Nucl. Phys. A **462**, 109 (1987).
20. J. Hubele et al., Phys. Rev. C **46**, R1577 (1992).
21. P. Desesquelles et al., Nucl. Phys. A **604**, 183 (1996).
22. P. Napolitani et al., Phys. Rev. C **70**, 054607 (2004).
23. J. Toke et al., Phys. Rev. C **72**, R031601 (2005).
24. W.A. Friedman, Phys. Rev. C **42**, 667 (1990).
25. A.S. Botvina, A.S. Iljinov, I.N. Mishustin, Sov. J. Nucl. Phys. **42**, 712 (1985).
26. J. Randrup, S.E. Koonin, Nucl. Phys. A **356**, 223 (1981); S.E. Koonin, J. Randrup, Nucl. Phys. A **471**, 355c (1987); J. Randrup, Comput. Phys. Commun. **77**, 153 (1993).
27. D.H.E. Gross et al., Z. Phys. A **309**, 41 (1982); X.Z. Zhang, D.H.E. Gross, S. Xu, Y.M. Zheng, Nucl. Phys. A **461**, 641 (1987).
28. J.P. Bondorf, R. Donangelo, I.N. Mishustin, C.J. Pethick, H. Schulz, K. Sneppen, Nucl. Phys. A **443**, 321 (1985); J.P. Bondorf, R. Donangelo, I.N. Mishustin, H. Schulz, Nucl. Phys. A **444**, 460 (1985).
29. A.S. Botvina et al., Nucl. Phys. A **584**, 737 (1995).
30. R.P. Scharenberg et al., Phys. Rev. C **64**, 054602 (2001).
31. L. Pienkowski et al., Phys. Rev. C **65**, 064606 (2002).
32. M. D'Agostino et al., Phys. Lett. B **371**, 175 (1996).
33. N. Bellaize et al., Nucl. Phys. A **709**, 367 (2002).
34. S.P. Avdeyev et al., Nucl. Phys. A **709**, 392 (2002).
35. J. Wang et al., Phys. Rev. C **71**, 054608 (2005).
36. J. Pochodzalla et al., Phys. Rev. Lett. **75**, 1040 (1995).
37. M. D'Agostino, A.S. Botvina, M. Bruno, A. Bonasera, J.P. Bondorf, I.N. Mishustin et al., Nucl. Phys. A **650**, 329 (1999).
38. Al.H. Raduta, Ad.R. Raduta, Phys. Rev. C **55**, 1344 (1997).
39. M.P. Tan et al., Phys. Rev. C **68**, 034609 (2003).
40. D. Durand, Nucl. Phys. A **541**, 266 (1992).
41. A.S. Botvina, I.N. Mishustin, Phys. Rev. C **63**, 061601(R) (2001).
42. A.Le Fevre et al., Nucl. Phys. A **735**, 219 (2004).
43. S. Das Gupta, A.Z. Mekjian, Phys. Rev. C **57**, 1361 (1998).
44. A.S. Parvan et al., Nucl. Phys. A **676**, 409 (2000).
45. A.S. Botvina et al., Z. Phys. A **345**, 297 (1993).
46. D. Hahn, H. Stoeker, Nucl. Phys. A **476**, 718 (1988).
47. H. Xi et al., Z. Phys. A **359**, 397 (1997).
48. Al.H. Raduta, Ad.R. Raduta, Phys. Rev. C **61**, 034611 (2000).
49. M.B. Tsang et al., Phys. Rev. C **64**, 054615 (2001).
50. A.S. Botvina, O.V. Lozhkin, W. Trautmann, Phys. Rev. C **65**, 044610 (2002).
51. J.P. Bondorf, D. Idier, I.N. Mishustin, Phys. Lett. B **359**, 261 (1995).
52. A.S. Botvina, I.N. Mishustin, Phys. Rev. Lett. **90**, 179201 (2003).
53. V.A. Karnaukhov et al., Phys. Rev. C **70**, 041601 (2004).
54. S. Fritz et al., Phys. Lett. B **461**, 315 (1999).
55. V.E. Viola et al., Phys. Rev. Lett. **93**, 132701 (2004).
56. A. Le Fevre et al., Phys. Rev. Lett. **94**, 162701 (2005).
57. D.V. Shetty et al., Phys. Rev. C **71**, 024602 (2005).
58. J. Iglio et al., Phys. Rev. C **74**, 024605 (2006).
59. S.K. Samaddar, J.N. De, A. Bonasera, Phys. Rev. C **71**, 011601 (2005).

60. T. Lefort *et al.*, Phys. Rev. C **62**, 031604 (2000).
61. C.B. Das, L. Shi, S. Das Gupta, Phys. Rev. C **70**, 064610 (2004).
62. F. Gulminelli, Ph. Chomaz, Nucl. Phys. A **734**, 581 (2004).
63. I.N. Mishustin, AIP Conf. Proc. **512**, 308 (2000).
64. W. Neubert, A.S. Botvina, Eur. Phys. J. A **17**, 559 (2003).
65. J.B. Natowitz *et al.*, Phys. Rev. C **65**, 034618 (2002).
66. N. Buyukcizmeci *et al.*, Eur. Phys. J. A **25**, 57 (2005).
67. B.K. Srivastava *et al.*, Phys. Rev. C **65**, 054617 (2002).
68. W. Bauer, A.S. Botvina, Phys. Rev. C **52**, R1760 (1995); M. Kleine Berkenbusch *et al.*, Phys. Rev. Lett. **88**, 022701 (2002).
69. J.B. Elliott *et al.*, Phys. Rev. Lett. **88**, 042701 (2002).
70. K.A. Bugaev *et al.*, Phys. Rev. C **62**, 044320 (2000).
71. A.S. Botvina, I.N. Mishustin, Phys. Lett. B **584**, 233 (2004).
72. H. Feldmeier, Nucl. Phys. A **681**, 398c (2001).
73. A. Ono, H. Horiuchi, Phys. Rev. C **53**, 2958 (1996).
74. V. Baran *et al.*, Nucl. Phys. A **703**, 603 (2002).
75. J.P. Bondorf, H. Feldmeier, I.N. Mishustin, G. Neergaard, Phys. Rev. C **65**, 017601 (2002).
76. H. Xu *et al.*, Phys. Rev. C **65**, 061602 (2002).
77. P.M. Milazzo *et al.*, Phys. Rev. C **62**, 041602 (2000).
78. T.X. Liu *et al.*, Phys. Rev. C **69**, 014603 (2004).
79. J. Toke *et al.*, Phys. Rev. C **63**, 024604 (2001).
80. D.H.E. Gross, Phys. Rep. **279**, 119 (1997).
81. Ph. Chomaz, F. Gulminelli, V. Duflot, Phys. Rev. E **64**, 046114 (2001).
82. M. D'Agostino *et al.*, Phys. Lett. B **473**, 219 (2000).
83. T.C. Sangster *et al.*, Phys. Rev. C **46**, 1404 (1992).
84. C. Volant *et al.*, Nucl. Phys. A **734**, 545 (2004).
85. K. Turzo *et al.*, Eur. Phys. J. A **21**, 293 (2004).
86. S. Agostinelli *et al.*, Nucl. Instrum. Methods A **506**, 250 (2003).
87. A.S. Botvina, I.N. Mishustin, Phys. Rev. C **72**, 048801 (2005).

Eur. Phys. J. A **30**, 129–139 (2006)
DOI 10.1140/epja/i2006-10111-0

THE EUROPEAN
PHYSICAL JOURNAL A

Comparisons of statistical multifragmentation and evaporation models for heavy-ion collisions

M.B. Tsang[1,a], R. Bougault[2], R. Charity[3], D. Durand[2], W.A. Friedman[4], F. Gulminelli[2], A. Le Fèvre[5], Al.H. Raduta[6,7], Ad.R. Raduta[7], S. Souza[8], W. Trautmann[5], and R. Wada[9]

[1] National Superconducting Cyclotron Laboratory and Department of Physics and Astronomy, Michigan State University, East Lansing, MI 48824, USA
[2] LPC/Ensicaen and University of Caen, 6 Bd du Maréchal Juin, F-14050 Caen cedex, France
[3] Chemistry Department, Washington University, St. Louis, MO 63130, USA
[4] Department of Physics, University of Wisconsin, Madison, WI 53706, USA
[5] Gesellschaft für Schwerionenforschung mbH, D-64291 Darmstadt, Germany
[6] Laboratori Nazionali del Sud and INFN, I-95123, Catania, Italy
[7] NIPNE, RO-76900 Bucharest-Magurele, Romania
[8] Instituto de Fisica, Universidade Federal do Rio de Janeiro, Cidade Universitária, CP 68528, 21945-970 Rio de Janeiro, Brazil
[9] Cyclotron Institute, Texas A&M University, College Station, TX 77843, USA

Received: 26 July 2006 /
Published online: 30 October 2006 – © Società Italiana di Fisica / Springer-Verlag 2006

Abstract. The results from ten statistical multifragmentation models have been compared with each other using selected experimental observables. Even though details in any single observable may differ, the general trends among models are similar. Thus, these models and similar ones are very good in providing important physics insights especially for general properties of the primary fragments and the multifragmentation process. Mean values and ratios of observables are also less sensitive to individual differences in the models. In addition to multifragmentation models, we have compared results from five commonly used evaporation codes. The fluctuations in isotope yield ratios are found to be a good indicator to evaluate the sequential decay implementation in the code. The systems and the observables studied here can be used as benchmarks for the development of statistical multifragmentation models and evaporation codes.

PACS. 25.70.Mn Projectile and target fragmentation – 25.70.Gh Compound nucleus – 25.70.Pq Multi-fragment emission and correlations

1 Introduction

During the later stages of a central collision between heavy nuclei at incident energies in excess of about $E/A = 50\,\mathrm{MeV}$, a rapid collective expansion of the combined system occurs [1]. Experimental evidence indicates that mixtures of intermediate-mass fragments (IMFs) with $3 \leq Z \leq 30$ and light charged particles (LCP, $Z \leq 2$) are emitted during this expansion stage. With increased nucleon collisions, the properties of the nuclear matter created can be described with equilibrium and statistical concepts [2–11]. Ultimately, one would like to describe nuclear collisions with a model that takes into account all the dynamics of nucleon-nucleon collisions. Until then, statistical models provide invaluable insight to the physics of multifragmentation of the last three decades, by reducing the intractable problem of time-dependent highly correlated interacting many-body fermion system to the much simpler picture of a system of non-interacting clusters [12].

Since most statistical multifragmentation codes have been developed to describe specific sets of data and nearly all of them have different assumptions, they are not equivalent [2–11,13–18]. One of the goals of this article is to examine the observables constructed with the isotope yields from different statistical multifragmentation models used in recent years. Even though the number of models we studied is limited, they represent the codes widely used in the heavy-ion community. The results show that all the statistical codes give similar general trends but different predictions to specific experimental observables. The conclusion is consistent with a recent study on models with different statistical assumptions [19]. We also find that the differences between models are much reduced for observables constructed with isotope yield ratios from different reactions.

---

[a] e-mail: `tsang@nscl.msu.edu`

Table 1. Summary of the different statistical multifragmentation models and evaporation codes studied in this article.

Code	Evaporation	User	Author	Ref.	(168,75)	(186,75)	(168,50)	Primary	Final
			Statistical Multifragmentation Models						
ISMM-c	MSU-decay	Tsang	Das Gupta	[2]	Y	Y		Y	Y
ISMM-m	MSU-decay	Souza	Souza	[13,14]	Y	Y		Y	Y
SMM95	own code	Bougault	Botvina	[4,9]	Y	Y		Y	Y
MMM1	own code	AH Raduta	AH Raduta	[15]	Y	Y	Y	Y	Y
MMM2	own code	AR Raduta	AR Raduta	[15]	Y	Y	Y	Y	Y
MMMC	own code	Le Fèvre	Gross	[5,16]	Y	Y	Y		Y
LGM	N/A	Regnard	Gulminelli	[17]	Y		Y	Y	
QSM	own code	Trautmann	Stöcker	[18]	Y	Y	Y		Y
EES	EES	Friedman	Friedman	[7,8]	Y	Y	Y	Y	Y
BNV-box	N/A	Colonna	Colonna	[24]	Y	Y		Y	
			Evaporation codes						
Gemini		Charity	Charity	[25]	Y	Y			Y
Gemini-w		Wada	Wada	[25–28]	Y	Y			Y
SIMON		Durand	Durand	[29]	Y	Y			Y
EES		Friedman	Friedman	[7,8]	Y	Y	Y		Y
MSU-decay		Tsang	Tan *et al.*	[14]	Y	Y			Y

The various codes and the benchmark systems which form the basis for comparison will be described in sect. 2. Comparisons of the statistical multifragmentation models are presented in sect. 3 and the results from the comparisons of five different evaporation codes are presented in sect. 4. Finally, we summarize our findings in sect. 5.

2 Benchmark systems

Nearly all statistical models assume that nucleons and fragments originate from a single emission source characterized by A_0 nucleons and Z_0 protons. The hot fragments then de-excite using evaporation models. To provide consistent comparisons between models, we have chosen the following source systems: 1) $A_0 = 168$, $Z_0 = 75$, $N_0/Z_0 = 1.24$, 2) $A_0 = 186$, $Z_0 = 75$, $N_0/Z_0 = 1.48$. These two systems have the same charge and are chosen to be 75% of the initial compound systems of ^{112}Sn + ^{112}Sn and ^{124}Sn + ^{124}Sn [20,21]. We also have calculations on system 3) $A_0 = 168$, $Z_0 = 84$, $N_0/Z_0 = 1.0$ which has the same mass but different charge from system 1. Even though most results of system 3 are not included in this article due to lack of space, they corroborate the conclusions. In each calculation, the same inputs are used. We require the source excitation energy, $E^\star$, to be 5 MeV per nucleon and the source density to be 1/6 of the normal nuclear-matter density.

At the time when this manuscript was prepared, we were able to get results from nine statistical multifragmentation model codes plus a hybrid dynamical-statistical code (BNV-box) and five evaporation codes. Table 1 lists all the codes, users (defined as the person who did the calculations shown in this paper) and the main authors of the codes. The users sent us the output files which contain mainly the neutron (N) and proton (Z) number and the yield of the hot fragments and/or the final fragments. All these output files can be found in the web: `http://groups.nscl.msu.edu/smodels/results.html`.

The statistical multifragmentation models studied here construct fragment yields from a maximum entropy principle, but they differ both in the degrees of freedom employed and in the chosen constraints. We have different versions of the Statistical Multifragmentation Model (SMM) [22]. All these models assume that the N-body source correlations are exhausted by clusterization and, therefore, describe the system as a collection of non-interacting clusters. (The Coulomb repulsion among fragments are approximately taken into account.) These codes differ in the freeze-out volume prescription, in the treatment of continuum states and in the numerical technique to span the phase space. The SMM95 code uses grand-canonical approximation [4,9] and Fermi-jet breakups for the de-excitation of hot fragments. The Improved Statistical Multifragmentation Model (ISMM) [14] uses experimental masses and level densities when available. When experimental information is not available, ISMM uses an improved algorithm to interpolate level densities for the hot fragments. It uses the MSU-decay code as an afterburner. ISMM-c [2] uses a canonical formalism, while ISMM-m [13] adopts a microcanonical approach. The sequential decay algorithm in ISMM [13] uses experimental masses and includes structure information for light fragments ($Z < 15$). The MMMC code uses a Metropolis-Monte Carlo method [5,16]. MMMC is the only model that can accommodate non-spherical sources but only neutrons are emitted in the sequential decays. We have two calculations using the microcanonical multifragmentation model MMM [15] with different freeze-out assumptions: 1) non-overlapping spherical fragments inside a spherical source and 2) free-volume approach. These two calculations are correspondingly denoted by MMM1 and MMM2. The Quantum Statistical Model (QSM) [18] is a simplified grand-canonical version of SMM models including only a limited number of light clusters ($A < 20$), which however are described with a detailed density of states accounting for all known discrete levels at the time when the code was written in the late eighties.

The Expanding Emitting-Source (EES) model [7,8] is an extended Weisskopf evaporation model [23] which couples the emission of fragments to the changing conditions, *i.e.*, density (volume), mass number, isospin, and entropy, of the source. The model assumes an equation of state for the source so that the thermal pressure and initial expansion determine the changes in the source due to emission. It is the only statistical model to account for the time dependence of the emission process. No specific density (volume) for emission is assumed, but the model predicts that the strongest emission often occurs from a dilute source during a narrow time period. (In this sense it is similar to the SMM.) Spectra are constructed by summing the contributions of emission from different times, with a switch from surface to volume emission at a low density of the source.

Finally, we have two microscopic model calculations. The Lattice Gas Model (LGM) [17] calculates the equilibrium configurations of a system of (semi)classical nucleons interacting via an Ising Hamiltonian. These configurations are generated in a given confining box by Monte Carlo. It is the only model that has no nuclear-physics input. The BNV-box model is based on the Boltzmann-Nordheim-Vlaslov (BNV) equations [24] and uses the effective Skyrme force augmented with a stochastic collision integral to calculate the equilibrium configurations which are generated via a dissipation dynamics in a box. In both models, the clusters have to be defined *a posteriori* via a clusterization algorithm.

Since de-excitation of the hot fragments is essential before comparison to experimental data, most codes have their own sequential decay algorithms. Ideally, one should compare the hot primary fragments and the decay fragments separately. Unfortunately, in some codes (*e.g.*, MMMC), the hot fragments cannot be extracted while in others (LGM and BNV-box) the hot fragments sent by the users have not undergone decay. This makes comparing the contributions from the evaporation portion of the code to the final fragments very difficult.

Since an "after-burner" or evaporation code is needed to allow the hot fragments to decay to ground states, codes that can be coupled to statistical and dynamical codes are very important. Thus, in addition to the fragmentation models, we also compare five different evaporation codes (listed in table 1) that have served the functions of "afterburners" to both statistical and dynamical codes. 1) The most widely used code is Gemini [25] which treats the physics of excited heavy residues very well. However, for the light fragments, it lacks complete structural information. 2) A modified code of an early version of Gemini [26, 27] has also been used extensively to de-excite hot frag-

ments generated in the Asymmetrized Molecular Dynamical (AMD) Model [28]. We labeled this modified version of Gemini as Gemini-w. 3) An event generator code called SIMON [29], based on Weisskopf emission rates [23], includes the narrowest discrete states for $Z \leq 9$ as well as in-flight evaporation. It has been used to de-excite fragments created in both BNV dynamical model [30] and a heavy-ion phase space model [31]. 4) The MSU-decay code [14] uses the Gemini code to decay heavy residues and includes much structural information such as the experimental masses, excited states with measured spin and parity for light fragments with $Z < 15$ in a table. This table also includes information of calculated states, which are not measured. 5) In principle, at very low excitation energy, the multifragmentation models can also be used as evaporation models. In this category, we have results from the EES model [7,8].

For the evaporation model comparison, the benchmark systems for the source are the same as the three systems used in the multifragmentation models, $(A_0, Z_0) = (168, 75)$, $(186, 75)$ and $(168, 84)$. The excitation energy is set to be $2\,\mathrm{MeV}$ per nucleon and the density is assumed to be the same as normal nuclear-matter density.

3 Results from multifragmentation models

In this section, we show results that illustrate the differences and similarities between calculations. Due to limited space, not all observables from the calculations are constructed or shown here. Since system 1 with $A_0 = 168$ and $Z_0 = 75$ have results from all the calculations, we tend to highlight this system. Some of the results on the ISMM-c calculations have been published in ref. [2]. If a choice has to be made between showing ISMM-c results or ISMM-m results due to lack of space, we choose to show the results of ISMM-m. For the LGM calculations [17,32], we have results using the micro-canonical approximations as well as results using canonical approximations. We show mainly the results with the microcanonical approximations. The differences between the microcanonical and canonical assumptions can be inferred from the results of ISMM-c and ISMM-m. The observables shown in sects. 3.1 to 3.5 are chosen for the relevance of the observables to the understanding of the multifragmentation process. More recently, the focus of heavy-ion collisions at intermediate energy has shifted to explore the isospin degree of freedom [2]. This is often done by studying two or more systems, which differ mainly in the isospin composition of the projectiles or targets. Isoscaling using isotope yield ratios is discussed in sect. 3.4. Instead of using isotope yield ratios for temperature, we use the fluctuations of different thermometers to determine how well the sequential decays in the code reproduce the observed fluctuations. The results will be described in sect. 3.5.

To provide some uniformity to the figures, we will try to use the same symbols for the results from the same code throughout this article. Where applicable, closed symbols often refer to the neutron-rich system ($A_0 = 186$, $Z_0 = 75$) and open symbols refer to the neutron-deficient system

$(A_0 = 168, Z_0 = 75)$. We also adopt the convention that the results are labeled with the user (who sent us the calculated results) and the code name. Even though comparison with data is not our goal, it is sometimes instructive to plot the data as reference points when appropriate. We have chosen the data from the central collisions of ^{124}Sn + ^{124}Sn and ^{112}Sn + ^{112}Sn [20,21,33] at 50 MeV per nucleon incident energy as represented by closed and open star symbols, respectively, mainly because this data set is readily available to the first author.

3.1 IMF multiplicities

The copious production of intermediate-mass fragments (IMFs) which are charged particles with $Z = 3$–20 is one signature of the multifragmentation process. The study of these fragments provides clues to the nuclear liquid-gas phase transition as they are considered as droplets formed from the condensation of nuclear gas and may provide information about the co-existence region. Figure 1 shows the mean multiplicities of IMFs produced by different models. Within errors, one cannot discern any dependence of the mean IMF multiplicity on the isospin composition of the initial sources by looking for systematic differences between solid and open symbols which represent the neutron-rich and neutron-deficient systems, respectively. If we compare the left and right panel of fig. 1, in general, sequential decays reduce the IMF multiplicities.

The two MMM calculations have different results due to different freeze-out assumptions used for the source. MMM1 which uses non-overlapping spherical fragments emits nearly two fragments less than MMM2. Only primary fragments before decay are available from the LGM and BNV-box calculations. For the MMMC and QSM models, we only have fragments after decay.

For the primary-fragment multiplicity (left panel), the BNV model emits slightly more primary fragments while the LGM model emits nearly a factor of two less fragments than the other models. For the final-fragment multiplicity (right panel), the QSM [18,34,35] emits many more IMFs. Indeed this model is not suited to predict absolute yields but rather should be used to compute relative yields of light isotopes, *e.g.* for thermometry purposes [35,36]. For comparisons, the data from the central collisions of Sn isotopes are represented by the star symbols in the right panel. The differences in the mean multiplicities between the ^{112}Sn + ^{112}Sn (open stars) and ^{124}Sn + ^{124}Sn (solid stars) [33] are much larger than those predicted by all the models after decay. The discrepancies between model predictions and data are not understood.

3.2 Mass distributions

Next, we examine the primary mass distributions of the $A_0 = 168$, $Z_0 = 75$ system. The steep drop of the light fragment ($A < 10$) multiplicity shown in fig. 2 are similar for nearly all the models but there are differences. Some of the differences (*e.g.* between MMM1 (upright triangles) and MMM2 (inverted triangles)) arise from differences in the freeze-out assumptions as described previously. The differences in the results from the two ISMM calculations may come from the difference between canonical and

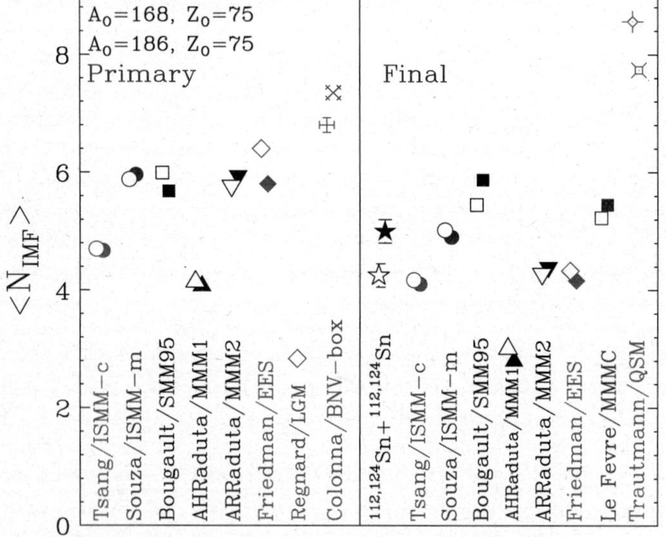

Fig. 1. Mean IMF multiplicity obtained from different statistical models listed in table 1 for primary fragments (left panel) and final fragments (right panel). At the bottom of the panels, the calculations are labeled by the name of the user and the name of the code. The open symbols refer to system 1 ($A_0 = 168$, $Z_0 = 75$) and the solid symbols refer to system 2 ($A_0 = 182$, $Z_0 = 75$). The open and solid stars in the right panel are data from the central collisions of ^{112}Sn + ^{112}Sn and ^{124}Sn + ^{124}Sn systems at $E/A = 50$ MeV [33].

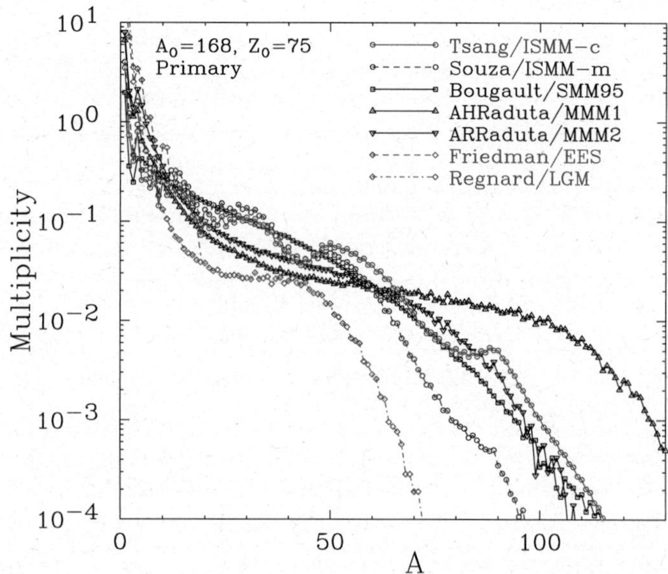

Fig. 2. Predicted primary-fragment mass distributions from the mutifragmentation of a source nucleus with $A_0 = 168$, $Z_0 = 75$ (system 1). See caption of fig. 1 for name convention.

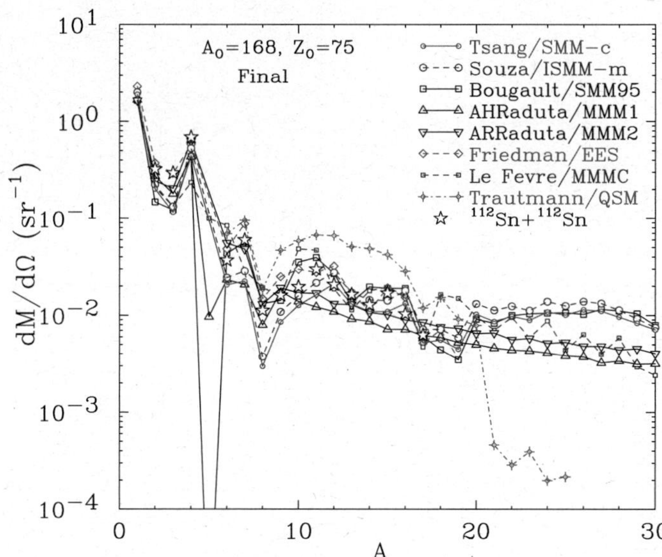

Fig. 3. Predicted final fragment mass distributions from the multifragmentation of a source nucleus with $A_0 = 168$, $Z_0 = 50$ (system 1). For comparison, data from the multifragmentation of central collisions of the ^{112}Sn + ^{112}Sn system [2,20] at $E/A = 50$ MeV are plotted as open stars.

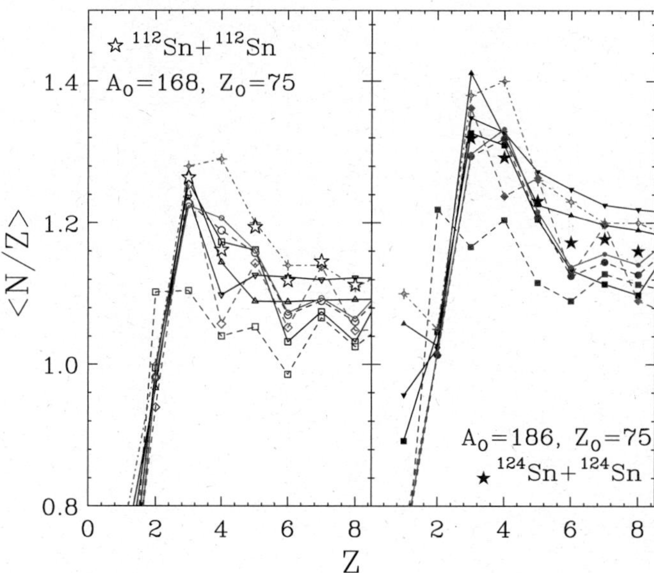

Fig. 4. The mean neutron to proton ratios as a function of the charge of the emitted fragment Z for system 1 (left panel) and system 2 (right panel). For comparison, results from the multifragmentation following central collisions of ^{112}Sn + ^{112}Sn and ^{124}Sn + ^{124}Sn are shown as open (left panel) and closed stars (right panel), data from ref. [20].

micro-canonical approximations used. (ISMM-c requires temperature instead of excitation energy as one property of the initial source.) SMM95 and the two MMM calculations have smooth distributions as the fragment masses are determined from liquid-drop mass formulae [37]. The LGM (the lowest curve with diamond symbols), which does not take into account the binding energies or nucleon masses shows a smooth dependence on mass but does not produce heavy residues.

In fig. 3, we have plotted the differential multiplicity of the final mass distributions for the same ($A_0 = 168$, $Z_0 = 75$) system in an expanded scale. Again, while the trends are similar for most calculations except the QSM model (crosses), there are significant differences in detailed comparisons. Most models do not have nuclei with mass 5 and cross-sections for mass 8 are much reduced in accordance to experimental observation. For reference, the data from the ^{112}Sn + ^{112}Sn system are plotted as open stars. The trends exhibited by most models are similar to those of the data. Primary and final fragments with $A \geq 20$ are ignored in the EES code. These heavy fragments are not included in the output files. The QSM does not produce fragments with $A > 20$. To conserve the total number of nucleons, more light charged fragments with $A \leq 20$ are produced, causing the over-production of IMFs seen in both figs. 1 and 3.

The charge distributions are similar to the mass distributions so they are not discussed here.

3.3 Isospin observables and isotope distributions

One observable to study the isospin degrees of freedom is the asymmetry, N/Z, of the fragments. Figure 4 shows

$\langle N/Z \rangle$ as a function of the fragment charge number Z predicted by different models. In this plot, the left panel shows results from the neutron-deficient system ($A_0 = 168$, $Z_0 = 75$) while the right panel contains results from the neutron-rich system ($A_0 = 186$, $Z_0 = 75$). Unlike the mass distributions shown in figs. 2 and 3, differences between different models are not very large, about 10%. (The zero of the vertical axis is suppressed in order to show the differences in greater details.) As expected, the $\langle N/Z \rangle$ of the fragments are larger for the more neutron-rich system. However, the $\langle N/Z \rangle$ values are much lower than the $\langle N_0/Z_0 \rangle$ of the initial system of 1.48 for the neutron-rich system. For the neutron-deficient system in the left panel, the initial $\langle N_0/Z_0 \rangle$ value is 1.24 which is only slightly larger than the fragment values. For reference, data from the central collisions of ^{124}Sn + ^{124}Sn (solid stars) and ^{112}Sn + ^{112}Sn systems (open stars) [20] are plotted in the left and right panels, respectively. Since the excited fragments in MMMC only emit neutrons [16], the fragment $\langle N/Z \rangle$ (squares) are lower than those derived from other models. All the other calculations exhibit similar trends as the data.

As the average values of the asymmetry of the fragments are determined from the isotope yields, it is instructive to examine the isotope distributions directly. Figure 5 shows the oxygen isotope distributions from different models before (left panels) and after (right panels) sequential decays. The upper panels indicate the isotopes from the neutron-rich ($A_0 = 186$, $Z_0 = 75$) system while fragments from the neutron-deficient ($A_0 = 168$, $Z_0 = 75$) system are plotted in the bottom panels. For reference, data [20] from the central collisions of ^{124}Sn + ^{124}Sn (solid stars)

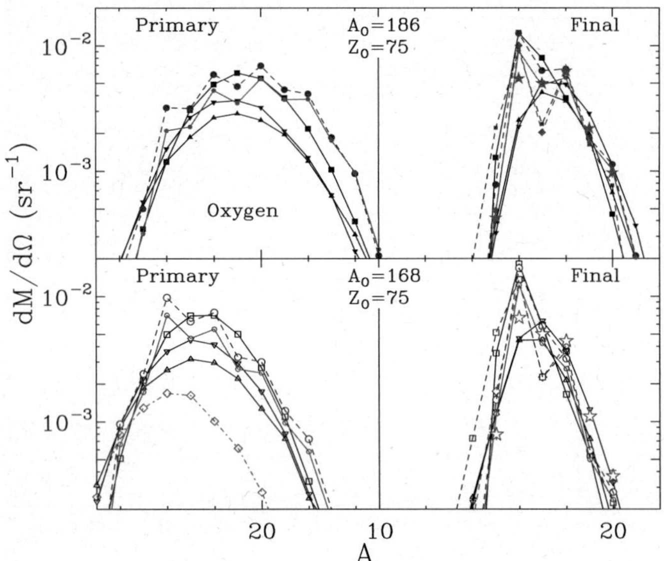

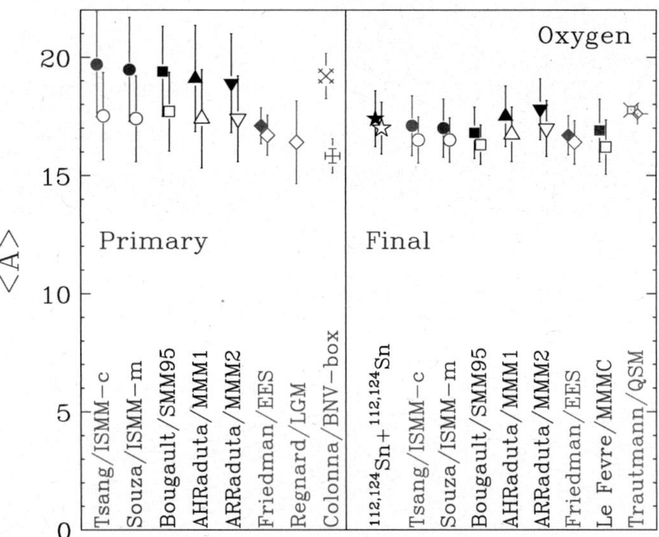

Fig. 5. Predicted isotope distributions for oxygen fragments from different models. Primary fragments are plotted in the left panels and final fragments in the right panels. The top panels contain results from the neutron-rich system 2 and the bottom panels contain distributions from the neutron-deficient system 1. The open (bottom right panel) and solid stars (top right panel) are data from ref. [20].

Fig. 6. Centroids and widths (variance) of the oxygen isotope distributions obtained from different models. Most of the distributions are shown in fig. 5.

and ^{112}Sn + ^{112}Sn systems (open stars) are plotted in the upper right panel and lower right panels, respectively.

The differences in the primary distributions between models (left panel) can be understood from the nuclear masses used in the different codes. Both the ISMM models (circle symbols) used experimental masses even for hot fragments [2,14], thus odd-even effects are evident in the primary mass distributions. The SMM95 (squares) [9] and the two MMM (upright and inverted triangles) [15,37] calculations use mass formulae resulting in smooth interpolations of isotope cross-sections. The deficiency of models like the LGM (open diamond symbols in the lower left panel), which do not include any nuclear-physics information, is obvious. The EES results are not presented here as the model ignores primary and secondary fragments with $A \geq 20$ and the oxygen isotope yields are not complete.

The isotope distributions from all models after decay (right panels) become much narrower and resemble that of the experimental data. The ISMM-c, ISMM-m and SMM95 models predict a peak at ^{16}O due to its large binding energy and the use of experimental masses in the decays. The ISMM calculations that incorporate the MSU-decay algorithms with experimental masses and structural information exhibit odd-even effects. In the decay code of the MMM calculations, fragment masses are derived from mass formulae [37]. As a result, the isotope distributions are rather smooth. The individual yields of oxygen isotopes are not available from the QSM output files, and the results of this model is not represented here.

In order to quantify the mean and the width of the distributions, we have plotted the mean mass number and the standard deviations of the oxygen distributions in fig. 6 for both the primary (left panel) and final (right panel) fragment distributions. The vertical bars represent the standard deviations of the isotope distributions. In general, all models produce much wider distributions for the primary isotopes and the widths are reduced by sequential decay effects. Sequential decays tend to move the centroids of the distributions towards the valley of stability and reduce the differences in the centroids of the isotope distributions between the neutron-rich and neutron-deficient systems.

3.4 Isoscaling

When isoscaling was first observed in experimental data [21,38], it was demonstrated through statistical model calculations that isoscaling could be preserved through sequential decays [38,39]. More importantly, statistical models relate the isoscaling phenomenon to the symmetry energy [38–41], which is of fundamental interest to general nuclear properties as well as astrophysics [42].

Isoscaling describes the exponential dependence on the isotope neutron (N) and proton (Z) number of the yield ratios from two different reactions,

$$R_{21} = \frac{Y_2(N,Z)}{Y_1(N,Z)} = Ce^{\alpha N + \beta Z}, \qquad (1)$$

where C, α, and β are the fitting parameters. In our specific examples of systems 1 and 2, $Y_2(N,Z)$ is the isotope yield emitted from the neutron-rich system $A_2 = 186$, $Z_2 = 75$ and $Y_1(N,Z)$ is the isotope yield emitted from the neutron-deficient system $A_1 = 168$, $Z_1 = 75$. Figure 7 shows that all statistical multifragmentation models exhibit good isoscaling behavior for the primary fragments. Each symbol corresponds to one element, $Z = 1$ (open triangles), $Z = 2$ (closed triangles), $Z = 3$ (open circles),

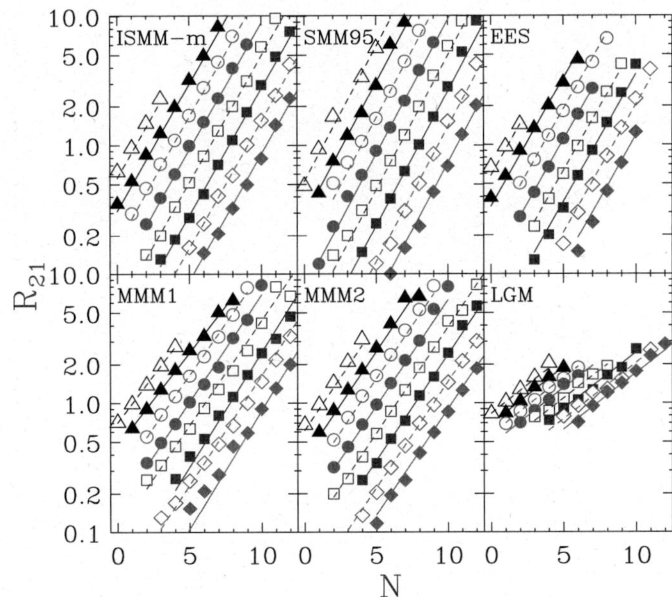

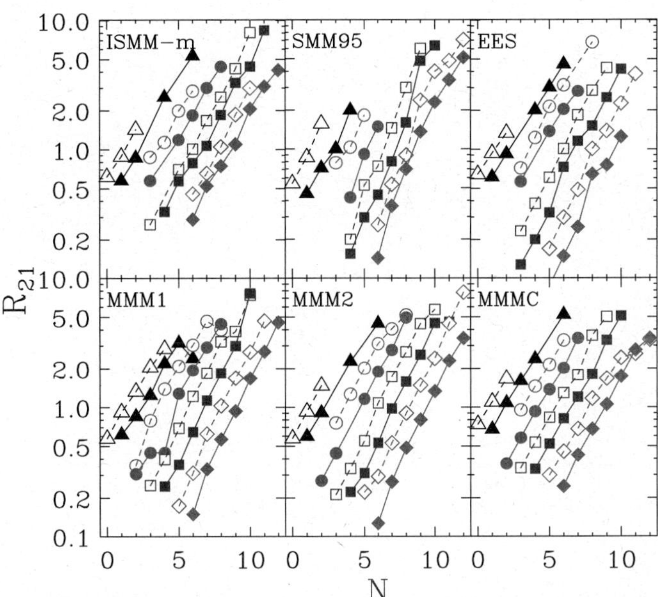

Fig. 7. Predicted yield ratios, $R_{21}(N,Z) = Y_2(N,Z)/Y_1(N,Z)$ from primary fragments for system 2 and system 1. (For the LGM, the calculations are for system 1 and system 3.) Each panel presents the results from one model calculation. The lines are best fits to the symbols according to eq. (1). Different lines correspond to atomic numbers $Z = 1$ to 8 starting with the leftmost line being $Z = 1$. Open points and dashed lines denote isotopes with odd Z while solid points and solid lines denote isotopes with even Z.

Fig. 8. Predicted yield ratios, $R_{21}(N,Z) = Y_2(N,Z)/Y_1(N,Z)$ from final fragments. The symbols have the same convention as in fig. 7. The lines are drawn to guide the eye.

$Z = 4$ (closed circles), $Z = 5$ (open squares), $Z = 6$ (closed squares), $Z = 7$ (open diamonds) and $Z = 8$ (closed diamonds). The solid and dashed lines are best fits from eq. (1). The slopes of the lines correspond to the neutron isoscaling parameter α and the distance between the lines corresponds to the isoscaling parameter β. All the models except LGM have similar slopes. The slope parameters from the two MMM models are slightly smaller. The LGM only has calculations on systems 1 and 3. Since the differences in the asymmetries between systems 1 and 2 and systems 1 and 3 are small, the LGM isoscaling slopes are expected to be slightly smaller but they are much smaller (lower left panel) than the other models. This is probably related to the lack of nuclear-physics input in such model.

An important contribution that statistical models make to the field of heavy-ion collision is the derivation that the isocaling parameter α is related to the symmetry energy coefficient, C_{sym}:

$$\alpha_{pri} = \frac{4C_{sym}}{T}\left[\left(\frac{Z_1}{A_1}\right)^2 - \left(\frac{Z_2}{A_2}\right)^2\right] = \frac{4C_{sym}}{T}\left[\left(\Delta\frac{Z}{A}\right)^2\right], \tag{2}$$

where α_{pri} is the isoscaling parameter extracted from the calculated yields of primary fragments, T is the temperature, Z_i/A_i is the proton fraction of the initial source with label i. To extract C_{sym} which is related to symmetry energy ($E_{sym} = C_{sym}I^2$) from data, it is important that

the sequential decays do not affect α, T and $[\Delta(Z/A)]^2$ significantly.

Figure 8 shows isoscaling plots constructed from final fragments after sequential decays. Isoscaling is no longer strictly observed over a large range of isotopes. Furthermore, the distances between elements are much less regular and the slopes vary from element to element. The distances between elements are related to the proton isoscaling parameter, β. Experimentally, the trends and magnitudes in both α and β are similar [21,43]. The irregular spacings between elements from the calculations is probably caused by the Coulomb treatment in different codes. Part of the lack of smoothness in the trends could come from the lack of statistics for primary isotopes with low cross-sections. By restricting the isoscaling analysis to the same set of isotopes measured in experiments, about 3 isotopes for each element [21,43], most models show that the effect from sequential decays on isoscaling is negligible as shown in the left panel of fig. 9. The solid points refer to the analysis of the systems with the same charge, system 1 and 2 in table 1, while the open points refer to the analysis of the systems with the same mass, system 1 and 3. In the two MMM calculations (triangles), the final fragments seem to retain more memory of the source than the other models as shown in fig. 6, resulting in the final isoscaling parameters being larger than the primary isoscaling parameters. By restricting the number of isotopes for fitting, the problems with statistics from fragment production may be minimized. On the other hand, such procedure may hide fundamental problems associated with the sequential decays.

All the statistical models except the LGM use the symmetry energy of stable nuclei in describing the mass of the fragments. Except for the EES model, the symme-

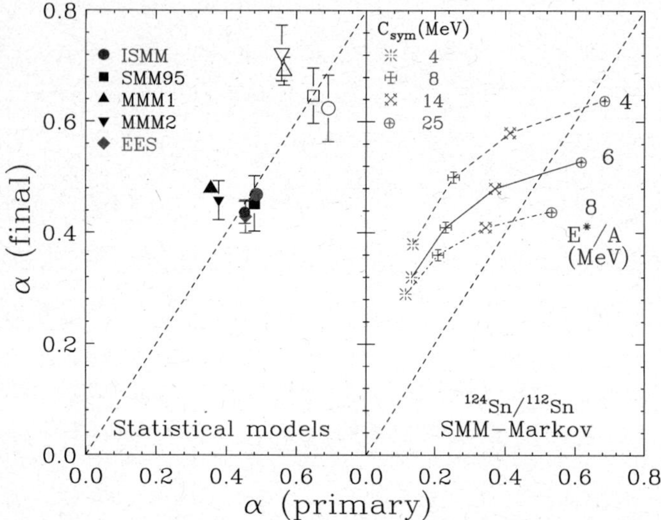

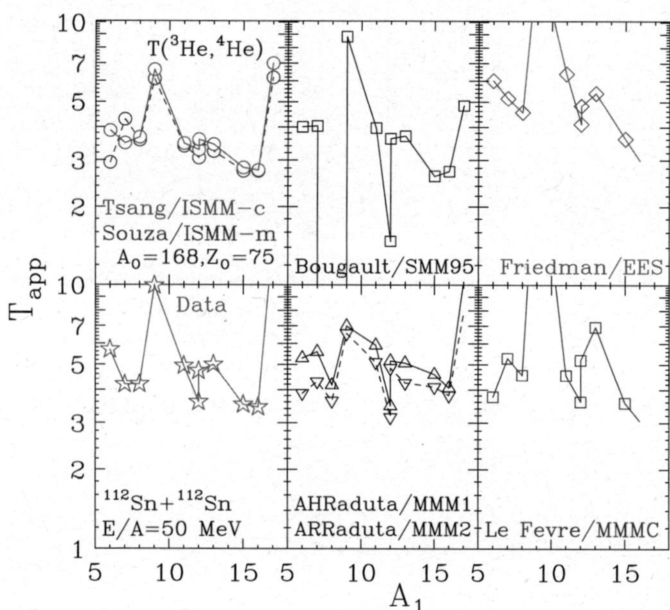

Fig. 9. Effect of sequential decays on the isoscaling parameter, α. Left panel shows the results of the statistical models studied in this work where the C_{sym} in the models assume a constant value of about 25 MeV and $E^*/A = 4$ MeV. Right panel shows the results from the microcanonical version of SMM with Markov chain where C_{sym} varies from 4 (burst symbols), 8 (crosses), 14 ($\times$ symbols) and 25 (circular symbols) MeV and E^*/A varies from 4 (dashed line), 6 (solid line) and 8 (dot-dashed line).

Fig. 10. Apparent isotope temperatures $T(^3\mathrm{He},{}^4\mathrm{He})$ constructed from different isotope pairs in the numerator of eq. (4) and the ratios of $Y(^3\mathrm{He})/Y(^4\mathrm{He})$ in the denominators are plotted as a function of A_1. The data [14] are plotted in the bottom left corner for reference. Models with similar decay codes such as ISMM-c and ISMM-m (top left panel) and MMM1 and MMM2 (bottom middle panel) are plotted together.

try energy coefficient, C_{sym}, remains constant throughout the reactions. Such prescription may not be realistic. In a recent study, when different (especially lower) values of C_{sym} are used in a Markov-chain version of the SMM95 code, the sequential decays effects are very different [44] as shown in the right panel of fig. 9. The lines denote the different excitation energy (4, 6, 8 MeV) used. For each excitation energy, calculations have been performed for $C_{sym} = 4$ (burst symbols), 8 (crosses), 14 ($\times$ symbols) and 25 (circular symbols) MeV. For $C_{sym} = 25$ MeV, the effect of sequential decays are similar to those shown in the left panel of fig. 9. However, for lower C_{sym} values, the α(final) are larger than α(primary). As discussed in [42], this trend is different from those observed in dynamical calculations. A detailed understanding of the effects of sequential decays on the isoscaling parameters α, the temperature T, and the proton fraction Z_i/A_i [45] is necessary before symmetry energy information can be extracted by applying eq. (2) to experimental data.

3.5 Fluctuations of isotope yield ratio temperatures

Ideally, a model should predict isotope cross-sections such as those shown in the right panels of fig. 5. All model comparisons involve the production of primary fragments and their decays. To disentangle the two parts of the calculations from the final fragments and to evaluate the sequential decay portion of the calculations, we need another observable that is mainly sensitive to the structural decay information, an important ingredient in sequential decay

models. It has been shown that the fluctuations observed in the isotope yield temperatures are sensitive to the sequential decay information [2,46].

The isotope yield ratio thermometer is defined as [47]

$$T = \frac{B}{\ln a \cdot R}, \tag{3}$$

where B is a binding energy parameter, a is the statistical factor that depends on statistical weights of the ground-state nuclear spins and R is the ground-state isotope yield ratio,

$$R = \frac{Y(A_1, Z_1)/Y(A_1 + 1, Z_1)}{Y(A_2, Z_2)/Y(A_2 + 1, Z_2)}. \tag{4}$$

In this section, our discussion is mainly focused on using T as a tool to evaluate the modeling of sequential decays. More details about using T as the temperature of the freeze-out source can be found in ref. [36, 46]. It is possible to construct many different thermometers from various combinations of the isotope yields using eqs. (3) and (4) [46]. In the grand-canonical approximation, if all fragments are produced directly in their ground states, these temperatures should all have the same value as the temperature of the initial system. Experimentally, we see large fluctuations of these isotope yield temperatures [46,48,49], *i.e.* T depends on the specific combinations of isotopes used in eq. (4). Without sequential decay corrections, the measured temperature is not the source temperature. Because of the fluctuations, the experimental measured temperatures are usually called T_{app}.

as denoted in fig. 10. As an example, we show $T(^3\mathrm{He}, ^4\mathrm{He})$ constructed with $Y(^3\mathrm{He})$ and $Y(^4\mathrm{He})$ yields in the denominators ($A_2 = 3$, $Z_2 = 2$) but different isotope pairs in the numerators of eq. (4). Specifically, we will examine eleven $T(^3\mathrm{He}, ^4\mathrm{He})$ thermometers constructed with the yields of the following isotope pairs in the numerators: $Y(^6\mathrm{Li})/Y(^7\mathrm{Li})$, $Y(^7\mathrm{Li})/Y(^8\mathrm{Li})$, $Y(^8\mathrm{Li})/Y(^9\mathrm{Li})$, $Y(^9\mathrm{Be})/Y(^{10}\mathrm{Be})$, $Y(^{11}\mathrm{B})/Y(^{12}\mathrm{B})$, $Y(^{12}\mathrm{B})/Y(^{13}\mathrm{B})$, $Y(^{12}\mathrm{C})/Y(^{13}\mathrm{C})$, $Y(^{13}\mathrm{C})/Y(^{14}\mathrm{C})$, $Y(^{15}\mathrm{N})/Y(^{16}\mathrm{N})$, $Y(^{16}\mathrm{O})/Y(^{17}\mathrm{O})$, and $Y(^{17}\mathrm{O})/Y(^{18}\mathrm{O})$.

These isotope pairs are chosen because the data for the central collisions of $^{112}\mathrm{Sn} + ^{112}\mathrm{Sn}$ at $E/A = 50\,\mathrm{MeV}$ are available [2, 20]. $T(^3\mathrm{He}, ^4\mathrm{He})$ are constructed from the values of a and B listed in ref. [14]. These temperatures are plotted as a function of A_1 in the lower left panel of fig. 10. To get a glimpse of how well different evaporation codes which are coupled to the statistical multifragmentation models listed in table 1 simulate sequential decays, $T(^3\mathrm{He}, ^4\mathrm{He})$ constructed with the final fragments produced from the different statistical models are plotted in the remaining panels of fig. 10. Instead of assuming a constant value, $T(^3\mathrm{He}, ^4\mathrm{He})$ fluctuates in all models. This suggests that decays to low-lying excited states occur. If a significant fraction of the particles de-excite to the gamma levels below the particle decay thresholds, the ground-state cross-sections are modified. Such contaminations may have caused the higher temperatures determined from the yields of $^9\mathrm{Be}$ ($A_1 = 9$, $Z_1 = 4$) and $^{18}\mathrm{O}$ ($A_1 + 1 = 18$, $Z_1 = 8$) which have several low-lying excited states below the neutron thresholds. Similar fluctuations have been observed in different reaction systems at different temperatures [14, 46, 48, 49]. They mainly originate from the detailed structure of the excited states. Thus the fluctuations in the isotope temperature provide a sensitive tool to evaluate whether proper decay levels have been taken into account in a code.

These fluctuations are mainly determined by the sequential decay portion of the code. Models with the same decay codes exhibit nearly the same fluctuations even though the primary IMF multiplicities and mass distributions are different. For example, different freeze-out assumptions used in the two MMM codes result in very different mean IMF multiplicities (fig. 1) and different residue distributions (fig. 2). However, the isotope yield ratio temperatures have the same trends (bottom middle panel) suggesting that sequential decays mask off some initial differences in the source. The fluctuations in ISMM-c and ISMM-m are similar (top left panel). Since the MSU-decay code incorporates the most structural information for the light fragments ($Z < 15$), $T(^3\mathrm{He}, ^4\mathrm{He})$ determined from the two ISMM codes that employ the MSU-decay as after-burners reproduce the trend of the experimental fluctuations the best (top left panel). However, $T(^3\mathrm{He}, ^4\mathrm{He})$ is lower than the input temperature of $4.7\,\mathrm{MeV}$ suggesting that the sequential decay effects on the initial temperature can be substantial. As $^9\mathrm{Li}$ isotopes are not produced in the SMM95 code, the temperatures involving this isotope drops (top middle panel). Individual temperature values do not agree among models even though the initial input

to the fragmenting source is the same. The differences in the isotope yield ratio temperatures probably reflect the difference in the decay codes.

4 Evaporation models

Before comparing calculated results with data, all hot fragments produced in any models must undergo decay. Unfortunately, the task to simulate sequential decays has proved to be rather difficult due to the lack of complete information on nuclear structures and level densities. In this section, we compare five sequential decay models (see table 1) that have been used in many studies. The benchmark systems are 1) $A_0 = 168$, $Z_0 = 75$ and 2) $A_0 = 186$, $Z_0 = 75$. The excitation energy is $2\,\mathrm{MeV}$ per nucleon. For brevity, we only discuss three observables, which illustrate the differences in the codes.

4.1 Mass distributions

Figure 11 shows the mass distributions from the decay of the $A_0 = 168$, $Z_0 = 75$ system (left panel) and $A_0 = 186$, $Z_0 = 75$ system (right panel). Contrary to the near exponential decrease of the production of fragments with increasing mass in multifragmentation processes (fig. 2), most evaporation models de-excite by emitting LCPs, leaving a residue. Fission is also a significant de-excitation mode in this mass region, resulting in a hump at about 10 mass units less than $A_0/2$. The inability of the EES model (symbols joined by dashed lines) to track fragments larger than $A = 20$ results in the artificial truncation of the mass distribution. Since the MSU-decay uses Gemini to decay fragments with $Z > 15$, results from Gemini (solid line)

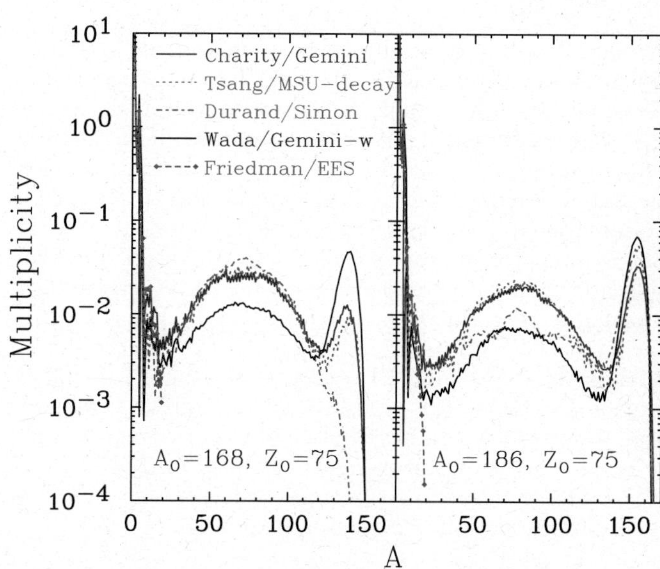

Fig. 11. Predicted mass distributions from the five evaporation codes listed in table 1 for the neutron-deficient system 1 (left panel) and neutron-rich system 2 (right panel).

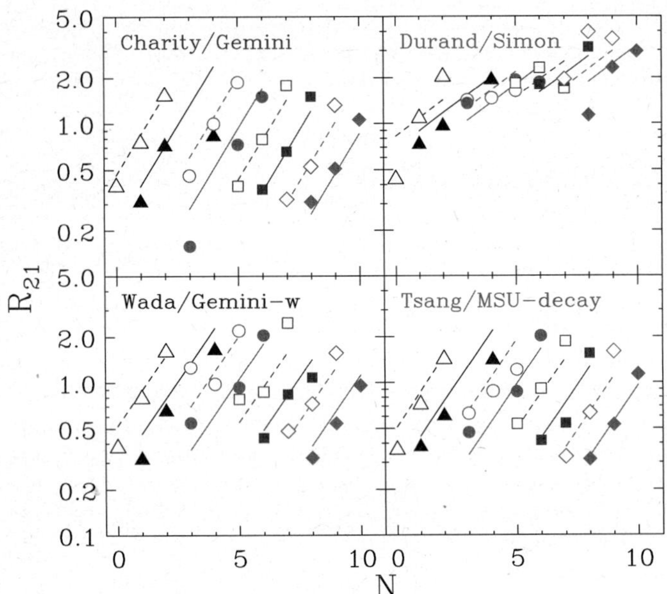

Fig. 12. Predicted ratios, $R_{21}(N,Z) = Y_2(N,Z)/Y_1(N,Z)$ of fragments evaporated from system 2 and system 1. Each panel presents the results from one model calculation. The results from the EES model are not plotted here as they are similar to those shown in fig. 8.

and MSU-decay models (dotted line) are very similar. In principle, Gemini-w (dashed line) should be the same as Gemini. However, an older version of Gemini was incorporated and Germini-w gives much larger residue cross-sections and correspondingly smaller fission fragment and IMF cross-sections. The event generator code, SIMON (dot-dashed line) has very different mass distributions than the other codes, e.g. it does not produce residues in the $A_0 = 168$, $Z_0 = 75$ system (left panel).

4.2 Isoscaling

Primary fragments produced from nearly all statistical multifragmentation codes observe isocaling, rigorously. However, isoscaling is not well observed over a large range of secondary fragments. For this reason, we limit the number of isotopes to three for each element, similar to those measured in experimental data. We use this observable to examine the differences between different models in fig. 12. The symbol convention of figs. 7 and 8 is used, i.e. symbols are yield ratios and lines are best fits. Isoscaling is reasonably reproduced except for SIMON. For the MSU-dacay and EES (not shown) decays, the results are similar to those of ISMM and EES calculations shown in fig. 8. Except for ^{6}He yield ratios, Gemini exhibits very good isoscaling. The isoscaling from Gemini-w fragments is not as good. The same problems that cause SIMON to produce different mass distributions could be the cause for the non-observation of isoscaling behavior.

4.3 Fluctuation of isotope yield ratio temperatures

In fig. 13, we show $T(^3\text{He}, ^4\text{He})$ constructed with $Y(^3\text{He})$ and $Y(^4\text{He})$ yields in the denominators ($A_2 = 3$, $Z_2 = 2$)

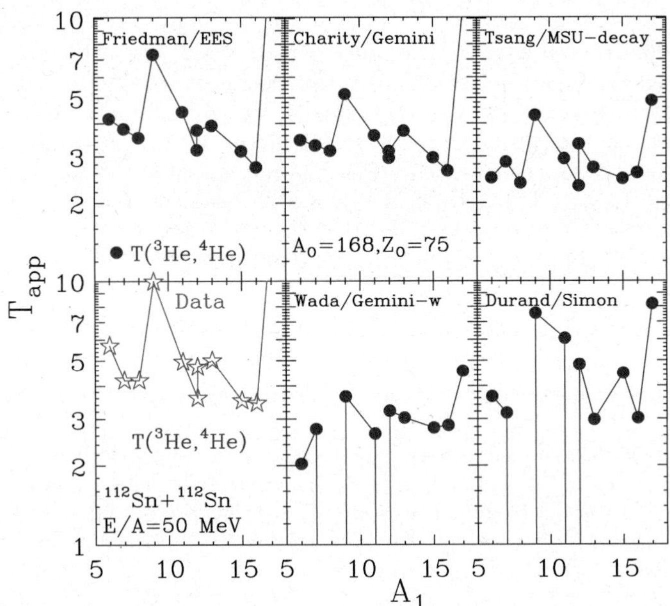

Fig. 13. Apparent isotope temperatures $T(^3\text{He}, ^4\text{He})$ plotted as a function of A_1. For reference, the data [14] are plotted in the bottom left corner.

but different isotope pairs in the numerators of eq. (4) as discussed in sect. 3.5. For reference, the Sn data are plotted in the lower left panel of fig. 13 as a function of A_1. As light-particle structure information has been included in EES, Gemini, and MSU-decay codes, they reproduce the fluctuations observed experimentally rather well as shown in the top three panels in fig. 13. On the other hand, SIMON and Gemini-w do not reproduce the general trends suggesting that the sequential decays are not properly taken into account in these codes.

Most of the isotope yield ratio temperatures from EES, Gemini and MSU-decay calculations are below 4 MeV, the input temperature of the source. Sequential decay effects are expected to reduce the initial temperature. It is interesting to note that of the three models that reproduce the fluctuations, the average temperature is the highest for the EES model and lowest for the MSU-decay model. This can be explained by the amount of structural information included in individual models. EES incorporates only a few low-lying excited states while the MSU-decay model incorporates most of the experimental level information for nuclei with $Z < 15$. The availability of a large number of decay levels in the latter code reduces the ground-state cross-sections more than in other calculations. This suggests that even at low excitation energy (2 MeV), sequential decays still significantly affect isotope yields.

5 Summary and conclusions

In summary, we have made comparisons of experimental observables using ten statistical multifragmentation codes. The general trends are similar among models suggesting that these models can provide important physical insights for the primary fragments and multifragmentation process. However, details in any single observable differ between models. The largest differences are observed in raw observables such as individual isotope yields, mass and charge distributions while the mean values of an observable such as IMF multiplicity, the mean fragment asymmetry $\langle N/Z \rangle$ or mean mass $\langle A \rangle$ of an element do not show as large differences. The effects of sequential decays on isoscaling parameters are not well understood.

As sequential decay codes are important to both dynamical and statistical models, we also compare five widely used codes. Relatively accurate structural information and experimental masses are required in evaporation models to reproduce the fluctuations of isotope yield temperatures. Such sensitivity allows one to evaluate the sequential decay properties of the evaporation codes.

The observables studied here are by no means an exhaustive list. However, these observables, which can be constructed easily from the isotope yields, provide important benchmarks to test any multifragmentation models or evaporation codes that describe sequential decays.

The authors wish to thank Dr. C. Bertulani for his help in the beginning of this project. MBT acknowledges the support of the National Science Foundation under Grant No. PHY-01-10253.

References

1. S. Das Gupta, A.Z. Mekjian, M.B. Tsang, Adv. Nucl. Phys. **26**, 91 (2001) and references therein.
2. C.B. Das, S. Das Gupta, W.G. Lynch, A.Z. Mekjian, M.B. Tsang, Phys. Rep. **406**, 1 (2005).
3. J. Randrup, S.E. Koonin, Nucl. Phys. A **356**, 223 (1981).
4. J.P. Bondorf, A.S. Botvina, A.S. Iljinov, I.N. Mishustin, K. Sneppen, Phys. Rep. **257**, 133 (1995) and references therein.
5. D.H.E. Gross, Phys. Rep. **279**, 119 (1997) and references therein.
6. W.A. Friedman, W.G. Lynch, Phys. Rev. C **28**, 16 (1983).
7. W.A. Friedman, Phys. Rev. Lett. **60**, 2125 (1988).
8. W.A. Friedman, Phys. Rev. C **42**, 667 (1990).
9. A.S. Botvina, A.S. Iljinov, I.N. Mishustin, J.P. Bondorf, R. Donangelo, K. Sneppen, Nucl. Phys. A **475**, 663 (1987).
10. J.P. Bondorf *et al.*, Nucl. Phys. A **443**, 321 (1985); **444**, 460 (1985); **448**, 753 (1986); K. Sneppen, Nucl. Phys. A **470**, 213 (1987).
11. A.S. Botvina, A.S. Iljinov, I.N. Mishustin, Sov. J. Nucl. Phys. **42**, 712 (1985).
12. G.F. Bertsch, Am. J. Phys. **72**, 983 (2004).
13. S.R. Souza, W.P. Tan, R. Donangelo, C.K. Gelbke, W.G. Lynch, M.B. Tsang, Phys. Rev. C **62**, 064607 (2000).
14. W.P. Tan, S.R. Souza, R.J. Charity, R. Donangelo, W.G. Lynch, M.B. Tsang, Phys. Rev. C **68**, 034609 (2003).
15. Al.H. Raduta, Ad.R. Raduta, Phys. Rev. C **65**, 054610 (2002).
16. A. Le Fèvre *et al.*, Nucl. Phys. A **735**, 219 (2004).
17. F. Gulminelli, Ph. Chomaz, Phys. Rev. Lett. **82**, 1402 (1999).
18. D. Hahn, H. Stöcker, Nucl. Phys. A **476**, 718 (1988).
19. C.E. Aguiar, R. Donangelo, S.R. Souza, Phys. Rev. C **73**, 024613 (2006).
20. T.X. Liu *et al.*, Phys. Rev. C **69**, 014603 (2004).
21. H.S. Xu *et al.*, Phys. Rev. Lett. **85**, 716 (2000).
22. See, *e.g.*, A.S. Botvina, I.N. Mishustin, this topical issue.
23. V.F. Weisskopf, D.H. Ewing, Phys. Rev. **57**, 472 (1940).
24. M. Colonna, private communications.
25. R.J. Charity *et al.*, Nucl. Phys. A **483**, 371 (1988), the code is available at `http://www.chemistry.wustl.edu/~rc/gemini_f77`.
26. K. Yuasa-Nakagawa *et al.*, Phys. Rev. C **53**, 997 (1996).
27. J. Cibor *et al.*, Phys. Rev. C **55**, 264 (1997).
28. R. Wada *et al.*, Phys. Rev. C **62**, 34601 (2000); **69**, 44610 (2004).
29. D. Durand, Nucl. Phys. A **541**, 266 (1992).
30. J.D. Frankland *et al.*, Nucl. Phys. A **689**, 940 (2001).
31. D. Lacroix, A. Van Lauwe, D. Durand, Phys. Rev. C **69**, 054604 (2004).
32. F. Gulminelli, Ph. Chomaz, Phys. Rev. C **71**, 054607 (2005).
33. G.J. Kunde *et al.*, Phys. Rev. Lett. **77**, 2897 (1996).
34. J. Konopka, H. Graf, H. Stöcker, W. Greiner, Phys. Rev. C **50**, 2085 (1994).
35. Hongfei Xi *et al.*, Z. Phys. A **359**, 397 (1997); Eur. Phys. J. A **1**, 235 (1998) (the second reference is an erratum to the first).
36. A. Kelić, J.B. Natowitz, K.-H. Schmidt, this topical issue.
37. W.D. Myers, W.J. Swiatecki, Nucl. Phys. **81**, 1 (1966); Ark. Fiz. **36**, 343 (1967).
38. M.B. Tsang, W.A. Friedman, C.K. Gelbke, W.G. Lynch, G. Verde, H.S. Xu, Phys. Rev. Lett. **86**, 5023 (2001).
39. M.B. Tsang *et al.*, Phys. Rev. C **64**, 054615 (2001).
40. A.S. Botvina, O.V. Lozhkin, W. Trautmann, Phys. Rev. C **65**, 044610 (2002).
41. A. Ono, P. Danielewicz, W.A. Friedman *et al.*, Phys. Rev. C **68**, 051601(R) (2003).
42. M. Colonna, M.B. Tsang, this topical issue.
43. E. Geraci *et al.*, Nucl. Phys. A **732**, 173 (2004).
44. A. Le Fèvre, G. Auger, M.L. Begemann-Blaich *et al.*, Phys. Rev. Lett. **94**, 162701 (2005).
45. A. Ono *et al.*, arXiv:nucl-ex/0507018.
46. M.B. Tsang, W.G. Lynch, H. Xi, W.A. Friedman, Phys. Rev. Lett. **78**, 3836 (1997).
47. S. Albergo *et al.*, Nuovo Cimento A **89**, 1 (1985).
48. M.J. Huang *et al.*, Phys. Rev. Lett. **78**, 1648 (1997).
49. H. Xi *et al.*, Phys. Lett. B **431**, 8 (1998).

Eur. Phys. J. A **30**, 141–151 (2006)
DOI 10.1140/epja/i2006-10112-y

THE EUROPEAN
PHYSICAL JOURNAL A

Instabilities in nuclear matter and finite nuclei

V. Baran[1,a] and J. Margueron[2]

[1] Physics Faculty, Bucharest University and IFIN-HH Bucharest, Romania
[2] Institut de Physique Nucléaire, Université Paris Sud F-91406 Orsay CEDEX, France

Received: 16 June 2006 /
Published online: 31 October 2006 – © Società Italiana di Fisica / Springer-Verlag 2006

Abstract. Spinodal instability in nuclear matter and finite nuclei is investigated. This instability occurs in the low-density region of the phase diagram. The thermodynamical and dynamical analysis is based on Landau theory of Fermi liquids. It is shown that asymmetric nuclear matter can be characterized by a unique spinodal region, defined by the instability against isoscalar-like fluctuation, as in symmetric nuclear matter. Everywhere in this density region the system is stable against isovector-like fluctuations related to the species separation tendency. Nevertheless, this instability in asymmetric nuclear matter induces isospin distillation leading to a more symmetric liquid phase and a more neutron-rich gas phase.

PACS. 21.65.+f Nuclear matter – 25.70.Pq Multifragment emission and correlations – 21.60.Ev Collective models

1 Introduction

The production of fragments represents an important dissipation mechanism in heavy-ion reactions at intermediate energies. A relevant phenomenon is the liquid-gas phase transition, very often invoked in discussing the nuclear multifragmentation. In this analogy, however, one should be aware also of the differences due to Coulomb, finite-size or quantum effects.

For phase transitions in macroscopic systems, the coexistence regions, corresponding to areas thermodynamically forbidden for one single phase, exhibit general features such as metastabilities or instabilities. At variance with the situation for macroscopic systems, where the observation time scale is much greater than the time scales of the microscopic processes that lead to drop (bubble) formation (even more exceptional is the study of critical points where the slowing-down phenomena require days of expectations for equilibration), in heavy-ions collisions the reaction times can be comparable to the fragment formation time which is of relevance for discussing about the kinetics of the phase transition. The violent collision and fast expansion may quench the system inside the instability region of the phase diagram. Moreover, a binary system, including asymmetric nuclear matter (ANM) (see [1]), manifest a richer thermodynamical behaviour, since it has to accommodate one more conservation law.

In this paper we will discuss first the nature of the instabilities and of the related fluctuations in such systems. Then, in sect. 3, we will describe the kinetics of

phase transition in ANM both in the linear and nonlinear regime. Finally, in the last section we will focus on the relevance of these results on nuclear multifragmentation and neck fragmentation in heavy-ion collisions at intermediate energies.

2 Instabilities and fluctuations in ANM

2.1 Thermodynamical approach

One-component systems may become unstable against density fluctuations as the result of the mean attractive interaction between constituents. In symmetric binary systems, like symmetric nuclear matter (SNM), one may encounter two kinds of density fluctuations: i) isoscalar, when the densities of the two components oscillate in phase with equal amplitude, ii) isovector when the two densities fluctuate still with equal amplitude but out of phase. Then mechanical instability is associated with instability against isoscalar fluctuations leading to cluster formation while chemical instability is related to instability against isovector fluctuations, leading to species separation. We will show in the following that in ANM, there is no longer a one-to-one correspondence between isoscalar (respectively isovector) fluctuations and mechanical (respectively chemical) instability. An appropriate framework for the study of instabilities is provided by the Fermi-liquid theory [2], which has been applied, for instance, to symmetric binary systems as SNM (the two components being protons and neutrons) [3], the liquid ^{3}He (spin-up

<hr>

[a] e-mail: `baran@lns.infn.it`

and spin-down components) [4,5] and proto-neutron stars to calculate neutrino propagation [6].

The starting point is an extension to the asymmetric case of the formalism introduced in [4]. The distribution functions for protons and neutrons are

$$f_q(\epsilon_p^q) = \Theta(\mu_q - \epsilon_p^q), \qquad q = n, p, \tag{1}$$

where μ_q are the corresponding chemical potentials. The nucleon interaction is characterized by the Landau parameters:

$$F^{q_1 q_2} = N_{q_1} V^2 \frac{\delta^2 H}{\delta f_{q_1} \delta f_{q_2}} = N_{q_1} \frac{\delta^2 H}{\delta \rho_{q_1} \delta \rho_{q_2}}, \tag{2}$$

$$N_q(T) = \int \frac{-2\, \mathrm{d}\mathbf{p}}{(2\pi\hbar)^3} \frac{\partial f_q(T)}{\partial \epsilon_p^q}, \tag{3}$$

where H is the energy density, V is the volume and N_q is the single-particle level density at the Fermi energy. At $T = 0$ this reduces to

$$N_q(0) = m p_{F,q}/(\pi^2 \hbar^3) = 3\rho_q/(2\epsilon_{F,q}),$$

where $p_{F,q}$ and $\epsilon_{F,q}$ are the Fermi momentum and Fermi energy of the q-component. Thermodynamical stability for $T = 0$ requires the energy of the system to be an absolute minimum for the undistorted distribution functions, so that the relation

$$\delta H - \mu_p \delta \rho_p - \mu_n \delta \rho_n > 0 \tag{4}$$

is satisfied when we deform proton and neutron Fermi seas.

Only monopolar deformations will be taken into account, since we consider here momentum-independent interactions, so that $F^{q_1 q_2}_{l=0}$ are the only non-zero Landau parameters. In fact, for momentum-independent interactions, all the information on all possible instabilities of the system is obtained just considering density variations. However, one should keep in mind that in the actual dynamical evolution of an unstable system in general one observes deformations of the Fermi sphere, hence the direction taken by the system in the dynamical evolution is not necessarily the most unstable one defined by the thermodynamical analysis.

Then, up to second order in the variations, the condition eq. (4) becomes

$$\delta H - \mu_p \delta \rho_p - \mu_n \delta \rho_n = \frac{1}{2}\left(a\delta \rho_p{}^2 + b\delta \rho_n{}^2 + c\delta \rho_p \delta \rho_n\right) > 0, \tag{5}$$

where

$$a = N_p(0)(1 + F_0^{pp}); \qquad b = N_n(0)(1 + F_0^{nn});$$
$$c = N_p(0)F_0^{pn} + N_n(0)F_0^{np} = 2N_p(0)F_0^{pn}. \tag{6}$$

The r.h.s. of eq. (5) is diagonalized by the following transformation:

$$u = \cos\beta\, \delta\rho_p + \sin\beta\, \delta\rho_n,$$
$$v = -\sin\beta\, \delta\rho_p + \cos\beta\, \delta\rho_n, \tag{7}$$

where the *mixing* angle $0 \leq \beta \leq \pi/2$ is given by

$$tg\, 2\beta = \frac{c}{a - b} = \frac{N_p(0)F_0^{pn} + N_n(0)F_0^{np}}{N_p(0)(1 + F_0^{pp}) - N_n(0)(1 + F_0^{nn})}. \tag{8}$$

Then eq. (5) takes the form

$$\delta H - \mu_p \delta \rho_p - \mu_n \delta \rho_n = Xu^2 + Yv^2 > 0, \tag{9}$$

where

$$X = \frac{1}{2}\left(a + b + \mathrm{sign}(c)\sqrt{(a - b)^2 + c^2}\right)$$
$$\equiv \frac{(N_p(0) + N_n(0))}{2}\left(1 + F_{0g}^s\right) \tag{10}$$

and

$$Y = \frac{1}{2}\left(a + b - \mathrm{sign}(c)\sqrt{(a - b)^2 + c^2}\right)$$
$$\equiv \frac{(N_p(0) + N_n(0))}{2}\left(1 + F_{0g}^a\right), \tag{11}$$

defining the new generalized Landau parameters $F_{0g}^{s,a}$.

Hence, thanks to the rotation eq. (7), it is possible to separate the total variation eq. (4) into two independent contributions, called the "normal" modes, and characterized by the "mixing angle" β, which depends on the density of states and the details of the interaction. Thus, the thermodynamical stability requires $X > 0$ and $Y > 0$. Equivalently, the following conditions have to be fulfilled:

$$1 + F_{0g}^s > 0 \quad \text{and} \quad 1 + F_{0g}^a > 0. \tag{12}$$

They represent Migdal-Pomeranchuk stability conditions extended to asymmetric binary systems.

The new stability conditions, eq. (12), are equivalent to mechanical and chemical stability of a thermodynamical state [7], *i.e.*

$$\left(\frac{\partial P}{\partial \rho}\right)_{T,y} > 0 \quad \text{and} \quad \left(\frac{\partial \mu_p}{\partial y}\right)_{T,P} > 0, \tag{13}$$

where P is the pressure and y the proton fraction. In fact, mechanical and chemical stability are very general conditions, deduced by requiring that the principal curvatures of thermodynamical potential surfaces, such as the free energy (or the entropy) with respect to the extensive variables are positive (negative).

It has been argued that the mechanical and chemical instability lead to very different phenomenons: the chemical instability with the concentration as order parameter and the mechanical instability for which the total density plays the role of a second-order parameter [8,9]. In the following, we will show that spinodal instability and phase transition in ANM should be instead discussed in terms of isoscalar- and isovector-like instabilities. In the case discussed here, it can be proved that [10]:

$$XY = N_p(0)N_n(0)\left[(1 + F_0^{nn})(1 + F_0^{pp}) - F_0^{np}F_0^{pn}\right]$$
$$= \frac{[N_p(0)N_n(0)]^2}{(1 - y)\rho^2}\left(\frac{\partial P}{\partial \rho}\right)_{T,y}\left(\frac{\partial \mu_p}{\partial y}\right)_{T,P} \tag{14}$$

and

$$\left(\frac{\partial P}{\partial \rho}\right)_{T,y} = \frac{\rho y (1-y)}{N_p(0) N_n(0)}\left(t\,a + \frac{1}{t}\,b + c\right)$$

$$\propto X\left(\sqrt{t}\cos\beta + \frac{1}{\sqrt{t}}\sin\beta\right)^2 + Y\left(\sqrt{t}\sin\beta - \frac{1}{\sqrt{t}}\cos\beta\right)^2$$

$$\text{with}\quad t = \frac{y}{1-y}\,\frac{N_n(0)}{N_p(0)}. \tag{15}$$

Let us assume that in the density range we are considering the quantities a and b remain positive. In this way one can study the effect of the interaction between the two components, given by c, on the instabilities of the mixture. If $c < 0$, *i.e.* for an attractive interaction between the two components, from eq. (11) one sees that the system is stable against isovector-like fluctuations. It becomes isoscalar unstable if $c < -2\sqrt{ab}$ (see eq. (10)). However thermodynamically this instability against isoscalar-like fluctuations will show up as a chemical instability if $(-ta - b/t) < c < -2\sqrt{ab}$ or as a mechanical instability if $c < (-ta - b/t) < -2\sqrt{ab}$ (see eq. (15)). This last observation is very interesting: it tells us that the nature of the thermodynamically instabilities can be related to the relative strength of the various interactions among the species. In other words, if it is possible to determine experimentally for a binary systems the signs of $(\frac{\partial P}{\partial \rho})_{T,y}$ and/or $(\frac{\partial \mu_p}{\partial y})_{T,P}$ we can learn about the inequalities, at a given density, between species interactions.

On the other hand, the distinction between the two kinds of instability (mechanical and chemical) is not really relevant regarding the nature of unstable fluctuations, being it essentially the same, *i.e.* isoscalar-like. The relevant instability region is defined in terms of instabilities against isoscalar fluctuations and we can speak, therefore, about a unique spinodal region. If $c > 0$, *i.e.* when the interaction between the components is repulsive, the thermodynamical state is always stable against isoscalar-like fluctuation, but can be isovector unstable if $c > 2\sqrt{ab}$. Since with our choices the system is mechanically stable $(a, b, c > 0$, see eq. (15)), the isovector instability is now always associated with chemical instability. Such situation will lead to a component separation of the liquid mixture. In this framework, a complete analysis of the instabilities of any binary system can be performed, in connection to signs, strengths and density dependence of the interactions.

2.2 Asymmetric nuclear-matter case

We show now quantitative calculations for asymmetric nuclear matter which illustrate the previous general discussion on instabilities. Let us consider a potential energy density of Skyrme type [11,12],

$$H_{pot}(\rho_n, \rho_p) = \frac{A}{2}\frac{(\rho_n + \rho_p)^2}{\rho_0} + \frac{B}{\alpha+2}\frac{(\rho_n + \rho_p)^{\alpha+2}}{\rho_0^{\alpha+1}}$$

$$+ \left(C_1 - C_2\left(\frac{\rho}{\rho_0}\right)^\alpha\right)\frac{(\rho_n - \rho_p)^2}{\rho_0}, \tag{16}$$

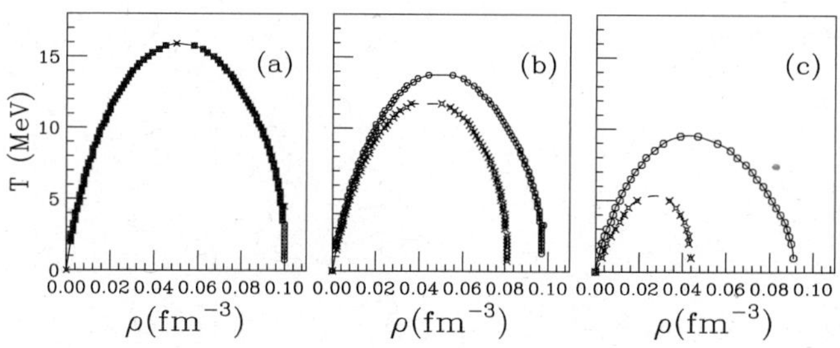

Fig. 1. Spinodal line corresponding to isoscalar-like instability of asymmetric nuclear matter (circles) and mechanical instability (crosses) for three proton fractions: $y = 0.5$ (a), $y = 0.25$ (b), $y = 0.1$ (c). The figure is taken from [10].

where $\rho_0 = 0.16\,\text{fm}^{-3}$ is the nuclear saturation density. The values of the parameters $A = -356.8\,\text{MeV}$, $B = 303.9\,\text{MeV}$, $\alpha = 1/6$, $C_1 = 125\,\text{MeV}$, $C_2 = 93.5\,\text{MeV}$ are adjusted to reproduce the saturation properties of symmetric nuclear matter and the symmetry energy coefficient.

We focus on the low-density region, where phase transitions of the liquid-gas type are expected to happen, in agreement with the experimental evidences of multifragmentation [13,14]. Since $a, b > 0$ and $c < 0$, we deal only with instability against isoscalar-like fluctuations, as for symmetric nuclear matter. In fig. 1 the circles represent the spinodal line corresponding to isoscalar-like instability, as defined above, for three values of the proton fraction. For asymmetric matter, $y < 0.5$, under this border one encounters either chemical instability, in the region between the two lines, or mechanical instability, under the inner line (crosses). The latter is defined by the set of values (ρ, T) for which $(\frac{\partial P}{\partial \rho})_{T,y} = 0$. We observe that the line defining chemical instability is more robust against the variation of the proton fraction in comparison to that defining mechanical instability: reducing the proton fraction makes it energetically less and less favorable for the system to break into clusters with the same initial asymmetry. However, we stress again the unique nature of the isoscalar-like instability. The change from the chemical to the mechanical character along this border line is not very meaningful and does not affect the properties of the system.

Let us now discuss the generality of the conclusions by comparing several models for the nuclear interaction. Indeed, the spinodal contours predicted by several models exhibit important differences (see fig. 2). In the case of SLy230a force (as well as SGII, D1P), the total density at which spinodal instability appears decreases when the asymmetry increases whereas for SIII (as well as D1, D1S) it increases up to large asymmetry and finally decreases. Despite the observed differences between the models, we observe that all forces which fulfill the global requirement that they reproduce the symmetric nuclear-matter (SNM) equation of state as well as the pure neutron matter calculations lead to the same curvature of the spinodal region [15].

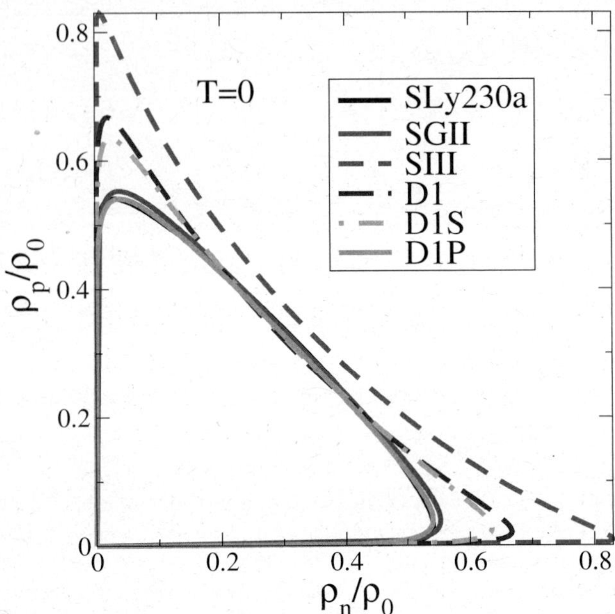

Fig. 2. Projection of the spinodal contour in the density plane for several effective interactions like Skyrme (SLy230a, SGII, SIII) and Gogny (D1, D1S, D1P). The figure is taken from [15].

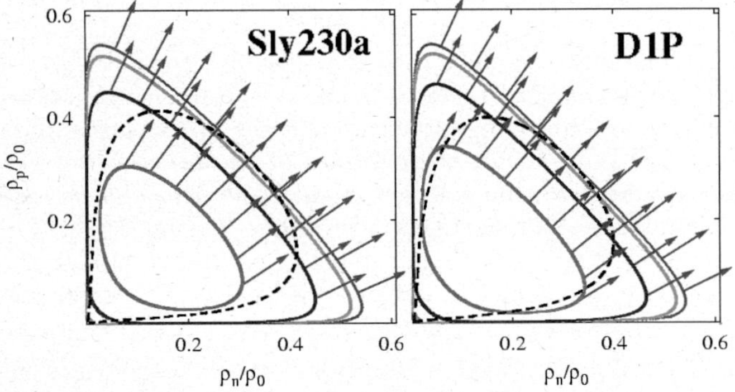

Fig. 3. Projection of the iso-eigen values on the density plane for Slya (left) and D1P (right). The arrows indicate the direction of instability. The mechanical instability is also indicated (dashed line). The figure is taken from [15].

Let us now focus on the direction of the instability. If the eigenvector associated with the unstable mode is along $y = \rho_p/\rho = \text{const}$ then the instability does not change the proton fraction. For symmetry reasons pure isoscalar and isovector modes appear only for SNM so it is interesting to introduce a generalization of isoscalar-like and isovector-like modes by considering if the protons and neutrons move in phase ($\delta\rho_n\delta\rho_p > 0$) or out of phase ($\delta\rho_n\delta\rho_p < 0$). Figure 3 shows the direction of instabilities along the spinodal border and some iso-instability lines. We observe that instability is always almost along the ρ-axis meaning that it is dominated by total density fluctuations even for large asymmetries. The instability direction is between the $y = \text{const}$ line and the ρ-direction. This shows that the unstable direction is of isoscalar nature as expected from the attractive proton-neutron interaction. The total density is, therefore, the dominant contribution to the order pa-

rameter showing that the transition is between two phases having different densities (*i.e.* liquid-gas phase transition). The angle with the ρ-axis is almost constant along a constant y line. This means that as the matter enters in the spinodal zone and then dives into it, there are no dramatic changes in the instability direction which remains essentially a density fluctuation. Moreover, the unstable eigenvector drives the dense phase (*i.e.* the liquid) toward a more symmetric point in the density plane. By particle conservation, the gas phase will be more asymmetric leading to the fractionation phenomenon.

We want to stress that those qualitative conclusions are very robust and have been reached for all the Skyrme and Gogny forces we have tested (SGII, SkM*, RATP, D1, D1S, D1P) including the most recent one (SLy230a, D1P) as well as the original one (like SIII, D1).

We eventually point out that also various relativistic mean-field hadron models were involved for the study of the phase transition from liquid to gas phases in ANM [8, 16, 17]. It was concluded that the largest differences between different parameterizations, regarding unstable behaviour in the low-density region, occur at finite temperature and in the high isospin asymmetry region.

3 The kinetics of phase transition in ANM

3.1 The linear response

The dynamical behaviour of a two-fluid system can be described, at the semi-classical level, by considering two Vlasov equations, for neutrons and protons in the nuclear-matter case [11, 12, 18, 19], coupled through the self-consistent nuclear field

$$\frac{\partial f_q(\mathbf{r},\mathbf{p},t)}{\partial t} + \frac{\mathbf{p}}{m}\frac{\partial f_q}{\partial \mathbf{r}} - \frac{\partial U_q(\mathbf{r},t)}{\partial \mathbf{r}}\frac{\partial f_q}{\partial \mathbf{p}} = 0, \qquad q = n, p. \quad (17)$$

For simplicity effective mass corrections are neglected. In fact, in the low-density region, of interest for our analysis of spinodal instabilities, effective mass corrections should not be large.

$U_q(\mathbf{r},t)$ is the self-consistent mean-field potential in a Skyrme-like form [11,12]:

$$U_q = \frac{\delta H_{pot}}{\delta \rho_q} = A\left(\frac{\rho}{\rho_0}\right) + B\left(\frac{\rho}{\rho_0}\right)^{\alpha+1} + C\left(\frac{\rho_3}{\rho_0}\right)\tau_q$$
$$+ \frac{1}{2}\frac{dC(\rho)}{d\rho}\frac{\rho_3^2}{\rho_0} - D\triangle\rho + D_3\tau_q\triangle\rho_3, \quad (18)$$

where

$$H_{pot}(\rho_n,\rho_p) = \frac{A}{2}\frac{\rho^2}{\rho_0} + \frac{B}{\alpha+2}\frac{\rho^{\alpha+2}}{\rho_0^{\alpha+1}}$$
$$+ \frac{C(\rho)}{2}\frac{\rho_3^2}{\rho_0} + \frac{D}{2}(\nabla\rho)^2 - \frac{D_3}{2}(\nabla\rho_3)^2 \quad (19)$$

is the potential energy density (see eq. (16)), where also surface terms are included; $\rho = \rho_n + \rho_p$ and $\rho_3 = \rho_n -$

ρ_p are, respectively, the total (isoscalar) and the relative (isovector) density; $\tau_q = +1$ ($q = n$), -1 ($q = p$).

The value of the parameter $D = 130\,\mathrm{MeV}\cdot\mathrm{fm}^5$ is adjusted to reproduce the surface energy coefficient in the Bethe-Weizsäcker mass formula $a_{surf} = 18.6\,\mathrm{MeV}$. The value $D_3 = 40\,\mathrm{MeV}\cdot\mathrm{fm}^5 \sim D/3$ is chosen according to ref. [20], and is also close to the value $D_3 = 34\,\mathrm{MeV}\cdot\mathrm{fm}^5$ given by the SKM* interaction [21].

Let us now discuss the linear response analysis to the Vlasov eqs. (17), corresponding to a semiclassical RPA approach. For a small amplitude perturbation of the distribution functions $f_q(\mathbf{r},\mathbf{p},t)$, periodic in time, $\delta f_q(\mathbf{r},\mathbf{p},t) \sim \exp(-i\omega t)$, eqs. (17) can be linearized leading to the following form:

$$-i\omega\delta f_q + \frac{\mathbf{p}}{m}\frac{\partial \delta f_q}{\partial \mathbf{r}} - \frac{\partial U_q^{(0)}}{\partial \mathbf{r}}\frac{\partial \delta f_q}{\partial \mathbf{p}} - \frac{\partial \delta U_q}{\partial \mathbf{r}}\frac{\partial f_q^{(0)}}{\partial \mathbf{p}} = 0, \quad (20)$$

where the superscript (0) labels stationary values and δU_q is the dynamical component of the mean-field potential. The unperturbed distribution function $f_q^{(0)}$ is a Fermi distribution at finite temperature

$$f_q^{(0)}\left(\epsilon_p^q\right) = \frac{1}{\exp\left(\epsilon_p^q - \mu_q\right)/T + 1}. \quad (21)$$

Since we are dealing with nuclear matter, $\nabla_r U_q^{(0)} = 0$ in eq. (20) and $\delta f_q \propto \exp(-i\omega t + i\mathbf{k}\mathbf{r})$. Following the standard Landau procedure [5,11], one can derive from eqs. (20) the following system of two equations for neutron and proton density perturbations:

$$\left[1 + F_0^{nn}\chi_n\right]\delta\rho_n + \left[F_0^{np}\chi_n\right]\delta\rho_p = 0, \quad (22)$$

$$\left[F_0^{pn}\chi_p\right]\delta\rho_n + \left[1 + F_0^{pp}\chi_p\right]\delta\rho_p = 0, \quad (23)$$

where

$$\chi_q(\omega,\mathbf{k}) = \frac{1}{N_q(T)}\int \frac{2\,\mathrm{d}\mathbf{p}}{(2\pi\hbar)^3}\frac{\mathbf{kv}}{\omega + i0 - \mathbf{kv}}\frac{\partial f_q^{(0)}}{\partial \epsilon_p^q}, \quad (24)$$

is the long-wavelength limit of the Lindhard function [5], $\mathbf{v} = \mathbf{p}/m$ and

$$F_0^{q_1 q_2}(k) = N_{q_1}(T)\frac{\delta U_{q_1}}{\delta \rho_{q_2}}, \qquad q_1 = n,p, \quad q_2 = n,p \quad (25)$$

are the usual zero-order Landau parameters, as already introduced in eq. (3), where now the k-dependence is due to the presence of space derivatives in the potentials (see eq. (18)). For the particular choice of potentials given by eq. (18), the Landau parameters are expressed as

$$F_0^{q_1 q_2}(k) = N_{q_1}(T)\Bigg[\frac{A}{\rho_0} + (\alpha+1)B\frac{\rho^\alpha}{\rho_0^{\alpha+1}} + Dk^2$$

$$+\left(\frac{C}{\rho_0} - D'k^2\right)\tau_{q_1}\tau_{q_2} + \frac{\mathrm{d}C}{\mathrm{d}\rho}\frac{\rho'}{\rho_0}(\tau_{q_1} + \tau_{q_2}) + \frac{\mathrm{d}^2 C}{\mathrm{d}\rho^2}\frac{\rho'^2}{2\rho_0}\Bigg]. \quad (26)$$

Multiplying the first equation by $N_n^{-1}\chi_p$ and the second one by $N_p^{-1}\chi_n$, we are led to define the following functions:

$$a(k,\omega) = N_p^{-1}\left(1 + F_0^{pp}\chi_p\right)\chi_n;$$

$$b(k,\omega) = N_n^{-1}\left(1 + F_0^{nn}\chi_n\right)\chi_p;$$

$$c(k,\omega) = \left(N_p^{-1}F_0^{pn} + N_n^{-1}F_0^{np}\right)\chi_n\chi_p =$$

$$2N_p^{-1}F_0^{pn}\chi_n\chi_p, \quad (27)$$

in some analogy with eqs. (6) and we obtain the following system of equations:

$$a\delta\rho_p + c/2\,\delta\rho_n = 0;$$
$$c/2\,\delta\rho_p + b\delta\rho_n = 0. \quad (28)$$

The system can be diagonalized with eigenvalues λ_s and λ_i, solutions of the equation:

$$(a - \lambda_{s,i})(b - \lambda_{s,i}) - c^2/4 = 0.$$

Formally we obtain for $\lambda_{s,i}$ the same expressions as given in eqs. (10), (11) for X and Y, but now a, b and c depend on ω. The unstable solutions for ω are obtained by solving the equations: $\lambda_s = 0$ (for isoscalar-like fluctuations), $\lambda_i = 0$ (for isovector-like fluctuations). This problem is completely equivalent to solve the equation: $c^2(\omega,k) = 4a(\omega,k)b(\omega,k)$, i.e. the dispersion relation

$$\left(1 + F_0^{nn}\chi_n\right)\left(1 + F_0^{pp}\chi_p\right) - F_0^{np}F_0^{pn}\chi_n\chi_p = 0, \quad (29)$$

that is also obtained directly by imposing the determinant of the system of eqs. (22), (23) equal to zero.

The dispersion relation is quadratic in ω and one finds two independent solutions (isoscalar-like and isovector-like solutions): ω_s^2 and ω_i^2. Then the structure of the eigenmodes can be determined and one finds

$$\delta\rho_p/\delta\rho_n = -2b(\omega_s,k)/c(\omega_s,k),$$

for the isoscalar-like modes and

$$\delta\rho_p/\delta\rho_n = -2b(\omega_i,k)/c(\omega_i,k),$$

for isovector-like oscillations. However, it is important to notice that the corresponding angles $\beta_{s,i}$ are not equal to the angle β determined in the thermodynamical analysis, eq. (8), because of the ω-dependence in a, b and c. They only coincide with β when $\omega = 0$ (and thus $\chi_{n,p} = 1$), i.e. at the border of the unstable region.

The dispersion relation, eq. (29), have been solved for various choices of the initial density, temperature and asymmetry of nuclear matter. Figure 4 reports the growth rate $\Gamma = \mathrm{Im}\,\omega(k)$ as a function of the wave vector k, for three situations inside the spinodal region. Results are shown for symmetric ($I = 0$) and asymmetric ($I = 0.5$) nuclear matter.

The growth rate has a maximum $\Gamma_0 = 0.01\text{--}0.03\,c/\mathrm{fm}$ corresponding to a wave vector value around $k_0 = 0.5\text{--}1\,\mathrm{fm}^{-1}$ and becomes equal to zero at $k \simeq 1.5k_0$, due to the k-dependence of the Landau parameters, as discussed above. One can see also that instabilities are reduced when increasing the temperature, an effect also present in the symmetric $N = Z$ case [22–24]. At larger initial asymmetry the development of the spinodal instabilities is slower, the maximum of the growth rate decreases. One should expect also an increase of the size of the produced fragments, decrease of the wave number corresponding to the maximum growth rate. From the long-dashed curves of fig. 4 we can predict the asymmetry effects to be more pronounced at higher temperature, when in fact the system is closer to the boundary of the spinodal region.

A full quantal investigation of spinodal instabilities and the related phase diagram was applied to finite nuclear systems, corresponding to Ca and Sn isotopes [25]. The frequencies and form factors of the unstable collective modes of an excited expanded system were obtained within the linearized time-dependent Hartree-Fock expansion, corresponding to RPA approximation. Dominant features are influenced by the quantum nature of the drop. So the first mode to become unstable is the low-lying octupole vibration. Diluted systems are unstable against low multipole deformations of the surface. It was shown that also in this case the instabilities are mostly of isoscalar nature, with an isovector component leading to isospin distillation, in agreement with the previous predictions for the nuclear-matter case [10].

3.2 Spinodal decomposition: numerical simulations

The previous analytical study is restricted to the onset of fragmentation, and related isospin distillation, in nuclear matter, in a linearized approach. Numerical calculations have been also performed in order to study all stages of

the fragment formation process [12,27]. We report on the results of ref. [12] where the same effective Skyrme interactions have been used.

In the numerical approach the dynamical response of nuclear matter is studied in a cubic box of size L imposing periodic boundary conditions. The Landau-Vlasov dynamics is simulated following a phase-space test particle method, using Gaussian wave packets [28–30]. The dynamics of nucleon-nucleon collisions is included by solving the Boltzmann-Nordheim collision integral using a Monte Carlo method [29]. The width of the Gaussians is chosen in order to correctly reproduce the surface energy value in finite systems. In this way, a cut-off appears in the short-wavelength unstable modes, preventing the formation of too small, unphysical, clusters [22]. The calculations are performed using 80 Gaussians per nucleon and the number of nucleons inside the box is fixed in order to reach the initial uniform density value. An initial temperature is introduced by distributing the test particle momenta according to a Fermi distribution.

We have followed the space-time evolution of test particles in a cubic box with side $L = 24$ fm for three values of the initial asymmetry $I = 0, 0.25$ and 0.5, at initial density $\rho^{(0)} = 0.06\,\mathrm{fm}^{-3} \simeq 0.4\rho_0$ and temperature $T = 5$ MeV. The initial density perturbation is created automatically due to the random choice of test particle positions.

The spinodal decomposition mechanism leads to a fast formation of the liquid (high density) and gaseous (low density) phases in the matter. Indeed this dynamical mechanism of clustering will roughly end when the variance saturates [31], *i.e.* around $250\,\mathrm{fm}/c$ in the asymmetric cases. We can also discuss the "chemistry" of the liquid-phase formation. In fig. 5 we report the time evolution of neutron (thick histogram in fig. 5a) and proton (thin histogram in fig. 5a) abundances and of asymmetry (fig. 5b) in various density bins. The dashed lines, respectively, shows the initial uniform density value $\rho \simeq 0.4\rho_0$ (fig. 5a) and the initial asymmetry $I = 0.5$ (fig. 5b). The drive to higher-density regions is clearly different for neutrons and protons: at the end of the dynamical clustering mechanism we have very different asymmetries in the liquid and gas phases (see the panel at $250\,\mathrm{fm}/c$ in fig. 5b).

It was shown in refs. [8,9,20], on the basis of thermodynamics, that the two phases should have different asymmetries, namely, $I_{gas} > I_{liquid}$, and actually a pure neutron gas was predicted at zero temperature if the initial global asymmetry is large enough ($I > 0.4$) [20]. Here we are studying this chemical effect in a non-equilibrium clustering process, on very short time scales, and we confirm the predictions of a linear response approach discussed before.

We can directly check the important result on the unique nature of the most unstable mode, independent of whether we start from a *mechanical* or from a *chemical* instability region. The isospin distillation dynamics presented in fig. 5 refers to the initial conditions of $T = 5$ MeV, average density $\rho = 0.06\,\mathrm{fm}^{-3}$ and asymmetry $I = 0.5$, *i.e.* we start from a point well inside the *mechanical* instability region of the used EOS, see fig. 1(b).

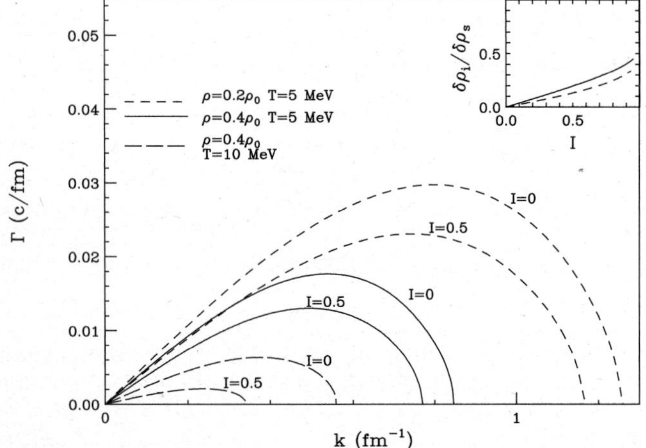

Fig. 4. Growth rates of instabilities as a function of the wave vector, as calculated from the dispersion relation eq. (29), for three situations inside the spinodal region. Lines are labeled with the asymmetry value I. The insert shows the asymmetry of the perturbation $\delta\rho_I/\delta\rho_S$, as a function of the asymmetry I of the initially uniform system, for the most unstable mode, in the case $\rho = 0.4\rho_0$, $T = 5$ MeV. The figure is taken from [26].

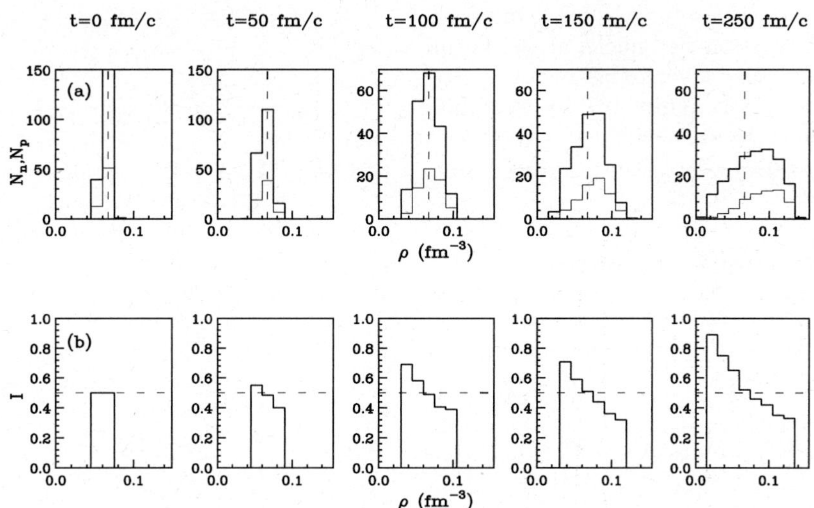

Fig. 5. Time evolution of neutron (thick lines) and proton (thin lines) abundances (a) and of asymmetry (b) in different density bins. The calculation refers to the case of $T = 5\,\mathrm{MeV}$, with initial average density $\rho = 0.06\,\mathrm{fm}^{-3}$ and asymmetry $I = 0.5$ (see the bottom left panel). The figure is taken from [12].

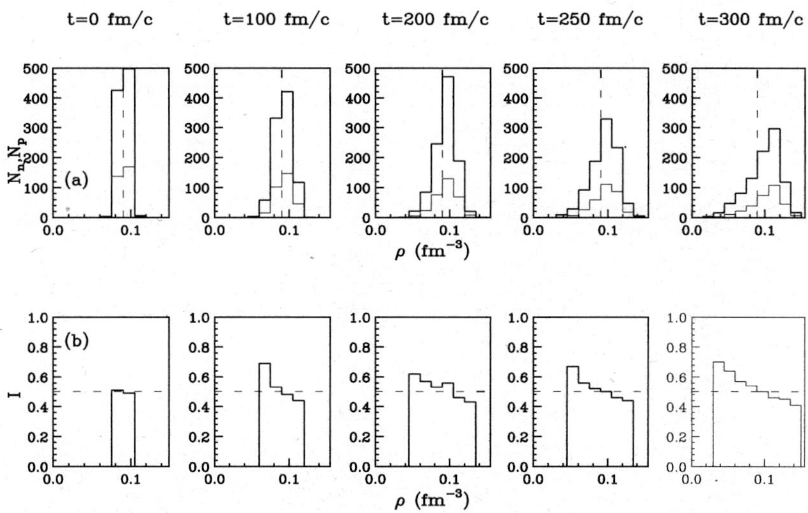

Fig. 6. Same calculation as in fig. 5 but with initial average density $\rho = 0.09\,\mathrm{fm}^{-3}$, inside the *chemical* instability region. The figure is taken from [26].

We can repeat the calculation at the same temperature and initial asymmetry, but starting from an initial average density $\rho = 0.09\,\mathrm{fm}^{-3}$, *i.e.* inside the *chemical* instability region of fig. 1(b). The results for the isospin distillation dynamics are shown in fig. 6. The trend is the same as in the previous fig. 5. This nicely shows the uniqueness of the unstable modes in the spinodal instability region, as discussed in detail in the previous subsection. Such result is due to gross properties of the n/p interaction, thus it should be not dependent on the use of a particular effective force. This has been clearly shown recently in the linear response frame [15], and in full transport simulations [27].

As intuitively expected, and as confirmed by the RPA analysis (see [12]), the isospin distillation effect becomes more important when increasing the initial asymmetry of the system. At the same time, the instability growth rates become smaller for the more asymmetric systems, see fig. 4.

Moreover, it is possible to observe a rather smooth and continuous transition from the trend observed at $\rho = 0.06\,\mathrm{fm}^{-3}$ (mechanical unstable region) to the trend observed at $\rho = 0.09\,\mathrm{fm}^{-3}$ (chemical unstable region), thus indicating that there is no qualitative change between the two kinds of instabilities. In fact they actually correspond to the same mechanism, the amplification of isoscalar-like fluctuations, with a significant chemical component (change of the concentration).

The conclusion is that the fast spinodal decomposition mechanism in neutron-rich matter will dynamically form more symmetric fragments surrounded by a less symmetric gas. Some recent experimental observations from fragmentation reactions with neutron-rich nuclei at the Fermi energies seem to be in agreement with this result on the fragment isotopic content: nearly symmetric intermediate mass fragments (IMF) have been detected in connection to very neutron-rich light ions [13,14].

4 From bulk to neck fragmentation

4.1 Multifragmentation

Since dynamical instabilities are playing an essential role in the reaction dynamics at Fermi energies it is essential to employ a stochastic transport theory. An approach has been adopted based on microscopic transport equations of Boltzmann-Nordheim-Vlasov (BNV) type [28,32–35] where asymmetry effects are suitably accounted for [36,37] and the dynamics of fluctuations is included [38,39].

The transport equations, with Pauli blocking consistently evaluated, are integrated following a test particle evolution on a lattice [35,40,41]. A parametrization of free NN cross-sections is used, with isospin, energy and angular dependence. The same symmetry term is utilized even in the initialization, *i.e.* in the ground-state construction of two colliding nuclei.

In particular, we report on a study of the 50 AMeV collisions of the systems ^{124}Sn + ^{124}Sn ^{112}Sn + ^{112}Sn and ^{124}Sn + ^{112}Sn, [42], where data are available from NSCL-MSU experiments for fragment production. One can identify quite generally three main stages of the collision, as observed also from the density contour plot of a typical event at $b = 2$ fm displayed in fig. 7: 1) in the early compression stage, during the first 40–50 fm/c, the density in the central region can reach values around 1.2–1.3 normal density; 2) the expansion phase, up to 110–120 fm/c, brings the system to a low-density state. The physical conditions of density and temperature reached during this stage correspond to an unstable nuclear-matter phase; 3) in the further expansion fragmentation is observed.

According to stochastic mean-field simulations, the fragmentation mechanism can be understood in terms of the growth of density fluctuations in the presence of instabilities. The volume instabilities have time to develop through spinodal decomposition leading to the formation of a liquid phase in the fragments and a gas of nucleons and light clusters. As seen in the figure, the fragment formation process typically takes place up to a *freeze-out* time (around 260–280 fm/c). This time is well defined in the simulations since it is the time of saturation of the average number of excited primary fragments. The clusters are rather far apart with a negligible nuclear interaction left among them.

Guided by the density contour plots we can investigate the behaviour of some characteristic quantities which give information on the isospin dynamics in fragment formation. In fig. 8, we report as a function of time:

(a) *The mass A* in the liquid phase (solid line and dots) and gas phase (solid line and squares).

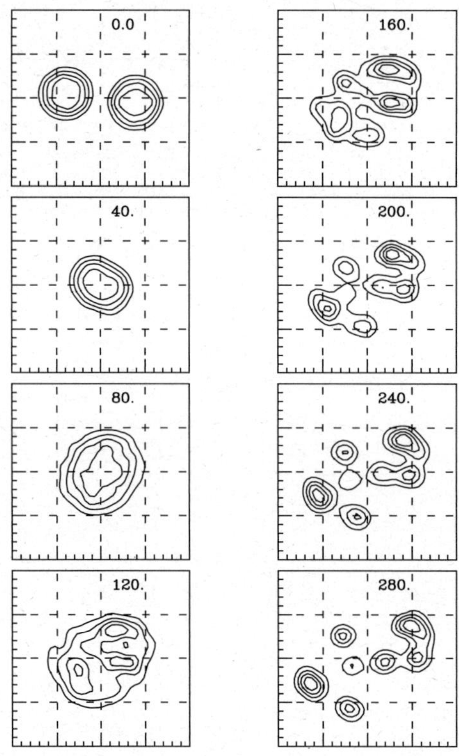

Fig. 7. Central $b = 2$ fm ^{124}Sn + ^{124}Sn collision at 50 AMeV: time evolution of the nucleon density projected on the reaction plane: approaching, compression and expansion phases. The times are written on each figure. The iso-density lines are plotted every 0.02 fm^{-3} starting from 0.02 fm^{-3}. The figure is taken from [42].

(b) *The asymmetry parameter $I = (N - Z)/(N + Z)$* in the gas "central" (solid line and squares), gas total (dashed+squares), liquid "central" (solid+circles) and IMFs (clusters with $3 < Z < 15$, stars). The horizontal line indicates the initial average asymmetry. "Central" means a box of linear dimension 20 fm around the center of mass of the total system.

(c) *The mean fragment multiplicity $Z \geq 3$* whose saturation defines the freeze-out time and configuration.

We also show some properties of the "primary" fragments in the *freeze-out configuration*:

(d) *The charge distribution probability $P(Z)$,*

(e) *The average asymmetry distribution $I_{av}(Z)$* and

(f) *The fragment multiplicity distribution $P(N)$* (normalized to 1).

For ^{124}Sn + ^{124}Sn we notice a neutron-dominated pre-equilibrium particle emission during the first 50 fm/c. The liquid phase becomes more symmetric during the compression and expansion. From the beginning of the fragment formation phase of the evolution, between 110 and 280 fm/c, we remark the peculiar trends of the liquid and gas phase asymmetry. In the "central region" the liquid asymmetry decreases while an *isospin burst* of the gas phase is observed. This behaviour is consistent with the

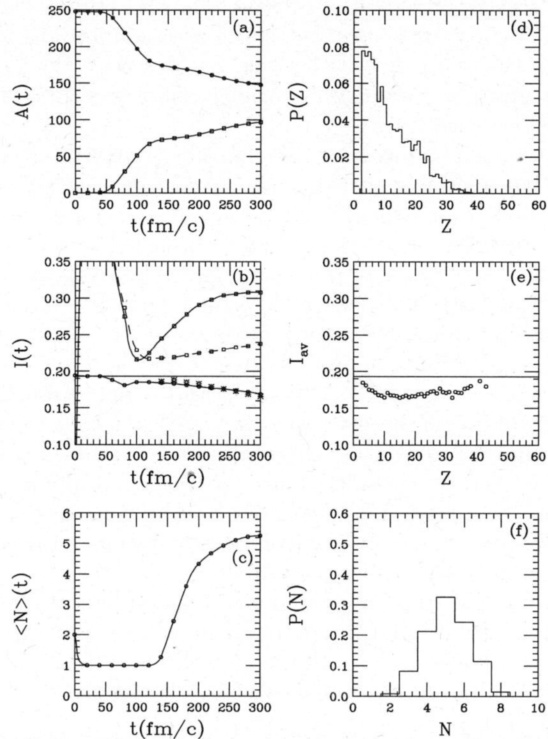

Fig. 8. The collision ^{124}Sn $+$ ^{124}Sn at $b = 2$ fm: time evolution (left) and freeze-out properties (right), ASY-STIFF EOS. The figure is taken from [42].

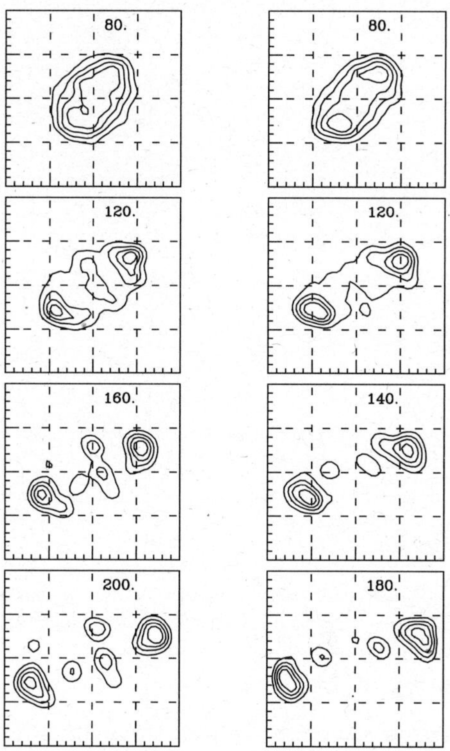

Fig. 9. ^{124}Sn $+$ ^{124}Sn collision at 50 AMeV: time evolution of the nucleon density projected on the reaction plane. Left column: $b = 4$ fm. Right column: $b = 6$ fm. The figure is taken from [42].

kinetic spinodal mechanism in dilute asymmetric nuclear matter leading to the *isospin distillation* between the liquid and the gas phase.

The effects of this process are clearly seen in the IMF isospin content, in both cases lower than at the beginning of the spinodal decomposition, fig. 8(e). Opposite trends for fragments with charge above and below $Z \approx 15$ can be observed. For heavier products the average asymmetry increases with the charge, a Coulomb related effect. However, the asymmetry rises again for lighter fragments. This can be a result of the differences in density and isospin between the regions in which the fragments grow, due to the fact that not all of them form simultaneously, as shown in the density contour plot. Let us also observe that the charge distribution of primary fragments has a rapidly decreasing trend, typical of a multifragmentation process.

4.2 Neck fragmentation

Summarizing the main experimental observations, we enumerate the following features of a "dynamical" IMF production mechanism in semi-peripheral collisions:

1. An enhanced emission is localized in the midrapidity region, intermediate between projectile-like fragment (PLF) and target-like fragments (TLF) sources, especially for IMFs with charge Z from 3 to 15 units.

2. The IMFs relative velocity distributions with respect to PLF (or TLF) cannot be explained in terms of a pure Coulomb repulsion following a statistical decay. A high degree of decoupling from the PLF (TLF) is also invoked.

3. Anisotropic IMFs angular distributions are indicating preferential emission directions and an alignment tendency.

4. For charge asymmetric systems the light particles and IMF emissions keep track of a neutron enrichment process that takes place in the neck region.

A fully consistent physical picture of the processes that can reproduce observed characteristics is still a matter of debate and several physical phenomena can be envisaged, ranging from the formation of a transient neck-like structure that would break-up due to Rayleigh instabilities or through a fission-like process, to the statistical decay of a hot source, triggered by the proximity with PLF and TLF [43–45].

The development of a neck structure in the overlap region of the two colliding nuclei is evidenced in fig. 9. During the interaction time this zone heats and expands but remains in contact with the denser and colder regions of PLF and/or TLF. The surface/volume instabilities of a cylindrically shaped neck region and the fast leading motion of the PLF and TLF will play an important role in the fragmentation dynamics. We notice the superimposed motion of the PL and TL pre-fragments linked to

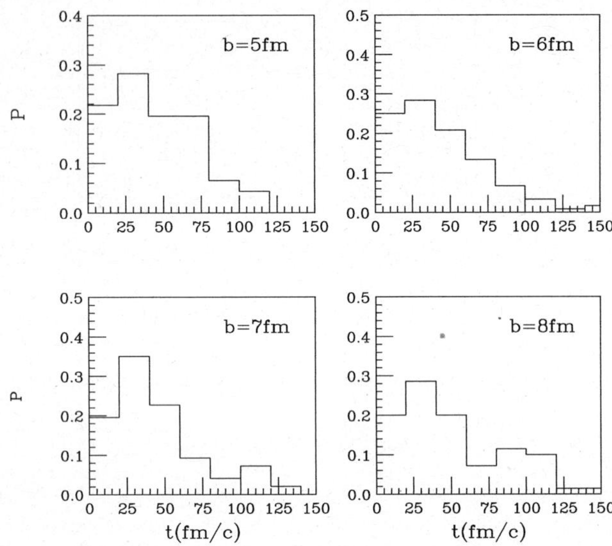

Fig. 10. The probability distribution of scission-to-scission time in the neck fragmentation for impact parameters from 5 to 8 fm. ^{124}Sn + ^{64}Ni at 35 AMeV and asystiff EOS. The figure is taken from [26].

the formation of a neck-like structure with a fast-changing geometry.

At the freeze-out time, with the neck rupture at about $140\,\mathrm{fm}/c$, intermediate mass fragments are produced in the mid-rapidity zone. In some events fragments form very early while, in others, they can remain for a longer time attached to the leading PLFs or TLFs. A transition behavior between multifragmentation and neck fragmentation is observed at $b = 4\,\mathrm{fm}$.

From the simulations we can extract an interesting information on the time scale of the Neck-IMF production. In fig. 10 we show, for different impact parameters, the probability distribution of the time interval between the instant of the first separation of the dinuclear system and the moment when a Neck-IMF is identified (scission-to-scission time). A large part of the Neck-IMFs are formed in short time intervals, within $50\,\mathrm{fm}/c$.

Finally, we would like to remark that the neck fragmentation shows a dependence on the nucleon-nucleon cross-sections and the EOS compressibility. The latter point is particularly interesting since it seems to indicate the relevance of volume instabilities even for the dynamics of the neck. This appears consistent with the short time scales shown before, see also the discussion in ref. [46].

5 Conclusions

In this work we investigated several properties of the asymmetric nuclear matter in the low-density region of phase diagram. The thermodynamical and dynamical analysis was based on Landau theory of Fermi liquid extended to binary systems. It was concluded that:

- at low densities, of interest for the nuclear liquid-gas phase transition, the asymmetric nuclear matter can

be characterized by a unique spinodal region, defined by the instability against isoscalar-like fluctuations; inside this we can identify the region where the system manifests mechanical instability and chemical instability, respectively;
- the physical meaning of thermodynamical chemical and mechanical instabilities should be related to the relative strengths of the interactions among the different species.
- everywhere in this density region the system is stable against the isovector-like fluctuations related to a tendency for species separation.
- at larger initial asymmetries the development of the spinodal instabilities is slower and a depletion of the maximum of the growth rate takes place. A decrease of the wave number corresponding to the maximum growth rate was deduced. Also the Coulomb force causes an overall decrease of growth rates. In this case the wave vector should exceed a threshold value in order to observe the instabilities.
- during the time development of the spinodal instabilities in ANM the fragment formation is accompanied by the isospin distillation leading to a more symmetric liquid phase and more neutron-rich gas phase.

We have made a connection of these features with isospin transport properties in simulations of fragmentation reactions based on stochastic BNV transport models. The presence and the role of the instabilities along the reaction dynamics in bulk fragmentation and neck fragmentation were discussed.

The results discussed here refer to the formation processes of primary fragments. *i.e.* at the freeze-out time. We explored the possibility that IMF appear as a result of a mechanism that initially started as spinodal decomposition triggered by isoscalar-like instabilities. These fragments are excited, and the subsequent statistical decay will certainly modify the signal. Therefore, it is important to search for various observables still keeping informations about the early stages of the fragments formation, for example those related to the kinematical properties (velocity distributions, angular distributions) and correlations between these observables and isospin content.

Moreover, the neck dynamics and corresponding isospin transport shows distinctive features related to the interplay between volume and surface instabilities. These should be better clarified in the future since they can contribute to a proper understanding of intermediate mass fragment production at Fermi energies.

V.B. acknowledges support of the Romanian Ministry for Education and Research for this work under the contract No. CEx-05-D10-02.

References

1. M. Barranco, J.R. Buchler, Phys. Rev. C **22**, 1729 (1980).
2. L.D. Landau, Sov. Phys. JETP **5**, 101 (1957).

3. A.B. Migdal, *Theory of finite Fermi systems and applications to atomic nuclei* (Wiley & Sons, N.Y., 1967).
4. G. Baym, C.J. Pethick, in *The Physics of Liquid and Solid Helium*, edited by K.H. Bennemann, J.B. Ketterson, Vol. **2** (Wiley, New-York, 1978) p. 1.
5. C.J. Pethick, D.G. Ravenhall, Ann. Phys. (N.Y.) **183**, 131 (1988).
6. N. Iwamoto, C.J. Pethick, Phys. Rev. D **25**, 313 (1982).
7. L.D. Landau, E.M. Lifshitz, *Statistical Physics* (Pergamonn Press, 1989) p. 288.
8. H. Müller, B.D. Serot, Phys. Rev. C **52**, 2072 (1995).
9. B.-A. Li, C.M. Ko, Nucl. Phys. A **618**, 498 (1997).
10. V. Baran, M. Colonna, M. Di Toro, V.G reco, Phys. Rev. Lett. **86**, 4492 (2001).
11. M. Colonna, M. Di Toro, A.B. Larionov, Phys. Lett. B **428**, 1 (1998).
12. V. Baran, M. Colonna, M. Di Toro, A.B. Larionov, Nucl. Phys. A **632**, 287 (1998).
13. H.S. Xu *et al.*, Phys. Rev. Lett. **85**, 716 (2000).
14. S.J. Yennello, *Proceedings of the International School-Seminar on Heavy Ion Physics*, edited by Yu.Ts. Oganessian (World Scientific, 1997).
15. J. Margueron, Ph. Chomaz, Phys. Rev. C **67**, 041602(R) (2003).
16. B. Liu, V. Greco, V. Baran, M. Colonna, M. Di Toro, Phys. Rev. C **65**, 045201 (2002).
17. S.S. Avancini, L. Brito, D.P. Menezes, C. Providencia, Phys. Rev. C **70**, 015203 (2004).
18. P. Haensel, Nucl. Phys. A **301**, 53 (1978).
19. F. Matera, V.Yu. Denisov, Phys. Rev. C **49**, 2816 (1994).
20. G. Baym, H.A. Bethe, C.J. Pethick, Nucl. Phys. A **175**, 225 (1971).
21. H. Krivine, J. Treiner, O. Bohigas, Nucl. Phys. A **336**, 155 (1990).
22. M. Colonna, Ph. Chomaz, Phys. Rev. C **49**, 1908 (1994).
23. M. Colonna, Ph. Chomaz, J. Randrup, Nucl. Phys. A **567**, 637 (1994).
24. Ph. Chomaz, M. Colonna, J. Randrup, Phys. Rep. **389**, 263 (2004).
25. M. Colonna, Ph. Chomaz, S. Ayik, V. Greco, Phys. Rev. Lett. **88**, 122701 (2002).
26. V. Baran, M. Colonna, V. Greco, M. Di Toro, Phys. Rep. **410**, 335 (2005).
27. B.-A. Li, A.T. Sustich, M. Tilley, B. Zhang, Nucl. Phys. A **699**, 493 (2002).
28. Ch. Grégoire *et al.*, Nucl. Phys. A **465**, 315 (1987).
29. A. Bonasera, F. Gulminelli, J. Molitoris, Phys. Rep. **243**, (1994).
30. V. Baran, A. Bonasera, M. Colonna, M. Di Toro, A. Guarnera, Prog. Part. Nucl. Phys. **38**, 263 (1997).
31. M. Colonna, M. Di Toro, A. Guarnera, Nucl. Phys. A **580**, 312 (1994).
32. G.F. Bertsch, S. Das Gupta, Phys. Rep. **160**, 189 (1988).
33. A. Bonasera *et al.*, Phys. Rep. **244**, 1 (1994).
34. A. Bonasera, G.F. Burgio, M. Di Toro, Phys. Lett. B **221**, 233 (1989).
35. TWINGO code: A. Guarnera PhD Thesis, University Caen (1996).
36. M. Colonna, M. Di Toro, G. Fabbri, S. Maccarone, Phys. Rev. C **57**, 1410 (1998).
37. L. Scalone, M. Colonna, M. Di Toro, Phys. Lett. B **461**, 9 (1999).
38. M. Di Toro *et al.*, Progr. Part. Nucl. Phys. **42**, 125 (1999).
39. M. Colonna *et al.*, Nucl. Phys. A **642**, 449 (1998).
40. A. Guarnera, M. Colonna, Ph. Chomaz, Phys. Lett. B **373**, 267 (1996).
41. V. Greco, Diploma Thesis (1997); V. Greco, A. Guarnera, M. Colonna, M. Di Toro, Phys. Rev. C **59**, 810 (1999).
42. V. Baran *et al.*, Nucl. Phys. A **703**, 603 (2002).
43. U. Brosa, S. Grossman, A. Muller, Phys. Rep. **197**, 167 (1990).
44. J. Lukasik *et al.*, Phys. Lett. B **566**, 76 (2003).
45. A.S. Botvina *et al.*, Phys. Rev. C **59**, 3444 (1999).
46. V. Baran, M. Colonna, M. Di Toro, Nucl. Phys. A **730**, 329 (2004).

Eur. Phys. J. A **30**, 153–163 (2006)

DOI 10.1140/epja/i2006-10113-x

THE EUROPEAN
PHYSICAL JOURNAL A

Isospin flows

M. Di Toro[1,a], S.J. Yennello[2], and B.-A. Li[3]

[1] Laboratori Nazionali del Sud INFN, Phys. Astron. Dept. Catania University, Via S. Sofia 62, I-95123 Catania, Italy
[2] Cyclotron Institute, Texas A&M University, College Station, TX 77843, USA
[3] Department of Chemistry and Physics, P.O. Box 419, Arkansas State University, AR 72467-0419, USA

Received: 18 May 2006 /
Published online: 24 October 2006 – © Società Italiana di Fisica / Springer-Verlag 2006

Abstract. In this report, we review the isospin dependence of various forms of the collective flow in heavy-ion reactions from Fermi to relativistic energies. The emphasis will be on suggested possible applications in directly exploring the underlying isovector potential and thus the Equation of State (EoS) of asymmetric nuclear matter, in particular in density regions far away from normal conditions. We also discuss forthcoming challenges and opportunities provided by high-energy radioactive beams.

PACS. 25.70.-z Low and intermediate energy heavy-ion reactions – 25.75.Ld Collective flow – 21.30.Fe Forces in hadronic systems and effective interactions – 21.65.+f Nuclear matter

1 Introduction

Nuclear collective flow is a motion characterized by space-momentum correlations of dynamical origins. It reveals itself in various forms in nuclear reactions. The study of several components of the collective flow in heavy-ion reactions has been found very useful for extracting information about the Equation of State (EoS) of symmetric nuclear matter [1–5]. The isospin flow refers to the dependence of the collective flow on the isospin asymmetry of the reaction system and/or of the reaction products. This isospin dependence of collective flow has been found useful for studying the isospin asymmetric part of the EoS, namely, the symmetry energy, of neutron-rich matter.

We begin by reviewing briefly our current understanding about the EoS of isospin-asymmetric matter. Several forms of the collective flow will then be introduced. Effects of the symmetry energy/potential on collective flows will be examined in the following sections, with an accurate analysis of the corresponding most sensitive observables. Particular attention will be given to the possibility of studying the symmetry term at high baryon density.

2 Equation of state of isospin-asymmetric nuclear matter

Here we shortly review the EoS of isospin-asymmetric matter and the related symmetry energy problem. In asymmetric matter the energy per nucleon, *i.e.* the equation of state, will be a functional of the total ($\rho = \rho_n + \rho_p$) and isospin ($\rho_3 = \rho_n - \rho_p$) densities. In the usual parabolic form in terms of the asymmetry parameter $I \equiv \rho_3/\rho = (N - Z)/A$ we can define a symmetry energy $\frac{E_{sym}}{A}(\rho)$:

$$\frac{E}{A}(\rho, I) = \frac{E}{A}(\rho) + \frac{E_{sym}}{A}(\rho)\, I^2. \tag{1}$$

The symmetry term gets a kinetic contribution directly from the basic Pauli correlations and a potential contribution from the properties of the isovector part of the effective nuclear interactions in the medium. Since the kinetic part can be exactly evaluated we can separate the two contributions, reducing the discussion just to a function $F(u)$ of the reduced density $u \equiv \rho/\rho_0$ linked to the interaction:

$$\epsilon_{sym} \equiv \frac{E_{sym}}{A}(\rho) \equiv \epsilon_{sym}(kin) + \epsilon_{sym}(pot)$$
$$= \frac{\epsilon_F(\rho)}{3} + \frac{C}{2}\, F(u), \tag{2}$$

with $F(1) = 1$, where ρ_0 is the saturation density and the parameter C is of the order $C \simeq 32\,\mathrm{MeV}$ to reproduce the a_4 term of the Bethe-Weizsäcker mass formula. The major uncertainties about the EoS and the symmetry energy are due to both our poor knowledge about the isospin dependence of nuclear effective interactions and the limitations of existing many-body techniques. Shown in fig. 1 are the density-dependent symmetry energies predicted by some of the most widely used microscopic many-body theories. It is seen that, at both sub-saturation and

a e-mail: `ditoro@lns.infn.it`

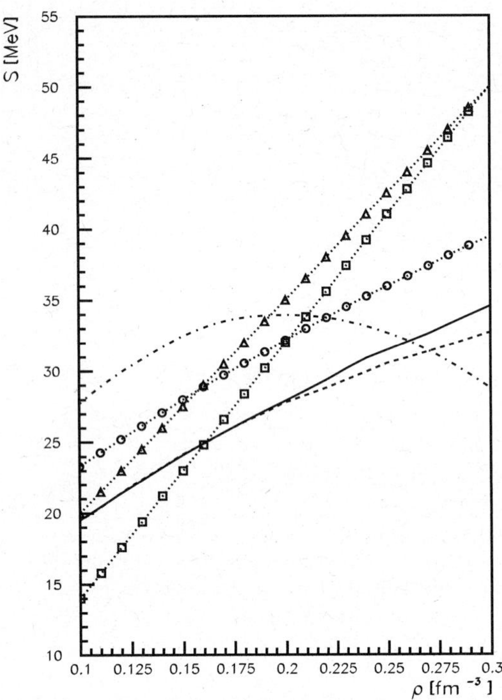

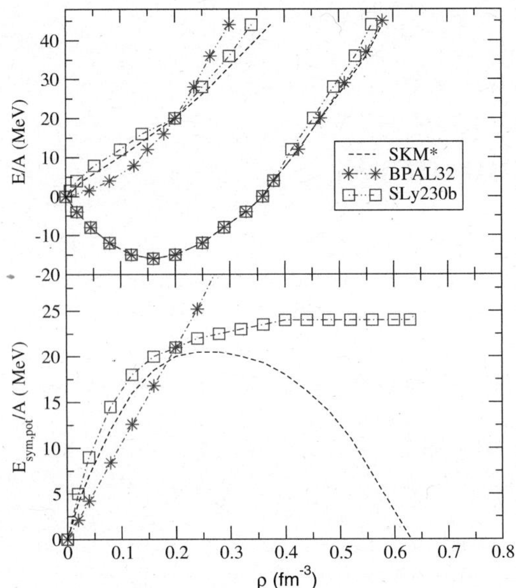

Fig. 2. EoS for various effective forces. Top: neutron matter (up), symmetric matter (down); Bottom: potential symmetry term. Taken from [23].

Fig. 1. Overview of several theoretical predictions for the symmetry energy S: Brueckner-Hartree-Fock with Reid93 potential (circles), self-consistent Green function theory with Reid93 potential (full line), variational calculation with Argonne Av14 potential (dashed line), Dirac-Brueckner-Hartree-Fock calculation (triangles), relativistic mean-field model (squares), effective field theory (dash-dotted line). Taken from [6].

supra-saturation densities, the predictions diverge very widely. We note that within each approach the prediction also depends on the two-body effective interaction used and whether/what three-body forces are included. To illustrate the dependence on the effective interactions used, we show in fig. 2 some typical EoSs obtained from Hartree-Fock calculations. It is necessary to stress that they all have *the same saturation properties for symmetric NM* (top): SKM* [7,8], SLy230b (SLy4) [9–11] and BPAL32 [12–15]. However, their predictions on the EoS of asymmetric matter, especially their contributions to the potential part of the symmetry energy, are very different. The major challenge is thus to constrain experimentally the potential part of the symmetry energy and the associated symmetry potential. The ultimate goal is to pin down the isospin dependence of nuclear effective interactions that is also responsible for the structure of rare isotopes.

In fig. 2 (bottom) the density dependence of the potential symmetry contribution for the three different effective interactions is reported. While all curves obviously cross at normal density ρ_0, quite large differences are present for values, slopes and curvatures in low-density and particularly in high-density regions. Moreover, even at the relatively well-known "crossing point" at normal density the

various effective forces are presenting controversial predictions for the momentum dependence of the fields acting on the nucleons and consequently for the splitting of the neutron/proton effective masses, of large interest for nuclear structure and dynamics. In recent years under the stimulating perspectives offered from nuclear astrophysics and from the new Radioactive Ion Beam (RIB) facilities a relevant activity has started in the field of the isospin degree of freedom in heavy-ion reactions, see for review refs. [20–23].

A traditional expansion to second order around normal density is used [18,24,25]

$$\epsilon_{sym} \equiv \frac{E_{sym}}{A}(\rho) = a_4 + \frac{L}{3}\left(\frac{\rho - \rho_0}{\rho_0}\right) + \frac{K_{sym}}{18}\left(\frac{\rho - \rho_0}{\rho_0}\right)^2,$$

$$(3)$$

in terms of a slope parameter

$$L \equiv 3\rho_0\left(\frac{\mathrm{d}\epsilon_{sym}}{\mathrm{d}\rho}\right)_{\rho=\rho_0} = \frac{3}{\rho_0}P_{sym}(\rho_0), \qquad (4)$$

which is simply related to the *symmetry pressure* $P_{sym} = \rho^2 \mathrm{d}\epsilon_{sym}/\mathrm{d}\rho$ at ρ_0, and a curvature parameter

$$K_{sym} \equiv 9\rho_0^2\left(\frac{\mathrm{d}^2\epsilon_{sym}}{\mathrm{d}^2\rho}\right)_{\rho=\rho_0}, \qquad (5)$$

a kind of symmetry compressibility. We remark that our present knowledge of these basic properties of the symmetry term around saturation is still very poor, see the analysis in ref. [26] and references therein. In particular, we note the uncertainty on the symmetry pressure at ρ_0, of large importance for structure calculations.

We have seen that asymmetry brings an extra pressure P_{sym}. For the collective flow discussion it is instructive to

Table 1. Symmetry term at saturation.

$F(u)$	L	K_{sym}	$[K_{sym} - 6L]$	$[K_{sym} + 6L]$
const $= 1$	$+25\,\text{MeV}$	$-25\,\text{MeV}$	$-175\,\text{MeV}$	$+125\,\text{MeV}$
$\sqrt{u}$	$+49\,\text{MeV}$	$-61\,\text{MeV}$	$-355\,\text{MeV}$	$+234\,\text{MeV}$
u	$+75\,\text{MeV}$	$-25\,\text{MeV}$	$-475\,\text{MeV}$	$+425\,\text{MeV}$
$u^2/(1+u)$	$+100\,\text{MeV}$	$+50\,\text{MeV}$	$-550\,\text{MeV}$	$+650\,\text{MeV}$

evaluate the density gradient of the *symmetry pressure* as a function of the slope and curvature of the symmetry term:

$$\frac{\mathrm{d}}{\mathrm{d}\rho}P_{sym} = \frac{1}{9}(K_{sym} + 6L), \qquad (6)$$

that around normal density gives

$$\frac{\mathrm{d}}{\mathrm{d}\rho}P_{sym} = \left(\frac{10}{27}\epsilon_F + C\left[\frac{\mathrm{d}}{\mathrm{d}u} + \frac{1}{2}\frac{\mathrm{d}^2}{\mathrm{d}u^2}\right]F(u)\Big|_{u=1}\right). \qquad (7)$$

The compressibility of the matter is also modified by the asymmetry [23,27]. For the compressibility shift *at equilibrium* we have, after some algebra,

$$\Delta K_{NM}(I) = 9\rho_0\left[\rho_0\frac{\mathrm{d}^2}{\mathrm{d}\rho^2} - 2\frac{\mathrm{d}}{\mathrm{d}\rho}\right]\epsilon_{sym}(\rho)\Big|_{\rho=\rho_0} I^2$$
$$= [K_{sym} - 6L]I^2 < 0. \qquad (8)$$

We note the different interplay between slope and curvature of the symmetry term for flows, eqs. (6), (7), and monopole, eq. (8), observables. In order to have a quantitative idea, we now show explicitly the influence on the L, K_{sym} parameters of a different density dependence in the potential part of the symmetry energy around saturation, *i.e.* of the function $F(u)$ of eq. (2):

$$L = \frac{2}{3}\epsilon_F + \frac{3}{2}C\frac{\mathrm{d}}{\mathrm{d}u}F(u)\Big|_{u=1},$$
$$K_{sym} = -\frac{2}{3}\epsilon_F + \frac{9}{2}C\frac{\mathrm{d}^2}{\mathrm{d}u^2}F(u)\Big|_{u=1}.$$

We obtain the rather instructive table 1 for various functional forms $F(u), u \equiv \rho/\rho_0$, around ρ_0. A stiffer symmetry term in general enhances the pressure gradient of asymmetric matter. We can expect direct effects on the nucleon emissions in the reaction dynamics, fast particles and collective flows. In particular, we will see larger flows in isospin-asymmetric collisions. Moreover, due to the different fields seen by neutrons and protons, we shall observe even specific isotopic effects.

In fig. 3 we report, for an asymmetry $(N - Z)/A = 0.2$ representative of ^{124}Sn, the density dependence of the symmetry contribution to the mean-field potential for the different effective interactions in the isovector channel. It is seen that in regions just off normal density the field "seen" by neutrons and protons in the three cases is very different. We thus expect important isospin effects on nucleon transport during reactions at intermediate energies (prompt particle emissions, collective flows, n/p interferometry) where the interacting asymmetric nuclear matter

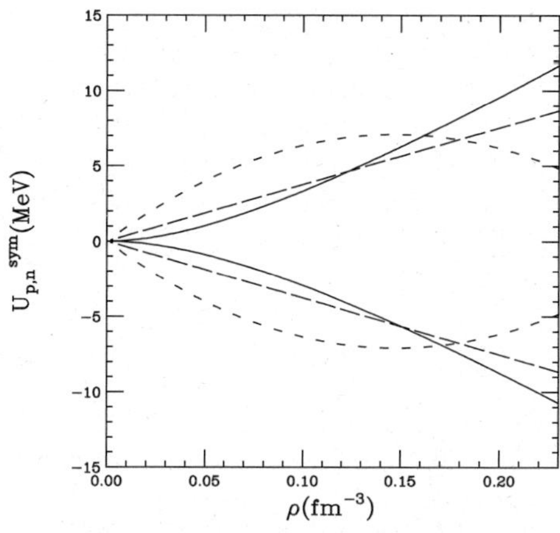

Fig. 3. Symmetry contribution to the mean field at $I = 0.2$ for neutrons (upper curves) and protons (lower curves): dashed lines "asy-soft", long-dashed lines "asy-stiff", solid lines "asy-superstiff". Taken from [23].

will experience compressed and expanding phases. These points have been analysed in some detail using isospin-dependent transport simulations for the reaction dynamics. We always compare results obtained with forces that have *the same saturation properties for symmetric NM*. We will refer to an *"asy-stiff"* EoS (*e.g.*, like BPAL32 of fig. 2). when we are considering a potential symmetry term linearly increasing with nuclear density and to a *"asy-soft"* EoS (*e.g.*, like SKM* of fig. 2) when the symmetry term shows a saturation and eventually a decrease above normal density. In some cases, in order to enhance the dynamical effects, we will consider also *"asy-superstiff"* behaviours, *i.e.* with a roughly parabolic increase of the symmetry term above normal density [15,28,29].

3 Collective flows: definitions

The collective motion can be characterized in several ways that pin down different space-momentum correlations that can be generated by the dynamics. The kind of collective flows that have been suggested and employed to get information on the equation of state can be divided into three categories: radial, sideward and elliptic. The sideward and elliptic flows have been and are currently useful tools for the study of the compressibility of symmetric nuclear matter. In the search for the density behaviour of the symmetry energy, similar concepts can be exploited but high-lightening the difference between neutrons and protons or light clusters with different isospin. We will define the different types of collective flow and we will discuss the current status of the effects expected due to different $E_{sym}(\rho)$, and related momentum dependence. We will see that first experimental results with stable beam already show hints of the effect of the symmetry energy. Thus future, more exclusive, experiments with radioactive beams should be able to set stringent constraints on the density dependence of the symmetry energy far from ground-state nuclear matter.

The sideward (transverse) flow is a deflection of forward- and backward-moving particles, within the reaction plane [30]. It is formed because for the compressed and excited matter it is easier to get out on one side of the beam axis than on the other. The sideward flow is often represented in terms of the average in-plane component of the transverse momentum at a given rapidity $\langle p_x(y)\rangle$:

$$F(y) \equiv \frac{1}{N(y)} \sum_{i=1}^{N(y)} p_{x_i} \equiv \langle p_x(y)\rangle. \tag{9}$$

The particular case in which the slope of the transverse flow is vanishing in a region around midrapidity is referred to as balance energy. It comes out from a balance between the attraction of the mean field and the repulsion of the two-body collisions.

The build up of sideward and elliptic flow is realized around the higher-density stage of the reaction and thus is a powerful tool for the search of the high-density behaviour of the symmetry energy. It represents a very general means of investigation, giving information on the dynamical response of excited nuclear matter in heavy-ion collisions, from the Fermi energies [1–5] up to the ultrarelativistic regime, in the search for a phase transition to QGP [31]. For the isospin effect the sum over the particles in eq. (9) is separated into protons and neutrons. In refs. [29,32] also the neutron-proton differential flow $F^{pn}(y)$ has been suggested as a very useful probe of the isovector part of the EoS since it appears rather insensitive to the isoscalar potential and to the in-medium nuclear cross-section and, as we will discuss, it combines the isospin distillation effects with the direct dynamical flow effect. The definition of the differential flow $F_{pn}(y)$ is

$$F_{pn}(y) \equiv \frac{1}{N(y)} \sum_{i=1}^{N(y)} p_{x_i}\tau_i \equiv \frac{N_n}{N(y)}F_n(y) - \frac{N_p}{N(y)}F_p(y),$$

$$\tag{10}$$

where $N(y)$ is the total number of free nucleons at rapidity y ($N_{n,p}$, neutron/proton multiplicities) and p_{x_i} is the transverse momentum of particle i in the reaction plane (τ_i is $+1$ and -1 for protons and neutrons). The flow observables can be seen, respectively, as the first and second coefficients from the Fourier expansion of the azimuthal distribution [33]:

$$\frac{dN}{d\phi}(y, p_t) \propto 1 + 2V_1\cos(\phi) + 2V_2\cos(2\phi),$$

where $p_t = \sqrt{p_x^2 + p_y^2}$ is the transverse momentum and y the rapidity along beam direction. The transverse flow can be also expressed as

$$V_1(y, p_t) = \left\langle \frac{p_x}{p_t} \right\rangle.$$

It provides information on the azimuthal anisotropy of the transverse nucleon emission and has been used to study the EoS and cross-section sensitivity of the balance energy [32].

The second coefficient of the expansion defines the elliptic flow v_2 that can be expressed as

$$V_2(y, p_t) = \left\langle \frac{p_x^2 - p_y^2}{p_t^2} \right\rangle.$$

It measures the competition between in-plane and out-of-plane emissions. The sign of V_2 indicates the azimuthal emission anisotropy: particles can be preferentially emitted either in the reaction plane ($V_2 > 0$) or out-of-plane (*squeeze-out*, $V_2 < 0$) [33,34]. The p_t-dependence of V_2, which has been recently investigated by various groups [5, 34–36], is very sensitive to the high-density behavior of the EoS since highly energetic particles ($p_t \geq 0.5$) originate from the initial compressed and out-of-equilibrium phase of the collision, see, *e.g.*, ref. [36]. Also at high energy it is allowing to get insight of the partonic stage and hadronization mechanism in ultrarelativistic heavy-ion collisions [31].

4 Collective flows at the Fermi energies: isospin effects around the balance energy

The Fermi energy range (roughly from 20 to 100 AMeV beam energies), transitional region from a mean field to a NN-collision dynamics with the related building up of density gradients, represents a kind of threshold for out-of-plane flows (radial and elliptic). Meanwhile the transverse flow shows the *balance* effect, *i.e.* it changes from negative to positive due to the competition between the attractive mean field and the repulsive NN collisions (plus Coulomb), see [3]:

$$\frac{dF(y)}{dy}(E_{bal})_{y=0} = 0.$$

Due to this delicate balance one would expect isospin effects on the mean field to be relevant.

The isospin dependence of the transverse collective flow near the balance energy was first pointed out in ref. [37], where it is stressed that the reactions involving neutron-rich nuclei should have a significantly stronger attractive flow and consequently a higher balance energy. Shown in fig. 4 is the impact parameter dependence of the flow parameter for the reaction of ^{58}Fe $+$ ^{58}Fe and ^{58}Ni $+$ ^{58}Ni at a beam energy of 55 MeV/nucleon from experiments done at MSU [38–40]. It is interesting to see that the flow parameter for the neutron-richer system is consistently higher and is in agreement with transport model predictions [37]. Pak *et al.* have also studied the

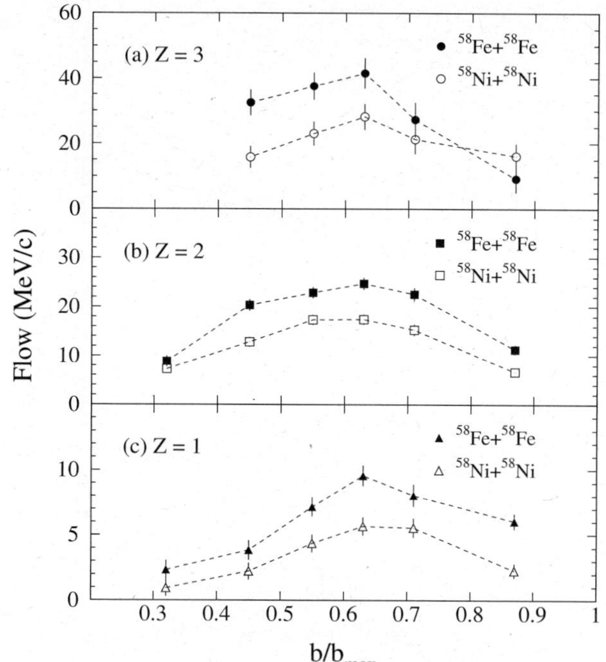

Fig. 4. Flow parameters for the reactions of ^{58}Fe + ^{58}Fe and ^{58}Ni + ^{58}Ni as a function of the reduced impact parameter at a beam energy of 55 MeV/nucleon. Taken from ref. [39].

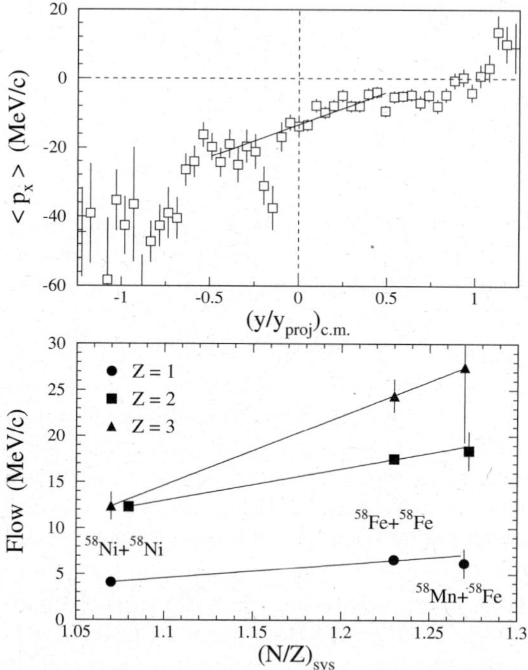

Fig. 5. Upper window: mean transverse momentum in the reaction plane *versus* the reduced c.m. rapidity for $Z = 2$ fragments from impact-parameter-inclusive ^{58}Mn + ^{58}Fe collisions at 55 MeV/nucleon. Lower window: isospin dependence of the flow parameter for inclusive collisions at a beam energy of 55 MeV/nucleon. Taken from ref. [39].

flow parameter as a function of the isotope ratio of the composite projectile plus target system for three different fragment types from three isotopic entrance channels. Shown in the upper window of fig. 5 is the mean transverse momentum in the reaction plane *versus* the reduced c.m. rapidity for $Z = 2$ fragments from impact-parameter-inclusive ^{58}Mn + ^{58}Fe collisions at 55 MeV/nucleon. The flow parameter extracted for inclusive events is plotted in the lower window of fig. 5 as a function of the ratio of neutrons to protons of the combined system $(N/Z)_{cs}$. The flow parameter increases linearly with the ratio $(N/Z)_{cs}$ for all three types of particles.

In spite of the low ^{58}F asymmetry ($I = 0.1$), in the iso-transport simulations of ref. [41] the shift of the balance energy is getting a noticeable contribution from the stiffness of the symmetry term. This is shown in fig. 6, where the flow slope at midrapidity *vs.* beam energy is reported: an *asy-stiff* behavior, more attractive for protons above normal density for the Fe asymmetric case, gives a clear shift in the balance energy as well as a larger (negative) flow at 55 AMeV, *i.e.* below the balance. Both effects are in agreement with the data and are disappearing in the *asy-soft* choice. Of course also the isospin and density dependence of the NN cross-sections is important (see the (c), (d) plots) but we note that a good sensitivity to the isovector part of the EoS is still present. In particular, we can see that the isospin dependence of the mean field is able to keep the transverse flow difference between protons in Fe-Fe and Ni-Ni. However a systematic study over different systems with more "exotic" isospin content is necessary to confirm this result. An important effect predicted by the simulations is the clear difference between

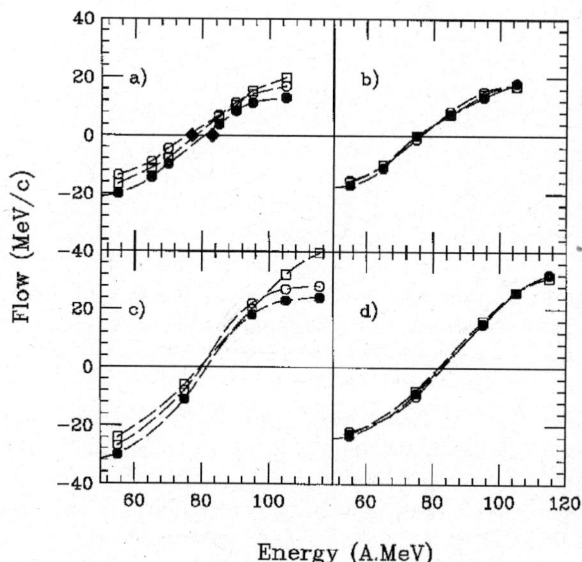

Fig. 6. Energy dependence of flows at $b_{red} = 0.45$ [42]: Fe + Fe protons (full circles); Ni + Ni protons (open circles); Fe + Fe neutrons (squares). (a) Asy-stiff; (b) asy-soft; (c), (d) same for $\sigma_{NN} = 2\,\mathrm{fm}^2$ no isospin dependent. The full diamonds in (a) represent the proton balance energy data of ref. [38] for the Fe + Fe (right) and Ni + Ni systems. Taken from [41].

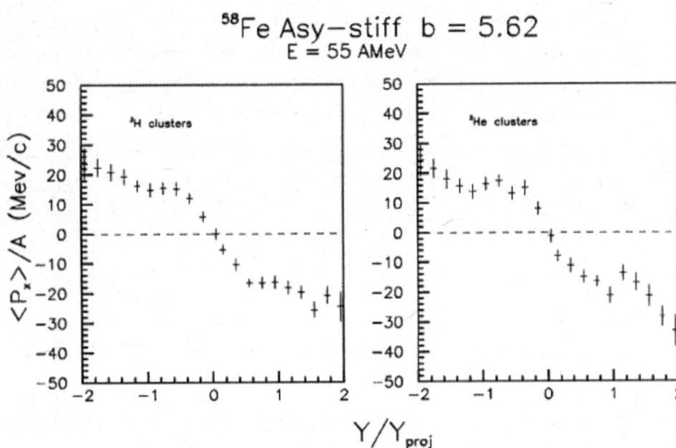

Fig. 7. Mean transverse momentum in the reaction plane *vs.* reduced rapidity for light ^{3}He-^{3}H isobars in the Fe + Fe collisions at 55 AMeV beam energy (*i.e.* below the balance) for semicentral impact parameter, $b_{red} = 0.6$. Asy-stiff parametrization. Taken from [41].

neutron and proton flows. Due to the difficulties in measuring neutrons this should be seen in a detailed study of light isobar flows. Moreover, we like to recall that clusters are better probing the higher-density regions. This point is quantitatively shown in fig. 7 where we present the transverse momentum *vs.* rapidity distributions for ^{3}He-triton clusters in semicentral Fe-Fe collisions at 55 AMeV, *i.e.* below the balance energy [41]. We can estimate a 20% larger (negative) flow for the ^{3}He ions, just opposite to what is expected from Coulomb effects. This appears to be a clear indication of the contribution of a much reduced (negative) neutron flow in the case of an *asy-stiff* force, *i.e.* a more repulsive symmetry term just above ρ_0. The effect would disappear in an *asy-soft* choice.

For heavier systems, with much larger Coulomb repulsion, the flow balance is at lower energies. The Iso-EoS effects are less evident for two main reasons: i) the smaller relative weight of symmetry *vs.* Coulomb contributions; ii) the reduced compression in the interacting region. This is clearly shown in fig. 8, from the iso-transport simulations of ref. [43], where the proton transverse flows for the ^{124}Sn + ^{124}Sn case at 50 AMeV (semicentral) are reported. There is no appreciable difference in the evaluations with two quite different density dependencies of the symmetry term, $F(u) = u^\gamma, u \equiv \rho/\rho_0$, $\gamma = 0.5$ (rather asy-soft) and $\gamma = 2$ (asy-superstiff).

Moreover, at the Fermi energies free nucleons can be emitted from various sources, from the early high-density stage as well as in the expansion phase, when fragments are formed (isofractionation or isodistillation) and finally from excited primary clusters. For the Iso-EoS studies more exclusive flow data are needed. In particular, a good selection for the source density could be based on the transverse momentum of the nucleons emitted at a given rapidity. The proton elliptic flow appears very sensitive to this analysis, see fig. 8 for the same Sn + Sn n-rich system [43]. At high p_t's the Iso-EoS differences are evident, with a reduced squeeze-out flow in the $\gamma = 2$ case. At this

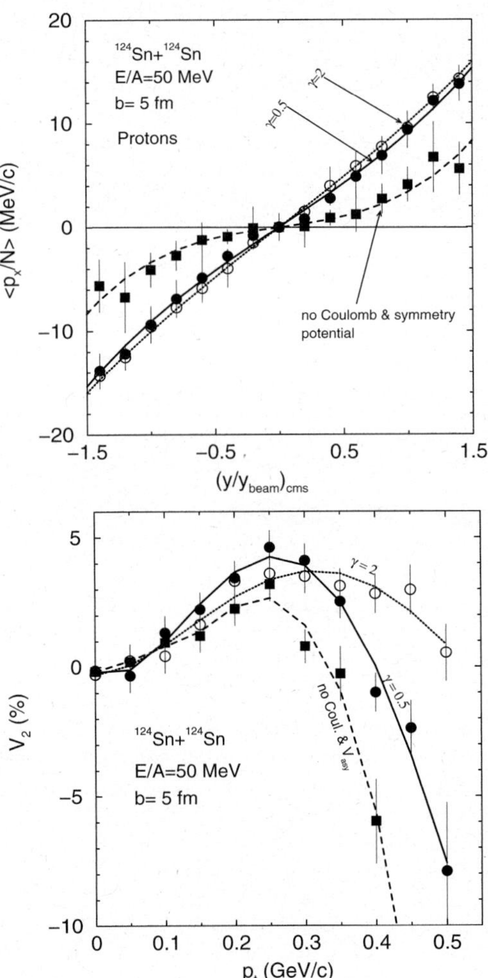

Fig. 8. Top: mean transverse momentum in the reaction plane *vs.* reduced rapidity for protons in the ^{124}Sn + ^{124}Sn collisions at 50 AMeV beam energy (*i.e.* just above the balance) for semicentral impact parameter, $b_{red} = 0.6$. Bottom: elliptic flow for midrapidity protons as a function of transverse momentum. Two different symmetry energy parametrizations are used (see text). From ref. [43].

low energy the less repulsive interactions (no Coulomb and symmetry potentials) give the largest squeeze-out, just opposite to what we will see at higher energies. Finally, we note that high-momentum particles will better probe the momentum dependence of the mean field, including its isospin-dependent part. This is the subject of the next section.

Despite the possible interpretation, in order to make the analysis of collective flow more sensitive to the symmetry potential, the *neutron-proton* differential flow, defined in eq. (10), has been introduced [32]. In such a way one combines constructively the difference in the neutron-proton collective flow and the difference in the number of protons and neutrons emitted. At the same time the influences of the isoscalar potential and the in-medium nucleon-nucleon cross-sections are also reduced. However, the measurement of such a differential flow demands not

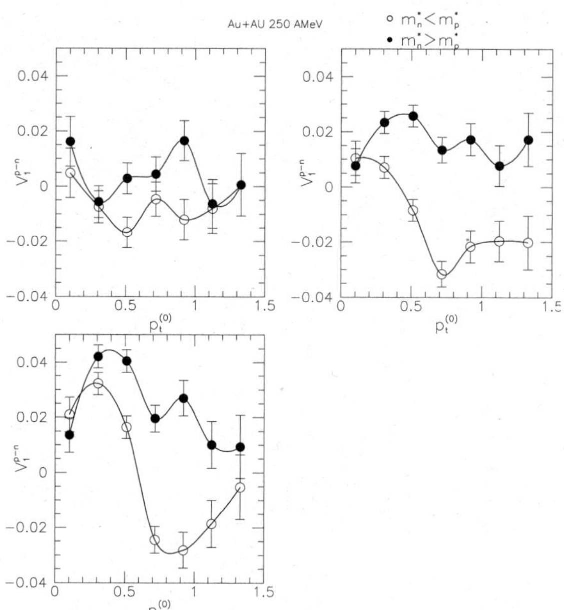

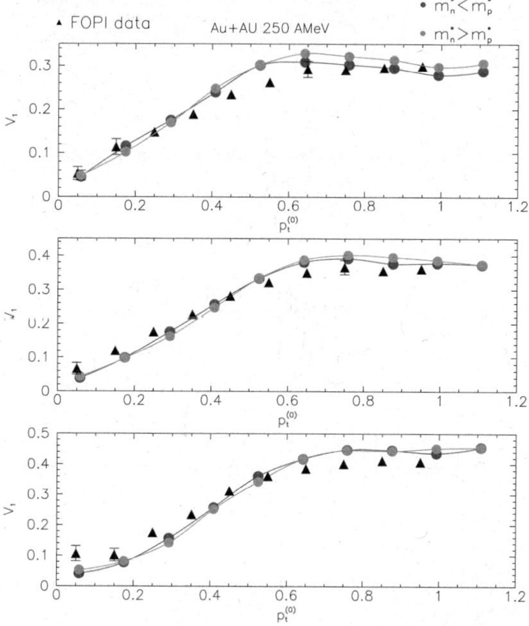

Fig. 9. Difference between proton and neutron V_1 flows in a semicentral reaction $\mathrm{Au} + \mathrm{Au}$ at $250\,A\mathrm{MeV}$ for three rapidity ranges. Upper left panel: $|y^{(0)}| \leq 0.3$; upper right: $0.3 \leq |y^{(0)}| \leq 0.7$; lower left: $0.6 \leq |y^{(0)}| \leq 0.9$. Taken from [47].

Fig. 10. Comparison of the V_1 proton flow with FOPI data [50] for three rapidity ranges. Top: $0.5 \leq |y^{(0)}| \leq 0.7$; center: $0.7 \leq |y^{(0)}| \leq 0.9$; bottom: $0.9 \leq |y^{(0)}| \leq 1.1$. Taken from [47].

only for, the measurement of neutron collective flow but also for a precise assessment of their number, which most likely is impossible. On the other hand, the idea to combine more than one isospin contribution in one observable is certainly important for elusive effects as those coming from the symmetry energy. In this respect, we would like to note that the usual problems caused by the neutrons can be overcome by looking at clusters. For example, one can use the definition of differential collective flow and apply it to the $^3\mathrm{H}$-$^3\mathrm{He}$ isospin doublet, see previous discussion.

5 Effective mass splitting and collective flows

The problem of momentum dependence (MD) in the isospin channel is still very controversial and it would be extremely important to get more definite experimental information, see the recent refs. [44–49]. Intermediate energies are important in order to have high-momentum particles and to test regions of high baryon (isoscalar) and isospin (isovector) density during the reaction dynamics. Now, we present some qualitative features of the dynamics in heavy-ion collisions in higher-energy regions, of large interest for the RIA facility, related to the splitting of nucleon effective masses.

Collective flows are very good candidates since they are expected to be very sensitive to the momentum dependence of the mean field, see [23,34] and references therein. We have then tested the isovector part of the momentum dependence just evaluating the *difference* of

neutron/proton transverse and elliptic flows

$$V_{1,2}^{(n-p)}(y, p_t) \equiv V_{1,2}^n(y, p_t) - V_{1,2}^p(y, p_t)$$

at various rapidities and transverse momenta in semicentral $(b/b_{max} = 0.5)$ $^{197}\mathrm{Au} + {}^{197}\mathrm{Au}$ collisons at $250\,A\mathrm{MeV}$, where some proton data are existing from the FOPI Collaboration at GSI [50,51].

We report here on expected effects of the isospin MD, studied by means of the Boltzmann-Nordheim-Vlasov transport code, refs. [52,53], implemented with a *BGBD-like* [54,55] mean field with a different (n,p) momentum dependence, see refs. [44,47], that allow to follow the dynamical effect of opposite n/p effective mass splitting while keeping the same density dependence of the symmetry energy.

Transverse flows

For the difference of nucleon transverse flows, see fig. 9, the mass splitting effect is evident at all rapidities, and nicely increasing at larger rapidities and transverse momenta, with more neutron flow when $m_n^* < m_p^*$. Just to show that our simulations give realistic results we compare in fig. 10 with the proton data of the FOPI Collaboration for similar selections of impact parameters rapidities and transverse momenta. The agreement is quite satisfactory. We see a slightly reduced proton flow at high transverse momenta in the $m_n^* < m_p^*$ choice, but the effect is too small to be seen from the data. Our suggestion of measuring just the difference of n/p flows looks much more promising. Similar calculations have been performed in ref. [45] for the $^{132}\mathrm{Sn}$-$^{124}\mathrm{Sn}$ system at $400\,A\mathrm{MeV}$ beam

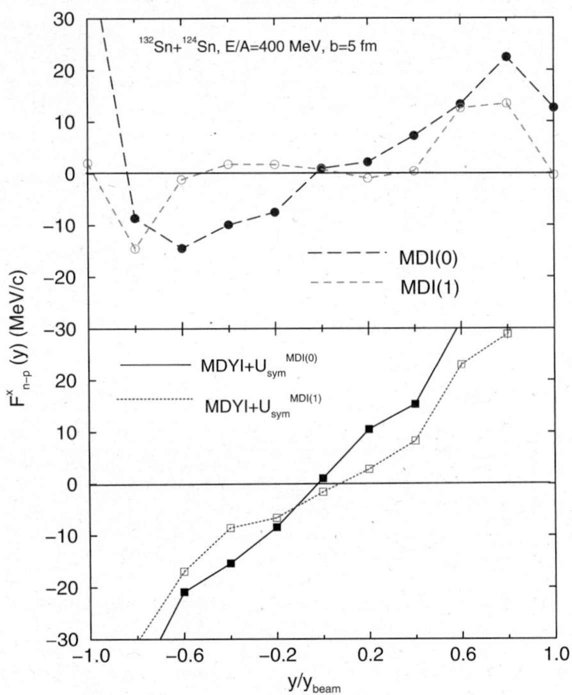

Fig. 11. Neutron-proton differential transverse flows *vs.* reduced rapidity with (upper panel) and without (lower panel) the momentum dependence of the symmetry potential for ^{132}Sn + ^{124}Sn collisions at $400\,A$MeV beam energy and semicentral impact parameter $b = 5\,$fm. Two different density parametrizations of the symmetry energy were used as indicated, see also text. Taken from [45].

energy. The *differential* transverse flow, eq. (10), is shown in fig. 11 with and without the *isospin-MD* of the mean field. The effect of the nucleon mass splitting is less evident. This could be related to the choice $m_n^* > m_p^*$ in this calculation, which tends to reduce symmetry effects on high momentum particles.

Elliptic flows

The same analysis has been performed for the difference of elliptic flows, see fig. 12. Again the mass splitting effects are more evident for different rapidity and transverse momentum selections. In particular, the differential elliptic flow becomes systematically negative at low rapidities when $m_n^* < m_p^*$. This is revealing a faster neutron emission from the high-density region and so a larger neutron squeeze out (more spectator shadowing) for high-energy collisions. In fig. 13 we also show a comparison with recent proton data from the FOPI Collaboration. The agreement is still satisfactory. As expected the proton flow is more negative (more proton squeeeze-out) when $m_n^* > m_p^*$. It is however difficult to draw definite conclusions only from proton data.

Again the measurement at least of a n/p flow difference appears essential. This could be in fact an experimental problem due to the difficulties in measuring neutrons. Our suggestion is to measure the difference between light

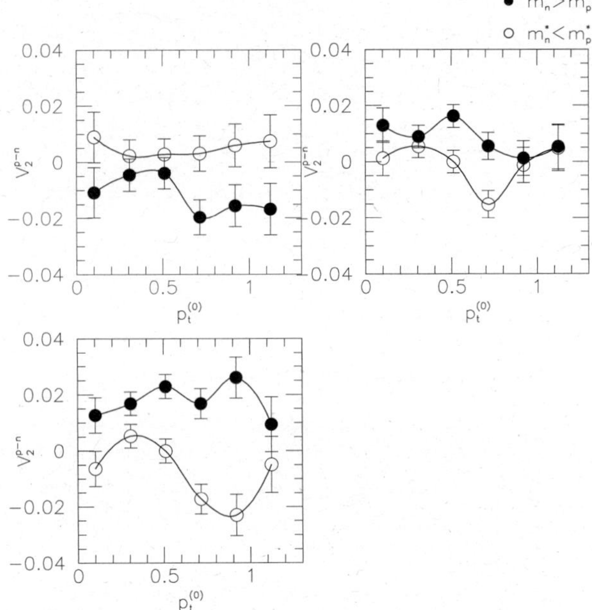

Fig. 12. Difference between proton and neutron elliptic flows for the same semicentral reaction Au + Au at 250 AMeV and rapidity ranges as in fig. 9. Taken from [47].

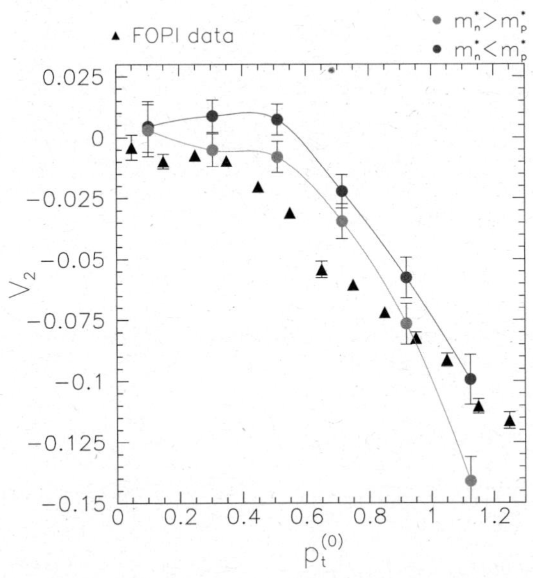

Fig. 13. Comparison of the elliptic proton flow with FOPI data [51] (M3 centrality bin, $|y^{(0)}| \leq 0.1$). Taken from [47].

isobar flows, like triton *vs.* ^{3}He and so on. We expect to clearly see the effective mass splitting effects, may be even enhanced due to larger overall flows shown by clusters, see [23,41].

6 Collective flows as probes of the high-density symmetry energy

Relativistic heavy-ion collisions open the unique possibility to explore the Equation of State (EoS) of nuclear matter far from saturation, in particular the density dependence of the symmetry energy [23]. The elliptic flows of nucleons and light isobars appear to be quite sensitive to the microscopic structure of the symmetry term, in particular, for particles with large transverse momenta, since they represent an earlier emission from a compressed source. Thus future, more exclusive, experiments with relativistic radioactive beams should be able to set stringent constraints on the density dependence of the symmetry energy far from ground-state nuclear matter. In recent years some efforts have been devoted to the effects of the scalar-isovector channel in finite nuclei. Such investigations have not shown a clear evidence for the δ-field and this can be understood considering that in finite nuclei one can test the interaction properties mainly below the normal density, where the effect of the δ-channel on symmetry energy and on the effective masses is indeed small [18] and eventually could be absorbed into nonlinear terms of the ρ-field. Moreover, even studies of the asymmetric nuclear matter by means of the Fermi-liquid theory [18] and a linear response analysis have concluded that some properties, like the borderline and the dynamical response inside the spinodal instability region, are not affected by the δ-field [19]. Here we show that heavy-ion collisions around 1 AGeV with radioactive beams can provide instead a unique opportunity to spot the presence of the scalar isovector channel [56]. In fact, due to the large counterstreaming nuclear currents one may exploit the different Lorentz nature of a scalar and a vector field. Oversimplifying the heavy-ion collision dynamics we consider locally neutrons and protons with the same γ factor (*i.e.* with the same speed). Then nucleon equations of motion can be expressed approximately by the following transparent form ($\rho_{S3} = \frac{M^*}{E^*}\rho_3$), [56]:

$$\frac{\mathrm{d}\mathbf{p}_p^*}{\mathrm{d}\tau} - \frac{\mathrm{d}\mathbf{p}_n^*}{\mathrm{d}\tau} \simeq 2\left[\gamma f_\rho - \frac{f_\delta}{\gamma}\right]\boldsymbol{\nabla}\rho_3, \qquad (11)$$

where γ is the Lorentz factor for the collective motion of a given ideal cell. Keeping in mind that $NL\rho\delta$ has a three times larger ρ-field [18], it is clear that dynamically the vector-isovector mean field acting during the heavy-ion collision is much greater than the one of the $NL\rho$, $NLD\rho$ cases ($NLD\rho$ is built with the same density dependence of the $NL\rho\delta$ symmetry energy, but without the δ coupling). Then the isospin effect is mostly caused by the different Lorentz structure of the "interaction" which results in a dynamical breaking of the balance between the ρ vector and δ scalar fields, present in nuclear matter at equilibrium. The Catania group has performed a set of relativistic transport simulations for the realistic ^{132}Sn $+$ ^{124}Sn reaction at 1.5 AGeV ($b = 6$ fm), that likely could be studied with the new planned radioactive beam facilities at intermediate energies. The transverse and elliptic differential flows are shown in fig. 14. The effect of the different structure of the isovector channel is quite clear. Particularly evident is the splitting in the high-p_t region of the elliptic flow. From fig. 14 we see that, in spite of the statistical errors, in the $(\rho + \delta)$ dynamics the high-p_t neutrons show a much larger *squeeze-out*. This is fully consistent

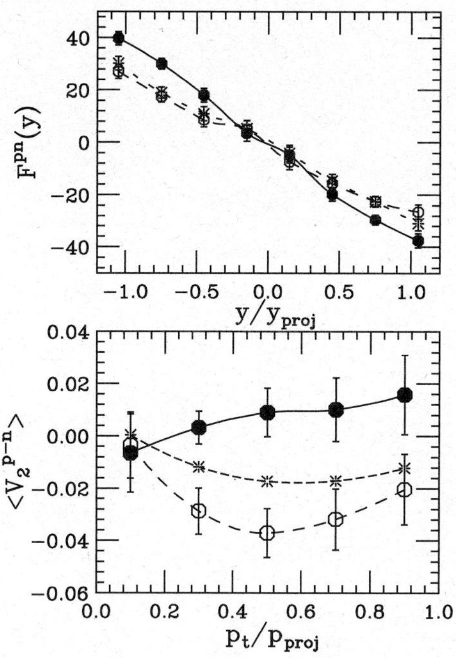

Fig. 14. Differential neutron-proton flows for the ^{132}Sn $+$ ^{124}Sn reaction at 1.5 AGeV ($b = 6$ fm) from the three different models for the isovector mean fields. Top: transverse flows. Bottom: elliptic flows. Full circles and solid line: $NL\rho\delta$. Open circles and dashed line: $NL\rho$. Stars and short dashed line: NL-$D\rho$. Error bars: see the text. Taken from [56].

with an early emission (more spectator shadowing) due to the larger repulsive ρ-field. We can expect this appreciable effect since the relativistic enhancement discussed above is relevant just at the first stage of the collision. The v_2 observable, which is a good *chronometer* of the reaction dynamics, appears to be particularly sensitive to the Lorentz structure of the effective interaction. We expect similar effects, even enhanced, from the measurements of differential flows for light isobars, like ^{3}H *vs.* ^{3}He.

Predictions have also been made with several other transport models. Shown in fig. 15 is the n-p differential flow for the reaction of ^{132}Sn $+$ ^{124}Sn at a beam energy of 400 MeV/nucleon and an impact parameter of 5 fm [57]. Effects of the symmetry energy are clearly revealed by changing the symmetry energy labeled with the parameter x in fig. 15. It is worth mentioning that the isospin dependence of radial flow at RIA energies has also been investigated very recently [58]. The difference in the radial flow velocity for neutrons and protons is the largest for the stiffest symmetry energy as one expects. As the symmetry energy becomes softer the difference disappears gradually. However, the overall effect of the symmetry energy on the radial flow is small, even for the stiffest symmetry energy with $x = -2$ the effect is only about 4%. This is because the pressure of the participant region is dominated by the kinetic contribution. Moreover, the compressional contribution to the pressure is overwhelmingly dominated by the isoscalar interactions. For protons, the radial flow

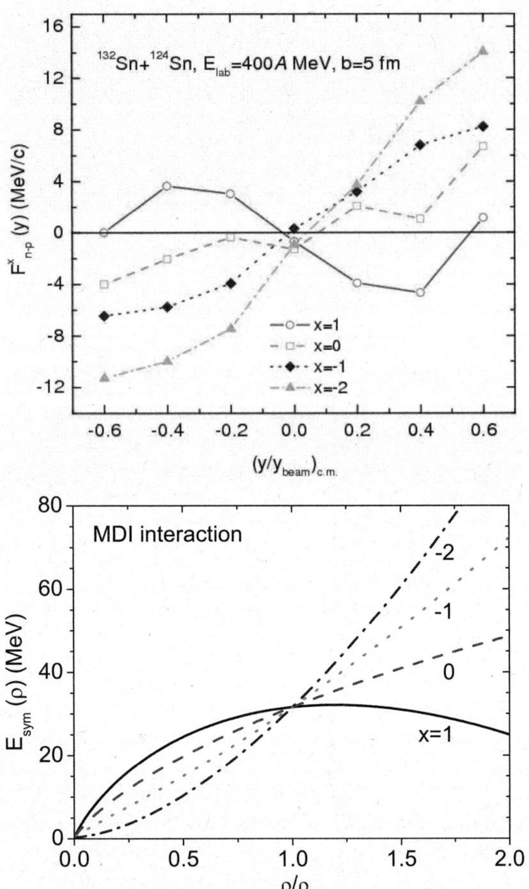

Fig. 15. Neutron-proton differential flow at RIA and GSI energies (top panel) and symmetry energy used in obtaining the above results (bottom panel). Taken from [57].

is affected much more by the Coulomb potential than the symmetry potential. In fact, the Coulomb potential almost cancels out the effect of the symmetry potential at $x = -2$. As the symmetry energy becomes softer, the radial flow for protons becomes higher than that for neutrons. The radial flow thus seems to be less useful for studying the EoS of neutron-rich matter.

7 Conclusions

The EoS of neutron-rich matter has been a long-standing fundamentally important topic in both nuclear physics and astrophysics. Nuclear reactions induced by neutron-rich nuclei provide a great opportunity to pin down the EoS of neutron-rich matter. In particular, the isospin dependence of various components/forms of nuclear collective flow is very useful for extracting interesting information about the EoS of neutron-rich matter. Some experimental evidence indicating the isospin dependence of collective flow has been obtained from heavy-ion reactions at the Fermi energies. In particular, it was shown both theoretically and experimentally that the flow strength of charged particles depends on the isospin asymmetry of the

reaction system. Moreover, the balance energy where the collective flow vanishes is also isospin dependent. A number of interesting predictions regarding the isospin flows have been made using isospin-dependent transport models. However, there are currently very few experimental data available to be compared with. Because of the fact that the isovector potential is rather small compared to the isoscalar potential during heavy-ion reactions, many of the sensitive observables use differences between neutrons and protons, such as the neutron-proton differential transverse and/or elliptic flow. They thus require the detection of neutrons simultaneously with charged particles. Although it is challenging to measure low-energy neutrons accurately, the transverse flow and squeeze-out of neutrons have been measured at both GSI and the Bevalac. In fact, neutron detectors have been built/planned at several radioactive beam facilities. One can thus expect to see high-quality neutron-proton differential flow data coming in the next few years. In the meantime, observables using differences of light isobaric nuclei can also provide some useful information albeit less sensitive than neutrons and protons.

While we have concentrated on the collective flow observables in this report, the readers are kindly reminded that there are many other equally useful observables, such as, the n/p ratio of pre-equilibrium nucleon emissions, π^-/π^+ and K^0/K^+ ratios as well as neutron-proton correlation functions, for studying the EoS of isospin-asymmetric matter. Correlations of multi-observables are critical for finally determining the EoS of neutron-rich matter. Based on transport model simulations, several interesting predictions were made in the literature. Our review here on the isospin flows serves as an example of a broad scope of interesting physics one can study with nuclear reactions induced by neutron-rich nuclei. With the construction of various radioactive beam facilities around the world, we expect that comparisons of theoretical predictions with future data will allow us to better understand the isospin dependence of in-medium nuclear effective interactions. In particular, high-energy radioactive beams being available at some of the facilities will provide us with a great opportunity to explore the EoS of dense neutron-rich matter which is of vast interest to astrophysics. Well-concerted collective actions by both experimentalists and theoreticians will certainly move this field forward quickly.

We would like to thank the organizers and participants of the past three WCI conferences in Catania, Smith College and Texas A&M University for their great efforts in generating the collective flow of excitement and interest in the study of isospin physics with heavy ions. The work of M. Di Toro was supported in part by INFN Italy. The work of S.J. Yennello was supported in part by the Robert A. Welch Foundation grant No. A-1266 and the DOE grant No. DE-FG03-93ER40773. The work of B.A. Li was supported in part by the US National Science Foundation under grant No. PHYS0354572, PHYS0456890 and by the NASA-Arkansas Space Grants Consortium award ASU15154.

References

1. H. Stöcker, W. Greiner, Phys. Rep. **137**, 277 (1986).
2. P. Danielewicz, Q. Pan, Phys. Rev. C **46**, 2002 (1992).
3. S. Das Gupta, G.W. Westfall, Phys. Today **46**, 34 (1993).
4. P. Danielewicz, Nucl. Phys. A **685**, 368c (2001).
5. P. Danielewicz, R. Lacey, W.G. Lynch, Science **298**, 1592 (2002) and references therein.
6. A.E.L. Dieperink, Y. Dewulf, D. Van Deck, M. Waroquier, V. Rodin, Phys. Rev. C **68**, 064307 (2003).
7. D. Vautherin, D.M. Brink, Phys. Rev. C **3**, 676 (1972).
8. H. Krivine, J. Treiner, O. Bohigas, Nucl. Phys. A **336**, 155 (1980).
9. E. Chabanat, P. Bonche, P. Haensel, J. Meyer, R. Schaeffer, Nucl. Phys. A **627**, 710 (1997).
10. E. Chabanat, P. Bonche, P. Haensel, J. Meyer, R. Schaeffer, Nucl. Phys. A **635**, 231 (1998).
11. F. Douchin, P. Haensel, J. Meyer, Nucl. Phys. A **665**, 419 (2000).
12. I. Bombaci, T.T.S. Kuo, U. Lombardo, Phys. Rep. **242**, 165 (1994).
13. I. Bombaci, Phys. Rev. C **55**, 1 (1997).
14. I. Bombaci, *EoS for isospin-asymmetric nuclear matter for astrophysical applications*, in *Isospin Physics in Heavy-ion Collisions at Intermediate Energies*, edited by Bao-An Li, W. Udo Schröder (Nova Science Publishers, New York, 2001) pp. 35-81 and references therein.
15. We remark here that also all relativistic mean-field approaches in the presently used approximation scheme are predicting a *stiff-like* symmetry term, proportional to the baryon density, coming from the ρ-meson field contribution, see ref. [16]. Actually a *superstiff-like* behaviour can be obtained when a coupling to the scalar charged meson δ is added, see refs. [17–19].
16. S. Yoshida, H. Sagawa, N. Takigawa, Phys. Rev. C **58**, 2796 (1998).
17. S. Kubis, M. Kutschera, Phys. Lett. B **399**, 191 (1997).
18. B. Liu, V. Greco, V. Baran, M. Colonna, M. Di Toro, Phys. Rev. C **65**, 045201 (2002).
19. V. Greco, M. Colonna, M. Di Toro, F. Matera, Phys. Rev. C **67**, 015203 (2003).
20. B.-A. Li, C.M. Ko, W. Bauer, Int. J. Mod. Phys. E **7**, 147 (1998).
21. Bao-An Li, W. Udo Schröder (Editors), *Isospin Physics in Heavy-ion Collisions at Intermediate Energies* (Nova Science Publishers, New York, 2001).
22. M. Di Toro, V. Baran, M. Colonna, V. Greco, S. Maccarone, M. Cabibbo, Eur. Phys. J. A **13**, 155 (2002).
23. V. Baran, M. Colonna, V. Greco, M. Di Toro, Phys. Rep. **410**, 335 (2005).
24. M. Lopez-Quelle, S. Marcos, R. Niembro, A. Bouyssy, N. Van Giai, Nucl. Phys. A **483**, 479 (1988).
25. Bao-An Li, Nucl. Phys. A **681**, 434c (2001).
26. R.J. Furnstahl, Nucl. Phys. A **706**, 85 (2002).
27. L.W. Chen, C.M. Ko, B.-A. Li, Phys. Rev. Lett. **94**, 03701 (2005).
28. M. Prakash *et al.*, Phys. Rep. **280**, 1 (1997).
29. B.-A. Li, Phys. Rev. Lett. **85**, 4221 (2000).
30. P. Danielewicz, G. Odyniec, Phys. Lett. B **157**, 146 (1985).
31. B. Zhang, M. Gyulassy, C.M. Ko, Phys. Lett. B **455**, 45 (1999); P. Kolb, J. Sollfrank, U. Heinz, Phys. Rev. C **62**, 054909 (2000); V. Greco, C.M. Ko, P. Levai, Phys. Rev. C **68**, 034904 (2003); D. Molnar, S. Voloshin, Phys. Rev. Lett. **91**, 092301 (2003).
32. B.-A. Li, A.T. Sustich, Phys. Rev. Lett. **82**, 5004 (1999).
33. J.Y. Ollitrault, Phys. Rev. D **46**, 229 (1992).
34. P. Danielewicz, Nucl. Phys. A **673**, 375 (2000).
35. A.B. Larionov, W. Cassing, C. Greiner, U. Mosel, Phys. Rev. C **62**, 064611 (2000).
36. T. Gaitanos, C. Fuchs, H.H. Wolter, A. Faessler, Eur. Phys. J. A **12**, 421 (2001).
37. B.-A. Li, Z. Ren, C.M. Ko, S.J. Yennello, Phys. Rev. Lett. **76**, 4492 (1996).
38. G. Westfall, Nucl. Phys. A **630**, 27c (1998).
39. R. Pak *et al.*, Phys. Rev. Lett. **78**, 1022 (1997).
40. R. Pak *et al.*, Phys. Rev. Lett. **78**, 1026 (1997).
41. L. Scalone, M. Colonna, M. Di Toro, Phys. Lett. B **461**, 9 (1999).
42. The reduced impact parameter is defined as $b_{red} \equiv b/b_{max}$, where b_{max} is the sum of the two nuclear radii.
43. B.-A. Li, A.T. Sustich, B. Zhang, Phys. Rev. C **64**, 054604 (2001).
44. J. Rizzo, M. Colonna, M. Di Toro, V. Greco, Nucl. Phys. A **732**, 202 (2004).
45. B.-A. Li, B. Das Champak, S. Das Gupta, C. Gale, Nucl. Phys. A **735**, 563 (2004); B.A. Li, L.W. Chen, Phys. Rev. C **72**, 064611 (2005).
46. Bao-An Li, Phys. Rev. C **69**, 064602 (2004).
47. M. Di Toro, M. Colonna, J. Rizzo, *On the splitting of nucleon effective masses at high isospin density: reaction observables*, *Argonne/MSU/JINA/RIA Workshop on Reaction Mechanisms for Rare Isotope Beams*, edited by B. Alex Brown, AIP Conf. Proc., Vol. **791** (AIP, 2005) pp. 70-82; J. Rizzo, M. Colonna, M. Di Toro, Phys. Rev. C **72**, 064609 (2005).
48. W. Zuo, L.G. Cao, B.-A. Li, U. Lombardo, C.W. Shen, Phys. Rev. C **72**, 014005 (2005).
49. E.N.E. van Dalen, C. Fuchs, A. Fässler, Phys. Rev. Lett. **95**, 022302 (2005).
50. FOPI Collaboration (A. Andronic *et al.*), Phys. Rev. C **67**, 034907 (2003).
51. FOPI Collaboration (A. Andronic *et al.*), Phys. Lett. B **612**, 173 (2005).
52. V. Greco, Diploma Thesis (1997); V. Greco, Á. Guarnera, M. Colonna, M. Di Toro, Phys. Rev. C **59**, 810 (1999); V. Greco, M. Colonna, M. Di Toro, A. Guarnera, Nuovo Cimento A **111**, 865 (1998).
53. P. Sapienza *et al.*, Phys. Rev. Lett. **87**, 2701 (2001).
54. C. Gale, G.F. Bertsch, S. Das Gupta, Phys. Rev. C **41**, 1545 (1990).
55. I. Bombaci *et al.*, Nucl. Phys. A **583**, 623 (1995).
56. V. Greco, V. Baran, M. Colonna, M. Di Toro, T. Gaitanos, H.H. Wolter, Phys. Lett. B **562**, 215 (2003); V. Greco, PhD Thesis (2002).
57. B.A. Li *et al.*, preprint nucl-th/0504069; AIP Conf. Proc. **791**, 22 (2005).
58. B.A. Li, G.C. Yong, W. Zuo, Phys. Rev. C **71**, 044604 (2005).

Eur. Phys. J. A **30**, 165–182 (2006)

DOI 10.1140/epja/i2006-10114-9

Isotopic compositions and scalings

M. Colonna[1,a] and M.B. Tsang[2]

[1] Laboratori Nazionali del Sud INFN via S. Sofia 62 and Dipartimento di Fisica e Astronomia, Universitá di Catania, I-95123 Catania, Italy

[2] National Superconducting Cyclotron Laboratory, Michigan State University, East Lansing, MI 48824, USA

Received: 2 March 2006 /
Published online: 23 October 2006 – © Società Italiana di Fisica / Springer-Verlag 2006

Abstract. We review experimental and theoretical studies devoted to extract information on the behaviour of the symmetry energy, in density regions different from the normal value, with charge-asymmetric reactions at Fermi energies. In particular, we focus on the analysis of fragmentation reactions and isotopic properties of the reaction products. Results concerning "isoscaling" properties and the N/Z equilibration among the reaction partners in semi-peripheral reactions are also discussed.

PACS. 21.30.Fe Forces in hadronic systems and effective interactions – 25.70.-z Low and intermediate energy heavy-ion reactions – 25.70.Lm Strongly damped collisions – 25.70.Pq Multifragment emission and correlations

1 Introduction

Heavy-ion collisions at Fermi energies offer the possibility to learn about the nuclear effective interaction in regions where the density and temperature are different from those of the stable nuclei. In particular, in charge-asymmetric systems, one can access information on the behaviour of the symmetry energy, E_{sym}, that is poorly known at low and high densities. Not only is the symmetry energy relevant for structure properties, being linked to the thickness of the neutron skin in heavy nuclei (see [1]), but this information is of interest also in the astrophysical context, providing constraints to the equation of state used in astrosphysical calculations [2,3]. Such information is essential for the understanding of the properties of supernovae and neutron stars [4–9].

In fig. 1 (bottom panel) we show the density dependence of the potential symmetry energy contribution, $E_{sym,pot}$ for three different effective interactions. While all curves cross around normal nuclear-matter density ρ_0, there are large differences, particularly in high-density regions. Even at the relatively well-known "crossing point" at normal density, various effective forces give controversial predictions for the momentum dependence of the fields acting on the nucleons and, consequently, for the splitting of the neutron/proton effective masses, which are important in nuclear structure and nuclear reaction dynamics. For discussion purpose, we will call interactions such as BPAL32 asy-stiff and such as SKM* asy-soft (fig. 1, ref. [10]).

---

[a] e-mail: colonna@lns.infn.it

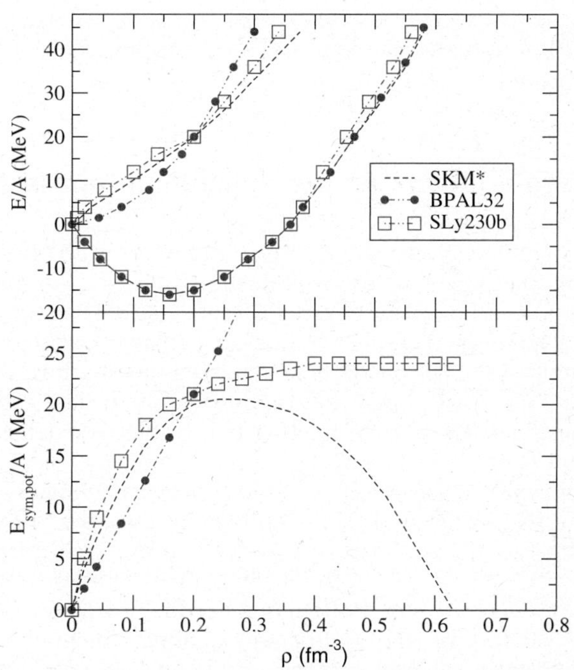

Fig. 1. Equation of state (EOS) for various effective forces. Top: neutron matter (upper curves), symmetric matter (lower curves). Bottom: potential symmetry energy. From [10].

Fragmentation mechanisms at Fermi energies can be used to study the symmetry energy at densities below and around the normal value. In violent collisions, where the full disassembly of the system into many fragments is observed, one can study new properties of liquid-gas phase transitions occurring in asymmetric matter. In neutron-rich matter, phase co-existence leads to different asymmetries in the liquid and in the gas phase: fragments (liquid) appear more symmetric with respect to the initial system, while light particles (gas) are more neutron rich [11–15]. This effect, caused by the decrease in the symmetry energy when the density gets lower, can be used to investigate the behaviour of the derivative of the symmetry energy with respect to density. The width of the isotopic distributions is more connected to the value of the symmetry energy. Information on the low-density properties of the symmetry energy can be obtained from fragmentation studies. Similarly, complementary information is obtained from the study of emitted nucleons and light particles (pre-equilibrium phase). The importance of the isotopic degree of freedom to obtain information about charge equilibration and its relation to the charge asymmetry dependent terms of the EOS has been recently pointed out [16]. We will review here experimental and theoretical results about isotopic properties of reaction products, with the aim of extracting information about the behaviour of the symmetry energy. The paper is organized as follows: we will first review results concerning the properties of pre-equilibrium emission and fragment isotopic content, iso-distillation, then we will discuss the relation of fragment isotopic distributions to the symmetry energy behaviour, focusing, in particular, on the recently introduced isoscaling analysis. Finally, we will discuss isospin transport mechanisms in mid-peripheral reactions and N/Z equilibration, before concluding.

2 Isospin effects on pre-equilibrium emission

Heavy-ion reactions, at energies larger than $30\,\text{MeV/A}$, are characterized by pre-equilibrium emission, fast particles emitted before and during thermalization, see ref. [17] and references therein. For nuclear collisions around the Fermi energy, fast particles are emitted mostly during the expansion phase, when the composite system has reached a density below normal making it possible to extract information on the behaviour of the symmetry energy at sub-normal densities. In collisions between neutron-rich nuclei, the N/Z of the pre-equilibrium emission directly reflect the value of the symmetry energy (being larger for larger values of E_{sym}). In dynamical models such as those based on the Boltzmann-Nordheim-Vlasov equations (BNV) or the Boltzmann-Uehling-Uhlenbeck approach (BUU) as well as stochastic mean-field (SMF) simulations, one observes different N/Z composition of pre-equilibrium emission depending on the asymmetric part of the nuclear equation of state (asy-EOS). At low density, the symmetry energy is larger in the asy-soft case, favoring neutron emission, than in the asy-stiff case. Similar results are obtained by molecular-dynamics studies, as discussed below [18,19].

We will define emitted particles in the "gas phase" as those particles localized in low-density regions ($\rho < \rho_0/3$

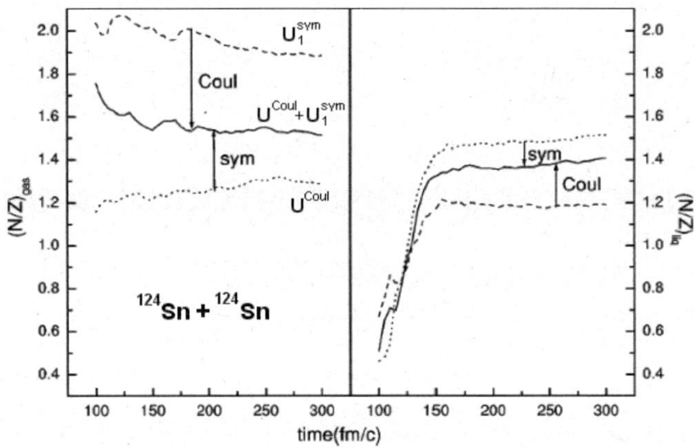

Fig. 2. The time evolution of $(N/Z)_{gas}$ (left panel) and $(N/Z)_{liq}$ (right panel) for three combinations of Coulomb and symmetry interactions, for the reaction $^{124}\text{Sn} + {}^{124}\text{Sn}$ at $50\,\text{MeV/A}$ and impact parameter $b = 2\,\text{fm}$, as obtained in IQMD.

for instance), while the remaining matter will be identified as "liquid phase". Pre-equilibrium emission can be associated with particles emitted (gas phase) at the first collisional stage, *i.e.* up to ≈ 100–$150\,\text{fm}/c$. Obviously, the properties of the "liquid phase" will be influenced by the characteristics of this particle emission. Figure 2 shows the time evolution of $(N/Z)_{gas}$ (left panel) and $(N/Z)_{liq}$ (right panel) for the reaction $^{124}\text{Sn} + {}^{124}\text{Sn}$ at $50\,\text{MeV/A}$, $b = 2\,\text{fm}$, obtained with the isospin quantum molecular dynamics (IQMD) model considering the full nucleon-nucleon interaction (full line), with Coulomb potential only (dashed line) and with symmetry potential only (dotted line). An asy-stiff parameterization has been used for the symmetry energy. The Coulomb interaction reduces the N/Z of the pre-equilibrium emission, while the symmetry energy enhances it. At $t \approx 150\,\text{fm}/c$ one obtains $(N/Z)_{gas} \approx 1.5$ while $(N/Z)_{liq} \approx 1.4$. Hence the gas phase is more neutron rich.

It is worthwhile to compare results obtained with the different transport models quantitatively. In ref. [20], the same reaction has been studied with the BUU code. Values of the $(N/Z)_{liq} = 1.44$ are obtained at $t = 100\,\text{fm}/c$ with an asy-stiff parameterization while an asy-soft parameterization leads to $(N/Z)_{liq} = 1.23$. The value compares rather well (within 3%) with the results of the IQMD model. The different time scales in the two models depend on the definition of the "liquid phase", that for IQMD corresponds to cluster and intermediate mass fragment (IMF) emission, while in the BUU model it is associated with a composite excited source. As shown in fig. 3, these results are also in agreement with stochastic mean-field (SMF) simulations. Indeed, one observes that, with an asy-stiff parameterization of the symmetry energy, after around $100\,\text{fm}/c$, the asymmetry $I = (N - Z)/A$ of the liquid phase equals 0.18, corresponding to $(N/Z)_{liq} = 1.44$.

The antisymmetrized molecular dynamics (AMD) simulations for different Ca isotopes at $35\,\text{MeV/A}$ are studied in ref. [18]. For the $^{48}\text{Ca} + {}^{48}\text{Ca}$ reaction, that has roughly

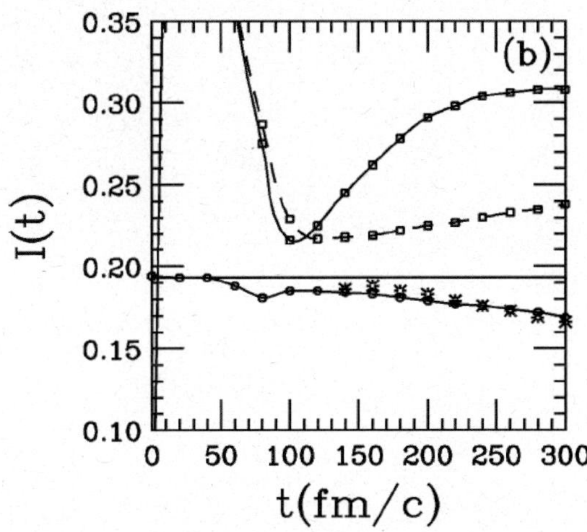

Fig. 3. Time evolution of the asymmetry of the liquid phase (circles, lower curve), gas phase (squares, dashed line) and gas phase in a central region (squares, full line), for the reaction ^{124}Sn + ^{124}Sn at 50 MeV/A, $b = 2$ fm, SMF calculations. An asy-stiff parameterization of the symmetry energy has been used.

the same asymmetry of the ^{124}Sn + ^{124}Sn reaction, one observes that after ≈ 80 fm/c, $(Z/A)_{liq} = 0.42$ corresponding to $(N/Z)_{liq} = 1.39$ (using an asy-stiff equation of state, Gogny-AS). However, at later times, $t \approx 300$ fm/c, the liquid phase appears to be more symmetric in the BUU calculations [18]. This indicates that the rate of neutron enrichment of the gas phase is not the same and may be related to the different evolution of the system in the two models. Indeed AMD calculations include clusters in the disassembly of the excited system, while in BUU calculations only a composite single excited source, that emits nucleons, not clusters, survives until late times. The isospin content of fragments formed in dissipative collisions at intermediate energies will be discussed in the next section.

It is interesting to look at the behaviour of pre-equilibrium emission in reactions at higher beam energies [21–23]. In this case, particles are emitted mostly from the high-density region (compression phase), allowing one to test the behaviour of the symmetry energy at densities above saturation. Pre-equilibrium emission is also sensitive to the momentum dependence of the isospin-dependent (isovector) part of the nuclear interaction. Typical BUU calculations are shown in fig. 4 for the reaction ^{132}Sn + ^{124}Sn at 400 MeV/A. The figure shows rapidity distributions of pre-equilibrium neutrons and protons, at $b = 5$ fm obtained with four interactions formed from the combinations of with (MDI) or without (MDYI) momentum dependence in the iso-vector part of the nuclear interaction with asy-soft (1) or asy-stiff (0) parameterizations of the density behaviour of the symmetry energy. With an asy-stiff parameterization (thick dashed and solid lines) more neutrons are emitted compared to the corresponding asy-soft parameterization (grey dashed and dotted line). This trend is the opposite to the low-density region as the symmetry energy for the asy-stiff parameterization is higher in

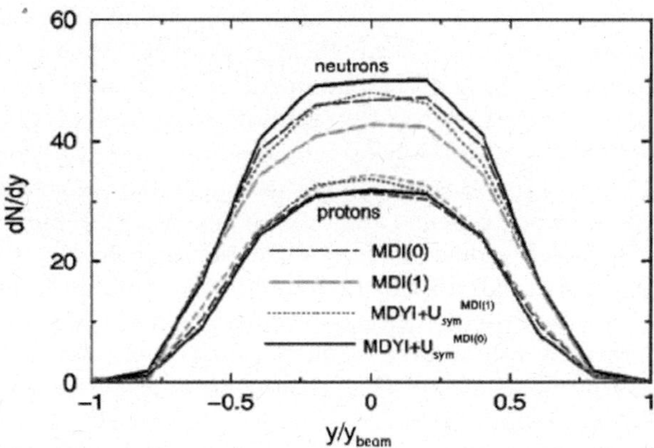

Fig. 4. Rapidity distributions of neutrons and protons, obtained in the reaction ^{132}Sn + ^{124}Sn at 400 MeV/A, $b = 5$ fm. The results of four interactions are presented: the MDI(0), asy-stiff (thick dashed line) and MDI(1), asy-soft (grey dashed line), that contain momentum dependence also in the isovector part of the interaction; the MDYI(0) (thick solid line) and MDYI(1) (dotted line), that are without iso-momentum dependence. From [21].

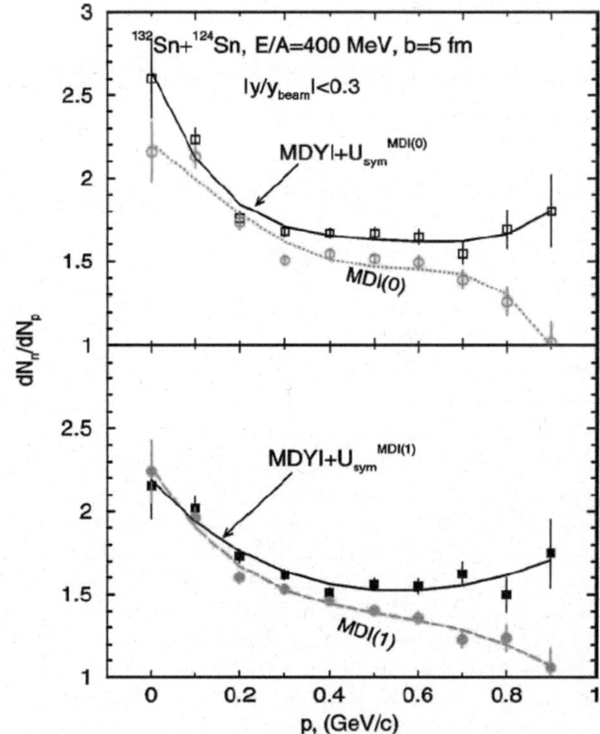

Fig. 5. Neutron to proton ratio, as a function of transverse momentum, for the same reaction and interactions of fig. 4. From [21].

the high-density phase. However, when the momentum dependence is implemented, less neutrons are emitted (compare the thick full line and grey dashed lines for instance). Indeed, for the interactions considered here, a splitting of neutron and proton effective masses with $m_n^* > m_p^*$ is obtained. This reduces the neutron repulsion. Therefore, we observe a kind of compensation between the effects due to the density dependence and to the momentum dependence of the symmetry potential. In fact, the thick dashed line (asy-stiff interaction with momentum dependence) and the dotted line (asy-soft interaction without momentum dependence) almost overlap. On the other hand, if the interactions have the opposite splitting, $m_n^* < m_p^*$, neutrons would be more repelled [22,24]. In fig. 5 the neutron to proton ratio is plotted as a function of transverse momentum, for particles in the mid-rapidity region. It is clear that the decrease of N/Z observed with the MDI interactions (see fig. 4) can be attributed to the high-momentum tail of the nucleon emission that is more sensitive to the momentum dependence.

In summary, the study of pre-equilibrium emission, as a function of the beam energy, can be considered as an interesting and promising tool to explore the reaction dynamics and to investigate the behaviour of the symmetry energy from low to high density. The isotopic composition, N/Z, of all emitted particles appears to be sensitive to the stiffness of the symmetry energy while the dependence of the isotopic content on rapidity, or transverse momentum, appears to be a good candidate to study the momentum dependence of the isovector part of the nuclear interaction.

3 Isotopic composition of fragments: the iso-distillation

As a consequence of the initial collisional shock, or thermal expansion effects, the excited nuclear system expands and enters the low-density (co-existence) region of the nuclear-matter phase diagram. Here a phase separation occurs and fragments are formed, surrounded by a neutron-rich gas. This process is often referred to as isospin distillation or fractionation [14,25]. The isotopic composition of nuclear-reaction products provides important information on the reaction dynamics and the possible occurrence of a phase transition in asymmetric nuclear matter [10–12], which leads to separation into a symmetric dense phase (fragments) and an asymmetric dilute phase (nucleons and light particles) [10,12,25]. Such a phase transition can be generated by fluctuations of density or concentration, leading to a coupling of different instability modes. This mechanism is predicted, for instance, by stochastic mean-field (SMF) simulations [26], where fragments are formed due to the development of spinodal (volume) instabilities. After the first stage of particle emission, the asymmetry of the liquid phase still decreases (fig. 3) while fragments are being formed. Thus fragmentation is accompanied by the iso-distillation process. The amplitude of the effect is strictly related to the derivative of the symmetry energy, as we will discuss more in detail in the section devoted to isospin transport, while the width of the isotopic distributions is more connected to the symmetry energy value. The distillation effect is also predicted by statistical multi-fragmentation models [27–29], where the partition of the system into fragments and light particles is determined according to the statistical weights, that depend on the cluster (free) energies and hence also on the symmetry energy coefficient.

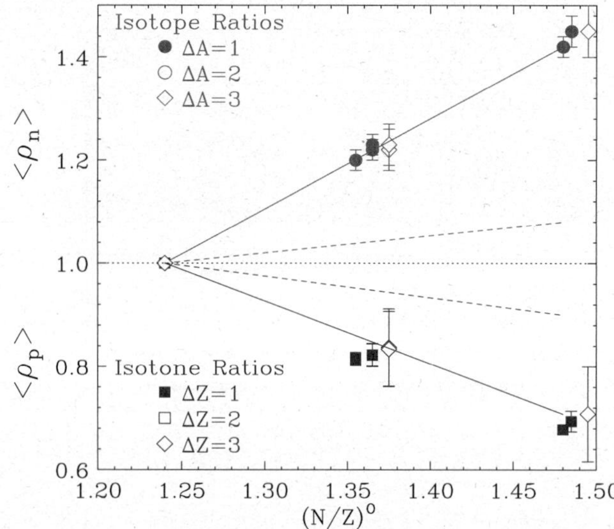

Fig. 6. The mean relative free neutron and free proton density as a function of $(N/Z)_0$. The dashed lines are the expected n-enrichment and p-depletion with the increase of isospin of the initial systems. The solid lines are drawn to guide the eye. From [30].

An experimental analysis of the N/Z of reaction products provides complementary information about the low-density dependence of the symmetry energy.

3.1 Experimental evidence

The N/Z degree of freedom has been studied experimentally with multifragmentation reactions [16,30,31]. Isotopically resolved data in the region $Z = 2\text{--}8$ have revealed systematic trends, which however are substantially affected by the decay of primary fragments.

In the study of the central collisions of four Sn systems at incident energy of 50 MeV per nucleon, the relative neutron and proton densities have been measured for the ^{112}Sn + ^{124}Sn, ^{124}Sn + ^{112}Sn, ^{124}Sn + ^{124}Sn, with respect to the ^{112}Sn + ^{112}Sn system [30]. The extracted relative neutron (ρ_n) and proton (ρ_p) densities are shown in fig. 6; ρ_n increases while ρ_p decreases with the $(N/Z)_0$ ratio of the total system. The increase of ρ_n is consistent with neutron enrichment in the gas phase while the decrease of ρ_p suggests proton depletion. The experimental trend (data points with the solid lines drawn to guide the eye) is much stronger than the trend expected if neutrons and protons were homogeneously mixed (dashed lines) in the breakup configuration. Adopting an equilibrium breakup model, the observation is consistent with isospin fractionation, a signal predicted in the liquid-gas phase transition. Since the isospin fractionation is governed by the symmetry energy of the neutron and proton, it is a more general property of heavy-ion reactions than the liquid-gas phase phenomenon. In fact, dynamical models also give predictions of isospin amplification, in qualitative agreement with the data [30]. In this analysis, complete cancelation of the sequential effects is not necessary as long as

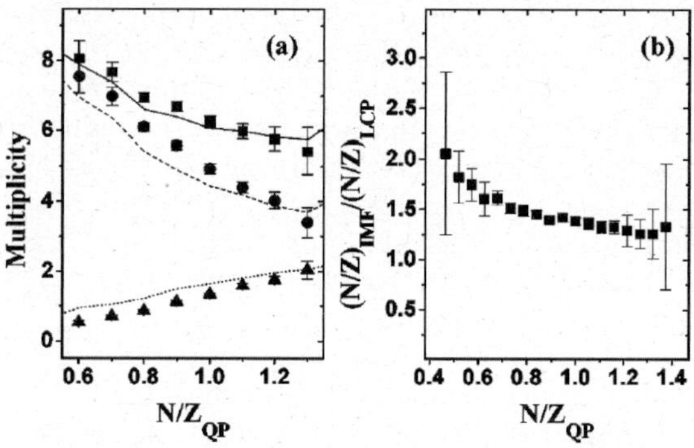

Fig. 7. a) Multiplicity of all charged particles (squares) as a function of the $(N/Z)_{QP}$ of the emitting PLF source, obtained in the reaction ^{28}Si + ^{112}Sn at 50 MeV/A. Circles and triangles represent the multiplicity of LCPs and IMFs, respectively. The lines are the results of model calculations. b) The ratio between the N/Z of IMFs and LCPs is plotted as a function of the $(N/Z)_{QP}$. From [32].

the final yield is related to the primary fragment yield by a multiplicative factor [30].

The composition of intermediate mass fragments in several combinations of isospin asymmetry and excitation energy of the fragmenting source has been studied by considering quasi-projectiles created via peripheral reactions of ^{28}Si + ^{112}Sn and ^{124}Sn at 30 and 50 MeV/A [32]. The quasi-projectiles have been reconstructed from isotopically identified fragments. It is observed that the dependence of the mean fragment multiplicity and the mean N/Z ratio of the fragments on the $(N/Z)_{QP}$ ratio of quasi-projectiles are different for light charged particles (LCPs) and intermediate mass fragments (IMFs). This is illustrated in fig. 7, for the reaction ^{28}Si + ^{112}Sn at 50 MeV/A. The squares represent the multiplicity of all charged particles. This is then broken down into the multiplicity of light charged particles (circles) and the multiplicity of intermediate mass fragments (triangles). We can see that the IMF multiplicity increases with the $(N/Z)_{QP}$ of the system. The lines represent the results of hybrid calculations, obtained by combining a description of transfer processes in deep-inelastic reactions (DIT model, [33]) to statistical calculations of fragment production (statistical multifragmentation model, SMM). They are in good agreement with the data. In [20,30], the multiplicity of IMFs increases as a function of the multiplicity of charged particles for the more neutron-rich systems. However, the decrease in multiplicity of LCPs for neutron-rich systems in fig. 7 is not observed in the multifragmentation of the Sn + Sn system [34].

This effect, however, weakens at higher energies [35]. This is consistent with the temperature dependence of the isospin distillation effect predicted by the lattice-gas model [36] or by dynamical calculations [26]. The ratio $(N/Z)_{IMF}/(N/Z)_{LCP}$ decreases as $(N/Z)_{QP}$ increases, as shown in fig. 7(b). As there are fewer neutrons available, the excess protons go into the smaller fragments rather than the larger ones. Neutron-poor quasi-projectiles prefer to break up into very neutron-deficient (proton rich)

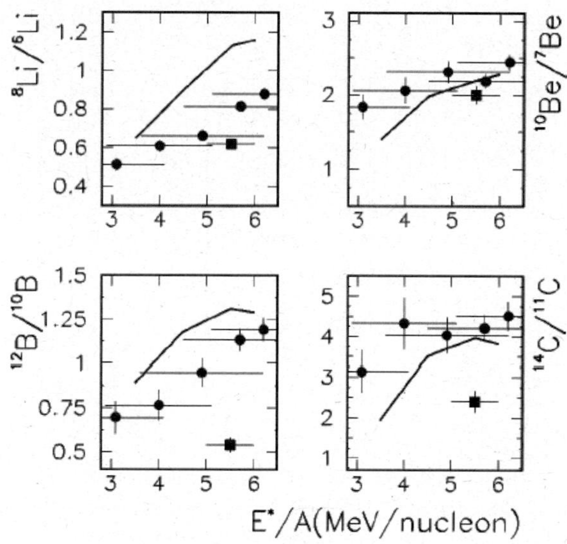

Fig. 8. Ratios of relative yields of neutron-rich to neutron-deficient isotopes, as a function of the excitation energy, as obtained in the fragmentation of PLF sources, from Au + Au at 35 AMeV (circles) or in central Xe + Cu collisions at 30 AMeV (squares). The lines are the predictions of SMM calculations. From [37].

LCPs and much more symmetric IMFs. On the contrary, neutron-rich quasi-projectiles break up into neutron-rich LCPs and more symmetric IMFs as a result of the distillation effect discussed before [12,14,25].

As for the evolution of the N/Z of fragments with the excitation energy of the fragmenting source, this is found to increase, as shown in ref. [37] for fragments emitted from excited PLF sources (see fig. 8). From the statistical point of view, this can be explained in terms of the larger amount of excitation energy available, that allows production of more exotic systems in a larger phase space. This effect is also compatible with the weakening of the distillation mechanism at high temperature, as discussed above.

4 Isoscaling in nuclear reactions

The availability of fragmentation data, obtained with good isotopic resolution for charge-asymmetric systems, makes it possible to examine systematic trends exhibited by isospin-dependent observables. In a series of recent papers, the scaling properties of cross-sections for fragment production with respect to the isotopic composition of the emitting systems were investigated [30,31]. The studied reactions include symmetric heavy-ion reactions at intermediate energy, leading to multifragment emission, as well as asymmetric reactions induced by α particles and ^{16}O projectiles at low to intermediate energies with fragment emission from excited heavy residues. To quantify the comparison of the isotope yields $Y(N, Z)$ obtained in reactions with different isospin asymmetry, the ratio $R_{21} = Y_2(N, Z)/Y_1(N, Z)$ is used. By convention, 2 denotes the more neutron-rich system.

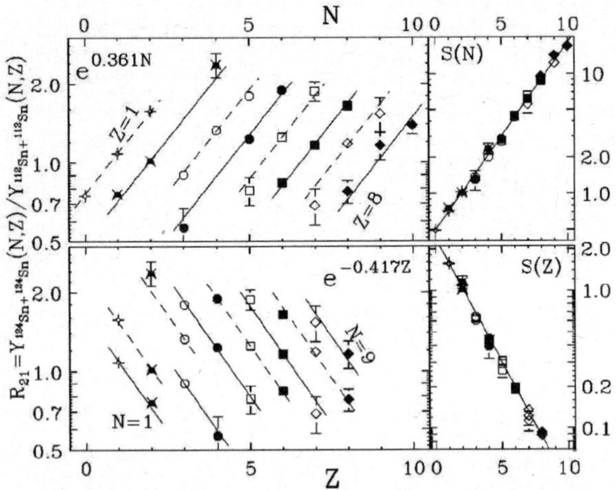

Fig. 9. The yield ratio R_{21} is plotted as a function of N (upper panel) or Z (lower panel). The central reactions considered are ^{124}Sn + ^{124}Sn and ^{112}Sn + ^{112}Sn at 50 MeV/A. From [38].

Figure 9 shows the isotope ratios, $R_{21}(N, Z)$, plotted as a function of N (upper panel) and Z (lower panel), for the central collisions of ^{124}Sn + ^{124}Sn and ^{112}Sn + ^{112}Sn, at 50 MeV/A. $R_{21}(N, Z)$ clearly exhibits an exponential dependence on N and Z, which is called the isoscaling relationship:

$$R_{21}(N, Z) = Y_2(N, Z)/Y_1(N, Z) = C \cdot \exp(N \cdot \alpha + Z \cdot \beta), \tag{1}$$

where C, α and β are fitting parameters. Since the introduction of isoscaling as an isospin observable, isoscaling has proved to be very robust and has been observed in many different types of reactions, such as multifragmentation, light ion-induced fragmentation, evaporation and deep-inelastic reactions [30,31,39,40]. There are also reports on the observation of isoscaling in spallation reactions [41] and, more recently, in fission [42], though the quality of the data is quite poor. Recent studies with realistic fission models [43] suggest that the behaviour of the isoscaling observed in fission is not the same as those observed in multifragmentation, *e.g.* the neutron isoscaling parameter α varies with the Z of the fission products. The most important aspect of isoscaling is its connection to symmetry energy and the temperature of the system, as will be discussed below.

4.1 Isoscaling in statistical models

Isoscaling arises very naturally within a statistical description of fragment production; it is the difference of the chemical potentials of systems with different (N_0/Z_0) ratio. In the grand-canonical statistical description of multifragmentation, the mean multiplicity of a fragment with mass number A and charge Z is given by

$$\langle N(A, Z)\rangle = g_{AZ}\frac{V_f}{\lambda_T^3}A^{3/2}\exp\left[-\frac{1}{T}(F_{AZ}(T,\rho)\right.$$
$$\left. -\mu_n(A - Z) - \mu_p Z)\right], \tag{2}$$

where T is the temperature of the fragmenting source, g_{AZ} is the degeneracy factor of the fragment, λ_T is the nucleon thermal wavelength, V_f is the "free" volume, F_{AZ} is the fragment free energy and μ_n and μ_p are the neutron and proton chemical potentials, respectively. It follows immediately that, for two systems 1 and 2 with different total mass and charge but with the same temperature and density, the ratio of fragment yields is given by eq. (1) with parameters $\alpha = \Delta\mu_n/T$ and $\beta = \Delta\mu_p/T$.

In ref. [44], the chemical potentials for ^{124}Sn and ^{112}Sn are calculated with the grand-canonical version of the statistical multifragmentation model. Despite a considerable variation of the individual potentials, their difference $\Delta\mu = \mu_{112} - \mu_{124}$ changes only slightly as a function of the temperature. At $T > 5$ MeV, the results are similar to that obtained with the Markov chain version of the statistical multifragmentation model (SMM), which takes a completely microcanonical approach. At lower temperature different results are obtained indicating that the exact conservation of charge, mass and energy makes important differences whether the grand-canonical or microcanonical approximation is adopted. Calculating the difference of chemical potentials within the grand-canonical approximation, it is possible to connect the isoscaling parameter α to the difference of asymmetry (Z/A) between the two systems considered and the values of symmetry energy and temperature, through the relation

$$\xi = \alpha/(4\Delta(Z/A)^2) = C_{sym}/T. \tag{3}$$

An analogous relation is derived for β. The symmetry coefficient C_{sym} is directly related to the symmetry energy (per nucleon) of a given fragment having asymmetry I, $E_{sym} = C_{sym}I^2$. Isoscaling is not limited to models within the grand-canonical approximation. As a matter of fact, eq. (3) was first derived in the expanding emitting-source EES model [45,46]. Isoscaling predictions have also been observed in different statistical multifragmentation models [29].

There is an alternative explanation within the statistical multifragmentation model why isoscaling should appear in finite systems. In most SMMs, a variant of the liquid-drop mass formula is used. Charge distribution of fragments with fixed mass numbers A, as well as mass distributions for fixed Z, are approximately Gaussian with average values and variances which are connected with the T and C_{sym} [44]. With a Gaussian distribution for the charge Z, for instance, we obtain, for fragments with a given mass A: $Y(Z) = \exp(-(Z-\langle Z\rangle)^2/2\sigma_Z^2)$. The ratio of this observable for two different systems is given by

$$Y_2(Z)/Y_1(Z) = c\exp\left(-\frac{Z^2}{2}(1/\sigma_2^2 - 1/\sigma_1^2)\right.$$
$$\left. +Z(\langle Z\rangle_2/\sigma_2^2 - \langle Z\rangle_1/\sigma_1^2)\right). \tag{4}$$

If the variances σ_1 and σ_2 are equal, then isoscaling is observed. This is not unlikely since, in the approximation $\sigma_Z \approx \sqrt{(AT/8C_{sym})}$, the variances depend only on the temperature and the symmetry-term coefficient. A similar

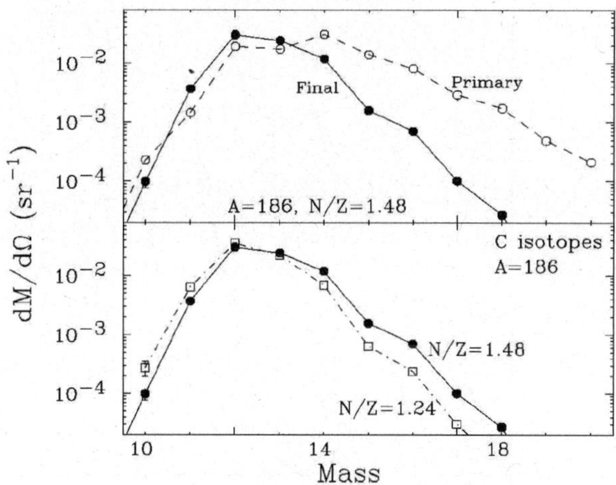

Fig. 10. Top: mass distribution of carbon isotopes, as obtained for the indicated systems, in SMM calculations, for primary hot fragments and final fragments. Bottom: predictions for final fragments are compared for two systems with different N/Z. From [46].

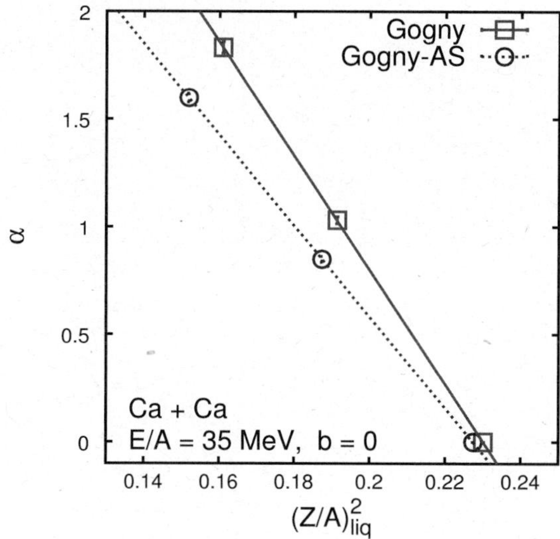

Fig. 11. Isoscaling parameter α, as a function of $(Z/A)^2_{liq}$ as obtained in AMD calculations with two different parameterizations of the symmetry energy.

expression for the mass distributions at given Z is also obtained [44].

In cases when the experimental masses are used, the isotope distributions are not strictly Gaussian. Figure 10 shows the hot carbon isotope distributions predicted by the canonical ISMM (open circles, top panel) from a source with $A_0 = 186, Z_0 = 75$ and $T = 5\,\mathrm{MeV}$, $\rho/\rho_0 = 1/6$. The solid circles correspond to the isotope distributions after sequential decays. Neither the primary nor the final distributions are Gaussian. Nonetheless, from two fragmenting sources with different (N_0/Z_0), one can derive the yield ratios and observe isoscaling [46]. In such calculations, isoscaling parameters are only slightly modified by the secondary-decay process, when the stable masses with the standard value of C_{sym} are used in the calculations. The effect of secondary decay is much larger if C_{sym} takes on lower values as will be discussed in sect. 4.4.2.

4.2 Origin of isoscaling in reaction dynamics

Isoscaling has been observed also in dynamical fragmentation models, such as the AMD model [18], QMD [47] and classical molecular dynamics (CMD) [19], as well as in SMF calculations [48] and in quasi-analytical calculations of the spinodal decomposition process [49]. For instance, fig. 11 shows the dependence of α on the charge to mass ratio, $(Z/A)^2_{liq}$, of the liquid phase for the collisions of Ca isotopes, at $35\,\mathrm{MeV/A}$, as predicted by the AMD model. A linear dependence of α on $(Z/A)^2_{liq}$ is observed. The two lines correspond to two different symmetry potentials used in the simulations (full line, asy-soft, Gogny; dotted line, asy-stiff, Gogny-AS). Thus, even in dynamical models, isoscaling is intimately related to the symmetry energy.

The study of isoscaling through dynamical simulations can elucidate the origin of this phenomenon. If chemical

equilibrium is reached during the fragmentation process, it is clear that one can apply the considerations outlined in the previous section and directly relate the isoscaling parameter to the value of the symmetry energy and the temperature.

However, the linear relation between α and $(Z/A)^2$ can be obtained in different conditions, without assuming statistical equilibrium. If the origin of fluctuations in a multi-fragmenting system can be considered a "white noise" source, then the probability to observe a given fluctuation of the isovector density $\delta\rho_i = \delta\rho_n - \delta\rho_p$, in a given volume V, can be expressed (for small amplitude fluctuations) as

$$P \approx \exp(-\delta\rho_i^2/2\sigma_{\rho_i}), \tag{5}$$

where the variance σ_{ρ_i} depends on the fragmentation mechanism. Then, for a fragment of volume V and mass A, the distribution $P(N - Z)$ can be written as

$$P(N - Z) \approx \exp(-[N - Z - (\bar{N} - \bar{Z})]^2/(\mathcal{F}\rho_{in}V)), \tag{6}$$

where $\bar{N}$ and $\bar{Z}$ are the average neutron and proton numbers in the volume V, and σ_{ρ_i} is proportional to ρ_{in}, the density of the fragmenting system [50]. $\mathcal{F}$ is a constant that depends on the symmetry energy and the temperature.

In spinodal decomposition, for instance, fragments are formed (and their density grows) due to the development of isoscalar-like unstable modes. Hence isoscalar density fluctuations grow while the isovector variance does not evolve and keeps the memory of the initial isovector fluctuations of the unstable diluted source. Therefore, we may expect reduced iso-vector fluctuations (and larger isoscaling parameters) with respect to the statistical case, where the isotopic content of the entire fragment mass may fluctuate.

As an example, the results of SMF, based on the spinodal decomposition scenario, are presented in fig. 12 [48].

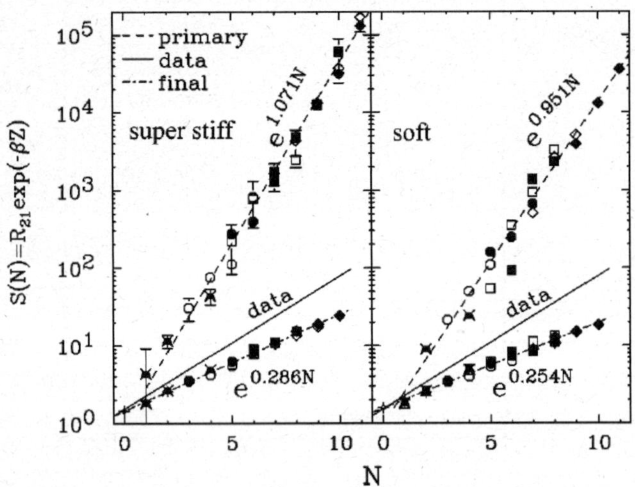

Fig. 12. Scaled isotope yield ratio, as a function of N, as obtained in SMF calculations of central reactions ^{124}Sn $+$ ^{124}Sn and ^{112}Sn $+$ ^{112}Sn, at 50 MeV/A, for primary and final fragments. Two parameterizations of the symmetry energy are considered. From [48].

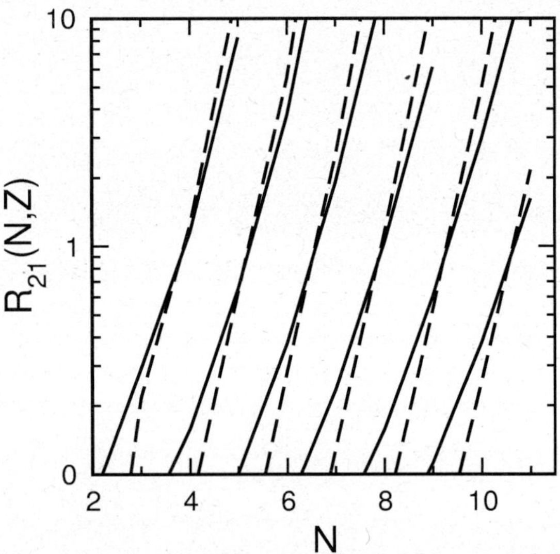

Fig. 13. Yield ratio $R_{21}(N, Z) = Y_{\alpha=0.2}(N, Z)/Y_{\alpha=0.1}(N, Z)$ calculated with the "super-stiff" symmetry term (solid lines) and with the asy-soft symmetry term (dashed lines). Lines correspond to different values of Z, $Z = 3$–8 from left to right. The system is prepared with density and temperature inside the spinodal region. The average values of the slope approximatively are 2.2 and 1.5 for the asy-soft case and the asy-super-stiff case, respectively. From [49].

Here large isoscaling parameters are observed for the primary fragments. However, they are significantly affected by the secondary decay and final values appear closer to the data. But the relative differences between the two parameterizations of the symmetry energy used are also much reduced.

The results obtained in a quasi-analytical description of spinodal decomposition, comparing fragment production in nuclear matter with asymmetry $I = 0.2$ and $I = 0.1$, are presented in fig. 13, for two parameterizations of the symmetry term [49].

The formula can be recast as follows:

$$P(Z, N) \approx \exp(-[(N - Z)^2/A - N(4(\bar{Z}/A)^2 - 1) - Z(4(\bar{N}/A)^2 - 1)]/[\mathcal{F}/\eta]), \qquad (7)$$

where η is the ratio between the fragment final density $\rho_{fin} = A/V$ and the initial density ρ_{in} (η larger or equal to 1).

In statistical models η is equal to 1 and $\mathcal{F}$ coincides with T/C_{sym}, while in the early spinodal decomposition process the variance of the fragment isotopic distribution, due to isovector fluctuations, is reduced with respect to the equilibrium value $\mathcal{F}$. However it should be noted that, within such a scenario, isoscalar-like modes also contribute to the variance, due to the beating of several unstable modes, that bear a different distillation effect [49].

If one assumes that $\bar{Z}/A$ (and $\bar{N}/A$) depends only slightly on A and can be related to the average distillation effect, that determines the average asymmetry of the formed fragments, then from eq. (7) the isoscaling parameters are equal to

$$\alpha = 4((Z_1/A_1)^2 - (Z_2/A_2)^2)/(\mathcal{F}/\eta),$$
$$\beta = 4((N_1/A_1)^2 - (N_2/A_2)^2)/(\mathcal{F}/\eta). \qquad (8)$$

Hence one gets the same formal expression as in statistical models, but with a more complex relation of the isoscaling parameters to the system properties. These parameters appear connected to the distillation effect, but also to the width of the isotopic distributions, that can in general differ from the predictions of statistical models.

The link between isoscaling parameters and symmetry energy depends on the way fragments are formed, while the observation of isoscaling and the relation to the $(Z/A)_{liq}$ value of the liquid phase appear as quite general properties and do not require the assumption of statistical equilibrium.

4.3 Temperature dependence of isoscaling

With the availability of models, we are able to explore the temperature dependence of isoscaling. All calculations (statistical or dynamical) show that the isoscaling parameters are inversely related to the temperature. So we would expect these parameters to decrease with increasing temperature, excitation energy or incident energy. Indeed, such phenomenon was observed in refs. [39,51]. Figure 14 shows the isoscaling parameter α as a function of the incident energy. In this study, isobars with mass $A = 58$ (^{58}Ni and ^{58}Fe) are used as target and projectile. Reaction 1 is taken to be the symmetric ^{58}Ni $+$ ^{58}Ni which has the initial $(N/Z)_{ini}$ value of 1.07. For the upper curve (solid points), reaction 2 is taken to be the symmetric system ^{58}Fe $+$ ^{58}Fe with $(N/Z)_{ini} = 1.23$ and, for the lower curve, the mixed system, ^{58}Fe $+$ ^{58}Ni, with $(N/Z)_{ini} = 1.15$ is used as reaction 2. This figure clearly shows that the α

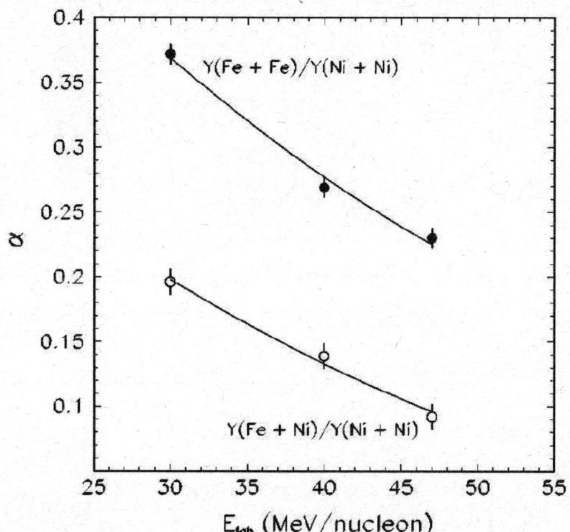

Fig. 14. Evolution of the isoscaling parameter α, as obtained in central collisions, *versus* the beam energy. From [52].

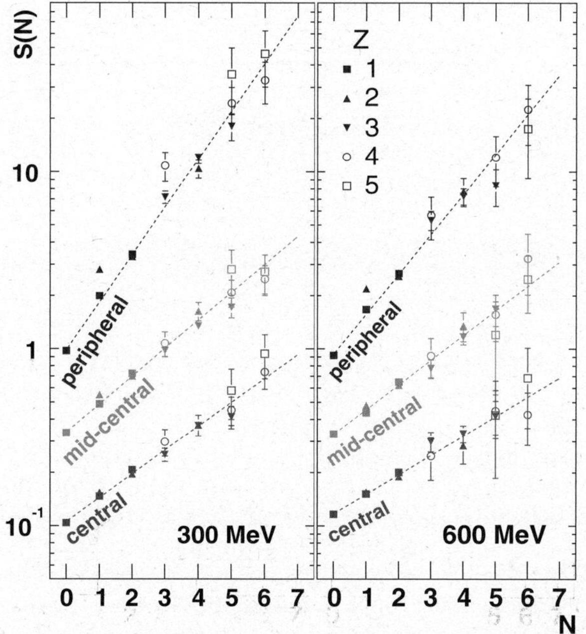

Fig. 15. Scaled isotopic ratios for ^{12}C $+$ ^{112,124}Sn at 300 (left) and 600 (right) MeV/A. Three different centrality bins are considered. From [53].

value decreases with incident energy from $E/A = 30$ MeV to 47 MeV. In addition, there is also a clear drop in the α values with the decrease of the $(N/Z)_{ini}$ values of the entrance channel. The latter has also been observed in the collisions of central Sn isotopes which first established the phenomenon of isoscaling [30].

Isoscaling has been studied in fragmentation processes of excited target residues, in the reactions ^{12}C $+$ ^{112,124}Sn at 300, 600 MeV/A [53]. Due to large uncertainties in the IMF isotope ratios, the isoscaling parameter α is mainly determined by the light charged particles, $Z = 1, 2$ (see fig. 15). The isoscaling parameter α is observed to decrease

with increasing centrality of the reaction and the beam energy. Even though the temperature (measured from isotope ratios) increases with centrality, the density of the emitting source in the peripheral collisions is expected to be larger than the density of the fragment formation created in the central collisions. Thus this experiment does not offer solid proof that α decreases with increasing temperature.

4.4 Extraction of symmetry energy from isoscaling analysis

In the past few years, there have been many attempts to extract information on the density behaviour of the symmetry energy from the isoscaling analysis of fragmentation reactions [44, 51–56]. In the following we will review recent developments in this field.

4.4.1 Insights from theoretical models

In the absence of equilibrium, the isoscaling parameter is not directly connected to the symmetry energy value. This is surely the case of SMF calculations, where large values of the isoscaling parameters are obtained for primary fragments, that do not reflect the low value of the symmetry energy observed in the model at the fragmentation stage.

If equilibrium is achieved as in statistical models, eq. (3) suggests that it is possible to extract information on the symmetry energy coefficient from the isoscaling analysis of the data. However, eq. (3) is strictly valid for primary fragments only. If experimental data are used, one must examine the effects of sequential decays on the variables α, $\Delta(Z/A)^2$, and T. To study the dependence of different parameters on the stiffness of the symmetry term in the nuclear EOS, it is useful to obtain some insights from model simulations.

The dependence of the α values as a function of stiffness parameters was first obtained in expanding emitting-source (EES) model calculation. In ref. [31], two sources with initial charge of 100 and initial mass of 224 for source 1 and 248 for source 2, initial thermal excitation energies of 9.5 MeV and collective radial expansion energies of 2.5 MeV (corresponding to the compound systems of ^{112}Sn $+$ ^{112}Sn and ^{124}Sn $+$ ^{124}Sn reactions at $E/A = 50$ MeV) are assumed in the calculations. In the model a power law dependence of $C_{sym} = 24.3(\rho/\rho_0)^\gamma$ is used to describe the fragmentation stage [31]. A nearly linear decrease of the isoscaling parameter α with γ is observed as shown in fig. 16. This suggests that the α value is larger for smaller γ, consistent with more neutrons emitted with an asy-soft interactions as discussed in sect. 3.

Unlike the EES and dynamical models, most statistical models assume that $C_{sym}(\approx 25$ MeV) is constant throughout the reaction. In most cases, this value is the same as the symmetry energy coefficient in the liquid-drop mass formula used in the model. Recently, the effects of the symmetry energy coefficient on fragment isotope distributions is studied with a microcanonical Markov-chain version of the statistical multifragmentation model [53]. The

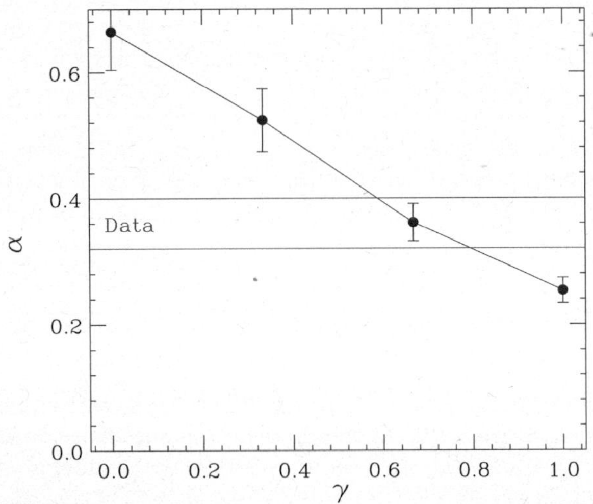

Fig. 16. Theoretical EES model predictions for the isoscaling parameter α extracted from the emitted $Z = 3$–6 fragments. From [31].

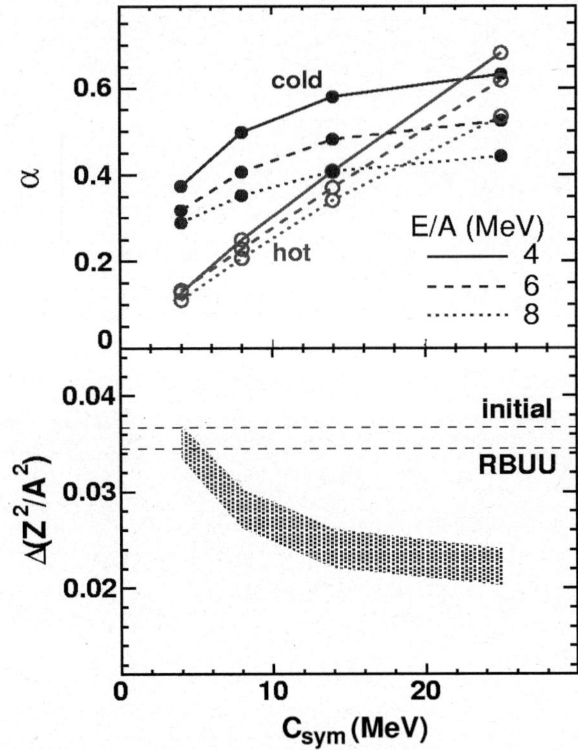

Fig. 17. The isoscaling parameter α, as obtained in SMM calculations for the same system as in fig. 15, as a function of the symmetry energy coefficient, for primary and final fragments. From [53].

target nuclei ^{112,124}Sn with excitation energies of 4,6 and 8 MeV/A were chosen as inputs and the symmetry energy C_{sym} was varied between 4 and 25 MeV in the model. The isoscaling coefficient α was determined from the calculated fragment yields before (hot fragments) and after (cold fragments) the sequential decay stage of the calculations. The hot fragments exhibit the linear increase of α values with increase in C_{sym}. In the low-density region, larger symmetry energy corresponds to asy-soft interactions. Thus the trend is consistent with that observed in the EES calculations.

With $C_{sym} \approx 25$ MeV, the sequential processes lower the α values by 10% to 20% [29]. However, at small values of C_{sym}, one gets surprisingly the opposite effect: the decay of the wings of the wider distributions of hot fragments, which is directed towards the valley of stability, leads to larger isoscaling coefficients (see fig. 17). Thus, the effect of sequential decays on α not only depends on the excitation energy but also depends on C_{sym}.

To further examine the sensitivity of the isoscaling parameters to the different steps of the reaction (pre-equilibrium emission, fragmentation stage, secondary de-excitation), the three quantities in eq. (3), α, $\Delta(Z/A)^2$ and $\xi = \alpha/(4\Delta(Z/A)^2)$ obtained from the AMD simulations of the central collisions of Ca isotopes are shown in fig. 18. ^{40}Ca + ^{40}Ca is chosen as reaction 1 in eq. (1) [57]. Reaction 2 of ^{48}Ca + ^{48}Ca and that of ^{60}Ca + ^{60}Ca are labeled in the figure. From the figure one can see that the isoscaling parameters are affected by the isospin-dependent pre-equilibrium emission and distillation effects which change the Z_0/A_0 of the source, and by the symmetry energy during the fragmentation stage, that influences the width of the isotopic distribution through the value of ξ (see eq. (8)). One may also note the compensation between the two effects: while ξ is larger in the asy-soft case (Gogny, closed points), $\Delta(Z/A)^2$ is larger in the asy-stiff case (Gogny-AS, open points).

As expected from isoscaling, α is nearly independent of the charge number of the isotopes. We note that while α (bottom panel), $\Delta(Z/A)^2$ (middle panel) depend on the reaction systems and the interacting potentials (Gogny (asy-soft, closed diamonds) and Gogny-AS (open squares)), $\xi = \alpha/(4\Delta(Z/A)^2)$ only depends on the interacting potentials. For asy-soft (Gogny) interaction, the symmetry energy is larger resulting in larger α and smaller $\Delta(Z/A)^2$ values (solid symbols) as compared to the corresponding values (open symbols) obtained if the asy-stiff (Gogny-AS) interaction is used. This trend is consistent with the results from the EES and SMM simulations.

To further examine the effects of sequential decays, the primary fragments emitted in the collisions of the Ca isotopes in AMD simulations are then allowed to decay and the corresponding quantities are plotted in fig. 19. Isoscaling is still preserved but the α values no longer show a difference between the two interactions. Furthermore, all the three quantities, α, $\Delta(Z/A)^2$ and ξ have very different values before and after sequential decays. To extract the correct information on the symmetry energy, ξ, (i.e. the value before decay), one must use the values of α and $\Delta(Z/A)^2$ from the primary fragments.

4.4.2 Experimental extraction of symmetry energy

One of the major objectives in heavy-ion collisions is to extract the density dependence of the symmetry energy

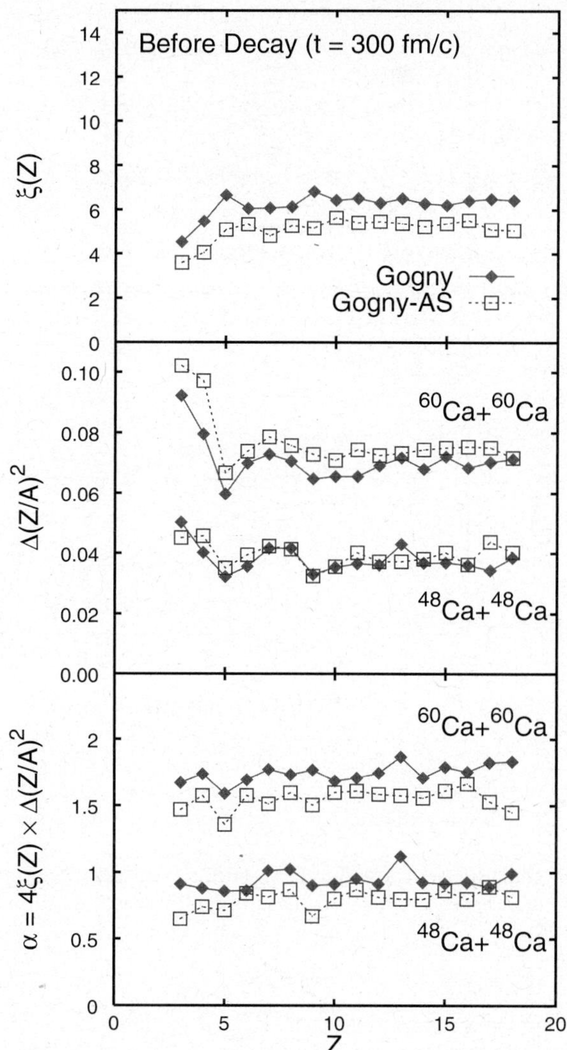

Fig. 18. The three quantities $\xi(Z)$ (top), $\Delta(Z/A)^2$ (middle) and α (bottom) for the primary fragments, as obtained in AMD calculations. The results with the Gogny and Gogny-AS forces are shown by filled diamonds and the open squares, respectively. From [57].

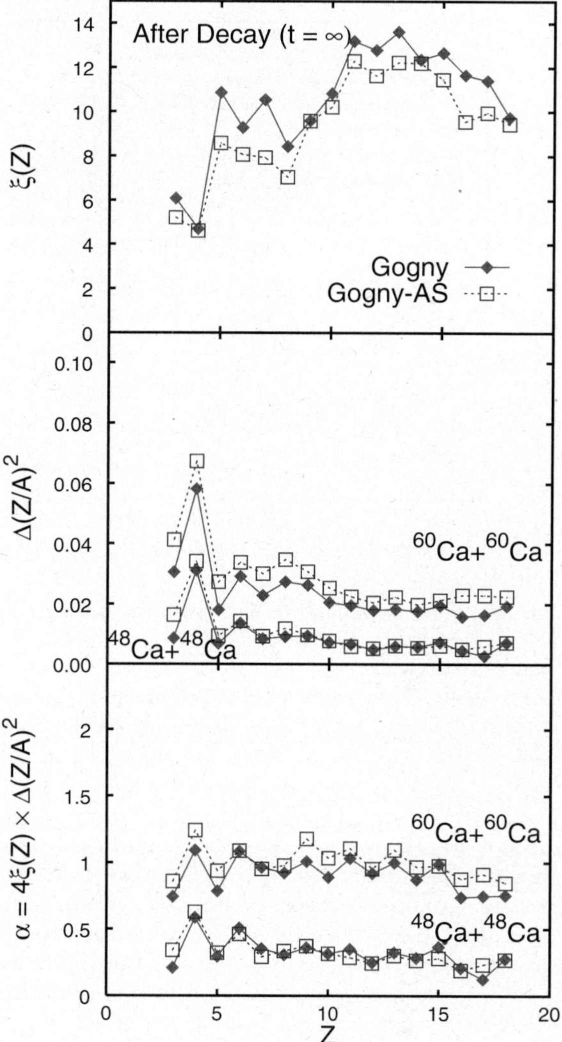

Fig. 19. The same as fig. 18 but for the final fragments. From [57].

in the nuclear equation of state. Since only ground-state particles are detected experimentally, theoretical models are needed to extrapolate the properties of the primary fragments or to simulate the effects of sequential decays. Unfortunately, theoretical developments in heavy-ion reactions have not reached the state that calculations can be done *a priori*. Nonetheless, we will review a few cases where isoscaling is used in the analysis.

In ref. [54], dynamical models are employed to deduce the $(Z/A)_{liq}$ of the liquid phase after pre-equilibrium emission, while the fragmentation stage is described by statistical models. Using this hybrid approach, it is concluded that an asy-stiff parameterization of C_{sym} leads to better agreement with the isoscaling analysis (^{112}Sn + ^{112}Sn and ^{124}Sn + ^{124}Sn at 50 MeV/A). The result, especially the trend, is in contradiction with the results from the same data analyzed with the expanding emitting-source model

as shown in fig. 16. To understand the inconsistency, we examine the contributions of each stage of the calculations. The BUU stage predicts a larger $\Delta(Z/A)^2$ value with asy-super-stiff interaction, this is similar to the predictions of EES, SMM and AMD discussed in last section. Accordingly, it is expected that the SMM fragmentation and decay stage would predict a smaller alpha value. However, the results from the hybrid calculation are just the opposite. One explanation is that inconsistent results are obtained if the dynamical stage (when symmetry energy evolves with density and time) is coupled with a statistical stage (when a fixed value of $C_{sym} \approx 25$ MeV is adopted).

The dependence of the isoscaling parameter on the symmetry interaction, simulated by the AMD model [18] has been used to analyze several systems at $T \approx$ 3.5 MeV [44,56], indicating that the data are more consistent with the Gogny-AS (asy-stiff) interaction, see fig. 20 [52]. To perform this comparison, an estimation of the $(Z/A)^2_{liq}$ is deduced, as a function of the $(Z/A)^2_{ini}$ of the initial system, by extrapolating the results of AMD

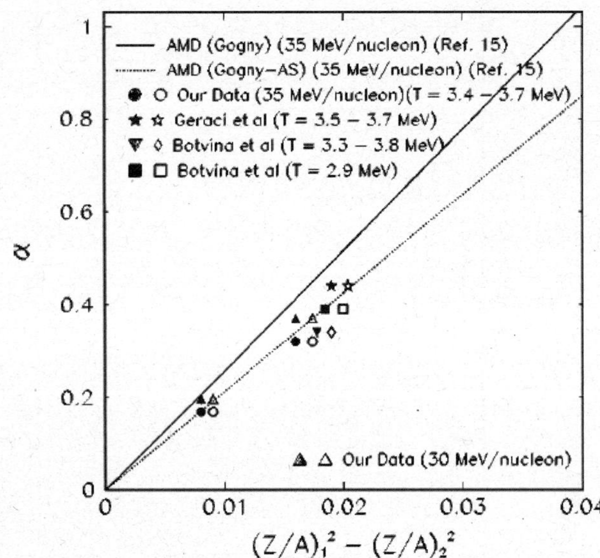

Fig. 20. Isoscaling parameter α, as a function of the difference $\Delta(Z/A)_{liq}^2$ between the two compared systems, as obtained in AMD calculations with two different parameterizations of the symmetry energy. The points refer to the isoscaling parameter extracted from several experimental data (see text for detail). From [52].

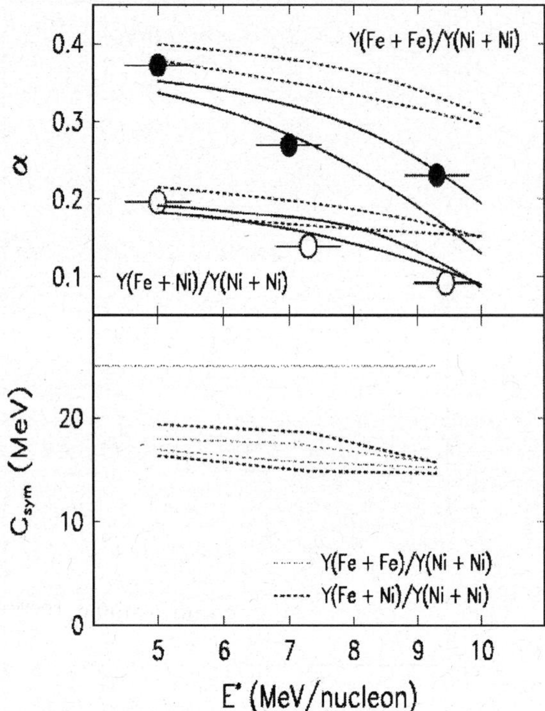

Fig. 21. Top: isoscaling parameter α *versus* the excitation energy, as obtained in the data (circles) and in SMM calculations for primary (dashed) or final (full lines) fragments. The band associated with calculations are due to uncertainty in the size of the fragmenting source. Bottom: symmetry energy coefficient, as extracted from α, assuming statistical equilibrium. From [51].

calculations for the Ca systems shown in fig. 11. Recent studies show that such extrapolations may not be valid outside the simulated range of impact parameters [58]. The isoscaling parameter deduced from the experimental data is compared with the results of the AMD model. However, this comparison is done without considering the secondary-decay effect which is large.

Other studies aimed to extract the symmetry energy value at the freeze-out configuration directly from the data assuming that thermal and chemical equilibrium have been reached. To do so, one needs to extract from the data the isoscaling parameter, as well as the temperature and the (Z_0/A_0) of the emitting source. In the study performed in ref. [51], the value of (Z_0/A_0), as provided by BNV calculations with a fixed value of C_{sym}, is used. The set of data has been analyzed with SMM calculations. The latter, that take into account the secondary decay, are in qualitatively good agreement with the experimental value of the isoscaling parameters. As it appears from the calculations, for a fixed value of the symmetry energy (around 25 MeV), the final isoscaling parameter, that can be compared to data, appears to be lower than the one obtained from the primary fragment yields, especially at high excitation energies. This observation is used to correct the experimental parameter, to try to remove the influence of the secondary de-excitation process. Then one can use eq. (3) to extract the value of the symmetry energy, once the temperature has been estimated. The results obtained for the symmetry energy are shown in fig. 21. Low symmetry energy values were obtained. However, it should be noted that the results of the analysis depend on the use of a hybrid model; the estimation of the (Z_0/A_0) of the

emitting sources is determined by the BNV model and sequential decay effects are obtained from the SMM. As discussed before, such approach has its own problems.

A similar analysis is performed in ref. [53], where fragmentation processes of excited target residues produced in the reactions $^{12}C + {}^{112,124}Sn$ at 300, 600 MeV/A are studied. The symmetry energy coefficient is extracted from the isoscaling parameters using eq. (3) and assuming that the Z_0/A_0 of the fragmenting sources is the same as the initial systems. The extracted C_{sym} value changes from 25 MeV to a value around 15 MeV (see fig. 22) when the centrality of the reaction is increased.

However, as stressed above, in this kind of analysis one has to use the isoscaling parameters as deduced from the primary fragments and not the final ones. So the authors try to go back to the properties of the primary fragments with the help of new SMM simulations, where the density dependence of the symmetry energy is partly taken into account by changing the symmetry energy constants used in the mass formula (see fig. 17).

This means that to get the observed isoscaling parameters for central reactions, C_{sym} should assume even lower values (4 MeV or less). These results would point to a strong reduction of the symmetry energy (and hence a strong dilution of the system) at the time of chemical freeze-out. It is clear, however, that any conclusion

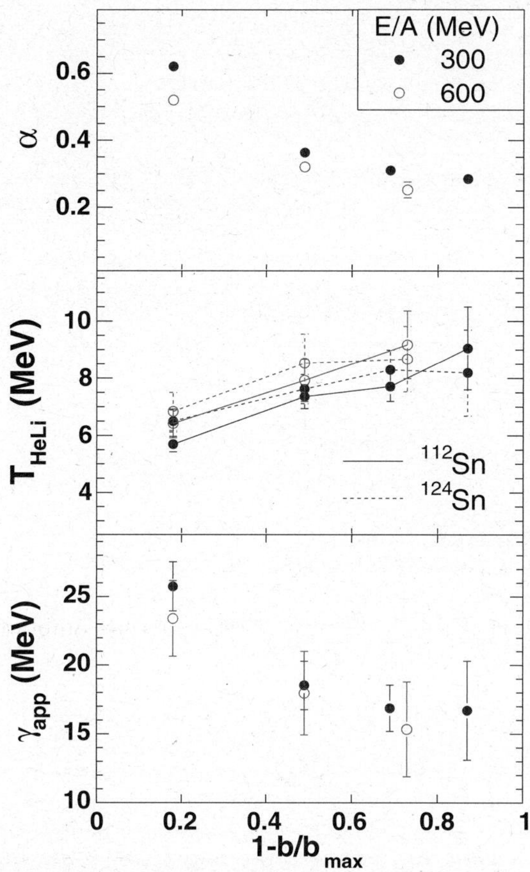

Fig. 22. Evolution of the isoscaling parameter *versus* the impact parameter, for the same reactions as in fig. 15 (top panel). The evolution of the temperature of the TLF sources and of the extracted apparent symmetry energy coefficient are shown in the middle and bottom panels, respectively. From [53].

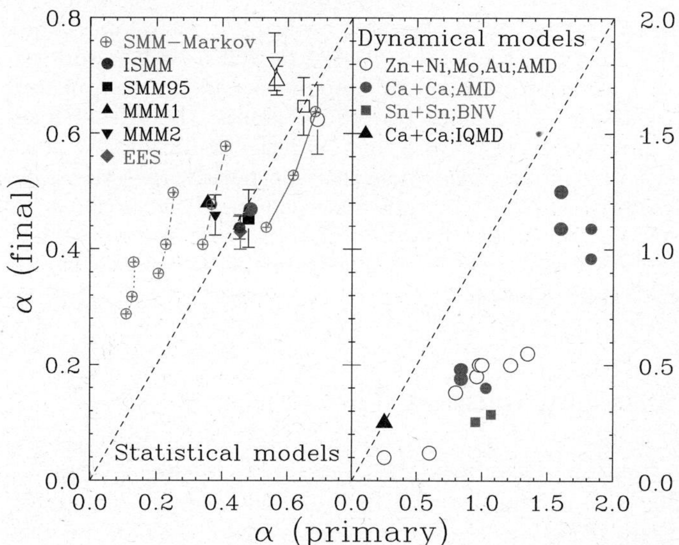

Fig. 23. Effect of sequential decays on the isoscaling parameter, α, for statistical models (left panel) and for dynamical models (right panel).

in this direction depends on the isotopic evolution of the multi-fragmenting system as it approaches the chemical freeze-out. Moreover, the results are only based on the predictions of the SMM and on a particular way to implement the density dependence of the symmetry energy in a fragmenting system.

While the effects of sequential decay on α and $\Delta(Z/A)^2$ and T are important in any analysis that involves using eq. (3) to extract the symmetry energy, such effects are not easy to evaluate and may depend on specific models. In fig. 23, we have compiled the α values before (horizontal axis) and after (vertical axis) sequential decays obtained for many systems using statistical models (left panel) and dynamical models (right panel). The statistical models used include statistical multifragmentaton models such as SMM95, I(mproved)SMM, microcanonical Markov-chain version of the SMM, microcanonical multi-fragmentation model (MMM), and the expanding emitting source (EES) [29]. The three dynamical models used are the asymmetrized molecular dynamical model, AMD, the Boltzmann-Nordheim-Vlasov model, BNV and the isospin quantum molecular dynamical models, IQMD. In general, the α values from primary fragments are much larger in

dynamical models such as SMF (fig. 13) and AMD (figs. 18 and 19) models than statistical multifragmentation models. (Note the difference in the vertical scale of the left and right panel of fig. 23.) Sequential decay effects reduce the α values in nearly all dynamical models and the effects are large, more than 50%. Such large effects obscure the differences in the isoscaling parameters from realistic interactions (see fig. 19).

The trend is not as clear in statistical models. For $C_{sym} \approx 25\,\text{MeV}$ (including results from SMM-Markov chain calculations joined by the solid lines), the sequential decay effects are small. However, if lower C_{sym} values are used, as in the simulations shown in fig. 17 [53], the α values increase after sequential decays (points joined by dot-dashed, dashed and dotted lines). It is unphysical to assume a constant C_{sym} throughout the expansion phase when fragments are formed. Furthermore, when the C_{sym} values are reduced, there will be a mismatch of the masses of the excited fragments and the ground-state fragments. More study is needed to understand the sequential-decay effects in isoscaling.

In conclusion, isoscaling remains a nice algorithm in data analysis. Unfortunately, the extraction of the symmetry energy using eq. (3) is not so transparent due to the uncertainties in the determination of the (Z_0/A_0) of the emitting source and the secondary-decay effects on the isoscaling parameter α of the emitting source. Our current understanding of the sequential-decay effects on primary fragments is not adequate to allow us to distinguish different realistic asy-EOS interactions. On the positive side, all studies seem to support that much lower values of the symmetry energy than that of normal nuclei are obtained in the low-density region when multifragmentation occurs. This is also in agreement with the results of AMD simulations [18]. Ideally, one should study the properties of the primary fragments directly from experiments. Alternatively, one could find an observable which cancels out the effects of sequential decays as well as other undesirable effects from Coulomb repulsion and pre-equilibrium emissions. The latter two effects mimic the effects from symmetry energy. From the theoretical point of view, it is crucial to follow the whole reaction path within the same model, including the same density dependence of the symmetry term in the equation of state from start to finish. Hybrid models definitely have drawbacks and should be used with caution.

5 Isospin transport

In peripheral collisions it is possible to identify projectile-like and target-like residues in model calculations, as well as in experiments. Calculations suggest that at incident energy above 30 MeV per nucleon and for charge-asymmetric reactions, the symmetry term of the nuclear EOS provides a significant driving force that speeds up the isospin equilibration between the two reaction partners. Thus peripheral collisions may allow one to measure the time scales for charge and mass transport and diffusion. The degree of equilibration, correlated to the interaction time, should provide some insights into transport properties of fermionic systems and, in particular, give information on transport coefficients of asymmetric nuclear matter [59].

5.1 Insight into isospin transport in nuclear reactions

The mechanisms responsible for isospin transport can be essentially related to the presence of isospin, but also to density gradients along the reaction path. Isospin diffusion bears information about the value of the symmetry energy at low density, while the drift (transport in the presence of density gradients) is more connected to the derivative of the symmetry energy.

In asymmetric systems, isospin transport can arise from isospin gradients (diffusion) and from density gradients (drift). Through the low-density neck region, density gradients may be present also in binary systems. The neutron-excess is pushed towards the low-density regions, because this situation is energetically more favorable. This mechanism can induce isospin transport even in reactions between nuclei with the same N/Z [60].

The role of the EOS in isospin transport mechanisms can be made more explicit by studying the response of nuclear matter, in the presence of neutron and proton density gradients. Since we are mostly facing situations where local thermal equilibrium is reached, we will discuss results obtained within the hydrodynamic limit, where the derivation of the isospin transport coefficients is more transparent.

In such a framework the proton and neutron migration is dictated by the spatial gradients of the corresponding chemical potentials $\mu_{p/n}(\rho_p, \rho_n, T)$, where ρ_p and ρ_n are

proton and neutron density and T denotes the temperature [10,61]. The currents of the two species can be expressed, in terms of the total density $\rho = \rho_n + \rho_p$ and asymmetry $I = (\rho_n - \rho_p)/\rho$, as follows:

$$j_n = D_n^\rho \nabla \rho - D_n^I \nabla I, \tag{9}$$

$$j_p = D_p^\rho \nabla \rho - D_p^I \nabla I, \tag{10}$$

where D_q^ρ and D_q^I are drift and diffusion coefficients due to density and isospin gradients, respectively:

$$D_q^\rho = ct \left(\frac{\partial \mu_q}{\partial \rho} \right)_{I,T}, \tag{11}$$

$$D_q^I = -ct \left(\frac{\partial \mu_q}{\partial I} \right)_{\rho,T}, \quad (q = n, p) \tag{12}$$

(ct is a negative constant).

They can be expressed as

$$D_q^\rho = ct \left[N^{-1} + \frac{\partial U}{\partial \rho} \pm 2I \frac{\partial C_{sym}}{\partial \rho} + O(I^2) \right], \tag{13}$$

$$D_q^I = \pm 2ct\, \rho \left[C_{sym} \pm I \left(\rho \frac{\partial C_{sym}}{\partial \rho} - C_{sym} \right) \right], \tag{14}$$

$$(+n, -p),$$

where N^{-1} is the level density of symmetric matter at the same density and temperature and $U(\rho)$ is the isoscalar part of the mean-field potential.

One can see that the isovector part of the nuclear interaction enters the coefficients D_q^ρ through the derivative of the total symmetry energy C_{sym}. On the other hand, the isospin diffusion coefficients D_q^I depend, in leading order, on C_{sym}. Moreover, it appears that the difference of neutron and proton drift coefficients, $D_n^\rho - D_p^\rho = \frac{\partial(\mu_n - \mu_p)}{\partial \rho}$, is equal to $4I\frac{\partial C_{sym}}{\partial \rho}$, as one can simply derive from the relation $\mu_n - \mu_p = 4C_{sym}I$.

In conclusion, the diffusion appears essentially related to the value of the symmetry energy, while the drift is connected to its derivative. From this study one can see more clearly that the isospin distillation effect, as discussed in sect. 3, which originates from the presence of density gradients in the fragmentation process, is sensitive essentially to the derivative of the symmetry energy at low density.

5.2 Experimental studies and comparison with calculations

Experimentally, one examines the isoscaling properties of the fragments originating from the (projectile) residues. Figure 24 shows the isoscaling phenomenon observed in the reaction of ^{124}Sn (projectile) + ^{112}Sn (target) and its inverse reaction ^{112}Sn (projectile) + ^{124}Sn (target) [16]. Unlike the central collision isoscaling data, the slope is larger for the reaction with neutron-rich projectile than the reaction with the proton-rich projectile. Differences in the asymmetric systems reflect the driving force that

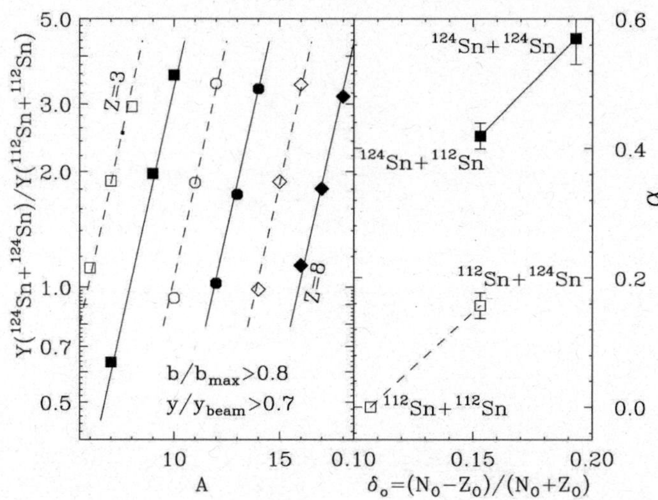

Fig. 24. Isotope yield ratios, as a function of the mass A, for fragments emitted from PLF in peripheral collisions (left). The isoscaling parameters, as obtained in different projectile-target combinations, are compared in the right part of the figure. From [16].

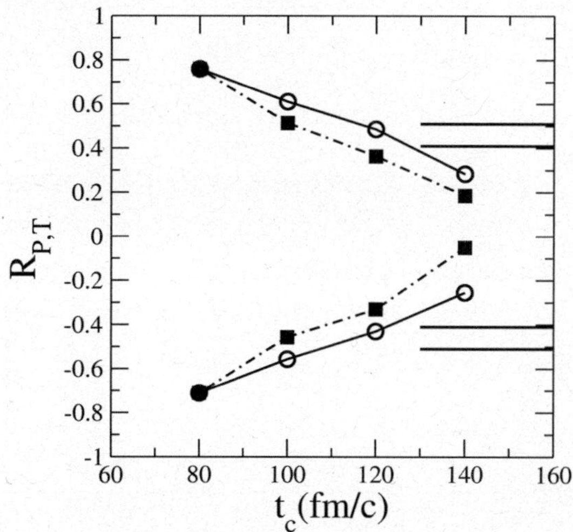

Fig. 25. The isospin transport ratio $R_{P,T}$, eq. (15), for the asy-soft (full squares) and asy-super-stiff (circles) interactions as a function of the interaction time t_c, corresponding to different impact parameter b, in the range 6–10 fm. The band between the two solid lines corresponds to the experimental data of [16].

causes isospin diffusion. As shown before, the force arises from the symmetry term in the equation of state. If the force is weak, one would not expect any isospin mixing to occur and the α value should resemble those of the projectiles in the symmetric systems. In order to quantify the transition from no isospin diffusion to complete mixing, the isospin transport ratio, R_i is used

$$R_i = \frac{2O_{PT} - O_{PP} - O_{TT}}{O_{PP} - O_{TT}}. \qquad (15)$$

Here $P, (T)$ stands for the projectile-like (target-like) fragment. The quantities O_i refer, in general, to any isospin-dependent observable, characterizing the fragments at separation time, for the mixed reaction (PT, $^{124}\mathrm{Sn} + ^{112}\mathrm{Sn}$ or $^{112}\mathrm{Sn} + ^{124}\mathrm{Sn}$), the reactions between the neutron-rich (PP, $^{124}\mathrm{Sn} + ^{124}\mathrm{Sn}$), and between the neutron-poor nuclei (TT, $^{112}\mathrm{Sn} + ^{112}\mathrm{Sn}$), respectively. Similar ratios constructed using free protons have been used as isospin tracer in central heavy-ion collisions [62], to check stopping and thermal equilibration. The insensitivity to systematic errors and the ability to calibrate the observables from the two symmetric systems, $^{124}\mathrm{Sn} + ^{124}\mathrm{Sn}$ and $^{112}\mathrm{Sn} + ^{112}\mathrm{Sn}$ to +1 and −1 offer many advantages. It has been shown that non-isospin effects such as effects from Coulomb force will be largely canceled using the isospin transport ratio [63]. Furthermore, in comparison with calculations that cannot predict the same experimental observables, due to model limitations, one can use another observable to construct the isospin transport ratios as long as both the experimental and theoretical observables are linearly related to each other. For example, in model calculations, one can use the asymmetry $I = (N - Z)/A$ of the emitting source, instead of isoscaling, to evaluate the transport (imbalance) ratio since, to the first order, α and I are linearly related. In fact, as we have seen previously,

α, as extracted from primary fragments, is related to the difference $\left(\frac{Z_1^2}{A_1^2} - \frac{Z_2^2}{A_2^2} \right)$, which for not too large asymmetries can be rewritten as $(I_2 - I_1)/2$. The linear relation between α and I has been confirmed experimentally [30,56], suggesting that it holds for the isoscaling parameters extracted from the final fragments. Hence, the asymmetry of projectile-like or target-like residues can be used as the observable O in eq. (15).

The dependence of isospin transport (and equilibration) on the centrality of the reaction has been investigated in ref. [64] using SMF simulations without momentum dependence. Figure 25 shows the isospin transport ratio as a function of the interaction time t_c among the two reaction partners, that is inversely related to the impact parameter, for two interactions: an asy-super-stiff (open symbols) and an asy-soft (SKM*) (solid symbols) interactions. A more detailed analysis shows that it is possible to explicitly estimate the effects of isospin transport and pre-equilibrium emission on the transport ratios R. The interplay between the two processes leads to a stronger equilibration for asy-soft EOS, as it is evidenced by the isospin transport ratio. Actually, in the asy-super-stiff case, a larger isospin transfer is observed in the calculations, due to the presence of density gradients, directed from PLF and TLF towards the neck region, in line with the analytical predictions illustrated above. Indeed, in the asy-stiff case, the derivative of the symmetry energy, just below normal density, acquires larger values. However, we observe a kind of compensation between the asymmetry of the matter transferred from projectile to target (I_{PT}) and from target to projectile(I_{TP}), so finally isospin equilibration is more effective in the asy-soft case.

In refs. [16,65] BUU calculations which use different density dependence of the symmetry terms in the equation of state are performed for the same system at $b = 6$ fm. In

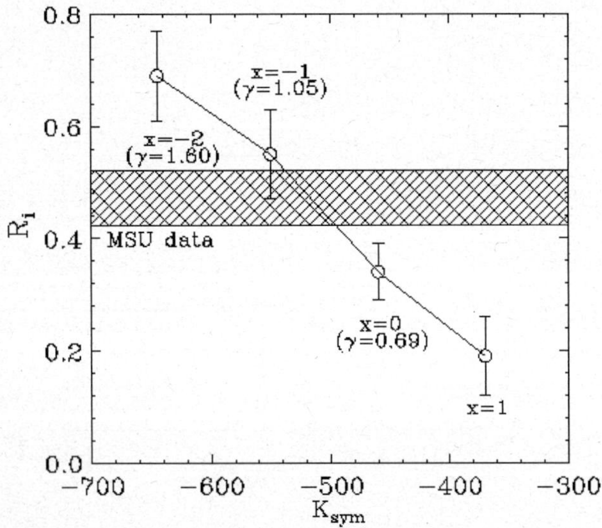

Fig. 26. Isospin transport ratio, relative to peripheral collisions of ^{124}Sn + ^{112}Sn, $E/A = 50$ MeV, as obtained in the data (hatched area) and in model calculations using different parameterizations of the symmetry energy.

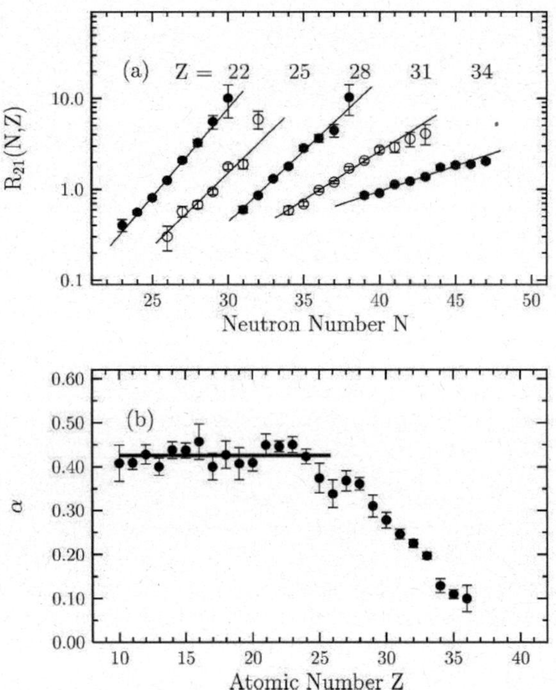

Fig. 27. a) Yield ratios $R_{21}(N, Z)$ of projectile residues from the reactions ^{86}Kr + ^{112,124}Sn at 25 MeV/A, with respect to N, for the Z's indicated. b) Isoscaling parameter α *versus* the charge of the fragment Z. The straight line is a constant value fit for the lighter fragments $Z = 10$–26. From [55].

fig. 26, the parameter x indicates the stiffness of the symmetry term, ranging from an asy-soft behaviour (larger x) to an asy-stiff behaviour (negative x). Momentum dependence is included in the calculations. Within error bars, the isospin transport ratio R_i decreases with the softness of the symmetry term suggesting that the driving force for isospin equilibration is much larger for the asy-soft interaction. Using the isoscaling fitting parameter α as the experimental isospin observable O in eq. (15), the isospin transport ratios of the two asymmetric systems shown in the right panel of fig. 24 appear as the two horizontal lines in fig. 25 and as the shaded region in fig. 26. Assuming that data can be directly compared with calculations performed at $b = 6$ fm, the best agreement is obtained in the range $x = -1, 0$. Without including momentum dependence in the calculations, a stiffer symmetry term would be needed to fit the data (see fig. 25 at 140 fm/c and [16]). This is due to the fact that the overall dynamics becomes more repulsive when the momentum dependence is included and isospin equilibration depends not only on the strength of the symmetry term, but also on the system interacting time.

It should be noted that this comparison is well suited only if one assumes that PLF and TLF fragments are near normal density, independent of the parameterization adopted for the symmetry energy. Otherwise, the isoscaling parameter cannot be simply related only to the (N_0/Z_0) of the source, but would also depend on the value of the symmetry energy, that is model dependent. It would be interesting to check this hypothesis using a model that includes fragmentation (such as AMD), and calculating the isospin transport ratio directly from the isotope yields, as done in the experiments. In this way one would better test the sensitivity of the results to the behaviour of the symmetry energy. In other words, as discussed in sect. 4.4,

it is essential to consider the density dependence of the symmetry energy in the fragmentation process.

The dependence of the $(N/Z)_{PLF}$ *versus* the excitation energy of the system has been studied for the INDRA reactions, Ni + Ni and Ni + Au at 52 and 74 MeV/A [66], looking at the reconstructed PLF, in binary collisions. In the symmetric case, $(N/Z)_{PLF}$ is seen to increase with centrality due to a proton-rich pre-equilibrium emission. For the Ni + Au case, it is possible to observe isospin equilibration among the reaction partners.

In the analysis presented in ref. [55], the goal was to extract the $(N/Z)_{PLF}$ of the PLF sources and hence to discuss isospin equilibration between projectile and target. The systems ^{86}Kr + ^{112,124}Sn at 25 MeV/A have been considered (fig. 27). The isoscaling parameter α, as a function of the charge Z of the projectile residue, exhibits a plateau, at $Z \leq 22$. This would suggest that fragments are emitted from equilibrated sources formed in more central collisions. (Any source characterization will require impact parameter selection, not available in this data set.) As Z increases, a decrease of α is observed, *i.e.* the strict isoscaling relation of eq. (1) is not obeyed suggesting a non-equilibrium process. One interpretation of this observation is that the heavier fragments coming from PLFs are formed in more peripheral reactions. In the extreme case when the fragment composition is close to that of the projectile with little transfer of nucleons from the targets, $\Delta(Z/A)^2$ becomes small resulting in small α values.

6 Conclusions and outlook

From the results discussed above, several indications on the density dependence of the symmetry potential can be extracted from the study of reactions with charge-asymmetric systems in the Fermi energy domain. The isotopic content of emitted light particles and IMFs, as well as the reconstructed degree of N/Z equilibration between reaction partners, bear important information on the symmetry energy. We summarize below the main conclusions, as well as improvements and new studies that can be envisaged.

- In model calculations, the isotopic content of pre-equilibrium emission appears quite sensitive to the stiffness of the symmetry term, and to its momentum dependence. A more neutron-rich pre-equilibrium emission is expected for higher values of the symmetry energy. Thus experiments to measure the ratios of free neutrons and free protons will complement the results obtained with fragments. In fragment formation the degree of isospin fractionation (*i.e.* the transfer of the neutron excess towards the gas phase) is sensitive to the slope of the symmetry energy at low density.
- The isoscaling analysis, based on the comparison of the isotopic content of fragments emitted from systems with different initial asymmetries could in principle be used to extract the value of the symmetry energy in situations where equilibrium is reached. However, it is not so easy to disentangle the predictions given by the different parameterizations. Indeed not only the isoscaling parameters reflect the width of the isotopic distributions, but they are quite sensitive to the difference $\Delta(Z/A)^2$ between the fragments asymmetries of the two compared systems, that is also largely influenced by the symmetry energy behaviour. Hence, compensation effects diminish the sensitivity of the final results to the asy-EOS. Moreover, as shown by both statistical and dynamical simulations, large secondary decays may reduce significantly the differences coming from different symmetry terms in the EOS used in the models. However, most experimental results seem to suggest that the symmetry energy of excited fragments at low density may be lower than the symmetry energy at normal nuclear-matter density. However, the extraction of the accurate value of symmetry energy and the differentiation of different interactions require theoretical models to simulate the sequential-decay effects and the reaction path. More work is needed in model development. Experiments that can extract the properties of primary fragments directly will also advance our understanding in this issue.
- It would be important to perform a cross-check of model predictions against several experimental observables, sensitive to the different phases of a reaction, from pre-equilibrium emission to fragmentation and de-excitation stage. This allows better identification of the isotopic content of the gas and liquid phases, that is essential for the analysis of the de-excitation process of the excited primary products and related observables (such as isoscaling). The use of models that can follow the whole path of the reaction is highly desirable. This would also allow to ascertain the fragmentation mechanism and the way the system approaches chemical freeze-out. The study of correlations between the fragment isotopic content and kinematical properties can be envisaged as a tool to learn about time scales for fragment formation and N/Z equilibration.
- In semi-peripheral collisions, it is interesting to compare the behaviour of reactions with different entrance channel asymmetries to investigate isospin exchange between projectile and target. Indeed the N/Z equilibration among the reaction partners gives information about the strength of the symmetry term. However, it should be noticed that the amount of isospin exchange is also strictly connected to the interaction time, when the two reaction partners are in contact. This is clearly influenced also by the isoscalar part of the nuclear interaction. Hence, the sensitivity of this observable to the value of the symmetry energy is not so transparent. The study of the interplay between isoscalar and isovector parts of the nuclear interaction in the reaction dynamics deserves further attention.

MBT acknowledges the support of the National Science Foundation under Grant No. PHY-01-10253.

References

1. It is important to put the properties of *neutron* matter at *subnuclear* densities on as firm a footing as possible, not only for astrophysical applications, but also for interpreting terrestrial experiments with coming radioactive beam facilities", C.J. Pethick, D.G. Ravenhall, in *The Lives of Neutron Stars*, NATO ASI Ser. C **450**, 59 (1995).
2. I. Bombaci, T.T.S. Kuo, U. Lombardo, Phys. Rep. **242**, 165 (1994); I. Bombaci, Phys. Rev. C **55**, 1587 (1997).
3. M. Prakash *et al.*, Phys. Rep. **280**, 1 (1997).
4. I.M. Irvine, *Neutron Stars* (Oxford University Press, 1978).
5. J.M. Lattimer *et al.*, Phys. Rev. Lett. **66**, 2701 (1991).
6. D. Pines, R. Tamagaki, S. Tsuruta, *Neutron Stars* (Addison-Wesley, New York, 1992).
7. K. Sumiyashi, H. Toki, Astrophys. J. **422**, 700 (1994).
8. C.J. Pethick, D.G. Ravenhall, *The Lives of Neutron Stars*, NATO ASI Ser. C **450**, 59 (1995).
9. C.H. Lee, Phys. Rep. **275**, 255 (1996).
10. V. Baran, M. Colonna, V. Greco, M. Di Toro, Phys. Rep. **410**, 335 (2005).
11. H. Müller, B.D. Serot, Phys. Rev. C **52**, 2072 (1995).
12. Bao-An Li, C.M. Ko, Nucl. Phys. A **618**, 498 (1997).
13. M. Colonna, M. Di Toro, A. Larionov, Phys. Lett. B **428**, 1 (1998).
14. V. Baran, M. Colonna, M. Di Toro, V. Greco, Phys. Rev. Lett. **86**, 4492 (2001).
15. M. Colonna, Ph. Chomaz, S. Ayik, Phys. Rev. Lett. **88**, 122701 (2002).
16. M.B. Tsang, T.X. Liu, L. Shi *et al.*, Phys. Rev. Lett. **92**, 062701 (2004).

17. P. Sapienza *et al.*, Phys. Rev. Lett. **87**, 072701 (2001) and references therein.

18. A. Ono, P. Danielewicz, W.A. Friedman *et al.*, Phys. Rev. C **68**, 051601(R) (2003); Phys. Rev. C **70**, 041604 (2004).

19. C.O. Dorso *et al.*, arXiv: nucl-th/0504036.

20. W.P. Tan, S.R. Souza, R.J. Charity *et al.*, Phys. Rev. C **68**, 034609 (2003).

21. B-A Li, C.B. Das, S. Das Gupta *et al.*, Nucl. Phys. A **735**, 563 (2004).

22. J. Rizzo *et al.*, Nucl. Phys. A **732**, 202 (2004).

23. J.Y. Liu, W.J. Guo, Y.Z. Xing *et al.*, High En. Phys. Nucl. **29**, 456 (2005).

24. J. Rizzo, M. Colonna, M. Di Toro, arXiv: nucl-th/0508008.

25. J. Margueron, Ph. Chomaz, Phys. Rev. C **67**, 041602 (2003).

26. V. Baran *et al.*, Nucl. Phys. A **703**, 603 (2002).

27. S. Das Gupta, A.Z. Mekjian, M.B. Tsang, Adv. Nucl. Phys. **26**, 91 (2001).

28. C.B. Das, S. Das Gupta, W.G. Lynch, A.Z. Mekjian, M.B. Tsang, Phys. Rep. **406**, 1 (2005).

29. M.B. Tsang *et al.*, this topical issue.

30. H.S. Xu *et al.*, Phys. Rev. Lett. **85**, 716 (2000).

31. M.B. Tsang, W.A. Friedman, C.K. Gelbke *et al.*, Phys. Rev. Lett. **86**, 5023 (2001).

32. M. Veselsky, R.W. Ibbotson, R. Laforest *et al.*, Phys. Rev. C **62**, 041605 (2000).

33. J. Randrup, Nucl. Phys. A **307**, 319 (1978).

34. G.J. Kunde *et al.*, Phys. Rev. Lett. **77**, 2897 (1996).

35. M.L. Miller *et al.*, Phys. Rev. Lett. **82**, 1399 (1999).

36. Ph. Chomaz, F. Gulminelli, Phys. Lett. B **447**, 221 (1999).

37. P. Milazzo *et al.*, Phys. Rev. C **62**, 041602 (2000).

38. M.B. Tsang *et al.*, Phys. Rev. C **64**, 041603 (2001).

39. D.V. Shetty, S.J. Yennello, E. Martin *et al.*, Phys. Rev. C **68**, 021602 (2003).

40. G.A. Souliotis, M. Veselsky, D.V. Shetty *et al.*, Nucl. Phys. A **746**, 526c (2004).

41. M.N. Andronenko, L.N. Andronenko, W. Neubert, Prog. Theor. Phys. Suppl. **146**, 538 (2002).

42. M. Veselsky, G.A. Souliotis, M. Jandel, Phys. Rev. C **69**, 044607 (2004).

43. W.A. Friedman, Phys. Rev. C **69**, 031601 (2004).

44. A.S. Botvina, O.V. Lozhkin, W. Trautmann, Phys. Rev. C **65**, 044610 (2002).

45. W.A. Friedman, Phys. Rev. C **42**, 667 (1990).

46. M.B. Tsang *et al.*, Phys. Rev. C **64**, 054615 (2001).

47. Qingfeng Li, Zhuxia Li, Horst Stöcker, Phys. Rev. C **73**, 051601 (2006).

48. T.X. Liu, M.J. van Goethem, X.D. Liu *et al.*, Phys. Rev. C **69**, 014603 (2004).

49. M. Colonna, F. Matera, Phys. Rev. C **71**, 064605 (2005).

50. M. Colonna *et al.*, Phys. Rev. C **47**, 1395 (1993).

51. D.V. Shetty, A.S. Botvina, S.J. Yennello *et al.*, Phys. Rev. C **71**, 024602 (2005); D.V. Shetty, A.S. Botvina, S.J. Yennello *et al.*, Phys. Rev. C **71**, 029903(E) (2005).

52. D.V. Shetty, S.J. Yennello, A.S. Botvina *et al.*, Phys. Rev. C **70**, 011601 (2004).

53. A. Le Fèvre, G. Auger, M.L. Begemann-Blaich *et al.*, Phys. Rev. Lett. **94**, 162701 (2005).

54. W.P. Tan, B.-A. Li, R. Donangelo *et al.*, Phys. Rev. C **64**, 051901 (2001).

55. G.A. Souliotis, M. Veselsky, D. Shetty *et al.*, Phys. Lett. B **588**, 35 (2004).

56. E. Geraci *et al.*, Nucl. Phys. A **732**, 173 (2004).

57. A. Ono *et al.*, *Proceedings for VI Latin American Symposium on Nuclear Physics and Applications, Iguazu, Argentina (2005)*, to be published in Acta Phys. Hung. A.

58. A. Ono *et al.*, arXiv: nucl-ex/0507018.

59. L. Shi, P. Danielewicz, Phys. Rev. C **68**, 064604 (2003).

60. R. Lionti, V. Baran, M. Colonna, M. Di Toro, Phys. Lett. B **625**, 33 (2005).

61. R. Balian, *From Microphysics to Macrophysics*, Vol **II** (Springer Verlag, Berlin, 1992).

62. F. Rami *et al.*, Phys. Rev. Lett. **84**, 1120 (2000).

63. M.B. Tsang, T.X. Liu, W.G. Lynch, in *Proceedings of the International Workshop on Multifragmentation IWM2005, Catania, Italy, 2005*, edited by R. Bougault *et al.*, Conf. Proc. Vol. **91** (Italian Physical Society, Bologna, 2006) p. 123.

64. V. Baran, M. Colonna, M. Di Toro *et al.*, Phys. Rev. C **72**, 064620 (2005) arXiv: nucl-th/0506078,

65. L.-W. Chen, C.-M. Ko, B.-A. Li, Phys. Rev. Lett. **94**, 032701 (2005).

66. E. Galichet *et al.*, submitted to Nucl. Phys. A.

Eur. Phys. J. A **30**, 183–202 (2006)
DOI 10.1140/epja/i2006-10116-7

THE EUROPEAN
PHYSICAL JOURNAL A

Evolution of the giant dipole resonance properties with excitation energy

D. Santonocito[1,a] and Y. Blumenfeld[2]

[1] INFN - Laboratori Nazionali del Sud, Via S. Sofia 62, I-95123 Catania, Italy
[2] Institut de Physique Nucléaire, IN2P3-CNRS, F-91406 Orsay, France

Received: 7 March 2006 /
Published online: 31 October 2006 – © Società Italiana di Fisica / Springer-Verlag 2006

Abstract. The studies of the evolution of the hot Giant Dipole Resonance (GDR) properties as a function of excitation energy are reviewed. The discussion will mainly focus on the $A \sim 100$–120 mass region where a large amount of data concerning the width and the strength evolution with excitation energy are available. Models proposed to interpret the main features and trends of the experimental results will be presented and compared to the available data in order to extract a coherent scenario on the limits of the development of the collective motion in nuclei at high excitation energy. Experimental results on the GDR built in hot nuclei in the mass region $A \sim 60$–70 will be also shown, allowing to investigate the mass dependence of the main GDR features. The comparison between limiting excitation energies for the collective motion and critical excitation energies extracted from caloric curve studies will suggest a possible link between the disappearance of collective motion and the liquid-gas phase transition.

PACS. 24.30.Cz Giant resonances – 25.70.Ef Resonances – 25.70.Gh Compound nucleus

1 Introduction

A well-established result of nuclear physics is the observation of giant resonances, small amplitude, high frequency, collective modes of excitation in nuclei. Among all possible modes of collective excitation, the Giant Dipole Resonance (GDR), a collective vibration of protons against neutrons with a dipole spatial pattern, has been widely investigated and is now considered a general feature of all nuclei.

The experiments performed over many years have shown that the GDR is an efficient tool to probe nuclear properties of the ground state as well as at finite temperature. In fact, the gamma-ray emission due to the GDR decay is sufficiently fast to compete with other decay modes with a sizable branching ratio and therefore to probe the characteristics of the nuclear system prevailing at that time. The resonance energy being proportional to the inverse of the nuclear radius, the investigation of the strength distribution gives access to the study of the nuclear deformations in the ground state but also to the shape evolution of nuclei as a function of spin and temperature of the system. Shape evolution and shape fluctuations are the main issues in the study of the GDR in nuclei populated at low excitation energy ($E^* < 100\,\mathrm{MeV}$) and spin up to the fission limit. This region has been extensively studied and the GDR properties, which are ex-

pected to be influenced by shell effects, are rather well understood [1,2] even if some recent results indicate an interesting discrepancy between data and theoretical models at temperatures $T \sim 1$–$1.5\,\mathrm{MeV}$ which deserves further investigation.

Conversely, populating nuclei at progressively higher thermal energies up to the limits of their existence, one can follow the evolution of the collective motion in extreme conditions up to its disappearance. The investigation of the GDR features at high excitation energy is particularly interesting because it also opens up the possibility to investigate the limits of validity of the standard statistical scenario in describing the decay properties of hot nuclei. The statistical model assumes, in fact, that the system reaches thermal equilibrium before it decays. Increasing the excitation energy, the compound nucleus lifetime decreases significantly and collective degrees of freedom might not reach equilibrium before the system decays. Therefore, the GDR strength distribution will reflect the relative influence of the different time scales which come into play, the population and decay time of the GDR on one hand and the equilibration and decay times of hot nuclei on the other. In the following the experimental results collected up to $E^* \sim 500\,\mathrm{MeV}$ will be presented and compared to statistical model calculations. The evidence in the gamma spectra of a vanishing of the GDR strength at high excitation energies relative to the standard statis-

---

a e-mail: santonocito@lns.infn.it

tical calculation led to the development of different theoretical models whose main features will be discussed in the text. The comparison between data and statistical calculations including different model prescriptions will allow us to draw some conclusions concerning the effects leading to the GDR disappearance. Eventually, the existence of a limiting excitation energy for the collective motion will be discussed and compared to the limiting excitation energies extracted from the caloric-curve studies in different mass regions. This will allow to investigate a link between the liquid-gas phase transition and the disappearance of collective motion.

2 GDR built on the ground state: general features

The GDR was first observed in 1947 by Baldwin and Klaiber in photo-absorption and photo-fission experiments [3,4]. They observed an increase of the absorption cross-section above $10\,\mathrm{MeV}$ in several nuclei with resonance energies between 16 and $30\,\mathrm{MeV}$.

The observed peak in the photo-absorption spectrum was interpreted by Goldhaber and Teller [5] as the excitation of a collective nuclear vibration in which all the protons in the nucleus move collectively against all the neutrons creating an electric dipole moment. Since then, the GDR has been extensively studied, and a broad systematics for almost all stable nuclei exists on the GDR built on ground states. Most of the information was extracted from photo-absorption experiments because of the high selectivity of this reaction to $E1$ transitions [6].

The shape of the resonance in the photo-absorption spectrum can be approximated, in the case of a spherical nucleus, by a single Lorentzian distribution [6,7]:

$$\sigma_{abs}(E_\gamma) = \frac{\sigma_0 E_\gamma^2 \Gamma_{GDR}^2}{(E_\gamma^2 - E_{GDR}^2)^2 + E_\gamma^2 \Gamma_{GDR}^2}, \qquad (1)$$

where σ_0, E_{GDR} and Γ are, respectively, the strength, the centroid energy and the width of the Giant Dipole Resonance. In nuclei with a static deformation, the GDR splits in two components corresponding to oscillations along and perpendicular to the symmetry axis, and the cross-section for photo-absorption can be well reproduced by the superposition of two Lorentzian distributions. This particular feature allows one to extract the nuclear deformation from the centroid energies of the two components and to distinguish, from the relative intensities, prolate from oblate deformations.

The systematics shows that the resonance energy decreases gradually with increasing mass number. This mass dependence can be reproduced by [6]:

$$E_{GDR} = 31.2A^{-1/3} + 20.6A^{-1/6} \qquad (2)$$

which is a linear combination of the mass dependencies predicted by Goldhaber-Teller and Steinwedel-Jensen macroscopic models for the energy of the GDR [5,7].

The width of the resonance is also strongly influenced by the shell structure of the nuclei. The systematics shows values ranging from about 4–$5\,\mathrm{MeV}$ for closed-shell nuclei up to about $8\,\mathrm{MeV}$ for nuclei between closed shells [6].

The collectivity of the excitation, which is related to the number of participating nucleons, can be estimated in terms of the Energy-Weighted Sum Rule (EWSR) for dipole radiation. This sum rule, also known as Thomas-Reiche-Kuhn (TRK) sum rule gives the total integrated cross-section for electric dipole photon absorption. It is given by:

$$\int_0^\infty \sigma_{abs}(E_\gamma)\mathrm{d}E_\gamma = \frac{2\pi^2 e^2 \hbar}{Mc}\frac{NZ}{A} = \\ 60\frac{NZ}{A}(\mathrm{MeV}\cdot\mathrm{mb}), \qquad (3)$$

where N, Z and A are, respectively, the neutron, the proton and the mass number and M is the nucleon mass [6]. The systematics shows that for nuclei of mass $A < 80$ the TRK sum rule is not exhausted by the data while in the region of mass $A \sim 100$ the photoneutron cross-section integrated up to about $30\,\mathrm{MeV}$ exhausts the TKR sum rule [6]. For heavier mass nuclei the experimental data exceed the TKR sum rule by about 20–30% [6].

3 GDR built on excited states: historical

The field of the study of Giant Resonances built on excited states was launched by Brink [8] who stated the hypothesis that Giant Resonances could be built on all nuclear states and that their characteristics, aside from the dependence on the shape, should not depend significantly on the nuclear state. This opened up the possibility of investigating nuclear shapes also in excited nuclei and to study the evolution of the properties of collective motion up to the limits of existence of nuclei. Indeed the disappearance of collective motion has been considered a further signature for a phase transition in nuclear matter.

Evidence in favor of the Brink hypothesis was extracted for the first time in 1974, in the study of the γ-ray spectrum emitted from spontaneous fission of $^{252}\mathrm{Cf}$ [9]. The enhancement observed in the γ spectrum above $10\,\mathrm{MeV}$ was, in fact, correctly attributed to the de-excitation of the GDR built on excited states of the fission products. The first evidence for the existence of the GDR built on an excited state using a reaction study emerged in a proton capture (p,γ) experiment on $^{11}\mathrm{B}$ where the GDR built on the first excited state of $^{12}\mathrm{C}$ was observed [10]. From subsequent (p,γ) and (n,γ) experiments on various other light nuclei emerged a coherent picture supporting the Brink hypothesis [11]. An important step further in the study of the GDR properties was made with the use of heavy-ion reactions which opened up the possibility to populate highly excited continuum states through the mechanism of complete fusion in a wide variety of nuclei. The first observation of the gamma-decay of the GDR built on highly excited states in nuclei formed in fusion

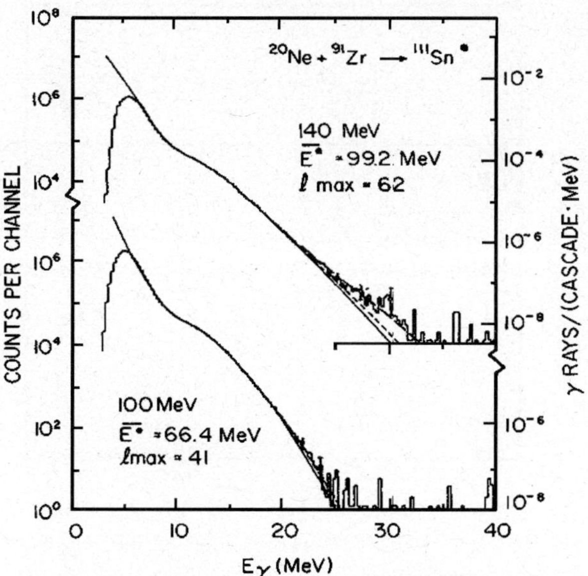

Fig. 1. Measured γ-ray spectra from the decay of ^{111}Sn at $E^* = 66$ and $E^* = 100$ MeV. Full lines represent the statistical-model calculations using a level density parameter $a = A/8$ while the short-dashed line is a similar calculation with $a = A/8.5$ and the long-dashed one is obtained including the decay from the giant quadrupole resonance [24].

reaction was made in 1981 studying ^{40}Ar-induced reactions on ^{82}Se, ^{110}Pd, and ^{124}Sn targets [12]. The importance of these measurements stems from the fact that they demonstrated the possibility to study the GDR in the γ-ray de-excitation spectra following fusion reactions where the statistical emission of high-energy gamma-rays occurs from an equilibrated system and in competition with particle evaporation indicating a sizable branching for gamma decay. The experiments performed since then have been focused on establishing the existence of the GDR built on excited states as a general feature of nuclei and on the evolution of the parameters governing the GDR properties as a function of the excitation energy, spin and mass.

4 GDR built on excited states: general features

When the GDR built on excited states is studied at increasing excitation energy the scenario becomes gradually more complex due to the opening of different decay channels. In heavy-ion collisions up to 5–$6A$ MeV the reaction dynamics are dominated by mean-field effects which lead, for central collisions, to a complete fusion of projectile and target nuclei. In this case a compound nucleus is formed with a well-defined excitation energy and a broad distribution in angular momentum. The equilibrated system will then undergo a statistical decay emitting light particles and gamma-rays according to their relative probabilities which can be very well accounted for in the framework of the statistical model. Gamma-rays can be emitted at

all steps during the decay sequence and the emission of a high-energy gamma-ray will be in competition with light-particle emission, driven by the ratio of the level densities between initial and final states for both decay channels. In general light-particle emission is much more probable than γ-decay, but the latter, which has a probability of the order of 10^{-3}, is a more useful probe of the GDR properties since the γ-ray carries all the energy of the resonance. The decay rate R_γ is given by

$$R_\gamma \mathrm{d}E_\gamma = \frac{\rho(E_2)}{\hbar\rho(E_1)} f_{GDR}(E_\gamma)\mathrm{d}E_\gamma, \qquad (4)$$

where $\rho(E_1)$ and $\rho(E_2)$ are, respectively, the level densities for the initial and final states which differ by an energy $E_\gamma = E_1 - E_2$ and $f_{GDR}(E_\gamma) \propto \sigma_{abs}E_\gamma^2$. It can be written as

$$f_{GDR}(E_\gamma) = \frac{4e^2}{3\pi\hbar mc^3} S_{GDR} \frac{NZ}{A}$$
$$\times \frac{\Gamma_{GDR}E_\gamma^4}{(E_\gamma^2 - E_{GDR}^2)^2 + E_\gamma^2\Gamma_{GDR}^2}, \qquad (5)$$

where σ_{abs} is the photo-absorption cross-section and S_{GDR} is the fraction of the exhausted sum rule. Comparing the above expressions with the neutron emission rate it follows that the GDR gamma yield is higher during the first steps of the decay cascade [1]. This means that, if the decay is statistical, the GDR γ-rays essentially reflect the GDR properties at the highest excitation energies. Besides, since the nuclear level density varies exponentially with the excitation energy the number of transition photons decreases exponentially with transition energy. This last argument together with the competition at all steps in the emission process reflects and explains the shape of the measured gamma-ray spectrum. A typical spectrum measured studying the decay of the GDR in hot nuclei populated in complete-fusion reactions between heavy ions at beam energies up to 5–$6A$ MeV is shown in fig. 1. Below $E_\gamma \sim 10$ MeV the spectrum is dominated by the statistical emission of gamma-rays from the equilibrated system at the end of the decay process. Above $E_\gamma \sim 10$ MeV a broad bump is observed which is a signature of the GDR decay.

In order to extract quantitative information on the GDR properties at different excitation energies from the spectrum we need to make a comparison with statistical calculations which take into account all the decay sequence. This kind of analysis is usually carried out using the statistical code CASCADE [13] which treats the statistical emission of neutrons, protons, alphas and γ-rays from a hot equilibrated system. In the code, the GDR is assumed to be Lorentzian in shape in analogy to the observation made on cold nuclei. The dipole emission is expected to dominate the spectrum above 10–12 MeV even if small contributions from quadrupole emission cannot be ruled out and are typically included in the calculation. All the results strongly depend on the assumptions made for the level densities.

In the following, the results concerning the GDR properties will be discussed for increasing excitation energy. We will generally assume that E^* and T can be related by the Fermi-gas formula $E^* = aT^2$ [14]. An extensive discussion of the determination of the level density parameter a can be found in ref. [15]. The main part of the discussion will be focused on mass region $A \sim 120$ due to the existing broad systematics. The discussion will be divided in two main sections, one for experiments up to $E^* \sim 200\,\mathrm{MeV}$ where the main issue is the increase of the GDR width while the strength retains its full collective character and a second one, above $E^* \sim 200$ where a progressive quenching of the GDR, is observed in all the experiments. A detailed analysis of this effect will be undertaken from the theoretical and the experimental point of view leading to some conclusions concerning the GDR properties up to the limits of its existence.

5 The evolution of the GDR at moderate excitation energies up to 200 MeV

Once the main features of GDR built on the ground state are well understood the question arises as to what happens to GDR properties built on the excited states. In this case the main aim is to probe the stability of collective motion in nuclei under increasing temperature and angular momentum. In particular, populating hot nuclei at increasing excitation energy and in different spin ranges one is able to follow the shape modifications and fluctuations associated to the weakening of shell effects which dominate the nuclear properties of the ground state. At the same time it is also possible to extract new information on the relative time scales involved in shape rearrangements. A further important issue in these studies is the evaluation of the relative influence of angular momentum and temperature effects on the evolution of the GDR parameters. In a typical fusion experiment the higher excitation energies are associated to large transfer of angular momentum. Recently, inelastic scattering has been used to populate nuclei in a wide range of excitation energies with little angular momentum transfer allowing to disentangle the relative contribution of angular momentum and temperature effect on the GDR features.

The existing hot GDR systematics can be reasonably well accounted for in the framework of the adiabatic thermal fluctuation model [16–19]. However, recent results on width measurements in the region of temperatures below about 1.5 MeV showed important discrepancies between predictions and data in different mass regions which remain hitherto unexplained [19–22]. In the following, we will concentrate on tin isotopes ($A \sim 110$) tracing an historically based overview of our understanding of the GDR features up to now. In this mass region the GDR built on the ground state is characterized by a resonance energy of about 15 MeV, a strength fulfilling 100% of the EWSR and a width of about 5 MeV. A significant modification of the GDR width was observed for the first time by Gaardhøje et al. [23] studying the gamma spectra emitted in the statistical decay of ^{108}Sn nuclei populated up to

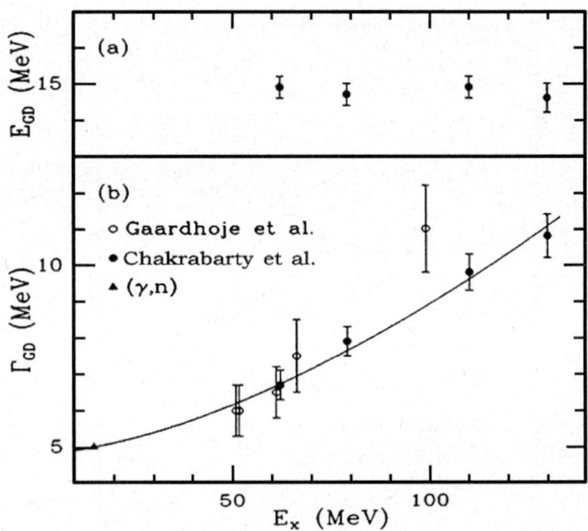

Fig. 2. The systematics for the energy (a) and width (b) of the GDR as a function of E^*. Open symbols are from refs. [23, 24] while full circles are from ref. [25]. The full line corresponds to the parametrization of the width given by eq. (6).

$E^* = 60\,\mathrm{MeV}$ and angular momenta up to $I \simeq 40\hbar$. Reproducing the data through a statistical calculation performed with the code CASCADE using a single Lorentzian function centered at $E_{GDR} = 15.5\,\mathrm{MeV}$ called for a width $\Gamma = 6\text{–}6.5\,\mathrm{MeV}$ for the three excitation energies investigated, values which were clearly in excess of the typical widths measured on the ground state.

Similar results were obtained in the study of the GDR decay from ^{111}Sn nuclei populated at $E^* = 66$ and $100\,\mathrm{MeV}$ excitation energies using the reaction ^{20}Ne $+ {}^{91}$Zr at $E_{beam} = 100$ and $140\,\mathrm{MeV}$ [24]. The measured γ spectra and the corresponding statistical calculation are shown in fig. 1. In this case a strength corresponding to 100% of the EWSR and widths of 7.5 and 11 MeV, respectively, were needed to reproduce the γ-ray spectra. Therefore, the results of these experiments, shown as open symbols in fig. 2, pointed to a progressive increase of the width with excitation energy at least up to $E^* = 100\,\mathrm{MeV}$. The authors suggested two possible interpretations for such an increase as due either to an increase of the GDR damping width with E^* and/or spin I or to a change in deformation.

The systematic study of the GDR properties in Sn isotopes was extended by the work of Chakrabarty et al. [25] at higher excitation energies ($E^* = 130\,\mathrm{MeV}$) and spin. The results concerning the centroid energies and widths are shown in fig. 2 as full circles. Within the experimental errors, the centroid energy of the GDR as extracted from best fits to experimental data are independent of excitation energy while the absolute value seems to be slightly lower than the one measured on the ground state. The width of the resonance was observed to increase with excitation energy although less strongly than reported in the previous work, the discrepancy being relevant only for $E^* = 100\,\mathrm{MeV}$. Different calculations were performed to

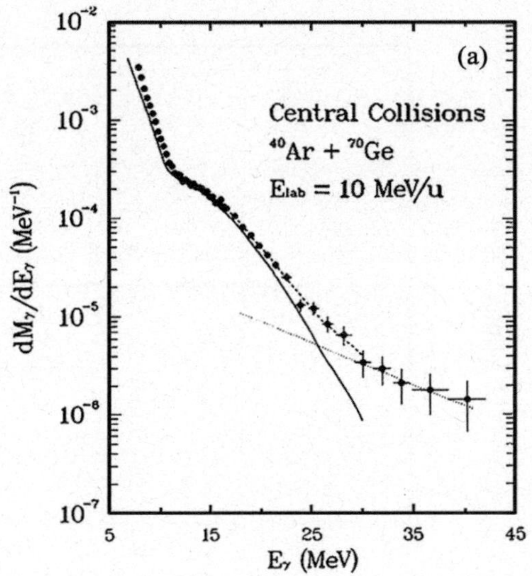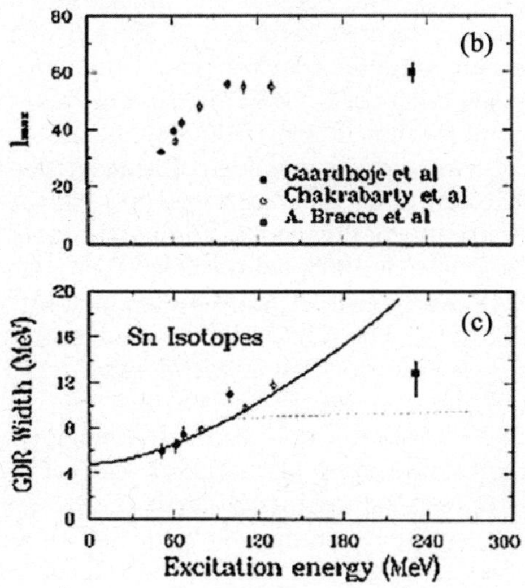

Fig. 3. a) Gamma spectrum measured in coincidence with fusion events from the reaction ^{40}Ar + ^{70}Ge at $E_{lab} = 10A$ MeV. The full line represents the statistical calculation, the dotted line indicates the bremsstrahlung contribution while the dashed line is the sum of the previous two. b) The trend of the maximum angular momentum populated in the different fusion reactions leading to Sn isotopes as a function of excitation energy. c) The systematics of the GDR width measured on $^{108-112}$Sn isotopes as a function of excitation energy. The width value at $E^* = 230$ MeV is the one measured by Bracco *et al.* [26] suggesting the width saturation. The full line corresponds to the parametrization of the width given by eq. (6).

investigate the sensitivity of the results to the level density parameter adopted. Small differences were observed and included in the estimated error bars. The average width trend, including the ground-state value, can be well accounted for by the relation

$$\Gamma = 4.8 + 0.0026E^{1.6} \text{ MeV}, \qquad (6)$$

where the first constant represents the ground-state value. Such a trend is in qualitative agreement with calculations performed on ^{108}Sn at high spin and temperatures which predict a width increase reflecting the increase of deformation at higher angular momenta and the progressive importance of thermal fluctuations of the nuclear shape [16]. In particular, Sn isotopes are predicted to evolve from spherical towards oblate shapes with increasing spin and temperature. Thermal fluctuations wash out the structures of the absorption strength function calculated for fixed deformation inducing a broadening and a smoothing of the strength function.

It is important to mention at this point that, while a clear trend as a function of excitation energy is observed for the width, each single statistical model calculation used as a comparison to estimate the GDR parameters was performed with a fixed width all along the decay chain. Therefore, the extracted parametrization reproduces the trend for the width averaged over the decay cascade. A proper treatment of the problem calls for the inclusion of the energy and spin dependencies of the width in the calculation. Chakrabarty and co-workers investigated these dependencies and found that a good fit to the data can be obtained assuming [25]

$$\Gamma = 4.5 + 0.0004E^2 + 0.003I^2. \qquad (7)$$

An abrupt change to the smooth increase of the GDR width with excitation energy observed up to 130 MeV was found in the study of the GDR structure at about $E^* = 230$ MeV. In this experiment performed using an ^{40}Ar beam at $10A$ MeV the GDR gamma decay was investigated in ^{110}Sn nuclei and a width similar to the one previously measured at 130 MeV was observed indicating the onset of a saturation effect above 130 MeV [26].

At $10A$ MeV beam energy the reaction dynamics are still dominated by the mean field which leads, for central collisions, mainly to complete-fusion events. However, modifications of the mean-field dynamics due to the effect of nucleon-nucleon collisions occur leading also to incomplete-fusion events characterized by a partial transfer of nucleons from the lightest to the heaviest partner of the collision which affects the final excitation energy and mass of the hot system produced. It is no longer straightforward to ascertain the excitation energy and spin distribution populated in the reaction. The higher bombarding energy also induces a new high-energy component in the γ-spectrum due to nucleon-nucleon bremsstrahlung in the first stages of the reaction. This component must be understood and subtracted before drawing conclusions on the GDR characteristics. A proper identification of the reaction mechanism and of the initial masses and excitation energies are needed in order to characterize the emitting source and follow the evolution of the GDR properties. Bracco and co-workers [26] used, in the experiment, two Parallel Plate Avalanche Counters (PPAC) to detect the reaction products in coincidence with γ-rays. Such a setup yields a measurement of the linear momentum transfer (LMT) from the projectile to the compound system. Complete-fusion events are characterized by 100% LMT.

Only such events were retained in the analysis and the gamma spectra were built accordingly. Figure 3 displays the γ-ray spectrum measured in ref. [26] which shows a clear bump associated to the GDR decay and, at higher energies, the contribution arising from bremsstrahlung γ-rays originating from nucleon-nucleon collisions in the first stage of the reaction. The full line represents the statistical model calculation performed assuming a Lorentzian shape for the GDR with 100% of the EWSR, a centroid energy of 16 MeV and a width of 13 MeV constant over the whole decay path. The dotted line is the estimate of bremsstrahlung contribution while the dashed one indicates the sum of the both statistical and bremsstrahlung contributions which nicely reproduces the whole spectrum. On the right side of fig. 3 the GDR width systematics for Sn isotopes is shown including the new result at $E^* = 230$ MeV. Its value is similar to the one extracted at $E^* = 130$ MeV suggesting a saturation of the effects which lead to the observed increase at lower excitation energy. Thermal fluctuations of the nuclear shape are expected to increase with the temperature of the emitting system and therefore the observation of a saturation suggests a different origin as the main contribution to the width increase. As already observed, angular momentum drives the nucleus towards shape modifications leading to prolate or oblate configurations which become stable at high spin. In fusion reactions the transferred angular momentum increases with beam energy reaching the maximum angular momentum a nucleus of mass $A \sim 110$ can sustain before fissioning at about $E^* \sim 100$ MeV. In fig. 3 the trend of maximum angular momenta populated in the reactions investigated is compared to the width increase in the same excitation energy region. The similarities observed in the two curves drove the authors to suggest that the angular momentum is the main effect for a width increase.

Evidence for a saturation of the width was also obtained by Enders *et al.* studying the GDR gamma decay in nuclei populated in deep inelastic reactions [27]. They studied the system ^{136}Xe $+$ ^{48}Ti at $18.5A$ MeV and measured the gamma-rays in coincidence with binary events. In order to investigate the excitation energy dependence of the GDR width, three different regions of excitation energy were selected and the gamma-ray spectra built accordingly. The results concerning the width show that a value of about 10 MeV reproduces the spectra at all excitation energies. Such a value is lower than the one measured by Bracco *et al.* [26] and the results seem to be insensitive to the particular choice of level density adopted in the statistical calculation. Further evidence for the width saturation came from the work of Hofmann *et al.* [28] who investigated the GDR properties using the 12.5 and $17.5A$ MeV ^{16}O beam impinging on the ^{118}Sn target. Assuming complete-fusion reactions, nuclei at excitation energies of 160 and 230 MeV, respectively, were populated [28]. The comparison of experimental spectra with statistical calculations indicated that a width value of 10.5–11 MeV led to a good reproduction of the data. Such values are in agreement, within the errors, with Enders' results. However, the systematics of momentum transfer

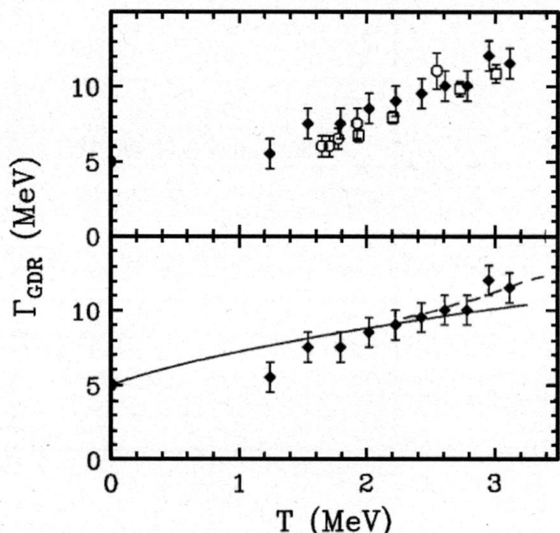

Fig. 4. Comparison of the GDR width extracted from $50A$ MeV α-particle inelastic-scattering experiment (full symbols) on ^{120}Sn [29] and from fusion reaction data (open symbols) on $^{108-112}$Sn nuclei [23–25]. The lower part shows the comparison of the α inelastic-scattering experiment results with adiabatic coupling calculations [32] shown as a full line. The dashed line includes the contribution to the width due to particle evaporation width [35].

indicates, for reactions at 12.5 and $17.5A$ MeV beam energy, an average value of 90% LMT. Calculations including corrections for incomplete momentum transfer led to differences of about 5% for the width, a value which did not affect the conclusions concerning the saturation [28]. However, this systematics of average momentum transfer was built measuring recoil velocities whose distribution becomes broader with increasing beam energy and they might not take properly into account pre-equilibrium emission which could affect the excitation energy and mass of the equilibrated system.

Once the systematics was established using fusion reactions, the next step was to attempt to disentangle the effects of the two parameters driving the width evolution, temperature and angular momentum.

A way to populate nuclei at well-determined temperatures and low angular momentum was proposed by Ramakrishnan *et al.* [29]. They used the inelastic scattering of α-particles at 40 and $50A$ MeV as a tool to populate ^{120}Sn nuclei in the excitation energies range of 30–130 MeV and low angular momentum states (about $15\hbar$ on the average) which allowed one for the first time to study the effects of large amplitude thermal fluctuations and angular momentum separately. The initial excitation energy of the target nuclei was determined from the energy loss of the scattered α-particle and the GDR evolution was followed gating on different windows of energy loss. Data analysis indicated a monotonic increase of GDR width with target excitation energy for both beam energies. Besides, the results were found in good agreement

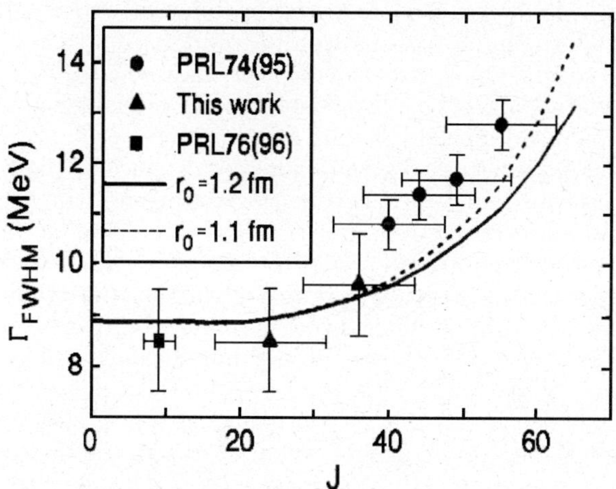

Fig. 5. Evolution of the width as a function of angular momentum at $T \sim 2$ MeV [30]. The solid line represents a calculation of the width evolution with spin assuming a moment of inertia of the nucleus equal to the rigid-rotor value while the dashed line is a similar calculation assuming a reduction of 16% in the rigid-rotor value [30].

with the existing systematics built on fusion data as shown in fig. 4. The agreement found up to $E^* = 130$ MeV in the data sets extracted using different reaction mechanisms which populate nuclei in rather different spin regions suggests that the increase in the width is mainly driven by temperature effects differently from what was previously suggested by Bracco *et al.* [26].

In order to evaluate the angular-momentum dependence of the GDR width at a fixed temperature fusion evaporation experiments were used to populate hot Sn nuclei at about $T \simeq 2$ MeV [30]. Differently from previously described fusion-evaporation experiments a multiplicity filter was used to select fusion events according to different average spin regions. The results, together with other exclusive measurements showed that the width measured at $T \simeq 2$ MeV is roughly constant up to spin $J \leq 35\hbar$ and then progressively increases up to the highest measured spin as shown in fig. 5 [30,31]. This trend is rather well reproduced by the calculations based on adiabatic theory of thermal shape fluctuations [19,32].

The results clearly suggest that the observed disagreement in the conclusions concerning the dominance of angular momentum and temperature effects on the width increase could be attributed to the different region of angular momentum investigated by the two types of experiments. In fact, the results of ref. [30] indicate that the influence of angular momentum on the width becomes really important only above $I \sim 35\hbar$, as is the case for the highest excitation-energy fusion data [26]. Such a conclusion finds a theoretical support in the results of adiabatic theory of thermal shape fluctuations which predicts the same effect.

Thermal shape fluctuation calculations give also a reasonable description of the overall GDR width increase as a function of the system temperature as shown in fig. 4.

The temperatures indicated are the initial temperatures of the compound nuclei. Taking into account a weighted average of the temperatures of nuclei contributing to the GDR gamma emission over the various decay steps would reduce these values by at most 0.6 MeV [33,34]. The inclusion of the evaporation width contribution [35] to the GDR width due to the finite width of both initial and final nuclear states involved in the GDR decay improves and extends the agreement between data and theory up to the highest E^* points (see fig. 4). However, the calculations are not able to reproduce the data trend below $T < 1.2$ MeV. The recent observation of a width close to the ground-state width at very low temperatures made the scenario a bit more confused casting some doubts on the validity of the calculations in the low-temperature region. Data from the ^{17}O inelastic scattering on ^{120}Sn [20] extracted at $T = 1$ MeV indicate that the GDR width in ^{120}Sn is 4 MeV, a value similar to the one extracted on the ground state compounding the difference with the calculation made in the framework of the standard theory of shape fluctuations [20]. This result cannot be currently explained in the framework of the thermal shape fluctuations [20,21]. Some new results were recently published also about the angular-momentum dependence of the width. In particular, an experiment on ^{86}Mo using fusion reactions did not show any dependence of the width on angular momentum which was measured to be constant up to $30\hbar$ at $T \simeq 1.3$ MeV [36]. The experimental evidence and theoretical framework discussed up to this point suggest that both angular momentum and temperature are effective in driving the nucleus towards more deformed or elongated shapes which influence the GDR width which becomes progressively broader with increasing excitation energy. At about $E^* \simeq 130$ MeV the system reaches the limiting angular momentum for a nucleus of mass $A \approx 120$ and this strongly affects the increase of the GDR width which seems to saturate. A smooth increase is instead predicted by thermal models due to the increase of the temperature effects which should lead to a $T^{1/2}$-dependence.

More recently some doubts were cast on the excitation energy determination in the fusion reactions. In particular it was pointed out that a proper determination of pre-equilibrium emission is mandatory in the estimate of the E^* of the system whose uncertainties could affect the conclusions concerning the width saturation. Recently, protons and α-particle pre-equilibrium emission has been established down to $E_{beam} = 7A$ MeV [37,38]. The measurements show that, on the average, the compound nucleus excitation energy is reduced by few percent at 7-$8A$ MeV and about $\simeq 20\%$ at $11A$ MeV using asymmetric reactions populating the $A \sim 115$–118 mass region [37,38]. At the same time, the mass of the compound system is reduced by a few units relative to complete fusion. The inclusion of the pre-equilibrium emission in the energy balance lowers the computed temperatures and increases the extracted GDR width and strength because of the lower excitation energy value used in statistical model calculations to reproduce the gamma-ray spectra. The ev-

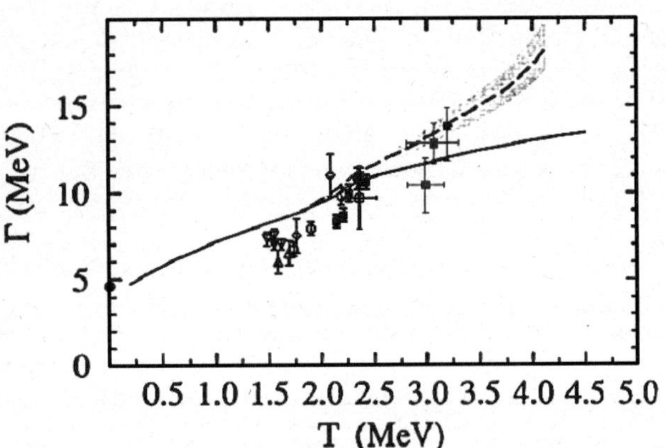

Fig. 6. Evolution of the width as a function of excitation energy applying the correction for pre-equilibrium emission to the highest E^* points [39] which were extracted in refs. [26,27].

idence of a pre-equilibrium emission already at $8A$ MeV suggested a re-analysis of the data taken above $10A$ MeV due to an overestimation of the initial excitation energy which, in some cases, was estimated assuming complete fusion. When excitation energies and temperature are re-computed including the pre-equilibrium emission the results of refs. [26,27] indicate that the width is still increasing up to temperatures $T \sim 3.2$ MeV as shown in fig. 6 [38,39]. The calculations based on adiabatic thermal shape fluctuations including the contribution to the GDR width coming from the evaporative decay support this conclusion (see fig. 6) up to $T \simeq 3$ MeV [19]. However, recent experimental findings showed a significant difference in pre-equilibrium emission between symmetric and asymmetric reactions which affects the final excitation energy of the system and, therefore, the conclusions concerning the GDR width saturation [40].

From the analysis of all the experimental findings up to an excitation energy of 200 MeV a scenario emerges where the strength of the GDR retains 100% of the EWSR and the width progressively increases due both to temperature and spin effects, the latter mainly playing a role above 35–40$\hbar$. The observed saturation of the width above about $E^* = 150$ MeV is due to limiting angular momentum from the opening of the fission channel. There is no strong evidence of a saturation of the width with increasing temperature at fixed angular momentum.

Dynamical effects start to set in at the highest end of the energy range discussed. They have non-negligible effects on the conclusions and are not completely under control. This problem will be exacerbated when moving to even higher energies in the next section.

6 Disappearance of the GDR above $E^* \sim 200$ MeV

The study of the GDR properties at high excitation energies was mainly focused in the Sn mass region where

broad systematics was collected in different experiments. The experimental investigation was undertaken by the different groups in a rather coherent way since in all experiments gamma-rays were detected in coincidence with reaction products. Typically, PPACs were used to identify incomplete-fusion events where only part of the light projectile is transferred to the target, through the simultaneous measurement of energy loss and time of flight of the residues. Broad distributions of recoil velocities were detected reflecting a range of momentum transfers leading to systems with different masses and excitation energies whose values can be estimated using a massive transfer model [41]. This is a very attractive feature of intermediate-energy heavy-ion collisions since, as long as the hot nuclei can be properly characterized, the broad excitation energy distribution measured can be used to follow the evolution of the gamma emission as a function of excitation energy in a single experiment. The recoil velocity distributions were sorted in bins corresponding to different average excitation energies and gamma spectra were built accordingly. The analysis of the gamma spectra was usually carried out using a statistical decay code which treats the statistical emission of γ-rays, neutrons, protons, alpha particles and in a few cases also a fourth particle (like a deuteron) from an equilibrated compound nucleus. The bremsstrahlung γ-ray contribution which dominates the high-energy part of the spectrum was estimated fitting the spectral shape with an exponential function above about 30 MeV. Since it gives also a sizeable contribution to the region below 30 MeV which is difficult to determine experimentally and which affects the estimate of the GDR yield, the exponential fit is extrapolated down to low energies. Eventually, this contribution was subtracted from the experimental spectrum in order to obtain the GDR gamma yield and to allow for a direct comparison with statistical calculations folded with detector response. However, even though the basic approach is the same, the authors followed, in the data analysis, different hypothesis concerning the GDR properties at high E^*, which lead, at least for some time, to controversial conclusions. In the following we will show the results of the different experiments, the procedure adopted in the analysis and eventually the comparison with theoretical models which provides the present understanding of the GDR behavior at very high temperature.

The first pioneering work to investigate the persistence of collective motion at very high excitation energies was performed by Gaardhøje *et al.* who studied the reaction ^{40}Ar + ^{70}Ge at 15 and $24A$ MeV beam energies [42]. Hot nuclei formed in incomplete-fusion reactions were populated at average excitation energies $E^* = 320$ MeV and $E^* = 600$ MeV for the two reactions. These estimates, based on average momentum transfer, did not take properly into account pre-equilibrium particle emission which affects significantly the excitation energy value in the case of the reaction at $24A$ MeV. However, even if corrections should be applied to extract a proper value of E^* for the emitting system, the general conclusions of this work remain the same, the excitation energy of the system

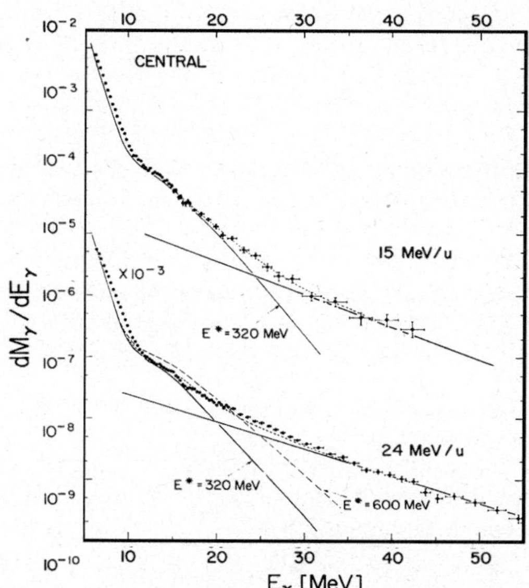

Fig. 7. Gamma spectra measured in the reactions ^{40}Ar + ^{70}Ge at 15 and $24A$ MeV. The full line represents the statistical model calculation performed at $E^* = 320$ MeV while the dashed line is a calculation assuming $E^* = 600$ MeV. Dotted lines are the sum of statistical plus bremsstrahlung contributions [42].

populated at $24A$ MeV is in any case much higher than 300 MeV.

The γ-ray spectra were reproduced assuming for the GDR a single Lorentzian shape with a centroid energy $E_{GDR} = 15.5$ MeV, a width $\Gamma = 15$ MeV and 100% of the EWSR. As it can be seen in fig. 7, the statistical calculation reproduces the gamma-ray spectrum measured at $15A$ MeV while a strong over-prediction of the GDR gamma yield is observed in the case of $24A$ MeV. These data indicated, for the first time, the existence of a suppression of the γ emission at high excitation energies [42] compared to the prediction of the statistical model which was interpreted as a loss of collectivity of the system. The spectrum measured at $24A$ MeV was found similar to the one measured at $15A$ MeV and could be reproduced assuming an excitation energy $E^* = 320$ MeV, a value much lower than the estimated one. Such an approach lead to the interpretation of the sudden disappearance of the GDR with increasing excitation energy. These observations suggested, for the first time, the existence of a limiting temperature $T \sim 4.5$ MeV for the collective motion.

Further evidence for the suppression of the γ yield at very high excitation energies was then found studying the reactions ^{40}Ar + ^{92}Mo at 21 and $26A$ MeV [43], ^{36}Ar + ^{90}Zr at $27A$ MeV [44] and ^{36}Ar + ^{98}Mo at $37A$ MeV [45]. These results could not be explained in the framework of statistical models because at higher excitation energies the number of emitted gamma-rays should increase due to the higher number of steps available for the GDR to compete with particle emission. Interest for this new problem spread through the theoretical community. Different approaches were proposed to explain the quenching of the GDR. The different ideas point to two main effects which could lead to a saturation of the GDR gamma yield at high excitation energy, either a suppression of the GDR or a rapid increase of the width. In the following section we will first describe the different theoretical models and then we will come back to a more detailed description of the experimental results.

6.1 Theoretical models: yield suppression

The statistical model used to reproduce the gamma-ray spectra emitted in the decay from a hot compound nucleus is based on the assumption that the nucleus survives long enough to reach thermal equilibrium before decaying. This hypothesis, valid for nuclei at low excitation energies, may not always be fulfilled at very high excitation energies where the time needed for the system to equilibrate the different degrees of freedom, in particular the collective ones, could become longer than the nucleus lifetime [46]. In this case the system will start to cool down by particle emission before being able to develop a collective oscillation.

The observation of a GDR quenching at high excitation energies has been interpreted by some theoreticians as a possible evidence of such pre-equilibrium effects. The time scale governing the GDR equilibration can be related to the GDR spreading width $\Gamma^\downarrow$. Since the particle evaporation width Γ_{ev} increases as a function of the temperature according to the statistical model predictions, as shown in fig. 8, the existence of the GDR above a certain excitation energy depends on the relative size of the spreading and evaporative widths [47]. The model suggested by Bortignon *et al.* [47] is based on the assumption

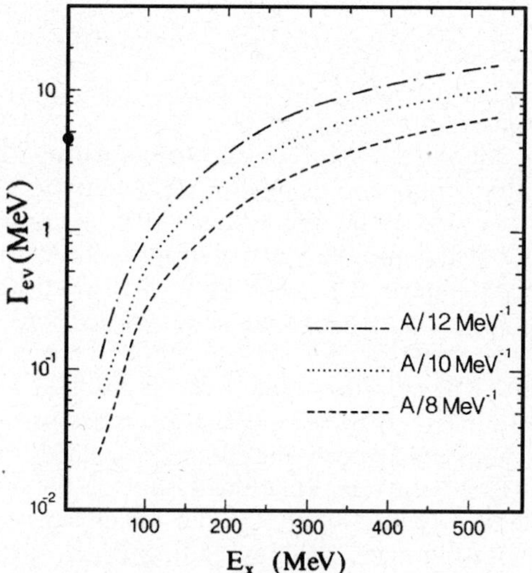

Fig. 8. Particle evaporation widths as a function of excitation energy estimated in the framework of the statistical model for three values of the level density parameter ranging from $a = A/8$ to $a = A/12$ [47]. The full symbol represents the value of the $\Gamma^\downarrow$ of the GDR measured on the ground state in ^{108}Sn.

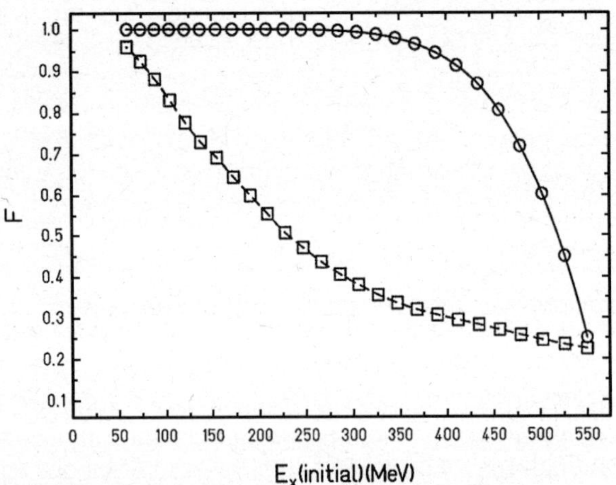

Fig. 9. Comparison of GDR suppression factors as a function of excitation energy for a system with $E^* = 550$ MeV. Circles indicate the results obtained estimating the suppression factor according to eq. (10) while squares are obtained using the relation given by eq. (8) [39].

that the compound nucleus states can exist in two different classes, with or without the GDR. Assuming that GDR states are not populated at the beginning of the reaction, the excitation energy at which the spreading width and evaporative width are comparable $\Gamma_{ev} \sim \Gamma^\downarrow$ defines a critical temperature for the existence of collective motion [47]. Above this temperature, in fact, the compound nucleus will evaporate particles before the GDR can be present in thermal equilibrium reducing its temperature. This affects the GDR yield which will be reduced by an amount related to the time needed to develop the collective oscillation relatively to the particle decay time. The model predicts a hindrance factor for the GDR emission dependent on excitation energy given by

$$F = \frac{\Gamma^\downarrow}{\Gamma^\downarrow + \Gamma_{ev}}, \qquad (8)$$

where Γ_{ev} increases rapidly with temperature. The fulfilment of the condition $\Gamma_{ev} \geq \Gamma^\downarrow$ relies on the temperature dependence of the spreading width as compared to the particle decay width. Since a suppression of the GDR was observed above $E^* \simeq 250$ MeV this was interpreted by Bortignon and co-workers as an indication that, at this excitation energy which corresponds to a temperature $T \sim 5$ MeV, the condition $\Gamma_{ev} \geq \Gamma^\downarrow$ is fulfilled. Comparing the value $\Gamma^\downarrow$ measured on the ground state which is about 4.5 MeV to Γ_{ev} calculated at $E^* = 250$ MeV which is ~ 5 MeV the authors concluded that $\Gamma^\downarrow$ is essentially independent of temperature. Similar conclusions concerning the independence of the spreading width of the temperature can be found in the theoretical work of Donati et al. [48].

It has been noted that the pre-equilibrium effects might be overestimated in the preceding model due to the hypothesis of a complete absence of population of GDR states at the beginning of the reaction on which the model

is based. In fact, the existence of some initial dipole oscillations in the fused system due to the long equilibration time of the charge degree of freedom has been theoretically investigated together with the excitation energy dependence of the spreading width. The calculations suggest that a suppression or an enhancement of the gamma emission could be observed depending on the initial conditions of the system out of equilibrium [49].

More recently the effect of the equilibration time for different degrees of freedom has been re-investigated [39]. Assuming that the equilibration of the collective vibration occurs with a probability given by

$$P(t) = 1 - \exp(-\mu_0 t), \qquad (9)$$

where $\mu_0 = \Gamma_0/\hbar$ is a characteristic mixing rate related to the spreading width and t is the time elapsed in the decay process, one can estimate the inhibition factor for the GDR decay et each step of the decay. At the first step the time to consider in eq. (9) will be the mean lifetime of the compound nucleus $t_{ev} = \hbar/\Gamma_{ev}$. For the n-th decay step the probability will be modified by the elapsed time which can be estimated as $t \sim \sum_{i=1}^{n} t_{ev}(i)$, where $t_{ev}(i)$ is the mean lifetime for the i-th decay step. Then the suppression factor is reduced and becomes [39]

$$F_n \sim 1 - \exp\left(-\Gamma_0 \sum_{i=1}^{n} \Gamma_{ev}(i)^{-1}\right). \qquad (10)$$

The comparison of this suppression factor with the one predicted by eq. (8) for a compound system with $E^* = 550$ MeV, mass $A = 110$, $\Gamma_0 = 4$ MeV and assuming $A/a = 11$ is shown in fig. 9. The different points in the figure are computed assuming an energy release per decay step given by $\Delta E = B_n + 2T$, where $B_n \approx 9$ MeV and $T = \sqrt{(E/a)}$ [39]. As it can be observed, the two suppression factors are similar at the first step but then, during the de-excitation process, eq. (10) predicts a rapid decrease of the suppression which becomes negligible already around $E^* = 300$–350 MeV while eq. (8) gives still a not negligible suppression at $E^* \sim 100$ MeV.

A different origin of the suppression of the GDR γ emission was suggested by Chomaz [50]. In his model the explanation of the quenching effect is again related to the different time scales which come into play in the emission process. Differently from the previous approach, he suggests to take also into account the period of one oscillation of the emitting system given by $T_{GDR} = 2\pi/E_{GDR}$. In fact, in order to be able to emit characteristic photons the system needs to make at least one full oscillation without perturbation of its dipole moment. Conversely, the associated spectrum cannot show a characteristic frequency. Since particle emission can induce fluctuations of the dipole moment the times which come into play and compete are the time between the sequential emission of two particles t_{ev} and the period of one collective oscillation T_{GDR}. The condition $t_{ev} \simeq T_{GDR}$ defines the threshold towards a chaotic regime where the collective oscillation is suppressed. The probability to make at least one oscillation can be computed and a GDR quenching factor can

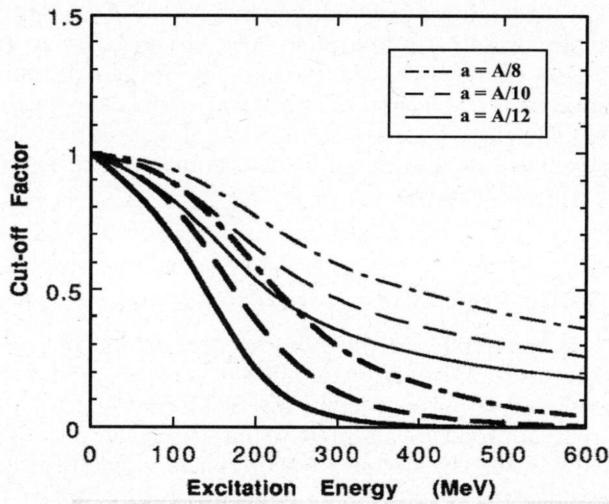

Fig. 10. Evolution of quenching factors predicted for Sn nuclei as a function of excitation energy for different values of the level density parameter. Thick lines show the reduction factor predicted in ref. [50] while thin lines are predictions made according to ref. [47].

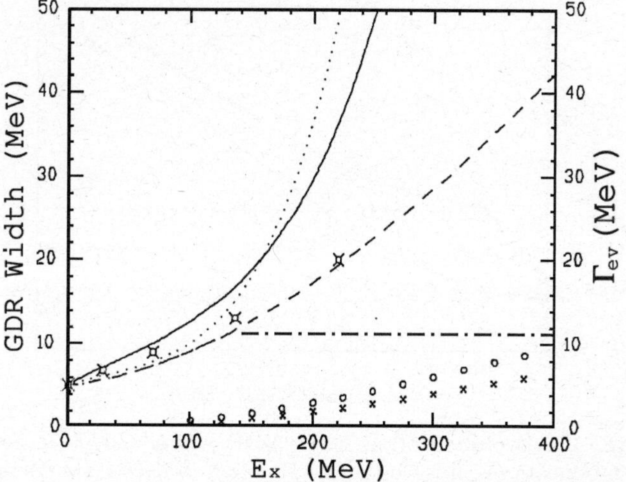

Fig. 11. Excitation energy dependence of the GDR width for Sn isotopes. Open square symbols represent the results of the calculations of Bonasera *et al.* [53]. The dashed line is the extrapolation to higher energies of the width parametrization given by eq. (6). The solid line is the estimate for the width increase according to Smerzi *et al.* [51], the dotted one is a fit of the width trend extracted from ref. [43] and the dot-dashed one is a parametrization assuming a constant width of 11 MeV above $E^* = 130$ MeV. In the same figure the Γ_{ev} values calculated for level density parameters $a = A/12$ (circles) and $a = A/10$ (crosses) are shown.

be extracted. It depends on the resonance energy and the evaporative width according to the relation

$$F = \exp\left(\frac{-2\pi\Gamma_{ev}}{E_{GDR}}\right). \qquad (11)$$

The excitation energy dependence of the GDR suppression factor is shown in fig. 10 for Sn nuclei. In the same figure the suppression factor proposed in ref. [47] is shown as a comparison. The effect of different values of the level density parameter on the suppression factor is also shown. Chomaz's approach to explain the GDR quenching leads to a suppression factor whose effects are much stronger than those predicted in ref. [47]. In particular, a sizeable quenching is predicted already between 150–200 MeV excitation energy, an excitation energy region where the GDR was measured to retain 100% of the EWSR.

6.2 Theoretical models: width increase

A completely different interpretation of the quenching effect was developed following the idea of a GDR width strongly increasing with the temperature. This argument is not in disagreement with the apparent saturation of the width at about 12–13 MeV observed above 250 MeV excitation energy in different experiments and, as we will see, its implication should give a clear signature in the gamma-ray spectrum which is not predicted by models which interpret the GDR quenching in terms of yield suppression. Such difference will become the key issue to segregate between the two theoretical interpretations of the GDR quenching.

In the attempt to reproduce the experimental data two different explanations leading to a rapid width increase at high excitation energy were put forward. Following a

semiclassical approach solving the Vlasov equation with a collision relaxation time, Smerzi *et al.* [51] studied the interplay between one- and two-body dissipation on the damping of collective motion. They evaluated the escape width $\Gamma^{\uparrow}$ and the spreading width $\Gamma^{\downarrow}$ contributions as a function of temperature. The escape width was found to be of the order of few hundred keV while a strong increase of the spreading width was observed as a function of the temperature [51]. Such an effect is due to two-body collisions which become increasingly important with temperature because of the suppression of Pauli blocking.

The excitation energy dependence of the GDR width for Sn isotopes is shown in fig. 11 as open squares and can be described reasonably well by the dashed line which is an extrapolation to higher energies of the Chakrabarty parametrization for the width found at lower excitation energies [52,53]. At about $E^* \simeq 230$ MeV the calculations predict a GDR spreading width of the order of the resonance energy and the contribution to the gamma-ray spectrum around the GDR energy becomes small. In fact, the γ-rays are spread out over a very large decay energy range. Therefore, the conclusion is that the GDR progressively disappears with excitation energy due to this broadening of the resonance. This interpretation should be able to explain the quenching of the γ yield and paradoxically is not in contradiction with the *apparent* width saturation. The analysis of the spectral shape in a region above the resonance should reveal the contributions not present at lower excitation energies. This part of the spectrum then

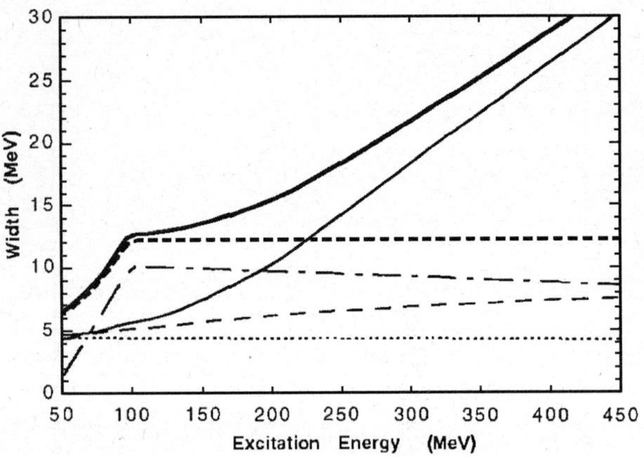

Fig. 12. Evolution of the total GDR width and its various components in Sn isotopes as a function of excitation energy. The thick solid line displays the total width predictions according to the parametrization suggested in ref. [54] including the particle evaporation width contribution [35]. The thick dashed line represents the standard prediction assuming the saturation at 12 MeV. The thin lines show the contribution to the width due to the various components. In particular, the thin solid line represents the intrinsic width including the particle evaporation width contribution while the dotted one shows the intrinsic width.

becomes of great importance to draw conclusions concerning the validity of the different models.

A different idea based on the width increase was proposed by Chomaz to explain the observed saturation of the yield [35,50]. The key issue is that each nuclear level involved in the GDR gamma decay has a finite lifetime τ due to particle evaporation. The value of the lifetime induces, according to the Heisenberg uncertainty principle, a width of each nuclear level of the order of $\hbar/\tau$. Therefore the transition energies between nuclear levels, like gamma-ray energies, cannot be determined to better than $2\hbar/\tau$. This indetermination affects the total width of the resonance but not the position of the centroid. Assuming for each nuclear level a width equal to the evaporation width of the compound system Γ_{ev} the total width of the GDR should contain the contributions coming from the spreading width and the natural width of the elementary gamma transition according to the relation [35]

$$\Gamma_{GDR} = \Gamma^{\downarrow} + 2\Gamma_{ev}. \tag{12}$$

While at low excitation energies the contribution arising from Γ_{ev} is negligible the statistical model predicts a strong increase of the particle evaporation width with excitation energy. In fig. 12 the evolution of the total GDR width calculated including the evaporation width effect is shown as a full thick line. In the same figure the contribution arising from the $2\Gamma_{ev}$ term is shown as full thin line. The comparison with the prediction assuming a saturating width according to the experimental observation shows that the new contribution starts to be significant in the region of $E^* \simeq 150$–200 MeV, becoming the dominant one above $E^* \sim 300$ MeV.

The strong increase predicted by both models could, in principle, explain the disappearance of the GDR at high excitation energies but the comparison of the experimental data with the results of statistical calculations including model prescriptions will show significant discrepancies which cannot be accounted for assuming a width increasing with temperature.

6.3 Evidence for the yield saturation

Now we come back to the experimental evidences for the yield saturation discussing in detail the results of different experiments performed in the Sn region together with the different approaches adopted to interpret the data. Historically, after the first evidence for the yield saturation observed by Gaardhøje, this issue was re-investigated at RIKEN by studying the gamma-ray spectra measured in coincidence with evaporation residues produced in the reactions ^{40}Ar + ^{92}Mo at 21 and $26A$ MeV [43]. At these bombarding energies incomplete fusion is the dominant reaction mechanism for central collisions and, therefore, the characterization of the emitting source becomes rather complex. Two methods were used to determine the excitation energy of the system: one based on the recoil velocities and the other on the measurement of neutron spectra. Gates on recoil velocity were applied to select nuclei with different average excitation energies whose values were estimated using a massive transfer model. Neutron and gamma-ray spectra were built accordingly. Neutron spectra were analyzed assuming the emission from two moving sources, one associated to the compound nucleus and the other to pre-equilibrium [55]. The results showed that both the temperature and the multiplicity of neutrons emitted from the compound nucleus source increase smoothly as a function of residue velocity [43,55] supporting the interpretation of a statistical emission from an equilibrated system formed at progressively higher excitation energy.

Gamma-ray spectra were extracted for both reactions and all velocity bins. The GDR gamma yield, integrated in the region $12 \leq E_{GDR} \leq 20$ MeV after bremsstrahlung subtraction, was observed to be almost constant, within the error bar, in the whole region above 250 MeV excitation energy [43,56], see fig. 13. The spectra were then analyzed using the standard statistical calculation assuming for the GDR a centroid energy $E_{GDR} = 15.5$ MeV, a width $\Gamma_{GDR} = 20$ MeV and full strength of the EWSR. The comparison clearly showed that the statistical calculation strongly overshoots the data in the GDR region.

In order to reproduce the spectra the authors proposed to include the energy dependence of the GDR width in the statistical calculation [43]. They showed that, taking into account the width variation at each step of the decay process, statistical model calculations were able to reproduce the γ-ray spectra at different E^* without introducing a reduction of the EWSR strength above a critical excitation energy. This was a really new approach since, traditionally, each single calculation was performed assuming a width constant during the de-excitation process. The inclusion of the excitation energy dependence

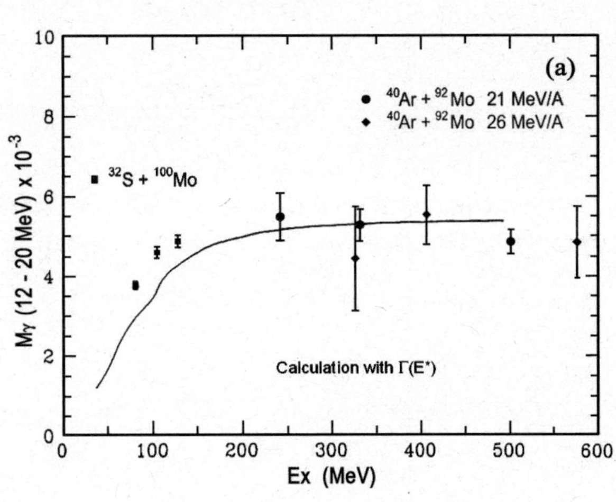
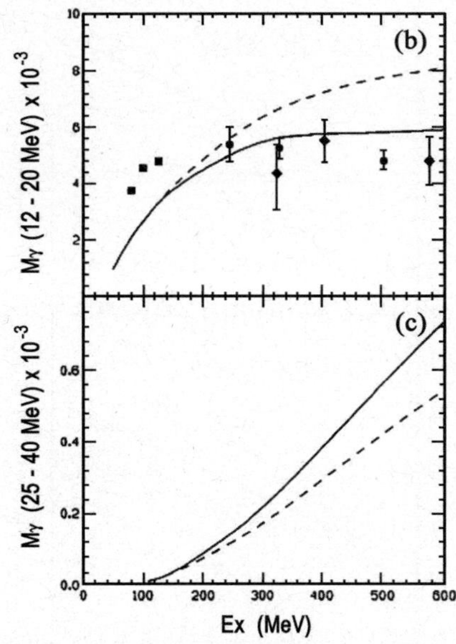

Fig. 13. a) Evolution of the gamma yield integrated in the region between 12 and 20 MeV as a function of excitation energy for different reactions. Circles correspond to $21A$ MeV data, diamonds to $26A$ MeV data while squares are from the reaction ^{32}S + ^{100}Mo at 150, 180 and 210 MeV beam energies [56]. The solid line represents the calculation of the γ yield, integrated in the same energy region, according to the parametrization adopted in ref. [43]. b) Comparison between the same set of data and the theoretical predictions according to Smerzi *et al.* [52], (solid line) and Bortignon *et al.* [47] (dashed line) c) Same model calculation as above but in the energy region 25–40 MeV.

of the width in the calculation produces a spread of the strength function outside the GDR peak region at high E^* leading, therefore, to a quenching of the gamma yield in the GDR region. Figure 13 shows, as a solid line, a calculation of the gamma yield integrated in the region $12 \leq E_\gamma \leq 20$ MeV including the energy dependence of the width. The data trend is rather well reproduced and the differences observed at low excitation energies can be ascribed to a different initial mass of the emitting system and to a different trigger adopted [56]. While a first analysis of the spectra suggested a strong dependence of the width on E^* [43], when the effect of equilibration time was taken into account in the calculation including the factor $\Gamma_{GDR}/(\Gamma_{GDR} + \Gamma_{ev})$ [47,57] a good fit was obtained with an energy dependence very similar to the one found by Chakrabarty at lower excitation energy [56,57].

The analysis in terms of a strongly increasing width found a significant theoretical support in a series of works where a strong width increase with temperature due to 2-body collisions was predicted [51–53]. Using this model the authors were able to reproduce the overall trend of the γ yield [52,53] shown as a solid line in the right panel of fig. 13. The saturation around $E^* = 250$–300 MeV is also reproduced leading to a corresponding limiting temperature $T \simeq 4$ MeV for the GDR in the mass region $A \sim 120$.

Therefore, while there was an agreement between Kasagi *et al.* and Gaardhøje *et al.* data on the GDR quenching and on the existence of a limiting temperature $T \simeq 4$ MeV for the collective motion, the different hypotheses adopted in the analysis led to controversial conclusions concerning the reasons of the GDR quench-

ing. The question how and why the GDR disappears was still open and the answers were found later, in the region of the spectrum above the resonance. In fact, the spread of the GDR strength function at high excitation energies predicted by a strong width increase affects the high-energy part of the spectrum where a sizeable difference in the spectral shape should be observed comparing constant width and increasing width prescriptions. In particular, a higher yield is predicted in the region $E_\gamma \geq 25$–40 MeV by calculations including a width increase as shown in the lower panel on the right of fig. 13. This region of the spectrum is rather difficult to analyze experimentally due to the presence of a significant contribution from np bremsstrahlung emission which dominates the γ-ray spectrum above 35 MeV. The bremsstrahlung contribution has to be evaluated and subtracted from the spectrum to allow for a proper determination of the GDR gamma multiplicity and to constrain different theoretical interpretations. The evaluation is typically done fitting with an exponential function the high-energy part of the spectrum ($E_\gamma \geq 30$–35 MeV) and then extrapolating the fit down to lower energies. High statistics is needed to allow for a precise determination of the slope of the bremsstrahlung component which is the crucial ingredient in the data analysis since it strongly affects the gamma yield determination. The limited statistics of the RIKEN data in the region above 25 MeV may have affected the proper determination of the bremsstrahlung contribution precluding a correct comparison of the spectral shape with statistical model calculations in this energy domain.

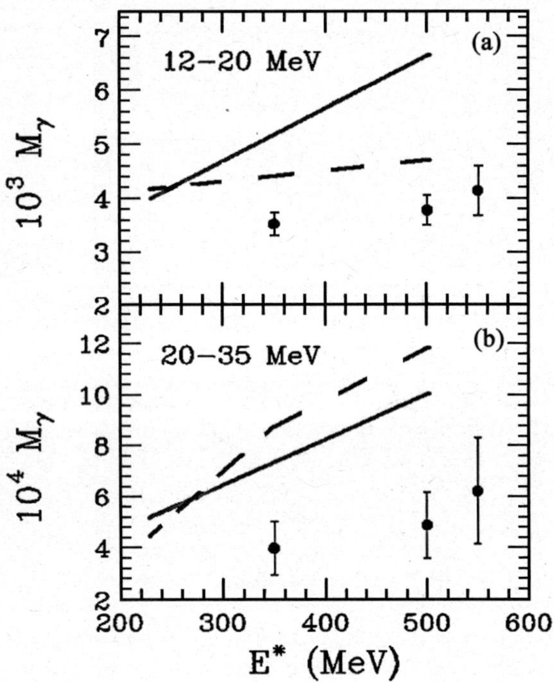

Fig. 14. a) Evolution of the gamma yield integrated in the region between 12 and 20 MeV as a function of excitation energy for the reaction ^{36}Ar $+$ ^{90}Zr at $27A$ MeV [61]. Full symbols are experimental results, the full line is the prediction of the standard statistical calculation while the dashed one is the prediction of a statistical calculation including a parametrization for the width given by eq. (6) [25,61]. b) Same comparison as above but in the energy region 20–35 MeV.

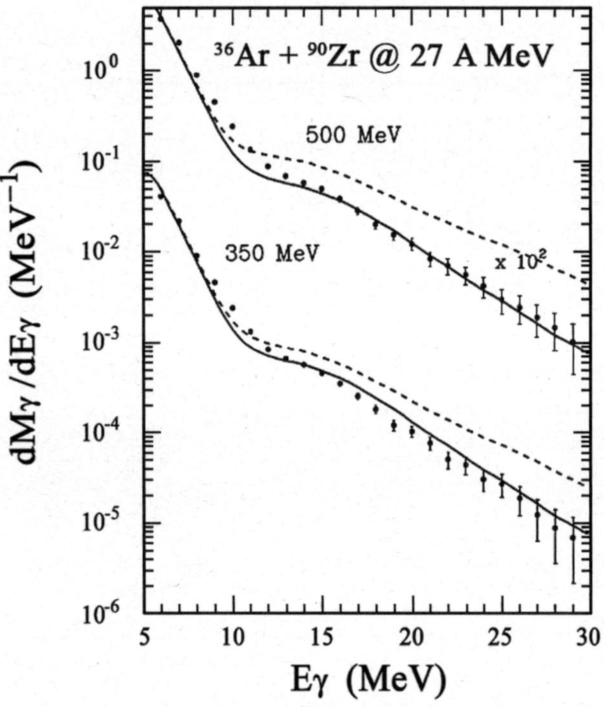

Fig. 15. Comparison of the gamma-ray spectra extracted for 350 and 500 MeV excitation energy bins after bremsstrahlung subtraction with statistical calculations [61]. The dashed lines represent the standard calculations while the full lines are the calculations including the suppression of the GDR emission above $E^* = 250$ MeV.

A clear answer to the open questions concerning the width and the strength behavior at high E^* was obtained in a set of experiments performed with the MEDEA detector [58] at GANIL and more recently at the LNS-Catania. In the GANIL experiments ^{36}Ar beams at 27 and $37A$ MeV impinging, respectively, on ^{90}Zr and ^{98}Mo targets were used to populate hot nuclei at excitation energies above 300 MeV [44,45]. The characterization of the hot nuclei was obtained through a complementary analysis of the recoil velocities and the study of light charged particle spectra [44,59]. The gamma spectra corresponding to the decay of systems with different average excitation energies were analyzed and the integrated gamma yield was observed to be almost constant within the error bar in the whole E^* region for each beam energy. The top panel in fig. 14 shows the gamma yield integrated in the region 12–20 MeV for $27A$ MeV data for the three excitation energy bins investigated. Slightly lower values were observed in the $37A$ MeV data.

The analysis of the spectra based on the comparison with standard statistical model calculations assuming a single Lorentzian shape for the GDR with centroid energy parametrized by $E_{GDR} = 76/A^{1/3}$, a constant width $\Gamma = 12$ MeV, a strength equal to 100% of the EWSR and a level density parameter dependent on the temperature [60] indicated a GDR quenching in both reactions. The sim-

plest way to reproduce the data was to introduce a sharp suppression of the gamma emission above a given excitation energy, the so called cut-off energy. In the analysis of the $27A$ MeV data the authors reproduced the spectra extracted at all the excitation energies using the same cut-off value of 250 MeV as shown in fig. 15 [61]. A slightly lower cut-off value was needed in the case of the $37A$ MeV data.

In order to constrain the different theoretical interpretations and find a definitive answer concerning how and why the GDR disappears, statistical calculations including the different model prescriptions were performed and compared to the spectra. The results of the calculation, shown in fig. 16 [61], clearly indicate that models including a continuously increasing width while leading to a decrease of the yield near the centroid of the resonance clearly fail to reproduce the high-energy part of the spectra both in yield and slope. Conversely the smooth cut-off prescription based on equilibration time effects suggested in ref. [47] gives a reasonable reproduction of the data. However, recently, Snover showed that this form of the cut-off while being valid at the first step of the decay process actually overestimates the inhibition over the entire decay chain [39]. The calculation including the modified smooth cut-off (see eq. (10)) led to a larger discrepancy between data and model [39].

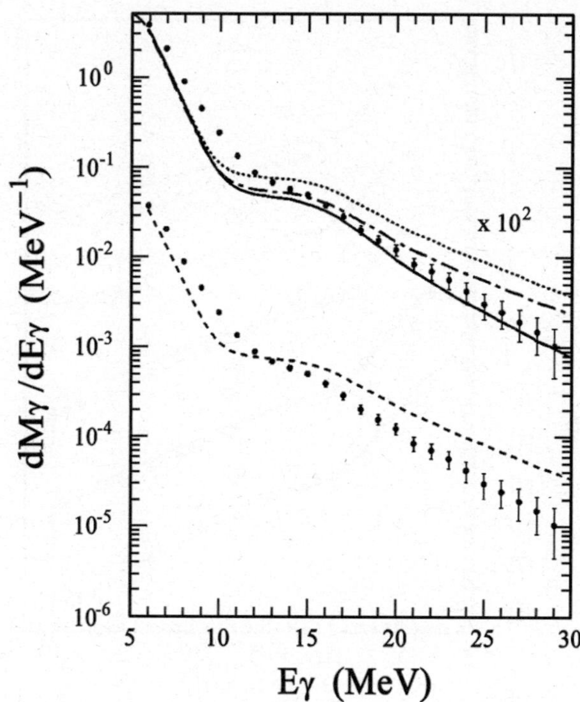

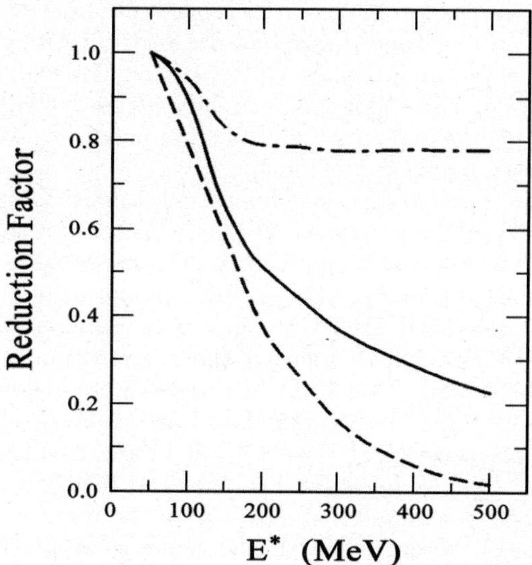

Fig. 17. Predictions for the GDR quenching factor as a function of E^* [61] according to the models of Bortignon *et al.* [47] (solid line), Smerzi *et al.* [51] (dot-dashed line) and Chomaz [50] (dashed line).

Fig. 16. Top: comparison between spectra extracted at 500 MeV excitation energy in the reaction ^{36}Ar + ^{90}Zr at $27A$ MeV [61] with statistical calculations including model prescriptions of Bortignon *et al.* [47] (solid line), Smerzi *et al.* [51] (dot-dashed line) and Chomaz [50] (dotted line). Bottom: same spectrum compared with the prediction of a statistical calculation including a width increasing with E^* according to eq. (6) shown as a dashed line.

The effect of the increasing width on the spectral shape can be evaluated in fig. 14 where the experimental integrated yield in the regions 12–20 MeV and 20–35 MeV are compared to the predictions of a statistical calculation including the parametrization given by eq. (6) (shown as a dashed line). As a reference, the yield according to the standard statistical calculation is also reported in the same figure as a solid line. The figure unambiguously shows that, while in the GDR peak region the calculation with an increasing width lies slightly above the data, this is no longer the case in the region 20–35 MeV where it predicts an increase even larger than the standard statistical calculation. Similar consideration holds for the slope of the spectrum calculated above 20 MeV after bremsstrahlung subtraction [61]. The reasons can be found in the statistical dipole emission rate formula (eq. (4)) where two ingredients contribute to the observed effect. The first is the level density ratio which is roughly proportional to $\exp\left(-E_\gamma/T\right)$ and with increasing temperature tends to increase the γ multiplicity at higher energies by decreasing the slope of the spectrum. The second is the factor E_γ^2 which multiplies the Lorentzian representing the GDR strength function. It shifts the γ yield to higher energies when the GDR width increases. Therefore, the overall effect, as already observed, is to induce a shift in the yield

rather then a quenching and this is not observed in the data. Such considerations hold for all models including a width increase in the calculation. The important conclusion of the work is that the GDR gamma-ray saturation is consistent with a disappearance of the GDR strength above E^* about 250 MeV. This led the authors to conclude that $E^*/A \sim 2.5$ MeV represents a limit for the existence of the dipole vibration for $A \sim 110$ nuclei [44].

Similar considerations hold also for $37A$ MeV data even if a slightly lower gamma multiplicity was observed. The comparison of the average multiplicity measured in the 27 and $37A$ MeV reactions with RIKEN data which were extracted in the same region of E^* but at lower beam energies indicates a significant decrease of the γ yield with bombarding energy suggesting the existence of a dynamical effect which could influence the equilibration time of the hot source and the development of collective motion [45]. However, a different pre-equilibrium emission among the reactions, not always properly evaluated, could lead to emitting sources with different average masses and charges, therefore affecting the emission probability which depends on $N \cdot Z/A$ of the emitting system. Besides, the comparison of the gamma yield between experiments performed using different experimental setups can be biased by the different response function of the detectors. These considerations may weaken the conclusion of a dependence of the γ-ray yield on beam energy.

A few other elements of this complicated puzzle remain still unexplained. In particular, the mechanism that suppresses the collective motion at high excitation energies is still unclear as well as the exact energy region where the quenching appears. In fact, in the GANIL experiments, all the systems were populated at excitation energies above the cut-off energy of 250 MeV, precluding a detailed study of the onset of the quenching. Besides, the introduction of a sharp cutoff approximation to reproduce the data, while pointing to a sudden disappearance of the GDR gamma emission does not preclude the existence of a progressive quenching of the GDR yield already below 300 MeV excitation energy which is actually predicted by different models (see fig. 17).

It thus appeared important to investigate a region of lower excitation energies where the saturation of the yield was expected to set in and to map the progressive disappearance of the GDR. The experiment performed at the LNS-Catania with MEDEA coupled to Superconductive Solenoid SOLE which focused the evaporation residues on the focal plane MACISTE [62] investigated the excitation energy region between 160 and 290 MeV through the study of the reactions ^{116}Sn + ^{12}C at 17 and 23A MeV and ^{116}Sn + ^{24}Mg at 17A MeV. The choice of reactions with a strong mass asymmetry was driven by the need to reduce the spread in momentum transfer which leads to a better determination of the excitation energy of the system. The reverse kinematics were used to better match the SOLE acceptance. A single velocity window centered around the center-of-mass velocity was selected for each reaction and gamma-ray spectra were built accordingly. The spectra were compared to standard statistical calculations assuming a fixed width $\Gamma = 12$ MeV, 100% of EWSR and a centroid energy $E_{GDR} = 76/A^{1/3}$ similarly to what was previously done for the reactions at 27 and 37A MeV.

The results shown in fig. 18 indicate that while the spectra up to $E^* = 200$ MeV are remarkably well reproduced by the calculation over almost six order of magnitude this is no longer the case for the spectrum at $E^* = 290$ MeV where the calculation slightly overshoots the data. In the same figure two spectra from the reaction at 37A MeV measured at $E^* = 350$ and 430 MeV are shown as a comparison together with the corresponding calculations. The overall set of data shows a clear evolution of the GDR yield with E^* from the low excitation energy domain where the statistical scenario provides a good description of the data to a region of excitation energies exceeding 300 MeV where the GDR quenching becomes progressively more pronounced suggesting that the critical region for the onset of the GDR quenching is between 200 and 290 MeV in nuclei of mass $A \sim 110$–130. All evidence collected points to a limiting excitation energy of about 250 MeV for the existence of collective motion which corresponds to a limiting $E^*/A \sim 2.5$ MeV. Above such a value a rather strong suppression of the GDR gamma emission is observed. This effect cannot be explained by a continuous increase of the width. The reason of the suppression has to be found in the competition between the development of collective motion and particle decay.

6.4 Hot GDR disappearance in nuclei of mass $A \sim 60$–70

Since Giant Dipole Resonances are a general feature of all nuclei it is important to investigate other mass regions to study the evolution of their main features. In the following, we will concentrate on the high-temperature region

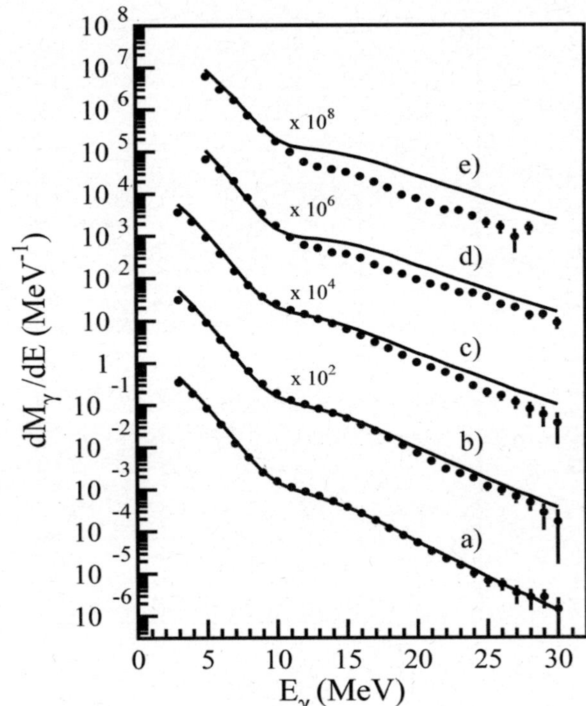

Fig. 18. Gamma-ray spectra measured at 160, 200, 290, 350 and 430 MeV excitation energy (a-e) in coincidence with evaporation residues. The spectra at $E^* = 160$ and 200 MeV are from the reactions ^{116}Sn + ^{12}C at 17 and 23A MeV, the one at $E^* = 290$ MeV is from the reaction ^{116}Sn + ^{24}Mg at 17A MeV while the spectra at higher excitation energies were measured in the reaction ^{36}Ar + ^{98}Mo at 37A MeV. Solid lines represent the corresponding CASCADE calculations performed assuming 100% of the EWSR and a constant width $\Gamma = 12$ MeV.

where further evidence for a saturation of the γ yield was recently observed in the mass region $A \sim 60$–70. The first information concerning the features of the GDR built on the ground state were collected in the early seventies, as in the case of the mass $A \sim 120$, through photo-neutron reaction studies [6].

The properties of the GDR built on excited states were then investigated in detail through the study of ^{59,63}Cu nuclei [63,64]. Different entrance channels and excitation energies were investigated in order to disentangle the effects driven by spin and temperature on the width and the energy of the resonance [63,64]. The collected systematics up to $E^* = 100$ MeV shows a centroid energy remarkably stable with temperature while the width increases from about 6–7 MeV in the ground state, depending on the isotope, up to about 15 MeV [1,63,64].

More recently the study of the reactions ^{40}Ca + ^{48}Ca and ^{40}Ca + ^{46}Ti at 25A MeV performed at the LNS-Catania with the TRASMA detector [65] demonstrated the existence of a limiting temperature for the collective motion in systems of mass $A \sim 60$ [66,67]. In this experiment pre-equilibrium γ-rays were also investigated [67] and a detailed description of this topic can be found in refs. [68–72]. Heavy residues populated at about $E^* =$

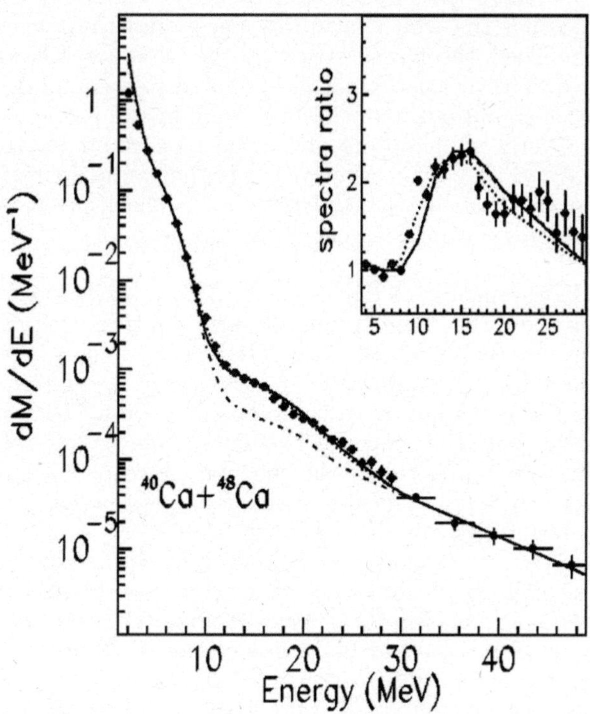

Fig. 19. Gamma-ray spectrum measured in the reaction ^{40}Ca + ^{48}Ca at $25A$ MeV in coincidence with evaporation residues. The full line is a calculation assuming a cut-off energy for the GDR emission at $E^* = 260$ MeV and a width $\Gamma = 15$ MeV. The dashed line is a calculation assuming a smooth cut-off expression according to [47]. The dotted line is instead obtained assuming again a cut-off energy at $E^* = 260$ MeV but a mass $A = 70$ for the emitting system. The dot-dashed line is a GDR zero strength calculation used to linearize the experimental data and calculations shown in the inset.

330–350 MeV were detected in coincidence with gamma-rays whose spectra were compared to statistical calculations assuming a centroid energy $E_{GDR} = 16.8$ MeV, 100% of EWSR and a width $\Gamma = 15$ MeV kept constant all along the decay process. The comparison provided evidence for a quenching of the yield similarly to what was previously observed in the mass region $A \sim 120$. In order to reproduce the data on ^{48}Ca the authors introduced a sharp suppression of the GDR gamma emission above $E^* = 260$ MeV corresponding to a $E_{cut\text{-}off}/A \simeq 4.7A$ MeV [66,67]. The statistical calculation shown as a solid line in fig. 19 nicely reproduced the whole spectrum. A smaller value for the cut-off energy was needed in the case of the ^{46}Ti target. The authors also investigated the effect on the cut-off of including a width dependent on excitation energy. A width increasing up to the saturation value of 15 MeV reached at $E^* = 100$ MeV, was used in the calculation and a cut-off energy of 240 MeV was extracted [67].

More refined calculations including the prescriptions of different smooth cut-offs were also performed. A good description of the data was obtained for both reactions adopting the smooth cut-off suggested by Bortignon *et al.* [47] which led to a slightly higher values of the cut-off energy compared to the sharp cut-off approximation. In particular, assuming a cut-off energy corresponding to $\Gamma^{\downarrow}/(\Gamma^{\downarrow} + \Gamma_{ev}) = 1/2$, values of about $5.4 \pm 0.5A$ MeV and $4.7 \pm 0.9A$ MeV were extracted for ^{48}Ca and ^{46}Ti targets. Figure 19 includes as a dashed line the statistical calculation performed using the smooth cut-off expression of ref. [47]. No difference with the calculation using a sharp cut-off approximation can be observed. Other prescriptions were investigated including the one assuming a width continuously increasing with excitation energy but a poorer agreement with data was found [67] confirming the results previously observed in the mass region $A \sim 110$–130. The important conclusion of this work concerns the first evidence for a limiting excitation energy for the GDR excitation in $A \sim 60$–70 nuclei. Its value of about 5 MeV/nucleon differs significantly from the one measured for nuclei in the mass region $A \sim 110$–130 and suggests the existence of a mass dependence of the limiting temperature for the excitation of collective motion.

7 Mass dependence of the limiting temperature

The study of the liquid-gas phase transition in nuclear matter represents an issue widely investigated during the last few years. It has been proposed that the presence of collective states can be a signature of the existence of a compound nucleus and that the disappearance of the GDR could be a further evidence for a phase transition in nuclei [49,73]. In particular, the GDR disappearance at high excitation energies gives access to the maximum excitation energy at which nuclei can still show a collective behavior. This energy can give complementary information to the caloric-curve studies which provide important information concerning the existence of a liquid-gas phase transition. Recently, the analysis of the nuclear caloric curve for nuclei in different mass regions has shown evidence for the existence of a plateau at high excitation energies which represents the region of the equilibrium phase coexistence between liquid and vapor. The limiting temperature represented by the plateau has been observed to decrease as a function of the nuclear mass [15]. This affects the excitation energy value at which the plateau appears which decreases with mass as shown in fig. 20 [15].

Interesting similarities with this trend were found studying the limiting excitation energy for the collective motion. In fact, the results indicate a decrease of the maximum excitation energy for the collective motion from about 5 MeV/nucleon for nuclei of mass $A \sim 60$–70 to about 2.5 MeV/nucleon for nuclei in the mass region $A \sim 110$. Moreover, the values of the excitation energies extracted in both mass regions are close to the energies where the plateau of the caloric curve appears (see fig. 20). This intriguing feature suggests the possible occurrence of a transition from order to chaos in nuclei for excitation energies close to the values where signals of a liquid-gas

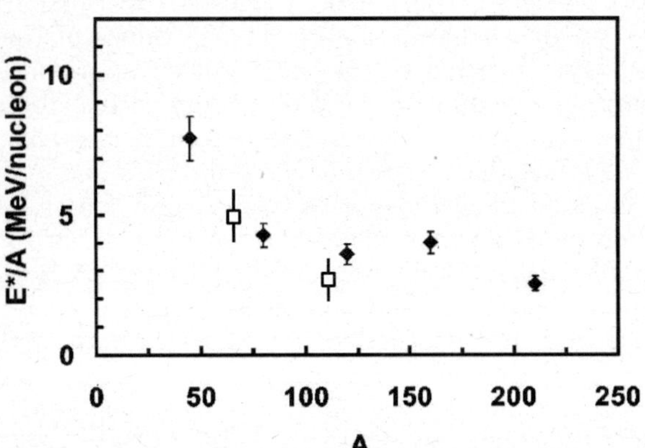

Fig. 20. Excitation energy per nucleon at which the limiting temperature is reached as a function of the system mass [15]. Open symbols are the limiting excitation energies per nucleon for the collective motion extracted in two mass regions.

phase transition were claimed to be present. The link between the two observations deserves further investigations.

8 Conclusions and perspectives

In the last twenty years of investigation, the main properties of the GDR built on excited states have been measured and understood. The most complete systematics was studied for medium mass nuclei around $A \sim 110$–120 for which measurements were performed for excitation energies between 10 and 500 MeV and spins up to $J \sim 60\hbar$ through both fusion reactions and inelastic scattering yielding an extraordinarily detailed picture of the GDR behavior.

No significant shift of the centroid energy with either temperature or angular momentum has been observed. The width increases both with excitation energy and spin, the latter becoming important only above about $35\hbar$. Inelastic-scattering experiments which populate a range of excitation energies at low spin and fusion experiments using a spin spectrometer setup have led to an experimental de-convolution of temperature and spin effects. It is now well established that the GDR width increases with spin up to $J \sim 60\hbar$ and with temperature up to at least $T \sim 3$ MeV. This behavior is satisfactorily accounted for by the adiabatic thermal fluctuation model.

Above these values many claims for saturation of the GDR width have been made in the literature. However above $60\hbar$ fission sets in as the main decay channel and increasing the angular momentum in the entrance channel does not probe higher spins. Many recent experiments have succeeded to reach temperatures above $T \sim 3$ MeV. In this region a saturation of the GDR gamma-ray yield is observed while statistical calculations predict a continuous increase with excitation energy. The results can be reproduced by a surprisingly sudden drop of the GDR strength at $E^* \sim 250$ MeV. The spectra are not compatible with a continuous increase of the width as was predicted in several theoretical papers. Nonetheless, it cannot be concluded that the width saturates since, once the strength has vanished, the characteristics of the GDR are no longer probed. The drop in strength is related to a competition between equilibration of collective motion and particle decay. Several models were developed to account for such effects and all predict a reduction of the gamma emission probability, albeit with different laws. The lack of data in the critical excitation energy region precludes us today from distinguishing between the different models.

Measurements for lighter nuclei at high excitation energies show the same trends but the limiting excitation energy for the existence of the GDR is $E^*_{lim}/A = 5$ MeV compared to approximately 2.5 MeV in the $A \sim 110$ mass region. It is intriguing to compare these values with the limiting excitation energies extracted from caloric-curve studies. A link between the disappearance of collective motion and a liquid-gas phase transition appears as a distinct possibility worthy of further studies.

Despite the global understanding of the characteristics of collective motion at high temperatures achieved over the past years several issues must still be elucidated. In particular, it would be of great interest to assess the sharpness of the disappearance of the GDR by measuring a more complete excitation function in this region. A slight bombarding-energy dependence of the GDR yield at a given excitation energy has been observed and remained hitherto unexplained. This effect, probably of dynamical nature, needs to be confirmed experimentally and understood theoretically. Finally, the link between the disappearance of the GDR and phase transition may be better understood by extending the systematics of high-energy GDR studies to heavier systems as, *e.g.*, in the lead region.

We would like to thank Drs P. Piattelli, P. Sapienza, R. Coniglione, R. Alba for many helpful discussions and valuable contributions in proofreading the manuscript. We would also thank all the participants in the MEDEA experiments at GANIL and LNS-Catania.

References

1. K.A. Snover, Annu. Rev. Nucl. Part. Sci. **36**, 545 (1986).
2. J.J. Gaardhøje, Annu. Rev. Nucl. Part. Sci. **42**, 483 (1992).
3. G.C. Baldwin, G.S. Klaiber, Phys. Rev. **71**, 3 (1947).
4. G.C. Baldwin, G.S. Klaiber, Phys. Rev. **73**, 1156 (1948).
5. M. Goldhaber, E. Teller, Phys. Rev. **74**, 1046 (1948).
6. B.L. Berman, S.C. Fultz, Rev. Mod. Phys. **47**, 713 (1975).
7. H. Steinwedel, J.H.D. Jensen, Z. Naturforsch A **5**, 413 (1950).
8. D.M. Brink, PhD Thesis, University of Oxford (1955).
9. F.S. Dietrich, J.C. Browne, W.J. O'Connell, M.J. Kay, Phys. Rev. C **10**, 795 (1974).
10. M.A. Kovash, S.L. Blatt, R.N. Boyd, T.R. Donoghue, H.J. Hausmann, A.D. Bacher, Phys. Rev. Lett. **42**, 700 (1979).
11. D.H. Dowell, G. Feldman, K.A. Snover, A.M. Sandorfi, M.T. Collins, Phys. Rev. Lett. **50**, 1191 (1983).
12. J.O. Newton, B. Herskind, R.M. Diamond, E.L. Dines, J.E. Draper, K.H. Lindenberger, C. Schück, S. Shih, F.S. Stephens, Phys. Rev. Lett. **46**, 1383 (1981).

13. F. Pühlhofer, Nucl. Phys. A **280**, 267 (1977).

14. A. Bohr, B.R. Mottelson, *Nuclear Structure*, Vol. **1** (Benjamin Inc., London, 1969).

15. J.B. Natowitz, R. Wada, K. Hagel, T. Keutgen, M. Murray, A. Makeev, L. Qin, P. Smith, C. Hamilton, Phys. Rev. C **65**, 034618 (2002) and references therein.

16. M. Gallardo, M. Diebel, T. Døssing, R.A. Broglia, Nucl. Phys. A **443**, 415 (1985).

17. J.M. Pacheco, C. Yannouleas, R.A. Broglia, Phys. Rev. Lett. **61**, 294 (1988).

18. Y. Alhassid, B. Bush, S. Levit, Phys. Rev. Lett. **71**, 1926 (1988).

19. D. Kusnezov, Y. Alhassid, K.A. Snover, Phys. Rev. Lett. **81**, 542 (1998) and references therein.

20. P. Heckman, D. Bazin, J.R. Beene, Y. Blumenfeld, M.J. Chromik, M.L. Halbert, J.F. Liang, E. Mohrmann, T. Nakamura, A. Navin, B.M. Sherrill, K.A. Snover, M. Thoennessen, E. Tryggestad, R.L. Varner, Phys. Lett. B **555**, 43 (2003).

21. M. Thoennessen, Nucl. Phys. A **731**, 131 (2004) and references therein.

22. F. Camera, A. Bracco, V. Nanal, M.P. Carpenter, F. Della Vedova, S. Leoni, B. Million, S. Mantovani, M. Pignanelli, O. Wieland, B.B. Back, A.M. Heinz, R.V.F. Janssens, D. Jenkins, T.L. Khoo, F.G. Kondev, T. Lauritsen, C.J. Lister, B. McClintock, S. Mitsuoka, E.F. Moore, D. Seweryniak, R.H. Siemssen, R.J. Van Swol, D. Hofmann, M. Thoennessen, K. Eisenman, P. Heckman, J. Seitz, R. Varner, M. Halbert, I. Dioszegi, A. Lopez-Martens, Phys. Lett. B **560**, 155 (2003).

23. J.J. Gaardhøje, C. Ellegaard, B. Herskind, S.G. Steadman, Phys. Rev. Lett. **53**, 148 (1984).

24. J.J. Gaardhøje, C. Ellegaard, B. Herskind, R.M. Diamond, M.A. Delaplanque, G. Dines, A.O. Macchiavelli, F.S. Stephens, Phys. Rev. Lett. **56**, 1783 (1986).

25. D.R. Chakrabarty, S. Sen, M. Thoennessen, N. Alamanos, P. Paul, R. Schicker, J. Stachel, J.J. Gaardhøje, Phys. Rev. C **36**, 1886 (1987).

26. A. Bracco, J.J. Gaardhøje, A.M. Bruce, J.D. Garrett, B. Herskind, M. Pignanelli, D. Barnéoud, H. Nifenecker, J.A. Pinston, C. Ristori, F. Schussler, J. Bacelar, H. Hofmann, Phys. Rev. Lett. **62**, 2080 (1989).

27. G. Enders, F.D. Berg, K. Hagel, W. Kühn, V. Metag, R. Novotny, M. Pfeiffer, O. Schwalb, R.J. Charity, A. Gobbi, R. Freifelder, W. Henning, K.D. Hildenbrand, R. Holzmann, R.S. Mayer, R.S. Simon, J.P. Wessels, G. Casini, A. Olmi, A.A. Stefanini, Phys. Rev. Lett. **69**, 249 (1992).

28. H.J. Hofmann, J.C. Bacelar, M.N. Harakeh, T.D. Poelhekken, A. van der Woude, Nucl. Phys. A **571**, 301 (1994).

29. E. Ramakrishnan, T. Baumann, A. Azhari, R.A. Kryger, R. Pfaff, M. Thoennessen, S. Yokoyama, J.R. Beene, M.L. Halbert, P.E. Mueller, D.W. Stracener, R.L. Varner, R.J. Charity, J.F. Dempsey, D.G. Sarantites, L.G. Sobotka, Phys. Rev. Lett. **76**, 2025 (1996).

30. M. Mattiuzzi, A. Bracco, F. Camera, W.E. Ormand, J.J. Gaardhøje, A. Maj, B. Million, M. Pignanelli, T. Tveter, Nucl. Phys. A **612**, 262 (1997).

31. A. Bracco, F. Camera, M. Mattiuzzi, B. Million, M. Pignanelli, J.J. Gaardhøje, A. Maj, T. Ramsøy, T. Tveter, Z. Zelazny, Phys. Rev. Lett. **74**, 3748 (1995).

32. W.E. Ormand, P.F. Bortignon, R.A. Broglia, Nucl. Phys. A **599**, 57c (1996).

33. G. Gervais, M. Thoennessen, W.E. Ormand, Nucl. Phys. A **649**, 173c (1999).

34. W.E. Ormand, Nucl. Phys. A **649**, 145c (1999).

35. P. Chomaz, Phys. Lett. B **347**, 1 (1995).

36. S.K. Rathi, D.R. Chakrabarty, V.M. Datar, Suresh Kumar, E.T. Mirgule, A. Mitra, V. Nanal, H.H. Oza, Phys. Rev. C **67**, 024603 (2003).

37. M.P. Kelly, J.F. Liang, A.A. Sonzogni, K.A. Snover, J.P.S. van Schagen, J.P. Lestone, Phys. Rev. C **56**, 3201 (1997).

38. M.P. Kelly, K.A. Snover, J.P.S. van Schagen, M. Kicińska-Habior, Z. Trznadel, Phys. Rev. Lett. **82**, 3404 (1999).

39. K.A. Snover, Nucl. Phys. A **687**, 337c (2001).

40. O. Wieland, S. Barlini, V.L. Kravchuk, F. Camera, F. Gramegna, A. Maj, G. Benzoni, N. Blasi, S. Brambilla, M. Brekiesz, M. Bruno, G. Casini, M. Chiari, E. Geraci, A. Giussani, M. Kmiecik, S. Leoni, A. Lanchais, P. Mastinu, B. Million, A. Moroni, A. Nannini, A. Ordine, G. Vannini, L. Vannucci, J. Phys. G: Nucl. Part. Phys. **31**, S1973 (2005).

41. H. Nifenecker, J. Blachot, J. Crançon, A. Gizon, A. Lleres, Nucl. Phys. A **447**, 533c (1985).

42. J.J. Gaardhøje, A.M. Bruce, J.D. Garrett, B. Herskind, D. Barnéoud, M. Maurel, H. Nifenecker, J.A. Pinston, P. Perrin, C. Ristori, F. Schussler, A. Bracco, M. Pignanelli, Phys. Rev. Lett. **59**, 1409 (1987).

43. K. Yoshida, J. Kasagi, H. Hama, M. Sakurai, M. Kodama, K. Furutaka, K. Ieki, W. Galster, T. Kubo, M. Ishihara, Phys. Lett. B **245**, 7 (1990).

44. J.H. Le Faou, T. Suomijärvi, Y. Blumenfeld, P. Piattelli, C. Agodi, N. Alamanos, R. Alba, F. Auger, G. Bellia, Ph. Chomaz, R. Coniglione, A. Del Zoppo, P. Finocchiaro, N. Frascaria, J.J. Gaardhøje, J.P. Garron, A. Gillibert, M. Lamehi-Racti, R. Liguori-Neto, C. Maiolino, E. Migneco, G. Russo, J.C. Roynette, D. Santonocito, P. Sapienza, J.A. Scarpaci, A. Smerzi, Phys. Rev. Lett. **72**, 3321 (1994).

45. P. Piattelli, T. Suomijärvi, Y. Blumenfeld, D. Santonocito, C. Agodi, N. Alamanos, R. Alba, F. Auger, G. Bellia, Ph. Chomaz, M. Colonna, R. Coniglione, A. Del Zoppo, P. Finocchiaro, N. Frascaria, A. Gillibert, J.H. Le Faou, K. Loukachine, C. Maiolino, E. Migneco, J.C. Roynette, P. Sapienza, J.A. Scarpaci, Nucl. Phys. A **649**, 181c (1999).

46. D.M. Brink, Nucl. Phys. A **519**, 3c (1990).

47. P.F. Bortignon, A. Bracco, D. Brink, R.A. Broglia, Phys. Rev. Lett. **67**, 3360 (1991).

48. P. Donati, N. Giovanardi, P.F. Bortignon, R.A. Broglia, Phys. Lett. B **383**, 15 (1996).

49. P. Chomaz, M. Di Toro, A. Smerzi, Nucl. Phys. A **563**, 509 (1993).

50. P. Chomaz, Nucl. Phys. A **569**, 203c (1994).

51. A. Smerzi, A. Bonasera, M. Di Toro, Phys. Rev. C **44**, 1713 (1991).

52. A. Smerzi, M. Di Toro, D.M. Brink, Phys. Lett. B **320**, 216 (1994).

53. A. Bonasera, M. Di Toro, A. Smerzi, D.M. Brink, Nucl. Phys. A **569**, 215c (1994).

54. R.A. Broglia, P.F. Bortignon, A. Bracco, Prog. Part. Nucl. Phys. **28**, 517 (1992).

55. K. Yoshida, J. Kasagi, H. Hama, M. Sakurai, M. Kodama, K. Furutaka, K. Ieki, W. Galster, T. Kubo, M. Ishihara, A. Galonsky, Phys. Rev. C **46**, 961 (1992).

56. J. Kasagi, K. Yoshida, Nucl. Phys. A **557**, 221c (1993).

57. J. Kasagi, K. Yoshida, Nucl. Phys. A **569**, 195c (1994).

58. E. Migneco, C. Agodi, R. Alba, G. Bellia, R. Coniglione, A. Del Zoppo, P. Finocchiaro, C. Maiolino, P. Piattelli, G. Raia, P. Sapienza, Nucl. Instrum. Methods Phys. Res. A **314**, 31 (1992).

59. D. Santonocito, P. Piattelli, Y. Blumenfeld, T. Suomijärvi, C. Agodi, N. Alamanos, R. Alba, F. Auger, G. Bellia, Ph. Chomaz, M. Colonna, R. Coniglione, A. Del Zoppo, P. Finocchiaro, N. Frascaria, A. Gillibert, J.H. Le Faou, K. Loukachine, C. Maiolino, E. Migneco, J.C. Roynette, P. Sapienza, J.A. Scarpaci, Phys. Rev. C **66**, 044619 (2002).

60. W.E. Ormand, P.F. Bortignon, A. Bracco, R.A. Broglia, Phys. Rev. C **40**, 1510 (1989).

61. T. Suomijärvi, Y. Blumenfeld, P. Piattelli, J.H. Le Faou, C. Agodi, N. Alamanos, R. Alba, F. Auger, G. Bellia, Ph. Chomaz, R. Coniglione, A. Del Zoppo, P. Finocchiaro, N. Frascaria, J.J. Gaardhøje, J.P. Garron, A. Gillibert, M. Lamehi-Racti, R. Liguori-Neto, C. Maiolino, E. Migneco, G. Russo, J.C. Roynette, D. Santonocito, P. Sapienza, J.A. Scarpaci, A. Smerzi, Phys. Rev. C **53**, 2258 (1996).

62. G. Bellia, P. Finocchiaro, K. Loukachine, C. Agodi, R. Alba, L. Calabretta, R. Coniglione, A. Del Zoppo, C. Maiolino, E. Migneco, P. Piattelli, G. Raciti, D. Rifugiato, D. Santonocito, P. Sapienza, IEEE Trans. Nucl. Sci. **43**, 1737 (1996).

63. M. Kicińska-Habior, K.A. Snover, C.A. Gossett, J.A. Behr, G. Feldman, H.K. Glatzel, J.H. Gundlach, E.F. Garman, Phys. Rev. C **36**, 612 (1987).

64. B. Fornal, F. Gramegna, G. Prete, R. Burch, G. D'Erasmo, E.M. Fiore, L. Fiore, A. Pantaleo, V. Paticchio, G. Viesti, P. Blasi, N. Gelli, F. Lucarelli, M. Anghinolfi, P. Corvisiero, M. Taiuti, A. Zucchiatti, P.F. Bortignon, D. Fabris, G. Nebbia, J.A. Ruiz, M. Gonin, J.B. Natowitz, Z. Phys. A **340**, 59 (1991).

65. A. Musumarra, G. Cardella, A. Di Pietro, S.L. Li, M. Papa, G. Pappalardo, F. Rizzo, S. Tudisco, J.P.S. van Schagen, Nucl. Instrum. Methods A **370**, 558 (1996).

66. S. Tudisco, G. Cardella, F. Amorini, A. Anzalone, A. Di Pietro, P. Figuera, F. Giustolisi, G. Lanzalone, Lu Jun, A. Musumarra, M. Papa, S. Pirrone, F. Rizzo, Europhys. Lett. **58**, 811 (2002).

67. F. Amorini, G. Cardella, A. Di Pietro, P. Figuera, G. Lanzalone, Lu Jun, A. Musumarra, M. Papa, S. Pirrone, F. Rizzo, W. Tian, S. Tudisco, Phys. Rev. C **69**, 014608 (2004).

68. M. Papa, contributed paper to the *WCI Texas A&M Conference, College Station, USA, 2005*.

69. M. Papa, A. Bonanno, F. Amorini, A. Bonasera, G. Cardella, A. Di Pietro, P. Figuera, T. Maruyama, G. Pappalardo, F. Rizzo, S. Tudisco, Phys. Rev. C **68**, 034606 (2003).

70. M. Papa, W. Tian, G. Giuliani, F. Amorini, G. Cardella, A. Di Pietro, P. Figuera, G. Lanzalone, S. Pirrone, F. Rizzo, D. Santonocito, Phys. Rev. C **72**, 064608 (2005).

71. D. Pierroutsakou, M. Di Toro, F. Amorini, V. Baran, A. Boiano, A. De Rosa, A. D'Onofrio, G. Inglima, M. La Commara, A. Ordine, N. Pellegriti, F. Rizzo, V. Roca, M. Romoli, M. Sandoli, M. Trotta, S. Tudisco, Eur. Phys. J. A **16**, 423 (2003).

72. D. Pierroutsakou, A. Boiano, A. De Rosa, M. Di Pietro, G. Inglima, M. La Commara, Ruhan Ming, B. Martin, R. Mordente, A. Ordine, F. Rizzo, V. Roca, M. Romoli, M. Sandoli, F. Soramel, L. Stroe, M. Trotta, E. Vardaci, Eur. Phys. J. A **17**, 71 (2003).

73. A. Bonasera, M. Bruno, P.F. Mastinu, C.A. Dorso, Riv. Nuovo Cimento **23**, no. 2 (2000).

Eur. Phys. J. A **30**, 203–213 (2006)

DOI 10.1140/epja/i2006-10117-6

THE EUROPEAN
PHYSICAL JOURNAL A

Nuclear thermometry

A. Kelić[1,a], J.B. Natowitz[2], and K.-H. Schmidt[1]

[1] GSI, Planckstr. 1, 64291 Darmstadt, Germany
[2] Department of Chemistry, Texas A&M University, College Station, TX 77842, USA

Received: 30 May 2006 /
Published online: 20 October 2006 – © Società Italiana di Fisica / Springer-Verlag 2006

Abstract. Different approaches for measuring nuclear temperatures are described. The quantitative results of different thermometer approaches are often not consistent. These differences are traced back to the different basic assumptions of the applied methods. Moreover, an overview of recent theoretical investigations is given, which study the quantitative influence of dynamical aspects of the nuclear-reaction process on the extracted apparent temperatures. The status of the present experimental and theoretical knowledge is reviewed. Guidelines for future investigations, especially concerning the properties of asymmetric nuclear matter, are given.

PACS. 24.60.-k Nuclear reaction: general: Statistical theory and fluctuations – 05.70.Fh Phase transitions: general studies – 25.70.-z Low and intermediate energy heavy-ion reactions – 21.10.Ma Level density

1 Introduction

The concept of a nuclear temperature was introduced some seventy years ago in pioneering works performed by Bethe [1] and Weisskopf [2]. The goal was to describe the formation and the decay of a compound nucleus formed in reactions induced by light projectiles, mostly neutrons. Later on, the concept of a nuclear temperature was extended to reactions involving high-energy projectiles and heavy ions [3]. These new studies were triggered by the quest for nuclear instabilities and a possible liquid-gas phase transition in nuclear matter [4,5]. To this goal, different experimental methods were developed and applied in order to extract information on thermal characteristics of highly excited nuclear systems (see, *e.g.*, [6] and references therein). Most of these "nuclear thermometers" rely on the application of thermodynamic relations to characterize the conditions at freeze-out. In general, the temperature of a system with fixed number of particles N_{part} at an energy E is defined according to statistical mechanics as

$$\frac{1}{T} = \frac{\partial S(E, N_{part})}{\partial E} = \frac{\partial \ln \rho(E, N_{part})}{\partial E}, \qquad (1)$$

where S is the entropy of the system, and ρ the density of states at energy E. In order to apply this formula to obtain a temperature, two conditions have to be fulfilled: Firstly, the system has to be in full statistical equilibrium, *i.e.* each of the states included in $\rho(E, N_{part})$ has to be populated with equal probability, and secondly the density of states

has to be known. For nuclear systems these two conditions can be critical. The degree to which the equilibrium is reached in high-energy heavy-ion collisions is not *a priori* known as the dynamical evolution of a nuclear system is still not fully understood. What concerns the nuclear state density, it is well known only at low energies. At high excitation energies, on the contrary, the knowledge of the nuclear state density is much poorer.

Apart from this, there are several other problems, which make the extraction of nuclear temperatures even more difficult:

– *The nucleus is a microscopic system.* External probes are not applicable. Consequently, information on temperature is obtained from the emission of (small) parts of the system itself assuming that the emitted clusters made part of the equilibrium and the density of states of the whole system before emission, and are, therefore, representative for the whole system.

– *The nucleus is an isolated system.* Due to the short range of the nuclear force, the nucleus cannot exchange its excitation energy with its external environment. Consequently, the nuclear system is defined by the conditions: $E = $ const, $N_{part} = $ const, and, therefore, the only appropriate statistical ensemble in case of the nucleus is the microcanonical ensemble used for isolated systems [6,7]. On the other hand, from the experiment it is not that easy to fix the value of energy, as the amount of deposited energy can vary strongly between different nuclear collisions, especially in cases where several different reaction mechanisms result in the emission of the same product.

---

a e-mail: `a.kelic@gsi.de`

– *The nucleus is a quantum fermionic system.* Nucleons inside the nucleus occupy different energy levels, and, moreover, due to the Pauli principle not all nucleons can participate in sharing the available energy. Consequently, the effective number of degrees of freedom depends on excitation energy, what is accounted for by the Fermi statistics. Moreover, the global properties of a nucleus change dynamically with energy (*e.g.*, the density of the nucleus reduces due to thermal expansion).

– *The nucleus is an electrically charged system.* The long-range Coulomb force between protons introduces instabilities [8] that could lead to a lowering of the critical temperature.

– *The nucleus heats up and cools down in a dynamical process.* Different signatures may correspond to different freeze-out conditions, or represent different stages in the dynamical evolution. Moreover, production during evaporation can contribute to the yields of light fragments, while expansion influences the kinetic energy of the fragments.

– *The thermodynamical parameters (e.g., pressure, volume, chemical potential) are not under control.* In the experiment one does not have direct access to thermodynamical parameters and is obliged to use model calculation in order to extract them.

– *Experimental signatures are modified by secondary decay.* Consequently, in most cases one needs robust signatures, which are least affected by secondary decay (*e.g.*, light IMFs).

2 Thermometer methods

In the literature, different thermometer methods have been applied. According to their approach they can be grouped as:

– *Population approaches.* Based on the grand-canonical concept. The value of the nuclear temperature is extracted from the yields of the produced clusters assuming a Boltzmann distribution: $Y_i \sim \exp(-E_i/T)$. The most often used methods are: Double ratios of isotopic yields [9,10], also called isotopic thermometer; Population of excited states (bound or unbound) [6,11–16]; Isobaric yields from a given source [17,18].

– *Kinetic approaches.* Based on the concept of a canonical ensemble. The value of the temperature is extracted from the slope of the measured particle kinetic-energy spectra; due to this, the method is named slope thermometer. Two processes are studied within this approach: Thermal evaporation from the compound nucleus [2] and sudden disintegration of an equilibrated source into observed nucleons and light nuclei [19–23] or gamma rays [24,25].

– *Thermal-energy approaches.* The excitation energy at the freeze-out is extracted by measuring the evaporation cascade from a thermalised source by variation of neutron-to-proton ratio N/Z. The temperature at freeze-out is then obtained from the deduced excitation energy. An example is the isospin thermometer [26,27].

2.1 Population approaches

2.1.1 Double ratios of isotopic yields

This method evaluates the temperature of equilibrated nuclear regions from which light fragments are emitted using the yields of different light nuclides [9]. The basic assumptions of the method are those of the grand-canonical approach.

During the cooling and expansion stage of a hot nuclear system, the interactions between the constituent particles take place until density and temperature become small enough so that the constituents do not longer interact. From this time on the particle composition remains unchanged (chemical freeze-out). As the system expands beyond this point the frozen particles escape. By detecting them one can obtain information on the freeze-out stage. The starting assumption of the method is that thermal equilibrium is established between free nucleons and composite fragments contained within a certain interaction volume V at a temperature T. In this case, the density of a particle (A, Z) is [9]

$$\rho(A, Z) = \frac{N_{part}}{V} = \frac{A^{3/2} \cdot \omega(A, Z)}{\lambda^3} \cdot \exp\left(\frac{\mu(A, Z)}{T}\right), \quad (2)$$

where ω is the internal partition function of the particle (A, Z): $\omega(A, Z) = \sum [2 \cdot s_j(A, Z) + 1] \cdot \exp[-E_j(A, Z)/T]$, λ is the thermal nucleon wavelength $\lambda = h/\sqrt{2 \cdot \pi m_N \cdot T}$, and μ is the chemical potential of the particle (A, Z).

In the next step, one imposes to the system also the condition of chemical equilibrium: $\mu(A, Z) = Z \cdot \mu_{pF} + (A - Z) \cdot \mu_{nF} + B(A, Z)$, B being the binding energy of the cluster (A, Z), and μ_{pF} and μ_{nF} the chemical potentials of free protons and neutrons, respectively.

Then for the ratio $Y(A, Z)/Y(A', Z')$ between the measured yields of two different emitted fragments one gets [9]:

$$\frac{Y(A, Z)}{Y(A', Z')} = \frac{\rho(A, Z)}{\rho(A', Z')} = \left(\frac{A}{A'}\right)^{3/2} \cdot \left(\frac{\lambda^3}{2}\right)^{A - A'}$$
$$\cdot \frac{\omega(A, Z)}{\omega(A', Z')} \cdot \rho_{pF}^{Z - Z'} \rho_{nF}^{(A - Z) - (A' - Z')}$$
$$\cdot \exp\left(\frac{B(A, Z) - B(A', Z')}{T}\right) \quad (3)$$

with ρ_{pF} and ρ_{nF} being, respectively, the densities of free protons and neutrons contained in the same interaction volume V at the temperature T as the cluster (A, Z). Using eq. (3) and two sets of the yields of two fragments differing only by one proton, one obtains the temperature of the emitting source at the moment of freeze-out [9]:

$$T = (\Delta B_1 - \Delta B_2)/\ln\left[\left(\frac{Y(A_1, Z_1)/Y(A_1 + 1, Z_1 + 1)}{Y(A_2, Z_2)/Y(A_2 + 1, Z_2 + 1)}\right)\right.$$
$$\cdot \left(\frac{(A_1 + 1) \cdot A_2}{A_1 \cdot (A_2 + 1)}\right)^{3/2}$$
$$\left. \cdot \left(\frac{\omega(A_1 + 1, Z_1 + 1) \cdot \omega(A_2, Z_2)}{\omega(A_1, Z_1) \cdot \omega(A_2 + 1, Z_2 + 1)}\right)\right], \quad (4)$$

where $\Delta B_i = B(A_i, Z_i) - B(A_i + 1, Z_i + 1)$, $i = 1, 2$. An analogous relation is obtained if one takes pairs of nuclei differing only by one neutron.

When applying this method to extract the value of the nuclear temperature, several precautions have to be taken. Firstly, as this method assumes that both thermal and chemical equilibrium at the freeze-out are reached, it is important to consider only those yields which can be attributed to the equilibrium component of the whole reaction mechanism. Secondly, one has to be sure that the studied light particles are emitted during the freeze-out and not as the product of secondary decay [28–30]. The side-feeding to the considered nuclides from secondary decay can result in a large spread of extracted temperature values. Finally, in order to obtain the value of nuclear temperature one needs to calculate the binding energies of observed fragments, see eq. (4). Although eq. (4) describes the situation at the freeze-out very often in its application the experimental binding energies have been used. One should not forget that a binding energy depends on the symmetry-energy coefficient used in the mass formula, which might depend on density and temperature, and, therefore, the use of experimental binding energies in order to describe the situation at the freeze-out could be questionable.

2.1.2 Population of excited states

This method has the same basic assumptions as the double-isotopic-ratio method. The departure point is that the population distribution of the excited states in a statistically equilibrated system should be given by the temperature of the system and by the spacing between the considered energy levels. The advantage of this method as compared to the double-isotopic-ratio method is that one can assume that isospin and dynamical aspects influencing the population of the two considered states are the same.

Following this picture, the ratio R of the populations of two states (if no feeding by particle decay takes place) is given, similarly to eq. (3), as

$$R = \frac{2 \cdot j_u + 1}{2 \cdot j_l + 1} \cdot \exp\left(-\frac{\Delta E}{T}\right), \qquad (5)$$

where j_u and j_l are the spins of the upper and lower state, respectively, and ΔE the energy difference between these two states. This energy difference limits the temperature that can be inferred by this method, as for temperatures higher than ΔE one reaches saturation, *i.e.* the ratio R approaches its asymptotic high-temperature value. The considered excited states can be either particle-bound or particle-unbound states. The advantage of taking particle-unbound states lies in the fact that for the unbound states ΔE has, generally, higher values than for the bound states, thus allowing for the measurement of higher temperatures. Moreover, the relative population between ground state and particle-bound state can be changed by the sequential decay of primary fragments produced in a particle-unbound state [11] or by the hadronic final-state interactions that occurs after emission from the equilibrated

system [31]. This is important, as in cases where the primary population ratio is strongly influenced by secondary decays the uncertainties in the extracted temperature can be large [11].

2.1.3 Isobaric yields from a given source

This thermometer is mostly applied in studies of the thermal properties of excited quasiprojectiles formed in heavy-ion reactions in the Fermi energy regime. It uses the model assumptions of the statistical multifragmentation model [5], according to which, in the grand-canonical picture, the ratio between yields of two observed fragments having the same ground-state spins and coming from the same source is given as [17]

$$\frac{Y(A_1, Z_1)}{Y(A_2, Z_2)} = \exp\left[-\frac{1}{T} \cdot (F_{A1,Z1}(T, V) - F_{A2,Z2}(T, V)\right.$$
$$\left. -\mu_n \cdot (N_1 - N_2) - \mu_p \cdot (Z_1 - Z_2)) \right] \qquad (6)$$

with $F(T, V)$ the internal free energy of the fragment, $N_i = A_i - Z_i$, T and V freeze-out temperature and volume, respectively. The internal free energy is calculated as given in [5]. The results of this thermometer using the $Y(^3\mathrm{H})/Y(^3\mathrm{He})$ ratio compares very well with results of double-isotopic-ratio methods using $^2\mathrm{H}$, $^3\mathrm{H}/^3\mathrm{He}$, $^4\mathrm{He}$ ratios [17]. The problems inherent to the previous two methods are also present in the isobaric-yields method.

2.2 Kinetic approaches

The method of the slope thermometer is based on fitting the exponential slope of measured particle spectra. The spectral distributions of particles emitted by an excited nucleus were firstly described by Weisskopf in case of neutron-induced reactions using the standard thermodynamic procedure [2]. The predicted spectra followed a Maxwell-Boltzmann distribution proportional to an energy-dependent pre-exponential factor and the Boltzmann function: $\mathrm{d}Y/\mathrm{d}E_{kin} = f(E_{kin}) \exp(-E_{kin}/T)$. The shape of the particle spectra was later discussed by Goldhaber who mostly concentrated on the form of the pre-exponential factor [32].

This method is applied to two processes:

– *Thermal evaporation from the compound nucleus.* Except at very low excitation energies, the decay of an excited nucleus proceeds through several de-excitation steps. Consequently, the mass and the temperature of the emitting source vary in time, and the observed spectra represent the convolution of all these different contributions. Therefore, for fitting the measured spectra dedicated models that properly describe the time evolution of the cooling process have to be applied (*e.g.* [33–38]).

– *Sudden disintegration.* One assumes a single freeze-out configuration from which nucleons and light particles are emitted. In this case, dynamical effects to be mentioned below if not properly described can lead to misleading results concerning the magnitude of the extracted nuclear temperature. Additional difficulties arise from the fact that the observed fragment can emerge from any location in the source, and that its Coulomb energy depends on the number and position of all the other created fragments [39]. Moreover, the Fermi motion of nucleons inside the projectile/target as well as inside the source has to be considered. The nucleonic Fermi motion within the colliding nuclei has been discussed by Goldhaber as the origin of the momenta of the produced fragments in fragmentation reactions [40]. He has also pointed out that the resulting behavior, *i.e.* the form of the fragment kinetic energies, is indistinguishable from that of a thermalised system with rather high temperature. Its relevance for the interpretation of the kinetic properties of nuclear decay products has been underlined by Westfall *et al.* [21]. In ref. [22], it was discussed that the slope temperature does not correspond to the thermal temperature at the freeze-out but rather reflects the intrinsic Fermi motion and, thus, the bulk density of the spectator system at the moment of break-up. This would suggest that it may be difficult to attribute the slope parameter of the energy spectra of the observed light fragments directly to the thermal characteristics of the decaying system. As in the case of surface emission, the temporal evolution of the emitting source [33,41,42] as well as the sequential decay of excited primary fragments [33,43] can complicate the interpretation of the measured kinetic-energy spectra. Recently, it was proposed to use the energy spectra of thermal Bremsstrahlung photons in order to extract the nuclear temperature at the freeze-out [24,25]. The advantage of using gamma rays instead of nucleons and light particles should lay in the following facts: minimal contribution from pre-equilibrium processes, absence of the reacceleration by the Coulomb field, sensitivity on the temperature of the system right after equilibration, and absence of final-state distortions [24].

The shape of the measured particle spectra can be influenced by collective dynamical effects —collective rotation [44,45], translatory motion [20,41] and collective expansion of the source [46,47]. Each of these effects can influence the spectra in a similar way as the changes in the temperature; for more details see ref. [6].

2.3 Thermal approaches

Thermal approaches are based on the assumption that the thermal energy after the freeze-out feeds an evaporation cascade. The excitation energy at the freeze-out is extracted by measuring the evaporation cascade from a thermalised source by detecting either final residues [26, 27] or light charged particles [48,49]. While for the other methods, the secondary decay is a disturbing effect, in thermal approaches the evaporation cascade is used to deduce the temperature at the freeze-out, and it is, therefore, also applicable to heavy reaction residues.

In the first approach —isospin thermometer— one gains information on the excitation energy and, conse-

quently, on the temperature at the freeze-out configuration by back-tracing the evaporation cascade [27]. This idea is the base of the "thermometer for peripheral nuclear collisions" [26], a method to deduce the temperature of nuclear systems from the isotopic distributions of the residues at the end of the evaporation cascade. The method consists of applying an evaporation code with the quite well-known ingredients of the statistical model in order to deduce the temperature at the beginning of the evaporation cascade. In this approach, the mean neutron-to-proton ratio of the final residues is calculated for different freeze-out temperatures, assuming that the N/Z ratio of fragments at the freeze-out is the same as that of the projectile. By obtaining agreement between measured and calculated N/Z ratios one deduces the value of nuclear temperature at the freeze-out. The assumption that the fragments enter the evaporation stage with the same N/Z as the projectile or, respectively, target nucleus is rather simplifying, since according to some descriptions of the nuclear break-up (*e.g.* [50,51]), the process of isospin fractionation should result in different isotopic compositions in case of heavy and light fragments (*i.e.* liquid and gas phase), leading to a more neutron-rich gas phase and a less neutron-rich liquid phase. While neglecting the isospin-fractionation process will likely introduce only a small uncertainty, details of the evaporation model especially at high excitation energy are important for the qualitative application of the isospin thermometer [52]. The isospin thermometer is mostly applied at relativistic energies as at Fermi energies the effect of isospin diffusion [53] can complicate the interpretation of this method.

In the second case [48,49], a correlation technique for the relative velocity between light charged particles and IMF is applied in order to extract multiplicities and velocity spectra of secondary evaporated particles. From this information the average size and average excitation energy of the primary hot fragments is reconstructed.

3 Corrections

One should not forget that one of the reasons for measuring the nuclear temperature is the possibility to reconstruct the nuclear caloric curve and to search for possible evidence of a liquid-gas phase transition. Very often, the predictions of different thermometers differ dramatically (see *e.g.* [54]), and it is, therefore, of prime interest to understand and apply all possible corrections that can influence the value of the obtained nuclear temperature.

Before we start a more detailed discussion on different corrections to be applied, we would like to express a word of caution —most methods mentioned above cannot result in the "correct" thermodynamical temperature of the nucleus, as they are all based either on the canonical or grand-canonical ensemble, but not on the microcanonical ensemble. Moreover, due to the basic difference between different methods (*e.g.* canonical *vs.* grand-canonical approach) one should not expect that the obtained, apparent, temperatures have the same values. One should also not forget that the measured quantity might reflect the temperatures on different stages (times) or different regions (positions) of the system, and this is also one of the reasons for different values of the apparent temperature. In connection with this, one can also pose the question if some of the basic assumptions of different methods, *i.e.* establishment of thermal and/or chemical equilibrium, are fulfilled in nuclear reactions. And if so, are the measured observables characteristic of the established equilibrium? An optimistic answer was given in ref. [55], where it was shown that caloric curves obtained using the above-mentioned thermometer methods can still carry the signal of the phase transition in a system with conserved energy.

If one assumes the validity of different thermometer methods, in order that they are applicable one has first to consider several corrections, and here we will discuss some of them: finite-size effects [39,56], emission time differences [57], multi-source emission [58], secondary decay [28–30,59] and recombination [60].

3.1 Finite-size effects

One of the consequences of applying the canonical or grand-canonical ensemble is that effects due to the finite size of the nucleus are neglected. In ref. [56] caloric curves obtained using different double-isotopic-ratio thermometers were compared with the results of microcanonical calculations [4,61]. Results of this comparison have shown that there are important differences between different double-isotopic-ratio temperatures themselves, as well as between double-isotopic-ratio temperatures and microcanonical temperatures. These deviations are especially important at higher excitation energies above $\sim 8\,\mathrm{MeV/nucleon}$.

The authors of ref. [56] proposed a method, independent of the size of the source, to "calibrate" the different thermometers using the microcanonical temperature. They applied this procedure to re-evaluate different experimental caloric curves (ALADIN [10], EOS [62], INDRA [63]). The re-evaluated caloric curves show the features of a liquid-gas phase transition, which were missing in the original experimental data.

3.2 Emission time differences

During a nuclear reaction, processes occurring on different time scales (*e.g.* fast break-up, pre-equilibrium emission, evaporation from the compound system) contribute to the production of the observed fragments and light particles. Fragments produced by these different mechanisms can have quite different characteristics (*e.g.* N/Z ratio, velocity, angular distribution), and already Albergo *et al.* have discussed the importance of selecting a proper subset of observed events [9]. The influence of the reaction dynamics on the observed yields of different isotopic thermometers was studied in ref. [57] in more detail. It was shown that the single ratios $Y(A, Z)/A(A + 1, Z)$ involving one nuclide with $N < Z$ have several times higher values at forward angles as compared to the backward angles, while

the single ratios including only $N \geq Z$ nuclides are approximately independent of the emission angle, the bombarding energy or the target-projectile system [57]. Based on the expanding-evaporating source model EES [47] these observations were interpreted as a consequence of differences in relative emission times of processes leading to the final fragments [57]. Similarly, Hudan *et al.* have found that in mid-peripheral and central collisions, isotopes with $N < Z$ have larger kinetic energies than heavier isotopes of the same element [58]. The same was observed by Liu *et al.* [64] for central collisions, and was explained by shorter emission times for neutron-deficient isotopes.

3.3 Multi-source emission

Production of light charged particles and intermediate-mass fragments is not only connected with different emission times, but also with different emitting sources. The composition and excitation energy of the emitting source can influence the size, composition and kinetic energy of the observed fragments [65,66], and, consequently, the value of the temperature extracted from yields or kinetic-energy spectra of fragments.

For example, it was shown in ref. [58] that fragments emitted from the mid-velocity source have broader peaks and higher tails in transverse-velocity distributions and are more neutron-rich as compared to fragments emitted from the projectile-like source. In ref. [67] a detailed study on the validity, accuracy and experimental limits of the excitation energy measurements in the Fermi-energy regime has been performed. There, it was shown that difficulties in separating particles coming from different sources, especially for mid-peripheral and central collisions, as well as different experimental thresholds and cuts can lead to uncertainties in the source reconstruction.

Therefore, it is very important to identify in an experiment all different sources that contribute to the production of the observed fragments and their characteristics. Otherwise, the extracted value of the nuclear temperature will represent an average over different processes and conditions.

3.4 Recombination

In statistical models based on the canonical or grand-canonical ensemble, the momentum distribution of fragments is Maxwellian at the corresponding temperature, and, consequently, there is a probability that some pairs of primary fragments come close enough to feel the nuclear force and may recombine to form an excited heavier fragment, which may also decay later. This question on the evolution of the primary fragments under the combined influence of Coulomb and nuclear fields was studied in refs. [60,68]. Samaddar *et al.* have shown that, while the calculations without recombination predict an increase in the temperature with excitation energy similar to the Fermi-gas model predictions, inclusion of the recombination effect resulted in a decrease of the nuclear temperature and a plateau in the caloric curve [60].

On the other hand, in models based on the microcanonical ensemble [4,61], the momenta and positions of the fragments are coupled, and the probability of having two fragments close in the freeze-out volume is strongly reduced by the Coulomb repulsion. Consequently, the effects of recombination may be reduced as compared to the above-mentioned results. Therefore, it would be very interesting to perform more detailed and dedicated calculations based on the microcanonical ensemble in order to quantitatively understand the recombination effect.

3.5 Secondary decay

The primary fragments produced at the freeze-out stage are usually highly excited and they can undergo secondary decays. Such decay is evidenced, for example, in refs. [48, 49] in which a method based on correlations between light charged particles and IMF was applied in order to extract multiplicities and velocity spectra of particles emitted during the evaporation from the primary hot fragments. Therefore, the measured yields used to extract the nuclear temperature are different from the primary distributions at the freeze-out stage. This question was studied on a theoretical basis by several authors, see *e.g.* [28–30,69].

Tsang *et al.* argued that the fluctuations observed in the value of the nuclear temperature applying different double-isotopic-ratio thermometers appear to originate from structure effects in the secondary-decay process and that each isotope ratio shows a characteristic behavior independently of the reaction [30]. Calculations performed by Xi *et al.* [69] indicated that due to strong feeding effects, the double-isotopic-ratio method is strongly influenced by secondary decays at temperatures above 6 MeV. Raduta and Raduta [29] have applied the sharp microcanonical multifragmentation model [56] with inclusion of secondary decay in order to evaluate the caloric curve from different isotopic thermometers for primary decay and asymptotic stages. In both stages, a dispersive character of the isotopic caloric curve increasing with the increase of the excitation energy was evidenced. The authors proposed a procedure to calibrate the isotopic thermometers on the microcanonical predictions independently of the source size and excitation energy [29].

A complex structure in the residue yields was recently evidenced in the fragmentation reaction ^{238}U+Ti at 1 A GeV [59]. From the light fragmentation residues, fully resolved in A and Z, an important even-odd staggering in the yields was observed. Using the statistical model of nuclear reactions, it was shown in ref. [59] that for all classes of nuclei except for $N = Z$ nuclei structural effects in nuclear binding and in the level density are responsible for the observed staggering. The chain of $N = Z$ nuclei appears as a special class of nuclei with increased enhancement in the production of even-even nuclei compared to other chains with $N - Z =$ even, and possible origins like the Wigner energy, alpha clustering, and neutron-proton pairing were discussed [59]. Therefore, when correcting for secondary decay, complex structure, as extremely strong even-odd staggering in $N = Z$ nuclei, must be considered.

4 Thermometer results

4.1 Nuclear caloric curves

As indicated in a number of previous reviews, measurements of nuclear temperatures, which have long been employed to explore excited nuclei, can also provide important information on the van der Waals-like nuclear equation of state and the postulated liquid-gas phase transition [3,28,70–73]. A large number of theoretical calculations have explored the nuclear equation of state and reported values for the critical temperature, T_C, of semi-infinite nuclear matter (nuclear matter with a surface). References [74–84] constitute a representative sampling of these calculations. The different nuclear interactions employed in the calculations lead to large differences in the critical temperatures derived from these interactions. Values from 13 to 24 MeV are reported in the cited references. For finite nuclei, early theoretical work by Bonche and collaborators explored the thermal properties and stability of highly excited nuclei by employing a temperature-dependent Hartree-Fock model with Skyrme interactions [8,85,86]. This work and later work with other models [87–92] predict the existence of "limiting temperatures". The temperatures at which the expanded nucleus reaches the limit of equilibrium phase coexistence between liquid and vapor were designated "Coulomb instability" temperatures [8,85].

In extensions of the work of refs. [85] and [8], Besprosvany and Levit mapped the limiting temperature surface as a function of N and Z [86]. The limiting temperatures that they calculated are shown in fig. 1. They are well below the critical temperature of nuclear matter. This reflects size effects, Coulomb effects and isospin asymmetry effects for the finite nuclei studied. It is important to note that such predictions are sensitive to both the chosen nuclear interaction and to the assumed temperature dependence of the surface energy [93]. One important goal of experimental measurements of temperatures of excited nuclei has been to derive information on T_C.

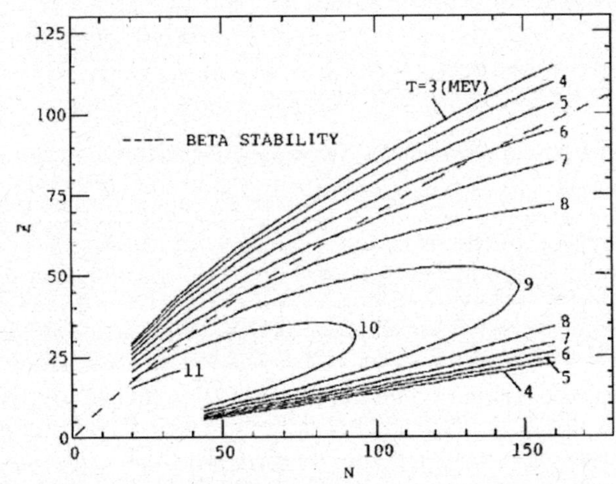

Fig. 1. Limiting temperatures predicted by Besprosvany and Levit [86].

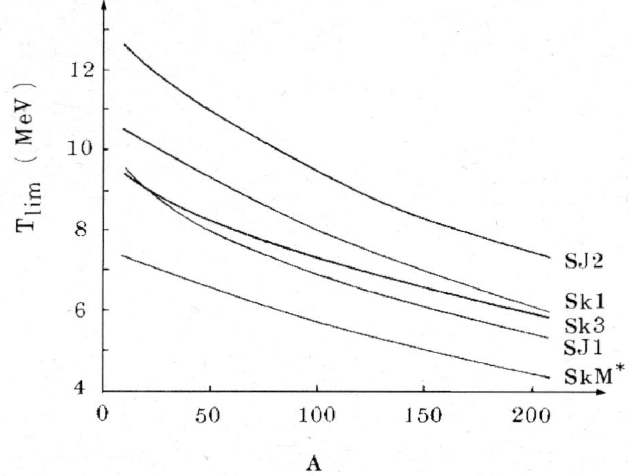

Fig. 2. Mass dependence of limiting temperatures for various Skyrme interactions [87].

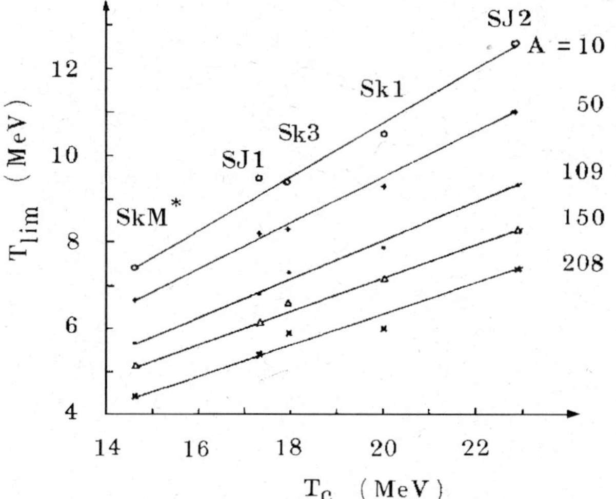

Fig. 3. Correlation between limiting temperature and T_C for nuclear matter [87].

Employing a variety of Skyrme interactions, Song and Su derived a mass-dependent scaling of the correlation of limiting temperatures with the critical temperature of nuclear matter [87]. Their results are shown in figs. 2 and 3. A similar scaling exists when other model interactions are employed [87–90]. The limiting temperatures are found to be quite sensitive to T_C and rather insensitive to the nuclear incompressibility, K_{inc}. These results, together with gathering experimental evidence of multi fragment disassembly modes at higher excitation energies, spurred the development of both statistical [4,5,94–96] and dynamic [97–104] models capable of exploring the multifragmentation process in much greater detail. Such models have made much more detailed predictions on the nature of multifragmentation processes and the excitation energy dependence of the temperature, *i.e.*, the nuclear caloric curve.

In statistical models of multifragmentation, increasing excitation energies lead to the onset of a plateau in the temperature. This plateau occurs at a "cracking energy" which may be associated with the Coulomb instability and leads to multiple fragment production [5,94,95]. Such plateaus are also observed within the framework of classical molecular dynamics calculations [104–106] and quantum molecular dynamics calculations [107,108].

Concurrently with these theoretical studies, many experimental investigations have resulted in the construction of caloric curves [14,109–120]. In ref. [121], a number of experimental caloric curves derived from charged particle observables were compared. The nature of the experimental collision dynamics encountered in the caloric-curve measurements is generally such that the masses of the excited nuclei that are produced in these experiments vary as the excitation energy varies. Although data from different experiments exhibit significant fluctuations, caloric curves may be constructed for different mass regions selected from the available data. Such curves, presented in fig. 4, are qualitatively similar and flatten into broad plateaus at higher excitation energies. Similar behavior is seen in a caloric curve derived using a very different technique, observation of "second chance" Bremsstrahlung gamma-ray emission for a series of reactions which span a wide range of mass [24,25,122].

Parameterized in terms of an inverse Fermi gas level density parameter, $k = T^2/(E_x/A)$, the data indicate that k initially increases from $k \sim 8$ to $k \sim 13$ as the excitation increases. Such behavior has been explained in models which take into account the change in effective nucleon mass with excitation energy [123–127]. Beyond excitation energies corresponding to the onset of the plateau, the derived values of k become progressively smaller reflecting the limiting temperature behavior seen in fig. 4. An analysis of this trend, carried out assuming a nondissipative uniform Fermi gas model, indicates a rapidly increasing expansion of the nuclei with increasing excitation energy above the excitation energy where the limiting temperatures are first reached [128]. Further evidence for this expansion is found in significant barrier lowering for ejected clusters [21,129,130] as well as in coalescence radius determinations [131]. Recent papers modeling the caloric curves assuming an expanding mononucleus are in generally good agreement with the experimental data [126,127]. Nevertheless, the effect of clustering on the level density of the system needs to be better understood. In models that include clusterization, the possible existence of negative heat capacities near the onset of the plateau has been extensively discussed [5,94,132–134] and some experimental evidences for observations of such negative heat capacities have been presented [135–137]. However, these interpretations have been subjected to some criticism [138,139]. It appears that, at present, the evidence for negative heat capacities is much more secure in analogous measurements of caloric curves for atomic clusters [140,141].

Although fig. 4 shows that there is a considerable spread in limiting temperature data from different measurements, the average temperatures in the plateau regions for each mass window have been employed to study the mass dependence of limiting temperatures. The lim-

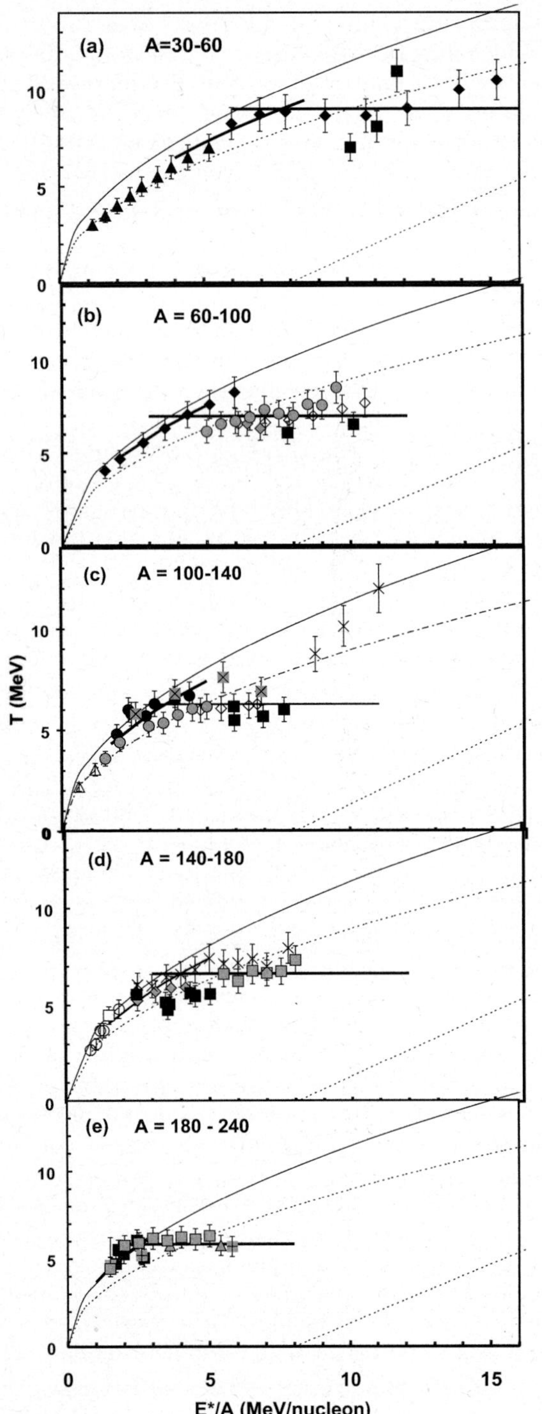

Fig. 4. Caloric curves for five selected regions of mass. See ref. [121].

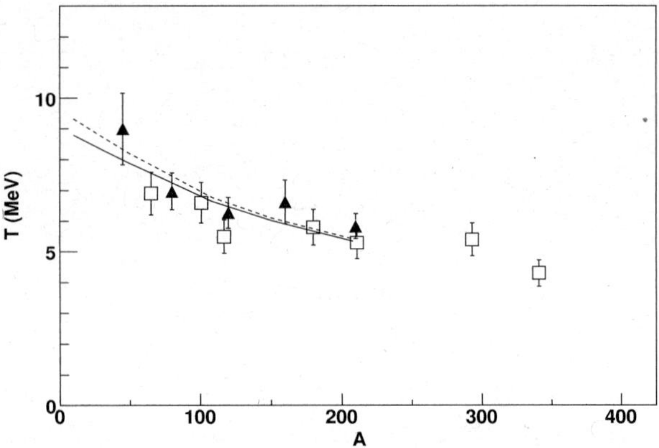

Fig. 5. Limiting values of temperature *vs.* mass. Temperatures derived from double isotope ratio measurements are indicated by solid diamonds. Temperatures derived from thermal Bremsstrahlung measurements are indicated by open squares. The lines represent the calculated limiting temperatures from references [90] (dashed line) and [82] (solid line).

ture is in good agreement with the limiting temperature deduced from the caloric curve. It appears that the point identified as the critical point by the droplet analysis is the point of initial flattening of the caloric curve.

4.2 Caloric curves and the nuclear equation of state

In ref. [143], the mean variation of T_{lim}/T_C with A determined from commonly used microscopic theoretical calculations has been used, together with the five experimental limiting temperatures reported in ref. [121], to extract a critical temperature of nuclear matter of $16.6 \pm 0.86\,\mathrm{MeV}$. Using a relationship between parameters used to describe nuclear matter suggested by Kapusta [144] and Lattimer and Swesty [145], both the incompressibility and the effective mass can be derived. The compressibility modulus for moderately excited nuclei, determined from the critical temperature in this manner is consistent with that determined from measurements of the nuclear Giant Monopole Resonance [146]. In attempts to derive the nuclear matter coexistence curve from Fisher scaling analysis nuclear matter critical temperatures of 10 to 14 MeV have been obtained [147]. These values are surprisingly close to the values derived for the finite systems studied [148]. Here, again, the temperature dependence of the surface energy plays an important role in the extrapolation to nuclear matter.

4.3 Temperature evolution

Both dynamic and thermodynamic considerations lead us to expect significant temperature changes as the reactions progress. Thus, probing the thermal evolution of the system can provide considerably more information on the history and degree of equilibration of the collisionally heated systems. In some recent measurements, the kinetic energy

iting temperatures characterizing these plateaus decrease with increasing nuclear mass (see fig. 5).

Both, the flattening of the caloric curve and the decrease of limiting temperature with increasing mass, are in agreement with a large number of theoretical calculations. Employing Fisher scaling analysis, Elliott *et al.* [142] concluded that the critical temperature for a nucleus with $A \sim 168$ at $E_x/A = 3.8\,\mathrm{MeV}$ is $6.7\,\mathrm{MeV}$. This tempera-

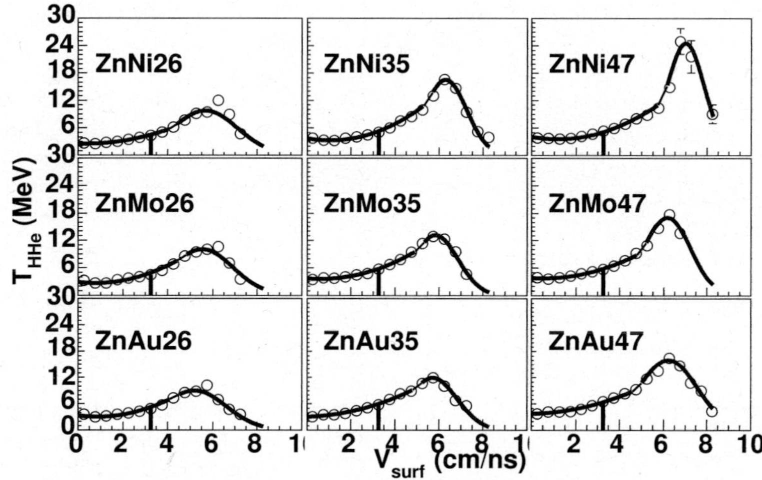

Fig. 6. T_{HHe} *vs.* surface velocity. See text. Horizontal bars are at 3–3.5 cm/ns corresponding to entry into the evaporation phase of the reaction. Solid lines indicate fits to data.

variation of emitted light clusters has been employed as a clock to explore the time evolution of the temperature for thermalizing composite systems in the reactions of $26A$, $35A$ and $47A$ MeV ^{64}Zn with ^{58}Ni, ^{92}Mo and ^{197}Au [149]. Figure 6 presents experimental results for the double isotope ratio temperatures as a function of velocity in the nucleon center-of-mass frame.

For the earliest stages of the collision, transport model calculations demonstrate a strong correlation of decreasing surface velocity with increasing time [149]. For each system investigated, the double isotope ratio temperature curve exhibits a high maximum apparent temperature, in the range of 10–25 MeV, at high ejectile velocity. These maximum values increase with increasing projectile energy and decrease with increasing target mass and are much higher than the limiting temperatures determined from caloric-curve measurements in similar reactions.

In each case, the temperature then decreases monotonically as the velocity decreases below the velocity at which the maximum is seen. The maxima in the temperature curves appear to signal the achievement of chemical equilibrium (a pre-requisite for employment of double isotope temperatures) at least on a local basis. They are quite comparable to those reported for QMD transport model calculations of the maximum and average temperatures and densities achieved in symmetric or near symmetric heavy ion collisions [150]. Those results strongly suggest the presence of an initial hot, locally equilibrated, participant zone surrounded by colder spectator matter. A similar picture is obtained in the AMD-V calculations of ref. [151]. For each different target, the subsequent cooling as the ejectile velocity decreases is quite similar. Temperatures comparable to those of limiting temperature systematics are reached at times when AMD-V transport model calculations predict entry into the final evaporative or fragmentation stage of de-excitation of the hot composite systems. Calibration of the time-scales using AMD-V calculations indicate that this occurs at times ranging from ~ 135 fm/c for the Ni target to ~ 165 fm/c for Au [149].

5 Conclusions

From all what was said, it is clear that it is not straightforward to determine the thermodynamical temperature T $(1/T = \partial S/\partial E)$ of a nuclear system. Important theoretical progress in understanding the conceptual differences in the apparent temperature values obtained from the different experimental methods has been made in the last ten years. Also on the experimental side, efforts have been made in order to obtain more information on the influence of the reaction dynamics on the apparent temperature values.

An enormous complexity of effects involved in the interpretation of apparent-temperature measurements has been evidenced. Understanding of these effects helped in approaching the results obtained using different thermometer methods. The question is whether we still have more complexity to expect. Le Fèvre *et al.* [55] have shown that apparent temperatures, even if uncertain in absolute value, seem to be surprisingly robust in showing signatures of phase transitions. In other words, caloric curves obtained using some of the above-mentioned thermometer methods can still carry the signal of the phase transition in a system with conserved energy.

On the other hand, for the systems already studied, the differences in the entrance channel isospins and in the first stage dynamics lead to some variation of the isospin of the fragmenting nuclei. However, the systematic uncertainties in the present measurements are such that sensitivity to this variable is not obvious. In the future, extension of caloric-curve measurements to nuclei far from stability should be very instructive. With the proposed radioactive beam facilities it will be possible to employ caloric-curve measurements to determine the critical parameters for quite asymmetric nuclei. In the future, determination of the nuclear level densities, of the limiting temperatures and of critical temperatures for asymmetric nuclear matter will play a significant role in providing a means to establish the isospin dependence of the nuclear equation of state and the nature of the phase transition in asymmetric nuclear matter.

References

1. H.A. Bethe, Rev. Mod. Phys. **9**, 69 (1937).
2. V.F. Weisskopf, Phys. Rev. **52**, 295 (1937).
3. E. Suraud, Ch. Grégoire, B. Tamain, Prog. Part. Nucl. Phys. **23**, 357 (1989).
4. D.H.E. Gross, Rep. Prog. Phys. **53**, 605 (1990).
5. J.P. Bondorf *et al.*, Phys. Rep. **257**, 133 (1995).
6. D.J. Morrissey, Annu. Rev. Nucl. Part. Sci. **44**, 27 (1994).
7. D.H.E. Gross, Entropy **6**, 158 (2004).
8. S. Levit, P. Bonche, Nucl. Phys. A **437**, 426 (1985).
9. S. Albergo *et al.*, Nuovo Cimento A **89**, 1 (1985).
10. J. Pochodzalla *et al.*, Phys. Rev. Lett. **75**, 1040 (1995).
11. D.J. Morrissey *et al.*, Phys. Lett. B **148**, 423 (1984).
12. D.J. Morrissey *et al.*, Phys. Rev. C **32**, 877 (1985).
13. J. Pochodzalla *et al.*, Phys. Rev. Lett. **55**, 177 (1985).
14. J. Pochodzalla *et al.*, Phys. Rev. C **35**, 35 (1987).
15. G.J. Kunde *et al.*, Phys. Lett. B **272**, 202 (1991).
16. V. Serfling *et al.*, Phys. Rev. Lett. **80**, 3928 (1998).
17. M. Veselsky *et al.*, Phys. Lett. B **497**, 1 (2001).
18. M. Veselsky *et al.*, Phys. Part. Nucl. **36**, 213 (2005).
19. G.D. Westfall, Phys. Lett. B **116**, 118 (1982).
20. B.V. Jacak *et al.*, Phys. Rev. Lett. **51**, 1846 (1983).
21. G.D. Westfall *et al.*, Phys. Rev. C **17**, 1368 (1978).
22. T. Odeh *et al.*, Phys. Rev. Lett. **84**, 4557 (2000).
23. J. Gosset *et al.*, Phys. Rev. C **16**, 629 (1977).
24. D.G. d'Enterria *et al.*, Phys. Lett. B **538**, 27 (2002).
25. R. Ortega *et al.*, Nucl. Phys. A **734**, 541 (2004).
26. K.-H. Schmidt *et al.*, Phys. Lett. B **300**, 313 (1993).
27. K.-H. Schmidt *et al.*, Nucl. Phys. A **710**, 157 (2002).
28. S. Shlomo *et al.*, Rep. Prog. Phys. **68**, 1 (2005).
29. Al.H. Raduta *et al.*, Nucl. Phys. A **671**, 609 (2000).
30. M.B. Tsang *et al.*, Phys. Rev. Lett. **78**, 3836 (1997).
31. D.H. Boal, Phys. Rev. C **30**, 749 (1984).
32. A.S. Goldhaber, Phys. Rev. C **17**, 2243 (1978).
33. W.A. Friedman, W.G. Lynch, Phys. Rev. C **28**, 16 (1983).
34. I. Dostrowsky, P. Robinowitz, P. Bivins, Phys. Rev. **111**, 1659 (1958).
35. R. Pühlhofer, Nucl. Phys. A **280**, 267 (1977).
36. A. Gavron, Phys. Rev. C **21**, 230 (1980).
37. M. Blann, Phys. Rev. C **23**, 205 (1981).
38. R.J. Charity *et al.*, Nucl. Phys. A **483**, 371 (1988).
39. S. Ban-Hao, D.H.E. Gross, Nucl. Phys. A **437**, 643 (1985).
40. A.S. Goldhaber, Phys. Lett. B **53**, 306 (1974).
41. D.J. Fields *et al.*, Phys. Rev. C **30**, 1912 (1984).
42. H. Stöcker *et al.*, Z. Phys. A **303**, 259 (1981).
43. W.A. Friedman, W.G. Lynch, Phys. Rev. C **28**, 950 (1983).
44. W.A. Friedman, Phys. Rev. C **37**, 976 (1988).
45. C.B. Chitwood *et al.*, Phys. Rev. C **34**, 858 (1986).
46. P.J. Siemens, J.O. Rasmussen, Phys. Rev. Lett. **42**, 880 (1979).
47. W.A. Friedmann, Phys. Rev. C **42**, 667 (1990).
48. N. Marie *et al.*, Phys. Rev. C **58**, 256 (1998).
49. S. Hudan *et al.*, Phys. Rev. C **67**, 064613 (2003).
50. H. Müller, B.D. Serot, Phys. Rev. C **52**, 2072 (1995).
51. V. Baran *et al.*, Nucl. Phys. A **703**, 603 (2002).
52. D. Henzlova, PhD Thesis, University of Prague (March 2006).
53. L. Shi, P. Danielewicz, Phys. Rev. C **68**, 064604 (2003).
54. H. Xi *et al.*, Phys. Lett. B **431**, 8 (1998).
55. A. Le Fèvre *et al.*, Nucl. Phys. A **657**, 446 (1999).
56. Al.H. Raduta *et al.*, Phys. Rev. C **59**, R1855 (1999).
57. V.E. Viola *et al.*, Phys. Rev. C **59**, 2660 (1999).
58. S. Hudan *et al.*, Phys. Rev. C **71**, 054604 (2005).
59. M.V. Ricciardi *et al.*, Nucl. Phys. A **733**, 299 (2004).
60. S.K. Samaddar *et al.*, Phys. Rev. C **71**, 011601 (2005).
61. D.H.E. Gross, Phys. Rep. **279**, 119 (1997).
62. J.A. Hauger *et al.*, Phys. Rev. Lett. **77**, 235 (1996).
63. Y.G. Ma *et al.*, Phys. Lett. B **390**, 41 (1997).
64. T.X. Liu *et al.*, Phys. Rev. C **69**, 014603 (2004).
65. M. Colonna *et al.*, Phys. Rev. Lett. **88**, 122701 (2002).
66. T. Sill *et al.*, Phys. Rev. C **69**, 014602 (2004).
67. E. Vient *et al.*, Nucl. Phys. A **700**, 555 (2002).
68. S. Pal *et al.*, Nucl. Phys. A **586**, 466 (1995).
69. H. Xi *et al.*, Phys. Rev. C **59**, 15667 (1999).
70. B. Tamain, D. Durand, University of Caen Report LPCC 96-16 (1996).
71. P. Chomaz, *Proceedings of the International Nuclear Physics Conference INPC2001, Berkeley CA, USA, 2001*, edited by E. Norman, L. Schröder, G. Wozniak, AIP Conf. Proc., Vol. **610** (Melville, New York, 2002) p. 167.
72. A. Bonasera *et al.*, Riv. Nuovo Cimento **23**, No. 2 (2000).
73. S. Das Gupta *et al.*, nucl-th/0009033.
74. J.P. Blaizot, Phys. Rep. **64**, 171 (1980).
75. H. Jaqaman *et al.*, Phys. Rev. C **27**, 2782 (1983); **29**, 2067 (1984).
76. W.D. Myers, W.J. Swiatecki, Phys. Rev. C **57**, 3020 (1998).
77. J. Zimanyi, S.A. Moszkowski, Phys. Rev. C **42**, 1416 (1990).
78. P.G. Reinhard *et al.*, Z. Phys. A **323**, 13 (1986).
79. J.P. Blaizot *et al.*, Nucl. Phys. A **591**, 435 (1995).
80. R.J. Furnstahl *et al.*, Nucl. Phys. A **615**, 441 (1997).
81. G. Lalazissis *et al.*, Phys. Rev. C **55**, 540 (1997).
82. M. Malheiro *et al.*, Phys. Rev. C **58**, 426 (1998).
83. E. Chabanat *et al.*, Nucl. Phys. A **627**, 710 (1997).
84. M. Farine *et al.*, Nucl. Phys. A **615**, 135 (1997).
85. P. Bonche *et al.*, Nucl. Phys. A **427**, 278 (1984).
86. J. Besprosvany, S. Levit, Phys. Lett. B **217**, 1 (1989).
87. H.Q. Song, R.K. Su, Phys. Rev. C **44**, 2505 (1991).
88. H.Q. Song *et al.*, Phys. Rev. C **47**, 2001 (1993); **49**, 2924 (1994).
89. Y. Zhang *et al.*, Phys. Rev. C **54**, 1137 (1996).
90. M. Baldo, L.S. Ferreira, Phys. Rev. C **59**, 682 (1999).
91. J.N. De *et al.*, Phys. Rev. C **55**, R1641 (1997).
92. P. Wang *et al.*, Nucl. Phys. A **748**, 226 (2005).
93. A.L. Goodman *et al.*, Phys. Rev. C **30**, 851 (1984).
94. W.A. Friedman, Phys. Rev. Lett. **60**, 2125 (1988).
95. J. Aichelin *et al.*, Phys. Rev. C **37**, 2451 (1988).
96. F. Gulminelli, D. Durand, Nucl. Phys. A **615**, 117 (1997).
97. J. Schnack, H. Feldmeier, Phys. Lett. B **409**, 6 (1997).
98. C. Fuchs *et al.*, Nucl. Phys. A **626**, 987 (1997).
99. H. Kruse *et al.*, Phys. Rev. C **31**, 170 (1985).
100. H. Horiuchi, Nucl. Phys. A **583**, 297c (1995).
101. A. Ono, Phys. Rev. C **59**, 853 (1999).
102. P. Chomaz *et al.*, Phys. Rev. Lett. **73**, 3512 (1994).
103. W. Nörenberg *et al.*, Eur. Phys. J. A **9**, 327 (2000); **14**, 43 (2002).
104. A. Chernomoretz *et al.*, Phys. Rev. C **64**, 044605 (2001).

105. A. Strachan, C.O. Dorso, Phys. Rev. C **58**, 632 (1998).

106. M.J. Ison *et al.*, Physica A **341**, 389 (2004).

107. Y. Sugawa, H. Horiuchi, Prog. Theor. Phys. **105**, 131 (2001).

108. T. Furuta, A. Ono, Prog. Theor. Phys. Suppl. **156**, 147 (2004).

109. K. Hagel *et al.*, Nucl. Phys. A **486**, 429 (1988).

110. R. Wada *et al.*, Phys. Rev. C **39**, 497 (1989).

111. D. Cussol *et al.*, Nucl. Phys. A **561**, 298 (1993).

112. M. Gonin *et al.*, Phys. Rev. C **42**, 2125 (1990).

113. T. Odeh, GSI Report Diss. 99-15, August 1999, and references therein.

114. J.A. Hauger *et al.*, Phys. Rev. C **62**, 024616 (2000).

115. R. Wada *et al.*, Phys. Lett. B **423**, 21 (1998).

116. K.B. Morley *et al.*, Phys. Rev. C **54**, 737; 749 (1996).

117. J. Cibor *et al.*, Phys. Lett. B **473**, 29 (2000).

118. K. Hagel *et al.*, Phys. Rev. C **62**, 034607-1 (2000).

119. A. Ruangma *et al.*, Phys. Rev. C **66**, 044603 (2002).

120. B. Borderie, *Proceedings of the Conference Bologna 2000, Bologna, Italy, 2000*, edited by G.C. Bonsignori *et al.*, Vol. **1**, *Nucleus-Nucleus Collisions* (World Scientific, Singapore, 2001) p. 187, preprint nucl-ex/0102016.

121. J.B. Natowitz *et al.*, Phys. Rev. C **65**, 034618 (2002).

122. D.G. d'Enterria *et al.*, Phys. Rev. Lett. **87**, 22701 (2002).

123. R. Hasse, P. Schuck, Phys. Lett. B **179**, 313 (1986).

124. S. Shlomo, J.B. Natowitz, Phys. Lett. B **252**, 187 (1990).

125. S. Shlomo, J.B. Natowitz, Phys. Rev. C **44**, 2878 (1991).

126. L.G. Sobotka *et al.*, Phys. Rev. Lett. **93**, 132702 (2004).

127. J.N. De *et al.*, Phys. Lett. B **638**, 160 (2006).

128. J.B. Natowitz *et al.*, Phys. Rev. C **66**, 031601 (2002).

129. A.S. Hirsch *et al.*, Phys. Rev. C **29**, 508 (1984).

130. V.E. Viola *et al.*, Phys. Rev. Lett. **93**, 132701 (2004).

131. J. Cibor *et al.*, in *Isospin Physics in Heavy-Ion Collisions at Intermediate Energies*, edited by Bao-An Li, W.U. Schröder (NOVA Science Publishers, Inc., New York, 2001) p. 283.

132. D.H.E. Gross *et al.*, Ann. Phys. **5**, 446 (1996).

133. P. Chomaz *et al.*, Phys. Rev. Lett. **85**, 3587 (2000).

134. P. Chomaz, F. Gulminelli, Phys. Lett. B **447**, 221 (1999).

135. D.H.E. Gross, J.F. Kenney, J. Chem. Phys. **122**, 224111 (2005).

136. A. Chbihi *et al.*, Eur. Phys. J. A **5**, 251 (1999).

137. M. D'Agostino *et al.*, Nucl. Phys. A **650**, 329 (1999).

138. W. Thirring *et al.*, Phys. Rev. Lett. **91**, 130601 (2003).

139. X. Campi *et al.*, Phys. Rev. C **71**, 041601 (2005).

140. M. Schmidt *et al.*, Phys. Rev. Lett. **87**, 203402 (2001).

141. D.H.E. Gross, P.A. Hervieux, Z. Phys. D **35**, 27 (1995).

142. J.B. Elliott *et al.*, Phys. Rev. Lett. **88**, 042701 (2002).

143. J.B. Natowitz *et al.*, Phys. Rev. Lett. **89**, 212701 (2002).

144. J. Kapusta, Phys. Rev. C **29**, 1735 (1984).

145. J.M. Lattimer, F.D. Swesty, Nucl. Phys. A **535**, 331 (1991).

146. D.H. Youngblood *et al.*, Phys. Rev. Lett. **82**, 691 (1999).

147. L.G. Moretto *et al.*, Phys. Rev. C **72**, 064605 (2005).

148. J.B. Elliott *et al.*, Phys. Rev. C **67**, 024609 (2003).

149. J. Wang *et al.*, Phys. Rev. C **72**, 024603 (2005).

150. A.D. Sood, R.K. Puri, Phys. Rev. C **70**, 034611 (2004).

151. R. Wada *et al.*, Phys. Rev. C **69**, 044610 (2004).

Eur. Phys. J. A **30**, 215–226 (2006)

DOI 10.1140/epja/i2006-10118-5

THE EUROPEAN
PHYSICAL JOURNAL A

Calorimetry

V.E. Viola[1,a] and R. Bougault[2]

[1] IUCF and Department of Chemistry, Indiana University, Bloomington, IN 47405, USA
[2] LPC/Ensicaen, 6 Bd du Maréchal Juin, 14050 Caen Cedex, France

Received: 12 June 2006 /
Published online: 25 October 2006 – © Società Italiana di Fisica / Springer-Verlag 2006

Abstract. Methods for determining the heat content E^*/A of hot nuclei formed in energetic nuclear reactions are discussed. The primary factors involved in converting raw data into thermal physics distributions include: 1) design of the detector array, 2) constraints imposed by the physics of the reaction mechanism, and 3) assumptions involved in converting the filtered data into E^*/A. The two primary sources of uncertainty in the calorimetry are the elimination of nonequilibrium emissions from the event components and accounting for the contribution of neutron emission to the excitation energy sum.

PACS. 25.40.Ve Other reactions above meson production thresholds (energies > 400 MeV) – 25.70.Pq Multifragment emission and correlations – 25.70.-z Low and intermediate energy heavy-ion reactions

1 Introduction

In order to describe the thermodynamic behavior of hot nuclear matter formed in energetic nuclear reactions, a knowledge of the heat content is fundamental. Stimulated by the caloric-curve measurements of the ALADiN group [1], extensive effort has been devoted to the determination of this energetic factor over the past decade. For hot nuclei the heat content is expressed in terms of the excitation energy E^*. Since nuclei are finite systems, the number of nucleons A is also necessary, so that the relevant thermodynamic quantity is E^*/A. This paper is devoted to the factors involved in evaluating E^*/A and the limitations imposed on the results due to experimental and physics constraints.

Ideally, the dynamics of the entrance channel lead to well-defined disintegrating ensembles and the calorimetric measurement of E^*/A requires an apparatus that collects the total kinetic energy (K), charge (Z) and mass (A) of all charged particles and neutrals that compose a given event. With this information each event can be reconstructed, permitting the calculation of E^* and A of the source, where

$$E^*_{source} = \sum_i K_{cp}(i) + \sum_j K_n(j) - Q(i,j) \qquad (1)$$

and

$$Z_{source} = \sum_i Z_{cp}(i), \qquad (2)$$

$$A_{source} = \sum_i A_{cp}(i) + \sum_j A_n(j). \qquad (3)$$

Here K_{cp} is the kinetic energy for all LCPs (H and He), IMFs $(3 \leq Z \lesssim 20)$ and heavy residues $(A \gtrsim 20)$. K_n is the neutron kinetic energy and energy of gammas, and the removal energy $(-Q)$ is the negative of the reaction Q-value. All kinetic energies should be calculated in the source frame. The charge and mass of the source are given by Z_{source} and A_{source}; the charge of the emitted charged particles is Z_{cp}, and A_{cp} and A_n are the mass numbers of the charged particles and neutrons, respectively.

However, no calorimeter is perfect and, moreover, the entrance channel dynamics may lead to several sources that produce particles. Thus, in order to extract E^*/A of a given source from data, one must construct a detector filter that converts the measured distributions into final data. Among the sources of energy, charge and mass loss or contamination are:

1. acceptance limitations imposed by the construction of the apparatus and the properties of its constituent detectors [2];
2. physics uncertainties, most importantly the criteria for accepting only studied source particles that are classified as "equilibrium-like"; *i.e.*, pre-equilibrium, midrapidity emissions and possible contamination from other source productions (target-like and fusion events if projectile-like events are under study) must be removed from the sums for eqs. (1)-(3).
3. Measurement uncertainties, most importantly particle characterization $(Z, A, \text{angles and energy})$ and the characterization of the source frame used for eq. (1).

a e-mail: viola@indiana.edu

Table 1. Systems and detectors reviewed for calorimetry measurements.

Reaction type	Detector/Collaboration	References
$\bar{p} + A$	Berlin n/cp Ball	[3,4]
$\bar{p}, p, \pi, {}^{3}\mathrm{He} + A$	ISiS	[5,6]
$p, {}^{4}\mathrm{He}, C + A$	FASA	[7]
$A + C$	EOS	[8,9]
$A + A$	ALADiN	[1,10]
$A + A$	INDRA	[11]
$A + A$	TAMU	[12,13]
$A + A$	Chimera	[14]
$A + A$	Laval array	[15]
$A + A$	Superball	[16]
$A + A$	Multics	[17–19]

4. Finally, since no two-detector arrays have the same acceptance, differences in procedures for converting the filtered data into E^*/A must be examined.

In the following sections, these issues are surveyed along with their inherent uncertainties. The analysis is drawn from those references in table 1, which are representative (but not complete) examples of the procedures currently employed in nuclear calorimetry.

2 The detection filter

The existing multifragmentation detector arrays are of various types. Most charged-particle detection involves some combination of silicon, gas ionization chamber and CsI scintillator telescopes for Z (and in some cases A) identification [3–7,11,12,14,15]. The EOS experiment [8, 9] employed a time projection chamber (TPC) and ALADiN utilized a magnetic spectrometer [1,10], both coupled to several detector arrays. Neutrons have been measured with large tanks of Gd-loaded scintillator liquid and via time-of-flight techniques [3,4,16]. Few experiments have been performed with simultaneous Z and A identification for the entire multifragmentation yield. For relativistic beams, the EOS TPC is well suited for complete Z and A identification in the forward laboratory hemisphere and the ALADiN experiments permit A detection over a significant mass range. Medium- and heavy-fragment identification by most other arrays relies on mass balance techniques and/or partial information (A or Z), with N/Z assumptions. In developing a reliable detector filter, several factors must be considered, as enumerated below. The filter must then be tested to ensure that it reproduces input from an appropriate simulation.

2.1 Solid-angle acceptance

In constructing any detector array, allowance must be made for beam entry/exit ports and any shadowing by the target. For light-ion reactions, for which the laboratory angular distributions are nearly isotropic, target shadowing must be treated carefully. For inverse kinematics or $A + A$ reactions, the effect of the exit port dead solid angle is projectile energy dependent. This effect may be controlled for high-energy beams by using a magnet. In any case the resulting geometric-acceptance factor must then be applied to all events, which due to fluctuations, may either over- or under-correct the data. For most detectors, geometric acceptance ranges from about 75% to the nearly complete acceptance for the EOS TPC in inverse kinematics. But those numbers should be reconsidered when speaking of real acceptance since the solid-angle acceptance depends on impact parameter and type of particle.

2.2 Detector granularity

Since the final states in multifragmentation reactions may involve large numbers of particles, high detector granularity ($N > 100$) is essential to minimize multiple-hit misidentification of fragments. In addition, angular information is required to test whether events classified as "equilibrium-like" meet the isotropic emission standard for a randomized system and is fundamental to rebuild the studied source velocity. The detector granularity and the angular resolution of a detector are technically different because they are related to different issues.

2.3 Detector characteristics —charged particles

The technical challenge of charged-particle detection is related to the large energetic range of particle detection and the necessity of identifying everything from light charged particles up to heavy residues.

- *Energy identification thresholds:*
  Ideally, for eqs. (1) and (2) Z and A identification of the products is required. Practically, this is almost done for light charged particles and light IMFs with an energetic threshold that depends, for example on the ΔE-E technique, on the thickness of the ΔE. Below the threshold and for all other charged particles either A (ΔE-E technique) or Z (time-of-flight technique) remains unknown and mean values are used in the calorimetry.
- *Energy thresholds:*
  The low-energy component of spectra measured with ΔE-E particle identification telescopes is constrained by the thickness of the ΔE element. Lowest thresholds are obtained with gas ionization chambers, essential for light-ion–induced reactions or excited target-like source reconstruction. The kinematic boost for fragments produced in heavy-ion reactions permits the use of higher stopping power, Si and CsI ΔE elements in the forward direction. Since the energy threshold depends on the detected species, this may affect the overall real detector acceptance. Ideally, corrections to the energy sum must be made for the missing part of the spectrum due to threshold effects.

— Detector resolution:
Si semiconductor detectors provide the highest energy resolution for the determination of energy loss and total energy K. For this reason Si-Si telescopes can provide both Z and A information for a significant range of the multifragmentation spectrum, limited by the minimum ΔE thickness and maximum E thickness. Si detectors are also used for time-of-flight A identification. Because of their minimum stopping power, gas ionization chambers are most effective as ΔE detectors for fragments with low kinetic energy per nucleon but usually do not yield both Z and A identification. Although CsI provides the poorest energy resolution, the ability to form very thick crystals makes it ideal for detecting the most energetic particles. Depending on the energy of the emitted particles, TPC measurements usually yield energy resolution intermediate between Si and CsI for IMFs and heavier fragments. In addition, plastic scintillators and pulse-shape discrimination have been employed, as well as silicon pulse-shape analysis, to identify fragments.

2.4 Detector characteristics —neutrons

The greatest experimental uncertainty in determining the total kinetic-energy sum is the contribution from neutron emission, for which multiplicities are greater than or comparable to charged particles. The energy associated with gamma rays is usually assumed to be small. The neutron kinetic-energy spectrum is measured via time-of-flight techniques, using fast plastic/liquid scintillators. Such measurements sample only a small fraction of 4π because of the spatial limitations imposed by the flight path. Hence, they yield only limited multiplicity information. Neutron multiplicities and charged-particle correlations have been determined with $\approx 4\pi$ tanks of Gd-loaded liquid scintillator [4,12,16]. Neutron detection must be corrected for energy-dependent efficiency losses, which contribute to the multiplicity uncertainty. When neutrons are not detected, mean values for neutron multiplicity and total neutron kinetic energy are assumed. In the case of full identification of the source charged products, the neutron multiplicity is accessible via a source N/Z hypothesis on an event-by-event basis. For the mean total neutron kinetic energy, several techniques based on proton characteristics or on experimental results or on average effective-temperature estimates are employed.

2.5 Replacement of unmeasured quantities

As has been described, there exist several methods to replace an unmeasured quantity in eqs. (1) and (2). Each method has its own advantage or disadvantage depending on the experimental context and the best choice depends on the goal of the measurement. Therefore, caution should be taken by keeping only conservative hypotheses, controlled with simulations taking into account the detector acceptance.

2.6 Statistics

The total number of events in the case of highly excited nuclei ($E^*/A \geq 2\,\mathrm{MeV}$) is a function of both the maximum event rate of the detector array and the availability of accelerator time. TPC and neutron tank measurements are limited in counting rate, so statistics are usually low. Detector array studies of $A+A$ reactions exhibit a wide range of statistics, depending on the number of systems studied in a finite amount of accelerator time. Most of the light-ion data have accumulated large numbers of events by using secondary beams over long (months) running time. Since the multifragmentation yield decreases with increasing excitation energy, the accumulation of high statistics is an important factor in determining reliable distributions.

3 Physics issues

In addition to mechanical and detector response contributions to the filter, several physics issues must be addressed, the most important of which is the selection of "equilibrium-like" events. The time evolution of nuclear reactions above the Fermi energy extends from the initial collision phase to an eventually randomized state that decays statistically. Particle emission occurs at all stages as translational projectile energy is converted into internal excitation energy. Selection of only those emissions that have a statistical origin is therefore a nontrivial problem. Other physics issues also come into play; *e.g.* neutron-proton multiplicity correlations, kinematics effects on the event reconstruction process in $A+A$ and inverse kinematics reactions, and the primary N/Z ratio of the emitted fragments. Below, these contributions to the filter are discussed.

3.1 Pre-equilibrium and mid-rapidity emission

Two essential first-order tests of a randomized system are the Maxwellian nature of its spectra and the forward-backward symmetry of its particles in the system frame. This is completely true when the collision dynamics lead to a unique fully equilibrated source of particles. It is not necessarily true for a deformed source or when Coulomb repulsion effects occur within the presence of another source of particles or in the presence of collective effects. Last but not least, the characterization of the source frame velocity is of prime importance for the angular symmetry test, where in some conditions this test may be used for determining the source velocity.

In figs. 1-5 spectra are shown for light-ion–induced reactions at GeV energies. These spectra best illustrate the prompt *vs.* statistical emission ambiguity, since there is only a single emitting source and the difference between the laboratory and center-of-mass velocities is small ($\approx 0.01\,c$). Figure 1 shows neutron spectra at a far backward angle for reactions of 1.2 GeV antiprotons on several targets [3,4]. Two components are present: a low-energy Maxwellian peak and an exponential high-energy tail. The

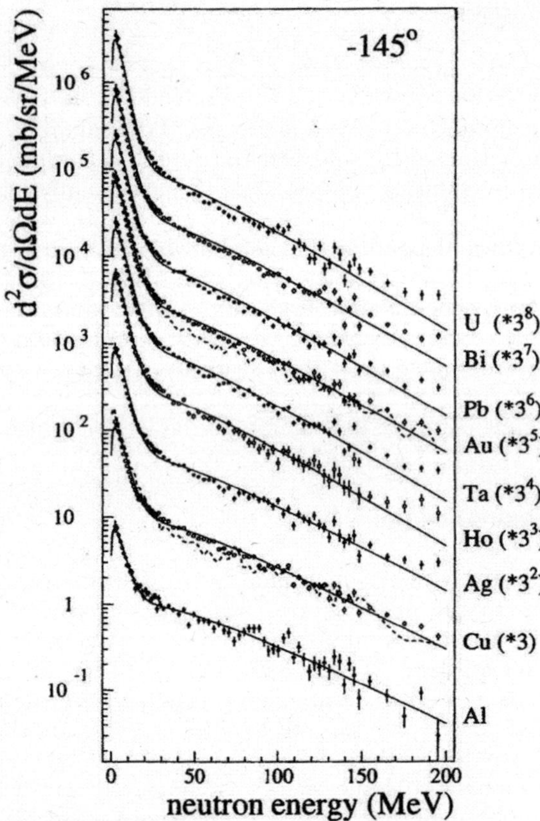

Fig. 1. Neutron kinetic-energy spectra at 145 deg for several target nuclei bombarded with 1.2 GeV antiprotons. Taken from ref. [20].

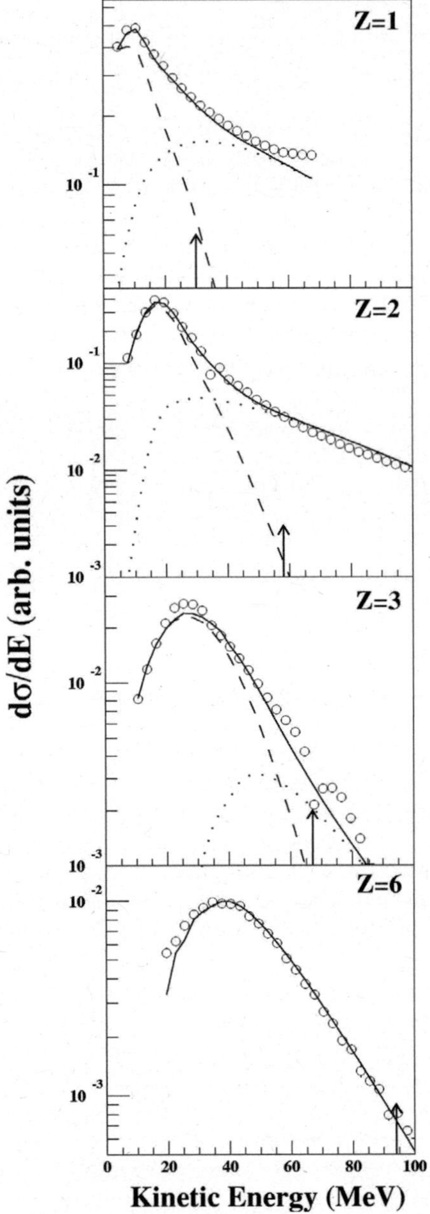

Fig. 2. Angle-integrated kinetic-energy spectra for $Z = 1, 2, 3$ and 6 particles observed in the $8.0\,\text{GeV}/c\ \pi^- + {}^{197}\text{Au}$ reaction. Lines are the result of a two-component moving-source fit composed of a thermal source (dashed line), nonequilibrium source (dotted line) and their sum (solid line). Taken from ref. [6].

former is associated with "equilibrium-like" behavior and the latter with pre-equilibrium emission. Separating these two components on an event-by-event basis is not a transparent procedure. Figure 2 shows inclusive spectra for LCPs and IMFs measured in the $8\,\text{GeV}/c\ \pi^- + {}^{197}\text{Au}$ reaction [6]. These spectra have been decomposed using a two-component moving-source model that assumes a statistical model for the low-energy component (dashed line) and an arbitrary Maxwellian function for the high-energy tail (dotted line). Pre-equilibrium emission is seen to be primarily important for LCPs and decreases in significance as the fragment charge increases.

In fig. 3 the angular dependence of the spectra is shown, along with the moving-source decomposition. Pre-equilibrium emission is forward-focused, whereas the statistical component (when integrated) is nearly isotropic in the lab system. By demanding forward-backward isotropy of the statistical component, the average source velocity can be determined, as well as the fragment energy at which the pre-equilibrium contribution is a negligible contribution to the total yield (cutoff energy). The average source velocity and cutoff energy can be determined from moving-source fits to the data. These are then incorporated into the filter, using a Z-dependent function for the cutoff energy.

As examples of how the separation between statistical and pre-equilibrium affects the determination of E^*/A,

fig. 4 compares the excitation energy distribution for the $\pi^- + {}^{197}\text{Au}$ reaction using both the EOS cutoff energy of $K_{cp}/A = 30\,\text{MeV}$ for all particles [8,9] and that employed by ISiS, $K = 30\,\text{MeV}$ for protons and $K_{cp} = 9Z + 30\,\text{MeV}$ for higher fragment charges [21]. The EOS prescription enhances the probability for high E^*/A values, leading to the difference of nearly 200 MeV at the 1% probability level. (In all other regards the $1\,\text{GeV}\ {}^{197}\text{Au} + {}^{12}\text{C}$ results from EOS are consistent with the $8\,\text{GeV}/c\ \pi^- + {}^{197}\text{Au}$ results from ISiS.)

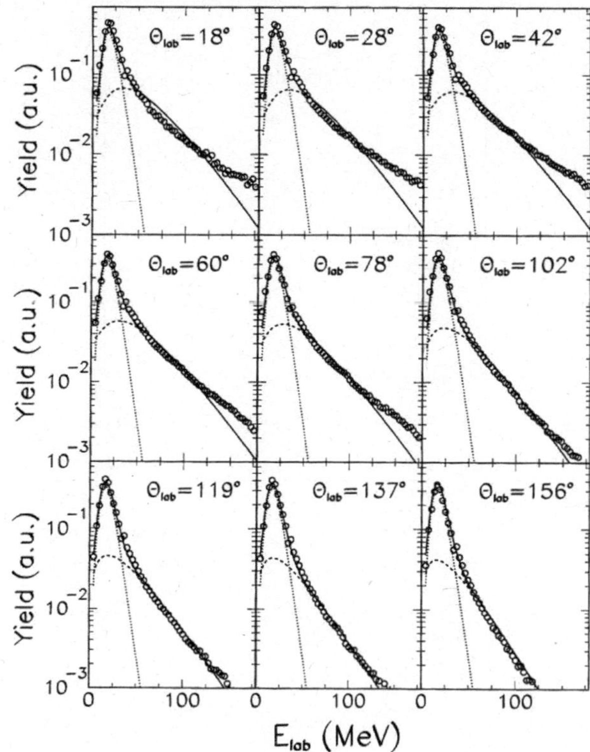

Fig. 3. Alpha-particle kinetic-energy spectra as a function of the angle, from ref. [6].

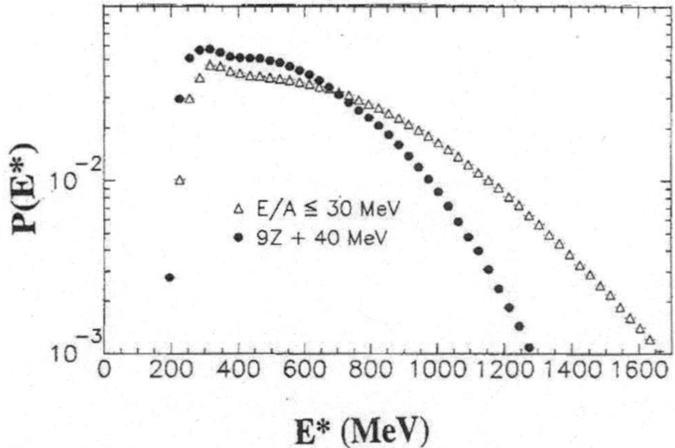

Fig. 4. Comparison of excitation energy distributions for the same experimental data set analyzed with the thermal cutoff energies of refs. [6] (circles) and [9] (triangles).

The pre-equilibrium/statistical separation process is further complicated by the evolution of the spectra with E^*/A as shown in fig. 5. Here H and He spectra, which dominate the pre-equilibrium yield, are shown for $E^*/A =$ 2–4, 4–6 and 6–9 MeV bins. The ISiS cutoff assumptions were derived from the lower-energy bin. However, as E^*/A increases, the spectra evolve into a single Maxwellian distribution, so that separation of the two components becomes more ambiguous.

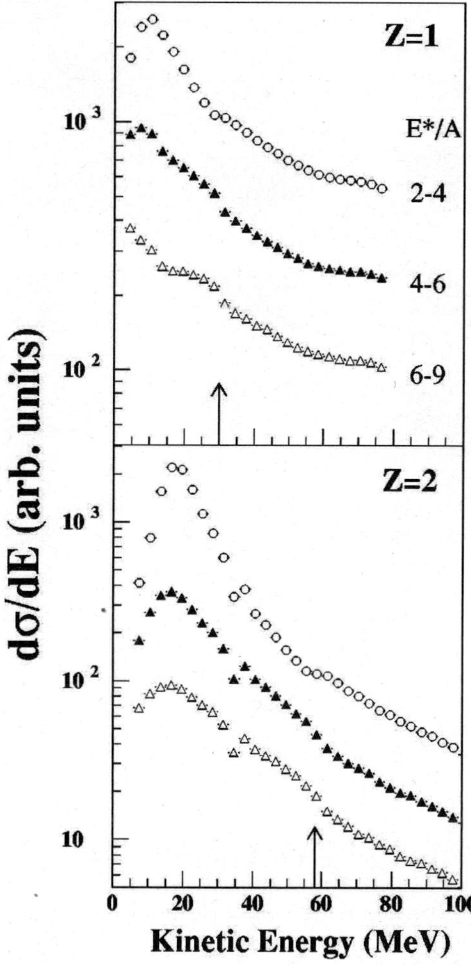

Fig. 5. Kinetic-energy spectra for $Z = 1$ and 2 nuclei as a function of excitation energy for the 8.0 GeV/u reaction. From ref. [6].

For $A + A$ reactions the situation is complicated by the existence of three sources: the projectile-like, target-like and mid-rapidity ones, each of which is then subjected to the same constraints as for light ions. The behavior of the three sources is illustrated in fig. 6, which shows invariant cross-section distributions for $Z = 3$, 6 and 9 fragments as a function of bombarding energy for peripheral ^{197}Au $+ ^{197}$Au data from the INDRA@GSI Collaboration [22]. The separation of the Coulomb rings for the projectile-like source (high y) from the target-like source (low y) becomes increasingly distinct as the bombarding energy increases. For $Z = 3$ the pre-equilibrium skewing of the spectra along the beam axis ($x = 0$) is apparent. For $Z = 6$ and 9 this contribution becomes less important. At lower bombarding energies, the mid-rapidity source masks the projectile-like and target-like statistical spectra, complicating their separation, a procedure that entails the same type of arbitrary assumptions that exist for the light-ion data.

The effect of assumptions about nonequilibrium emission is presented in fig. 7 for peripheral ^{197}Au $+ ^{197}$Au re-

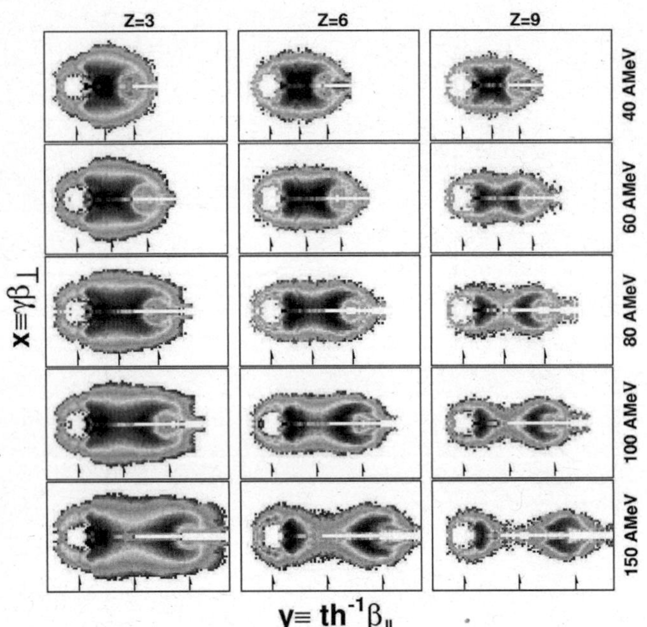

Fig. 6. Invariant cross-section plots for $Z = 3$, 6 and 9 fragments as a function of bombarding energy for the ^{197}Au $+$ ^{197}Au reaction [22].

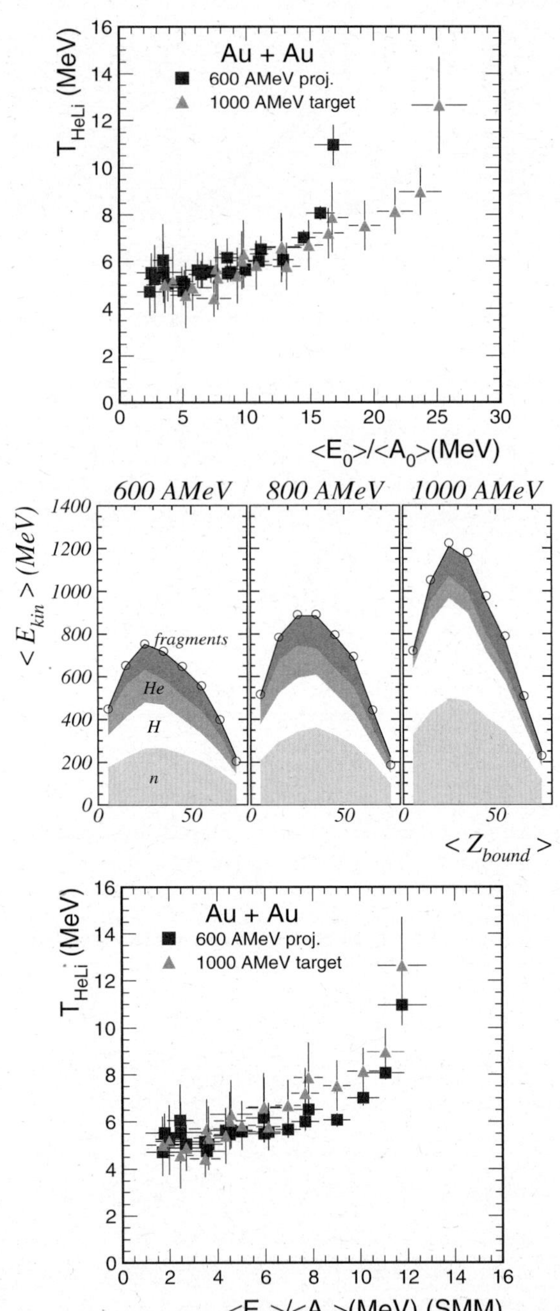

Fig. 7. Top panel: caloric-curve comparison for $E/A = 600$ and 1000 MeV ^{197}Au $+$ ^{197}Au reaction, uncorrected for nonthermal effects. Center panel: average fragment kinetic energies for $E/A = 600$, 800 and 1000 MeV as a function of Z_{bound}. Bottom panel: caloric-curve comparison when corrected for nonthermal effects [23].

action energies of 600, 800 and 1000 GeV, data obtained by the ALADiN group [23]. In the caloric curve shown in the top panel, the E^*/A distribution extends up to 25 MeV for the 1000 A MeV data, and the two caloric curves are not consistent. In the central panel, the relative contributions of neutrons, LCPs and IMFs indicate that the nonequilibrium contributions to the spectra grow significantly between 600 A and 1000 A MeV. As shown in the bottom panel, when corrections are made to eliminate nonequilibrium components, the caloric curves overlap, with maximum E^*/A values reduced to $E^*/A \approx 12$ MeV for both bombarding energies.

For $A + A$ central collisions at lower energies (see for example [24]), a single statistical source can be identified and concerning light charged particles, cuts are applied in order to take into account particle emission at different stages. In general the more massive particles are assumed to originate from a single source, even though their angular distributions in the source frame present some anisotropy. This so-called source deformation depends on bombarding energy. Sophisticated event selections based on isotropy criteria and only operatives in case of quasi-complete detection are also used in order to extract from the central events almost fully equilibrated events. Even in this case, cuts are applied to light charged particles for excitation energy measurement because of pre-equilibrium emission.

3.2 Neutrons

The evaluation of E^*/A via eqs. (1) and (2) requires a knowledge of the kinetic energy and multiplicity of the neutrons in an event. Because of the inherent difficulties in measuring neutrons, as discussed previously, only few measurements exist that measure neutrons and charged particles simultaneously. Important examples are the studies carried out with the Berlin Ball [3,4], the Rochester Superball [16], the ORION detector [25] and the Texas A&M NIMROD [12] systems. For those ar-

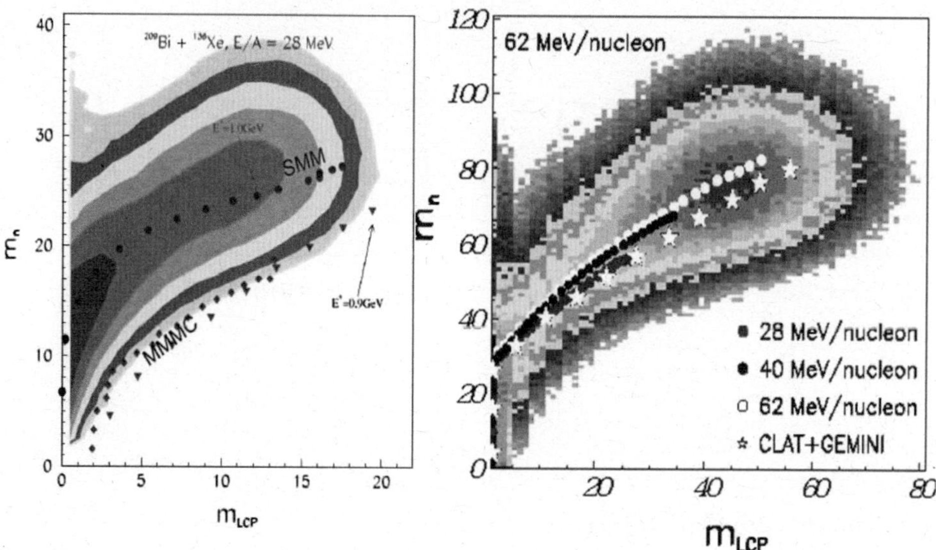

Fig. 8. Experimental correlations between neutron and light-charged-particle multiplicities from the ^{209}Bi + ^{136}Xe reactions [26]. The left panel shows a correlation compared with SMM and MMMC predictions; the right panel compares results for different bombarding energies.

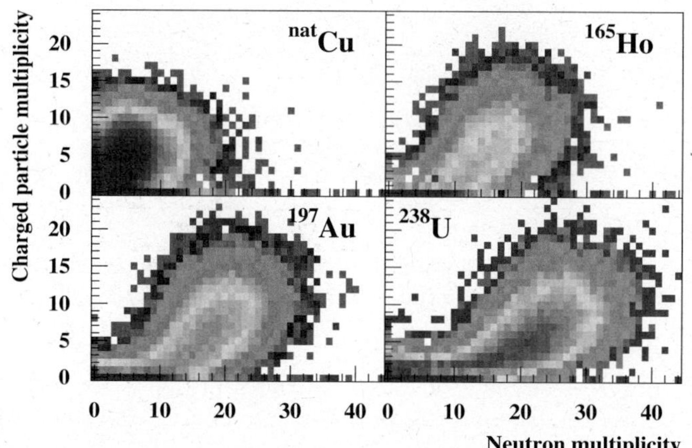

Fig. 9. Experimental neutron *vs.* charged-particle multiplicity for 1.2 GeV antiprotons on several targets. Taken from ref. [4].

rays that detect only charged particles, the existing measurements of neutron-proton multiplicity correlations and spectra must be relied upon to estimate the missing neutron contribution to E^*/A. Neutron spectra were shown in fig. 1 which emphasizes the fact that neutrons may also originate from pre-equilibrium emission.

In figs. 8 and 9 the neutron-light-charged particle multiplicity correlations are compared for $A + A$ and light-ion reactions. Two systems are shown: $28\,\text{MeV}/A$ Xe + Bi [26] and $1.2\,\text{GeV}\ \bar{p} +$ several targets [3,4], respectively. Both cases behave similarly, with a target mass-dependence (fig. 9) that favors an increasing growth in the n/LCP ratio with increasing target mass. The total particle multiplicity is known to be strongly correlated with excitation energy. For heavy targets the neutron multiplicity increases rapidly with excitation energy up to

$E^*/A \approx 2\,\text{MeV}$, while charged-particle emission remains low due to Coulomb inhibition. At higher excitation energies, the probability for additional neutron emission is approximately balanced by LCP emission. The left part of fig. 8 compares the Xe + Bi multiplicity correlation with the average predicted by two multifragmentation models, SMM [27] and MMMC [28]. It was assumed that neutrons and light charged particles are emitted from excited projectile- and target-like sources containing total excitation energy of 0.9 GeV or 1.0 GeV. Without entering into details of models and data comparisons, we see that the mean trend of the correlation can be understood within the framework of an equilibration scenario. The effect of increased bombarding energy on the n/LCP correlation is indicated for the Xe + Bi case in the right part of fig. 8. No strong dependence is observed and here again a statistical de-excitation model (GEMINI [29]) is able to reproduce the mean trend.

For heavy targets the relative insensitivity of the n/LCP ratio to colliding system or bombarding energy, as well as the general agreement with models, provides guidance in accounting for the missing neutron fraction of E^*/A in arrays that measure only charged particles. Various approaches have been followed: use of model calculations calibrated to the LCP multiplicity, or direct use of the experimental multiplicity correlation centroids. An alternative is to employ a mass balance approach, as determined from the experimental event structure. In this latter event-by-event method the neutron multiplicity is determined by mass conservation assuming the N/Z ratio of the studied source. This requires a reliable determination of total Z.

Figure 10 illustrates the effectiveness of such techniques, using the $1.2\,\text{GeV}\ \bar{p} + {}^{197}$Au results [3,4]. SMM (dotted curve) and SIMON-evaporation [30] (dashed line), both statistical models, provide a reasonable description

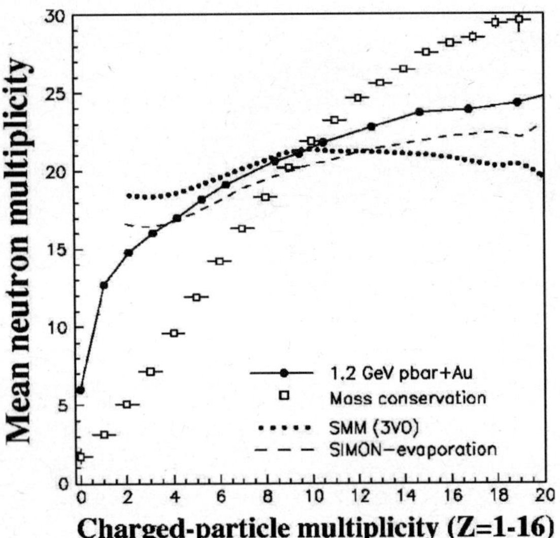

Fig. 10. Comparison of neutron *vs.* charged-particle multiplicity for the 1.2 GeV antiproton + ^{197}Au reaction with SMM, SIMON-evaporation and mass balance assumptions as indicated in the legend. Taken from ref. [6].

of the data for $M_{\mathrm{LCP}} \geq 3$ ($E^*/A \approx 2\,\mathrm{MeV}$), considering that statistical models do not describe the pre-equilibrium particle emission. Mass conservation (open squares) is not satisfactory in this case, although in some instances the nature of the data may provide a more satisfactory fit.

Finally, even in case of neutron detection, the primary difficulty in the determination of the neutron contribution to E^* and the source mass A_{source} is that the neutron tanks provide multiplicity information (good for A_{source}) but not neutron energies, while the time-of-flight method provides energies (good for K_n) but only limited multiplicity data. By use of LCP-calibrated models, it is possible to obtain a reasonable approximation to the total excitation energy contributed by neutrons. However, in doing so, one is employing averages that fail to introduce fluctuations in M_n and K_n.

3.3 Additional factors

While non-equilibrium emission and neutron emission constitute the major sources of uncertainty in the determination of E^*/A, several other factors must be taken into account, as discussed in the following.

– *Source reconstruction:*
  Among the various multifragmentation programs, calculations of the properties of the emitting source —charge, mass and velocity— are usually detector-array–dependent. For light ion + A reactions, for which there is only one source, the reconstruction depends on the acceptance of the array. The EOS TPC measurements provide nearly complete charged-particle detection [9], as shown in the left part of fig. 11, from which the statistical component of an event can be extracted.

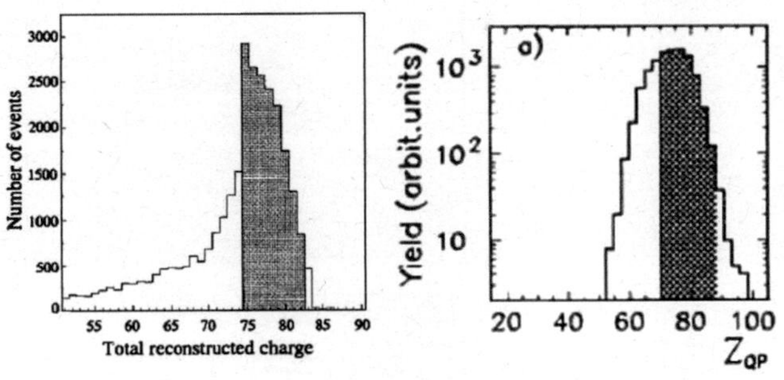

Fig. 11. Quasiprojectile reconstructed charge distributions for EOS (left part, taken from ref. [9]), and for MULTICS-MINIBALL (right part, taken from ref. [17]). In both cases the initial projectile charge is $Z_P = 79$. The shaded areas give the events used for the calorimetric and thermodynamic analyses.

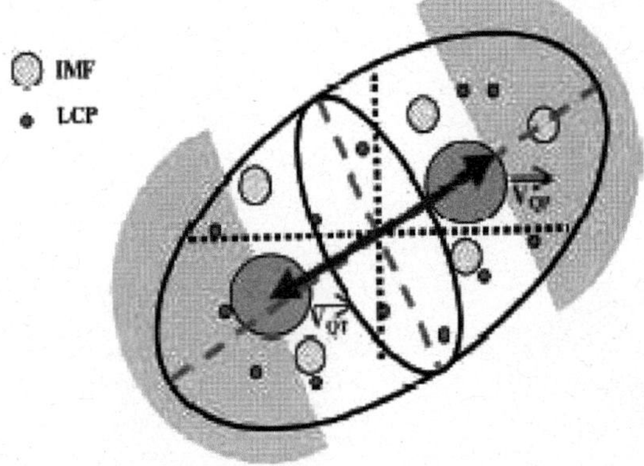

Fig. 12. Schematic representation of the momentum tensor analysis to reconstruct the reaction plane (see text).

For $A + A$ reactions some authors utilize a center-of-mass tensor analysis in order to reconstruct the reaction plane, schematically shown in fig. 12. This procedure, only adopted in the case of a nearly perfect detection, is used to isolate the fragments of the studied source. For the schematic example of fig. 12, the forward center-of-mass emitted fragments are assumed to originate from the excited projectile-like source, and only light charged particles emitted in the forward hemisphere of the source are taken into account for calorimetry because of pre-equilibrium effects at midrapidity. Going back to fig. 6, we can see that midrapidity emission concerns also light fragments and thus, depending on the reaction and the experimental apparatus, additional criteria and/or checks are used for accepting only studied source fragments that are classified as "equilibrium-like". As an example, total-charge results for projectile-like fragments from ^{197}Au + ^{197}Au studies of the MULTICS/MINIBALL are presented in the right part of fig. 11 [17]. In nearly all cases, the emitting-source mass is determined from the A/Z ratio of the heavy collision partner(s). The velocity of the statistical source is determined as the vectorial sum of fragment velocities.

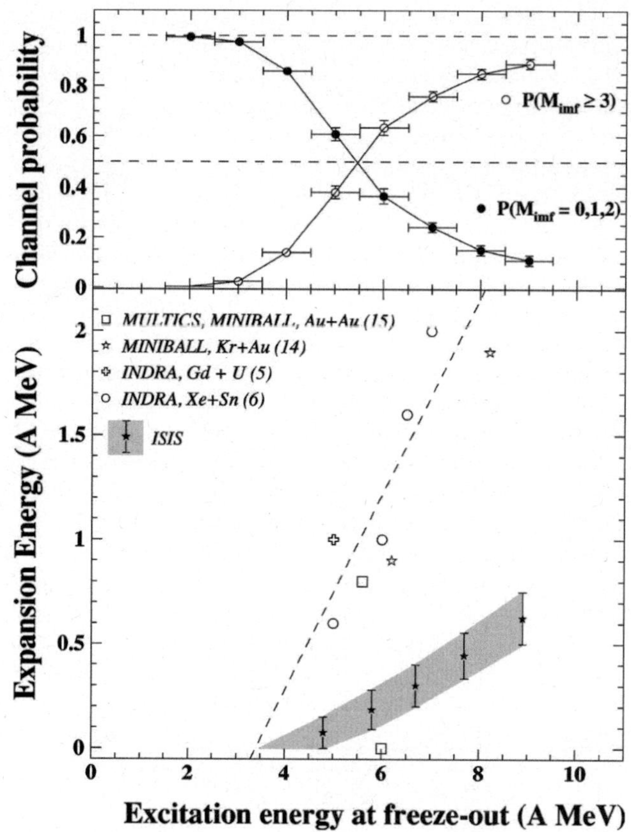

Fig. 13. Upper panel: decay channel probability for multiplicities $M \leq 2$ and $M \leq 3$. Lower panel: extra radial expansion energy for light-ion–induced reaction (yellow band) and several heavy-ion reactions, as indicated in the legend. Taken from ref. [31].

Representative protocols for evaluating the source properties are described in the next section.

– *Primary-fragment N/Z ratio:*
In the framework of statistical multifragmentation, to trace back the freeze-out stage of an event, it is necessary to know the mass and the charge of primary fragments before secondary decay. This information is not necessary for calorimetry but it is relevant to measure, for example, the mass and volume of the source. In most cases, only the fragment charge is measured for all but the lightest elements. Various approaches have addressed the conversion of the data to primary yields. One is to use the N/Z ratio of the cold fragments, the composite system, or some combination of the two. Another is to use the N/Z ratio of IMFs emitted in reactions of protons with heavy nuclei at energies below 500 MeV, where secondary emission should be small.

– *Collective energy:*
Finally, any internal energy used to expand/deform or rotate the hot source, must be subtracted from the excitation energy sum of eq. (1) in order to access thermodynamic properties of nuclear systems. As shown in fig. 13, for light-ion reactions this is a small, but non-

negligible contribution at high excitation energies [31]. For $A + A$ reactions, compression effects produce considerable expansion and therefore can contribute a significant amount to the E^* sum at high excitation energies. This correction also accounts in part for the bombarding energy dependence of E^* shown in fig. 7. The amount of collective energy is generally deduced from data-model comparisons. In the case of expansion, the comparison is based on kinetic properties of the fragments and thus is dependent on model assumptions about the source volume since Coulomb repulsion is acting [19]. A and Z fragment identification over a wide range, as well as correlation measurements, could disentangle this problem.

4 E^*/A protocols

The procedures for converting measured data to E^*/A differ for every multifragmentation experiment. Once detector calibration and filter development are complete, the salient variables can be applied to eqs. (1) and (2). In this section, several methods are described that are representative of the approaches that have been employed.

4.1 Model-based calorimetry

In the $\bar{p}$ studies of the Berlin Neutron/Silicon Ball [3, 4], E^* is determined by comparison of the light-charged particle multiplicity with that predicted by the evaporation code GEMINI [28] at a given excitation energy. Since IMF multiplicities are rarely greater than unity in this experiment, the use of an evaporation code is appropriate. For the higher-energy LCP + A studies employed by the FASA group, an empirical parameter α is obtained from the comparison of observed charged-particle multiplicities with the multiplicity distribution predicted by a hybrid RC + SMM model [32]. The excitation energy is then taken to be a function α times the predicted excitation energy. For experiments with good fragment detection, the element distribution (or the distribution of the biggest fragment) is used to deduce the mean excitation energy and size of the source. Genuine distributions of source characteristics (size, excitation energy, volume, ...) of collected event ensembles are accessible via backtracing procedures. In both cases this is done with data-model comparisons and event-by-event information is not accessible.

Below, we summarize the calorimetry procedures used in several representative systems that calculate E^*/A on an event-by-event basis.

4.2 Calorimetric protocols

In this section we schematically review different assumptions that have been employed in the literature to perform the calorimetric measurement. They include both corrections for the incomplete detection (neutron multiplicity and energy, masses of heavy products) and techniques to separate pre-equilibrium and source mixing contaminations, as we have discussed in the previous sections. For more details about the procedures and their justification, we refer the reader to the original publications of the different collaborations. As a general statement, the validity of a calorimetric protocol is usually tested from an extensive use of statistical as well as dynamical simulations using realistic models [11,19]. For example in the case of the MULTICS detector, the input energy of SMM simulations, filtered using the same calorimetric protocol as data, is reproduced within 10% in the multifragmentation regime [19]. A similar performance can be associated to the INDRA detector [11].

– *EOS*:
In the analyses of the data obtained with the EOS detection system [8,9], the total charge of the studied source is estimated from the initial projectile charge by subtracting pre-equilibrium particles and fragments identified according to a cutoff energy prescription (see sect. 3.1). A number M_{cp}^{neq} of nonequilibrium charged products of charge Z_i^{neq} is thus defined event by event, leading to

$$Z_{source} = Z_{proj} - \sum_i Z_i^{neq}, \qquad (4)$$

$$A_{source} = A_{proj} - \left(\sum_i Z_i^{neq} + 1.7 M_{cp}^{neq} \right). \qquad (5)$$

Once the source size is known, the neutron multiplicity, M_n, can be inferred from mass conservation, while the average neutron energy, K_n, is estimated via an effective temperature T obtained within a Fermi-gas ansatz with a level density parameter $a = A/13\,\mathrm{MeV}^{-1}$:

$$M_n = A_{source} - \sum_i A_i^{thermal}, \qquad (6)$$

$$\langle K_n \rangle = M_n \cdot \frac{3}{2} T. \qquad (7)$$

– *ISiS*:
In the case of the ISiS detector [5,6] the procedure is similar to that of EOS, as discussed above, except that excited target-like residues are studied:

$$Z_{source} = Z_{target} - \sum_i Z_i^{neq}, \qquad (8)$$

$$A_{source} = A_{target} - \left(\sum_i Z_i^{neq} + 1.93 M_{cp}^{neq} \right). \qquad (9)$$

Since heavy residues are not detected, it is assumed that all missing charge resides in a single fragment, an assumption that is in good agreement with the EOS data and SMM simulations. Neutron multiplicities have been calibrated by the measured neutron-charged particle correlations of ref. [4] and kinetic energies were based on both Fermi gas and model simulation results. The IMF mass A_{IMF} is estimated based on the data of ref. [33], with the assumption that no charged-particle decay of IMFs has occurred. An overall geometrical efficiency correction is also applied.

– *INDRA and MULTICS quasiprojectile*:
Such a correction is not applied to INDRA data since the complete Z identification over 4π allows the selection of events with complete charge. Completeness conditions can vary from 70% to 90% in the different analyses and are often complemented by completeness condition of linear momentum. For MULTICS data the completeness condition is 90% and good Z identification of all emitted products is effective in the forward direction (up to about 30 degrees in the laboratory frame). We have already seen in fig. 6 that at low incident energies in the Fermi-energy range the kinematic distinction between different emission sources is blurred. The contamination from non-quasiprojectile sources is minimized by selecting as the QP i) only forward-emitted fragments (IMFs and heavy residues) via a tensor analysis, and ii) only light charged particles forward emitted in the source frame:

$$Z_{source} = \sum_i Z_i^{\mathrm{IMF+HR}} + 2 \sum_i Z_i^{\mathrm{LCP}}, \qquad (10)$$

$$A_{source} = (A/Z)_{proj} \cdot Z_{source}, \qquad (11)$$

$$M_n = A_{source} - \sum_i A_i^{\mathrm{IMF+HR}} - 2 \sum_i A_i^{\mathrm{LCP}}. \qquad (12)$$

In these expressions, all sums over IMFs and HR are restricted to the forward hemisphere in the center-of-mass frame, while the sums over LCPs are restricted to the forward hemisphere in the reconstructed QP frame. Concerning the neutron kinetic energies, three different prescriptions have been shown to give comparable results [11]:

$$\langle K_n \rangle = bT, \qquad (13)$$

$$\langle K_n \rangle = \langle K_{Z=1} \rangle - 3.5\,\mathrm{MeV}, \qquad (14)$$

$$\langle K_n \rangle = M_n \cdot \frac{3}{2} T. \qquad (15)$$

For the first K_n prescription the b parameter varies from $b = 1$ to $b = 2$ depending on excitation energy.

– *INDRA central*:
For INDRA symmetric central collisions the same QP-prescription has been used, but i) all detected IMFs and heavy residues have been attributed to the source, ii) the retained LCPs are those emitted between 60 degrees and 120 degrees in the center-of-mass frame and the A/Z ratio is that of the total entrance channel.

– *TAMU central*:
For asymmetric systems central collisions detected by TAMU experiments cited above [12,13], the studied source corresponds to target-like sources. The selected events correspond to central events (multiplicity cut). Three sources are present: the projectile-like, target-like and a hypothetical source whose velocity corresponds to the velocity of the nucleon-nucleon collision frame. The data rely on identification of p, d, t, ^{3}He,

^{4}He and of $3 < Z < 14$ elements. The neutron multiplicity is measured. A three-source fitting procedure is applied to energy spectra (with efficiency corrections) to obtain the size of the target-like source and the equilibrated target-like neutron multiplicity (M_n) out of the total neutron multiplicity:

$$\langle K_n \rangle = M_n \cdot \frac{3}{2} T; \qquad T = 4 - 4.7 \, \text{MeV}. \qquad (16)$$

– *ALADiN*:

For ALADiN experiments cited above, the collision of the relativistic projectile and the target leads to a projectile spectator, a target spectator and a fireball whose size is increasing with centrality. The studied source is the forward-emitted projectile spectator whose size is decreasing with centrality. The selection among detected source particles is based on a rapidity (y) cut and a bombarding energy-dependent angular cut. Neutrons are detected by the apparatus but $Z = 1$ particles are not. The He isotopes are not identified over the full solid angle. Therefore ALADiN does not use eqs. (1) and (2) on an event-by-event basis but rather uses them with mean assumptions for unknown quantities to extract a mean value. The data are divided into Z_{bound} (sum of selected Z from He up to projectile size) ensembles and for each ensemble a mean calorimetry is applied.

– *Alternative method*:

An alternative method relative to eq. (1) to estimate the excitation energy is to measure the source velocity. Its validity is based on the reaction mechanism which is at the origin of the source excitation. Complete or incomplete fusion of asymmetric systems (mass tranfer) or a pure binary collision mechanism provide a link between E^* and the source velocity [14]. Because precise measurements of particle angles, masses and energies are needed, this method is used as a check after using eqs. (1)-(3) [17].

5 Summary and conclusions

From the analysis of the filtered multifragmentation data, all three terms in eq. (1), charged-particle kinetic energies, neutron kinetic energies and removal energy ($-Q$) are found to have significant weights in the excitation energy sum. In the upper panel of fig. 14 the relative kinetic-energy percentages are shown for LCPs, neutrons and IMFs as a function of E^*/A for ISiS data [6]. Neutrons and LCPs are roughly equivalent, each accounting for 20–30% over the entire E^*/A range. IMFs do not become significant until about $E^*/A \approx 3$–4 MeV, reaching a maximum of $\approx 10\%$ near $E^*/A \approx 6$ MeV. Above $E^*/A \approx 6$ MeV, all three percentages remain nearly constant. As is apparent from the previous discussions, these percentages vary, depending on assumptions about nonequilibrium emission, neutrons, etc.

The bottom frame of fig. 14 compares the percentage of the E^* sum for total kinetic-energy release with that

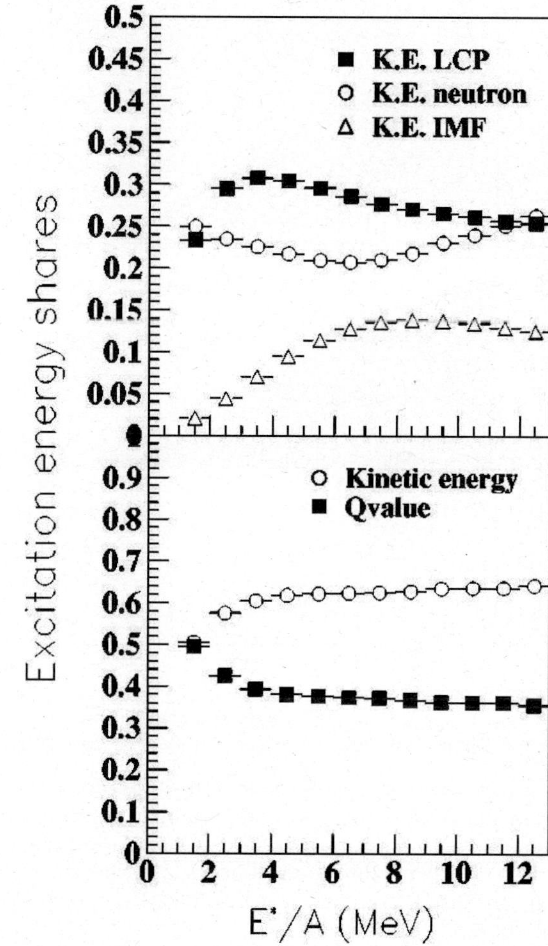

Fig. 14. Upper panel: relative kinetic-energy contributions to the excitation energy for neutrons, LCPs and IMFs as a function of excitation energy. Lower panel: relative contributions to the excitation energy from total kinetic energy and removal energy, as indicated in the legend. Taken from ref. [6].

for the removal energy derived from event reconstruction. For low excitation energies, the kinetic-energy sum and removal energy are roughly equivalent. One factor that tends to stabilize eq. (1) with respect to input assumptions is that some of the uncertainties are self-compensating. If, for example, the neutron multiplicity and/or energy input to the filter is too high, the separation energy decreases, and vice versa. Another factor that must be kept in mind is that many of the assumptions that are involved in the filter are averages, and therefore do not adequately account for fluctuations in the distributions. Because of the exponential decrease in yield with increasing E^*/A, fluctuations skew the distribution toward lower excitation energies. This effect is demonstrated in fig. 15. The upper frame shows the average yield as a function of E^*/A bin size (heavy solid line). Superimposed on each bin is a Gaussian approximation to the fluctuation widths which are assumed to increase with excitation energy (light lines). The effect on the E^*/A distribution is demonstrated in the middle frame of fig. 15, showing the yield (yellow on-

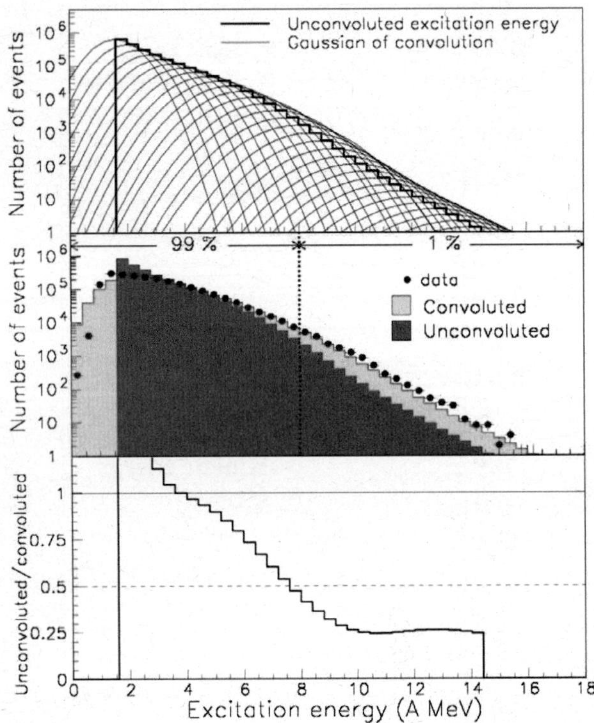

Fig. 15. Upper panel: Gaussian decomposition of the excitation energy distribution. Middle panel: comparison between the excitation energy distribution derived from calorimetry procedures and the deconvoluted distribution. Bottom panel: ratio of the deconvoluted distribution to original distribution. Taken from ref. [6].

line) for the data (solid points) and that for the deconvoluted distribution (red on-line). Over the range up to $E^*/A \approx 8$ MeV there is relative agreement between the two distributions. Above this energy, the most probable E^*/A value increasingly falls below that of the average.

From examination of the existing analyses, it is estimated that as a thermodynamic variable, all of the results are self-consistent over about a 20% range in E^*/A. Given this uncertainty, however, there is general agreement among all of the data sets. In the range $E^*/A \approx 4$–5 MeV, a distinct change occurs in multifragmentation observables, indicating a change in the reaction mechanism. Within a phase transition scenario, this excitation energy would represent the liquid-gas transition energy. The consistency of the measurements is perhaps best illustrated by the caloric-curve analysis of Natowitz [34], in which all of the caloric-curve measurements are decomposed

as a function of source mass. When this decomposition is performed, a systematic behavior is revealed that lends greater credence to the caloric-curve behavior in hot nuclear systems. To go beyond and thus relate quantitatively the experimental results to the nuclear equation of state, Z and A identification over a wide range would be needed.

References

1. J. Pochodzalla *et al.*, Phys. Rev. Lett. **75**, 1040 (1995).
2. R.T. de Souza *et al.*, contribution VI.1, this topical issue.
3. L. Pienkowski *et al.*, Phys. Lett. B **336**, 147 (1994).
4. F. Goldenbaum *et al.*, Phys. Rev. Lett. **77**, 1230 (1996).
5. K. Kwiatkowski *et al.*, Phys. Lett. B **433**, 21 (1998).
6. T. Lefort *et al.*, Phys. Rev. C **64**, 064603 (2001).
7. V. Radionov *et al.*, Nucl. Phys. A **700**, 457 (2002).
8. J.A. Hauger *et al.*, Phys. Rev. Lett. **62**, 024616 (2000).
9. J.A. Hauger *et al.*, Phys. Rev. C **57**, 764 (1998).
10. A. Schüttauf *et al.*, Nucl. Phys. A **607**, 457 (1996).
11. J.C. Steckmeyer *et al.*, Nucl. Phys. A **686**, 537 (2001); E. Vient *et al.*, Nucl. Phys. A **700**, 555 (2002).
12. R. Wada *et al.*, Phys. Rev. C **55**, 227 (1997); K. Hagel *et al.*, Phys. Rev. C **62**, 034607 (2000).
13. Y.G. Ma *et al.*, Nucl. Phys. A **749**, 106 (2005).
14. CHIMERA Collaboration (E. Galichet), private communication.
15. L. Beaulieu *et al.*, Phys. Rev. Lett. **77**, 462 (1996).
16. B. Djerroud *et al.*, Phys. Rev. C **64**, 034603 (2001).
17. M. D'Agostino *et al.*, Nucl. Phys. A **650**, 329 (1999).
18. M. D'Agostino *et al.*, Phys. Lett. B **473**, 219 (2000).
19. M. D'Agostino *et al.*, Nucl. Phys. A **699**, 795 (2002).
20. T. von Egidy *et al.*, Eur. Phys. J. A **8**, 197 (2000).
21. K.B. Morley *et al.*, Phys. Rev. C **54**, 737 (1996).
22. J. Łukasik *et al.*, Phys. Lett. B **566**, 76 (2003).
23. ALADiN Collaboration (W. Trautmann), private communication.
24. M.F. Rivet *et al.*, Phys. Lett. B **423**, 217 (1998); P. Desquelles *et al.*, Phys. Rev. C **62**, 024614 (2000).
25. Y. Perier *et al.*, Nucl. Instrum. Methods A **413**, 32 (1998).
26. J. Toke *et al.*, Phys. Rev. Lett. **75**, 2920 (1995).
27. A. Botvina, A.S. Ilijinov, I.N. Mishustin, Nucl. Phys. A **507**, 649 (1990).
28. D.H.E. Gross, Rep. Prog. Phys. **53**, 605 (1990).
29. R. Charity *et al.*, Nucl. Phys. A **483**, 371 (1998).
30. D. Durand, Nucl. Phys. A **541**, 266 (1992).
31. T. Lefort *et al.*, Phys. Rev. C **62**, 031604R (2000).
32. S.P. Avdeyev *et al.*, Nucl. Phys. A **709**, 392 (2002).
33. R.E.L. Green *et al.*, Phys. Rev. C **29**, 1806 (1984).
34. J.B. Natowitz *et al.*, Phys. Rev. C **65**, 034618 (2002); see also A. Kelić *et al.*, contribution V.2, this topical issue.

Eur. Phys. J. A **30**, 227–242 (2006)

DOI 10.1140/epja/i2006-10119-4

THE EUROPEAN
PHYSICAL JOURNAL A

Moment analysis and Zipf law

Y.G. Ma[a]

Shanghai Institute of Applied Physics, Chinese Academy of Sciences, Shanghai 201800, PRC

Received: 26 April 2006 /
Published online: 23 October 2006 – © Società Italiana di Fisica / Springer-Verlag 2006

Abstract. The moment analysis method and nuclear Zipf's law of fragment size distributions are reviewed to study nuclear disassembly. In this report, we present a compilation of both theoretical and experimental studies on moment analysis and Zipf law performed so far. The relationship of both methods to a possible critical behavior or phase transition of nuclear disassembly is discussed. In addition, scaled factorial moments and intermittency are reviewed.

PACS. 05.70.Jk Critical point phenomena – 64.60.Fr Equilibrium properties near critical points, critical exponents – 25.70.Mn Projectile and target fragmentation – 25.70.Pq Multifragment emission and correlations

1 Introduction

Hot nuclei can be formed in energetic heavy ion collisions (HIC) and de-excite by different decay modes, such as evaporation and multifragmentation. Experimentally, multifragment emission was observed to evolve with excitation energy. The multiplicity, N_{imf}, of intermediate mass fragment (IMF) rises with the beam energy, reaches a maximum, and finally falls to a lower value. The onset of multifragmentation may indicate the coexistence of liquid and gas phases [1]. Phenomenologically, the mass (charge) distribution of IMF distribution can be expressed as a power law with parameter τ_{eff}, and a minimum τ_{min} of τ_{eff} emerges around the onset point, which suggests that a kind of critical behavior may take place. In the framework of Fisher's droplet model, the mass distribution can be described by a power law with a critical exponent of $\tau \sim 2.3$ when the system is in the vicinity of the critical point [2].

On the other hand, the caloric-curve measurement can also provide useful information on the liquid-gas phase transition [3–7]. The analysis of other independent critical exponents provides additional indications of critical behavior of finite nuclear systems [8–13]. In addition, more observables have been proposed to sign the liquid-gas phase transition or critical behavior of nuclei [14–19]. Some reviews can be found in this topical issue [20–24].

In this report, we shall review the moment analysis method and Zipf law of fragment size distribution. The phenomenological basis of moment analysis is introduced in sect. 2. Finite-size effects are discussed in sect. 3. Section 4 gives the application of moment analysis

to multifragmention and its relation to critical behavior. Scaled factorial moments and intermittency are discussed in sect. 5. In sect. 6 Zipf law is introduced for the nuclear fragment distribution and the corresponding simulations are given; some experimental indications of nuclear Zipf law are presented in sect. 7; finally, the summary and outlook are given in sect. 8.

2 Phenomenological basis of moment analysis

Campi [25,26] and Bauer [27,28] *et al.* first suggested that the methods used in percolation studies may be applied to nuclear multifragmentation data. In percolation theory the moments of the cluster distribution contain a signature of critical behavior [29]. The method of moment analysis has been experimentally used to search for evidence of the critical behavior in multifragmentation. The definition of the k moments of the cluster size distribution for each event is

$$M_k = \sum_{A \neq A_{max}} A^k n_A, \qquad (1)$$

where A is the fragment mass, and n_A is the number of charged fragments whose charge is Z and mass is A. The sum runs over all masses A in the event including neutrons except the heaviest fragment (A_{max}). This quantity was taken as a basic tool in extracting critical exponents in Au + C data [9]. It has been argued that there should be an enhancement in the critical region of the moment M_k, for $k > \tau - 1$, with a critical exponent $\tau > 2$ [25,26].

In experimental analyses, events are sorted by different conditions. In this case, so-called conditional moments are used to describe the fragment distribution. Usually

---

[a] e-mail: ygma@sinap.ac.cn

the mean value of $M_k(m)$ for events with given control parameters, *e.g.* the moment M_k for events with a given multiplicity m, or total bound charge number Z_{bound}, or excitation energy E^*, is called conditional moment.

More insight in the shape of the fragment size distribution is obtained by looking at a combination of moments M_k. For example, the quantity

$$\gamma_2 = \frac{M_2 M_0}{M_1^2} = \frac{\sigma^2}{\langle s \rangle^2} + 1, \qquad (2)$$

has been used, where M_1 and M_2 are the first and second moments of the mass distribution and M_0 is the total multiplicity including neutrons. σ^2 is the variance of the fragment distribution and $\langle s \rangle = M_1/M_0$ represents the mean fragment size. γ_2 takes the value $\gamma_2 = 2$ for a pure exponential distribution, $N(s) \sim \exp(-\alpha s)$ regardless of the value of α, but $\gamma_2 \gg 2$ for a power law distribution, $N(s) \sim s^{-\tau}$ when $\tau > 2$. In the percolation model, the position of the maximum γ_2 value defines the critical point, where the fluctuations in the fragment size distribution are the largest. In principle, a genuine critical behavior requires the peak value of γ_2 to be larger than 2 [25,26]. However, due to finite-size effects, this is not always true when the system size decreases, as we will see in the following sections.

Campi also suggested to use the single event (j) moment, *i.e.*

$$M_k^{(j)} = \sum_{A \neq A_{max}} A^k n^{(j)} \qquad (3)$$

to investigate the shape of fragment size distribution. Also normalized moments [25]

$$S_k^{(j)} = M_k^{(j)} / M_1^{(j)} \qquad (4)$$

can be defined. It was suggested to use the event-by-event scatter-plots of the natural log of the size (A_{max}) or charge number (Z_{max}) of the largest cluster, $\ln A_{max}$ or $\ln Z_{max}$ *vs.* the natural log of the second moment, $\ln M_2$, or the normalized moment $\ln S_2$ to search for the largest fluctuation point. Some examples will be given in the following sections.

In the percolation model, the cluster size distribution for infinite systems near a critical point can be expressed by

$$n(s) \sim s^{-\tau} f(\epsilon s^{\sigma}), \qquad (5)$$

where s is the size of finite clusters, τ and σ two critical exponents and ϵ a variable that characterizes the state of the system. In thermal phase transitions, $\epsilon = T - T_c$ is the distance to the critical temperature T_c. In percolation, $\epsilon = p_c - p$ is the distance to the critical fraction of active bonds or occupied sites p_c. The scaling function $f(\epsilon s^{\sigma})$ satisfies $f(0) = 1$, decaying rapidly (exponentially) for large values of $|\epsilon|$. In addition, theory predicts that when $\epsilon < 0$ one infinite cluster (liquid or gel) is present in the system while no such cluster exists when $\epsilon > 0$ (only droplets or n-mers). In finite systems a similar behavior is observed, especially when the largest cluster is counted separately.

The moment analysis method is useful to obtain some information about the possible occurrence of a critical behavior. In general, critical exponents can be defined according to the standard procedure followed in condensed-matter physics [30]. For example,

$$M_k(\epsilon) = \sum_A = A^k n_A(\epsilon) \sim |\epsilon|^{\frac{\tau - k - 1}{\sigma}} \quad (\epsilon \to 0), \qquad (6)$$

where τ and σ are the critical exponents. For the percolation phase transition and the critical point in the Fisher droplet model, the exponent τ satisfies $2 < \tau < 3$ and thus the second and high moments diverge at the critical point. In contrast, the lower moments M_0 and M_1, which correspond to the number of fragments and the total mass, do not diverge.

Based upon the scaling relation eq. (5), there exists the following relationship between critical exponents and moments:

$$\begin{aligned} M_0 &\sim |\epsilon|^{2-\alpha}, \\ M_1 &\sim |\epsilon|^{\beta}, \\ M_2 &\sim |\epsilon|^{-\gamma}, \end{aligned} \qquad (7)$$

where β and γ are two other critical exponents. Some relationships among critical exponents exist (hyperscaling relations), for instance

$$2\beta + \gamma = \frac{\tau - 1}{\sigma} = 2 - \alpha. \qquad (8)$$

In finite systems transitions are smooth, but it is still possible to determine some critical exponents, as we will discuss in the next section. By analogy with the infinite-system behavior, one says that these moments exhibit a critical behavior also for finite systems. In particular, in the Fisher model, the thermal critical point is also a critical point for moments of the fragment size distribution.

In order to illustrate the application of moment analysis, we show the EOS data and NIMROD data as examples in sect. 4.

3 Finite-size effects

Since the nucleus is a finite-size system, the macroscopic thermal limit cannot be applied. Therefore finite-size effects on phase transition behavior should be checked. In this section, we give some examples to illustrate this problem.

A percolation on a cubic lattice of linear size L containing L^3 sites, for $L = 4$ to 10, where all sites are occupied and bonds are assumed to exist between neighbouring sites with bond probability p, has been considered [31]. Sites that are connected together by such bonds are said to belong to the same cluster. It is well known that in such a model there exists a critical (or threshold) probability p_c such that for $p > p_c$ there is a large cluster that percolates throughout the lattice from end to end whereas for $p < p_c$ no such cluster exists and all the sites belong to small

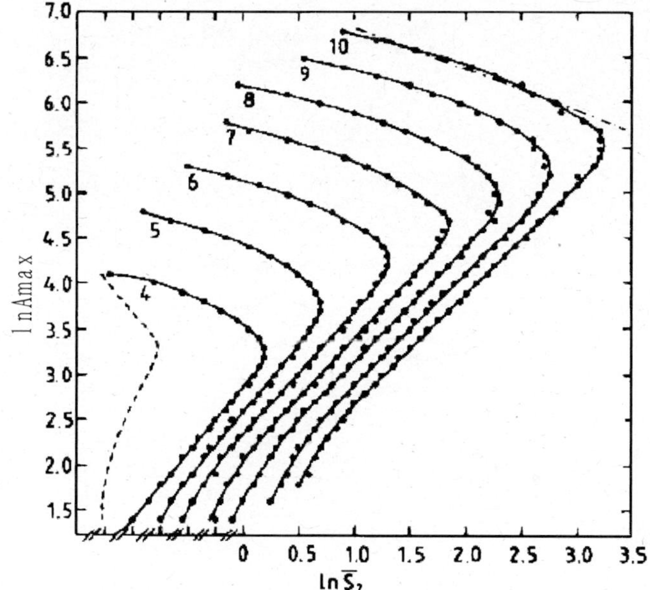

Fig. 1. The logarithm of the largest fragment size A_{max} as a function of the logarithm of the corresponding average normalized second moment $\overline{S_2}$ for bond percolation on simple cubic lattices of linear size ranging as $L = 4$–10 sites. The dots represent the actual calculation results and the curves drawn are just to guide the eye. The number next to each curve gives the value of the linear size L. Note that the $\ln \overline{S_2}$ scale given corresponds to the $L = 10$ curve. The other curves are successively shifted to the left with respect to each other by a distance of 0.25. The dashed curve and the dot-dashed straight line are explained in the text. The figure is taken from ref. [31].

clusters (including isolated sites, *i.e.* singlet or clusters of size 1). As $L \to \infty$, the transition becomes sharper and p_c approaches a limiting value which for bond percolation on a cubic lattice is $p_c = 0.249$ [29]. For finite systems the threshold percolation probability is not so sharply defined.

In order to quantitatively illustrate finite-size effects on critical behavior, the average of normalized second moment S_2 $(\overline{S_2})$ over all events belonging to the same value of $\ln(A_{max})$ was calculated [31]. The results obtained by such averaging are presented by the dots shown in fig. 1 for various cubic lattices with linear dimension $L = 4$–10 sites [31]. The location of the maximum value of $\overline{S_2}$ is now defined as corresponding to the location of the critical point, which is a standard way of determining the percolation threshold [32]. The slope of the lower branches of the curves in fig. 1 can also be calculated. This slope is expected to be $1 + \beta/\gamma$ which for percolation in three dimensions is equal to 1.23. For comparison the slopes of the straight lines by a lest-squares fit to the lower branches of the $L = 4$ to 10 curves are found, in ascending order of L, to have the values 1.582 ± 0.036, 1.503 ± 0.029, 1.375 ± 0.017, 1.355 ± 0.021, 1.260 ± 0.007, 1.258 ± 0.014 and 1.242 ± 0.015 [31]. This indicates that these slopes rapidly approach the value expected in the thermodynamic limit. In calculating these slopes one has excluded the points near the bottom of the branch in the region where the

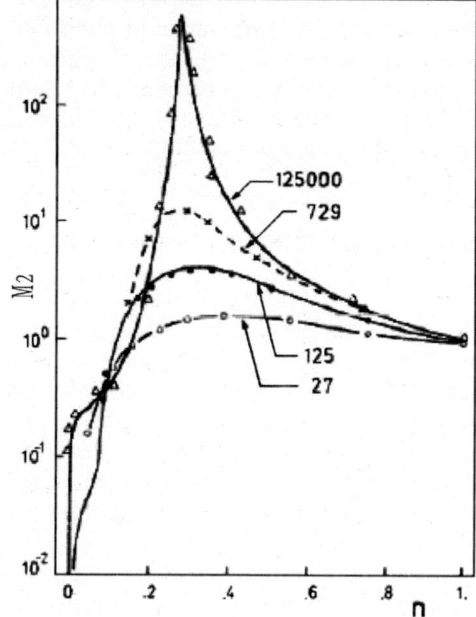

Fig. 2. The conditional moments $M_2(n)$ for percolation in a cubic lattice of linear size $L = 3, 5, 9$ and 50 (the corresponding cubic lattice is L^3 which is shown as the number in the insert). The figure is taken from ref. [26].

curves in fig. 1 deviate noticeably from a straight line. These points correspond to events that are far from the critical region.

Similarly to the analysis for the correlation of $\overline{S_2}$ and $\ln(A_{max})$, finite-size effects have been also investigated for M_2 by Campi [26]. This is shown in fig. 2, where $M_2(n)$ is plotted for various system sizes (50^3, 9^3, 5^3 and 3^3) in a percolation model. We see clearly the critical behavior for the largest system, namely a well-defined peak, and how this peak is smoothed when decreasing the size [26].

4 Application of the moment analysis method

4.1 EOS data

4.1.1 Experimental description

The reverse kinematic EOS experiment was performed with 1 A GeV ^{197}Au, ^{139}La, and ^{84}Kr beams on carbon targets. The experiment was done with the EOS Time Project Chamber (TPC) and multiple sampling ionization chamber (MUSIC II). The excellent charge resolution of this detector permitted the identification of all detected fragments. The fully reconstructed multifragmentation events for which the total charge of the system was taken as $79 \le Z \le 83$, $54 \le Z \le 60$, $33 \le Z \le 39$ for Au, La, and Kr, respectively [33–36] were analyzed. The remnant refers to the equilibrated nucleus formed after the emission of prompt particles. The charge and mass of the remnant were obtained by removing for each event the total charge of the prompt particles. The excitation energy of the remnant E^* was based on an energy balance between the excited remnant and the final stage of the fragments for each event [37]. The thermal excitation energy E_{th}^* of the remnant was obtained as the difference between E^* and E_x which is a nonthermal component, namely an expansion energy [33–36,38].

4.1.2 Determination of critical point and exponent in terms of moment analysis

The determination of the critical point and the associated exponents in the multifragmentation of gold nuclei was first attempted by the EOS Collaboration [9]. In their early publication [9], they use the multiplicity m, as a control variable for the collision violence and assume that m is a linear measure of the distance from the critical point. Then the critical exponents β, γ and τ, can be determined according to eqs. (7), (8) above. They find that these exponents are close to the nominal liquid-gas universality class values. However, this method is very delicate. In particular, due to the small size of the system, an important rounding of the transition is expected which may distort considerably the determined critical exponents. For a review of this debate, see the arguments between Bauer [39] and Gilkes [40].

A different analysis was also proposed by the EOS Collaboration [41]. In this work, thermal excitation energy has been taken as a control variable, which is believed to be more suitable to characterize the collision violence.

The γ_2 analysis is shown in fig. 3 for all three systems. The position of the maximum γ_2 value defines the critical excitation energy E_c^*, which corresponds to the largest fluctuation point in the fragment size distribution. The peak in γ_2 is well defined for La and Au. For Kr, the peak is very broad and the value γ_2 is less than 2.

Figure 3 also shows a γ_2 calculation using the statistical multifragmentation model (SMM). The fission contribution to γ_2 has been removed both from the data and SMM. In the case of Au, the γ_2 value remains above two for most of the excitation energy range both in data and SMM. The E_{th}^* width over which $\gamma_2 > 2$ is smaller for La and disappears for Kr. The decrease in γ_2 with decreasing system size is also seen in 3D percolation studies and these differences have been attributed to finite-size effects [41–43].

The exponent τ can be obtained if the second moment M_2 and the third moment M_3 of the fragment mass distributions are known. A plot of $\ln(M_3)$ vs. $\ln(M_2)$ should give a straight line with a slope given by

$$ S = \frac{\Delta \ln(M_3)}{\Delta \ln(M_2)} = \frac{\tau - 4}{\tau - 3}. \tag{9} $$

Figure 4 shows a scatter-plot of $\ln(M_3)$ vs. $\ln(M_2)$ for the three systems constructed with data above the critical excitation energy E_c^* (see fig. 3) and with SMM simulations. A linear fit to $\ln(M_3)$ vs. $\ln(M_2)$ gives the value of τ. The fitted τ values are 2.16 ± 0.08, 2.10 ± 0.06 and 1.88 ± 0.08, respectively. The former two are very close to

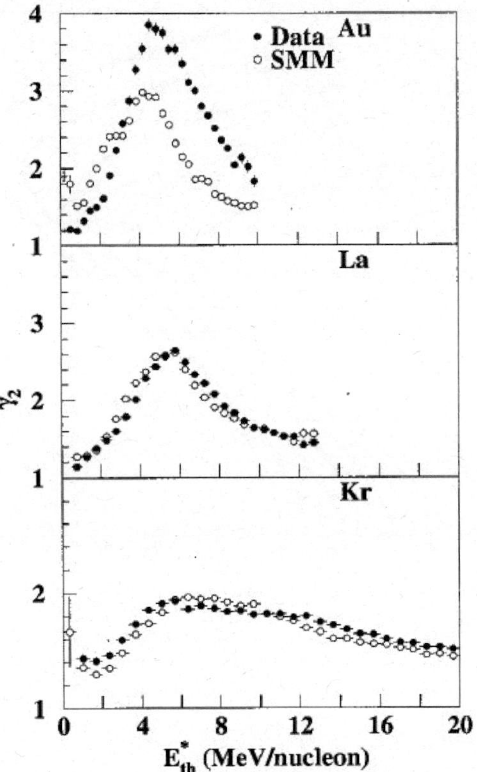

Fig. 3. γ_2 as a function of E_{th}^* for all three systems of 1 A GeV Au, La, and Kr collisions with C target and SMM calculations. The figure is taken from ref. [33].

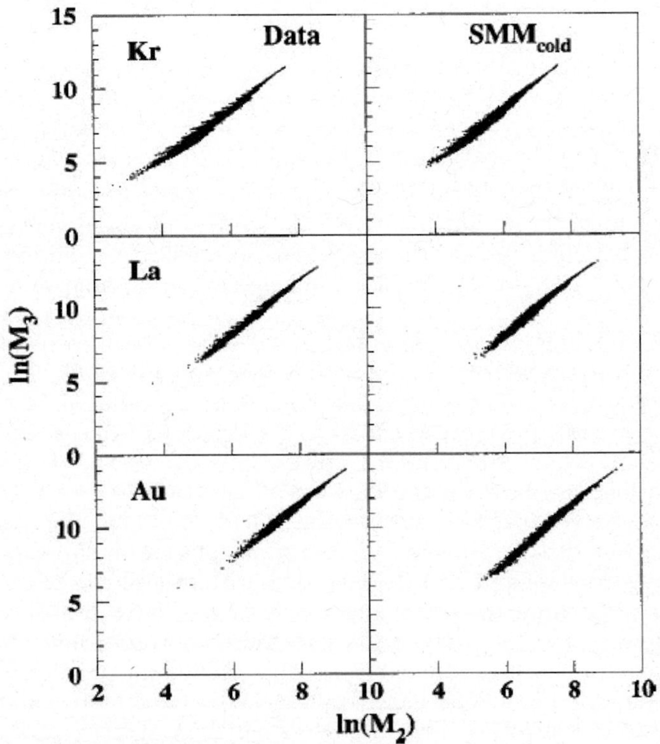

Fig. 4. $\ln(M_3)$ vs. $\ln(M_2)$ for Au, La, and Kr above the critical energy. The figure is taken from ref. [33].

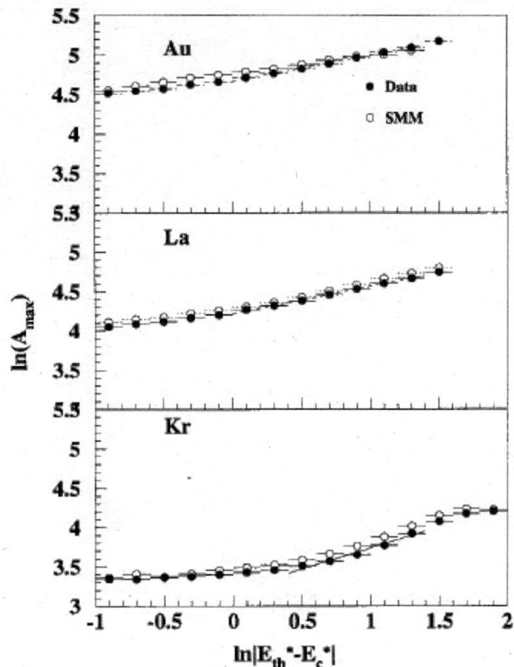

Fig. 5. $\ln(A_{max})$ *vs.* $\ln|E_{th}^* - E_C^*|$ for Au, La, and Kr below the critical energy for exponent β determination. The figure is taken from ref. [33].

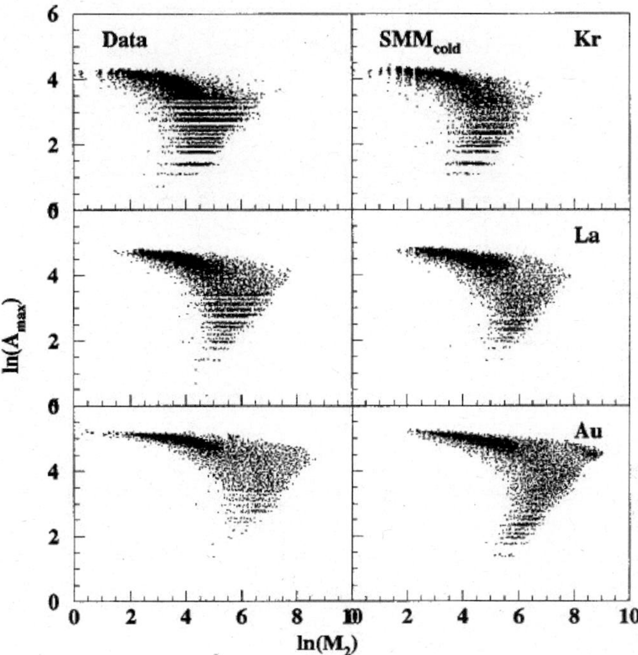

Fig. 6. Scatter-plots of $\ln(A_{max})$ *vs.* $\ln(M_2)$ from the data for Au, La, and Kr. Left panel: EOS data; right panel: SMM simulation. The figure is taken from ref. [33].

the critical exponents $\tau \sim 2.3$ of the liquid-gas universal class.

The exponent β can be obtained for the multifragmentation data by the relation

$$A_{max} \sim |\epsilon|^{\beta}, \qquad (10)$$

where $\epsilon = p - p_c$ and $\epsilon > 0$. In the multifragmentation case of this work p and p_c have been replaced by E_{th}^* and E_c^*. In an infinite system, the finite cluster exists only on the liquid side of p_c. In a finite system, a largest cluster is present on both sides of the critical point, but the above equation holds only on the liquid side. Figure 5 shows a plot of $\ln(A_{max})$ *vs.* $\ln|E_{th}^* - E_c^*|$ for Au, La, and Kr. The values of β extracted for Au and La are 0.32 ± 0.02 and 0.34 ± 0.02, respectively, which are close to the value of 0.33 predicted for a liquid-gas phase transition. On the other hand, the value of $\beta = 0.53 \pm 0.05$ for Kr is much higher than that of Au and La.

As shown in sect. 2, Campi also suggested that the correlation between the size of the biggest fragment A_{max} and the moments in each event, *i.e.* the scatter-plot, can measure the critical behavior in nuclei. Figure 6 depicts a scatter-plot with logarithmic scale for Au, La, and Kr of EOS data. The two branches corresponding to the sub-critical (upper branch) and overcritical (lower branch) events are clearly seen for Au and La. The scatter-plot is very broad for Kr and fills most of the available space. The sub- and over-critical branches seem to overlap and are not well separated. Studies on SMM show a similar behavior. If one knows the location of the critical point from some other methods, then the scatter-plot can be

used to calculate the ratio of critical exponents β/γ from the slope of the sub-critical branch. In EOS data, the position of the largest γ_2 was used to define the critical point, which corresponds to the largest fluctuation of the fragment distribution. In this context, β/γ values for Au, La and Kr can be extracted from the linear fit to the upper branch; they are 0.22 ± 0.03, 0.25 ± 0.01 and 0.50 ± 0.01, respectively. β/γ values of Au and La are close to the value 0.26 expected for the liquid-gas universality class.

To summarize the critical exponent analysis of the EOS data, the experimental results in conjunction with SMM provide some indications on the order of the phase transition in Au, La and Kr. The values of the critical exponents τ, β, and γ, which are close to the values of a liquid-gas system, along with nearly zero latent heat (this subject is beyond the discussion topics in this review, but the interested reader is reported to refs. [33,34]) have been interpreted by the authors as a continuous phase transition in Au and La. However, the analysis of Kr leads to very different critical exponents. A recent analysis based on the shape of SMM microcanonical caloric curve indicates a first-order phase transition for the multifragmentation of Kr [33,34].

4.2 NIMROD data

4.2.1 Experimental set-up and analysis details

Using the TAMU NIMROD (Neutron Ion Multidetector for Reaction Oriented Dynamics) and beams from the TAMU K500 super-conducting cyclotron, we have probed

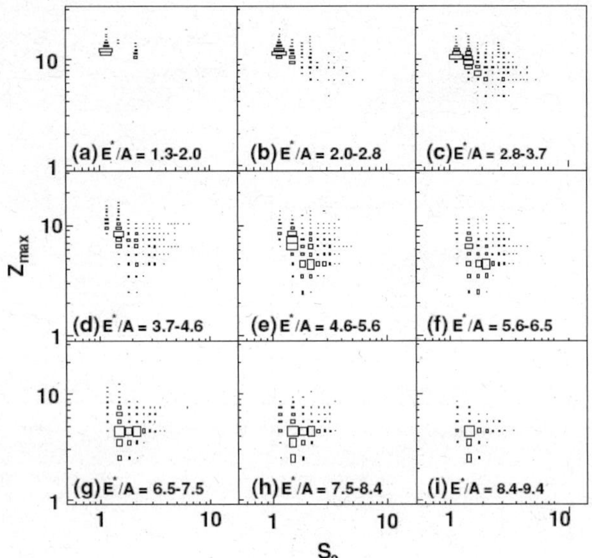

Fig. 7. Campi plots for nine intervals of excitation energy for the QP formed in ^{40}Ar $+$ ^{58}Ni. The figure is taken from ref. [14].

the properties of excited projectile-like fragments produced in the reactions of 47 MeV/nucleon ^{40}Ar $+$ ^{27}Al, ^{48}Ti and ^{58}Ni. The charged-particle detector array of NIMROD, which is set inside a neutron ball, includes 166 individual CsI detectors arranged in 12 rings in polar angles from $\sim 3°$ to $\sim 170°$. The detailed description for the experiment can be found in [14]. The correlation of the charged-particle multiplicity (M_{cp}) and the neutron multiplicity (M_n) was used to sort the event violence. After the reconstruction of the quasi-projectile (QP) particle source, the excitation energy was deduced event by event using the energy balance equation [37].

4.2.2 Critical-point determination via moment analysis

In fig. 7 we present Campi scatter-plots for the nine selected excitation energy bins. In the low excitation energy bins of $E^*/A \leq 3.7$ MeV/u, the upper (liquid phase) branch is strongly dominant while at $E^*/A \geq 7.5$ MeV/u, the lower Z_{max} (gas phase) branch is strongly dominant. In the region of intermediate E^*/A of 4.6–6.5 MeV/u, the transition from the liquid-dominated branch to the vapor branch occurs, indicating that the region of maximal fluctuations is to be found in that range.

The excitation energy dependence of the average values of γ_2 obtained in an event-by-event analysis of our data are shown in fig. 8. γ_2 reaches its maximum in the 5–6 MeV excitation energy range. In contrast to observations for heavier systems of Au and La [33,41], there is no well-defined peak in γ_2 for our very light system and γ_2 is relatively constant at higher excitation energies. This is similar to the case of Kr of EOS data. We note also that the peak value of γ_2 is lower than 2 which is the expected smallest value for critical behavior in large systems. However, 3D percolation studies indicate that finite-size effects

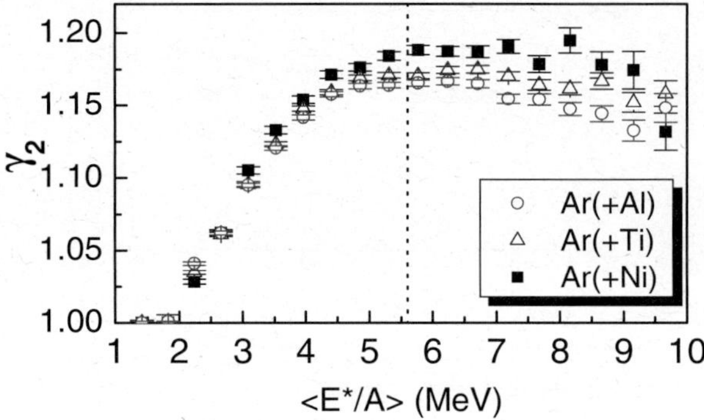

Fig. 8. γ_2 of the QP systems formed in Ar $+$ Al (open circles), Ti (open triangles) and Ni (solid squares) as a function of excitation energy.

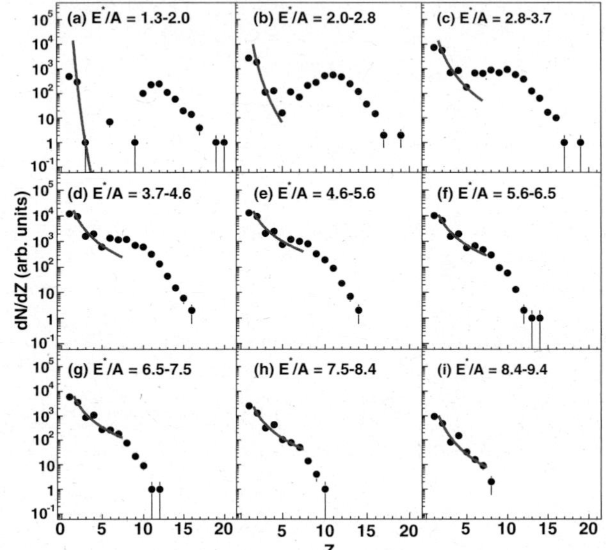

Fig. 9. Charge distribution of QP in different E^*/A window for the reaction ^{40}Ar $+$ ^{58}Ni. Lines represent fits. The figure is taken from ref. [14].

can lead to a decrease of γ_2 with system size [42,43]. For a percolation system with 64 sites, peaks in γ_2 under two are observed. Therefore, the criterion $\gamma_2 > 2$ alone is not sufficient to discriminate whether or not the critical point is reached.

In the Fisher droplet model, the critical exponent τ can be deduced from the cluster distribution near the critical point. To quantitatively pin down the possible phase transition point, we use a power law fit to the QP charge distribution in the range of $Z = 2$–7 (fig. 9) to extract the effective Fisher-law parameter τ_{eff} by

$$dN/dZ \sim Z^{-\tau_{eff}}. \tag{11}$$

Figure 10(a) shows the effective Fisher-law parameter τ_{eff} as a function of excitation energy. A minimum with $\tau_{eff} \sim 2.3$ is seen to occur in the E^*/A range of 5 to 6 MeV/u [44]. This value is close to the critical exponent of the liquid-gas phase transition universality class [2].

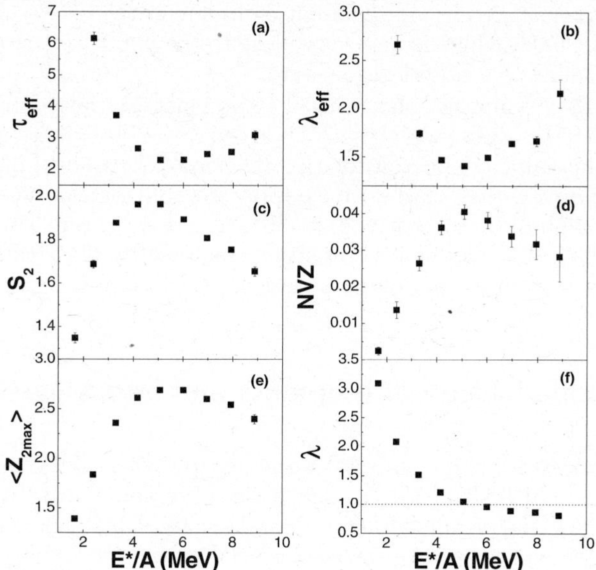

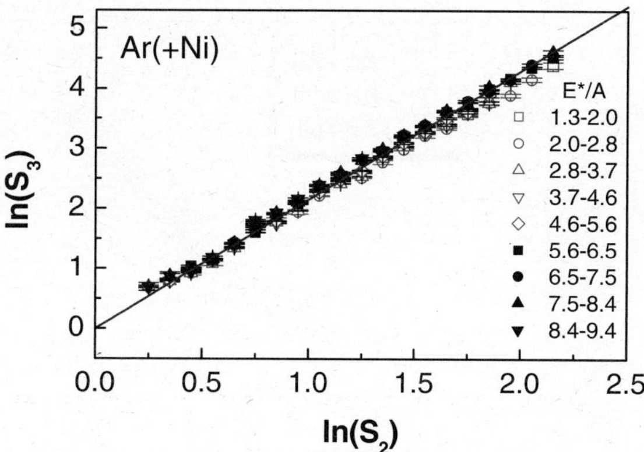

Fig. 11. The correlation between $\ln(S_3)$ $vs.$ $\ln(S_2)$ and a linear fit.

Fig. 10. The effective Fisher-law parameter (τ_{eff}) (a), the effective exponential law parameter (λ_{eff}) (b), $\langle S_2 \rangle$ (c), NVZ fluctuation (d), the mean charge number of the second largest fragment $\langle Z_{2max} \rangle$ (e), the Zipf-law parameter λ (f). See details in the text. The figure is taken from ref. [44].

Assuming that the heaviest cluster in each event represents the liquid phase, we have attempted to isolate the gas phase by event-by-event removal of the heaviest cluster from the charge distributions. We find that the resultant distributions are better described with an exponential form $\exp^{-\lambda_{eff}Z}$. The fitting parameter λ_{eff} was derived and is plotted against excitation energy in fig. 10(b). A minimum is seen in the same region where τ_{eff} shows a minimum. To further explore this region we have investigated other proposed observables commonly related to fluctuations and critical behavior. Figure 10(c) shows the mean normalized second moment, $\langle S_2 \rangle$ as a function of excitation energy. A peak is seen around $5.6\,\mathrm{MeV/u}$, it indicates that the fluctuation of the fragment distribution is the largest in this excitation energy region. Similarly, the normalized variance in Z_{max}/Z_{QP} distribution ($i.e.$ $\mathrm{NVZ} = \frac{\sigma^2_{Z_{max}/Z_{QP}}}{\langle Z_{max}/Z_{QP} \rangle}$) [45] shows a maximum in the same excitation energy region (fig. 10(d)), which illustrates the maximal fluctuation for the largest fragment is reached around $E^*/A = 5.6\,\mathrm{MeV}$. The second largest fragment shows a behavior similar to the one of the largest fragment. Figure 10(e) shows a broad peak of $\langle Z_{2max} \rangle$ —the average atomic number of the second largest fragment— also occurring in the same excitation energy range around $5.6\,\mathrm{MeV/u}$.

More variables have been collected to support the determination of the critical point around $5.6\,\mathrm{MeV/u}$ of excitation energy for our system [14], such as Δ-scaling [46] or energy fluctuations [23]. In addition, the measurement of the caloric curve [14] gives the temperature $T_c \sim 8.3\,\mathrm{MeV}$ around $E^*/A = 5.6\,\mathrm{MeV}$. The value of the critical tem-

perature is needed for the determination of the critical exponents, as explained in the following subsection.

4.2.3 Determination of critical exponents based on moment analysis

In terms of the scaling theory, τ can also be deduced from eq. (9). Since the value of $T_c = 8.3\,\mathrm{MeV}$ has been determined from our caloric-curve measurements [14], we can explore the correlation of S_2 and S_3 in two ranges of excitation energy (see fig. 11). The moments were calculated by excluding the species with Z_{max} for the "liquid" phase but including it in the "vapor" phase. The slopes were determined from linear fits to the "vapor" and "liquid" regions, respectively, and then averaged. In this way, we obtained a value of $\tau = 2.13 \pm 0.1$.

Other critical exponents can also be related to other moments of the cluster distribution, M_k. Using our caloric-curve measurements [14], we can use temperature as a control parameter for such determinations. Then the critical exponent β can be extracted from the relation

$$Z_{max} \propto \left(1 - \frac{T}{T_c}\right)^\beta, \tag{12}$$

and the critical exponent γ can be extracted from the second moment via

$$M_2 \propto \left|1 - \frac{T}{T_c}\right|^{-\gamma}. \tag{13}$$

In both equations, $|1 - \frac{T}{T_c}|$ is the parameter which measures the distance from the critical point.

The upper panel of fig. 12 explores the dependence of Z_{max} on $(1 - \frac{T}{T_c})$. A dramatic change of Z_{max} around the critical temperature T_c is observed. Lattice-gas model (LGM) calculations also predict that the slope of Z_{max} $vs.$ T will change at the liquid-gas phase transition [47]. Using the liquid side points, we can deduce the critical exponent β by $\ln(Z_{max})$ $vs.$ $\ln|1 - T/T_c|$. Figure 12(a) shows the extraction of β using eq. (12). An excellent fit was obtained

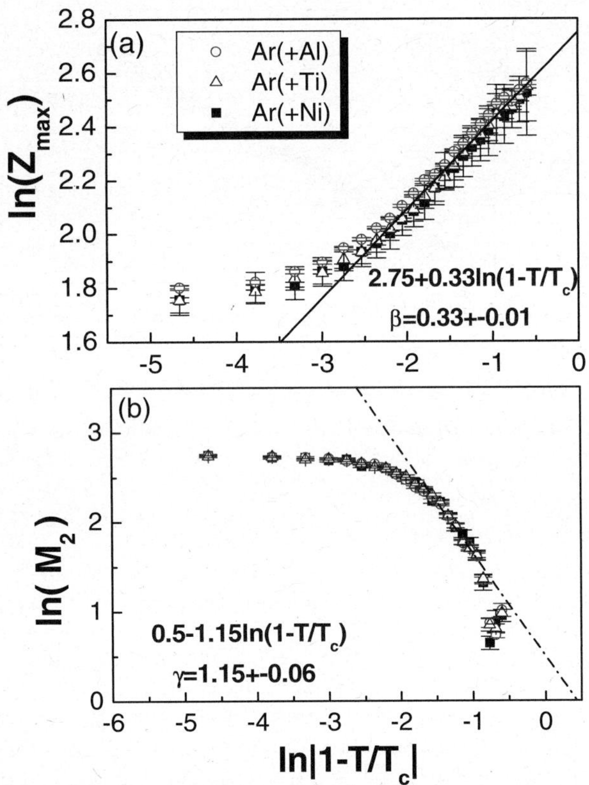

Fig. 12. The extraction of the critical exponent β (a) and γ (b). See text for details.

Table 1. Comparison of the critical exponents.

Exponents	3D percolation	Liquid-gas	NIMROD
τ	2.18	2.21	2.13 ± 0.10
β	0.41	0.33	0.33 ± 0.01
γ	1.8	1.23	1.15 ± 0.06
σ	0.45	0.64	0.68 ± 0.04

in the region away from the critical point, which indicates a critical exponent $\beta = 0.33 \pm 0.01$. Near the critical point, finite-size effects become stronger so that the scaling law is violated. The extracted value of β is that expected for a liquid-gas transition (see table 1) [29].

To extract the critical exponent γ, we take M_2 on the liquid side without Z_{max}. Figure 12(b) shows $\ln(M_2)$ as a function of $\ln(|1 - \frac{T}{T_c}|)$. We center our fit to eq. (13) about the center of the range of $(1 - T/T_c)$ which leads to the linear fit and extraction of β as represented in fig. 12. We obtain a critical exponent $\gamma = 1.15 \pm 0.06$. This value of γ is also close to the value expected for the liquid-gas universality class (see table 1). It is seen that the selected region has a good power law dependence.

Since we have the critical exponents β and γ, we can use the scaling relation

$$\sigma = \frac{1}{\beta + \gamma} \qquad (14)$$

to derive the critical exponent σ. In such way, we get $\sigma = 0.68 \pm 0.04$, which is also very close to the expected critical exponent of a liquid-gas system.

To summarize the critical exponents extracted from NIMROD data, we present the results in table 1 as well as the values expected for the 3D percolation and liquid-gas universality classes. It is apparent that our values for this light system with $A \sim 36$ are closer to the values of the liquid-gas phase transition universality class rather than to the 3D percolation class.

5 Scaled factorial moments and intermittency

Intermittency is related to the existence of large non-statistical fluctuations and is a signal of self-similarity of the fluctuation distribution at all scales. This signal can be deduced from the scaled factorial moments [48],

$$F_k(\delta) = \frac{\Sigma_{i=1}^{X_{max}/\delta} \langle n_i(n_i - 1)(n_i - 2) \ldots (n_i - k + 1) \rangle}{\Sigma_{i=1}^{X_{max}/\delta} \langle n_i \rangle^k}, \qquad (15)$$

where X_{max} is an upper characteristic value of the system (*i.e.* total mass or charge, maximum transverse energy or momentum, etc.) and k is the order of the moment. The total interval 0–X_{max} (1–A_{max}, Z_{max} in the case of mass or charge distributions) is divided into X_{max}/δ bins of the size δ, n_i is the number of particles in the i-th bin for an event, and the ensemble average $\langle \rangle$ is performed over all events. The concept of intermittency was originally developed in the field of fluid dynamics to study the fluctuations occurring in turbulent flows [49,50]. Its presence in the velocity and temperature distributions is established by the existence of large non-statistical fluctuations which exhibit scale invariance. Intermittency in physical systems is studied by examining the scaling properties of the moments of the distributions of relevant variables over a range of scales [51]. The concept of intermittency was first introduced for the study of dynamical fluctuations in the density distribution of particles produced in high-energy collisions by Bialas and Peschanski [48]. It soon led to the discovery of a characteristic power law dependence of the factorial moments, F_k, of an order k on the resolution scale, δ: $F_k \propto (1/\delta)^{f(k)}$. The specific properties of the intermittency exponent, $f(k)$, can be associated either with a random production process [48, 52] or with a second-order phase transition [52–54] depending on the values obtained. Thus an analysis of the factorial moments may provide important information on the dynamical properties of the system. Ploszajczak and Tucholski were the first to suggest searching for intermittency patterns in the mass and charge distributions of the fragments produced in energetic collisions [55]. Since then many studies show that an intermittency pattern of fluctuations in the fragmentation charge distributions has been observed in many data and models. Much effort has been devoted to find the relation between fragmentation, a possible critical behavior, and intermittency [45,56–60].

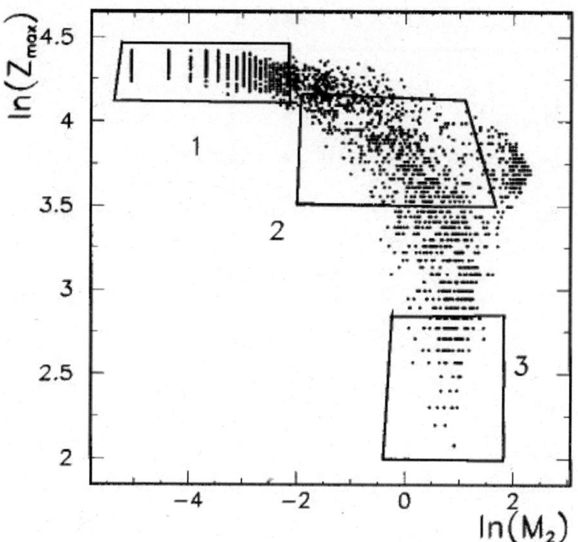

Fig. 13. Experimental Campi scatter-plots from ref. [61]. Three cuts are employed to selected the upper branch (1), the lower branch (3), and the central region (2).

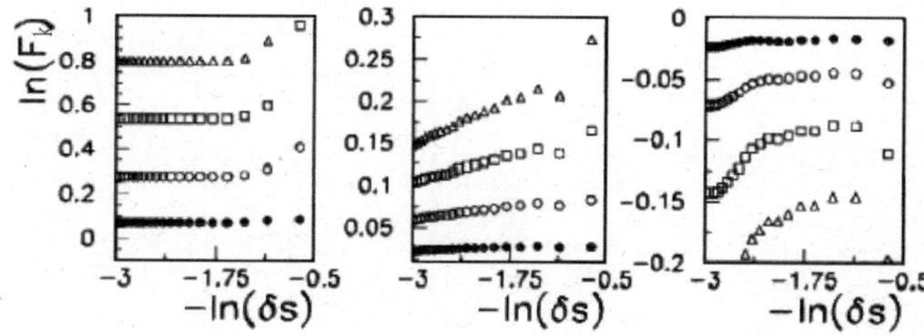

Fig. 14. Experimental results from ref. [61]. Scaled factorial moments $\ln(F_k)$ $vs.$ $-\ln(\delta_s)$ for the three cuts made in fig. 13: left part cut 1, central part cut 2, and right part cut 3. Solid circles represent the SFM of order $k = 2$, open circles $k = 3$, open squares $k = 4$, and open triangles $k = 5$. The figure is taken from ref. [61].

Intermittency is defined by the relation

$$F_k(\delta') \equiv F_k(a\delta) = a^{-f(k)}F_k(\delta), \qquad (16)$$

between factorial moments $F_k(\delta')$ and $F_k(\delta_s)$ obtained for two different binning parameters δ and $\delta' = a\delta$. Intermittency implies a linear relationship in the double logarithmic plot of $\ln F_k$ $vs.$ $-\ln\delta$.

The fractal intermittency exponent, $f(k)$, is related to the factorial dimension d_k by

$$f(k) = \frac{d_k}{k-1} > 0. \qquad (17)$$

Different processes seem to give a different behavior of these anomalous fractal dimension d_k: 1) d_k = constant corresponds to a monofractal, second-order phase transition in the Ising model and in the Feynman-Wilson fluid [53,54]. It has been also demonstrated that in the case of a second-order phase transition in the Ginzburg-Landau description one gets $d_k = d_2(k-1)^{\mu-1}$ with $\mu = 1.304$ [53]. 2) $d_k \propto k$ corresponds to multifractal, cascading processes [48]. Therefore, a study of the anomalous fractal dimensions can give useful information about the evolution of the system.

Several models have been introduced to study the intermittency signal. One of the simplest models, widely used in the analysis of experimental data and which gives intermittency, is the percolation model. Percolation models predict a phase transition corrected for finite-size effects and produce, at the critical point for this phase transition, a mass distribution following a power law and obeying scaling properties.

An intermittency analysis has been performed on many heavy-ion collision data as well as emulsion data. Here we give an example of the multifragmentation data of Au + Au collisions at 35 MeV/u which was performed at NSCL by the Multics-Miniball Collaboration [61]. A power law charge distribution, $A^{-\tau}$ with $\tau \simeq 2.2$ and an intermittency signal has been observed for the events selected in the region of the Campi scatter-plot where "critical" behavior is expected. As shown in fig. 13, three cuts have been tested. The upper branch is mostly related to the liquid branch and the lower branch to the gas branch, while the central cut (2) is expected to belong to a region where critical behavior takes place. Actually the resultant charge distribution of cut (2) shows a power law distribution with $\tau \simeq 2.2$ which is close to the droplet model predicted if the liquid-gas critical point is explored. The scaled factorial moments are shown in fig. 14 for the different cuts of fig. 13. For cut 3, the logarithm of the scaled factorial moment is always negative and almost independent of $-\ln\delta$; there is no intermittency signal. The situation is different for cut 2 (the central part). The logarithm of the scaled factorial moments is positive and almost linearly increasing as a function of $-\ln\delta$, and an intermittency has been observed. Cut 1 gives a zero slope, no intermittency signal again.

It has been argued that the interpretation of this experimentally observed intermittency signal may, however, be problematic due to an ensemble average effect [56]. Since cut 2 involves a large range of impact parameters, the observed intermittency signal could be an artifact of ensemble averaging, and cannot be seen as a definite evidence of large fluctuation driven by a critical behavior.

Actually, several criticisms have been raised about the role of the intermittency signal in nuclear fragmentation. For instance, Elattari $et\ al.$ showed that an intermittency signal can be obtained even for a simple fragmentation generator model by the random population of mass bins with a power law distribution in which the only non-statistical source of fluctuations is the mass conservation law [57]. It has also been shown that the intermittency signal is washed out when events of fixed total multiplicity are selected [45,60] or when the size of the system tends to infinity in the percolation model in which the fluctuations are of nontrivial origin [60]. Moreover, the intermittency signal is not observed in the narrow excitation energy region where the phase transition occurs in the framework

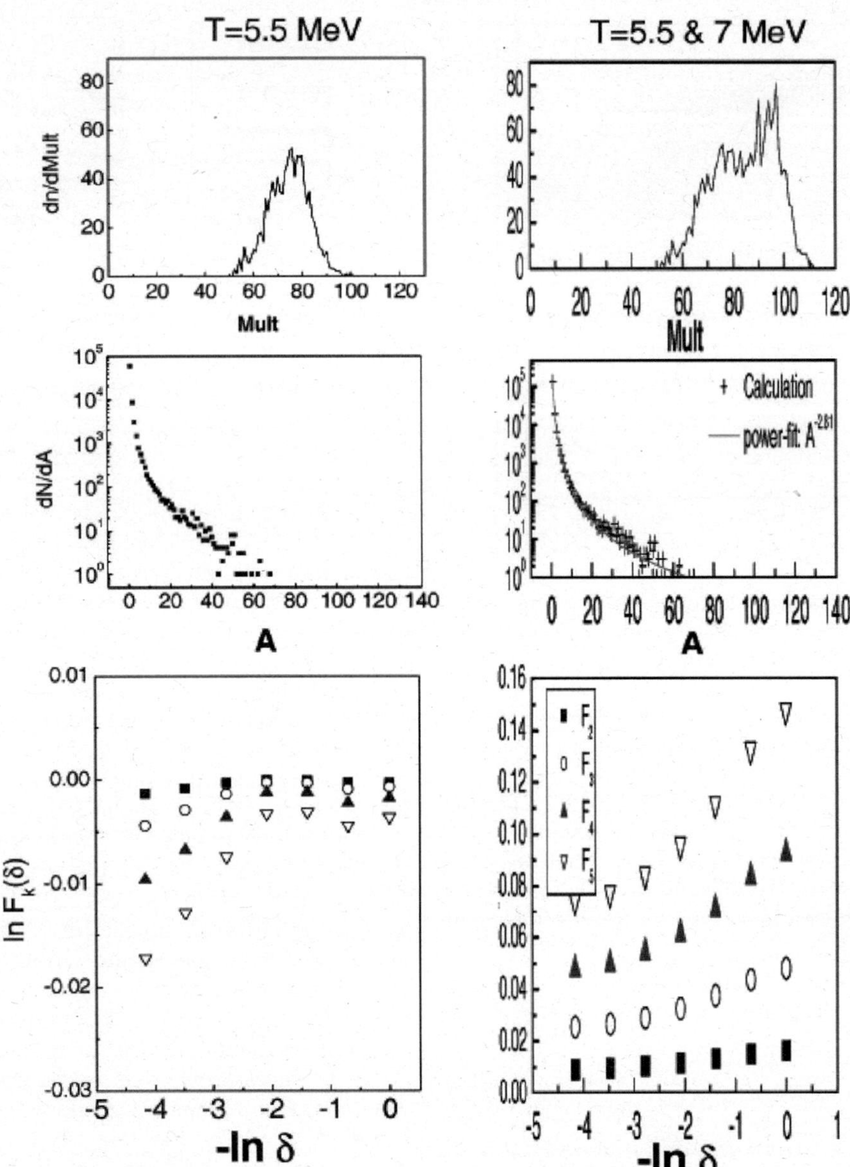

Fig. 15. Left panels: the multiplicity distribution (upper panel), the mass distribution (middle panel), the scaled factorial moments (bottom panel) with the multiplicity restriction for ^{129}Xe in the lattice gas model calculation. Right panel: same as the left panel but for the events mixed with $T = 5.5$ MeV and $T = 7$ MeV. The figure is taken from ref. [62].

of the well-known Copenhagen statistical multifragmentation model [58] or in the data of 35–110 MeV/nucleon ^{36}Ar + ^{197}Au when the effects of impact parameter averaging are reduced by some appropriate cuts [56]. However, it is important to notice that there is no reason to expect intermittency if the phase transition is first order.

As an example, we check the intermittency behavior [62] in the lattice gas model for the disassembly of the system ^{129}Xe at $0.38\rho_0$ in the framework of LGM (for the details of the model description, please see the following section). At a temperature $T = 5.5$ MeV, the mass distribution shows a power law distribution with an effective power law parameter $\tau = 2.43$. In a previous work with the same model, it was shown that the liquid-gas phase transition occurs near 5.5 MeV for this system in the

LGM [63,64]. The $\ln F_k$ shows slight negative values with slightly positive slopes $vs. -\ln \delta$. However, this kind of the positive slopes with a moment less than unity may be of trivial origin and does not demonstrate the appearance of intermittency which is characteristic of systems exhibiting larger than Poisson fluctuations (*i.e.* the moment should be larger than unity). In order to check the event mixture effect on the scaled factorial moment, we mixed all the events at $T = 4$ MeV and $T = 7$ MeV and also used the multiplicity cuts ($29 \leq M \leq 101$) and ($M < 29$ or $M > 101$) to see if an intermittency behavior can be found in such mixed events. Figure 15 shows these results. Even though all the $\ln F_k$ values are positive, they are flat, *i.e.* there is no intermittency signal. In these cases, the fluctuation is large enough but the mass distribution shows no power law distribution. Hence, intermittency is absent. However, intermittency emerges when the moments were calculated from the mixed events of $T = 5.5\,\mathrm{MeV}$ and $T = 7\,\mathrm{MeV}$ (fig. 15). In this case, the mass distribution shows a quite good power law distribution and fluctuations are also large enough to induce intermittency.

From the above discussions, the apparent signals of intermittency which emerge in many experimental data are not easy to understand since many experimental conditions bring some complexities to the pure signal of intermittency, such as event mixing. More precise experimental measurements in the future are needed to probe the intermittency signal, which then may be taken as a signal of true critical behavior.

6 Phenomenological basis of nuclear Zipf law and model simulation

In the above sections, we have focussed on the moment analysis, namely the behavior of the moments of the fragment size distribution, or of the scaled factorial moments. Both are related to the fluctuations of some physical observables. In this section, we would like to emphasize the topological structure of the fragment size distribution, *i.e.* how the fragments distribute from the largest to the smallest in nuclear fragmentation. To this end, we introduce the Zipf-type plot, *i.e.* rank-ordering plot, in the fragment size distribution as well as Zipf's law which will be illustrated in the following [64,65].

The original Zipf's law [66] has been used for the diagnosis of nuclear liquid-gas phase transition and as such we have called it the nuclear Zipf's law. Zipf's law has been known as a statistical phenomenon concerning the relation between English words and their frequency in the literature in the field of linguistics [66]. The law states that, when we list the words in the order of decreasing population, the frequency of a word is inversely proportional to its rank [66]. This relation was found not only in linguistics but also in other fields of sciences. For instance, the law appeared in distributions of populations in cities, distributions of income of corporations, distributions of areas of lakes and cluster-size distribution in percolation processes [67,68]. The details for the proposal of nuclear Zipf's law can been found in refs. [64,65]. In this report, we firstly define the nuclear Zipf plot for the fragment mass (charge) distribution and nuclear Zipf's law in the simulation with the help of the lattice gas model. Then we show some experimental evidences for the nuclear Zipf law as well as some remarks.

The tools we will use here are the isospin-dependent lattice gas model (LGM) and molecular dynamical model (MD). The lattice gas model was developed to describe the liquid-gas phase transition for atomic systems by Lee and Yang [69]. The same model has already been applied to nuclear physics for isospin symmetrical systems in the grand-canonical ensemble [70] with a sampling of the canonical ensemble [63,71–76], and also for isospin asymmetrical nuclear matter in the mean-field approximation [77]. In

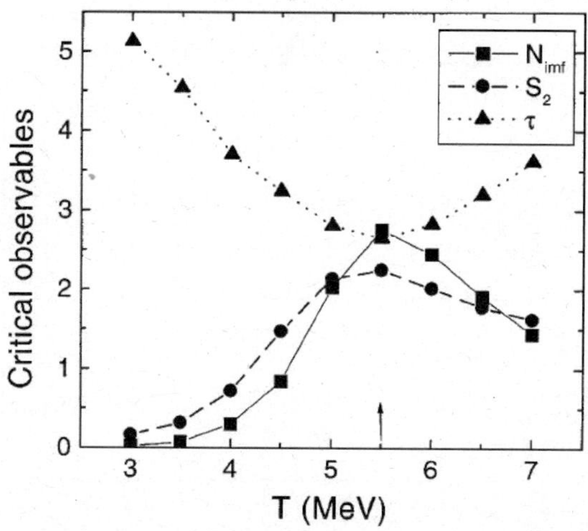

Fig. 16. Effective power law parameter, τ, second moment of the cluster distribution, S_2, and multiplicity of intermediate mass fragments, N_{imf} as a function of temperature for the disassembly of ^{129}Xe at $\rho_f \sim 0.38\rho_0$ in I-LGM. The arrow represents the estimated temperature of the phase transition. The figure is taken from [65].

addition, a classical molecular dynamical model is used to compare its results with the results of the lattice gas model.

In the lattice gas model, $A \ (= N + Z)$ nucleons with an occupation number s which is defined $s = 1 \ (-1)$ for a proton (neutron) or $s = 0$ for a vacancy, are placed on the L sites of the lattice. Nucleons in the nearest-neighboring sites interact with an energy $\epsilon_{s_i s_j}$. The Hamiltonian is written as $E = \sum_{i=1}^{A} \frac{P_i^2}{2m} - \sum_{i<j} \epsilon_{s_i s_j} s_i s_j$. A three-dimension cubic lattice with L sites is used. The freeze-out density of disassembling system is assumed to be $\rho_f = \frac{A}{L}\rho_0$, where ρ_0 is the normal nuclear density. The disassembly of the system is to be calculated at ρ_f, beyond which nucleons are too far apart to interact. Nucleons are put into lattice by Monte Carlo Metropolis sampling. Once the nucleons have been placed we also ascribe to each of them a momentum by Monte Carlo samplings of a Maxwell-Boltzmann distribution. Once this is done, the LGM immediately gives the cluster distribution using the rule that two nucleons are part of the same cluster if $P_r^2/2\mu - \epsilon_{s_i s_j} s_i s_j < 0$. This method is similar to the Coniglio-Klein prescription [78] in condensed-matter physics and was shown to be valid in LGM [71,72,74,76]. In addition, to calculate clusters using MD we propagate the particles from the initial configuration for a long time under the influence of the chosen force. The form of the force is chosen to compare with the results of LGM. The system evolves with the potential. At asymptotic times the clusters are easily recognized. Observables based on the cluster distribution in both models can now be compared. In the case of proton-proton interactions, the Coulomb interaction can also be added separately and it can be compared with the case without Coulomb effects.

In order to check the phase transition behavior in the I-LGM, we will first show the calculations of some physical

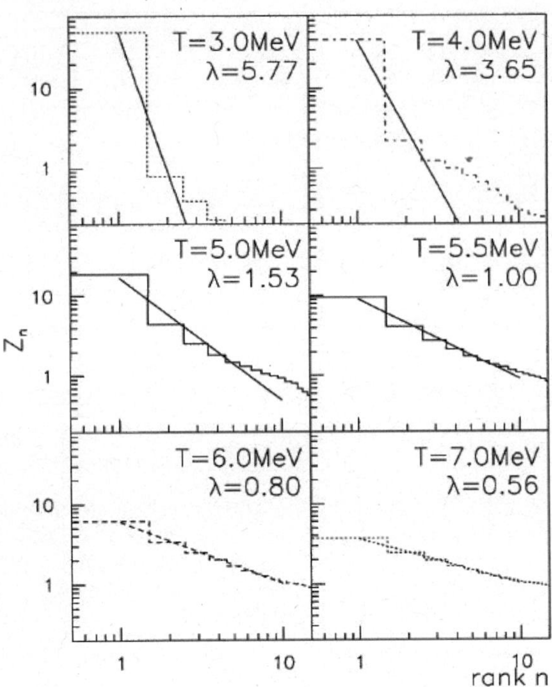

Fig. 17. Average charge Z_n with rank n as a function of n for ^{129}Xe $\rho_f \sim 0.38\rho_0$ in I-LGM. The histograms are the calculation results and the straight lines are their fits with $Z_n \propto n^{-\lambda}$. The figure is taken from [65].

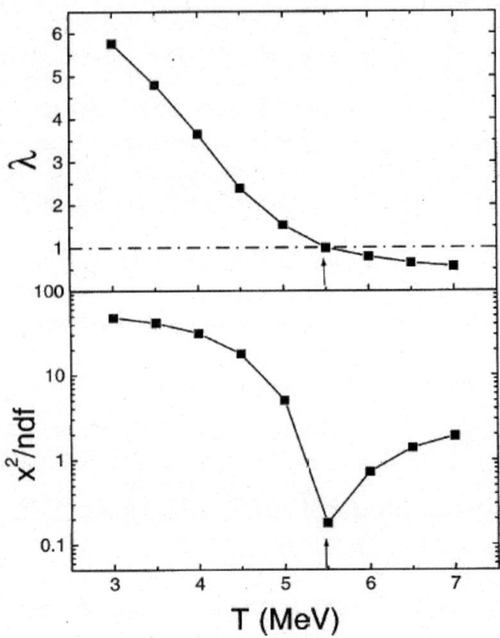

Fig. 18. Slope parameter λ of Z_n to n (top) and χ^2 test for Zipf's law (bottom) as a function of temperature for ^{129}Xe at $\rho_f \sim 0.38\rho_0$. The arrow represents the estimated temperature of the phase transition. The figure is taken from [65].

observables in fig. 16, namely the effective power law parameter, τ, the second moment of the cluster distribution, S_2 [60], and the multiplicity of intermediate mass fragments, N_{imf}, for the disassembly of ^{129}Xe at the freeze-out density $\rho_f \sim 0.38\rho_0$. These observables have been successfully employed in previous works to probe the liquid-gas phase transition, as shown in refs. [63,65,75]. The valley of τ, the peaks of N_{imf} and S_2 are located around $T \sim 5.5$ MeV which is the signature of the onset of the phase transition. Because of the exact mapping between the LGM and the Ising model, we know that at this point the transition is first order.

Now we present the results for testing Zipf's law in the charge distribution of clusters. The law states that the relation between the sizes and their ranks is described by $Z_n = c/n$ $(n = 1, 2, 3, \ldots)$, where c is a constant and Z_n (or A_n) is the average charge (or mass) of rank n in a charge (or mass) list when we arrange the clusters in the order of decreasing size. For instance, the charge Z_2 of the second largest cluster with rank $n = 2$ is one-half of the charge Z_1 of the largest cluster, the charge Z_3 of the third largest cluster with rank $n = 3$ is one-third of the charge Z_1 of the largest cluster, and so on. In the simulations of this work, we averaged the charges for each rank in charge lists of the events: we averaged the charges for the largest clusters in each event, averaged them for the second largest clusters, averaged them for the third largest clusters, and so on. From the averaged charges, we examined the relation between the charges Z_n and their

ranks n. Figure 17 shows such relations of Z_n and n for Xe with different temperatures. The histogram is the simulated results and the straight lines represent the fit with $Z_n \propto n^{-\lambda}$ in the range of $1 \leq n \leq 10$, where λ is the slope parameter. λ is 5.77 at $T = 3$ MeV. Then we increased the temperature and examined the same relation and obtained $\lambda = 3.65$ and 1.53 at $T = 4$ and 5 MeV, respectively. Up to $T = 5.5$ MeV, $\lambda = 1.00$, $i.e.$, at this temperature the relation is satisfied to Zipf's law: $Z_n \propto n^{-1}$. When the temperature increases, λ decreases; for instance, $\lambda = 0.80$ at $T = 6$ MeV and $\lambda = 0.56$ at $T = 7$. The temperature at which Zipf's law emerges is consistent with the phase transition temperature obtained in fig. 16, illustrating that Zipf's law is also an additional signal to determine the location of a phase transition. From a statistical point of view, Zipf's law could also be related to a critical phenomenon [2,29]. The upper panel of fig. 18 summarizes the parameter λ as a function of temperature.

In order to further illustrate that Zipf's law is most probably fulfilled in phase transition points, we directly reproduce the histograms with Zipf's law: $Z_n = c/n$. In this case, c is the only parameter, but what we are interested in is to check the hypothesis of Zipf's law through a χ^2 test. The bottom panel of fig. 18 shows the χ^2/ndf for the Z_n-n relations at different T. As expected, the minimum χ^2/ndf is observed around the phase transition temperature, which further indicates that Zipf's law of the fragment distribution occurs around the liquid-gas phase transition point.

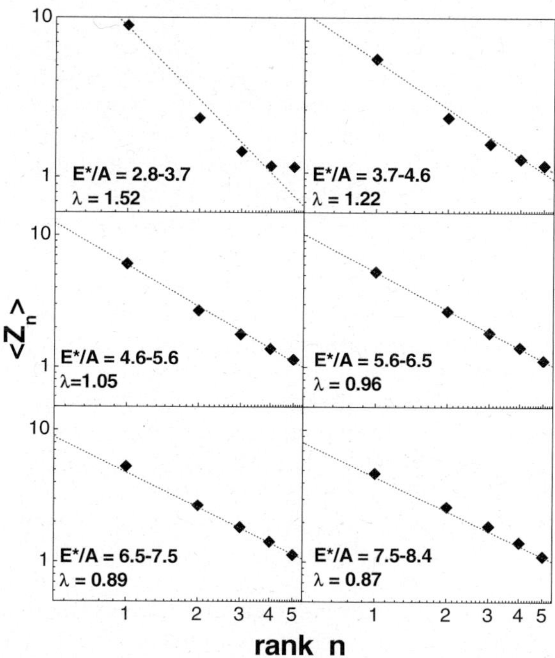

Fig. 19. Zipf plots in six different excitation energy bins for the QP formed in ^{40}Ar + ^{58}Ni. The dots are data and the lines are Zipf-law fits. The statistical error is smaller than the size of the symbols.

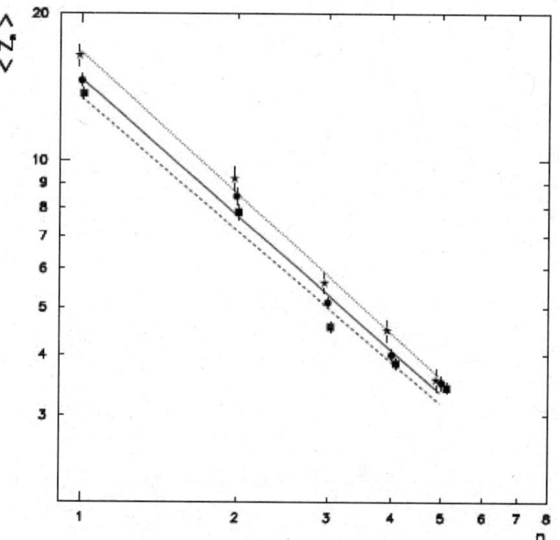

Fig. 20. Zipf-law fit to the dependences of the mean charge of the fragment on its rank. The different symbols represent the multifragmentation data of different beams with an emulsion target. Circles and solid line represent Pb beam at 158 A GeV, squares and dashed line represent Au beam at 10.6 A GeV, star and dotted line represent Au beam at 0.64 A GeV. Data are taken from ref. [84].

7 Experimental evidences of nuclear Zipf law

7.1 NIMROD results

In sect. 4.2, we gave some information on critical behaviors for the Texas A&M NIRMOD data based on the moment analysis technique. Different signals of critical behavior coherently pointing to the same excitation energy interval have been shown. In this section, we will further show the significance of the 5–6 MeV region in NIMROD data using a Zipf's law analysis. In fig. 19 we present Zipf plots for rank-ordered average Z in six different energy bins. The lines in the figure are fits to the power law expression $\langle Z_n \rangle \propto n^{-\lambda}$. Figure 10(f) shows the fitted Zipf exponent, λ parameter, as a function of excitation energy. As shown in fig. 19, this rank ordering of the observation probability of fragments of a given atomic number, from the largest to the smallest, does indeed lead to a Zipf's power law parameter $\lambda = 1$ in the 5–6 MeV/nucleon range. Around this excitation energy, the mean size of the second largest fragment is 1/2 of that of the largest fragment; that of the third largest fragment is 1/3 of the largest fragment's one, etc. This is a special kind of size topology of fragment distributions, which is very different from the equal-size fragment distribution expected if fragments are formed through a spinodal instability inside the phase coexistence region [22,79–83]. This shows the relevance of using Zipf plots to explore the fragment size topology.

7.2 CERN emulsion experiment

The nuclear Zipf-type plot has been also applied in the analysis of CERN emulsion or plastic data of Pb + Pb or plastic at 158 A GeV following Ma's proposal on Zipf law, and it was found that the nuclear Zipf law is satisfied in coincidence with other proposed signals of phase transition [84,85].

Dabrowska *et al.* have extended these studies to the multifragmentation of lead projectiles at an energy of 158 A GeV [84]. The analyzed data were obtained from the CERN EMU13 experiment in which emulsion chambers, composed of nuclear target foils and thin emulsion plates interleaved with spacers, allow for precise measurements of emission angles and charges of all projectile fragments emitted from Pb-nucleus interactions. The results on fragment multiplicities, charge distributions and angular correlations are analyzed for multifragmentation of the Pb projectile after an interaction with heavy (Pb) and light (plastic-$C_5H_4O_2$) targets. A detailed description of the emulsion experiment can be found in ref. [84].

Figure 20 shows the Zipf-type plot for charged fragments heavier than helium emitted in multifragmentation events of Au or Pb projectile at different beam energies. The values of λ exponents from fits $\langle Z_n \rangle \sim n^{-\lambda}$ are 0.92 ± 0.03, 0.90 ± 0.02 and 0.96 ± 0.04 for beam energies of 158, 10.6 and 0.64 A GeV, respectively. Within the statistical errors, the values of the λ coefficient are the same in the studied energy interval (< 1–158) A GeV and do not differ significantly from unity [84].

Dabrowska *et al.* also studied the dependence of the power law exponent λ on the control parameter m, the normalized multiplicity with respect to the total charge of spectator particles [85]. In fig. 21(a) are shown the mean multiplicity $\langle N_f \rangle$ of fragments with $Z \geq 3$ and the mean

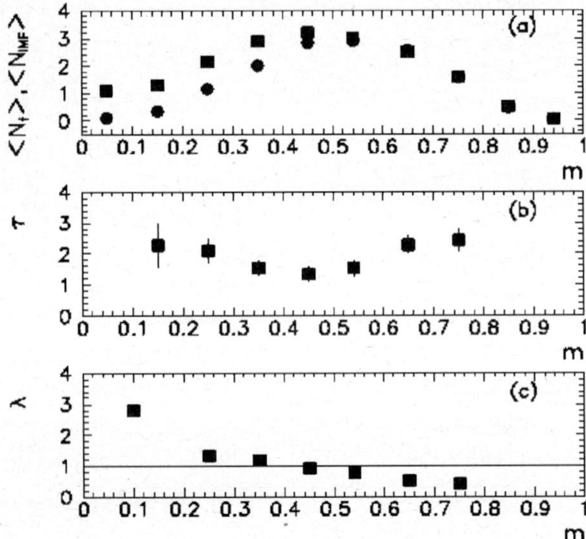

Fig. 21. (a) Mean number, $\langle N_f \rangle$, of fragments (squares) and mean number, $\langle N_{IMF} \rangle$, of intermediate mass fragments (circles) as a function of the normalized multiplicity m. Error bars are smaller than the size of the squares and circles. (b) Power law exponent, τ, of the charge distribution of fragments in different intervals of m. (c) Power law exponent, λ, in Zipf's law (see text) in different intervals of m. Error bars are smaller than data points. The data is taken from ref. [85].

number $\langle N_{IMF} \rangle$ of the intermediate fragments. The latter are usually defined as fragments with $3 \leq Z \leq 30$. In fig. 21(b) the dependence of the exponent τ of the power fits to the charge distribution of fragments, performed at different ranges of m, is also given. In this analysis, the fits are restricted to fragment charges smaller than $Z = 16$. At small values of m, a system has few light fragments and the power law is steep; at large values of m there are basically only many light fragments leading again to a steep power law. At the moderate excitation energies where heavier fragments appear and where we expect the phase transition, the exponent τ has its lowest value. As can be seen from fig. 21(b), the minimum τ occurs for m values between 0.35 and 0.55. In fig. 21(c) the dependence of λ obtained from the fits $\langle Z_n \rangle \sim n^{-\lambda}$, as a function of m is depicted. The exponent λ decreases with increasing m. Between $m \approx 0.3$ and $m \approx 0.5$ the value of λ is close to unity and Zipf's law is satisfied. This suggests that at this value of m the liquid-gas phase transition might occur. It has been checked that $\lambda \sim 1$ occurs in the same region of m, irrespectively of the mass of the target [85]. This means that the liquid-gas phase transition occurs when a given amount of energy is deposited into the nucleus and does not depend on the mass of the target. As expected, in the case of a liquid-gas phase transition, the previously shown maxima in frequency distributions of multiply charged fragments (fig. 21(a)) as well as a minimum of the power law parameter τ (fig. 21(b)), all occur at the same values of m, where Zipf's law emerges.

7.3 Some remarks on Zipf law

Campi *et al.* pointed out that for an infinite system, Zipf's law is a mathematical consequence of a power law cluster size distribution with exponent $\tau \simeq 2$ [86]. More precisely, both Zipf law exponent λ and Fisher scaling power law exponent τ are connected through the formula $\lambda = 1/(\tau - 1)$ in *an infinite system* assuming that the cluster size distribution is a power law distribution. They argued that such distributions appear at the critical point with $\tau \simeq 2$ of many theories, *e.g.* various theories of cluster formation but also in the super-critical region of the lattice-gas and realistic Lennard-Jones fluids [87]. However, the experimental fragment size distribution is mostly neither power law distribution nor exponential distribution except for some special situations. Also, the nuclear system is always a finite system, which means that the relationship between λ and τ mentioned above is not strictly valid. To account for finite-size effects, Bauer *et al.* [88] have evaluated the fragment probabilities as a function of their rank at the critical point for a finite system with fragment distributions obeying a finite-size scaling ansatz. From this analytical evaluation, where however the assumption is made that all fragments including the largest are much smaller than the source, they suggest to extend the simple Zipf's law to a more general Zipf-Mandelbrot distribution [89, 90], $\langle A_r \rangle = c(r + k)^{-\lambda}$, where the offset k is an additional constant that one has to introduce, and λ is asymptotically approximated as a function of the critical exponent τ, $\lambda = 1/(\tau - 1)$ of the *infinite system*.

In any case, the Zipf-type plot is a direct observable allowing to characterize the fragment hierarchy in nuclear disassembly, and as such it is a useful signal of phase transition or critical behavior.

8 Summary and outlook

In summary, the moment analysis method has been introduced and some applications to nuclear multifragmentation have been presented. Since we are dealing with a finite nucleus rather than infinite nuclear matter, finite-size effects must always be discussed in the model calculations and data analysis. Experimentally, the critical behavior of nuclear disassembly can be investigated with the help of moment analysis. The occurrence of a fluctuation peak which can be extracted from the moment analysis method can be interpreted as a signal of critical behavior. Using the same analysis method as for the percolation model, the liquid-gas universality class exponents are approximately obtained in nuclear multifragmentation, such as in EOS data and NIMROD data. This would point to the observation of the liquid-gas critical point or second-order phase transition. However, when we think about the system size dependence of critical exponents and we consider some results using lattice gas model simulations and other related different analysis methods, it appears that some open questions still remain concerning the order of the phase transition. For instance, EOS Collaboration claimed that there are continuous phase transitions for heavier systems, namely Au and La and first-order phase transitions for lighter systems, namely Kr. On the other hand, NIMROD data show critical behavior, corresponding to a continuous phase transition, for the light system Quasi-Ar. Different conclusions are then reached for similar light systems. Recent systematic analyses of caloric curves [6,14,20] and configurational energy fluctuations [23], indicate that heavier systems may undergo a first-order phase transition while lighter systems can probably sustain a higher temperature, possibly even above the critical point, which would make the first-order phase transition observed in heavy nuclei become a crossover in lighter systems. Concerning configurational energy fluctuations, a well-pronounced peak at an excitation energy around $5\,\mathrm{MeV}$ was shown in Multics, Indra, Isis and NIMROD data [23]. However, this fluctuation appears monotonically decreasing in EOS data [33]. Thus it deserves further investigations.

Scaled factorial moments and intermittency have also been reviewed and some examples given to show the apparent intermittency in nuclear fragmentation. However, some complex ingredients in experimental measurements, such as mixtures of event multiplicities or temperature fluctuations in the data can induce spurious intermittency-like behavior which implies that the apparent "intermittency" cannot be taken as a unique signal of the critical behavior. Without 4π detector upgrades allowing better data sorting, it remains difficult to take apparent "intermittency" behaviors as a signal of critical behavior in nuclear multifragmentation.

Finally, nuclear Zipf-type plots are introduced and Zipf's law is proposed to be related to a phase transition or a critical behavior of nuclei. Around the transition point, the cluster mass (charge) shows inversely to its rank, *i.e.* Zipf's law appears. Even though the criterion is phenomenological, it is a simple and practicable tool to characterize the fragment hierarchy in nuclear disassembly. The 4π multifragmentation data of heavy-ion collision at Texas A&M University and the CERN emulsion/plastic data exhibit the Zipf law around the same excitation energy deposit. The satisfaction of the Zipf law for the cluster distributions illustrates that the clusters obey at this point a particular rank ordering distribution very different from the equal-size fragment distribution which may occur due to spinodal instability inside the liquid-gas coexistence region. To conclude, we should mention that all these transition signals, such as the fluctuation peak, critical exponents, Fisher scaling as well as Zipf's law, etc. may not be very robust individually since we are facing a transient finite charged nuclear system. A unique signal cannot give any definite information as to whether the system is in a critical point or is undergoing a phase transition. Only many coherent signals emerging together can corroborate the observation of a phase transition or a critical behavior in finite nuclei.

This work was supported in part by the Shanghai Development Foundation for Science and Technology under Grant Number 05XD14021, the National Natural Science Foundation of China under Grant Nos. 19725521, 10328259, 10135030, 10535010 and the Major State Basic Research Development Program under Contract No. G200077404.

References

1. C.A. Ogilvie *et al.*, Phys. Rev. Lett. **67**, 1214 (1991); M.B. Tsang *et al.*, Phys. Rev. Lett. **71**, 1502 (1993); Y.G. Ma, W.Q. Shen, Phys. Rev. C **51**, 710 (1995).
2. M.E. Fisher, Physics (N.Y.) **3**, 255 (1967).
3. S. Albergo, S. Costa, E. Costanzo, A. Rubbino, Nuovo Cimento A **89**, 1 (1985).
4. J. Pochodzalla *et al.*, Phys. Rev. Lett. **75**, 1040 (1995).
5. Y.G. Ma *et al.*, Phys. Lett. B **390**, 41 (1997).
6. J.B. Natowitz *et al.*, Phys. Rev. C **65**, 034618 (2002).
7. J.B. Natowitz, K. Hagel, Y.G. Ma, M. Murray, L. Qin, R. Wada, J. Wang, Phys. Rev. Lett. **89**, 212701 (2002).
8. W. Bauer, Phys. Rev. C **38**, 1297 (1988).
9. M.L. Gilkes *et al.*, Phys. Rev. Lett. **73**, 1590 (1994).
10. M. D'Agostino *et al.*, Nucl. Phys. A **650**, 329 (1999).
11. J.B. Elliott *et al.*, Phys. Rev. C **55**, 1319 (1997).
12. J.B. Elliott *et al.*, Phys. Rev. C **49**, 3185 (1994).
13. M. Kleine Berkenbusch, W. Bauer, K. Dillman, S. Pratt, L. Beaulieu, K. Kwiatkowski, T. Lefort, W.C. Hsi, V.E. Viola, S.J. Yennello, R.G. Korteling, H. Breuer, Phys. Rev. Lett. **88**, 022701 (2002).
14. NIMROD Collaboration (Y.G. Ma *et al.*), Phys. Rev. C **69**, 031604(R) (2004); **71**, 054606 (2005).
15. J. Richert, P. Wagner, Phys. Rep. **350**, 1 (2001).
16. S. Das Gupta, A.Z. Mekjian, M.B. Tsang, Adv. Nucl. Phys. **26**, 89 (2001).
17. A. Bonasera, M. Bruno, P.F. Mastinu, C.O. Dorso, Riv. Nuovo Cimento **23**, No. 2 (2000).
18. Ph. Chomaz, *Proceedings of the International Nuclear Physics Conference INPC2001, Berkeley CA, USA, 2001*, edited by E. Norman, L. Schroeder, G. Wozniak, AIP Conf. Proc., Vol. **610** (Melville, New York, 2002) p. 167.
19. L.G. Moretto, J.B. Elliott, L. Phair, G.J. Wozniak, C.M. Mader, A. Chappars, in [18] p. 182.
20. A. Kelic, J.B. Natowitz, K.-H. Schmidt, this topical issue.
21. V.E. Viola, R. Bougault, this topical issue.
22. B. Borderie, P. Désesquelles, this topical issue.
23. F. Gulminelli, M. D'Agostino, this topical issue.
24. O. Lopez, M.F. Rivet, this topical issue.
25. X. Campi, J. Phys. A **19**, L 917 (1986).
26. X. Campi, Phys. Lett. B **208**, 351 (1988).
27. W. Bauer *et al.*, Phys. Lett. B **150**, 53 (1985).
28. W. Bauer *et al.*, Nucl. Phys. A **452**, 699 (1986).
29. D. Stauffer, *Introduction to Percolation Theory* (Taylor and Francis, London, 1985).
30. K. Binder, *Monte Carlo Methods in Statistical Mechanics*, 2nd ed. (Springer-Verlag, Berlin, 1986).
31. H.R. Jaqaman, D.H.E. Gross, Nucl. Phys. A **524**, 321 (1991).
32. N. Tan *et al.*, Phys. Rev. B **29**, 6354 (1984).
33. B.K. Srivastava *et al.*, Phys. Rev. C **64**, 041605 (2001); **65**, 054617 (2002).
34. R.P. Scharenberg *et al.*, Phys. Rev. C **64**, 054602 (2001).
35. J.A. Hauger *et al.*, Phys. Rev. C **57**, 764 (1998).
36. J.A. Hauger *et al.*, Phys. Rev. C **62**, 024616 (2000).
37. D. Cussol *et al.*, Nucl. Phys. A **561**, 298 (1993).
38. J. Lauret *et al.*, Phys. Rev. C **62**, R1051 (1998).

39. W. Bauer, W.A. Friedman, Phys. Rev. Lett. **77**, 767c (1995).
40. M.L. Gilkes *et al.*, Phys. Rev. Lett. **77**, 768c (1995).
41. J.B. Elliott *et al.*, Phys. Rev. C **67**, 024609 (2003).
42. X. Campi, H. Krivine, Nucl. Phys. A **545**, 161c (1992).
43. X. Campi, H. Krivine, Z. Phys. A **344**, 81 (1992).
44. Y.G. Ma *et al.*, Nucl. Phys. A **749**, 106c (2005).
45. C.O. Dorso, V.C. Latora, A. Bonasera, Phys. Rev. C **60**, 034606 (1999).
46. R. Botet, M. Ploszajczak, A. Chbihi, B. Borderie, D. Durand, J. Frankland, Phys. Rev. Lett. **86**, 3514 (2001).
47. Y.G. Ma, J. Phys. G **27**, 2455 (2001).
48. A. Bialas, R. Peschanski, Nucl. Phys. B **273**, 703 (1986).
49. B. Mandelbrot, J. Fluid Mech. **62**, 331 (1974); U. Frisch, P. Sulem, M. Nelkin, J. Fluid Mech. **87**, 719 (1978).
50. Ya. B. Zeldovich *et al.*, Sov. Phys. Usp. **30**, 353 (1987).
51. G. Paladin, V. Vulpiani, Phys. Rep. **156**, 147 (1987).
52. A. Bialias, R.C. Hwa, Phys. Lett. B **207**, 59 (1988).
53. R.C. Hwa, M.T. Nazirov, Phys. Rev. Lett. **69**, 741 (1992).
54. H. Satz, Nucl. Phys. B **326**, 613 (1989).
55. M. Ploszajczak, A. Tucholski, Phys. Rev. Lett. **65**, 1539 (1999).
56. L. Phair *et al.*, Phys. Rev. Lett. **79**, 3538 (1996); Phys. Lett. B **291**, 7 (1992).
57. B. Elattari, J. Richert, P. Wagner, Phys. Rev. Lett. **69**, 45 (1992); Nucl. Phys. A **560**, 603 (1993).
58. H.W. Barz *et al.*, Phys. Rev. C **45**, R2541 (1992).
59. V. Latora, M. Belkacem, A. Bonesera, Phys. Rev. Lett. **73**, 1765 (1994); M. Belkacem, V. Latora, A. Bonasera, Phys. Rev. C **52**, 271 (1995).
60. X. Campi, H. Krivine, Nucl. Phys. A **589**, 505 (1995).
61. P.F. Mastinu *et al.*, Phys. Rev. Lett. **76**, 2646 (1996).
62. Y.G. Ma, unpublished.
63. Y.G. Ma *et al.*, Phys. Rev. C **60**, 024607 (1999).
64. Y.G. Ma, Phys. Rev. Lett. **83**, 3617 (1999).
65. Y.G. Ma, Eur. Phys. J. A **6**, 367 (1999).
66. G.K. Zipf, *Human Behavior and the Principle of Least Effort* (Addisson-Wesley Press, Cambridge, MA, 1949).
67. D.L. Turcotte, Rep. Prog. Phys. **62**, 1377 (1999).
68. M. Watanabe, Phys. Rev. E **53**, 4187 (1996).
69. T.D. Lee, C.N. Yang, Phys. Rev. **87**, 410 (1952).
70. T.S. Biro *et al.*, Nucl. Phys. A **459**, 692 (1986); S.K. Samaddar, J. Richert, Phys. Lett. B **218**, 381 (1989); Z. Phys. A **332**, 443 (1989); J.M. Carmona *et al.*, Nucl. Phys. A **643**, 115 (1998).
71. X. Campi, H. Krivine, Nucl. Phys. A **620**, 46 (1997).
72. J. Pan, S. Das Gupta, Phys. Rev. C **53**, 1319 (1996).
73. W.F.J. Müller, Phys. Rev. C **56**, 2873 (1997).
74. J. Pan, S. Das Gupta, Phys. Lett. B **344**, 29 (1995); Phys. Rev. C **51**, 1384 (1995); Phys. Rev. Lett. **80**, 1182 (1998); S. Das Gupta *et al.*, Nucl. Phys. A **621**, 897 (1997).
75. J. Pan, S. Das Gupta, Phys. Rev. C **57**, 1839 (1998).
76. F. Gulminelli, Ph. Chomaz, Phys. Rev. Lett. **82**, 1402 (1999).
77. S. Ray *et al.*, Phys. Lett. B **392**, 7 (1997).
78. A. Coniglio, E. Klein, J. Phys. A **13**, 2775 (1980).
79. G.F. Bertsch, P.J. Siemens, Phys. Lett. B **126**, 9 (1983).
80. L.G. Moretto *et al.*, Phys. Rev. Lett. **77**, 2634 (1996).
81. L. Beaulieu *et al.*, Phys. Rev. Lett. **84**, 5971 (2000).
82. M. Colonna *et al.*, Phys. Rev. Lett. **88**, 122701 (2002).
83. B. Borderie *et al.*, Phys. Rev. Lett. **86**, 003252 (2001).
84. A. Dabrowska, M. Szarska, A. Trzupek, W. Wolter, B. Bosiek, Acta Phys. Pol. B **32**, 3099 (2001).
85. A. Dabrowska, M. Szarska, A. Trzupek, W. Wolter, B. Bosiek, Acta Phys. Pol. B **35**, 2109 (2004).
86. X. Campi, H. Krivine, Phys. Rev. C **72**, 057602 (2005).
87. N. Sator, Phys. Rep. **376**, 1 (2003).
88. W. Bauer, B. Alleman, S. Pratt, nucl-th/0512101.
89. B. Mandelbrot, *An informational theory of the statistical structure of language*, in *Communication Theory*, edited by W. Jackson (Betterworths, 1953).
90. B. Mandelbrot, *The Fractal Geometry of Nature* (Freeman, 1982).

Eur. Phys. J. A **30**, 243–251 (2006)
DOI 10.1140/epja/i2006-10120-y

THE EUROPEAN
PHYSICAL JOURNAL A

Many-fragment correlations and possible signature of spinodal fragmentation

B. Borderie[1,a] and P. Désesquelles[2]

[1] Institut de Physique Nucléaire, IN2P3-CNRS, F-91406 Orsay cedex, France
[2] Centre de Spectrométrie Nucléaire et de Spectrométrie de Masse, IN2P3 et Université, F-91406 Orsay cedex, France

Received: 3 February 2006 /
Published online: 25 October 2006 – © Società Italiana di Fisica / Springer-Verlag 2006

Abstract. Abnormal production of events with almost equal-sized fragments was theoretically proposed as a signature of spinodal instabilities responsible for nuclear multifragmentation. Many-fragment correlations can be used to enlighten any extra production of events with specific fragment partitions. The high sensitivity of such correlation methods makes it particularly appropriate to look for small numbers of events as those expected to have kept a memory of spinodal decomposition properties and to reveal the dynamics of a first-order phase transition for nuclear matter and nuclei. This paper summarizes results obtained so far for both experimental and dynamical simulations data.

PACS. 25.70.Pq Multifragment emission and correlations – 24.60.Ky Fluctuation phenomena

1 Introduction

Thermodynamics describes phase transitions in terms of static conditions. Information on the existence of phases and coexistence of phases is derived depending on thermodynamical parameters (temperature, pressure...). How to pass from a phase to another? What is the time needed? To answer these questions, dynamics of phase transitions must be studied. Therefore the aim of this paper is to discuss signals which could be related to the dynamics of phase transition involved in hot unstable nuclei produced in nucleus-nucleus collisions. Of particular relevance is the possible occurrence of spinodal instabilities. Indeed, during a collision, a wide zone of the nuclear-matter phase diagram may be explored and the nuclear system may enter the coexistence region (at low density) and even more precisely the unstable spinodal region (domain of negative compressibility). Thus, a possible origin of multifragmentation may be found through the growth of density fluctuations in this unstable region. Within this theoretical scenario a breakup into nearly equal-sized "primitive" fragments should be favoured. Hence many fragment correlations have been analyzed to investigate this possible scenario. They were applied on selected central collision events produced in experiments and also in 3D stochastic mean-field simulations of head-on collisions.

2 Spinodal instabilities for nuclear matter and nuclei

In the last fifteen years a big theoretical effort has been realized to understand and learn about spinodal decomposition in the nuclear context. A review can be found in ref. [1].

2.1 Nuclear matter

We shall first briefly discuss what are the specificities of spinodal decomposition as far as infinite nuclear matter is concerned. Associated to negative compressibility the mechanically unstable spinodal region can be investigated by studying the propagation of small density perturbations [2,3]. To do that the linear response framework is used to solve the RPA equations. In the spinodal region some modes do not oscillate but are amplified because of the instability. They have an imaginary eigenfrequency, this frequency being the inverse of the instability growth time. Figure 1 presents an example of nuclear dispersion relation at 3 MeV temperature for two different densities $\rho_0/2$ and $\rho_0/3$. Imaginary RPA frequencies are reported as a function of the wave number k of the considered perturbation. This dispersion relation exhibits a strong maximum at a given wave number followed by a cut-off at large k values. This cut-off reflects the fact that fluctuations with wavelength smaller than the range of the force cannot be amplified. The most unstable modes cor-

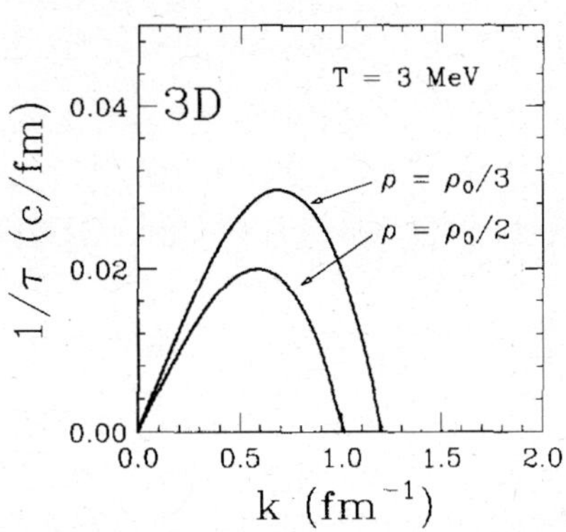

Fig. 1. Nuclear-matter dispersion relation at 3 MeV temperature for two different densities; ρ_0 is the normal density. (From [2].)

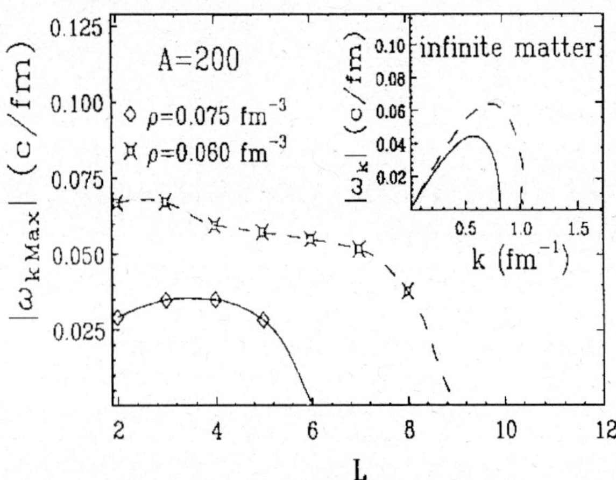

Fig. 2. Growth rates of the most unstable modes for a spherical source with 200 nucleons as a function of the multipolarity L and for two different central densities. (From [5].)

respond to wavelengths lying around $\lambda \approx 10\,\mathrm{fm}$ and associated characteristic times are almost identical, around 30–$50\,\mathrm{fm}/c$, depending on density ($\rho_0/2$–$\rho_0/8$) and temperature (0–$9\,\mathrm{MeV}$) [2,4]. A direct consequence of the dispersion relation is the production of "primitive" fragments with size $\lambda/2 \approx 5\,\mathrm{fm}$ which correspond to $Z \approx 10$. However, this simple and rather academic picture is expected to be largely blurred by several effects. We do not have a single unstable mode and consequently the beating of different modes occurs. Coalescence effects due to the residual interaction between fragments before the complete disassembly are also expected [2].

2.2 Finite systems

Does the signal discussed for nuclear matter survive (in final fragment partitions experimentally measured) if we consider the case of a hot expanding nucleus formed in heavy-ion collisions which undergoes multifragmentation? First of all, the fused system produced has to stay long enough in the spinodal region (≈ 3 characteristic time: 100–$150\,\mathrm{fm}/c$) to allow an important amplification of the initial fluctuations. Second, the presence of a surface introduces an explicit breaking of the translational symmetry. Figure 2 shows the growth rates of the most unstable modes for a spherical source of $A = 200$ with a Fermi shape profile and for two different central densities [5]. The growth rates are nearly the same for different multipolarities L up to a maximum multipolarity L_{max} (see also [6]). This result indicates that the unstable finite system breaks into different channels with nearly equal probabilities. Depending on multipolarity L, equal-sized "primitive" fragments are expected to be produced with sizes in the range $A_F/2$–A_F/L_{max}; A_F being the part of the system leading to fragments during the spinodal decomposition. One can also note that the Coulomb potential has a very small effect on the growth rates of unstable collective modes except close to the border of the spinodal zone where it stabilizes very long-wavelength unstable modes [7].

On the other hand, for a finite system, Coulomb interaction reduces the freeze-out time and enhances the chance to keep a memory of the dynamical instabilities; a similar comment can be made if collective expansion of the system is present. Both effects push the "primitive" fragments apart from each other and reduce the time of their mutual interaction.

3 Selected central collision events

Central collisions between medium or heavy nuclei leading to "fused" systems are very appropriate in the incident energy range 20–$50\,\mathrm{AMeV}$ to produce well-defined pieces of excited nuclear matter for which one could expect that bulk effects related to spinodal instabilities can occur. Such collisions represent a small (a few percent) part of cross-sections and corresponding events have been selected using global variables such as the total transverse energy E_t ($^{129}\mathrm{Xe} + {}^{nat}\mathrm{Cu}$, $^{129}\mathrm{Xe} + {}^{197}\mathrm{Au}$ and $^{36}\mathrm{Ar} + {}^{197}\mathrm{Au}$ at $50\,\mathrm{AMeV}$) [9] or the flow angle ($^{129}\mathrm{Xe} + {}^{nat}\mathrm{Sn}$ for the incident energy domain 32–$50\,\mathrm{AMeV}$) [10] and the discriminant analysis method ($^{58}\mathrm{Ni} + {}^{58}\mathrm{Ni}$ and $^{58}\mathrm{Ni} + {}^{197}\mathrm{Au}$ for the range 32–$52\,\mathrm{AMeV}$) [11–13].

4 Multi-fragment correlation functions and production of events with nearly equal-sized fragments

Following early studies related to nearly equal-sized fragment partitions [8], ten years ago a method called higher-order charge correlations [9] was proposed to enlighten any extra production of events with specific fragment partitions. The high sensitivity of the method makes it particularly appropriate to look for small numbers of events as those expected to have kept a memory of spinodal decomposition properties. Thus, such a charge correlation method allows to examine model-independent signatures that would indicate a preferred decay into a number of equal-sized fragments in events from experimental data or from simulations.

4.1 Methods

The classical two-fragment charge correlation method considers the coincidence yield $Y(Z_1, Z_2)$ of two fragments of atomic numbers $Z_{1,2}$, in the events of multiplicity M_f of a sample. A background yield $Y'(Z_1, Z_2)$ is constructed by mixing, at random, fragments from different coincidence events selected by the same cut on M_f. The two-particle correlation function is given by the ratio of these yields. When searching for enhanced production of events which break into equal-sized fragments, the higher-order correlation method appears much more sensitive. All fragments of one event with fragment multiplicity $M_f = M = \sum_Z n_Z$, where n_Z is the number of fragments with charge Z in the partition, are taken into account. By means of the normalized first-order

$$\langle Z \rangle = \frac{1}{M} \sum_Z n_Z Z \tag{1}$$

and second-order

$$\sigma_Z^2 = \frac{1}{M} \sum_Z n_Z (Z - \langle Z \rangle)^2 \tag{2}$$

moments of the fragment charge distribution in the event, one may define the higher-order charge correlation function as

$$1 + R(\sigma_Z, \langle Z \rangle) = \left. \frac{Y(\sigma_Z, \langle Z \rangle)}{Y'(\sigma_Z, \langle Z \rangle)} \right|_M . \tag{3}$$

Here, the numerator $Y(\sigma_Z, \langle Z \rangle)$ is the yield of events with given $\langle Z \rangle$ and σ_Z values. Because the measurement of the charge belonging to a given event is not subject to statistical fluctuations, one can use expression (2) rather than the "nonbiased estimator" of the variance, $\frac{1}{M-1} \sum_Z n_Z (Z - \langle Z \rangle)^2$, as proposed in [9] and also used in [14]. Note that this choice has no qualitative influence on the forthcoming conclusions. The denominator $Y'(\sigma_Z, \langle Z \rangle)$, which represents the uncorrelated yield of pseudo-events, was built in [9], as for classical correlation methods, by taking fragments at random in different events of the selected sample of a certain fragment multiplicity; this way to evaluate the denominator will be denoted as Fragments at Random Method (FRM) in what follows. This Monte Carlo generation of the denominator $Y'(\sigma_Z, \langle Z \rangle)$ can be replaced by a fast algebraic calculation which is equivalent to the sampling of an infinite number of pseudo-events [15]. Its contribution to the statistical error of the correlation function is thus eliminated. However, owing to the way the denominator was constructed, only the fragment charge distribution $\mathrm{d}M/\mathrm{d}Z$ of the parent sample is reproduced but the constraints imposed by charge conservation are not taken into account. This has,

in particular, a strong effect on the charge bound in fragments $\mathrm{d}M/\mathrm{d}Z_{bound}$ distribution. This fact makes the denominator yield distributions as a function of $\langle Z \rangle$ wider and flatter than those of the numerator [16]. Consequently, even in the absence of a physical correlation signal, the ratio (3) is not a constant equal to one. The correlations induced by the finite size of the system (charge conservation) distorts the amplitude, or may even cancel other less trivial correlations. Therefore, a new method for the evaluation of the denominator [15], based on the "intrinsic probability" of emission of a given charge, was proposed. It minimizes these effects and replicates all features of the partitions of the numerator, except those (of interest) due to other reasons than charge conservation. The principle of the method is to take into account, in a combinatorial way, the trivial correlations due to charge conservation. If there is no correlation between the charges, each charge can be fully described by an emission probability referred to as intrinsic probability. This new method to build the denominator will be denoted as the Intrinsic Probability Method (IPM) in what follows. However, the explicit calculation of the intrinsic probabilities may not be the only method for building a denominator including only the correlations induced by charge conservation. Another procedure was also proposed in [17]: the denominator is built by mixing events through random exchanges of two fragments between two events under the constraint that the sum of the two exchanged fragments is conserved, which satisfies Z_{bound} conservation (see also sect. 6 for a comparison with the IPM method). This last method will be denoted as the Random Exchange of Two-Fragment Method (RETFM) in what follows.

4.2 Stochastic mean-field simulations and spinodal instabilities

Dynamical stochastic mean-field simulations have been proposed for a long time to describe processes involving instabilities like those leading to spinodal decomposition [18–20]. In this approach, spinodal decomposition of hot and dilute finite nuclear systems can be mimicked through the Brownian One-Body (BOB) dynamics [21–23], which consists in employing a Brownian force in the kinetic equations. Simulations have been performed for head-on ^{129}Xe on ^{119}Sn collisions at 32 AMeV. The ingredients of the simulations can be found in [23] as well as a detailed comparison between filtered simulated events (to account for the experimental device) and experimental data. A good agreement between both is revealed.

To refine the comparison, higher-order charge correlations have been calculated for the simulated events [10], keeping the compact presentation proposed in [14]: charge correlation functions are built for all events, whatever their multiplicity, by summing the correlated yields for all M and by replacing the variable $\langle Z \rangle$ by $Z_{bound} = M \times \langle Z \rangle = \sum_Z Z n_Z$. Uncorrelated events are constructed and weighted in proportion to real events of each multiplicity. This presentation is based on the experimental observation that the peaks observed independently for each

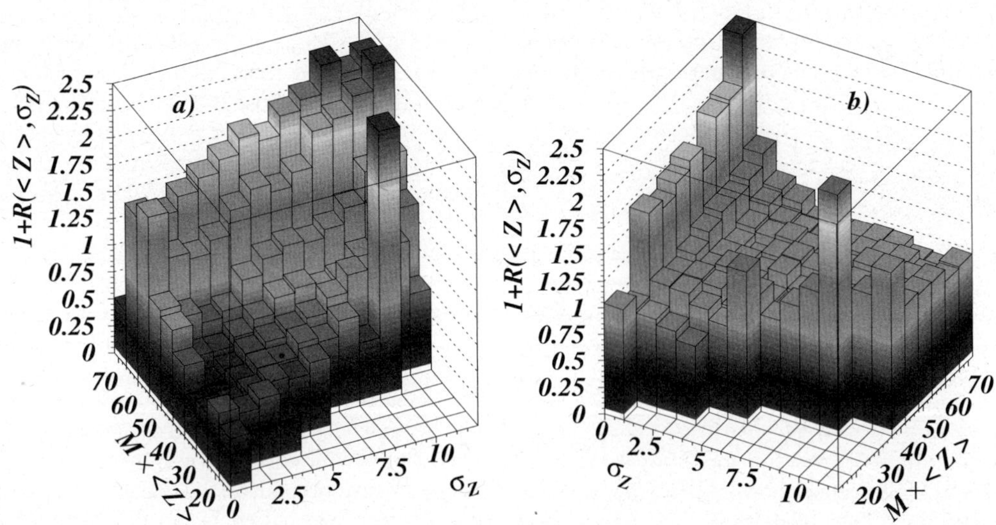

Fig. 3. Correlation functions for events with $M_f = 3$ to 6, simulated with the Brownian One-Body (BOB) model for 32 AMeV ^{129}Xe + ^{nat}Sn collisions: a) with an analytical denominator provided by FRM; b) with a denominator calculated with the IPM. The orientations of a) and b) are different for a better visualisation of the landscapes. (From [10].)

fragment multiplicity correspond to the same Z_{bound} region [14]. The variance bin was chosen equal to one charge unit. One can recall that in the considered domain of excitation energy, around 3 MeV per nucleon [24,23], secondary evaporation leads to fragments one charge unit smaller, on average, than the primary $Z \approx 10$–20 ones, with a standard deviation around one [16]. If a weak enhanced production of exactly equal-sized fragments exists, peaks are expected to appear in the interval $\sigma_Z = 0$–1, because of secondary evaporation. This interval in σ_Z is hence the *minimum* value which must be chosen to look for nearly equal-sized fragments. Any (unknown) intrinsic spread in the fragment size coming from the break-up process itself may enlarge the σ_Z interval of interest. Here, only events with $\sigma_Z < 1$ were considered, which corresponds to differences of at most two units between the fragment atomic numbers in one event.

Figure 3 shows the correlation function calculated using the analytical denominator of FRM (a) or the denominator given by the IPM (b). Both functions are drawn *versus* the variables $Z_{bound} = M \times \langle Z \rangle$ and σ_Z. In fig. 3a, the equal-sized fragment correlations in the first bin are superimposed over trivial correlations due to the finite size of the system. For this reason, the ratio (3) is generally different from one and smoothly varies with the variables Z_{bound} and σ_Z. For each bin in Z_{bound} (fixed at 6 atomic number units), an exponential evolution of the correlation function is observed from $\sigma_Z = 7$–8 down to $\sigma_Z = 2$–3. This exponential evolution was thus taken as a "background" empirically extrapolated down to the first bin $\sigma_Z = 0$–1. The amplitude of the correlation function in the domain $Z_{bound} = 36$–60 is well above the background, with a confidence level higher than 90%, proving thus a statistically significant enhancement of equal-sized

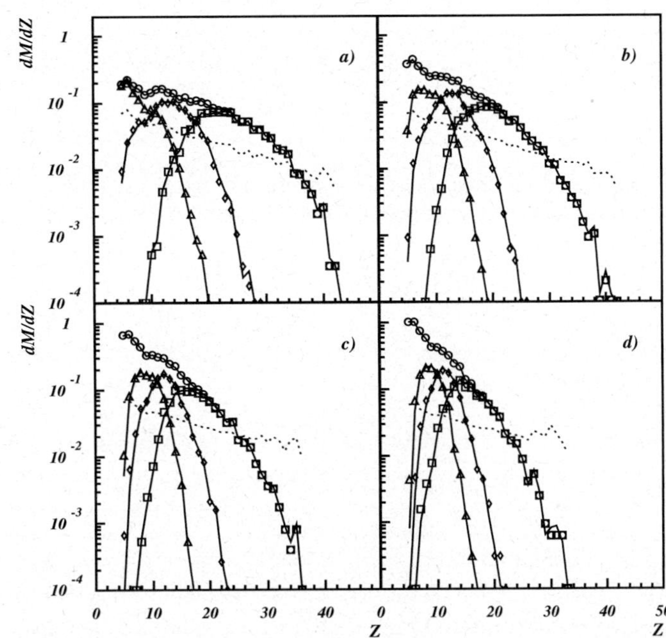

Fig. 4. Experimental differential charge multiplicity distributions (circles) for the single source formed in central 39 AMeV ^{129}Xe on ^{nat}Sn collisions. Parts a), b), c) and d) refer, respectively, to fragment multiplicities 3, 4, 5, 6. The Z distributions for the first (squares), second (diamonds) and third (triangles) heaviest fragments are presented too. The lines correspond to the results obtained with IPM. The dashed lines display the intrinsic probabilities. (From [10].)

fragment partitions. Of the 1% of events having $\sigma_Z < 1$, (0.13 ± 0.02)% (called extra-events from now on) are in excess of the background. In fig. 3b, as one could expect,

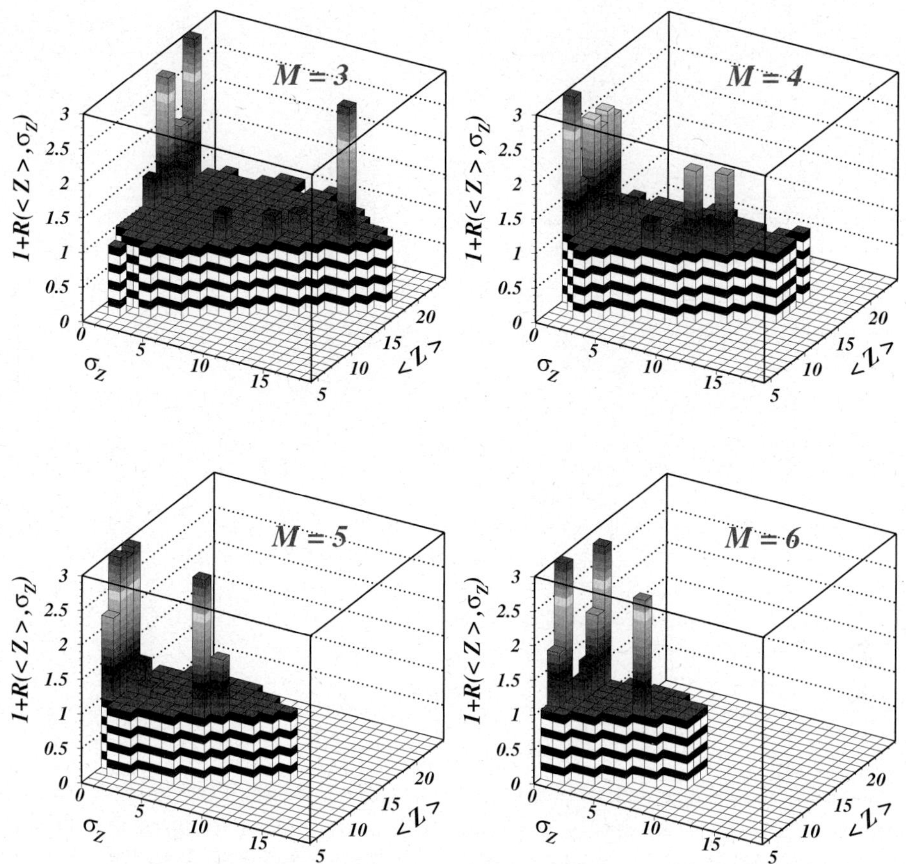

Fig. 5. Experimental higher-order charge correlations for selected events formed in central 39 AMeV ^{129}Xe on ^{nat}Sn collisions, for fragment multiplicities 3 to 6. The maximum value of the scale of the correlation function is limited to 3 on the picture. (From [10].)

all correlations due to the charge conservation are suppressed and the correlation function is equal to 1 (within statistical fluctuations) wherever no additional correlation is present. Again peaks for $\sigma_Z < 1$ are observed. The percentage of extra-events is $0.36 \pm 0.03\%$, higher than the one obtained with the previous method. Moreover, with this method, peaks also appear at the maximum values of σ_Z for a given Z_{bound}. They correspond to events composed of one big (a heavy residue) and several lighter fragments (sequentially emitted from the big one). In that case fusion-multifragmention does not occur and the peaks reveal the small proportion (0.15%) of events which undergo the fusion-evaporation process.

Note that very recently higher-order charge correlations were also studied for central Ni + Ni collisions simulated using the LATINO semiclassical model [25]. A single source at 4.75 AMeV excitation energy was measured, which de-excites with an abnormal production of four equal-sized fragments.

To conclude this part one can say that, although all events in the simulation arise from spinodal decomposition, only a very small fraction of the final partitions have nearly equal-sized fragments. Let us recall again the different effects: beating of different modes, coalescence of nascent fragments, secondary decay of the excited fragments and, above all, finite-size effects are responsible for this fact [5,2]. The signature of spinodal decomposition can only reveal itself as a "fossil" signal.

4.3 Experimental results

As an example, higher-order charge correlations for selected experimental events concerning ^{129}Xe on ^{nat}Sn collisions at 39 AMeV incident energy [10] are presented. This is in the framework of the IPM for the denominator. The first step consists in determining the intrinsic probabilities of fragments for each multiplicity. These probabilities are obtained by a recursive procedure of minimization. The minimization criterion is the normalized χ^2 between experimental and combinatorial fragment partition probabilities. Charge distributions experimentally observed for the different fragment multiplicities are shown in fig. 4. Dashed lines refer to the intrinsic probabilities calculated with IPM and the corresponding charge distributions are the full lines. One can note the excellent agreement between calculations and data. The contributions to the Z distribution of the three heaviest fragments of each par-

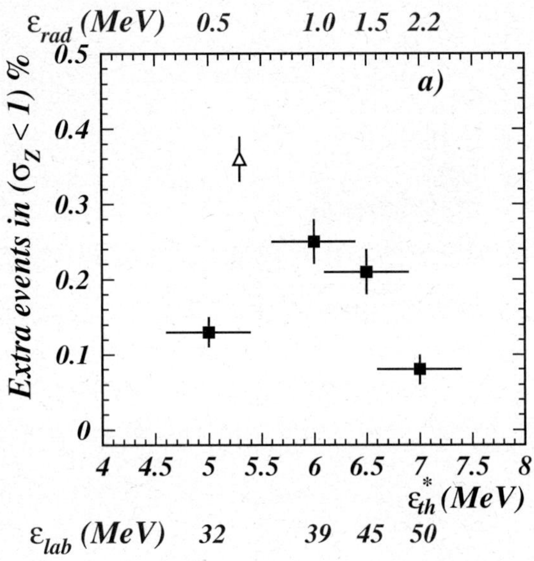
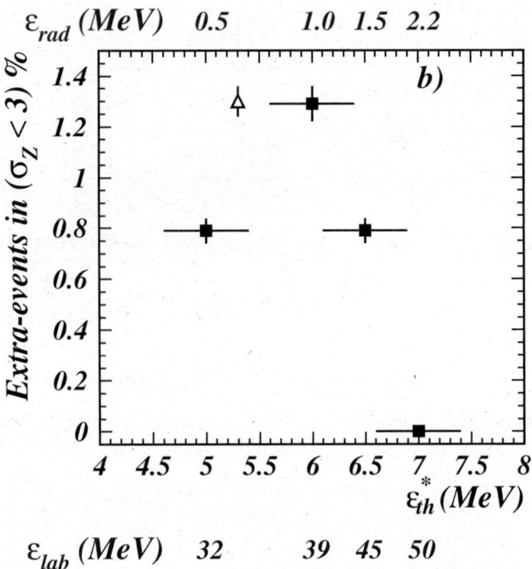

Fig. 6. Abnormal production of events with nearly equal-sized fragments (a) $\sigma_Z < 1$ and b) $\sigma_Z < 3$) as a function of thermal excitation energy (full points); the incident and radial energy scales are also indicated. ϵ_{th}^* and ϵ_{rad} are deduced from comparisons with SMM. The open point refers to the result from BOB simulations; the average thermal excitation energy is used. Vertical bars correspond to statistical errors and horizontal bars refer to estimated uncertainties on the backtraced quantity, ϵ_{th}^*. (From [10].)

tition are also well described and the charges bound in fragments (not shown) are perfectly reproduced.

Figure 5 illustrates the higher-order correlation functions measured for the different fragment multiplicities. To make the effects more visible, peaks with confidence level lower than 80% were flattened out. We observe significant peaks in the bin $\sigma = 0$–1 for each fragment multiplicity. For $M = 6$, peaks are essentially located in the bin $\sigma_Z = 1$–2. As observed in simulations, peaks corresponding to events composed of a heavy residue and light fragments (σ_Z in the region 5–10 associated with low $\langle Z \rangle$) are also visible. At 32 AMeV incident energy, for the same system, similar results were obtained using FRM or IPM methods and well compare to those obtained with events from dynamical simulations, BOB [14,10]. On the other hand, no abnormal production of events with nearly equal-sized fragments was obtained using the RETFM [17].

Moreover, using the IPM method, a rise and fall of the percentage of "fossil partitions" from spinodal decomposition is measured over the incident energy range 32–50 AMeV (see fig. 6) The percentages of events with $\sigma_Z < 3$ are also reported. The conclusions are the same: while more events have small values of σ_Z when the incident energy increases, the percentage of extra-events shows a maximum at 39 AMeV but vanishes at 50 AMeV. Figure 6 also reveals some difference between the experimental (full symbols) and simulated events (open symbols): the experimental percentages of extra-events are closer to the simulated ones in fig. 6b than in fig. 6a. This means that the charge distributions inside an event are slightly narrower in the simulation than in the experiment either because of the primary intrinsic spread, or because the width due to evaporation is underestimated. For the

considered system, incident energies around 35–40 AMeV could appear as the most favourable to induce spinodal decomposition; it corresponds to about 5.5–6 AMeV thermal excitation energy associated to a very gentle expansion energy around 0.5–1 AMeV. The qualitative explanation for those numbers can be well understood in terms of a necessary compromise between two times. On one hand, the fused systems have to stay in the spinodal region $\approx$ 100–150 fm/c [2,4,26], to allow an important amplification of the initial fluctuations and thus permit spinodal decomposition; this requires a not too high incident energy, high enough however for multifragmentation to occur. On the other hand, for a finite system, Coulomb interaction and collective expansion push the "primitive" fragments apart and reduce the time of their mutual interaction, which is efficient to keep a memory of "primitive" size properties. Note that such an explanation cannot be derived using the RETFM for which no abnormal production of events with nearly equal-sized fragments was measured, neither in BOB simulations [27,28] nor in experimental data [17].

Table 1 summarizes all the results concerning charge correlation studies performed up to now.

5 Observation of correlated signals

The concept of spinodal instability applies in general to macroscopically uniform systems that are suddenly brought into the coexistence region of their phase diagram. This instability occurs when the entropy function for the uniform system has a local convexity. Then the system splits into two independent subsystems (spinodal

Table 1. World-wide results on fragment correlations. DA refers to the Discriminant Analysis method. Percentage of extra events refers to the extra-percentage of events with nearly equal-sized fragments which correspond to $\sigma_Z < 1$.

System	Energy (AMeV)	Source selection	Detection (% of Zsyst)	Correlation method	$\langle Z \rangle$ range	Percentage of extra events	Ref.
$^{129}\mathrm{Xe} + {}^{nat}\mathrm{Cu}$	50	top 5% E_t	–	FRM	–	no events	[9]
$^{129}\mathrm{Xe} + {}^{197}\mathrm{Au}$	50	top 5% E_t	–	FRM	–	no events	[9]
$^{36}\mathrm{Ar} + {}^{197}\mathrm{Au}$	50	top 5% E_t	–	FRM	–	no events	[9]
$^{129}\mathrm{Xe} + {}^{119}\mathrm{Sn}$	32	BOB $(b=0)$	INDRA filter	FRM	10–19	0.13	[10]
$^{129}\mathrm{Xe} + {}^{119}\mathrm{Sn}$	32	BOB $(b=0)$	INDRA filter	IPM	8–20	0.36	[10]
$^{129}\mathrm{Xe} + {}^{119}\mathrm{Sn}$	32	BOB $(b=0)$	INDRA filter	RETFM	–	no events	[27,28]
$^{129}\mathrm{Xe} + {}^{nat}\mathrm{Sn}$	32	$\theta_{flow} > 60°$	> 80%	FRM	10–19	0.10	[14]
$^{129}\mathrm{Xe} + {}^{nat}\mathrm{Sn}$	32	$\theta_{flow} > 60°$	> 80%	IPM	11–21	0.13	[10]
$^{129}\mathrm{Xe} + {}^{nat}\mathrm{Sn}$	32	$\theta_{flow} > 60°$	> 80%	RETFM	–	no events	[17]
$^{129}\mathrm{Xe} + {}^{nat}\mathrm{Sn}$	39	$\theta_{flow} > 60°$	> 80%	IPM	6–20	0.25	[10]
$^{129}\mathrm{Xe} + {}^{nat}\mathrm{Sn}$	39	$\theta_{flow} > 60°$	> 80%	RETFM	–	no events	[17]
$^{129}\mathrm{Xe} + {}^{nat}\mathrm{Sn}$	45	$\theta_{flow} > 60°$	> 80%	IPM	6–18	0.21	[10]
$^{129}\mathrm{Xe} + {}^{nat}\mathrm{Sn}$	45	$\theta_{flow} > 60°$	> 80%	RETFM	–	no events	[17]
$^{129}\mathrm{Xe} + {}^{nat}\mathrm{Sn}$	50	$\theta_{flow} > 60°$	> 80%	IPM	7–9	0.08	[10]
$^{129}\mathrm{Xe} + {}^{nat}\mathrm{Sn}$	50	$\theta_{flow} > 60°$	> 80%	RETFM	–	no events	[17]
$^{58}\mathrm{Ni} + {}^{197}\mathrm{Au}$	32	DA (SIMON training)	> 60%	IPM	–	no events	[12]
$^{58}\mathrm{Ni} + {}^{197}\mathrm{Au}$	52	DA (SIMON training)	> 60%	IPM	7–15	not given	[12]
$^{58}\mathrm{Ni} + {}^{58}\mathrm{Ni}$	32	DA (data training)	> 80%	IPM	–	no events	[13]
$^{58}\mathrm{Ni} + {}^{58}\mathrm{Ni}$	40	DA (data training)	> 80%	IPM	5	events ?	[13]
$^{58}\mathrm{Ni} + {}^{58}\mathrm{Ni}$	52	DA (data training)	> 80%	IPM	4–8	0.85	[13]

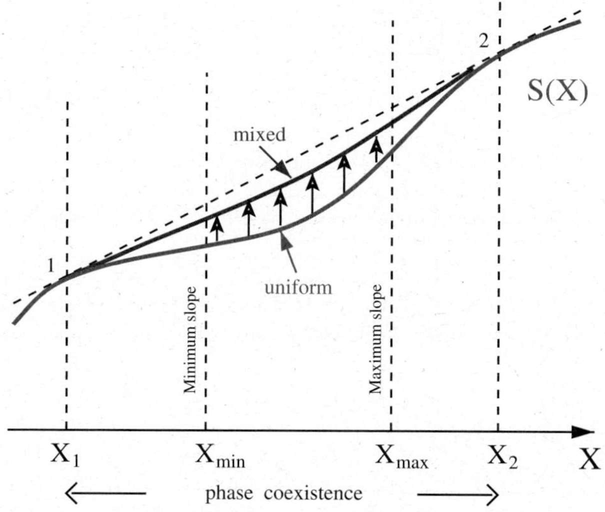

Fig. 7. Isolated finite system. The entropy function for a uniform system (lower curve) has a convexity region and the system gains entropy reorganizing itself into two subsystems but the resulting equilibrium entropy function (upper curve) will always lie below the common tangent (dashed line). From [1].

spond to the sum of the individual subsystem entropies: the Maxwell construction is no more valid as illustrated in fig. 7. Thus one can stress an important fact related to finite systems. It concerns the sign of the heat capacity in the spinodal region: if spinodal decomposition is observed, one must measure correlatively a negative microcanonical heat capacity related to the resulting equilibrium entropy function with local convexity.

Both signals (spinodal decomposition and negative microcanonical heat capacity) have been simultaneously studied on different fused systems which undergo multifragmentation [10,12,13,29]. Results are summarized in table 2. For the different systems we have also indicated the associated thermal and radial collective energies derived from data. We generally observe a correlation between the two signals. They are present when a total (thermal+radial) energy in the range 5.5–8.0 AMeV is measured. Note that the effect of a very gentle compression phase leading to 0.5–1.0 AMeV radial expansion energy seems to play the same role as a slightly higher thermal energy (Ni + Au system at 52 AMeV). This can be understood in terms of a required threshold for expansion energy; in the latter case this threshold should be reached by thermal expansion only.

decomposition) to increase entropy; in the thermodynamical limit the entropy is additive and the Maxwell construction operates. For finite systems these features are no more correct. Interfaces between coexisting phases are no longer negligible and the entropy at equilibrium does not corre-

6 Correlation methods and confidence level

As we have seen in sect. 4, very different results are obtained for the $^{129}\mathrm{Xe} + {}^{nat}\mathrm{Sn}$ system using IPM or RETFM

Table 2. Summary of the findings for phase transition signals.

System	Ni + Au	Ni + Ni	Xe + Sn	Xe + Sn	Ni + Au	Ni + Ni	Xe + Sn	Xe + Sn	Ni + Ni
Incident Energy AMeV	32.0	32.0	32.0	39.0	52.0	40.0	45.0	50.0	52.0
Thermal energy AMeV	5.0 ± 1.0	5.0 ± 1.0	5.0 ± 0.5	6.0 ± 0.5	6.0 ± 1.0	6.3 ± 1.0	6.5 ± 0.5	7.0 ± 0.5	8.0 ± 1.0
Radial energy AMeV	0.0	0.8 ± 0.5	0.5 ± 0.2	1.0 ± 0.3	0.0	1.7 ± 0.5	1.5 ± 0.4	2.2 ± 0.4	2.4 ± 0.5
Spinodal decomposition	no	no	yes	yes	yes	yes?	yes	no	yes
Negative microcanonical heat capacity	no	yes	yes	yes	yes	yes?	yes?	no	no

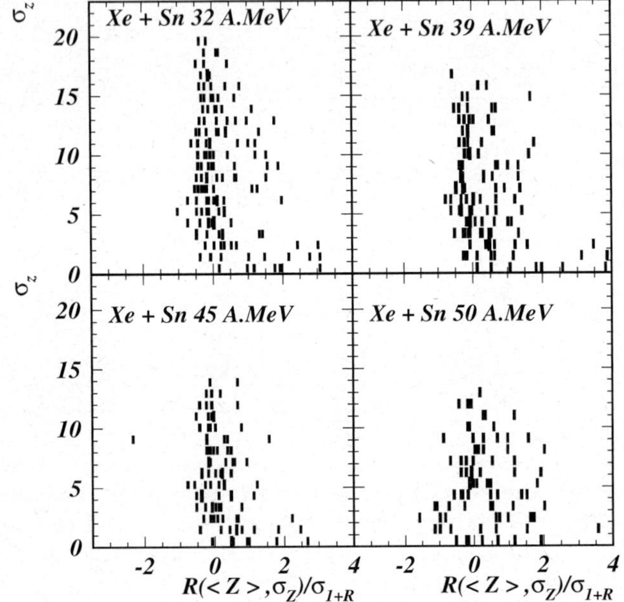

Fig. 8. Deviations from 1 of the correlation functions divided by the statistical errors in abcissa, for the different values of σ_Z. Correlation functions calculated by the IPM. (From [10].)

Fig. 9. Deviations from 1 of the correlation functions divided by the statistical errors in abcissa, for the different values of σ_Z. Correlation functions calculated by the RETFM. (From [10].)

for the calculation of the denominator. The sensitivity of the two methods was tested [10] by building correlation functions $1 + R(\sigma_Z, \langle Z \rangle)$ with σ_Z being calculated from eq. (2). The results of the two methods are compared in fig. 8 and fig. 9, displaying for each bin of the plane $(\sigma_Z, M \times \langle Z \rangle)$, the deviation from 1, $R(\sigma_Z, \langle Z \rangle)$, of the correlation function, normalized to its statistical error bar, $\sigma_{1+R(\sigma_Z, \langle Z \rangle)}$, calculated from the numerator only.

The greater sensitivity of the IPM appears in fig. 8: the values of the correlation function are closer to one ($R = 0$) except at low σ_Z where one observes correlations with significant confidence level (2 to 4 σ_{1+R}). Conversely, the exchange method, fig. 9, leads to a large dispersion of the values of $R(\sigma_Z, \langle Z \rangle)/\sigma_{1+R(\sigma_Z, \langle Z \rangle)}$, ~ 1.6 times broader than with the IPM at 32, 39 and 45 AMeV. This observation may be related to the fact that in the IPM one adjusts the partition probabilities. Thus all experimental charge distributions are reproduced (see fig. 4) whereas in the RETFM only the distribution of the total charge emitted in fragments, Z_{bound}, is constrained. So the IPM approach should appear more suitable to reveal weak correlations.

7 Perspectives

Charge correlation functions for compact fused systems which undergo multifragmentation have been investigated, as a function of the incident energy, from 30 to 50 AMeV. Enhanced production of events with almost equal-sized fragments at the level of 0.1–0.8% were possibly revealed. Supported by theoretical simulations, this abnormal enhancement can be interpreted as a signature of spinodal instabilities as the origin of multifragmentation in the Fermi energy domain. This fossil signal seems to culminate for incident energy around 35–40 AMeV. Microcanonical heat capacities observed in correlation with fossil signals also plead for spinodal decomposition to describe the dynamics of a phase transition for hot nuclei.

However, confidence levels, lower than 5 σ_{1+R}, observed for charge correlations prevent any definitive conclusion.

To firmly assess or not the validity of this fossil signal new studies must be performed:

– by achieving new experiments with higher statistics to reach relevant confidence levels,

– by performing dynamical simulations at higher incident energies and for different impact parameters,

– by increasing, on same event samples, the cross-check of different signals predicted to be correlated.

Moreover, more direct experimental determinations of thermal and radial energies of fragment sources are required to better determine the domain where "fossil partitions" are produced and preserved.

References

1. P. Chomaz, M. Colonna, J. Randrup, Phys. Rep. **389**, 263 (2004).
2. M. Colonna, P. Chomaz, A. Guarnera, Nucl. Phys. A **613**, 165 (1997).
3. S. Ayik, M. Colonna, P. Chomaz, Phys. Lett. B **353**, 417 (1995).
4. D. Idier, M. Farine, B. Remaud, F. Sébille, Ann. Phys. (Paris) **19**, 159 (1994).
5. B. Jacquot, S. Ayik, P. Chomaz, M. Colonna, Phys. Lett. B **383**, 247 (1996).
6. W. Nörenberg, G. Papp, P. Rozmej, Eur. Phys. J. A **9**, 327 (2000).
7. B. Jacquot, thèse de doctorat, Université de Caen (1996) GANIL T 96 05.
8. M. Bruno et al., Phys. Lett. B **292**, 251 (1992); Nucl. Phys. A **576**, 138 (1994).
9. L.G. Moretto, T. Rubehn, L. Phair, N. Colonna, G.J. Wozniak et al., Phys. Rev. Lett. **77**, 2634 (1996).
10. G. Tăbăcaru, B. Borderie, P. Désesquelles, M. Pârlog, M.F. Rivet et al., Eur. Phys. J. A **18**, 103 (2003).
11. INDRA Collaboration (P. Désesquelles, A.M. Maskay, P. Lautesse, A. Demeyer, E. Gerlic et al.), Phys. Rev. C **62**, 024614 (2000).
12. B. Guiot, thèse de doctorat, Université de Caen (2002) http://tel.ccsd.cnrs.fr/documents/archives0/00/00/37/53/.
13. R. Moustabchir, thèse de doctorat, Université Claude Bernard - Lyon I et Université Laval Québec (2004) http://tel.ccsd.cnrs.fr/documents/archives0/00/00/86/54/.
14. INDRA Collaboration (B. Borderie, G. Tăbăcaru, P. Chomaz, M. Colonna, A. Guarnera et al.), Phys. Rev. Lett. **86**, 3252 (2001).
15. P. Désesquelles, Phys. Rev. C **65**, 034604 (2002).
16. G. Tăbăcaru, thèse de doctorat, Université Paris-XI Orsay (2000) http://tel.ccsd.cnrs.fr/documents/archives0/00/00/79/12/.
17. J.L. Charvet, R. Dayras, D. Durand, O. Lopez, D. Cussol et al., Nucl. Phys. A **730**, 431 (2004).
18. J. Randrup, B. Remaud, Nucl. Phys. A **514**, 339 (1990).
19. P. Chomaz, G.F. Burgio, J. Randrup, Phys. Lett. B **254**, 340 (1991).
20. G.F. Burgio, P. Chomaz, J. Randrup, Nucl. Phys. A **529**, 157 (1991).
21. P. Chomaz, M. Colonna, A. Guarnera, J. Randrup, Phys. Rev. Lett. **73**, 3512 (1994).
22. A. Guarnera, P. Chomaz, M. Colonna, J. Randrup, Phys. Lett. B **403**, 191 (1997).
23. INDRA Collaboration (J.D. Frankland, B. Borderie, M. Colonna, M.F. Rivet, C.O. Bacri et al.), Nucl. Phys. A **689**, 940 (2001).
24. INDRA Collaboration (N. Marie, A. Chbihi, J. Natowitz, A. Le Fèvre, S. Salou et al.), Phys. Rev. C **58**, 256 (1998).
25. A. Barranon, J.A. Lopez, nucl-th/0503070 (2005).
26. W. Nörenberg, G. Papp, P. Rozmej, Eur. Phys. J. A **14**, 43 (2002).
27. INDRA Collaboration (M.F. Rivet et al.), Proceedings of the International Workshop on Multifragmentation 2003, GANIL, Caen, France, edited by G. Agnello, A. Pagano, S. Pirrone (2003) p. 22, nucl-ex/0501008.
28. INDRA and ALADIN Collaborations (M.F. Rivet, N. Le Neindre, J. Wieleczko, B. Borderie, R. Bougault et al.), Nucl. Phys. A **749**, 73 (2005).
29. N. Le Neindre, thèse de doctorat, Université de Caen (1999) http://tel.ccsd.cnrs.fr/documents/archives0/00/00/37/41/.

Eur. Phys. J. A **30**, 253–262 (2006)

DOI 10.1140/epja/i2006-10121-x

THE EUROPEAN
PHYSICAL JOURNAL A

Fluctuations of fragment observables

F. Gulminelli[1,a] and M. D'Agostino[2]

[1] LPC Caen (IN2P3-CNRS/Ensicaen et Université), F-14050 Caen Cédex, France
[2] Dipartimento di Fisica and INFN, Bologna, Italy

Received: 1 March 2006 /
Published online: 23 October 2006 – © Società Italiana di Fisica / Springer-Verlag 2006

Abstract. This contribution presents a review of our present theoretical as well as experimental knowledge of different fluctuation observables relevant to nuclear multifragmentation. The possible connection between the presence of a fluctuation peak and the occurrence of a phase transition or a critical phenomenon is critically analyzed. Many different phenomena can lead both to the creation and to the suppression of a fluctuation peak. In particular, the role of constraints due to conservation laws and to data sorting is shown to be essential. From the experimental point of view, a comparison of the available fragmentation data reveals that there is a good agreement between different data sets of basic fluctuation observables, if the fragmenting source is of comparable size. This compatibility suggests that the fragmentation process is largely independent of the reaction mechanism (central *vs.* peripheral collisions, symmetric *vs.* asymmetric systems, light ions *vs.* heavy-ion–induced reactions). Configurational energy fluctuations, that may give important information on the heat capacity of the fragmenting system at the freeze-out stage, are not fully compatible among different data sets and require further analysis to properly account for Coulomb effects and secondary decays. Some basic theoretical questions, concerning the interplay between the dynamics of the collision and the fragmentation process, and the cluster definition in dense and hot media, are still open and are addressed at the end of the paper. A comparison with realistic models and/or a quantitative analysis of the fluctuation properties will be needed to clarify in the next future the nature of the transition observed from compound nucleus evaporation to multi-fragment production.

PACS. 24.10.Pa Thermal and statistical models – 24.60.Ky Fluctuation phenomena – 25.70.Pq Multi-fragment emission and correlations – 68.35.Rh Phase transitions and critical phenomena

1 Fluctuations and phase transitions

Since the first inclusive heavy-ion experiments, multifragmentation has been tentatively associated with a phase transition or a critical phenomenon. This expectation was triggered by the first pioneering theoretical studies of the nuclear phase diagram [1] which contains a coexistence region delimited, at each temperature below an upper critical value, by two critical points at different asymmetries [2,3].

Even more important, the first exclusive multifragmentation studies have shown that multifragmentation is a threshold process occurring at a relatively well-defined deposited energy [4–7]. The wide variation of possible fragment partitions naturally leads to important fluctuations of the associated partition sizes and energies.

Different observables have been proposed to measure such fluctuations. Using the general definition of the n-th moment as

$$M_n = \sum_{Z_i \neq Z_{max}} Z_i^n \cdot n_i(Z_i), \tag{1}$$

the variance of the charge distribution is measured by the second moment M_2 or by the normalized quantity [8]

$$\gamma_2 = \frac{M_2 M_0}{M_1^2}. \tag{2}$$

The root mean-square fluctuation per particle

$$\sigma_m = \sqrt{\langle (Z_m/Z_0 - \langle Z_m/Z_0 \rangle)^2 \rangle} \tag{3}$$

of the distribution of the largest fragment Z_m detected in each event completes the information. We will also consider the total fluctuation

$$\Sigma_m^2 = \langle Z_0 \rangle \sigma_m^2 \tag{4}$$

and the fluctuation

$$\sigma_k^2 = \langle (E_p/A_0 - \langle E_p/A_0 \rangle)^2 \rangle \tag{5}$$

[a] e-mail: `gulminelli@lpccaen.in2p3.fr`

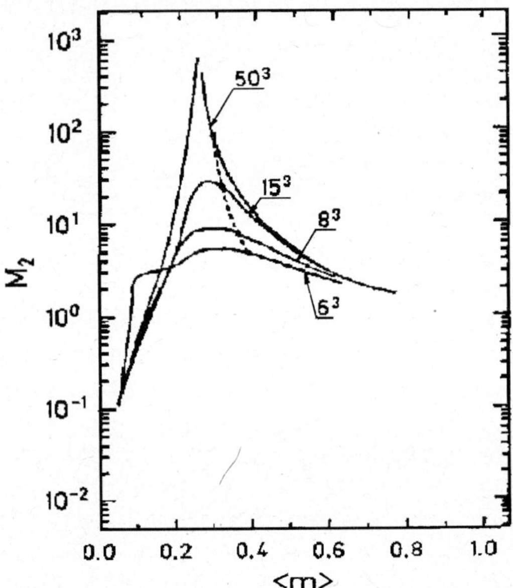

Fig. 1. Second moment of the size distribution (see [10] for a precise definition) as a function of the average cluster multiplicity for the three-dimensional percolation model for different lattice sizes. The figure is taken from [8].

of the configurational energy per particle associated with each fragment partition (k)

$$E_p^{(k)} = \sum_{i=1}^{m_k} (BE)_i + \alpha^2 \sum_{i,j=1}^{m_k} \frac{Z_i Z_j}{\langle |\mathbf{r}_i - \mathbf{r}_j| \rangle}, \qquad (6)$$

where m_k is the multiplicity of event k, BE is the ground-state binding energy of each fragment, and $\langle |\mathbf{r}_i - \mathbf{r}_j| \rangle$ is the average interfragment distance at the formation time. The quantities A_0, Z_0 in eqs. (3), (5) represent the reconstructed charge and mass of the fragmenting system, $Z_0 = \sum_{i=1}^{m_k} Z_i$, $A_0 = \sum_{i=1}^{m_k} A_i$.

In a simple statistical picture the fluctuation of any observable can be related to the associated generalized susceptibility by

$$\chi = -\frac{\partial \langle A \rangle}{\partial \lambda} = \langle A^2 \rangle - \langle A \rangle^2, \qquad (7)$$

where λ is the intensive variable associated with the generic observable A. Since the intensive variable associated with a particle density N/V is the susceptibility $\chi = \partial \langle N \rangle / \partial \mu$, then the large variance of the charge distribution observed in multifragmentation experiments could be connected to the diverging critical point fluctuation which would signal a diverging susceptibility and a diverging density correlation length. The apparent self-similar behavior and scaling properties of fragment yields [9] tend to support this intuitive picture.

1.1 Finite-size effects

Many different effects can, however, blur this simple connection. First of all, since fragmenting sources cannot ex-

ceed a few hundred nucleons, we have certainly to expect finite-size rounding effects, which smooth the fluctuation signal [8]. Not only the transition point is expected to be loosely defined and shifted in the finite system as shown in the three-dimensional percolation model in fig. 1, but also the signal is qualitatively the same for a critical point, a first-order transition or even a continuous change or crossover.

Finite-size effects have other consequences on the distribution than the simple smoothing of the transition. It has been shown in different model calculations that the presence of conservation constraints as well as the use of different event sorting procedures can sensibly distort the fluctuation observables. To give a simple example, the presence of a peak in the largest fragment's size fluctuation as a function of the energy deposit is trivially produced by the baryon number conservation constraint which forces this fluctuation to decrease with increasing average multiplicity [9]. In the case of a genuine critical behavior as for the percolation model, the fact of sorting events according to the percolation parameter p or according to some other correlated observable, as for instance the total cluster multiplicity, modifies [9,5] the behavior of m_2, γ_2, and all other related moments [10] measuring the fluctuation properties of the system. All these effects can be understood in the general framework of the non-equivalence of statistical ensembles for finite systems, which we will discuss in the next section.

1.2 Thermal invariance properties

Another problem when trying to connect a fluctuation peak to a phase transition or a critical behavior in a finite system is given by the possible existence of thermodynamic ambiguities. It has been observed by different independent works that in the framework of equilibrium fragmentation models the fluctuation behavior is qualitatively independent of the break-up density [11–14]. An example is given in fig. 2, which gives the second moment of the charge ($S_2 = M_2 - M_1^2$) and of the energy (C_v) distribution as a function of temperature in the lattice gas model for different break-up densities in the subcritical regime.

A peak in the fluctuation observables can be seen at all densities, at a temperature which is systematically below the critical temperature of the system and close to the first-order transition temperature in the thermodynamic limit. A similar behavior has been observed in different fluctuation observables and also at supercritical densities along the Kertesz percolation line, where the system does not present any phase transition. Table 1 gives, as a function of the lattice size, the inverse temperature at which the variable S_2 shows a maximum in the three-dimensional IMFM model [12] at different densities. As a general statement, the fluctuation peak as well as the global scaling properties of the size distribution [14,15] in these models can be found along a curve in the $T(\rho)$ diagram passing through the thermodynamic critical point but extending in the subcritical as well as supercritical region [16]. The

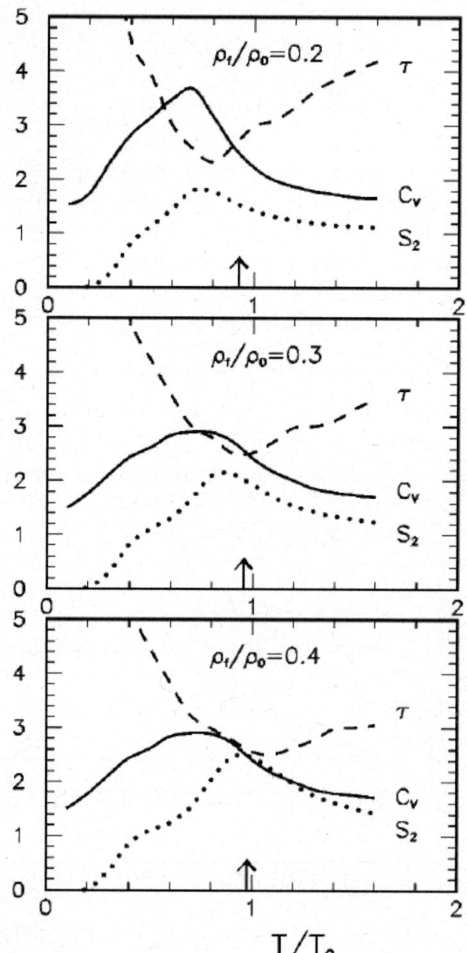

Fig. 2. Second moment of the charge (S_2) and of the energy (C_v) distribution as a function of temperature in the lattice gas model for different densities for a system of linear dimension $L = 7$. Arrows: first-order transition temperature in the thermodynamic limit. The figure is taken from ref. [13].

subcritical behavior can be understood as a finite-size effect, when the correlation length, close to the first-order transition point, becomes comparable to the linear size of the system, while the supercritical behavior is linked to the definition of clusters in dense and hot media [16]. For the subcritical region, a clusterization algorithm has been suggested to eliminate such behaviors in Ising simulations [17]. The possible pertinence of all these observations to experimental data is still a subject of debate, and essentially depends on the relationship between the measured clusters and the cluster definitions of the models.

Last but not least, the presence of different time scales in the reaction [18,19] and the dynamics of the fragmentation process may have important effects in the quantitative value of charge partition fluctuations [20], as we will discuss in the last section.

For all these reasons, it is clear that the well-documented presence of a fluctuation peak in the measured charge distributions [7] cannot be taken as such as a proof of a critical behavior and/or phase transition. In

Table 1. Inverse temperature at which the second moment $S_2 = M_2 - M_1^2$ is maximal for different densities and lattice sizes in the three-dimensional IMFM model. Taken from ref. [12].

L	$\beta_c(\rho = 0.3)$	$\beta_c(\rho = 0.5)$	$\beta_c(\rho = 0.7)$
10	0.2560(5)	0.225(3)	0.194(2)
16	0.2440(2)	0.2230(5)	0.1984(2)
20	0.23960(10)	0.2227(4)	0.1990(6)
24	0.2367(3)	0.2227(2)	0.2005(6)

order to connect the fluctuation behaviour to a phase transition and to conclude on its order, it is indispensable to compare with models and/or to quantify the fluctuation peak.

2 Theory

2.1 Fluctuations and constraints

It is clear that fluctuations on a given observable A will be suppressed if a constraint is applied to a variable correlated to A. This trivial fact has a deep thermodynamic meaning and is linked to the non-equivalence of statistical ensembles in finite systems [21]. Indeed, the basic statistical relation between a fluctuation and the associated susceptibility eq. (7) is only valid in the ensemble in which the fluctuations of A are such as to maximize the total entropy under the constraint of $\langle A \rangle$ ("canonical" ensemble). The thermodynamics in the ensemble where the generic observable A is controlled event by event ("microcanonical" ensemble), or in the ensemble where σ_A is externally fixed ("Gaussian" ensemble [22]) is a perfectly defined statistical problem, but the thermodynamic relationships have to be explicitly worked out [23]. As an example we show in fig. 3 the correlation between the size of the largest cluster A_{big} and the total energy in the isobar lattice gas model [23] at the transition temperature. The presence of two energy solutions at the same temperature and pressure clearly shows that the transition is in this case first order [24]. The A_{big} fluctuation properties are very different in the canonical ensemble (left part) and in the microcanonical ensemble (right part) at the same (average) total energy. Because of the important correlation between the total energy and the fragmentation partition, fragment size fluctuations can be compared only for samples with comparable widths of the energy distribution.

From the experimental viewpoint, different constraints apply to fragmentation data and have to be taken into account. Apart from the sorting conditions [7], the collisional dynamics can also give important constraints to the fragmentation pattern (e.g. flows, deformation in r-space and p-space). This means that fluctuations have to be compared with calculations performed in the statistical ensemble corresponding to the pertinent experimental constraints [25].

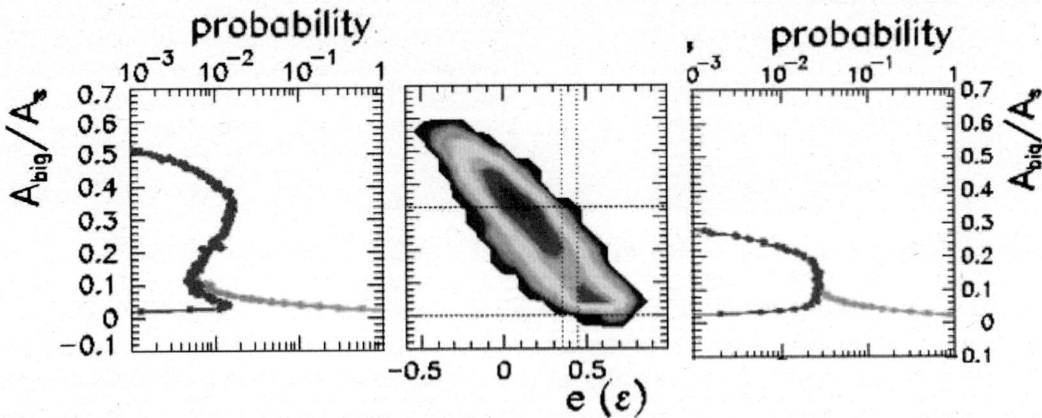

Fig. 3. Center: correlation between the largest fragment's size and the total energy in the isobar lattice gas model close to the transition temperature, for a system of 216 particles. Left side: projection over the A_{big} direction. Right side: same as the left side, but only events within a narrow energy interval around the average energy have been retained.

2.2 Fluctuations and susceptibilities

In the last subsection we have stated that a connection between a fluctuation and the associated susceptibility can always be in principle worked out if the constraints acting on the observable are known. In the case of sharp constraints (*e.g.* fixed total mass, charge, deposited energy), the connection between the fluctuation on a variable correlated to the constraint (*e.g.* size or charge of the largest fragment, configurational energy) and the associated susceptibility are in many cases analytical [26–28]. If a conservation constraint $A = A_1 + A_2 = $ const applies and the system can be split into two statistically independent components such that $W(A) = W(A_1)W(A_2)$, then the partial fluctuations are linked to the total susceptibility by

$$\frac{\chi_1}{\chi} = 1 - \frac{\sigma_1^2}{\sigma_{ref}^2}, \tag{8}$$

where $\chi_i^{-1} = \partial_{A_i}^2 W_i$, σ_{ref}^2 is the fluctuation of A_1 in the ensemble where only the average value $\langle A \rangle$ is constrained, and we have approximated the distribution of A_1 with a Gaussian [23]. The case of the total energy constraint has been particularly studied in the literature. Indeed the total energy deposit can be (approximately [29]) measured event by event in 4π experiments, allowing to experimentally construct a microcanonical ensemble by sorting. For classical systems with momentum-independent interactions the potential energy fluctuation σ_I^2 at a fixed total energy is linked to the total microcanonical heat capacity by

$$\frac{C_k}{C} = 1 - \frac{\sigma_k^2}{\sigma_{can}^2}, \tag{9}$$

where C_k, C are the kinetic and total heat capacity, $\sigma_k^2 = \sigma_I^2$ and $\sigma_{can}^2 = c_k T^2$ is the kinetic energy fluctuation in the canonical ensemble. Apart from the microstate equi-probability inherent to all statistical calculations, the above formula is obtained in the saddle point approximation for the partial energy distributions. The contribution of non-Gaussian tails can be also analytically worked

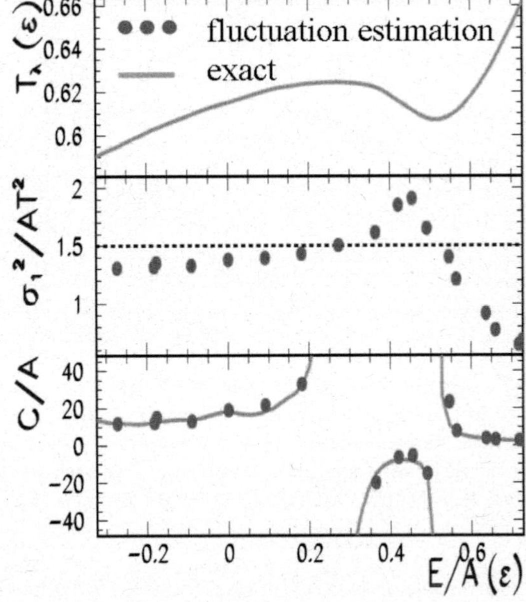

Fig. 4. Temperature, normalized binding energy fluctuation and heat capacity in the microcanonical isobar lattice gas model as a function of the total energy for a system of 108 particles. In the lower panel the heat capacity estimated from fluctuations via eq. (9) (dots) is compared to the exact expression from the entropy curvature (line). The figure is taken from [30].

out [27] and has been found to be negligible in all theoretical as well as experimental data samples analyzed so far [23]. An exemple of the quality of the approximation is given in fig. 4 which gives the temperature, normalized potential energy fluctuation and heat capacity in the isobar lattice gas model for a system of 108 particles.

Table 2. Maximum γ_2 (columns 2–5) and σ_m (columns 6, 7) values measured in the break-up of an Au system within different data sets sorted in Z_{bound}, total multiplicity (m) or calorimetric excitation energy (ϵ^*). Different values for the same case denote different bombarding energies. Values taken from refs. [5, 9, 31–35].

	γ_2 [5]	[31–33]	[34, 9]	[35]	σ_m [34]	[35]
Z_{bound}	1.4	1.3				
m		1.85	3.2	2.23		0.15
		1.85				
		2.5				
ϵ^*			3.7	2.5	0.12	0.14

Table 3. Maximum γ_2, Σ_m^2 and σ_m values measured within different data sets for various system sizes Z_0. Different values for the same case denote different targets. Values taken from refs. [34, 36, 35].

$\langle Z_0 \rangle$	γ_2 [34, 36, 35]	Σ_m^2 [34, 36, 35]	σ_m [34, 36, 35]
76	2.5	1.49	0.14
59	3.7	0.85	0.12
43	2.4	0.73	0.13
27	1.75	0.39	0.125
16	1.19	0.22	0.114
	1.17	0.22	0.114
	1.16	0.22	0.114

3 Experiment

3.1 Effect of the sorting variable

In this section we turn to compare different sets of experimental data available in the literature. Special attention has been paid by different collaborations to the largest fragment fluctuation σ_m eq. (3) and to the γ_2 observable eq. (2) [5, 31–35]. For all data sets of comparable total size these observables, as well as the others we will show in the next subsections, show a well-defined peak at comparable values of the chosen sorting variable. This is an important and non-trivial result considering that data are taken with different apparata and the multifragmenting systems are obtained with very different reaction mechanisms. The effect of the sorting variable is explored in table 2, that gives the maximum value of γ_2 and σ_m with different data sets sorted in bins of the total measured bound charge Z_{bound}, total measured charged particles multiplicity m, or calorimetric excitation energy [29]. Even if the systematics should certainly be completed and errors should definitely be evaluated, we can observe from table 2 that different data sets show a reasonable agreement when the same sorting is employed.

We can also note that a higher γ_2 is systematically obtained when data are analyzed in bins of total charge multiplicity, with respect to a sorting in Z_{bound}. This can be qualitatively understood if we recall that γ_2 measures the variance of the charge bound in fragments, and this quantity is obviously strongly correlated with Z_{bound} and loosely correlated with m. The calorimetric excitation energy sorting leads to results comparable to the multiplicity sorting. The value of γ_2 is slightly increased, which may be explained by a reduced correlation of ϵ^* with respect to m with the total fragment charge, since the excitation energy contains the extra information of the kinetic energy of the fragments. However, the effect goes in the opposite direction as the fluctuation of Z_m is concerned. A detailed study of the correlation coefficient between the considered observables and the sorting variables is needed to fully understand these trends. It is also possible that the fluctuations obtained with these two sortings may be compatible within error bars, which stresses the importance of an analysis of errors.

The fluctuation values appear to be largely independent of the reaction mechanism and incident energy [5, 31, 33]. The only exception is the value $\gamma_2 \approx 2.5$ obtained from emulsion data in ref. [32], which is significantly higher than the values obtained at the other bombarding energies for the same system. Such anomaly might be due to the presence of fission events that have been excluded in the other analyses [31, 33]. The independence on the incident energy tends to show that the fragmentation process is essentially statistical.

3.2 Effect of the system size

The effect of the system size is further analyzed in table 3. All presented data are sorted in bins of calorimetric excitation energy.

The fluctuation properties of quasi-projectile decay appear to be largely independent of the target. This well-known behavior at relativistic energy [5] appears confirmed in the case of the NIMROD experiment [36] which was performed with a beam energy as low as $47\,\mathrm{MeV}/A$. This suggests that a quasi-projectile emission source can be extracted [7] in spite of the important midrapidity contribution in the Fermi energy regime [19].

From table 3 we can also see that Σ_m^2 decrease monotonically with the system mass. The evolution with the system size, at least in the size range analyzed, appears as a simple scaling behavior as shown by the fact that the normalization to the source size in σ_m makes the fluctuation almost independent of the size. Similar conclusions can be drawn concerning the γ_2 observable, even if the behavior for the heaviest sources is less clear. This interesting scaling behavior should be confirmed using hyperscaling techniques [10].

To conclude, we have seen that fluctuations can vary by a factor of two when changing the sorting variable. This stresses the need of confronting the experimental data with statistical predictions containing the same constraints, *i.e.* performed in the adapted statistical ensemble. Interesting enough, when the same sorting is adopted the different available data sets agree within $\approx 15\%$, both in the value of the peak and in the position where the peak is observed. More data are needed to confirm these trends.

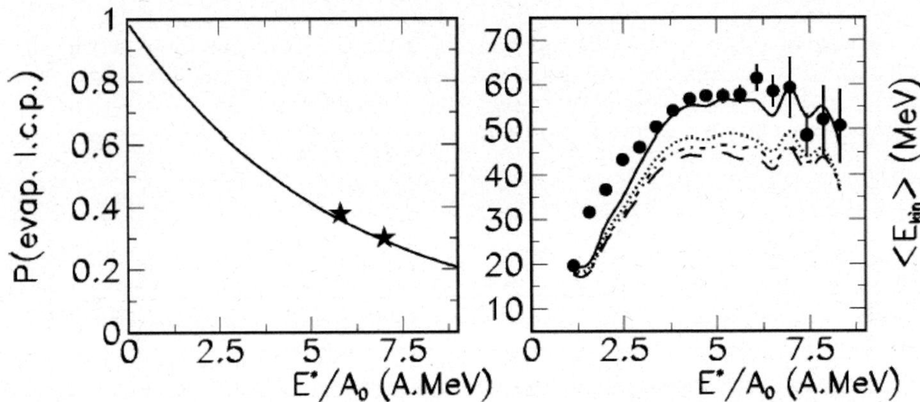

Fig. 5. Left part: percentage of secondarily emitted light charged particles taken from correlation function measurements (see ref. [37]). Right part: total measured fragment kinetic energy (points) compared with Coulomb trajectory calculations where the volume is changed from $6V_0$ to $3V_0$. Both quantities are plotted as a function of the calorimetric excitation energy. The figure is taken from [38].

3.3 Configurational energy fluctuations

One of the most interesting aspects of studying fluctuation observables, is their possible connection with a susceptibility or a heat capacity via eq. (9). Configurational energy fluctuations have been studied at length by the Multics Collaboration [38–40] and by the INDRA Collaboration [38,41–43] on Au sources. The observable used in these studies is an estimation of the energy stored in the configurational degrees of freedom at the time of fragment formation, defined as follows:

$$E_I = \sum_{i=1}^{N_{imf}} Q\left(Z_i^p, A_i^p\right)$$
$$+ \sum_{i=n,p,d,t,^3He,^4He} Q\left(Z_i, A_i\right) M_i^p \qquad (10)$$
$$+ V_{coul}\left(\{Z_i^p\}, V_{FO}\right),$$

where Q indicates the mass defects and V_{coul} the Coulomb energy. The measured fragment charges Z_i and lcp multiplicities M_i are corrected in each event to approximately account for secondary decay,

$$Z_i^p = Z_i + \langle M_H^{ev} + 2M_{He}^{ev}\rangle \frac{Z_i}{\sum_{i=1}^{N_{imf}} Z_i}, \qquad (11)$$

$$M_i^p = M_i - \langle M_i^{ev}\rangle, \qquad (12)$$

where $\langle M_i^{ev}\rangle$ is the estimated multiplicity of secondary emitted light charged particles for each calorimetric excitation energy bin.

Three quantities need to be estimated in each excitation energy bin to compute E_I:

1) The freeze-out volume V_{FO} which determines the total Coulomb energy. Its average value is deduced from the measured fragment kinetic energies through Coulomb trajectories calculations (see fig. 5, right part).
2) The average multiplicities of secondarily emitted particles $\langle M_{lcp}^{ev}\rangle$ to account for side-feeding effects. They

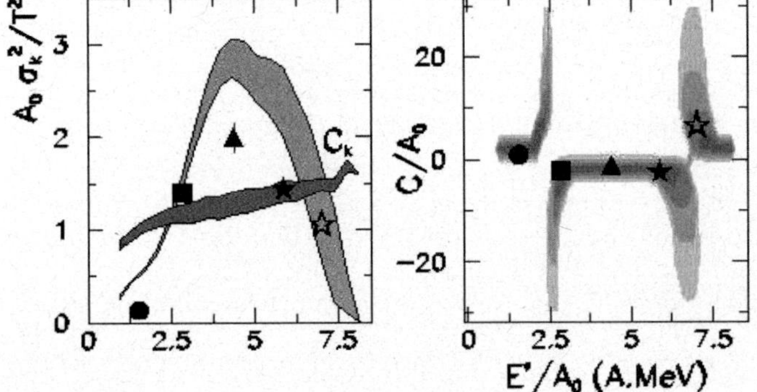

Fig. 6. Left side: normalized fluctuation of E_I and estimated C_k (see text) as a function of the calorimetric excitation energy. Grey zone: peripheral $35\,A$ MeV Au + Au collisions. Symbols: central Au + C, Au + Cu, Au + Au at 25 and 35 A MeV. Right side: heat capacity from eq. (9). The figure is taken from [40].

are deduced from fragment-particle correlation functions (see fig. 5, left part).

3) The isotopic content A_i^p/Z_i^p of primary fragments. It is assumed that it is equal to the isotopic content of the fragmenting system. This quantity allows in turn to determine the number of free neutrons at freeze-out from baryon number conservation.

A general protocol has been proposed to minimize the spurious fluctuations due to the implementation of this missing information [38]. The resulting fluctuation of E_I $\sigma_I^2 = \sigma_k^2$ is shown for different Multics data in fig. 6. The temperature has been estimated alternatively using isotopic thermometers or solving the kinetic equation of state and comes out to be in good agreement [40] with the general temperature systematics [44] (around 4.5 MeV in the fragmentation region). Similar to the other fluctuation observables, configurational energy fluctuations show a well-pronounced peak at an excitation energy around $5\,A$ MeV. This general feature is apparent in Multics [40], INDRA [43], Isis [45] and NIMROD [36] data. The only

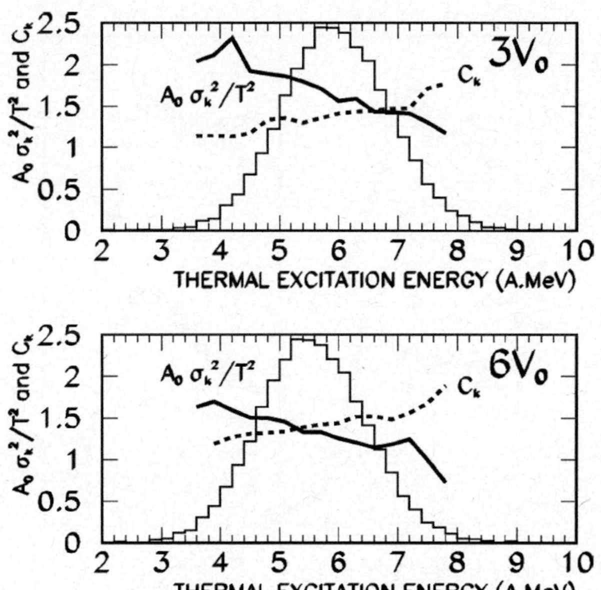

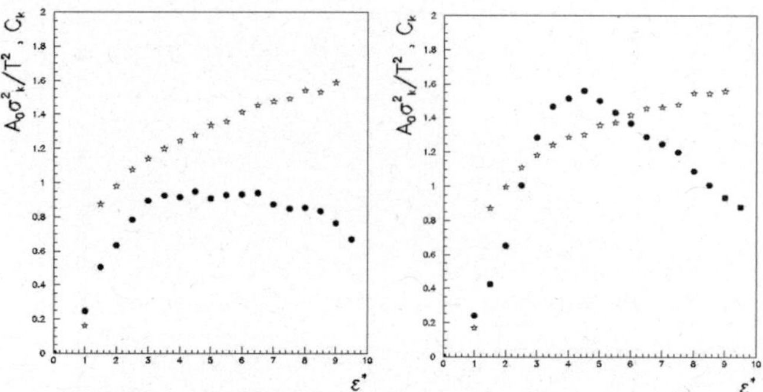

Fig. 8. Normalized fluctuation and kinetic heat capacity (stars) for $80\,A$ MeV Au + Au peripheral collisions measured by the INDRA and ALADIN Collaborations as a function of the calorimetric excitation energy, for all quasi-projectile events (left side) and after subtraction of events elongated along the beam axis (right side). The figure is taken from [43].

Fig. 7. Normalized fluctuation and kinetic heat capacity (dashed lines) for $32\,A$ MeV Xe + Sn central collisions measured by the INDRA Collaboration as a function of the calorimetric excitation energy with two different hypothesis on the freeze-out volume. The histogram gives the event distribution. The figure is taken from [38, 41].

exception is EOS data [46] where this fluctuation appears monotonically decreasing.

In the hypothesis of thermal equilibrium at the freeze-out configuration this fluctuation is a measure of the heat capacity according to eq. (9). The value expected for this fluctuation in the canonical ensemble can be written as $\sigma_{can}^2 = c_k T^2$. The kinetic heat capacity c_k is calculated from the measured fragment yields [38]. We can see that the fluctuation peak overcomes the upper classical limit $c_k = 3/2$ suggesting a negative heat capacity as expected in a first-order phase transition analyzed in the micro-canonical ensemble [47, 48].

The same analysis performed on INDRA data of central Xe + Sn collisions at different bombarding energies leads to compatible temperatures and volumes and a fluctuation estimation that agrees within 25% with the presented Multics results [38], as shown for the $32\,A$ MeV data in fig. 7 (upper part). In the absence of isotopic resolution for fragments, Coulomb repulsion cannot be distinguished from a radial collective expansion due to a possible initial compression. If an important radial flow component is assumed for these central collisions, data can also be compatible with a bigger freeze-out volume (lower part of the figure) leading to a shift of the abnormal fluctuation behavior towards lower energy. This volume/flow ambiguity in central collisions can only be solved with third-generation multidetectors [49].

INDRA data on a source of the same size as the Au quasi-projectile analyzed by the Multics Collaboration lead to a fluctuation measurement about 40% lower, see fig. 8. This difference is tentatively explained as an

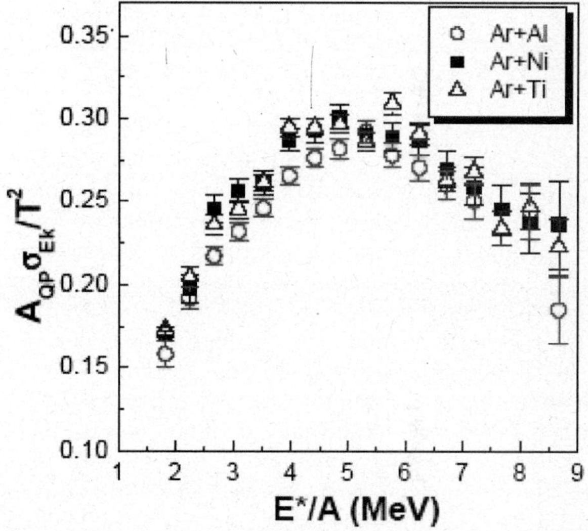

Fig. 9. Normalized fluctuation and kinetic heat capacity for $47\,A$ MeV argon quasi-projectiles on different targets measured by the NIMROD Collaboration as a function of the calorimetric excitation energy. The figure is taken from [36].

effect of emission from the neck which leads to a reduced occupation of the available phase space [43].

Recent NIMROD data [36] on the fragmentation of a much lighter system show a similar value for the energy corresponding to the fluctuation peak, but an absolute value for the fluctuation of a factor 10 lower than for Multics data, as shown in fig. 9. If we consider the global fluctuation $\langle A_0 \rangle \sigma_k^2$ without the normalization to the estimated temperature, this factor is reduced to about a factor 4. These results go in the same direction as the general behavior of Σ_m^2 that we have analyzed in sect. 3.2. Recall that the fluctuation of the biggest fragment for the quasi-Au source [35] is a factor 6.8 higher than for the quasi-Ar one [36]. This fluctuation reduction seems then to be a general feature of light-system fragmentation and has been tentatively explained as an effect of the

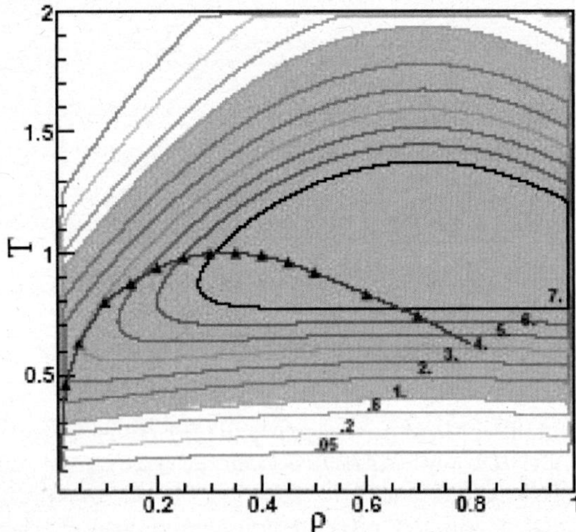

Fig. 10. Phase diagram of 64 Lennard-Jones particles confined in a box. Filled triangles give the coexistence border. Isocontours give values of normalized fluctuations $\sigma_k^2/\sigma_{can}^2$ calculated from the ground-state Q value of clusters defined with the Hill algorithm. The figure is taken from [50].

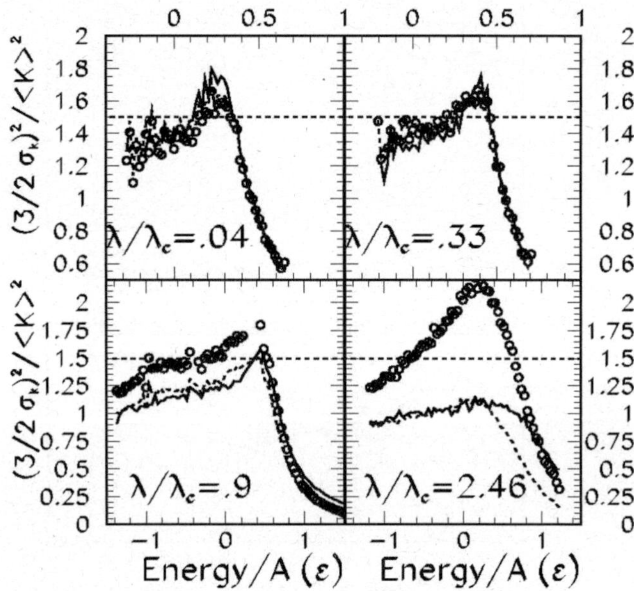

Fig. 11. Normalized fluctuations σ_k^2/T^2 as a function of energy for a system of 216 lattice gas particles in the isobar ensemble at different pressures. Full lines: exact results. Symbols: estimation from the ground-state Q value of Coniglio-Klein clusters. Dashed lines: as the symbols, but data are sampled in bins of energy reconstructed from cluster kinetic energies and sizes. λ_c gives the critical pressure. The figure is taken from [51].

higher temperature that light systems can sustain [36]. In this interpretation, a higher temperature region of the phase diagram, possibly above the critical point, is explored in the fragmentation of light systems, and the first-order phase transition observed in heavy nuclei becomes a smooth crossover.

As a general remark, the configurational energy fluctuation signal is a very interesting one due to its possible connection with a heat capacity, but it is also a very indirect and fragile experimental signal which needs precise calorimetric measurements, a careful data analysis, extensive simulations to assess the effect of the different hypotheses in the event sorting and reconstruction procedure. Moreover, the different techniques to exclude or minimize pre-equilibrium and neck emission seem to have a strong influence in the absolute value of fluctuations.

The evaluation of systematic errors in fluctuation measurements is necessary to achieve a quantitative estimation of fluctuations: some first encouraging results in this direction have been presented in ref. [40]. The confirmation (or infirmation) of the fluctuation enhancement is certainly one of the most important challenges of the field in the next years with third-generations multidetectors.

4 Open questions

The possibility of accessing a thermodynamic information on the nuclear phase diagram from measured fragment properties entirely relies on the representation of the system at the freeze-out stage as an ideal gas of fragments [25] in thermal equilibrium. This is true for fluctuation observables as well as for all other thermodynamic analyses [34,44]. This is an important conceptual point which is presently largely debated in the heavy-ion community.

A first open question concerns the structure of the systems at the freeze-out stage, *i.e.* at the time when fragments decouple from each other. Contrary to the ultrarelativistic regime [52], we do not expect much difference between the chemical and kinetic decoupling times due to the small collective motions implied in these low-energy collisions. We can, therefore, speak at least in a first approximation of a single freeze-out time. If at this time the system is still relatively dense, the cluster properties may be very different from the ones asymptotically measured, and the question arises [16] as to whether the energetic information measured on ground-state properties can be taken backward in time up to the freeze-out. Calculations from classical molecular dynamics [50] show that the ground-state Q-value is a very bad approximation of the interaction energy of Hill clusters in dense systems. This is due both to the deformation of clusters when recognized in a dense medium through the Hill algorithm, and to the interaction energy among clusters in dense configurations where cluster surfaces touch. As a consequence, comparable fluctuations are obtained in the subcritical and supercritical region of the Lennard-Jones phase diagram. This result is shown in fig. 10. Calculations in a similar model, the lattice gas model, show that even in the supercritical regime the correct fluctuation behavior can be obtained if both the total energy and the interaction energy are consistently estimated with the same approximate algorithm as is done in the experimental data analysis [51]. Indeed, the high value of the estimated configurational energy Q fluctuations is essentially due to the spurious fluctuation of the total energy $E_K + E_I$ obtained when E_I is esti-

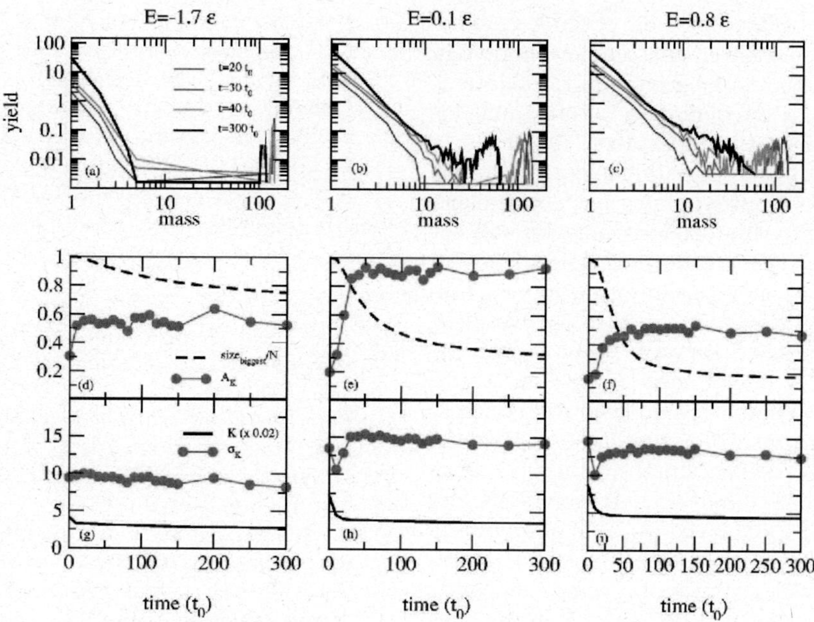

Fig. 12. Time evolution of a Lennard-Jones system initially confined in a dense supercritical configuration and freely expanding in the vacuum at different total energies. Upper part: minimum-spanning-tree (MST) fragment size distribution at different times. Lower part: average kinetic-energy (full lines), total (lower symbols) and normalized (upper symbols) kinetic-energy fluctuations, and size of the largest MST cluster (dashed lines). Abnormal fluctuations in these units correspond to $A_k \gtrsim 0.7$. The figure is taken from [53].

mated through Q; such an effect is eliminated if data are analyzed in bins of $E_K + Q$. This calculation is shown in fig. 11.

A second related question which needs further work is the relevance of the equilibrium assumption at freeze-out. Molecular dynamics models applied to study the time evolution of the reaction [20,53–55] predict that the decoupling between fragment degrees of freedom (freeze-out) occurs very rapidly during the reaction. At this stage, however, the configuration is considerably diluted due to the early presence of collective motions [20]. An example taken from classical molecular dynamics for an initially equilibrated compact configuration freely evolving in the vacuum is shown in fig. 12. At this reaction stage cluster energies may be well approximated (within a side-feeding correction) by their asymptotically measured values, but it is not clear whether this configuration can correspond to an equilibrium, more precisely whether the hypothesis of equiprobability of the different charge partitions holds.

5 Conclusions and outlooks

In this paper we have presented a short review of the experimental as well as theoretical studies of fluctuation observables of fragments produced in a multifragmentation heavy-ion reaction. The aim of these studies is the understanding of the nature of the nuclear fragmentation transition as well as the thermodynamic characterization of the finite-temperature nuclear phase diagram. This vast and ambitious program is still in its infancy. Many promising

results already exist, but the analyses are not yet conclusive and need to be intensively pursued in the future.

The nuclear fragmentation phenomenon, well documented by a series of independent experiments [7], presents many features compatible with a critical phenomenon [10] or a phase transition [9,24,34]. Only a careful study of fluctuation properties will allow to discriminate between the different scenarii. Even more important, the phase diagram of finite nuclei is theoretically expected to present an anomalous thermodynamics [47,48] which should be characteristic of any non-extensive system undergoing first-order phase transitions in the thermodynamic limit. Once the difficulties linked to the imperfect detection and sorting ambiguities will be overcome, fluctuation observables will be a unique tool to quantitatively study this new thermodynamics with its interdisciplinary applications [47,48,56].

From the theoretical point of view, the theoretical connections between fluctuations and susceptibilities in the different statistical ensembles are well established, and the different experimental constraints can be consistently adressed by the theory. However, the evaluation of a thermodynamics for a clusterized system opens the difficult theoretical problem of cluster definition in dense quantum media. To produce quantitative estimations of measurable fluctuation observables, the pertinence of classical models has to be checked through detailed comparisons with microscopic [54] and macroscopic [25] nuclear models.

On the experimental side, multiplicities and size fluctuations agree reasonably well if comparable size fragmenting systems are studied, even if the effect of the system size has to be clarified. Configurational energy fluctuations are especially interesting because of their possible connection with a heat capacity measurement. The methodology to extract such fluctuations from fragmentation data is presently under debate; in particular, a careful analysis of systematic errors is presently undertaken [40]. From a more conceptual point of view, the influence of the different time scales in the reaction dynamics has to be clarified. Configurational energy fluctuations may be subject to strong ambiguities since they use information from all the particles of the event, and this information is integrated over the whole reaction dynamics. In this respect, an interesting complementary observable may be given by fluctuations of the heaviest cluster size [24,28,57].

To solve the existing ambiguities we need full comparisons with a well-defined protocol and consistency checks between different data sets. The simultaneous measurement of fragment mass and charge on a 4π geometry [49] will be essential to measure the basic variable of any thermodynamic study, namely the deposited energy. No definitive conclusion about the occurrence of a thermodynamic phase transition and its order can be drawn without this detection upgrade.

References

1. G. Bertsch, P.J. Siemens, Phys. Lett. B **126**, 9 (1983).
2. H. Müller, B.D. Serot, Phys. Rev. C **52**, 2072 (1995).
3. C. Ducoin et al., nucl-th/0512029.
4. G. Bizard et al., Phys. Lett. B **208**, 162 (1993).
5. A. Schüttauf et al., Nucl. Phys. A **607**, 457 (1996).
6. T. Beaulieu et al., Phys. Rev. C **64**, 064604 (2001).
7. B. Tamain, this topical issue.
8. X. Campi, Phys. Lett. B **208**, 351 (1988).
9. J.B. Elliott et al., Phys. Rev. C **62**, 064603 (2000).
10. Y.G. Ma, this topical issue.
11. Al.H. Raduta et al., Phys. Rev. C **65**, 034606 (2002).
12. J. Carmona et al., Nucl. Phys. A **643**, 115 (1998).
13. J. Pan et al., Phys. Rev. Lett. **80**, 1182 (1998).
14. Ph. Chomaz, F. Gulminelli, Phys. Rev. Lett. **82**, 1402 (1999).
15. F. Gulminelli et al., Phys. Rev. C **65**, 051601 (2002).
16. N. Sator, Phys. Rep. **376**, 1 (2003).
17. L.G. Moretto et al., Phys. Rev. Lett. **94**, 202701 (2005).
18. A. Bonasera et al., this topical issue.
19. M. Di Toro et al., Neck dynamics, this topical issue.
20. P. Balenzuela et al., Phys. Rev. C **66**, 024613 (2002).
21. F. Gulminelli et al., Phys. Rev. E **68**, 026120 (2003).
22. M.S. Challa, J.H. Hetherington, Phys. Rev. Lett. **60**, 77 (1988).
23. F. Gulminelli, Ann. Phys. (Paris) **29**, 6 (2004).
24. O. Lopez, M.F. Rivet, this topical issue.
25. A.S. Botvina, I.N. Mishustin, this topical issue.
26. J.L. Lebowitz et al., Phys. Rev. **153**, 250 (1967).
27. F. Gulminelli, Ph. Chomaz, Nucl. Phys. A **647**, 153 (1999).
28. F. Gulminelli, Ph. Chomaz, Phys. Rev. C **71**, 054607 (2005).
29. V.E. Viola, R. Bougault, this topical issue.
30. F. Gulminelli, Ph. Chomaz, V. Duflot, Europhys. Lett. **50**, 434 (2000).
31. P.L. Jain et al., Phys. Rev. C **50**, 1085 (1994).
32. M.I. Adamovitch et al., Eur. Phys. J. A **1**, 77 (1998).
33. D. Kudzia et al., Phys. Rev. C **68**, 054903 (2003).
34. J.B. Elliott et al., Phys. Rev. C **67**, 024609 (2003).
35. A. Bonasera et al., Riv. Nuovo Cimento **23**, No. 2 (2000); M. D'Agostino, private communication.
36. Y.G. Ma et al., Nucl. Phys. A **749**, 106 (2005); Y.G. Ma, private communication.
37. G. Verde et al., this topical issue.
38. M. D'Agostino et al., Nucl. Phys. A **699**, 795 (2002).
39. M. D'Agostino et al., Phys. Lett. B **473**, 219 (2000).
40. M. D'Agostino et al., Nucl. Phys. A **734**, 512 (2004).
41. N. Leneindre, PhD Thesis, `http://tel.ccsd.cnrs.fr/tel-0003741`.
42. M.F. Rivet et al., Nucl. Phys. A **749**, 73 (2005).
43. M. Pichon et al., nucl-ex/0602003; to be published in Nucl. Phys. A.
44. A. Kelić, J.B. Natowitz, K-.H. Schmidt, this topical issue.
45. T. Lefort, V.E. Viola, private communication.
46. B.K. Srivastava et al., Phys. Rev. C **65**, 054617 (2002).
47. Ph. Chomaz, F. Gulminelli, this topical issue.
48. D.H.E. Gross, this topical issue.
49. R.T. de Souza et al., this topical issue.
50. X. Campi et al., Phys. Rev. C **71**, 41601 (2005).
51. F. Gulminelli, Ph. Chomaz, M. D'Agostino, Phys. Rev. C **72**, 064618 (2005).
52. I.N. Mishustin, this topical issue.
53. A. Chernomoretz et al., Phys. Rev. C **69**, 034610 (2004).
54. A. Ono, H. Horiuchi, Progr. Part. Nucl. Phys. **53**, 501 (2004).
55. A.D. Sood, R.K. Puri, J. Aichelin, Phys. Lett. **594**, 260 (2004).
56. H. Behringer et al., J. Phys. A **38**, 973 (2005).
57. J.D. Frankland et al., Phys. Rev. C **71**, 034607 (2005).

Eur. Phys. J. A **30**, 263–274 (2006)
DOI 10.1140/epja/i2006-10122-9

Bimodalities: A survey of experimental data and models

O. Lopez[1],[a] and M.F. Rivet[2]

[1] Laboratoire de Physique Corpusculaire, IN2P3-CNRS/ENSICAEN/Université, F-14050 Caen cedex, France
[2] Institut de Physique Nucléaire, IN2P3-CNRS, F-91406 Orsay cedex, France

Received: 13 February 2006 /
Published online: 25 October 2006 – © Società Italiana di Fisica / Springer-Verlag 2006

Abstract. Bimodal distributions of some chosen variables measured in nuclear collisions were recently proposed as a non-ambiguous signature of a first-order phase transition in nuclei. This section presents a compilation of both theoretical and experimental studies on bimodalities performed so far, in relation with the liquid-gas phase transition in nuclear matter.

PACS. 25.70.Pq Multifragment emission and correlations – 24.60.-k Statistical theory and fluctuations – 64.60.Cn Order-disorder transformations; statistical mechanics of model systems – 25.70.-z Low and intermediate energy heavy-ion reactions

After a formulation of the theoretical bases of bimodality, world-wide experimental results will be reviewed and discussed, as well as the occurrence of some kind of bimodality in models. Finally, conclusions on the perspectives of such analyses in the near future and the possible connections to other proposed signals of the liquid-gas phase transition in nuclear matter will be given.

1 Theoretical bases

1.1 Definition

Bimodality is a property of finite systems undergoing a first-order phase transition [1–3]. It is thus a generic feature which concerns not only nuclear physics but a broad domain of physics such as astrophysics, or soft-matter physics. Bimodality means that *the probability distribution of an order parameter of the considered system at phase transition exhibits two peaks separated by a minimum.* Indeed, if the system is in a pure phase, the order parameter distribution consists in one peak and can be characterized by its mean value and its variance. By contrast, if the system is in the coexistence region, the distribution presents two peaks, well separated, whose properties are related to the two different phases of the system [2]. Bimodality is then one of the signals associated to a first-order phase transition [3], beside others such as scaling laws, critical exponents or negative heat capacities.

In the following, the term "bimodality" will abbreviate "the probability distribution of some variable, in a given region of the phase diagram of the system, is bimodal".

1.2 Pioneering studies

Bimodality and its relationship to phase transition has been studied since the '80s. Figure 1 shows an Ising model simulation of a ferromagnet studied by Binder and Landau [4]. In this analysis, the authors studied the magnetization M of the system as a function of the applied magnetic field H. When the magnetic field comes close to the critical value H_c, the spontaneous magnetization of the ferromagnet presents a sudden change; in this case the probability distribution of the magnetization is never bimodal, as the system "jumps" suddenly from the negative value $-M_{sp}$ to the positive one $+M_{sp}$: the transition between the two regimes is sharp at the thermodynamical limit.

By contrast, when the size of the system is finite (and defined by the number of sites L), the step function is replaced by a smooth curve in fig. 1, with a slope proportional to L^d —where d is the dimensionality of the system. Consequently, in the vicinity of H_c, the magnetization M exhibits a bimodal structure, as shown in the bottom panel of fig. 1.

1.3 Link with phase transition in thermodynamics

It was recently demonstrated by Chomaz and Gulminelli that bimodality of the probability distribution of the order parameter is equivalent to the other definitions of phase transition proposed up to now [5].

1.3.1 Relationship to the Yang-Lee theorem

The Yang-Lee theorem [6] is considered as the standard definition of first-order phase transitions at the thermo-

---

[a] e-mail: `lopezo@lpccaen.in2p3.fr`

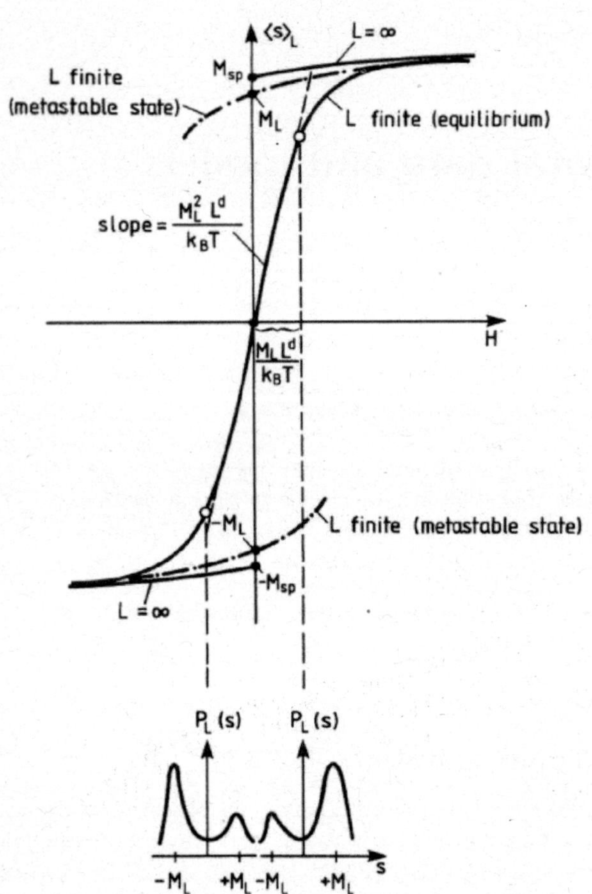

Fig. 1. Evolution of the magnetization M, as a function of the applied magnetic field H, in the Ising model for a lattice defined by the size L. The bottom panel presents a schematic probability distribution of the magnetization between $-M_L$ and $+M_L$ around the critical field value H_c. Taken from [4].

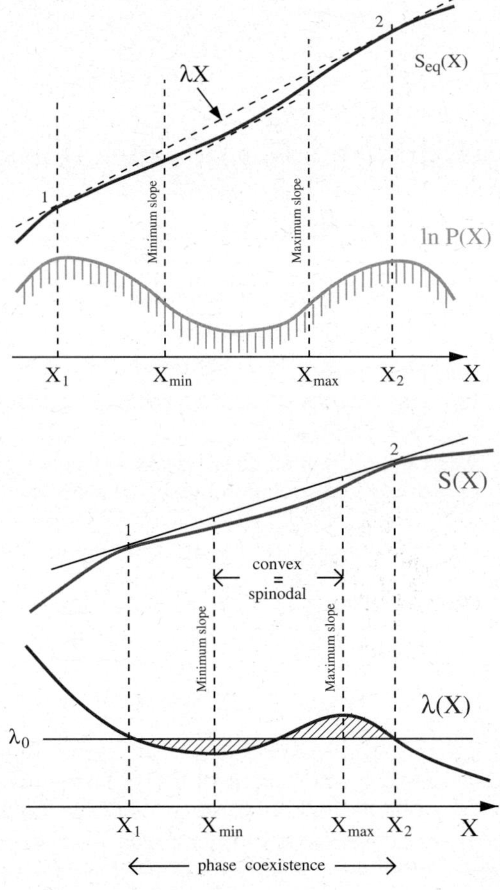

Fig. 2. Entropy S of the system as a function of an order parameter X of the phase transition. The relation is made between the convex intruder of S, the bimodal distribution in X (top) and the abnormal fluctuations of X in the phase coexistence region (bottom). λ is the intensive variable associated with X. Taken from [7].

dynamic limit. As demonstrated in [5], bimodality is a necessary and sufficient condition for zeroes of the partition sum in the control intensive variable complex plane to be distributed on a line perpendicular to the real axis.

1.3.2 Anomaly of thermodynamical potentials

A first-order phase transition is characterized by an *inverted curvature* of the relevant thermodynamical potential (entropy, free energy) [7,8]. This feature is also equivalent to a bimodality in the event probability of the given order parameter X as displayed in the upper part of fig. 2.

1.3.3 Negative derivatives of the thermodynamical potentials

A first-order phase transition was also related to a backbending in the equation of state of the system [7], characterized by a negative second derivative of the thermo-

dynamical potential, as, for example, the heat capacity if the energy is the order parameter, fig. 2 bottom.

1.4 Microcanonical vs. canonical ensemble

Among the observables signing a phase transition, the heat capacity is related to the fluctuations of the partial energy of the system and needs to be studied in the microcanonical ensemble, while bimodality can only be observed when the system is free to fluctuate in terms of the associated extensive variable (*i.e.* energy or volume). This case corresponds to canonical or isobar ensembles. In other words, events must be selected without constraint on the extensive variable in order to study bimodality. However, in nuclear-physics experiments, the two colliding nuclei form an isolated system: it seems thus natural to work in a microcanonical ensemble, and cuts can be applied on the energy of the system, determined, for instance, by calorimetry. It seems conversely out of reach

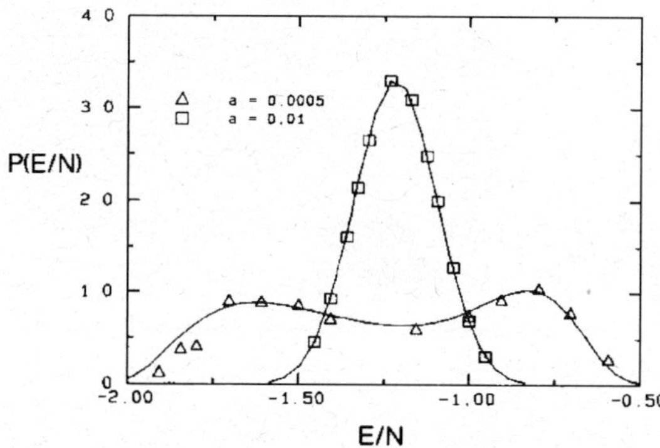

Fig. 3. Energy distributions obtained in the Gaussian ensemble for different $a = N/N'$. Taken from [9].

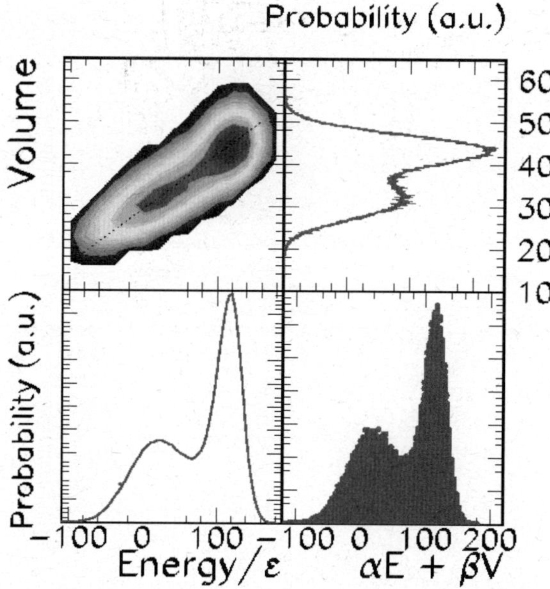

Fig. 4. Probability distributions of the energy E, volume V and a combination of the two variables coming from a lattice-gas simulation in the canonical ensemble. Taken from [2].

to be in a canonical framework which would require the existence of a large heat bath.

The situation is not hopeless, as it was shown some years ago that properties of phase transitions can be observed even if the working ensemble is not strictly microcanonical or canonical, but is an interpolating ensemble. In Gaussian ensembles, for instance, it is supposed that N particles are in contact with a system of N' particles acting as a heat bath at temperature T. When N' varies from 0 to ∞, the working ensemble mimicks the transition between microcanonical and canonical [9]. Figure 3 presents the results of such a simulation, where it is clearly seen that the probability distribution of the energy —in the transition region— presents a bimodal shape only when N/N' is small enough ($< 1/1000$), while for larger N/N', the situation is that of the microcanonical case with only one peak in the distribution.

1.5 Liquid-gas phase transition

Since nuclei are supposed to undergo a liquid-gas phase transition, specific studies of this peculiar transition were undertaken through lattice-gas calculations. In liquid-gas phase transitions, volume as well as energy are order parameters. The bimodality of the event probability distribution in the first-order phase transition region is evident in fig. 4 which shows the location of events in the volume *vs.* energy plane (top left). The projections along the axes (E, V) also display the expected bimodality, as does a linear combination of these two order parameters (bottom right). In this framework (lattice-gas model), bimodality is evidenced if we are able to select (sort) events in a canonical way (or as close as possible, see previous section), and plot the event probability distribution of the energy or volume, or any observable directly related to them.

2 Experimental observations

Since bimodality was proposed as a signature of liquid-gas phase transition, it was extensively searched for in event samples resulting from nuclear collisions; studies were made for central collisions, where the liquid-gas phase transition is clearly evidenced by previous analyses (see sect. "Signals of phase transition", this topical issue) as well as for peripheral collisions, where a large range of excitation energy can be explored.

2.1 Central collisions: systems with mass ~ 250

Systems with total mass close to 250 were studied with the INDRA array using two entrance channels, an asymmetric one, Ni + Au, and an almost symmetric one, Xe + Sn. In both cases, in the incident energy range scanned, it was shown that a fused system was formed in central collisions.

Bellaize *et al.* [10] have reported the observation of bimodality of the size asymmetry of the two largest fragments in central events for the Ni + Au system at $32\,A$, $52\,A$ and $90\,A$ MeV. It was associated with two fragmentation patterns (see first row of fig. 5), one similar to residue-evaporation (one large fragment with few small ones, zone 1 in fig. 5), the other to multifragmentation (fragments of nearly equal size, zone 2). A variable built with the charges of the three largest fragments, Z_1, Z_2, Z_3 in decreasing order,

$$Z_1 - 3(Z_2 + Z_3), \qquad (1)$$

also has a bimodal distribution at $32\,A$ and $52\,A$ MeV, as shown in the bottom row of fig. 5, but no longer at $90\,A$ MeV. This fact is compatible with the location of the system in the coexistence region below $52\,A$ MeV, where it can experience a first-order phase transition by exploring different densities and temperatures. For higher energies (here $90\,A$ MeV), the system passes directly through the

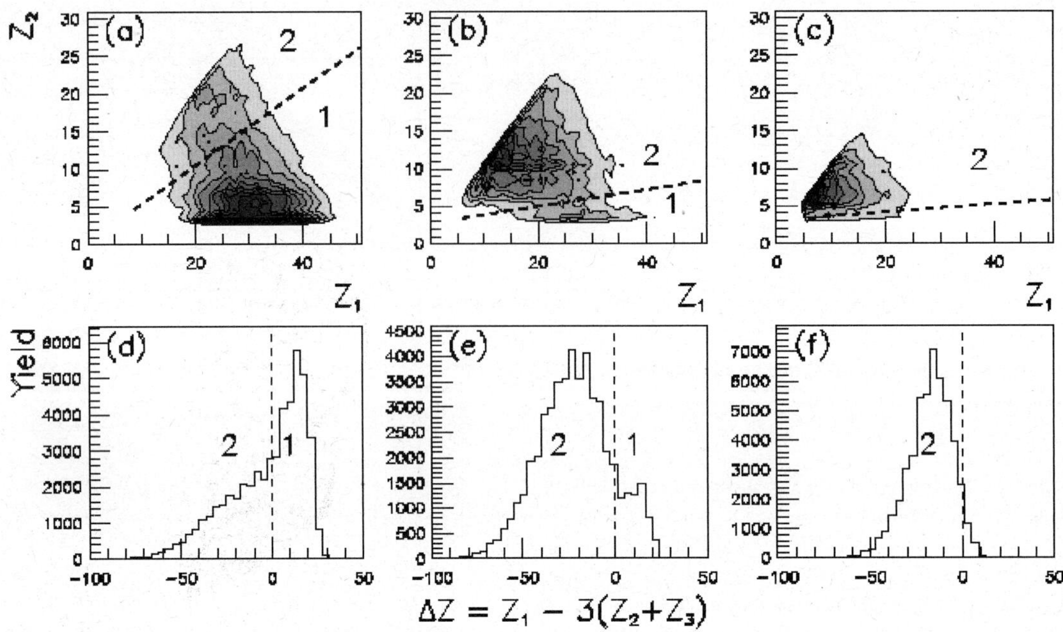

Fig. 5. Correlation between the two largest fragments, Z_1 and Z_2 obtained in central collisions for the Ni + Au system at $32\,A$ (left), $52\,A$ (middle) and $90\,A$ MeV (right). The bottom row shows an asymmetry variable built as a linear combination of the atomic number of the three largest fragments. Taken from [10].

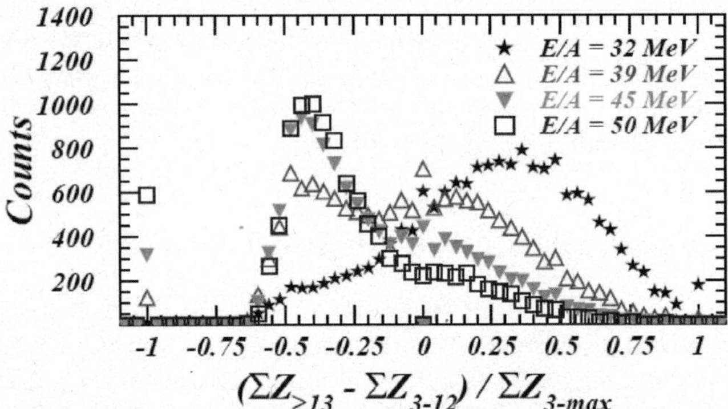

Fig. 6. Probability distributions of the charge asymmetry between light ($Z = 3$–12) and heavy fragments ($Z \geq 12$) for fused events in the Xe + Sn system at $32\,A$, $39\,A$, $45\,A$ and $50\,A$ MeV. Taken from [11].

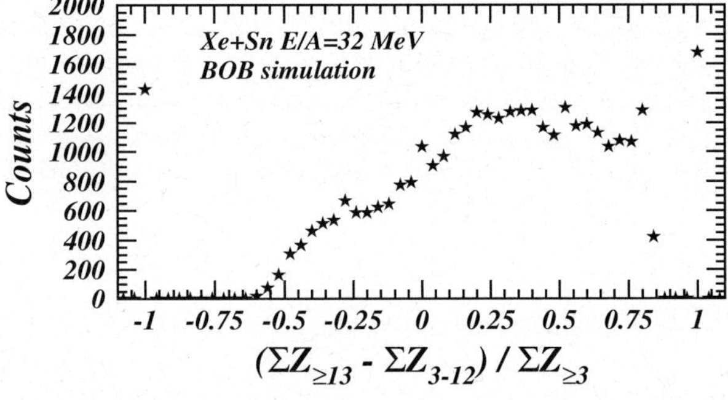

Fig. 7. Charge asymmetry obtained by using a stochastic mean-field simulation (BOB [12]) for central events of the Xe + Sn system at $32\,A$ MeV. Unpublished results from the authors of [14].

coexistence region and we observe only the presence of the multifragmentation regime, which could indicate that the system explores only the low-density part of the phase diagram.

Figure 6 shows the distributions obtained when looking at the asymmetry ratio between heavy, ($Z \geq 13$), and light, ($Z = 3$–12), fragments

$$\left(\sum Z_{\geq 13} - \sum Z_{3-12} \right) / \sum Z_{\geq 3} \qquad (2)$$

for single-source events produced in central Xe + Sn collisions between $32\,A$ and $50\,A$ MeV [11]. Bimodality is present at all energies, with dominant "liquid-type" events at $32\,A$ MeV, and a dominance of "gas-like" events at and

above $45\,A$ MeV; the two types of events are in roughly equal number at $39\,A$ MeV, where other phase transition signals have been already observed (see contribution V.5, *Many-fragment correlations and possible signature of spinodal fragmentation*, this topical issue). The authors of [11] relate the chosen asymmetry variable to the density difference between the coexisting liquid and gas phases of nuclear matter. The same variable was built for the events resulting from a stochastic mean-field simulation [12] of head-on collisions between Xe and Sn at $32\,A$ MeV. In this simulation, which was shown to well reproduce many experimental features, single variable distributions as well as different correlations [13–15] (see contribution V.5, *Many-fragment correlations and possible signature of spinodal fragmentation*, this topical issue), the system enters the coexistence region and multifragments through spinodal decomposition. The equivalent of fig. 6 for simulated events is shown in fig. 7; the picture is very similar to the experimental data at the same energy (black stars in fig. 6), a bimodal behaviour appears with a dominance of events of liquid type.

2.2 Central collisions: systems with mass ~ 100

Central collisions between two ^{58}Ni nuclei were studied at incident energies between 32 and 90 A MeV; event selection was made through a discriminant factorial analysis trained, at variance with ref. [16], on the complete experimental events. A bimodal distribution of the largest fragment was observed at 52 A MeV, intermediate between the Gaussian distributions measured at lower energies and the asymmetric distributions found from 74 A MeV upwards [17], see fig. 8. The minimum is rather shallow (about 80% of the peak value); at 64 A MeV a bimodal distribution persists, but now the peak on the more fragmented side is dominant. Conversely, the distributions of the fragments of higher rank (not shown) are monotonous. To our knowledge, this is the only direct observation of bimodality on the largest fragment.

2.2.1 Going further

Central collisions allowed to study and evidence a bimodal behaviour of some asymmetry variables, which can be connected to the density difference between a liquid and a gas phase; in that sense they would be good candidates for being order parameters of a liquid-gas–type transition. Nevertheless, several drawbacks can be pointed out; firstly it was shown that the lighter fragments exhibit a pre-equilibrium component in Ni + Au [10], while radial-flow effects were recognised in symmetric systems, Xe + Sn [18–20] and Ni + Ni [17]. But above all, the sorting of central events selects a rather narrow region in excitation energy for each incident energy (about 1–2 A MeV at half-maximum of the distribution). This is closer to a microcanonical working ensemble and may prevent a very clear observation of bimodality.

2.3 Quasi-projectiles in peripheral collisions

Analyses of quasi-projectiles formed in peripheral and semi-peripheral reactions are thus mandatory, as they allow to overcome some of the abovementioned problems. In particular a broad excitation energy distribution of quasi-projectiles (QP) can be accessed. Exchanges of energy and particles with the quasi-target (QT), while it lies in the neighbourhood of the QP and especially when it is heavy, mimick a small heat bath and an almost canonical sorting can be envisaged. Whenever the incident energy is high enough, the different components (the QT and the QP, and the pre-equilibrium or neck part) can be better disentangled, or at least the uncertainties caused by their existence can be circumvented.

Most of the studies on quasi-projectiles arise from Au on Au collisions at various energies. Extensive results concerning a very light nucleus, close to argon were also recently proposed. Several variables are used for sorting events as a function of the violence of the collisions; among the most commonly employed one can cite multiplicities and the transverse energy (relative to the beam axis) of charged products, either all of them or only light charged

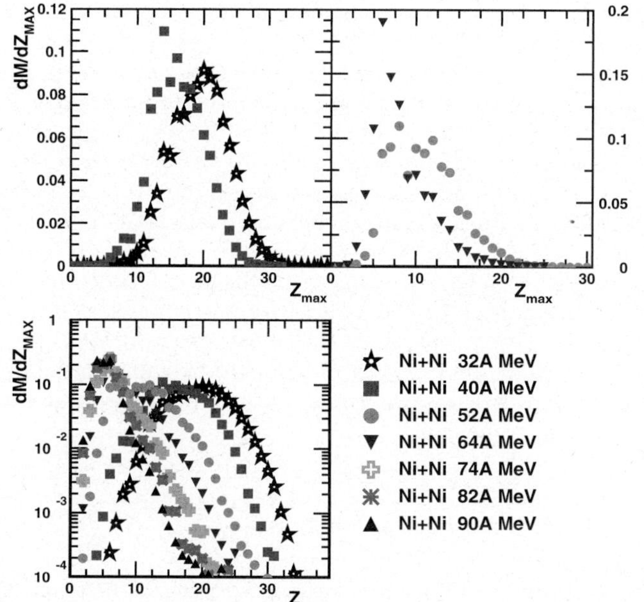

Fig. 8. Distributions of the largest fragment for central Ni + Ni collisions from 32 to 90 A MeV (bottom). The same distributions at the four lowest energies are displayed in linear scale in the top panels. Taken from [17].

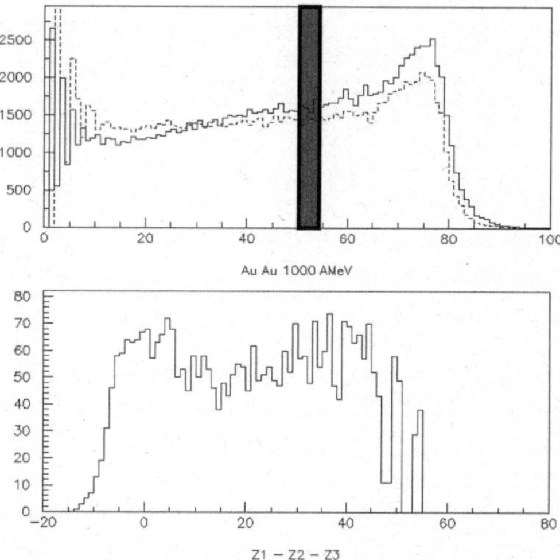

Fig. 9. Z_{bound} (top) and charge asymmetry distributions (bottom) for the Au + Au system at 1 A GeV. The bottom panel corresponds to the Z_{bound} selection displayed by the highlighted area in the top panel. Taken from [27].

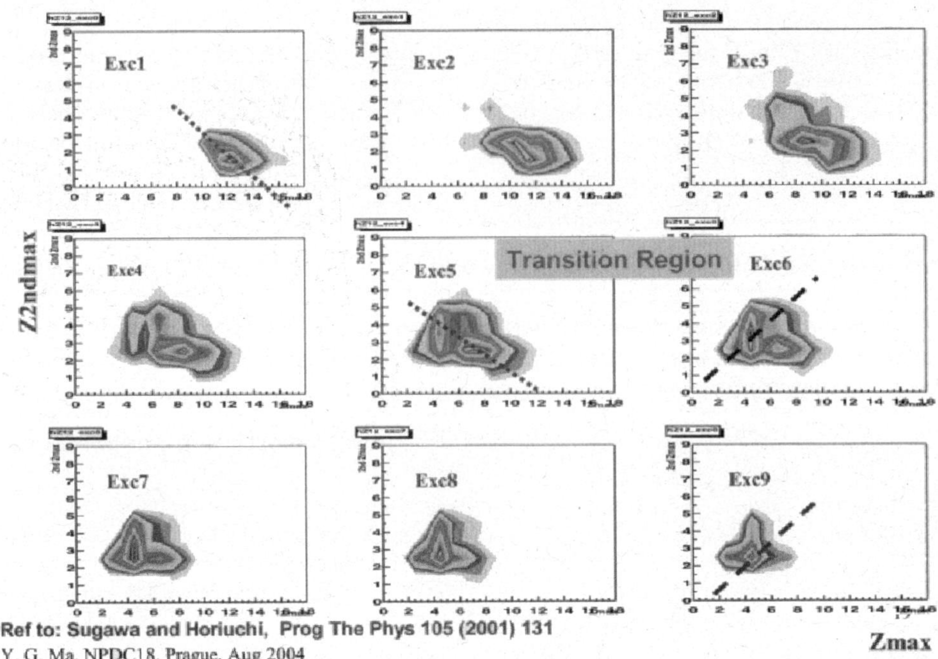

Fig. 10. Correlation between the largest charge (Z_{max}) and the second largest ($Z_{2nd\ max}$) for the QP in peripheral Ar + Ni collisions at $47\,A$ MeV. Panels from Exc1 to Exc9 correspond to a selection in increasing excitation energy (see text). Taken from [26].

particles ($Z = 1, 2$) [21–24]. Other sortings are based on Z_{bound} (the sum of charges for fragments, $Z > 2$), as proposed by the ALADIN Collaboration [25], or on the excitation energy (NIMROD Collaboration) [26].

2.3.1 Au quasi-projectiles at relativistic energies

The ALADIN Collaboration reported the presence of bimodality for peripheral Au + Au reactions at $1\,A$ GeV [27]. Figure 9 shows the Z_{bound} distribution (top panel) where the selected region, $Z_{bound} = 53$–55, is highlighted for which was drawn the charge asymmetry between the three largest fragments,

$$Z_1 - Z_2 - Z_3\,, \qquad (3)$$

in the bottom panel. The charge asymmetry exhibits two components, the first one centered at low values (close to 0), which is associated to multifragmentation events, and the second one located at values around 40, which is more likely due to an evaporation residue of charge Z close to Z_{bound}. It is worth saying that a percolation simulation was able to reproduce this bimodality in the charge asymmetry at the transition point. In this case, this is a second-order phase transition. This point will be discussed below in the section "Pending questions".

2.3.2 A smaller system with mass ~ 40

In a very complete analysis, Ma *et al.* [26] scrutinized data collected with the NIMROD array. They were able to reconstruct, from their emitted particles and fragments, the quasi-projectiles formed in $47\,A$ MeV Ar + Al, Ti, Ni collisions. The method used consisted in tagging the particles with the help of a three-moving-source fit (QP, QT and mid-rapidity) and then attributing to each of them, event per event, a probability to be emitted by one of these sources. Completeness of quasi-projectiles, ($Z_{QP} \geq 12$), from semi-peripheral collisions was further required; QP excitation energy was determined using the energy balance equation. The distributions of excitation energy so obtained for the three targets superimpose, showing that the QP excitation energy calculation is under control.

Plots of the charge of the second largest fragment *vs.* the largest one are shown in fig. 10. As for heavier systems, the topology evolves from residue-evaporation to multifragmentation with increasing excitation energy. An equipartition of events between two topologies is observed for $E^*/A = 5.5$ MeV, where at the same time fluctuations on the size of the largest fragment are the largest, the power law exponent for the charge distribution is minimum, and scaling laws are present. Here again, bimodality is observed at the same time as other possible indicators of a phase transition.

2.3.3 Toward a canonical event sorting?

In the previous cases the sorting for peripheral reactions uses properties of the studied source itself (here the QP) and is then probably more akin to a microcanonical than a canonical sorting. Indeed the bimodal character of the distribution is not very marked, as expected if the experimental sorting constrains strongly the excitation energy [28].

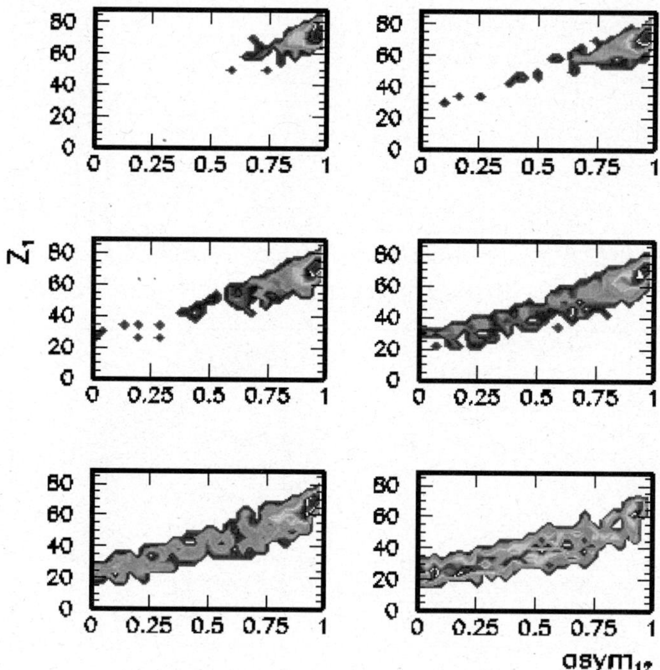

Fig. 11. Correlation between the charge of the largest fragment (Z_1) and the charge asymmetry (asym$_{12}$) between the two largest fragments for peripheral events of the Au + Au system at 35 A MeV. The panels correspond to a selection in increasing transverse energy of particles coming from the QT side from top left to bottom right. Taken from [29].

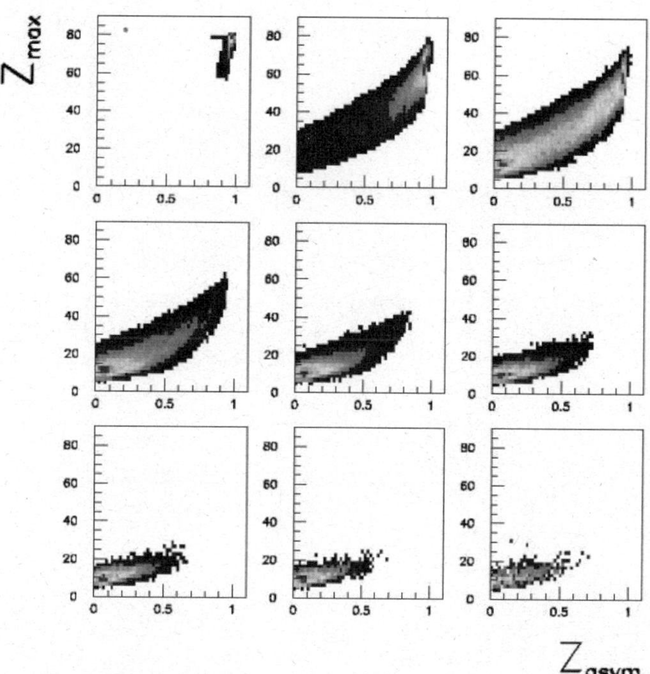

Fig. 12. Same as for fig. 11 but for Au + Au at 80 A MeV. The panels corresponds to the same cuts in QT transverse energy. Taken from [30].

To attempt a true canonical sorting one must discriminate the studied system from some heat bath. A first tentative in that aim was the study of Au quasi-projectiles through a sorting performed on the transverse energy of the particles of the Au quasi-target (as the system is symmetric, this amounts to particles emitted backward in the c.m.). This sorting is illustrated by the results presented hereafter.

2.3.4 Au-like nuclei in a "canonical" sorting

Au quasi-projectiles from Au + Au collisions at various incident energies were widely studied. Two examples are given here, at 35 A MeV —results from the MULTICS-MINIBALL Collaboration [29]— and at 80 A MeV, data from the INDRA/ALADIN Collaboration [30]. In both cases data were sorted *vs.* the transverse energy of the QT light charged particles. The charge of the largest fragment in each event is plotted in figs. 11, 12 *vs.* the charge asymmetry of the two largest fragments,

$$(Z_1 - Z_2)/(Z_1 + Z_2). \tag{4}$$

Whatever the incident energy, the picture evolves from an evaporation residue to a multifragmentation configuration, passing through a zone where the two topologies coexist, separated by a neat minimum; in this zone (last one at 35 A MeV, third one at 80 A MeV) the distributions

present a bimodal behaviour. Note that the bimodal character is not very strong when one projects the bidimensional figures on either Z_{max} or on the asymmetry. This is attributed in [30] to the presence of pre-equilibrium effects, and some remaining aligned momentum which tend to shallow the minimum of a bimodal distribution.

3 Bimodality in models

Different statistical as well as dynamical models explicitly or implicitly contain a phase transition. They predict the occurrence of bimodal distributions for selected variables around some transition energy. Examples are given in this section.

3.1 SMM: Statistical Multifragmentation Model

Buyukcizmeci, Ogul and Botvina [31] analyzed SMM simulations for heavy nuclei of various sizes, with excitation energy ranging from 2 to 20 MeV/nucleon. They found that all nuclei exhibit the same caloric curve, depicted in the top panel of fig. 13, with the well-known "plateau" between 4 and 7 MeV/nucleon (note in passing that the common temperature at plateau whatever the mass of the considered nucleus is in contradiction with the experimental results analyzed in ref. [32]). In the same energy interval as that of the plateau, the fluctuations of A_{max} (not shown) and of the temperature (panel (b) of fig. 13) are maximum. The authors sorted the events following the size of the largest fragment, A_{max}. They defined two

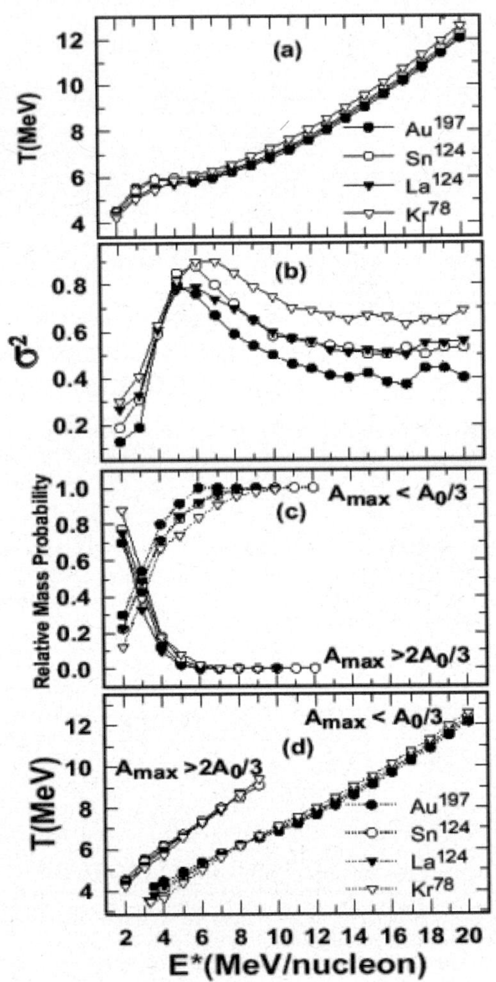

Fig. 13. Temperature average value (a) and variance (b), probability of events selected on A_{max} (c) and average temperature for these events (d) *vs.* E^* for Kr, La, Sn and Au nuclei (SMM simulations). Taken from [31].

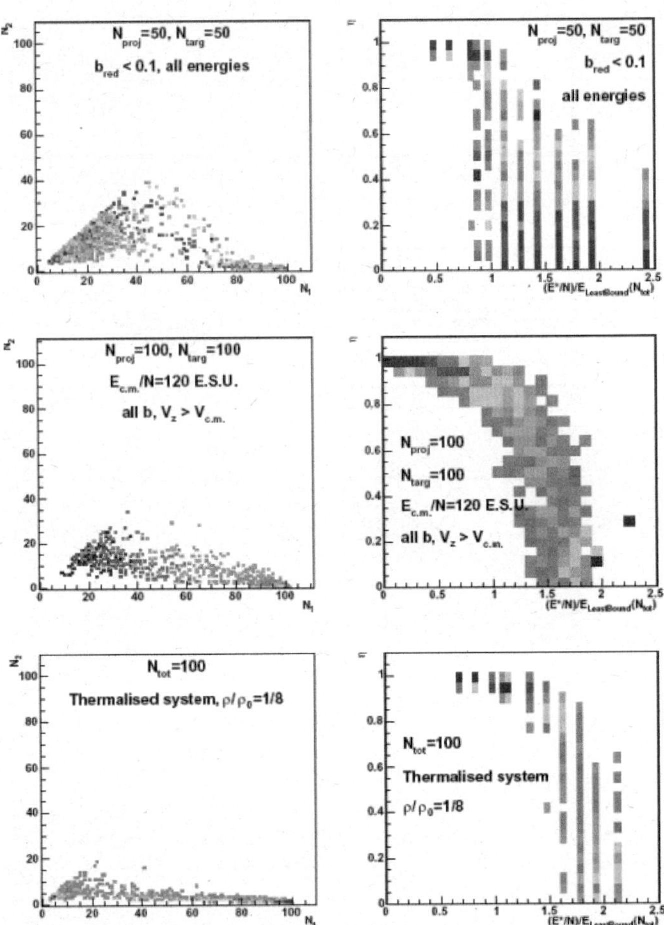

Fig. 14. Mass correlation between the two largest fragments (left) and mass asymmetry, η, as a function of the excitation energy (right), for collisions of LJ droplets. The top panel is associated to central collisions; the middle to quasi-projectiles from peripheral collisions and the bottom panel to "thermalized" systems (see text) [34].

3.2 CMD: Classical Molecular Dynamics

Signals of phase transition were searched for in dynamical models. A simple example is a Classical Molecular Dynamics model with a Lennard-Jones potential implemented by Cussol [33]. With such a potential, analogous to the van der Waals interaction for fluids, the model includes a liquid-gas phase transition. Symmetric collisions of LJ droplets with sizes of $50 + 50$ and $100 + 100$ are analyzed. Systems were prepared in three different conditions:

- central collisions (small impact parameters),
- peripheral collisions (all impact parameters but looking at the forward zone, "quasi-projectiles"),
- "thermalized" systems (particles are placed in a box of volume $V/V_0 = 8$ and released after a time sufficient to reach thermal equilibrium).

Two variables were scrutinized, the size asymmetry between the two largest fragments, η (eq. (4)), and the mass

event classes, one with $A_{max} \geq 2A_0/3$, representative of the residue-evaporation channel and the other one with $A_{max} \leq A_0/3$, characterizing multifragmentation events —A_0 being the total system size. Panel (c) of fig. 13 shows that the probability of the first group decreases rapidly in the excitation energy range 2–6 MeV/nucleon, while that of the second one increases. The temperatures T associated to each class are different, as appears on the related caloric curves: the residue-evaporation class shows a Fermi-gas behaviour (proportional to T squared), while the multifragmentation class is associated to a classical gas (linear in T). The combination of these two behaviours gives rise to the plateau zone in the total caloric curve and explains the inflexion point of this curve. One is thus dealing with a direct bimodal behaviour, with two excitation energies associated with one temperature in the transition region. This behaviour is an intrinsic feature of the phase space population in the SMM.

correlation between these same fragments [34]. The results, for systems comprising 100 droplets, are presented in fig. 14 (top: central collisions, middle: quasi-projectiles and bottom: thermalized system). Excitation energies are expressed in ESU, ratio between the excitation energy per particle and the binding energy of the least bound particle.

Bimodality —the occurrence of two fragmentation patterns in a given energy zone— is present in all situations, but at different excitation energies: 1 ESU for central collisions, 1.5 ESU for quasi-projectiles, and ~ 1.8 ESU for the thermalized system. It is however worth to mention that if the thermalized system is prepared at higher densities ($\rho/\rho_0 = 1$–1.5) the transition between the fragmentation patterns also occurs but at lower excitation energy, namely < 1 ESU [34]. Cussol attributes the differences in the transition energy to the lack of complete thermalization of any source produced in nuclear collisions, whatever the impact parameter. One can conversely argue that this study proves that bimodality is a robust signature of phase transition, as it survives even if the system is not fully thermalized, although the apparent transition energy is displaced. This point will be developed later.

3.3 HIPSE: Heavy-Ion Phase Space Exploration

The Heavy-Ion Phase Space Exploration model (HIPSE) comprises a full (classical) treatment of the entrance channel (nucleus-nucleus potential, NN collisions). It is followed by a random sampling of nucleons in the participant zone from Thomas-Fermi distributions of the two colliding nuclei to form fragments in the dense zone [35]. Excitation energy is shared among all products, taking into account the total energy constraint. Finally a statistical de-excitation (SIMON code [36]) of the fragments, including QP and QT —if they are still present— is performed.

Simulations were done for all impact parameters, to mimick a real $50\,A\,\mathrm{MeV}$ Xe + Sn experiment, then the same analysis as in [30] was performed by Lopez *et al.* [37]; a bimodal structure was observed in the correlation between Z_{max} and the charge asymmetry of the two largest fragments (eq. (4)). In a model however one can go further and track the origin of the bimodal behaviour: is it due to the entrance channel (dynamical effect) or to the de-excitation step? The first hypothesis was ruled out, as no discontinuity was found in the evolution of the size of the hot largest fragment with the impact parameter: the bimodality was clearly attributed to the statistical de-excitation of the QP. A deeper analysis of the de-excitation stage was then achieved through the simulated statistical de-excitation of xenon nuclei of different excitation energies and spins with the SIMON code [36]. This is depicted in fig. 15, where the distributions of the asymmetry variable (eq. (4)) are plotted for several initial conditions. Increasing the excitation energy does decrease the average charge asymmetry, but never down to the small values observed in the data. Conversely, if more spin is given to the nucleus, the asymmetry variable displays a sharp transition around 60–70$\hbar$, which corresponds indeed to the an-

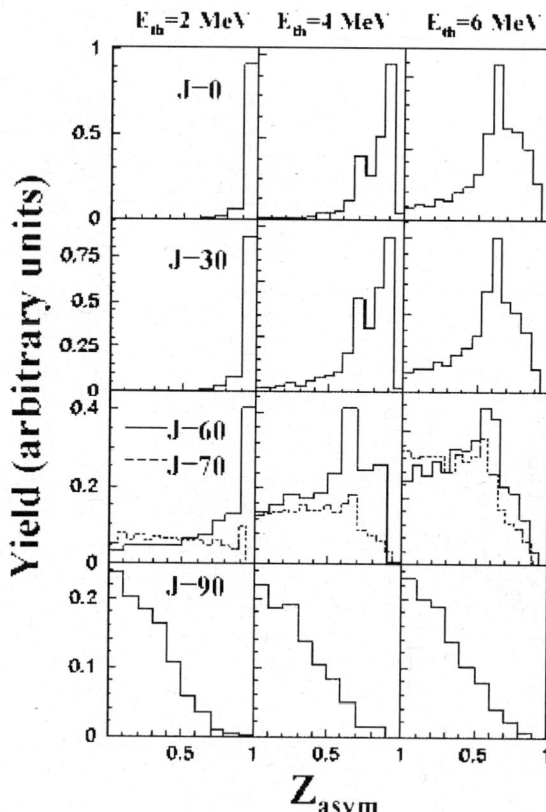

Fig. 15. Charge asymmetry distributions resulting from the de-excitation of hot Sn nuclei with different initial excitation energies (columns) and spin (rows), with the SIMON code. Taken from [37].

gular momentum for which the symmetric fission barrier vanishes.

The authors of [37] conclude that, in the HIPSE model, the observed bimodality found its origin in the spin rather than in the excitation energy transferred to the QP, being still a phase transition but not of the liquid-gas type. It is worth mentioning that using the SMM model for the de-excitation stage, the authors also observe bimodality in the size of the largest fragment. This is not surprising in view of the abovementioned study with the SMM. However, this raises the important issue —still under debate— of the order parameters (and then the type of the phase transition) which govern the bimodality. This point will be discussed below in the section "Perspectives".

4 Pending questions

As seen in the previous sections, bimodality is a very common feature in nuclear collisions at intermediate energies. It is present in central as well as in peripheral collisions. It takes place for a large range of masses, $A = 40$–200. It was however mentioned in the course of the text that it is experimentally difficult to isolate a source, because of dynamical effects leading to a mixture of pre-equilibrium

Table 1. World-wide experimental results on bimodality recorded in July 2005.

Results from	Reaction centrality	Source size	Bimodal variable
INDRA	Central	~ 200	$Z_1 - Z_2$ (eqs. (1), (4))
INDRA	Central	~ 200	$Z_{liq} - Z_{gas}$ (eq. (2))
INDRA	Central	~ 100	Z_{max}
INDRA	Peripheral	160–180	$Z_1 - Z_2$ (eq. (4))
MULTICS/ MINIBALL	Peripheral	~ 180	$Z_1 - Z_2$ (eq. (4))
ALADIN	Peripheral	~ 130	$Z_1 - Z_2 - Z_3$ (eq. (3))
NIMROD	Peripheral	24–40	$Z_{liq} - Z_{gas}$

products and of QP/QT de-excitation particles. Even if a source can be properly defined, one has to verify its degree of thermalization. Indeed radial flow was found, particularly in central collisions, and transparency effects were also evidenced [38]. It seems however from both experimental [30] and theoretical [37] studies that bimodality is not mainly driven by dynamical effects. Ambiguities remain in the type of phase transition observed, and consequently on the definition of a true order parameter. Some of these questions were addressed recently and are presented in the following.

Table 1 gathers all experimental results on bimodality found so far. A glance at the table indicates that bimodality was essentially found in charge asymmetry variables comprising the two or three largest fragments of each event. Such variables can in some sense be related to the density difference between a dense (liquid) and a dilute (gas) phase; in some models, for instance the Fisher droplet model, the largest fragment is assimilated to the liquid while all the other form the gas.

4.1 Are Z_{max}, A_{max}, or the asymmetry order parameters?

Simulations were performed in different frameworks to test whether the observables Z_{max}, A_{max}, or the asymmetry, reliably sign a phase transition. Let us recall that a bimodality of an order parameter signs the occurrence of a first-order phase transition in a finite system. Figure 16 shows the outcomes of three simulations in the transition region —when it exists; there is no phase transition in the random partitions calculation, while percolation has a second-order transition and lattice gas a first-order one [39]. The distributions of the largest fragment A_{max} evidence that A_{max} only presents a bimodal distribution for the canonical lattice-gas calculation. This means that A_{max} is indeed an order parameter of the first-order phase transition of the lattice gas. The distribution presents a wide plateau, as expected, in the case of a continuous transition (percolation). By contrast, the mass asymmetry, A_{asy}, defined in a similar way as the charge asymmetry (eq. (4)), also displays a bimodality (although with a less marked minimum) for simulations which have a

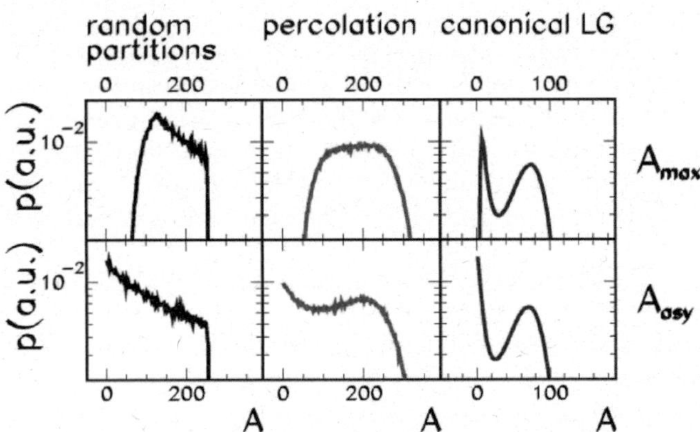

Fig. 16. Largest fragment A_{max} (top) and mass asymmetry A_{asy} (bottom) distributions for three simulations. Left: random partitions (no phase transition), middle: percolation (2nd-order phase transition) and right: canonical lattice gas (1st-order phase transition). Taken from [39].

2nd-order phase transition (percolation, middle column). The conclusion of this study is that both A_{max} and A_{asy} clearly signal a phase transition —note that none of them presents bimodality in a model without phase transition— but A_{max} is the only unambiguous signature of the order of the transition.

4.2 Order parameters of the liquid-gas phase transition

If nuclei undergo a liquid-gas–type phase transition, then the order parameters are known: energy, volume. In some of the experimental studies cited above, the authors try to push the analysis beyond the single observation of bimodality on the asymmetry variable. As a first attempt, in central collisions between Ni and Au at $52\,A\,$MeV [10], the excitation energies (experimentally deduced from the energy balance equation) associated to the two fragmentation patterns were found slightly different (by $1\,A\,$MeV) [40]. This bimodality of the excitation energy is an indication in favour of the liquid-gas type of the phase transition observed.

Studies of Au quasi-projectiles were deepened by the authors of ref. [30]: a test of the reliability of the canonical picture was accomplished by estimating the apparent temperatures of the two types of events, from the slope of the emitted proton spectra for residue-like events, and from double isotope ratios in the multifragmentation regime. As seen in fig. 17 both temperatures are close enough in the region where bimodality is present (Etrans = 0.8–1.2 $A\,$MeV), while the excitation energies, calculated with the energy balance equation, are different. This is expected if bimodality has a thermal origin and validates the sorting as close to a canonical one.

4.3 Does bimodality survive out-of-equilibrium effects?

The influence of non-equilibrium effects on signals of phase transition was studied in [41] in the case of incompletely

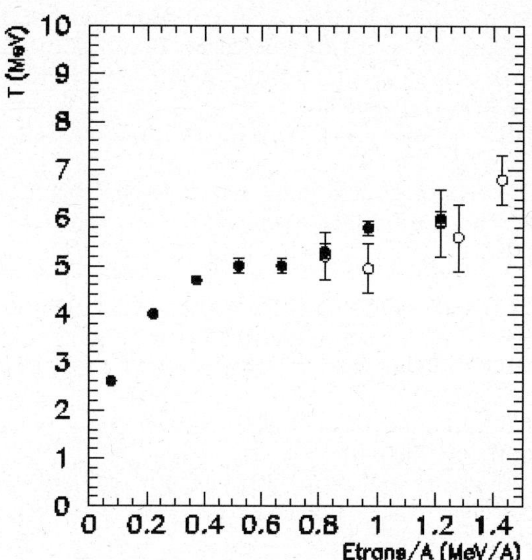

Fig. 17. Apparent temperatures of Au quasi-projectiles as a function of the normalized transverse energy for residue (filled symbols) and multifragmentation events (open symbols). Taken from [30].

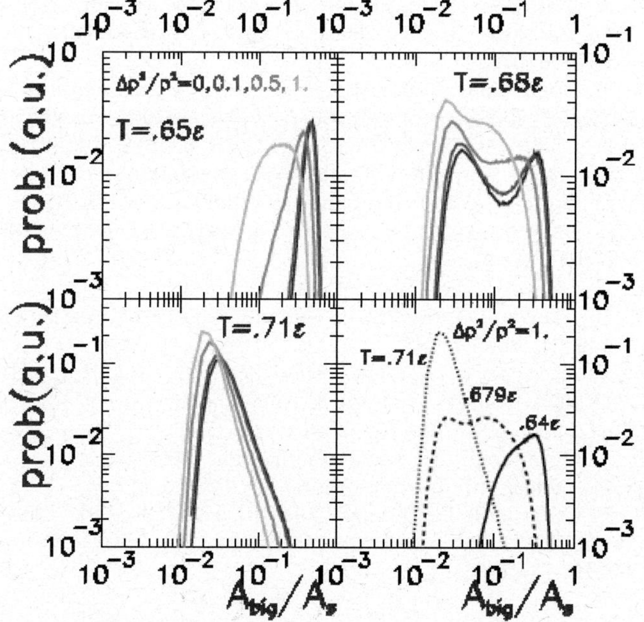

Fig. 18. Canonical lattice-gas simulations for different temperatures T around the critical one $T_c = 0.68\varepsilon$ for the distributions of the largest fragment. Simulations are performed by adding an extra radial-flow energy $\Delta p^2/p^2$ between 0 and 1. Taken from [2].

relaxed incoming momentum (transparency) and of self-similar radial flow. Both effects were indeed recognized in experimental data. Figure 18 displays results of (canonical) lattice-gas simulations with different radial-flow energies; $\Delta p^2/p^2$ is equivalent to the ratio $\varepsilon_{flow}/\varepsilon_{th}$ and is varied from 0 to 1. At the transition temperature, $T = 0.68\varepsilon$, a bimodality of A_{max} is clearly seen in the absence of flow,

and is still visible even when the flow energy is as important as the thermal energy (top right panel). The authors state thus that radial flow does disturb the signal, partially filling the gap between the two components, but does not destroy it as long as the flow does not dominate the global energetics. Similar conclusions were drawn in this paper in the presence of longitudinal flow (transparency effects).

These two examples illustrate the robustness of bimodality *vs.* external (and realistic) constraints due to the dynamics of the collision; similar conclusions can also be derived from CMD simulations (see above).

In experimental data on Au quasi-projectiles [30], refined treatments aiming at better isolating quasi-projectiles from the mid-rapidity contribution and keeping only events where this contribution was smaller were tempted. In all cases the bimodal picture comes out better, although it occurs for a lower value of the sorting variable (smaller dissipation), for a given incident energy. This is again an evidence of the robustness of bimodality against non-equilibrium effects.

5 Perspectives

Bimodality is a very promising signature of first-order phase transition because of its simplicity and robustness against dynamical constraints. It was shown in this contribution that the signal is quite common for the decay of hot nuclei and can be observed in rather different experimental conditions (central/peripheral collisions, small/large source sizes).

Nevertheless, some open questions need to be answered in order to firmly assess the validity of this signal. Several strategies can be envisaged in order to progress in this direction:

- cross the observation of the bimodality signal with that of all the other proposed signals for the phase transition such as critical exponents, scalings (Delta-scaling, Fisher scaling, Zipf law), negative heat capacities, or space-time correlations (emission times and correlation functions). Obviously, when possible, all signals should be studied on the same sample of events to minimise biases due to sorting. Such cross controls were started by the INDRA [42] and NIMROD [26] Collaborations. One must solve however the problem of the non-equivalence of statistical ensembles in some cases.
- Test the effect of sorting. Indeed different ways of sorting were proposed (impact parameter selectors, compact shape events, source selection). The robustness of any signal will be established if its observation is not drastically dependent on the chosen sorting for a given centrality, for instance.
- Compare the results of different entrance channels for nuclear collisions; by using asymmetrical reactions such as light ions impinging on heavy targets, or nucleon/pion-nucleus reactions, one may hope to disentangle the different effects which could possibly govern bimodality. By using these very different entrance

channels reactions, the pre-equilibrium/neck contributions can be evaluated and even subtracted. Moreover, the effects of large collective motions such as radial flow (for central collisions) or spin (angular-momentum transfer in semi-peripheral reactions) can also be measured. This will possibly help to answer the fundamental question of the type of phase transition which is experienced by hot nuclei.

We thank all the nuclear physicists around the world who send us their results —published or not.

References

1. T.L. Hill, *Thermodynamics of Small Systems* (Benjamin, New York, 1963).
2. Ph. Chomaz, F. Gulminelli, V. Duflot, Phys. Rev. E **64**, 046114 (2001).
3. K.C. Lee, Phys. Rev. E **53**, 6558 (1996).
4. K. Binder, D.P. Landau, Phys. Rev. B **30**, 1477 (1984).
5. Ph. Chomaz, F. Gulminelli, Physica A **330**, 451 (2003).
6. C.N. Yang, T.D. Lee, Phys. Rev. **87**, 404 (1952).
7. Ph. Chomaz, M. Colonna, J. Randrup, Phys. Rep. **389**, 263 (2004).
8. D.H.E. Gross, *Lecture Notes in Physics*, Vol. **602** (Springer-Verlag, 2002) p. 23.
9. M.S. Challa, J.H. Hetherington, Phys. Rev. Lett. **60**, 77 (1988).
10. INDRA Collaboration (N. Bellaize et al.), Nucl. Phys. A **709**, 367 (2002).
11. B. Borderie, J. Phys. G: Nucl. Part. Phys. **28**, R217 (2002).
12. Ph. Chomaz, G.F. Burgio, J. Randrup, Phys. Lett. B **254**, 340 (1991); A. Guarnera, Ph. Chomaz, M. Colonna, J. Randrup, Phys. Lett. B **403**, 191 (1997).
13. INDRA Collaboration (J.D. Frankland et al.), Nucl. Phys. A **689**, 940 (2001).
14. INDRA Collaboration (G. Tăbăcaru et al.), Eur. Phys. J. A **18**, 103 (2003).
15. INDRA Collaboration (G. Tăbăcaru et al.), Nucl. Phys. A **764**, 371 (2006).
16. INDRA Collaboration (P. Lautesse et al.), Phys. Rev. C **71**, 034602 (2005).
17. P. Lautesse, Habilitation à diriger des recherches, Université Claude Bernard Lyon I (2006); R. Moustabchir, thèse de doctorat, Québec/Lyon (2005), `http://tel.ccsd.cnrs.fr/tel-00008654`.
18. INDRA Collaboration (N. Marie et al.), Phys. Lett. B **391**, 15 (1997).
19. N. Le Neindre, thèse de doctorat, Caen (1999), `http://tel.ccsd.cnrs.fr/tel-00003741`.
20. INDRA Collaboration (B. Borderie et al.), Nucl. Phys. A **734**, 495 (2004).
21. W.J. Llope et al., Phys. Rev. C **51**, 1325 (1995).
22. J. Péter et al., Nucl. Phys. A **593**, 95 (1995).
23. INDRA Collaboration (J. Lukasik et al.), Phys. Rev. C **55**, 1906 (1997).
24. INDRA Collaboration (J.D. Frankland et al.), Nucl. Phys. A **689**, 905 (2001).
25. A. Schüttauf et al., Nucl. Phys. A **607**, 457 (1996).
26. NIMROD Collaboration (Y.G. Ma et al.), nucl-ex/0410018.
27. ALADIN Collaboration (W. Trautmann et al.), private communication (2005).
28. F. Gulminelli, Ph. Chomaz, Phys. Rev. C **71**, 054607 (2005).
29. MULTICS-MINIBALL Collaboration (M. d'Agostino et al.), private communication (2004).
30. INDRA Collaboration (M. Pichon et al.), Nucl. Phys. A **749**, 93c (2004); nucl-ex/0602003; M. Pichon, thèse de doctorat, Caen (2004), `http://tel.ccsd.cnrs.fr/tel-00007451`.
31. N. Buyukcizmeci, R. Ogul, A.S. Botvina, Eur. Phys. J. A **25**, 57 (2005).
32. J.B. Natowitz et al., Phys. Rev. C **65**, 034618 (2002).
33. D. Cussol, Phys. Rev. C **65**, 054614 (2002).
34. D. Cussol, private communication (2005).
35. D. Lacroix, A. Van Lauwe, D. Durand, Phys. Rev. C **69**, 054604 (2004).
36. D. Durand, Nucl. Phys. A **541**, 266 (1992).
37. O. Lopez, D. Lacroix, E. Vient, Phys. Rev. Lett. **95**, 242701 (2005).
38. W. Reisdorf et al., Phys. Rev. Lett. **92**, 232301 (2004); INDRA Collaboration (C. Escaño-Rodriguez et al.), nucl-ex/0503007.
39. F. Gulminelli, private communication (2005).
40. O. Lopez, unpublished results.
41. F. Gulminelli, P. Chomaz, Nucl. Phys. A **734**, 581 (2004).
42. INDRA Collaboration (M.F. Rivet et al.), Nucl. Phys. A **749**, 73 (2005).

Eur. Phys. J. A **30**, 275–291 (2006)
DOI 10.1140/epja/i2006-10123-8

THE EUROPEAN
PHYSICAL JOURNAL A

Detection

R.T. de Souza[1], N. Le Neindre[2,a], A. Pagano[3], and K.-H. Schmidt[4]

[1] Indiana University Cyclotron Facility, Indiana University, Bloomington, IN 47405, USA
[2] Institut de Physique Nucléaire d'Orsay, CNRS-IN2P3, F-91406 Orsay, France
[3] Istituto Nazionale di Fisica Nucleare INFN and Università di Catania, I-95123 Catania, Italy
[4] GSI mbH, D-64291 Darmstadt, Germany

Received: 4 April 2006 /
Published online: 24 October 2006 – © Società Italiana di Fisica / Springer-Verlag 2006

Abstract. This review on second- and third-generation multidetectors devoted to heavy-ion collisions aims to cover the last twenty years. The presented list of devices is not exhaustive but regroups most of the techniques used during this period for nuclear reactions at intermediate energy ($\simeq 10\,A$ MeV to $1\,A$ GeV), both for charged-particle and neutron detection. The main part will be devoted to 4π multidetectors, projectile decay fragmentation, high-resolution magnetic spectrometers, auxiliary detectors and neutron detection. The last part will present the progress in electronics and detection in view of the construction of future-generation detectors.

PACS. 29.30.Aj Charged-particle spectrometers: electric and magnetic – 29.40.Cs Gas-filled counters: ionization chambers, proportional, and avalanche counters – 29.40.Mc Scintillation detectors – 29.40.Wk Solid-state detectors

1 Introduction

The empirical knowledge of the dynamics and the thermodynamics with nuclear degrees of freedom emerges from experimental studies of nuclear collisions. For this purpose, powerful experimental installations have been developed at many accelerator facilities with the aim of registering as many observables of the reaction products as possible with the best possible resolution. The key properties of these installations are large solid-angle coverage, high granularity and low detection thresholds. Generally, one strives for observation and full identification of all reaction products. For charged particles, the determination of the atomic number Z has become standard, while isotopic identification is still limited to the lighter products in most installations.

Full isotopic identification of all residues has become an important issue, and strong efforts are being made to achieve this goal. The attempts follow two roads: On the one hand, detector telescopes are being developed with ToF or pulse-shape analysis to extend the mass range where isotopic resolution can be achieved. On the other hand, full identification over the whole mass range has been obtained by using powerful magnetic spectrometers.

Motivated by specific characteristics of the reaction dynamics, experiments are performed from the Fermi-energy regime up to the GeV-per-nucleon range, in accordance

with the capabilities of the corresponding accelerators. Traditionally, most experiments have been performed in the Fermi-energy regime, where the partly overlapping Fermi spheres of projectile and target lead to very specific features like isospin diffusion. Reactions at energies around $1\,A$ GeV, where the Fermi spheres of projectile and target are well separated, are rather goverened by a clear distinction of participant and spectator nucleons.

Different requirements are imposed on the detector equipment for these different energy regimes. Experiments at low energies need large angular coverage. Higher energies lead to a strong kinematical focussing of projectile-like reaction products in the forward direction. They also facilitate full identification of heavy residues in mass and atomic number and high-precision measurements of their kinematic properties as well as the simultaneous detection of neutrons and charged particles.

2 Heavy ions

This section presents the most important 4π multidetectors, *i.e.* devices covering as much of the solid angle as possible, used during the last twenty years all over the world. With the evidencing of multifragmentation in the early 1980s, multidetectors able to detect most of the products coming from reactions between heavy-ion collisions became essential. Following the progress made in microelectronics, detection and in understanding the new phe-

------

[a] e-mail: `leneindre@ipno.in2p3.fr`

nomena, always more and more powerful devices appeared with always better and better granularity, better angular coverage, lower detection thresholds, and better identification (including isotopic resolution).

2.1 NAUTILUS

The NAUTILUS multidetector installed at Ganil was constituted by four multidetectors, DELF, XYZt, MUR and TONNEAU [1].

2.1.1 General characteristics

- DELF and XYZt:
- position-sensitive gas telescope consisting of a parallel-plate avalanche detector and an ionization chamber;
- full fragment efficiency for $Z \geq 8$;
- angular coverage 3°–150°;
- DELF: detection thresholds $0.13\,A$ MeV, angular resolution $\Delta\theta = 0.5°$, velocity resolution $\Delta v/v = 4\%$;
- XYZt: detection thresholds $2.0\,A$ MeV, angular resolution $\Delta\theta = 0.1°$, velocity resolution $\Delta v/v = 7\%$.
- MUR and TONNEAU:
- plastic scintillators for light-charged-particle detection $Z = 1$–2;
- energy thresholds $1\,A$ MeV;
- angular coverage 3°–150°.
- Complete device: 35% of 4π.

Fragments and particles produced in the heavy-ion collisions were detected by the four multidetectors of NAUTILUS. DELF and XYZt are gaseous multidetectors in which each module was constituted by a parallel-plate avalanche detector for localisation and an ionization chamber. They detected with full efficiency fragments of charge greater than eight and their angular coverage was between 3° and 150°, which corresponds to almost 2π. For DELF, energy detection thresholds were $0.13\,A$ MeV, the angular resolution $\Delta\theta = 0.5°$ and the velocity precision $\Delta v/v = 4\%$. For XYZt, energy detection thresholds were $2\,A$ MeV, the angular resolution $\Delta\theta = 0.1°$ and the velocity precision $\Delta v/v = 7\%$. The MUR and TONNEAU were constituted by scintillation plastics which detected light charged particles (essentially $Z = 1$–2), with an energy threshold of about $1\,A$ MeV. They covered an angular range between 3° and 150° representing 70% of 4π.

When XYZt and DELF were mounted inside the reaction chamber they shadowed the MUR and TONNEAU multidetectors reducing thus their detection efficiency for light charged particles to 35% of 4π. Before being detected, particles lose energy in the fragment detectors. Therefore, detection thresholds were higher, of the order of $4\,A$ MeV, and there was degradation in the quality of time-of-flight measurements (light-particle velocities). This was mainly the case for particles emitted by the quasi target. Nevertheless, in spite of these drawbacks on light charged particles, this device was perfectly adapted for fission studies. The very low-energy thresholds allowed the detection of fragments emitted by the target side both for residues and fission fragments and projectile fragments which benefited from the recoil energy of the emitting nucleus to largely overcome the thresholds.

2.2 MSU 4π

Michigan State University 4π multidetector.

2.2.1 General characteristics

- Soccer ball geometry.
- $18° \leq \theta \leq 162°$.
- 215 fast (3 mm) - slow plastic (25 cm).
- 55 Bragg curve detectors.
- Thresholds:
- $17\,A$ MeV for plastic only;
- few A MeV with Bragg curve detectors.
- p, d, t, He, Li, Be, B, C with $E/A \leq 200$ MeV.

2.2.2 Main results

Thanks to the measurements of the momentum in an event, the MSU 4π device allowed the determination of the "balance energy" at which the nuclear potential changes from attractive to repulsive. The momentum dependence of the mean field is one of the basic ingredients of transport models of any kind. The analysis performed with such a multidetector has permitted to extract the evolution of the balance energy with beam energy as a function of the reduced impact parameter, bringing thus important constraints for the models.

2.3 MSU MINIBALL

Michigan State University MINIBALL [2].

2.3.1 General characteristics

- 188 detectors: $9° \leq \theta \leq 160°$.
- Large angular coverage 89% of 4π.
- Thin fast plastic $(40\,\mu\text{m})$ - CsI(Tl) (2 cm).
- Thresholds: $1.5\,A$ MeV for ^{4}He up to $\simeq 3\,A$ MeV around Ca.
- Charge identification for $Z \leq 18$ and mass separation for $Z = 1$–2.

2.3.2 Main results

The fragment detection with low thresholds and good coverage allowed the investigation of the dependence of fragment number on total charge multiplicity (dissipation) and incident beam energy [3]. Thanks to the reasonable energy and angular resolution, IMF-IMF correlations were performed giving information on the mean time between IMF emissions. Times as short as 100–200 fm/c were found depending on the relative momentum between the two partners [3].

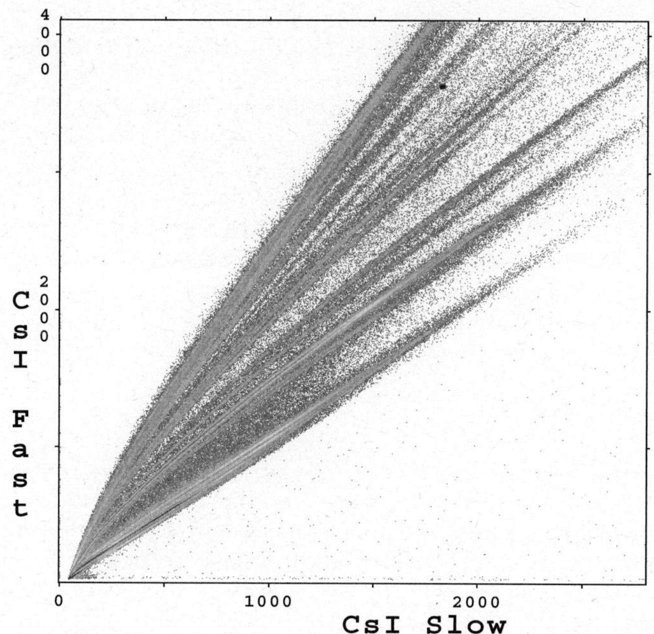

Fig. 1. Fast-slow INDRA CsI matrix obtained by accumulating high-energy light charged particles coming from two reactions, ^{124}Xe + ^{112}Sn and ^{136}Xe + ^{124}Sn at $45\,A$ MeV, during the fifth INDRA campaign at Ganil. Isotopic identification is obtained for H, He, Li, Be and B produced during the collisions.

2.4 INDRA

Identification de Noyaux et Détection avec Résolution Accrues [4].

2.4.1 Main characteristics

- Large angular coverage: 90% of 4π.
- High granularity provided by 336 telescopes distributed on 17 rings from $2°$ to $176°$.
- Phoswich or Si-CsI telescopes: $2°$ to $3°$.
- Ionization chamber-Si-CsI telescopes: $3°$ to $45°$.
- Ionization chamber-CsI telescopes: $45°$ to $176°$.
- Low-energy thresholds, around $0.8\,A$ MeV for $Z \leq 12$ and $\simeq 1.3\,A$ MeV above.
- Charge identification for all Z.
- Mass identification for $1 \leq Z \leq 4$, see fig. 1 (to be extended to $Z \simeq 10$ with optimum electronic gain and software, in progress).

2.4.2 Main advantages

INDRA is one of the first 4π multidetectors of the second generation dedicated to multifragmentation studies. The large solid angular coverage of INDRA, 90% of 4π, can be seen in fig. 2. Alpha-particle centre-of-mass $v_{parallel}$-$v_{perpendicular}$ plots are shown for different dissipation regimes. Starting from peripheral collisions (upper-left panel), we clearly see Coulomb rings caracterisctic

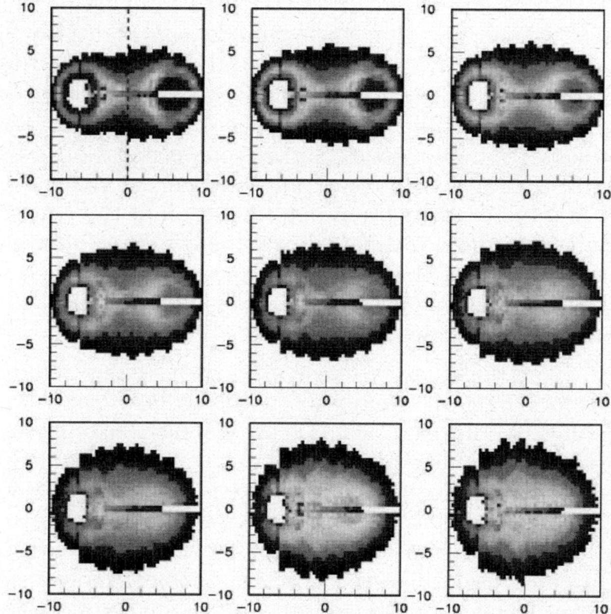

Fig. 2. Alpha-particle centre-of-mass $v_{parallel}$-$v_{perpendicular}$ plots for different dissipation regimes measured with the INDRA multidetector in Xe + Sn collisions at $80\,A$ MeV. Peripheral to central collisions are displayed from upper-left to bottom-right panel (M. Pichon, PhD Thesis, Caen University (2004)).

of an evaporation regime. At the opposite, for more central collisions (bottom-right panel), a quasi-homogeneous emission is seen in the centre of mass.

Over more than ten years, INDRA has proved its reliability during four campaigns of measurements at Ganil-France, plus a campaign conducted at GSI-Germany. It has also proved its versatility, as it was coupled with the first CHIMERA ring in 1997, it was associated to other telescopes for time-of-flight measurements, position-sensitive telescopes for crystal-blocking experiments on fission lifetimes of super-heavy elements, and it is scheduled to be moved and coupled to the Vamos spectrometer at Ganil for residue identification of compound nuclei. Symmetric as well as asymmetric systems, in both reverse and direct kinematics, for projectile energies as small as $\simeq 5\,A$ MeV and as large as $1\,A$ GeV (for C beams at GSI) have been recorded since 1993. We have learnt, during all these experiments, that if INDRA is well suited for studying central collisions of symmetric systems, nevertheless this detector has some drawbacks concerning asymmetric systems where the centre-of-mass velocity is small (Ni + Au $32\,A$ MeV, C + Sn). In these cases the lack of time-of-flight measurements (*i.e.* very low-energy thresholds) for very slow fragments leads to a reduced detection quality (especially at backward angles).

2.4.3 Main INDRA results

INDRA is able to perform physics analyses as different as the study of vaporization of a quasi projectile (Ar + Ni), studies of fission products (U + U, Ta + Au, Au + Au), central collisions leading to fusion-like sources (Xe + Sn, Ni + Au, Ni + Ni), de-excitation studies on a large scale of dissipation for hot nuclei (mainly quasi projectile), evidence of mid-rapidity emission, neck, ... Because of its good detection properties, intra-event correlations were studied leading to various results in many domains: impact parameter estimation, determination of the reaction plane for spin measurements, calorimetry event by event, reconstruction of hot primary fragments. With such an efficient tool all main subjects of heavy-ion collisions at intermediate energies were covered:

– Phase transition: with temperature measurements (caloric curve), studies of negative heat capacities, abnormal fluctuations (Delta scaling), scaling laws (Fisher's scaling, critical exponents, ...), bimodality of an order parameter, spinodal decomposition, phase diagram, random break-up, etc.
– Reaction mechanism: mid-rapidity emission, neck formation, fusion-like events, fission, momentum transfer, frag-ment-formation mechanism, isospin equilibration, chrono-meter of the fragmentation process, ...
– De-excitation of hot nuclei: from evaporation to vaporization, passing through fission, multifragmentation of any kind, ...
– Comparison with models: statistical models (MMMC, SMM, MMM, Simon, Gemini), dynamical models (QMD, AQMD, CNBD, HIPSE, BNV, BUU, stochastic mean-field approach (BOB)), lattice gas, EES, etc.

2.4.4 To go forward

With the increasing availability in the near future of radioactive nuclear beams (Spiral, Spiral II, Eurisol), the role of the isospin degree of freedom in nuclear reactions will be studied. Such a study will require a new generation of 4π multidetectors keeping all the qualities of the present generation (low-energy thresholds, granularity) and representing a step forward in terms of isotopic identification (both charge and mass for nuclei up to $Z \simeq 30$), always a better granularity for more precise intra-event correlation functions and also a coupling with neutron detectors (see recommendations from NuPECC long-range plan 2004). A French-Italian group has been working on this subject during the last four years. Its main goal is an R&D program on the feasibility of such a new detector. The name of this project is FAZIA (Four π A-Z Identification Array). For this goal, NTD silicon detectors, CsI scintillators and associated digital electronics are being tested. A report on the feasibility of such a new project is in progress.

2.5 CHIMERA

Charged Heavy Ion Mass and Energy Resolving Array [5].

2.5.1 General characteristics

– 9×2 rings in the range $1° \leq \theta \leq 30°$.
– The sphere: 17 rings in the range $30° \leq \theta \leq 176°$.
– For an amount of 1192 Si-Csi(Tl) telescopes with time-of-flight measurements.
– Very high granularity and efficiency: 94% of 4π.
– Mass identification at very low energy $< 0.3\,A$ MeV for heavy ion (TOF).
– Z and A identification by ΔE-E for $Z \leq 9$.
– Z and A identification for high-energy light charged particles (using fast-slow components on CsI(Tl) signals) $Z \geq 5$.
– Z identification up to beam charge for $Z > 9$ (ΔE-E).

2.5.2 Physics goals

CHIMERA is the last example of second-generation 4π detector. The high granularity of the detector, its high solid angular coverage, its low-energy thresholds allow to work in different experimental conditions, reverse or direct kinematics, symmetric or asymmetric systems, small or heavy nuclei (target-projectile).

Thanks to its good capabilities, many aspects of physics in heavy-ions collisions can be studied [6]: starting from dynamical aspects (fragment formation from projectile, target or neck emission, time scale in neck fragmentation, mid-velocity emission, pre-equilibrium effects, isospin equilibration in time, ...), thermodynamical characteristics (exploration of the phase diagram in temperature, excitation energy, density, volume, isospin, signals of phase transition in hot nuclear matter, isoscaling in multifragmenting sources, ...) as well as prospective experiments (Bose condensates, search for alpha-particle condensates in hot diluted nuclei).

2.5.3 CHIMERA-PS —upgrading

In the near future (2006) an upgrade of the CHIMERA electronics will appear. The new method consists in the rise time measurement for pulse-height application. It results in a charge identification up to $Z \simeq 10$ with a $\simeq 4\,A$ MeV energy threshold for particles stopped in silicon detectors, see fig. 3. It will be coupled with time-of-flight identification that gives both A and Z for low-energy light fragments.

2.6 GARFIELD

General ARray for Fragment Identification and for Emitted Light particles in Dissipative collisions [7].

2.6.1 General characteristics

– High granularity (400 ΔE-E telescopes $\theta \simeq 4°$–150°).
– Low-energy thresholds (ionization chambers as ΔE).
– A and Z identification ($1 \leq Z \leq 8$) up to $\theta \simeq 90°$.
– Digital electronics for pulse-shape discrimination.

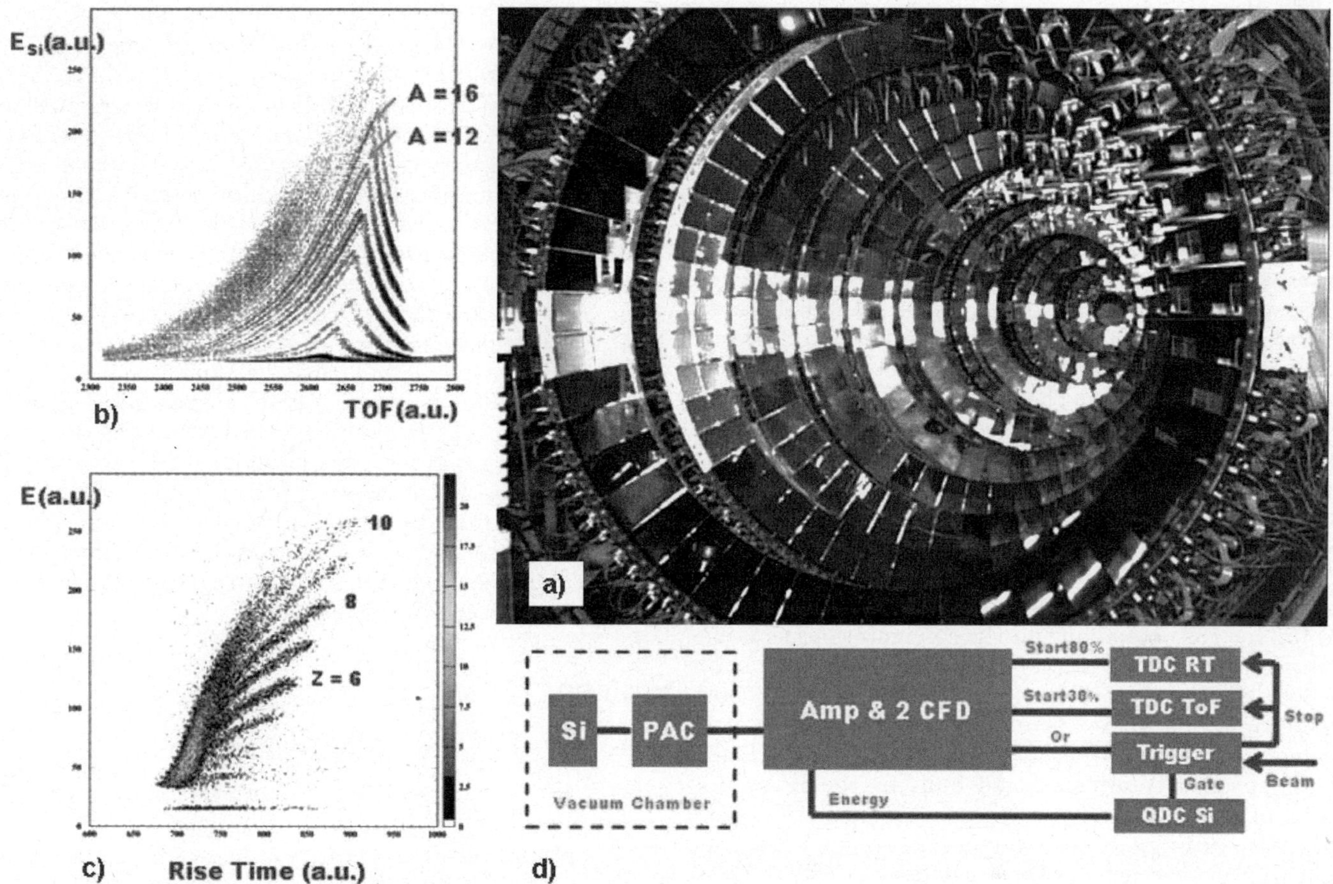

Fig. 3. a) Forward part of the CHIMERA detector (as seen by the target). b) Energy plot from silicon detector *vs.* time of flight for reaction products of Ne + Al at $20\,A$ MeV (detected at $\theta = 12°$ and at $180\,\mathrm{cm}$ from the target). Charged particles stopped (low kinetic energy) in the first stage silicon detector or (alternatively) stopped in the second stage CsI(Tl) crystal are seen in the plot. c) Energy plot *vs.* signal rise time showing the charge identification obtained with the PSD only for charged particles that are stopped in the silicon detector. d) Upgraded electronics chain of CHIMERA for the PSD analysis with silicon detectors.

2.6.2 The GARFIELD drift chamber

- 180 double-stage E (CsI(Tl))-ΔE (Multi segmented gas chamber) telescopes.
- Angular coverage: $(30°–85°, 95°–150°)$.
- Charge resolution from proton to heavy ions, with $\Delta Z/Z = 1/28$.
- Angular resolution ($\Delta\theta = 1°$, $\Delta\phi = 7.5°$).

2.6.3 Scientific goals

Installed at the LNL Legnaro Italy, GARFIELD is used by the Nucl-ex Collaboration whose main interest is the study of the dynamics and thermodynamics of reactions at low-to-medium energy. It is designed to detect and identify with low-energy thresholds both light charged particles and heavy fragments. It is based on a gas drift chamber which conveys primary ionization electrons on gas microstrip devices where multiplication occurs and the energy-loss signals are generated. Silicon detectors or CsI(Tl) crystals operate as residual-energy detectors. This detector can be coupled to other systems like the MULTICS phoswich scintillators [8] or Hector (8 large BaF$_2$ detectors) according to the physic (mechanism) one wants to study.

2.7 FIASCO

Florentine Initiative After Superconducting Cyclotron Opening [9].

2.7.1 General characteristics

- 24 positive position-sensitive gas detectors, Parallel-Plate Avalanche Detectors (PPADs), for velocity-vector determination of heavy fragments $A > 20$.
- 96 silicon telescopes for the measurements of projectile-like fragment products (energy and charge).

- 158 phoswich modules for light charged particles and small fragments $3 \leq Z \leq 20$.
- Time-of-flight measurements.

2.7.2 Main goals and topics

The FIASCO multidetector is a low-threshold apparatus, optimized for the investigation of peripheral to semicentral collisions in heavy-ion reactions at Fermi energies. It consists of three types of detectors. The first detector layer is a shell of 24 position-sensitive Parallel-Plate Avalanche Detectors PPADs, covering about 70% of the forward hemisphere, which measure the velocity-vectors of heavy ($Z \geq 10$) reaction products. Below and around the grazing angle, behind the most forward PPADs, there are 96 ΔE-E silicon telescopes (with thickness of 200 and $500 \, \mu$m, respectively); they are mainly used to measure the energy of the projectile-like fragment and to identify its charge and, via the time of flight of the PPADs, also its mass. Finally, behind most of the PPADs there are 158 (or 182, depending on the configuration) scintillation detectors, mostly of the phoswich type, which cover 25–30% of the forward hemisphere; they identify both light charged particles ($Z = 1$–2) and intermediate-mass fragments ($3 \leq Z \leq 20$) measuring also their time of flight. It was specifically designed and built by the Heavy-Ion Group of the INFN and the Department of Physics of the University of Florence for studying non-central collisions in heavy-ion reactions at Fermi energies (15–40 A MeV) with the beam delivered by the superconducting cyclotron of the Laboratori Nazionali del Sud, LNS in Catania, Italy.

2.8 NIMROD

Neutron Ion Multidetector for Reaction Oriented Dynamics [10].

2.8.1 General characteristics

The charged-particle detector setup is composed of:

- 176 CsI(Tl) arranged in 13 rings:
  - 6 rings of 12 detectors in the range $3.5°$–$27.8°$;
  - 2 rings of 24 detectors in the range $28.7°$–$45°$;
  - 2 rings of 16 detectors in the range $48.3°$–$75°$;
  - 1 rings of 8 detectors in the range $75°$–$105°$;
  - 1 rings of 8 detectors in the range $105°$–$135°$;
  - 1 rings of 8 detectors in the range $135°$–$170°$.
- For the four forward rings, 20 silicon telescopes composed for each ring by
  - $2 \times (150 + 500$ microns) (Super Telescopes) isotopic resolution for $Z \leq 8$;
  - 1×150 microns (Z identification for all particles ΔE-E technique);
  - 2×300 microns (Z identification for all particles ΔE-E technique)
  placed in front of 5 CsI scintillators in each ring.

- 96 ionization chambers.

NIMROD is a 4π neutron and charged-particle detection system built at Texas A&M to study reaction mechanisms in heavy-ion reactions. It is used to select collisions according to their violence. Dynamic and thermodynamic information is derived from the observed multiplicities, energies and angular distributions of particles and fragments produced in the nuclear reaction. Neutrons are detected using a liquid scintillator which is contained in vessels around the target. The charged-particle detectors are composed of ionization chambers, silicon telescopes and CsI(Tl) scintillators covering angles between 3 and 170 degrees. These charged-particle detectors are placed in a cavity inside the revamped TAMU (Texas A&M University) neutron ball. To minimize the cost of the detector, the "INDRA geometry" was adopted in the forward direction. The forward rings of detectors thus cover polar angles from 3.6 to 45 degrees in 8 concentric rings of 12 CsI(Tl) detectors each. This arrangement completes the already existing TAMU CsI Ball array which is used almost "as is" from 45–170 degrees.

2.9 FOPI

FOur PI (π) detector for charged particles [11].

2.9.1 General characteristics

- Superconducting solenoid.
- Forward plastic wall.
- Forward drift chamber.
- Central drift chamber.
- Plastic barrel.

The FOPI detector has documented the investigation of the fragments and the particles produced in central heavy-ion collisions. This detector detects, identifies, and determines the momentum of all charged particles emitted in a heavy-ion reaction. Based on the modular combination of different detector systems, the total assembly achieves its goal of covering the complete range of beam energies (0.1–2 GeV per nucleon) made available by GSI's SIS heavy-ion synchrotron. The target is located within a superconducting coil, which produces a magnetic field of 0.6 tesla. Charged particles within this field travel along curved paths before passing through the drift chambers. These chambers register both the particle track and the energy loss suffered by the particle in its passage through the detector gas. In addition, lines have been drawn through connected hits, which have been recognized by the automatic track-recognition of the event. Further on through the detector, the majority of the particles land in scintillation counters, with which their flight time from the target can be determined. The combination of different measurements enables the unambiguous identification of the particles.

The FOPI detector's potential ability to identify particles is, however, not restricted to charged particles. Although neutral particles do not themselves leave signals in the detector system, some of them can be recognized and reconstructed thanks to their decay. This interesting class of particles includes neutral particles containing a strange quark, such as the neutral kaon and the Λ-particle. Due to their strangeness, they are relatively long-lived, decaying after a flight of several centimeters into charged particles, which leave tracks in the detectors. As the momenta and particle types are already known, the calculation of the invariant masses and comparison with the rest masses of the candidate particles enable the direct determination of which tracks belong together.

2.9.2 Main results

The FOPI detector can simultaneously detect all the charged and some of the neutral particles produced in a heavy-ion reaction. Global correlations among the particles are thus possible. Also known as collective effects, these correlations provide meaningful signatures for the properties of nuclear matter. Main results concern: collective expansion, stopping and directed sidewards flow at the highest energies, the latter of which provides interesting information, particularly in the case of strange particles, on interactions with the surrounding matter (in-medium effect) [12].

Collective expansion
An interpretation of the FOPI results leads to the conclusion that the observed fragments arise from an expanding flow of matter, from which they all carry the same velocity component into the detector. Although the density is not directly accessible from the measurements, comparison of the spectra and the rapidity-density distribution has been used with the predictions of so-called transport models to determine the change in the density over time and the maximum density reached. For heavy systems, a beam energy of 1 GeV per nucleon produces approximately the double density of the ground state. Variation of the beam energy and of the projectile-target combination enables the region between one and 2.5 times normal density to be covered.

Directed sidewards flow
The question of how generated particles, in particular vector mesons and kaons, behave under conditions of high temperature and density was investigated. It would appear that the observed probabilities of production of the antikaons in particular can currently only be described theoretically if it is assumed that particle masses are lower in-medium. Another window on possible "in-medium properties" of kaons is founded upon the directed sidewards flow. As nuclear matter is extremely difficult to compress, the particles attempt to get out of the way. Theoretically one expects a change of hadron properties in hot and dense nuclear matter. Thus, strange particles are ideal probes for in-medium effects. FOPI capabilities in this detection domain have brought strong information in this field.

Stopping
Rise and fall of the stopping properties in nuclear matter

was investigated both with incident energy and its size dependence. It was found that stopping is maximal around $\simeq 400\,A$ MeV and decreases toward higher beam energies. Stopping increases also with system size (Au + Au as compared to Ca + Ca), and systems exhibit always transparency, which increases with higher beam energy. It was also shown that stopping in nuclear matter is correlated with flow and pressure measurements.

2.10 MEDEA

Multi Element DEtector Array for γ-rays and light-charged-particles detection [13].

2.10.1 General characteristics

- 180 BaF2 scintillator crystal detectors.
- 120 plastic phoswich detectors.
- Spherical geometry around the target.
- Angular coverage: 90% of 4π.
- Gamma-ray detection up to $\simeq 300\,$MeV.
- + Light-charged-particle detection.

MEDEA's basic configuration consists of 180 barium fluoride scintillator crystals, arranged in the shape of a ball, plus a forward-angle wall of 120 phoswich detectors. The inner radius of the ball (22 cm) and the distance of the wall from the target (55 cm) allow the placement of other detectors in the inner volume. MEDEA was first installed at Ganil (France) in 1989-1993 and then moved to the Laboratori Nazionali del Sud in Catania (Italy). Coupled with other detectors like Multics [8], MEDEA has given its contribution to intermediate-energy physics, both in the field of fragment production and in pre-equilibrium production of particles and γ-rays. The hot giant dipole resonance GDR has been also investigated taking advantage of the coupling with the superconducting solenoid SOLE and its focal-plane detector MACISTE (Mass And Charge Identification Spectroscopy with TElescope (gas chamber for ΔE-wire chamber for x-y position and TOF-plastic scintillator for E)) [13] installed at the LNS-Catania.

3 Auxiliary detectors

3.1 LASSA

Large-Area Silicon-Strip Array.

3.1.1 General characteristics

- Highly segmented Si(65 μm)-Si(0.5–1.5 mm)-Csi(Tl) 4–6 cm read out by photodiode.
- Each silicon segmented in 16–32 strips covered by 4 CsI.
- 10% thickness variation in ΔE.

– Segmentation helps both particle identification and angular resolution.
– Angular resolution $\Delta\theta \simeq 0.8°$.
– 9 to 20 telescopes.
– Thresholds: $2\,A$ MeV for ^{4}He, $4\,A$ MeV for ^{12}C.
– Isotopic identification for $Z \leq 9$.

LASSA consists of 9 individual telescopes which may be arranged in a variety of geometries. This array was built to provide isotopic identification of fragments ($Z < 10$) produced in low- and intermediate-energy heavy-ion reactions. In addition to good isotopic resolution, it was essential to provide a low threshold for particle identification as many of the fragments emitted in these reactions are low in energy.

Each LASSA telescope is composed of a stack of two silicon strip detectors followed by 4 CsI(Tl) crystals. The silicon which faces the target is 65 microns thick while the second silicon is 500 microns thick. Both silicons are ion-implanted passivated detectors, Si(IP). The 65 micron silicon wafer is segmented into 16 strips which are read out individually. The 500 micron wafer is segmented into 16 strips on its junction (front) side while the Ohmic surface (rear) is segmented into 16 strips in the orthogonal direction. Collection of holes and electrons in orthogonal directions provides two-dimensional position sensitivity from this detector alone. The additional position information from the 65 micron detector is used as a redundancy check. The pitch of the detector is nominally 3 mm with a 100 micron inter-strip gap. Behind the silicon detectors are 4 independent 6 cm CsI(Tl) crystals to stop penetrating particles. Scintillation caused by ionizing particles impinging on these scintillators is detected by 2 cm × 2 cm photodiodes (PD). The signals from the PD are amplified by pre-amplifiers mounted in the detector housing.

Each detector has 16 strips and an area of 5 cm by 5 cm. The detector set consists of a variety of ΔE and E detectors. The real ΔE detectors are 65 microns thick and are all one-sided for the readout. A set of 9 detectors 500 microns thick are double sided in readout. Finally, a set of 6 detectors are 1000 microns thick and are one sided. In the high-energy applications the 65 and 500 microns detectors are used backed with thick CsI(Tl) scintillators. For the Gammasphere applications only four of the 65 and 1000 microns telescopes are used. LASSA has been used in experiments at NSCL (Michigan State University), ATLAS (Argonne National Laboratory), and at the Cyclotron Institute (Texas A&M University).

3.2 HiRA

High Resolution Array [14].

3.2.1 General characteristics

– 20 telescopes Si($65\,\mu$m)-Si(1.5 mm)-CsI(Tl) read out by photodiode.
– For an amount of 1920 strips.

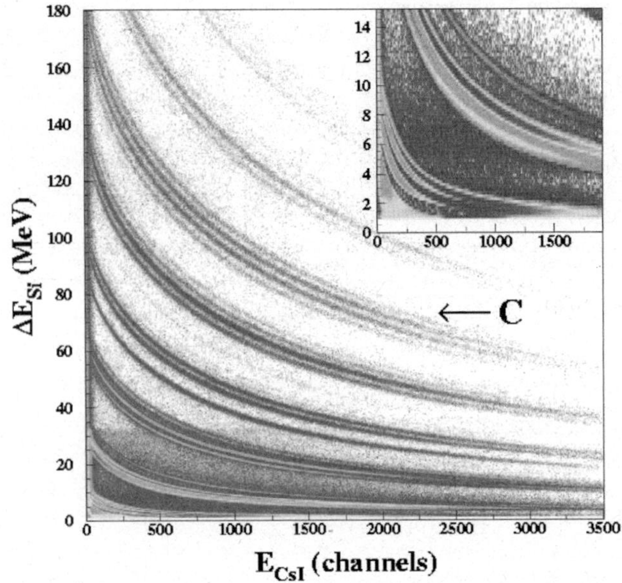

Fig. 4. An example of a ΔE-E matrix obtained with the LASSA-HIRA setup. The resolution for $Z = 1, 2$ is shown in the insert.

– Highly configurable for different experiments.

HiRA is an array of 20 telescopes with approximately three times the geometric efficiency of LASSA. This array is capable of addressing a wide range of physics goals including resonance spectroscopy, transfer reactions with radioactive ("exotic") beams, and reaction dynamics related to the equation of state. It was designed to develop a large solid-angle (high-efficiency) array of Si-Si-CsI(Tl) telescopes with high angular and energy resolution for radioactive-beam studies. The principal physics objectives of HiRA are elastic- and inelastic-scattering measurements with radioactive beams (important measurements in astrophysics), resonance spectroscopy, isospin dependence of the nuclear equation of state, and studies of nuclear-reaction dynamics. The design and construction of HiRA was built upon the experience in the design, construction, and performance of the LASSA array.

HiRA consists of 20 telescopes based upon a Si-Si-CsI(Tl) stack for identifying charged particles in Z and A by the ΔE-E technique. An example of a ΔE-E matrix for the Hira setup is presented in fig. 4. High isotopic resolution is obtained from H up to F. Each telescope is constructed from a 65 micron strip Si(IP) detector backed by a 1.5 mm silicon strip detector. The 65 micron detector is single-sided (32 strips on the junction (front) side only), while the 1.5 mm detector is segmented in 32 strips on the junction side with 32 orthogonal strips on the Ohmic (rear) surface. Thus the 1.5 mm will provide two-dimensional position information. The detectors have a 2 mm pitch with a 25 or 40 micron interstrip gap (junction and Ohmic surfaces, respectively).

4 Light ions

Bombarding light relativistic projectiles (π, p, C, ...) on much heavier nuclei allows to investigate the multifragment decay of a heavy target. This kind of study gives complementary information to that obtained from heavy-ion collisions and the comparison allows one to extract the influence of compression and rotation on the multifragment decay. Indeed for such collisions few compressional effects are expected as regards to central heavy-ion collisions. We therefore hope to explore a different portion of the phase diagram of hot nuclear matter.

4.1 ISIS

Indiana SIlicon Sphere.

4.1.1 General characteristics

- 162 individual telescopes covering 74% of 4π.
- Gas ionization chamber, $500\,\mu$m Si - 28 mm thick CsI(Tl) read out by photodiode.
- Each telescope measures Z, A, E and θ.
- Charge identification for Z with $0.6 \leq E/A \leq 96$ MeV.
- Isotopic separation for $Z \leq 4$ for $E/A \geq 8$ MeV.

The Indiana Silicon Sphere detector array is based on a spherical geometry, designed primarily for the study of light-ion–induced reactions. It consists of 162 triple telescopes, 90 in the forward hemisphere and 72 in the backward hemisphere, covering the angular ranges from 14° to 86.5° and 93.5° to 166°. The design consists of eight rings, each composed of 18 truncated-pyramid telescope housings. To increase granularity for the most forward angles, the ring nearest to 0° is segmented into two components. Each telescope is composed of: a gas-ionization chamber operated at 25–40 torr of CF4 or 12–20 torr of C3F8; a $500\,\mu$m ion-implanted passivated silicon detector, Si(IP), and a 28 mm thick CsI(Tl) crystal with light guide and photodiode readout. The telescope dynamic range permits measurement of $Z = 1$–20 fragments with discrete charge resolution over the dynamic range $0.6 \leq E/A \leq 96$ MeV. The Si(IP)/CsI(Tl) telescopes also provide particle identification (Z and A) for energetic H, He, Li and Be isotopes (E/A 8 MeV). The Si(IP) detectors constitute a critical component of the array in that they provide both excellent energy resolution and reliable energy calibration for the gas ionization chamber and CsI(Tl) elements.

4.2 FASA

4.2.1 General characteristics

- 55 thin scintillator crystals CsI(Tl).
- 5 time-of-flight telescopes.
- A large-area position-sensitive parallel-plate avalanche detector (PPAD).

The FASA setup [15], installed at the JINR synchrophasotron providing light ion beams with energies up to 3.65 GeV/nucleon, is a fragment multiplicity detector, consisting of 55 scintillation counters made of thin CsI(Tl) films, five time-of-flight telescopes and a large-area position-sensitive parallel-plate avalanche chamber. The basic aim of the device is to determine with high precision the energy, mass, and velocity of the fragments detected in the time-of-flight telescopes (TOF) while for the other fragments global multiplicity information is obtained. Therefore, the TOF telescopes serve as a trigger. In addition, angular correlations and distributions and relative velocity correlations for coincident fragments can be measured. The FASA setup was also upgraded to FASA-2. Light-charged-particle (LCP) multiplicity detectors have been developed (64 plastic scintillator counters in the forward hemisphere). This gives the possibility to select the events according to impact parameter. A new detector system (25 ΔE(gas)-E(Si) telescopes) was made. It gives the better possibility for measuring IMF-IMF angular (and relative velocity) correlations, which are important for the time-scale study. This system is used also for triggering the FASA setup.

4.2.2 Main physics goals

In the FASA project the light relativistic projectiles from protons to carbon are used to investigate the multifragment decay of a heavy target. The study gives complementary information to that obtained from heavy-ion collisions, and the comparison allows one to extract the influence of compression and rotation on the multifragment decay.

5 Projectile decay fragmentation

Thanks to the forward focusing of the product emission in the laboratory frame (even better with increasing incident energy and reverse kinematic), many studies were performed on the decay of hot projectiles. Indeed in this case the complete angular coverage is not necessary as only the forward part is fired, reducing the number of detectors (and even their type as detection thresholds are not critical here) and associated electronics and thus the cost.

5.1 ALADIN

5.1.1 General characteristics

- Magnet.
- TP music.
- Time-of-flight wall.
- + coupling with LAND (Large Area Neutron Detector) [16].

5.1.2 Main ALADIN results

The ALADIN spectrometer, coupled with other detectors like the MINIBALL/wall [2], LAND [16] or Hodoscopes (for example, the Catania SIS Hodoscopes), was optimized for inverse-kinematics studies, namely projectile fragmentation at energies between 100 and 1000 A MeV: Au + C, Al, Cu, Pb at 600 A MeV, Au + Au at 100-250-400-1000 A MeV, Xe, Au, U + Be at 600-800-1000 A MeV [17]. Many aspect of multifragmenting projectiles were revealed, like the rise and fall in the fragment emission multiplicity with dissipation, collective expansion in central collisions, the universality behaviour of spectator fragmentation, the caloric curve of hot nuclei (T $vs.$ E^* which shows the transition from a Fermi gas at low energy to a Boltzmann gas at higher energy), temperature measurements in exploding nuclei, break-up densities, ...

5.2 MULTICS

5.2.1 General characteristics

- 3 layers telescopes.
- Silicon 500 μm position sensitive.
- CsI(Tl) + photodiode.
- Angular coverage $3° \leq \theta \leq 25°$.
- Energy threshold $\simeq 1.5\,A$ MeV.
- Z identification up to the beam charge.

5.2.2 Main MULTICS results

Coupled with the MSU MINIBALL detector [2], experiments performed with the MULTICS-MINIBALL had a geometrical acceptance of 87% of 4π. Light charged particles and fragments with charge up to the beam charge were detected at θ_{lab} from 3° to 25° by the MULTICS array [8], with an energy threshold of about 1.5 A MeV, nearly independent of fragment charge. Light charged particles, $Z = 1$ and 2 isotopes and fragments with charge up to $Z \simeq 20$ were fully identified by 160 phoswich detector elements of the MSU MINIBALL, covering the angular range from 25° to 160°. The charge identification thresholds were about 2, 3, 4 A MeV for $Z = 3$, 10 and 18, respectively.

This setup device was mainly devoted to thermodynamical studies on the Au + Au system at 35 A MeV [18] and the search for signals of phase transition in nuclear matter: critical behaviour inside the coexistence region (critical exponent), caloric curve, exploration of the phase diagram of hot nuclear matter, negative heat capacity (associated to abnormal kinetic-energy fluctuations), scaling laws (Fisher's scaling), ...

5.3 FAUST

Forward Array Using Silicon Technology [19].

5.3.1 General characteristics

- 68 detectors: $1.6° \leq \theta \leq 33.6°$.
- 300 μm Si - 3 cm CsI(Tl) read out by photodiodes.
- Silicon are single area edge mounted.
- Isotopic resolution for $Z \leq 6$.
- Pulse-shape discrimination in CsI for light charged particles.

The FAUST detector was used for studying fragmentation of projectiles excited via peripheral interactions with heavy targets. This requires a forward array with relatively high granularity and good solid-angle coverage. Isotopic resolution of light fragments is also an additional dimension for these studies. FAUST is an array of 68 detector telescopes (Si-Csi(Tl) read out by photodiodes) arranged in five rings that are squares projected onto spherical surfaces. Ring 1 contains 8 telescopes and covers laboratory angles in the range $1.6° \leq \theta \leq 4.5°$. Ring 2 contains 12 telescopes and covers the range $4.6° \leq \theta \leq 8.7°$. Ring 3-5 each contain 16 detectors and cover angles in the ranges $8.8° \leq \theta \leq 14.1°$, $14.3° \leq \theta \leq 22.4°$ and $22.6° \leq \theta \leq 33.6°$. The mounting structure of each ring is hidden behind the active area of the ring in front of it. This combination provides a very good geometric coverage. The solid angular coverage from 2.3° to 33.6° is 89.7%. Isotopic resolution is obtained from $Z = 1$ up to charge $Z \simeq 6$ thanks to different identification techniques, namely ΔE-E for $Z \geq 2$ and pulse-shape discrimination applied to the signal coming from the CsI(Tl) for $Z \leq 3$.

5.3.2 Main FAUST results

The fragments resulting from the fragmentation have a source with a different N/Z than the initial beam. The N/Z present in the fragmenting system is on average not equal to the N/Z of the initial beam; there is a shift toward the valley of stability. Moreover there is a distribution in the N/Z of the fragmenting system, which appears to be isospin dependent. The overall dependence of the excitation energy on $(N/Z)_{QP}$ increases as the N/Z of the beam increases. Production of neutron-rich nuclides decreases with increasing excitation energy [20].

5.4 FIRST

Forward Indiana Ring Silicon Telescopes.

5.4.1 General characteristics

FIRST is a set of three annular Si-CsI(Tl) telescopes for the detection of the projectile-like fragment and forward-focused charged products. This array coupled to existing arrays such as LASSA is being built to better understand neck fragmentation and the projectile-fragmentation of radioactive ("exotic") beams. FIRST is intended to study the processes of neck and projectile fragmentation that

Table 1. Parameters of the three types of telescopes used in the FIRST array.

Designation	First element	Second element	Third element	No. of concentric rings	No. of pies
T1: 2.0°–7.0°	200–220 μm Si(IP)	1 mm Si(IP)	2 cm CsI(Tl)-PD	48	16
T2: 7.3°–14.5°	300 μm Si(IP)	2 cm CsI(Tl)/PD		16	16
T3: 15.0°–28.0°	300 μm Si(IP)	2 cm CsI(Tl)/PD		16	16

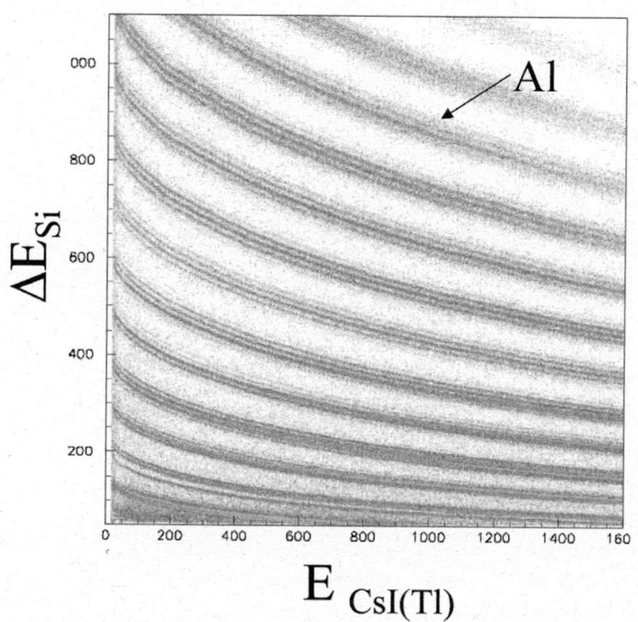

Fig. 5. A ΔE-E matrix for a FIRST telescope. Isotopic identification is performed up to Si, at least.

occurs for peripheral and mid-central collisions between two heavy ions.

FIRST consists of 3 telescopes based upon either a Si-Si-CsI(Tl) stack or just a Si-CsI(Tl) stack for identifying charged particles in Z and A by the ΔE-E technique. An example for the good isotopic resolution obtained with a First telescope over a wide range of elements is given in fig. 5. The most forward telescope is designated T1 and the more backward telescopes are T2 and T3, respectively. The parameters of the three telescopes are summarized in table 1.

5.5 ARGOS

5.5.1 General characteristics

The ARGOS multidetector [21] consists of 111 elements. Each element is a BaF2 crystal, 5 or 10 cm thick and of hexagonal shape, with a surface of 25 cm^2. An upgrade can be obtained just putting in front of the crystal a foil of plastic scintillator. In detecting a particle, the photo-multiplier signal is charge-integrated by two different gates, 20 and 300 ns wide, respectively. These fast and total (or slow) components together with the time-of-flight information allow charge identification for all detected particles, $Z < 30$, and mass identification for light

charged particles and light ions. Due to the fast light response, timing characteristics of the detector are excellent, and time resolutions as low as 300 ps have recently been obtained, comprehensive of the beam burst width. Therefore, the neutron detection is also possible, with a measured efficiency between 5% and 20%, depending on the crystal thickness, neutron energy and electronic threshold. The elements can be assembled in different ways. In a typical geometry, the ARGOS "eyes" are distributed: 60 in a forward and 36 in a backward wall, both honeycomb shaped. As an example, for the first wall and for a distance of 2.35 m, all the particles in the angular range between 0.75° and 7° are detected. At the same time, for the backward wall at a distance of 50 cm, the angular range between 160° and 177° is covered.

6 High-resolution magnetic spectrometers

Full isotopic identification of the reaction residues over the whole mass range can be achieved by the use of high-resolution magnetic spectrometers. Experimental programs on nuclear-reaction dynamics have been performed at essentially three facilities, which will be listed in the following subsections. All three spectrometers have a common basic design, consisting of two stages with an intermediate dispersive and a final achromatic image plane. The projectile energies range from around 20 A MeV to 1 A GeV.

6.1 MARS recoil separator at Texas A&M University

6.1.1 General characteristics

The K500 superconducting cyclotron delivers beams with maximum energies from 20 to 50 A MeV, depending on the projectile mass. The MARS magnetic spectrometer [22] has an energy acceptance of ±9% and an angular acceptance of 9 msr. It can be positioned at angles between 0° and 30° with respect to the beam axis. Nuclide identification is performed by energy loss, residual energy, time of flight, and magnetic rigidity. ToF and position are measured by 2 PPACS, ΔE and E by a Si detector telescope. Important part of the ions with $Z > 12$ are not completely stripped, and thus the ionic-charge-state distribution has to be determined. The angular distribution up to 30° and the full momentum distribution of the reaction products can be determined by combining different measurements with different positions of the spectrometer and different magnetic fields.

6.1.2 Main results

The experiments with the MARS separator contribute to the institute's program in heavy-ion reaction dynamics and thermodynamics, investigating the properties and the decay modes of nuclear systems from low energy up to the limits of thermal and rotational stability, testing theories of many-body systems, chaotic-regime dynamics and the statistical mechanics of strongly interacting, finite quantum systems [23]. Systematic measurements in intermediate-energy heavy-ion collisions establish the degree of thermal, chemical and isospin equilibration, the mechanism of nuclear disassembly, the caloric curve and the mass and isospin dependence of limiting temperatures.

6.2 A1900 fragment separator at MSU, East Lansing

6.2.1 General characteristics

The coupled K500-K1200 cyclotrons deliver beams with maximum energies from 100 to 200 A MeV, depending on the projectile mass. The A1900 fragment separator [24] has a momentum acceptance of 5.5% and an angular acceptance of 8 msr. The heavier residues are fully transmitted. Nuclide identification is performed by Brho, ToF, and ΔE, which are measured by a scintillator, a PPAC, and a silicon detector telescope. Typically, reaction products with $Z < 30$ are fully stripped.

6.2.2 Main results

Systematic measurements of projectile fragments with full isotopic resolution for projectiles in the mass range around 50 to 60 have been performed, and their momentum distributions have been determined with the A1900 magnetic system [25]. The experiments are analyzed for testing and adapting dedicated model calculations of the nuclear-reaction process. A major aim is to optimize the production of exotic nuclei. Previously, a number of similar investigations had been performed with the A1200 system [26] at lower energies.

6.3 FRS magnetic spectrometer at GSI, Darmstadt

6.3.1 General characteristics

The heavy-ion synchrotron SIS18 delivers beams with maximum energies from 1 to 2 A GeV, depending on the projectile mass. The FRS magnetic spectrometer [27] has a momentum acceptance of $\pm 2.5\%$ and an angular acceptance of 15 mrad around the beam axis. This is adapted to full transmission of all heavier residues. For the lightest residues and fission fragments, the full momentum range can be obtained by combining different measurements, while the angular range is generally covered only up to 15 mrad. Nuclide identification is performed by Brho, ToF, and ΔE, which are measured by scintillation detectors and an ionization chamber. In figs. 6 and 7 we can see the good isotopic identification for all reaction products. Typically, they are fully stripped.

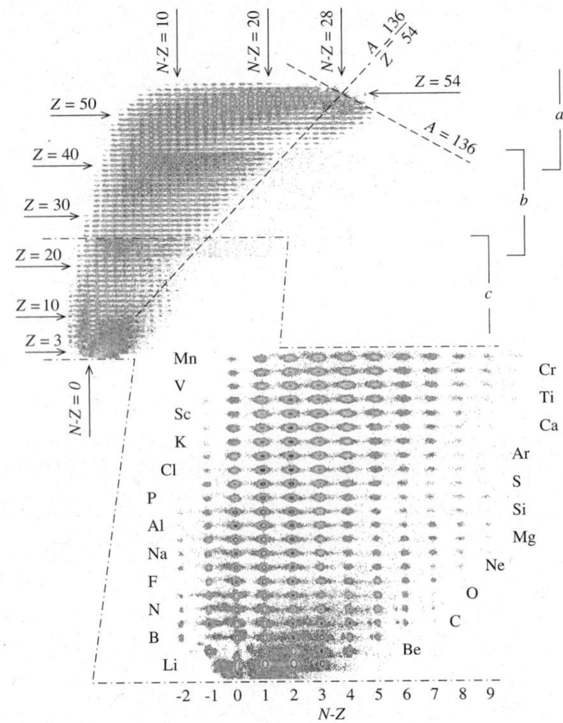

Fig. 6. Composition of all identified events measured with the FRS spectrometer with a ^{136}Xe beam at 1 A GeV on a hydrogen target. Three overlapping bands, a, b, c, correspond to the three groups of magnetic settings for the central isotopes ^{120}Ag, ^{69}Zn and ^{24}Al, respectively. The band c collecting light nuclides is enlarged in order to show the isotopic resolution (P. Napolitani, PhD Thesis, University Paris XI (2004)).

6.3.2 Main results

Dedicated experiments were performed on the identification of the projectile-like fragments and on their momentum distributions. The shift in the N/Z content during the evaporation cascade has been studied to determine the initial excitation energy introduced in the abrasion process [28] and to deduce the freeze-out temperature after thermal nuclear break-up [29]. The kinematical properties of projectile fragments produced in mid-peripheral collisions were introduced as a new access to study the nuclear equation of state, in particular to pin down the momentum dependence of the nuclear mean field [30]. This approach is complementary to the analysis of the flow pattern performed with 4π detectors up to now. A detailed analysis of the shape of the momentum distributions of the projectile fragments has been related to the decay characteristics of the excited spectators [31]. The charge-pickup channels provide information on the in-medium nucleon-nucleon cross-sections and the excitation of the Delta resonance [32].

6.3.3 Future projects

As the most prominent examples for next-generation magnetic systems, we concentrate on some new installa-

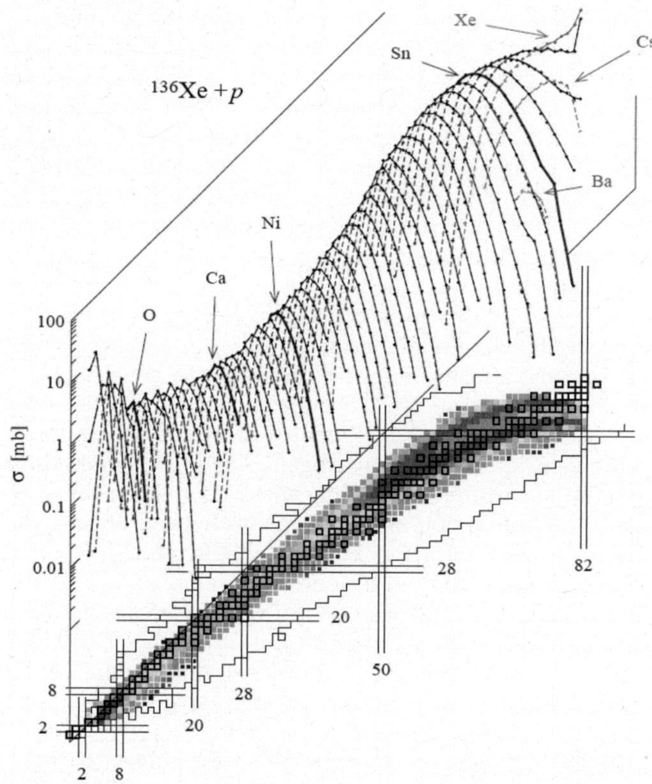

Fig. 7. Isotopic production cross-sections shown on a chart of nucleides for the reactions ^{136}Xe + p at 1 A GeV obtained with the FRS spectrometer (P. Napolitani, PhD Thesis, University Paris XI (2004)).

tions planned for the FAIR project [33]. The FAIR project currently represents a major step in the development of a new experimental installation, including more powerful magnetic spectrometers. The R3B project [34] aims for a large-acceptance dipole magnet with high resolution, which will combine isotopic resolution of all residues with simultaneous detection of light reaction products. Another important project is the Super-FRS [35], which follows the same ion-optical design as the present FRS with essentially larger acceptances in angle and momentum. Moreover, a pre-separator will facilitate the installation of additional detectors in the target area.

7 Neutron detection

In parallel to an always better charged-particle identification, it became rapidly evident that a lot of information was lost if neutron detection was missing. As neutron determination involves a different technique essentially incompatible with 4π detectors, since the coverage by silicon or scintillation detectors, plus their pre-amplifier, cooling system and so on, screen everything, it is very difficult to couple both charged particles and neutron detectors. Nonetheless, some experiments have tried to mix both techniques allowing compromise either on efficiency, granularity or particle identification and sometimes on intra-

event correlation. In many cases part of the neutron information was obtained on their energy and/or their multiplicity. The combination ALADIN-LAND is an exception, since the magnet deflects the charged particles from the direction of the LAND detectors.

7.1 Rochester SUPERBALL

7.1.1 General characteristics

- Gd-doped liquid scintillator.
- Geometrical flexibility.
- Neutron multiplicity measured.
- Angular distribution given by 5 segments.
- Possibility of 4π charged particles in coincidence.

The University of Rochester SUPERBALL is a five-segment, 16000 liter gadolinium-loaded liquid scintillator 4π neutron calorimeter surrounding a vacuum scattering chamber. It delivers, through its 52 photomultiplier tubes, trains of electronic pulses in response to multi-neutron events. These trains contain information on neutron multiplicity (number of pulses in the train, excluding the prompt train head signal) and total kinetic energy of the injected neutrons (intensity of the prompt head signal). By its geometrical design, the SUPERBALL delivers additionally event-by-event angular information on the neutron yield. Its overall efficiency for neutrons from a 252-Cf fission-neutron source is 90% for typical threshold settings. The SUPERBALL measures the multiplicity of neutrons very efficiently. In addition, it measures the kinetic energies of neutrons and their emission directions. It surrounds the reaction chamber from all sides with a 1–1.5 m thick layer of scintillating liquid.

7.2 The National Superconducting Cyclotron Laboratory (NSCL) neutron walls

7.2.1 General characteristics

The neutron walls [36] are two large-area (2 m × 2 m), high-efficiency, position-sensitive neutron detectors. Each wall consists of a stack of 25 glass cells filled with the scintillator liquid NE213, with which one can distinguish neutron from gamma-ray pulses by pulse-shape analysis. Each cell is two meters long and has phototubes at its ends. Light from an interaction in the liquid reaches the phototubes via total internal reflection. Each wall has its own carriage and can be positioned independently of the other. It is mainly devoted to extend our knowledge of the structure of exotic nuclei.

7.3 LAND

Large Area Neutron Detector [16].

7.3.1 General characteristics

- Overall dimension $2\,\mathrm{m} \times 2\,\mathrm{m} \times 1\,\mathrm{m}$ deep.
- 200 paddles of $200\,\mathrm{cm} \times 10\,\mathrm{cm}$ area $\times\, 10\,\mathrm{cm}$ deep.
- Each paddle 11 sheets of Fe $2.5\,\mathrm{mm}/5\,\mathrm{mm}$ thick.
- Each paddle 10 sheets of $5\,\mathrm{mm}$ thick plastic scintillator.
- Alternating planes in perpendicular dimensions.
- Position resolution $\pm 1\,\mathrm{cm}$.

The LAND detector can be coupled to other devices installed at GSI-Germany in order to study different aspects of heavy-ion collisions.

Multifragmentation
The measurement of neutrons emerging from such a process allows to determine the excitation energy of the pre-fragments. For that purpose, LAND was installed in combination with a number of fragment detectors installed at the magnetic spectrometer ALADIN [17].

Collective flow of nuclear matter
The flow of neutral nuclear matter in (semi-)central heavy-ion collisions at high energy ($\lesssim 1\,A$ GeV) is studied employing LAND, operated in conjunction with the FOPI [11] charged-particle spectrometers. Directed flow ("squeeze-out") was observed in close resemblance to corresponding effects found for charged particles.

7.4 ORION

ORganic Interceptor Of Neutrons [37].

The ORION detector allows excitation-energy measurements of hot nuclei by observing the evaporated neutrons. It is constituted of 4200 liters of Gd-doped (0.3% of the weight) liquid scintillator (NE343). Light produced is collected by 22 photomultipliers. ORION is composed of separated modules of $1.60\,\mathrm{m}$ diameter covering the full solid angle and surrounding a large ($1.30\,\mathrm{m} \times 60\,\mathrm{cm}$ diameter) reaction chamber. For the amplification-transformation of the primary signals, photomultipliers are placed at the periphery of each module containing the liquid scintillator. The coincidence of at least 2 phototube-signals is required in order to reduce the influence of their intrinsic noise. The efficiency of the neutron detection is a function of the energy. The efficiency is quite high for neutrons of energy lower than $20\,\mathrm{MeV}$ which is the main region of interest for ORION measuring evaporation neutrons from sources of small velocity.

ORION is used for studies of:

- The properties of hot nuclei produced in heavy-ion–induced reactions or spallation reactions induced by light hadrons of several GeV.
- The influence of an existing halo of neutrons in a nucleus on the reaction mechanisms.
- The characteristics of spallation neutron sources.

7.5 DEMON: neutron wall

DÉtecteur MOdulaire de Neutrons [38].

DEMON is issued from a Belgian-French collaboration. It consists of a hundred individual large-size liquid-scintillator cells whose characteristics allow to accede to the angular and energy distributions of the emitted neutrons over a large energy range. DEMON is usually associated to master detectors. It has been conceived essentially to study the reaction mechanisms. But it can also be used in many other domains like in measurements of neutron halo, nuclear interferometry, or neutron cross-section for transmutation of nuclear waste. DEMON's modularity allows to adapt the geometry of the setup to particular needs. It has already been mounted in a cylindrical, a spherical and a wall configuration. DEMON consists of 100 individual large-size NE213 liquid-scintillator cells. Each cell is $20\,\mathrm{cm}$ long, has a diameter of $16\,\mathrm{cm}$ and contains 4.5 liters of liquid rich in hydrogen. A neutron coming into the scintillator interacts mainly with the protons which results in the ionization of the atoms of the scintillator. The subsequent de-excitation induces a light emission which has two components: a fast one and a slow one. The outcoming light is transformed into an electrical signal by a specially designed XP4512B photomultiplier tube. The shape analysis of this pulse allows the discrimination between incident neutrons and γ which have different slow components: a fast one and a slow one.

A proton rejection system (SYREP) is used to avoid proton contamination in the neutron spectra. DEMON operates in the atmosphere. Therefore, slow charged particles do not reach the scintillators. To avoid the contamination of the neutron spectra by very energetic protons which may reach the neutron detectors, DEMON disposes of 24 NE102 scintillators ($3\,\mathrm{mm}$ thin) coupled to photomultipliers which can be mounted in front of the DEMON cells. These plastic scintillators are nearly 100% efficient to the protons and less than 0.5% to the neutrons. Thus, the protons can be rejected in the analysis of the data by an anti-coincidence between the two signals delivered by the two photomultipliers.

The n-γ discrimination is obtained by a pulse-shape analysis by comparison of the slow component of the charge to the total one. A very good discrimination is obtained down to $150\,\mathrm{keV}$ which corresponds to an energy of less than $1\,\mathrm{MeV}$. DEMON has a high intrinsic efficiency over a large energy range: 50% for a neutron of $10\,\mathrm{MeV}$ and still 30% to 40% at $50\,\mathrm{MeV}$. The energy of the neutrons is obtained by the measured time of flight with a resolution of $1.2\,\mathrm{ns}$. The time of flight is corrected for the interaction distance inside the $20\,\mathrm{cm}$ deep cell. The geometrical acceptance of DEMON is of about 4–5% when the scintillators are at a distance of around 1.8 meters. The cross-talk of DEMON is very low and becomes negligible when the distance between two adjacent cells is of $16\,\mathrm{cm}$.

8 Future detectors

With the availability of future radioactive beams (Spiral I-II, Eurisol, RIA, ...) it becomes clear that a complete charge and mass identification of all products coming from

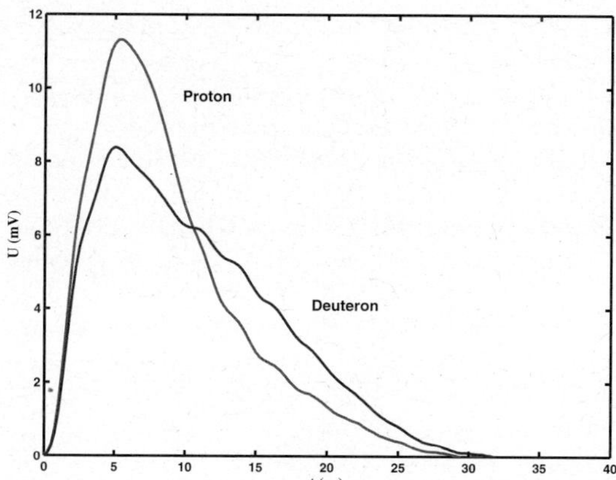

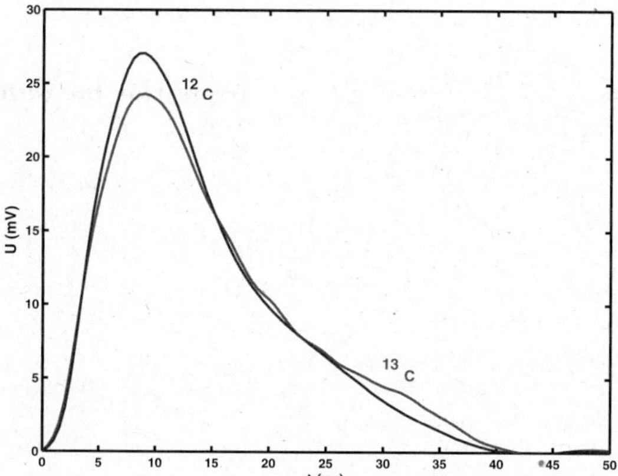

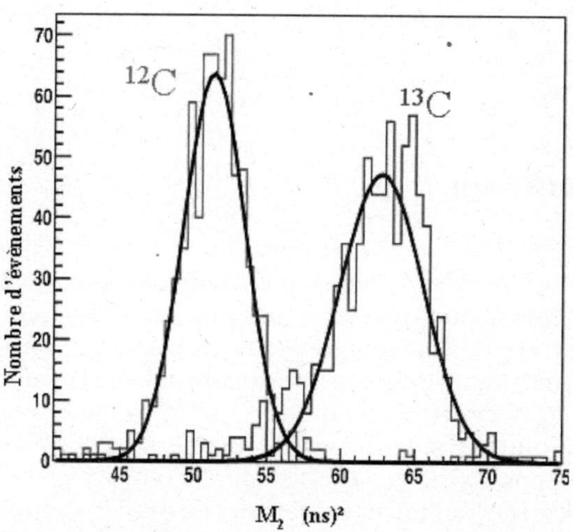

Fig. 8. Example of a mean experimental current signal sampling for two kind of particles, proton and deuteron, at the same incident energy of 5 MeV [39].

Fig. 9. Example of a mean experimental current signal sampling for two carbon isotopes, ^{12}C and ^{13}C, at the same incident energy of 80 MeV. Isotopic determination is well established with the second-moment values M_2 of digitized current pulse, see ref. [39] and fig. 10.

Fig. 10. Discrimination of carbon isotopes. Histograms correspond to the distributions of the second-moment values M_2 of the pulses and full lines to Gaussian fits, see [39].

Many groups all over the world are involved in research and developments on these topics. The final aim is to cover, by varying different techniques (ΔE-E, time of flight, pulse shape, ...), the widest isotopic range as possible.

Neutron detection (and light charged particles, $Z = 1$–2, of very high energy) are not forgotten in this context. Indeed when we are dealing with radioactive nuclei, neutron emission could be of very strong importance (chemical equilibration in time evolving processing, ...). Thus, more than ever their determination event by event is of first importance, including energy and multiplicity.

Morever we have learnt during the last decades that intra-event correlations are fundamental tools to extract precise information on the reaction process, either on time evolution, time emission, re-construction of hot fragments and thus calorimetry, volume determination, ... Therefore, still higher and higher granularity is requested, increasing angular resolution, reducing dead zones and multi-hits. Segmentation could be an issue, but for 4π detector it is a challenge.

What we have learnt so far with 4π multidetectors is the huge calibration time requested by such experiments. Months (even years) are usually necessary for complete and satisfactory results both on particle determination and their energy measurements. With a new generation of multidetectors the number of exit channels will explode. CHIMERA is at present constituted by 1192 telescopes Si-Csi(Tl). For the European project FAZIA (Four π A-Z Identification Array), 5000 to 6000 modules are envisaged, composed by a ΔE_1 (300 μm silicon), ΔE_2 (700 μm silicon) and a 4 cm crystal scintillator (CsI(Tl)). Moreover, pixel detectors are also still considered in the R&D project. It results, with such an amount of detectors, that calibration becomes one of the biggest challenges of such a program. Automatic procedures (neural networks, auto lines-recognition, ...) allowing to sample directly the particle characteristics (and not only their rough signals) are

heavy-ions reactions will be necessary. This goal has only been reached with high-resolution spectrometers, which, however, are rather limited in their angular coverage and their momentum acceptance. Up to now, the maximum isotopic determination in multidetector systems is reached for $Z \simeq 10$ thanks to a ΔE-E method, usually Si-Si or Si-CsI. It is insufficient. New technological advances are necessary for the next generation of multidetector. This may be achieved either with the help of fast digital electronics coupled to appropriate detectors and a careful analysis of the pulse shape of the signals released by a particle crossing the detection material, see [39]. Figures 8 and 9 present different mean experimental current signal samplings for particles at the same energy, proton and deuteron at 5 MeV and two different carbon isotopes ^{12}C and ^{13}C at 80 MeV. If for light-particles discrimination is clearly observed, see fig. 8, for light fragments the use of the second-moment values M_2 is necessary to disentangle the two isotopes, see fig. 10.

envisaged. Those procedures are far from being reached but appear like a crucial goal, a key point of major importance.

9 Conclusion

Approaches based on multidetector systems have yielded many excellent and important results. Anyway the need for an always high-precision information is commonly recognized. Since the requirements to achieve large acceptance and high mass resolution are difficult in many aspects, for forward-focused heavy projectile-like fragments high-resolution magnetic spectrometers have been introduced as powerful tools. This approach provides full identification in A and Z of forward-focus reaction products as well as high-precision measurements of their longitudinal momentum with a resolution not reachable by multidetector systems alone. Nevertheless, complete event-by-event particle information is missing. The importance of isotopic identification of heavy residues being accepted, the combination of full-acceptance detectors for neutrons and light charged particles and intermediate mass fragment with high-resolution magnetic spectrometers for the heavier projectile-like fragments might be one way to combine the advantages of these two systems for certain reaction studies. For the others, a complete knowledge of the full intra-event correlations is necessary and only reachable with complete 4π multidetectors of a new generation.

References

1. DELF: R. Bougault *et al.*, Nucl. Instrum. Methods A **259**, 473 (1987); XYZt: G. Rudolf *et al.*, Nucl. Instrum. Methods A **307**, 325 (1991); MUR: G. Bizard *et al.*, Nucl. Instrum. Methods A **244**, 489 (1986); TONNEAU: A. Peghaire *et al.*, Nucl. Instrum. Methods A **295**, 365 (1990).
2. R.T. DeSouza *et al.*, Nucl. Instrum. Methods A **295**, 109 (1990).
3. D.R. Bowman *et al.*, Phys. Rev. Lett. **67**, 1527 (1991); Y.D. Kim *et al.*, Phys. Rev. Lett. **67**, 14 (1991).
4. J. Pouthas *et al.*, Nucl. Instrum. Methods A **357**, 418 (1995).
5. S. Aiello *et al.*, Nucl. Phys. A **583**, 561 (1995); S. Aiello *et al.*, Nucl. Instrum. Methods A **369**, 50 (1996); S. Aiello *et al.*, Nucl. Instrum. Methods A **400**, 469 (1997); M. Alderighi *et al.*, Nucl. Instrum. Methods A **489**, 257 (2002); N. Le Neindre *et al.*, Nucl. Instrum. Methods A **490**, 251 (2002).
6. A. Pagano *et al.*, Nucl. Phys. A **681**, 331c (2001); E. Geraci *et al.*, Nucl. Phys. A **732**, 173 (2004); A. Pagano *et al.*, Nucl. Phys. A **734**, 504 (2004); E. De Filippo *et al.*, Phys. Rev. C **71**, 044602 (2005); E. De Filippo *et al.*, Phys. Rev. C **71**, 064604 (2005).
7. F. Gramegna *et al.*, Nucl. Instrum. Methods A **389**, 474 (1997); U. Abbondanno *et al.*, Nucl. Instrum. Methods A **488**, 604 (2002).
8. I. Iori *et al.*, Nucl. Instrum. Methods A **325**, 458 (1993); M. Bruno *et al.*, Nucl. Instrum. Methods A **311**, 189 (1992); N. Colonna *et al.*, Nucl. Instrum. Methods A **321**, 529 (1992); P.F. Mastinu *et al.*, Nucl. Instrum. Methods A **338**, 419 (1994).
9. M. Bini *et al.*, Nucl. Instrum. Methods A **515**, 497 (2003).
10. See `http://cyclotron.tamu.edu/nimrod/`.
11. A. Gobbi *et al.*, Nucl. Instrum. Methods A **324**, 156 (1993).
12. W. Reisdorf *et al.*, Nucl. Phys. A **612**, 493 (1997); F. Rami *et al.*, Phys. Rev. Lett. **84**, 1120 (2000); P. Crochet *et al.*, Phys. Lett. B **486**, 6 (2000); W. Reisdorf *et al.*, Phys. Rev. Lett. **92**, 232301 (2004).
13. E. Migneco *et al.*, Nucl. Instrum. Methods A **314**, 31 (1992); MACISTE, IEEE Trans. Nucl. Sci. **43**, 1737 (1996).
14. A. Wagner *et al.*, Nucl. Instrum. Methods A **456**, 290 (2000); B. Gavin *et al.*, Nucl. Instrum. Methods A **473**, 301 (2001).
15. S.P. Avdeyev *et al.*, Nucl. Instrum. Methods A **332**, 149 (1993).
16. Th. Blaich *et al.*, Nucl. Instrum. Methods A **314**, 136 (1992).
17. C.A. Ogilvie *et al.*, Phys. Rev. Lett. **67**, 1214 (1991); W.C. Hsi *et al.*, Phys. Rev. Lett. **73**, 3367 (1994); A. Schüttauf *et al.*, Nucl. Phys. A **607**, 457 (1996); J. Pochodzalla *et al.*, Phys. Rev. Lett. **75**, 1040 (1995); V. Serfling *et al.*, Phys. Rev. Lett. **80**, 3928 (1998); S. Fritz *et al.*, Phys. Lett. B **461**, 315 (1999); T. Odeh *et al.*, Phys. Rev. Lett. **84**, 4557 (2000).
18. M. D'Agostino *et al.*, Nucl. Phys. A **650**, 329 (1999); M. D'Agostino *et al.*, Phys. Lett. B **473**, 219 (2000); M. D'Agostino *et al.*, Nucl. Phys. A **699**, 795 (2002); M. D'Agostino *et al.*, Nucl. Phys. A **724**, 455 (2003); M. D'Agostino *et al.*, Nucl. Phys. A **734**, 512 (2004).
19. F. Gimeno-Nogues *et al.*, Nucl. Instrum. Methods A **399**, 94 (1997); R. Laforest *et al.*, Nucl. Instrum. Methods A **404**, 470 (1998).
20. D.J. Rowland *et al.*, Phys. Rev. C **67**, 064602 (2003).
21. G. Lanzano *et al.*, Nucl. Instrum. Methods A **323**, 694 (1992); G. Lanzano *et al.*, Nucl. Phys. A **683**, 566 (2001).
22. R.E. Tribble *et al.*, Nucl. Instrum. Methods A **285**, 441 (1989).
23. G.A. Souliotis *et al.*, Phys. Rev. C **68**, 024605 (2003); G.A. Souliotis *et al.*, Phys. Lett. B **588**, 35 (2004).
24. D.J. Morrissey, NSCL Staff, Nucl. Instrum. Methods B **26**, 316 (1997); D.J. Morrissey *et al.*, Nucl. Instrum. Methods B **204**, 90 (2003).
25. M. Mocko *et al.*, Nucl. Phys. A **734**, 532 (2004); A. Stolzet *et al.*, Phys. Lett. B **627**, 32 (2005).
26. R. Pfaff *et al.*, Phys. Rev. C **51**, 1348 (1995); K.A. Hanold *et al.*, Phys. Rev. C **52**, 1462 (1995); R. Pfaff *et al.*, Phys. Rev. C **53**, 1753 (1996); G.A. Souliotis *et al.*, Phys. Rev. C **57**, 3129 (1998); G.A. Souliotis *et al.*, Nucl. Phys. A **705**, 279 (2002).
27. H. Geissel *et al.*, Nucl. Instrum. Methods B **70**, 286 (1992).
28. K.-H. Schmidt *et al.*, Phys. Lett. B **300**, 313 (1993).
29. K.-H. Schmidt *et al.*, Nucl. Phys. A **710**, 157 (2002).
30. M.V. Ricciardi *et al.*, Phys. Rev. Lett. **90**, 212302 (2003).
31. P. Napolitani *et al.*, Phys. Rev. C **70**, 054607 (2004).
32. A. Kelic *et al.*, Phys. Rev. C **70**, 064608 (2004).
33. `http://www.gsi.de/fair/index.html`.
34. `http://www-land.gsi.de/r3b/index.html`.
35. H. Geissel *et al.*, Nucl. Instrum. Methods B **204**, 71 (2003).
36. P.D. Zecher *et al.*, Nucl. Instrum. Methods A **401**, 329 (1997).

37. Y. Périer *et al.*, Nucl. Instrum. Methods A **413**, 321 (1998).
38. G. Bizard *et al.*, Nucl. Phys. News **1**, No. 5, 15 (1991); M. Moszynski *et al.*, Nucl. Instrum. Methods A **307**, 97 (1991); M. Moszynski *et al.*, Nucl. Instrum. Methods A **317**, 262 (1992); M. Moszynski *et al.*, Nucl. Instrum. Methods A **343**, 563 (1994); M. Moszynski *et al.*, Nucl. Instrum. Methods A **350**, 226 (1994); S. Mouatassim *et al.*, Nucl. Instrum. Methods A **359**, 530 (1995); I. Tilquin *et al.*, Nucl. Instrum. Methods A **365**, 446 (1995).
39. H. Hamrita *et al.*, Nucl. Instrum Methods A **531**, 607 (2004) and PhD Thesis, University Paris VI (2005).

Eur. Phys. J. A **30**, 293–302 (2006)

DOI 10.1140/epja/i2005-10317-6

Nuclear multifragmentation, its relation to general physics

A rich test ground of the fundamentals of statistical mechanics

D.H.E. Gross[a]

Hahn-Meitner Institute Glienickerstr. 100, 14109 Berlin, Germany and
Freie Universität Berlin, Fachbereich Physik, Berlin, Germany

Received: 22 November 2005 /
Published online: 2 November 2006 – © Società Italiana di Fisica / Springer-Verlag 2006

Abstract. Heat can flow from cold to hot at any phase separation even in macroscopic systems. Therefore also Lynden-Bell's famous gravo-thermal catastrophe must be reconsidered. In contrast to traditional canonical Boltzmann-Gibbs statistics this is correctly described only by microcanonical statistics. Systems studied in chemical thermodynamics (ChTh) by using canonical statistics consist of several *homogeneous macroscopic* phases. Evidently, macroscopic statistics as in chemistry cannot and should not be applied to non-extensive or inhomogeneous systems like nuclei or galaxies. Nuclei are *small and inhomogeneous*. Multifragmented nuclei are even more inhomogeneous and the fragments even smaller. Phase transitions of first order and especially phase separations therefore cannot be described by a (*homogeneous*) canonical ensemble. Taking this serious, fascinating perspectives open for statistical nuclear fragmentation as test ground for the basic principles of statistical mechanics, especially of phase transitions, *without the use of the thermodynamic limit*. Moreover, there is also a lot of similarity between the accessible phase space of fragmenting nuclei and inhomogeneous multistellar systems. This underlines the fundamental significance for statistical physics in general.

PACS. 04.40.-b Self-gravitating systems; continuous media and classical fields in curved spacetime – 05.20.Gg Classical ensemble theory – 25.70.Pq Multifragment emission and correlations – 64.60.-i General studies of phase transitions

1 Introduction

In 1981 Randrup and Koonin [1] proposed the statistical (grand-canonical) decay of an excited nucleus into several light fragments. As the grand-canonical ensemble fixes the mean mass by an intensive control parameter, the chemical potential μ, but has no information about the total mass M_t of the decaying nucleus, this works only for fragment masses $M_i \ll M_t$. This touches already the central point of the discussion to follow, the difference between intensive parameters (fields) used in canonical statistics in contrast to the mechanical extensive control parameters used in microcanonical statistics.

The statistical *multifragmentation* of a hot nucleus simultaneously into larger fragments was introduced by [2,3] (details of the historical development of the theory of statistical multifragmentation is discussed in appendix A of [4].) Of course the finiteness of the total mass and charge is then crucial. Meanwhile statistical multifragmentation developed to a powerful and successful description even of sophisticated correlations seen in nuclear

multifragmentation, cf. also [5–8]. A presentation of its far-reaching implications for the fundamental understanding of statistical mechanics in general is now demanding.

Here I will give mainly the motivation. In sect. 2 I address the general basis of statistical mechanics without invoking the thermodynamic limit. Then I give the physical definition of entropy S, I show how phase-separation is necessarily linked to convexities of $S(E)$ and negative heat capacities. In [9] I discussed in detail the general topology of the entropy surface $S(E)$ indicating phase transitions in general. In sect. 3 I present shortly the application to three characteristic phenomena: Nuclear multifragmentation, the fragmentation of small atomic clusters and finally the fragmentation of stellar objects under large angular momentum.

In sect. 3.1 I only discuss the implications of the new formalism for statistical nuclear fragmentation. In this topical issue there will be many contributions that compare detailed experimental data to the predictions of the different models for statistical multifragmentation of hot nuclei. Here I will put the new statistics of nuclear multifragmentation into a more general perspective: I show how, similar to nuclear fragmentation, also atomic clus-

---

[a] e-mail: `gross@hmi.de`;
`http://www.hmi.de/people/gross/`.

ters fragment with rising excitation into more and more medium-sized fragments. In close similarity to nuclear multifragmentation also the accessible phase space of self-gravitating astro-physical systems splits under rising energy and/or angular momentum into various inhomogeneous phases of single stars, rotating multi-star systems, and sometimes even more exotic configurations as ring systems and others.

2 Fundamentals of thermostatistics without thermodynamic limit

Since the beginning of thermodynamics in the first half of the 19th century its original motivation was the description of steam engines and the liquid-to-gas transition of water. Here water becomes inhomogeneous and develops a separation of the gas phase from the liquid, *i.e.* water boils. Thus, *phase separations* were in the focus some 170 years ago. Every child realizes phase separation by the inter-phase surface. And every child distinguishes a solid crystal from a liquid by the hard surface of the latter. It is an irony of the history of statistical mechanics that phase transitions of first order can only be signaled indirectly by the academic construct of a Yang-Lee singularity [10]. There is no information about the necessary and characteristic inter-phase surface. Of course this is because of the use of the thermodynamic limit and the use of intensive Lagrange parameters as control parameters.

A little later statistical mechanics was proposed by Boltzmann [11,12] to explain the microscopic mechanical basis of thermodynamics. Up to now it is generally believed that this is given by the Boltzmann-Gibbs canonical statistics. As traditional canonical statistics works only for *homogeneous, infinite* systems, *phase separations* remain outside standard Boltzmann-Gibbs thermostatistics, which, consequently, signal phase transitions of first order by Yang-Lee singularities.

It is amusing that this fact that is essential for the original purpose of thermodynamics to describe steam engines was never treated completely in the past 150 years. The system must be somewhat artificially split into (still macroscopic and homogeneous) pieces of each individual phase [13]. The most interesting configurations like two co-existing phases cannot be described by a *single* canonical ensemble. Important *inter-phase fluctuations* remain outside the picture, etc. Of course these are essential for the fragmentation process. These *inter-phase fluctuations* are also responsible for the negative heat capacity [14]. This is all hidden due to the restriction to homogeneous systems in the thermodynamic limit and the use of intensive control parameters like temperature, pressure, chemical potentials etc. What may be more surprising is the fact that the curvature of $S(E)$ can stay convex even at the thermodynamic limit. The leading volume term of $S(E)$ follows the Maxwell double tangent (concave hull) and has curvature 0. In the intermediate energy range between the liquid and the gas the surface contribution $\delta^2 S_{surf} \propto N^{2/3} > 0$ is the *dominant curvature*. It leads to a deep intruder in $S(E)$ *also in the thermodynamic limit.*

Also the second law can rigorously be formulated only microcanonically. Already Clausius [15–17] distinguished between external and internal entropy generating mechanisms. The second law is only related to the latter mechanism [18], the internal entropy generation. Again, canonical Boltzmann-Gibbs statistics is insensitive to this important difference.

For this purpose, and also to describe small systems like fragmenting nuclei or non-extensive ones like self-gravitating very large systems, we need a new and deeper definition of statistical mechanics and at the heart of it, of entropy.

2.1 What is entropy?

Entropy, S, is *the* characteristic entity of thermodynamics and statistics. Its use distinguishes thermodynamics from all other physics; therefore, its proper understanding is essential. The understanding of entropy is sometimes obscured by frequent use of the Boltzmann-Gibbs canonical ensemble, and the thermodynamic limit. Also its relationship to the second law is often beset with confusion between external transfers of entropy d_eS and its internal production d_iS.

The main source of the confusion is of course the lack of a clear *microscopic* and *mechanical* understanding of the fundamental quantities of thermodynamics like heat, external *vs.* internal work, temperature, and last but not least entropy, at the times of Clausius and possibly even today.

Clausius [15,16] defined a quantity which he first called the *"value of metamorphosis"*, in German *"Wert der Verwandlung"* in [16]. Eleven years later he [17] gave it the name "entropy" S:

$$S_b - S_a = \int_a^b \frac{\mathrm{d}E}{T}\,, \tag{1}$$

where T is the absolute temperature of the body when the momentary change is done, and $\mathrm{d}E$ is the increment (positive, respectively, negative) of all different forms of energy (heat and potential) put into, respectively, taken out of the system. (Later, however, we will learn that care must be taken of additional constraints on other control parameters like, *e.g.*, the volume, see below.)

From the observation that heat does not flow from cold to hot (see, however, sect. 2.2) he went on to enunciate the second law as

$$\Delta S = \oint \frac{\mathrm{d}E}{T} \geq 0, \tag{2}$$

which Clausius called the *"uncompensated metamorphosis"*. As will be worked out later, the second law as presented by eq. (2) remains valid even in cases where heat (energy) flows during relaxation from low to higher temperatures.

Prigogine [18], cf. [13], quite clearly stated that the variation of S with time is determined by two, crucially different, mechanisms of its changes: the flow of entropy d_eS to or from the system under consideration, and its internal production d_iS. While the first type of entropy change d_eS (that effected by exchange of heat d_eQ with its surroundings) can be positive, negative or zero, the second type of entropy change d_iS is fundamentally related to the spontaneous internal evolution ("Verwandlungen", "metamorphosis" [15]) of the system, and states the universal irreversibility of spontaneous transitions. It can be only positive or zero in any spontaneous transformation.

Clausius gives an illuminating example in [16]: When an ideal gas suddenly streams under insulating conditions from a small vessel with volume V_1 into a larger one ($V_2 > V_1$), neither its internal energy U, nor its temperature changes, nor external work done, but its internal (Boltzmann) entropy S_i, eq. (3), rises by $\Delta S = N \ln (V_2/V_1)$. Only by compressing the gas (*e.g.*, isentropically) and creating heat $\Delta E = E_1[(V_2/V_1)^{2/3} - 1]$ (which must be finally drained) it can be brought back into its initial state. Then, however, the entropy production in the cycle, as expressed by integral (2), is positive ($= N \ln (V_2/V_1)$). This is also a clear example for a microcanonical situation where the entropy change by an irreversible metamorphosis of the system is absolutely internal. It occurs during the first part of the cycle, the expansion, where there is no heat exchange with the environment and no work done, and consequently no contribution to the integral (2). The construction by eq. (2) is correct though artificial. After completing the cycle the Boltzmann entropy of the gas is of course the same as initially. All this will become much more clear by Boltzmann's microscopic definition of entropy, which will moreover clarify its real *statistical* nature.

Boltzmann [11,12] later defined the entropy of an isolated system (for which the energy exchange with the environment $d_eQ \equiv 0$) in terms of the sum of possible configurations, W, which the system can assume consistent with its constraints of given energy and volume:

$$\boxed{S = k * \ln W} \tag{3}$$

as written on Boltzmann's tombstone, with

$$W(E, N, V) = \int \frac{\mathrm{d}^{3N}\vec{p}\,\mathrm{d}^{3N}\vec{q}}{N!(2\pi\hbar)^{3N}} \epsilon_0 \, \delta(E - H\{\vec{q}, \vec{p}\}) \tag{4}$$

in semi-classical approximation. E is the total energy, N is the number of particles and V the volume. Or, more appropriate for a finite quantum-mechanical system:

$$W(E, N, V) = \mathrm{Tr}[\mathcal{P}_E] \tag{5}$$

$$= \sum \begin{array}{l} \text{all eigenstates } n \text{ of } H \text{ with given } N, V, \\ \text{and } E < E_n \leq E + \epsilon_0 \end{array}$$

and $\epsilon_0 \approx$ the macroscopic energy resolution. This is still up to day the deepest, most fundamental, and most simple definition of entropy. There is no need of the thermodynamic limit, no need of concavity, extensivity and homogeneity. In its semi-classical approximation, eq. (4), $W(E, N, V, \cdots)$ simply measures the area of the sub-manifold of points in the $6N$-dimensional phase space (Γ-space)

with prescribed energy E, particle number N, volume V, and some other time invariant constraints which are here suppressed for simplicity. Because it was Planck who coined it in this mathematical form, I will call it the Boltzmann-Planck principle.

The Boltzmann-Planck formula has a simple but deep physical interpretation: W or S measure our ignorance about the complete set of initial values for all $6N$ microscopic degrees of freedom which are needed to specify the N-body system unambiguously [19]. To have complete knowledge of the system we would need to know (within its semi-classical approximation (4)) the initial positions and velocities of all N particles in the system, which means we would need to know a total of $6N$ values. Then W would be equal to one and the entropy, S, would be zero. However, we usually only know the value of a few parameters that change slowly with time, such as the energy, number of particles, volume and so on. We generally know very little about the positions and velocities of the particles. The manifold of all these points in the $6N$-dimensional phase space, consistent with the given macroscopic constraints of $E, N, V, \cdots$, is the microcanonical ensemble, which has a well-defined geometrical size W and, by eq. (3), a non-vanishing entropy, $S(E, N, V, \cdots)$. The dependence of $S(E, N, V, \cdots)$ on its arguments determines completely thermostatics and equilibrium thermodynamics.

Clearly, Hamiltonian (Liouvillean) dynamics of the system cannot create the missing information about the initial values —*i.e.* the entropy $S(E, N, V, \cdots)$ cannot decrease. As has been further worked out in [20] and more recently in [21] the inherent finite resolution of the macroscopic description implies an increase of W or S with time when an external constraint is relaxed. This is a statement of the second law of thermodynamics, which requires that the *internal* production of entropy be positive or zero for every spontaneous process. The analysis of the consequences of the second law by the microcanonical ensemble is appropriate because, in an isolated system (which is the one relevant for the microcanonical ensemble), the changes in total entropy must represent the *internal* production of entropy, see above, and there are no additional uncontrolled fluctuating energy exchanges with the environment.

2.2 The zeroth law in conventional extensive thermodynamics

In conventional (extensive) thermodynamics thermal equilibrium of two systems (1 and 2) is established by bringing them into thermal contact which allows free energy exchange. Equilibrium is established when the total entropy

$$S_{1+2}(E, E_1) = S_1(E_1) + S_2(E - E_1) \tag{6}$$

is maximal:

$$\mathrm{d}S_{1+2}(E, E_1)|_E = \mathrm{d}S_1(E_1) + \mathrm{d}S_2(E - E_1) = 0. \tag{7}$$

Under an energy flux $\Delta E_{2\to 1}$ from $2 \to 1$ the total entropy changes to lowest order in ΔE by

$$\Delta S_{1+2}|_E = (\beta_1 - \beta_2)\Delta E_{2\to 1}, \qquad (8)$$

$$\beta = \mathrm{d}S/\mathrm{d}E = \frac{1}{T}. \qquad (9)$$

Consequently, a maximum of $S_{total}(E = E_1 + E_2, E_1)|_E \geq S_{1+2}$ will be approached when

$$\mathrm{sign}(\Delta S_{total}) = \mathrm{sign}(T_2 - T_1)\,\mathrm{sign}(\Delta E_{2\to 1}) > 0. \qquad (10)$$

From here Clausius' first formulation of the second law follows: "Heat always flows from hot to cold". Essential for this conclusion is the *additivity* of S under the split (eq. (6)). There are no correlations which are destroyed when an extensive system is split. Temperature is an appropriate control parameter for extensive systems.

It is further easy to see that the heat capacity of an extensive system with $S(E,N) = Ns(e = E/N) = 2S(E/2, N/2)$ is necessarily non-negative:

$$C_V(E) = \partial E/\partial T = -\frac{(\partial S/\partial E)^2}{\partial^2 S/\partial E^2} \geq 0. \qquad (11)$$

The combination of two pieces of $N/2$ particles each, one at the specific energy $e_a = e_2 - \Delta e/2$ and a second at $e_b = e_2 + \Delta e/2$, must lead to $S(E_2, N) \geq S(E_a/2, N/2) + S(E_b/2, N/2)$, the simple algebraic sum of the individual entropies, because by combining the two pieces one normally loses information. This, however, is for extensive systems equal to $[S(E_a, N) + S(E_b, N)]/2$, thus $S(E_2, N) \geq [S(E_a, N) + S(E_b, N)]/2$. I.e., the *entropy $S(E, N)$ of an extensive system is necessarily nonconvex, $\partial^2 S/\partial E^2 \leq 0$* and eq. (11) follows. In the next subsection we will see that therefore *extensive systems cannot have phase transitions of first order*.

2.3 No phase separation, no boiling water, without a convex, non-extensive S(E)

At phase separation the weight $e^{S(E)-E/T}$ of the configurations with energy E in the definition of the canonical partition sum

$$Z(T) = \int_0^\infty e^{S(E)-E/T}\mathrm{d}E \qquad (12)$$

becomes here *bimodal*: at the transition temperature it has two peaks, the liquid and the gas configurations which are separated in energy by the latent heat. Consequently, $S(E)$ must be convex ($\partial^2 S/\partial E^2 > 0$, like $y = x^2$) and the weight in (12) has a minimum between the two pure phases. Of course, the minimum can only be seen in the microcanonical ensemble where the energy is controlled and its fluctuations forbidden. Otherwise, the system would fluctuate between the two pure phases by an, for macroscopic systems even macroscopic, energy $\Delta E \sim E_{lat} \propto N$ of the order of the latent heat. Canonically, phase separations are unstable, however, not

microcanonically, and of course not in real nature. The heat capacity is

$$C_V(E) = \partial E/\partial T = -\frac{(\partial S/\partial E)^2}{\partial^2 S/\partial E^2} < 0. \qquad (13)$$

I.e., the convexity of $S(E)$ and the negative heat capacity are the generic and necessary signals of phase separation [4]. It is amusing that this fact that is essential for the original purpose of thermodynamics to describe steam engines and boiling water seems never been really recognized in the past 150 years. However, such macroscopic energy fluctuations and the resulting negative specific heat are already early discussed in high-energy physics by Carlitz [22].

2.3.1 Physical origin of positive curvature, the surface tension

For short-range forces the depth of the convex intruder into $S(E)$ is linked to the inter-phase surface tension. This is demonstrated by fig. 1 which shows an MMMC simulation of the entropy per atom of a cluster of 1000 sodium atoms.

At the energy $e \leq e_1$ the system is in the pure liquid phase and at $e \geq e_3$ in the pure gas phase, of course with fluctuations. The latent heat per atom is $q_{lat} = e_3 - e_1$. *Attention:* the curve $s(e)$ is artifically sheared by subtracting a linear function $25 + e * 11.5$ in order to make the convex intruder visible. *$s(e)$ is always a steeply monotonic rising function.* We clearly see the global concave (downwards bending) nature of $s(e)$ and its convex intruder. Its depth is the entropy loss due to additional correlations by the interfaces. It scales $\propto N^{-1/3}$. From this one can calculate the surface tension per surface atom $\sigma_{surf}/T_{tr} = \Delta s_{surf} * N_0/N_{surf}$. This quantity, as well as other relevant parameters of the transition, is given in table 1. The double tangent (Gibbs construction) is the concave hull of $s(e)$. Its derivative gives the Maxwell line

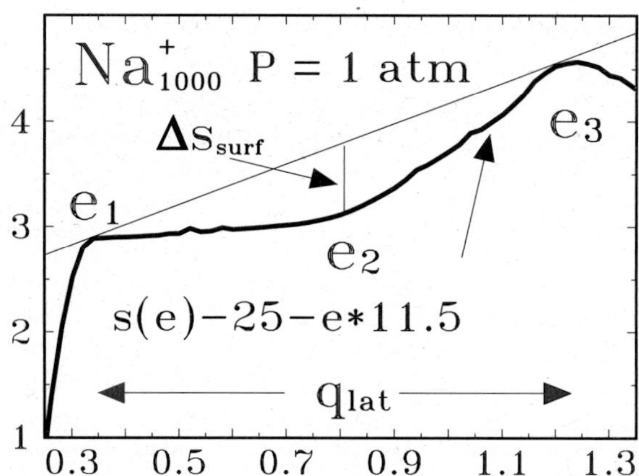

Fig. 1. MMMC [4] simulation of the entropy $s(e)$ per atom (e in eV per atom) of a system of $N_0 = 1000$ sodium atoms at an external pressure of 1 atm.

Table 1. Parameters of the liquid-gas transition of small sodium clusters (MMMC calculation [4]) in comparison with the bulk for a rising number N_0 of atoms, N_{surf} is the average number of surface atoms (estimated here as $\sum N_{cluster}^{2/3}$) of all clusters with $N_i \geq 2$ together. $\sigma/T_{tr} = \Delta s_{surf} * N_0/N_{surf}$ corresponds to the surface tension. Its bulk value is adjusted to agree with the experimental values of the a_s parameter which we used in the liquid-drop formula for the binding energies of small clusters, cf. Brechignac *et al.* [23], and which are used in this calculation [4] for the individual clusters.

	N_0	200	1000	3000	Bulk
	T_{tr} [K]	940	990	1095	1156
	q_{lat} [eV]	0.82	0.91	0.94	0.923
Na	s_{boil}	10.1	10.7	9.9	9.267
	Δs_{surf}	0.55	0.56	0.44	
	N_{surf}	39.94	98.53	186.6	∞
	σ/T_{tr}	2.75	5.68	7.07	7.41

in the caloric curve $T(e)$ at T_{tr}. In the thermodynamic limit the intruder would disappear and $s(e)$ would approach the double tangent from below. Nevertheless, even there, the probability of configurations with phase separations are suppressed by the (infinitesimal small) factor $e^{-N^{2/3}}$ relative to the pure phases and the distribution remains *strictly bimodal in the canonical ensemble*. The region $e_1 < e < e_3$ of phase separation gets lost. Consequently, the intruder can only be seen when the system is insolated (thermo-flask) and the energy can be controlled. *I.e., in the microcanonical situation.*

The existence of the negative heat capacity at phase separation has a surprising but fundamental consequence: Combining two systems with negative heat capacity they will relax with a flow of energy from the lower to the higher temperature! This is consistent with the naive picture of an *energy equilibration*. Thus *Clausius' "energy flows always from hot to cold", i.e. the dominant control-role of the temperature in thermostatistics, as emphasized by Hertz [24], is violated.* Of course this shows quite clearly that *unlike to extensive thermodynamics the temperature is not the appropriate control parameter in non-extensive situations like, e.g., at phase separations, nuclear fragmentation, or stellar systems* [25].

2.3.2 Lynden-Bell's paradox

By the same reason the well-known paradox of Antonov in astro-physics due to the occurrence of negative heat capacities must be reconsidered: Lynden-Bell [26] uses standard arguments from extensive thermodynamics that a system a with negative heat capacity $C_a < 0$ in gravitational contact with another b with positive heat capacity $C_b > 0$ will be unstable: If initially $T_a > T_b$ the hotter system a transfers energy to the colder b and by this both become even hotter! If $C_b > -C_a$, T_a rises faster than T_b and this will go for ever. This is wrong because just the opposite happens,

the hotter a even *absorbs* energy from the colder b and both system come to equilibrium at the same intermediate temperature, cf. [25, 27]. Negative heat capacity can only occur in the microcanonical ensemble. Temperature is *not* controlling the direction of energy (heat) flow when the heat capacity is negative. This is controlled by *entropy* according to the second law. Isothermal self-gravitating systems appear somehow paradoxical. Moreover, one cannot argue, as for extensive systems, that $S_{1+2} = S_1 + S_2$ and $E_{1+2} = E_1 + E_2$ as discussed above. There are far-reaching correlations between the two systems due to long-ranged gravity.

In the thermodynamic limit $N \to \infty$ of a system with short-range coupling the depth of the convex intruder $\Delta S_{surf} \sim N^{2/3}$, i.e. $\Delta S_{surf}/N = \Delta s_{surf} \propto N^{-1/3}$ must go to 0 due to van Hove's theorem. Of course it is only the *specific* surface entropy $\Delta S_{surf}/N$ which disappears. As phase separation exists also in the thermodynamic limit, by the same arguments as above, *the curvature of $S(E)$ remains convex, $\partial^2 S/(\partial E)^2 > 0$. Consequently, the negative heat capacity at phase separation should also be seen in ordinary macroscopic systems in chemistry!*

Searching for example in Guggenheims book [13] one finds some cryptic notes in § 3 that the heat capacity of steam at saturation is negative. No notice that *this is the generic effect at any phase separation!*

It is interesting to notice that, if ordinary macroscopic thermodynamics is used in describing finite systems, artificial unphysical effects need to be invoked to obtain negative heat capacities at first-order phase transitions [28]. Therefore, let me recapitulate in the next subsection how chemists treat phase separation of macroscopic systems and then point out why this does not work in non-extensive systems like fragmenting nuclei, at phase separation in normal macroscopic systems, or large astronomical systems.

2.4 Macroscopic systems in chemistry

Systems studied in chemical thermodynamics consist of several *homogeneous macroscopic* phases $\alpha_1, \alpha_2, \cdots$ cf. [13]. Their mutual equilibrium must be explicitly constructed from outside.

Each of these phases are assumed to be homogeneous and macroscopic (in the "thermodynamic limit" ($N_\alpha \to \infty|_{\rho_\alpha = \text{const}}$)). There is no *common* canonical ensemble for the entire system of the coexisting phases. Only the canonical ensemble of *each* phase separately becomes equivalent in the limit to its microcanonical counterpart.

The canonical partition sum of *each* phase α is defined as the Laplace transform of the underlying microcanonical sum of states $W(E)_\alpha = e^{S_\alpha(E)}$ [29,30]

$$Z_\alpha(T) = \int_0^\infty e^{S_\alpha(E) - E/T_\alpha} dE. \qquad (14)$$

The mean canonical energy is

$$\langle E_\alpha(T_\alpha) \rangle = -\partial \ln Z_\alpha(T_\alpha)/\partial \beta_\alpha,$$
$$\beta_\alpha = \frac{1}{T_\alpha}. \qquad (15)$$

In chemical situations proper the assumption of homogeneous macroscopic individual phases is of course acceptable. In the thermodynamic limit $(N_\alpha \to \infty|_{\rho_\alpha = \mathrm{const}})$ of a *homogeneous* phase α, the canonical energy $\langle E_\alpha(T_\alpha)\rangle$ becomes identical to the microcanonical energy E_α when the temperature is determined by

$$T_\alpha^{-1} = \beta_\alpha = \left.\frac{\partial S_\alpha(E, V_\alpha)}{\partial E}\right|_{E_\alpha}. \qquad (16)$$

The relative width of the canonical energy is

$$\Delta E(T)_\alpha = \frac{\sqrt{\langle E_\alpha^2\rangle_T - \langle E_\alpha\rangle_T^2}}{\langle E_\alpha\rangle_T} \propto \frac{1}{\sqrt{N_\alpha}}. \qquad (17)$$

The heat capacity at constant volume is (care must be taken about the constraints (!))

$$C_\alpha|_{V_\alpha} = \frac{\partial\langle E_\alpha(T_\alpha, V_\alpha)\rangle}{\partial T_\alpha} \qquad (18)$$

$$= \frac{\langle E_\alpha^2\rangle_{T_\alpha} - \langle E_\alpha\rangle_{T_\alpha}^2}{T_\alpha^2} \geq 0. \qquad (19)$$

Only in the thermodynamic limit $(N_\alpha \to \infty|_{\rho_\alpha = \mathrm{const}})$ does the relative energy uncertainty $\Delta E_\alpha \to 0$, and the canonical and the microcanonical ensembles for each homogeneous phase (α) become equivalent. This equivalence is the *only* justification of the canonical ensemble controlled by intensive temperature T, or chemical potential μ, or pressure P. I do not know of any microscopic foundation of the canonical ensemble and intensive control parameters apart from the limit. This is also the reason why, *e.g.*, the Clausius-Clapeyron equation as an equation between intensive variables is *not* applicable away from the thermodynamic limit, *e.g.* in nuclei.

The positiveness of any canonical $C_V(T)$ or $C_P(T)$ is of course the reason why the inhomogeneous system of several coexisting phases (α_1 and α_2) with an overall *negative* heat capacity cannot be described by a *single common* canonical distribution [4,31]. The inter-phase fluctuations are ignored.

This new fundamental interpretation of thermo statistics was introduced to the chemistry community in [32,33].

2.5 A remark on "non-equilibrium" thermodynamics of small sytems

Prigogine quite clearly gives a short introduction into the logical foundations of non-equilibrium thermodynamics in [34]. The system is assumed to be composed by small subsystems internally in thermodynamic equilibrium. Each one is itself macroscopic and homogeneous that the conventional canonical Boltzmann-Gibbs statistics applies. However, the individual subsystems are not assumed to be in mutual thermodynamic equilibrium. There are temperature and/or pressure gradients, there may be a flow of the subsystems etc. Hydrodynamics or heat conductivity are examples. Clearly, this is certainly not possible in small systems like atomic nuclei or

atomic clusters. Therefore, attempts to transfer macro-thermo-dynamic concepts like temperature or Gibbs-free energy $G(T, P)$ to nano-objects [35] like single biological molecules and the exploration of Jarzynski's equality [36] must be considered with reservation cf. [37,38]. Temperature, and pressure are ill defined in such small objects [39].

3 Statistical fragmentation

3.1 Nuclear fragmentation

The new lesson to be learned is that if one defines the phases by individual peaks in $e^{S(E)-E/T}$ in (12), then there exist also *inhomogeneous phases* like in fragmented nuclei or stellar systems. The general concept of thermo-statistics becomes enormously widened.

However, before applying the microcanonical thermo-statistics to nuclear collisions, a clarification is necessary: Nuclear collisions are *transient* phenomena. Thus, a theory of statistical nuclear fragmentation is an *approximation* to a *dynamical* process. This is well known and applies as well to the old Weisskopf theory of the statistical decay of a compound nucleus. The scenario one has in mind is that the emissions of fragments over the barrier is so slow that all accessible exit channels are tested. This is the open phase space *at or on top of the exit barrier*. In the statistical fragmentation model MMMC [4] this is taken care of by sampling all fragments under non-overlapping conditions inside a "freeze-out" volume corresponding to $\sim 5 \times V_0$ the volume of the nucleus in its ground state. The average distance between neighboring fragments is then about 2 fm. The experimental discovery of nuclear *multifragmentation* by [40] and its theoretical interpretation by [3] was the clear recognition that within a time of $\ll 10^{-21}$ s *several* medium-sized fragments can cross the decay barrier. This is much shorter than the time the fragments need to come out of mutual Coulomb fields, a fact discussed in detail in Chapt. 5.2.1 of my book [4].

At this point a clarifying comment must be made on the paper *Information theory of open fragmenting systems* and its relevance for nuclear fragmentation [41] especially to the established statistical fragmentation models. The authors write on p. 2: "*More important, to specify the density matrix, the projector P_S (which projects on the given boundary condition (my explanation)) has to be exactly known and this is in fact impossible. The nature of P_S is intrinsically different from the usual global observables A_l. Not only it is a many-body operator, but P_S requires the exact knowledge of each point of the boundary surface while no or few parameters are sufficient to define the A_l. This infinity of points corresponds to an infinite amount of information to be known to define the density matrix. . . the same is true for the standard (N, E, V) ensembles when dealing with finite unbound unconfined systems.*" The (N, E, V) ensemble of a bound system is the most fundamental ensemble of statistical mechanics. The treatment of the ideal gas in a box is one of the most elementary exercises in statistical mechanics that can be solved analytically. It is *the* paradigm of statistical mechanics and we should keep close to it as much as we can,

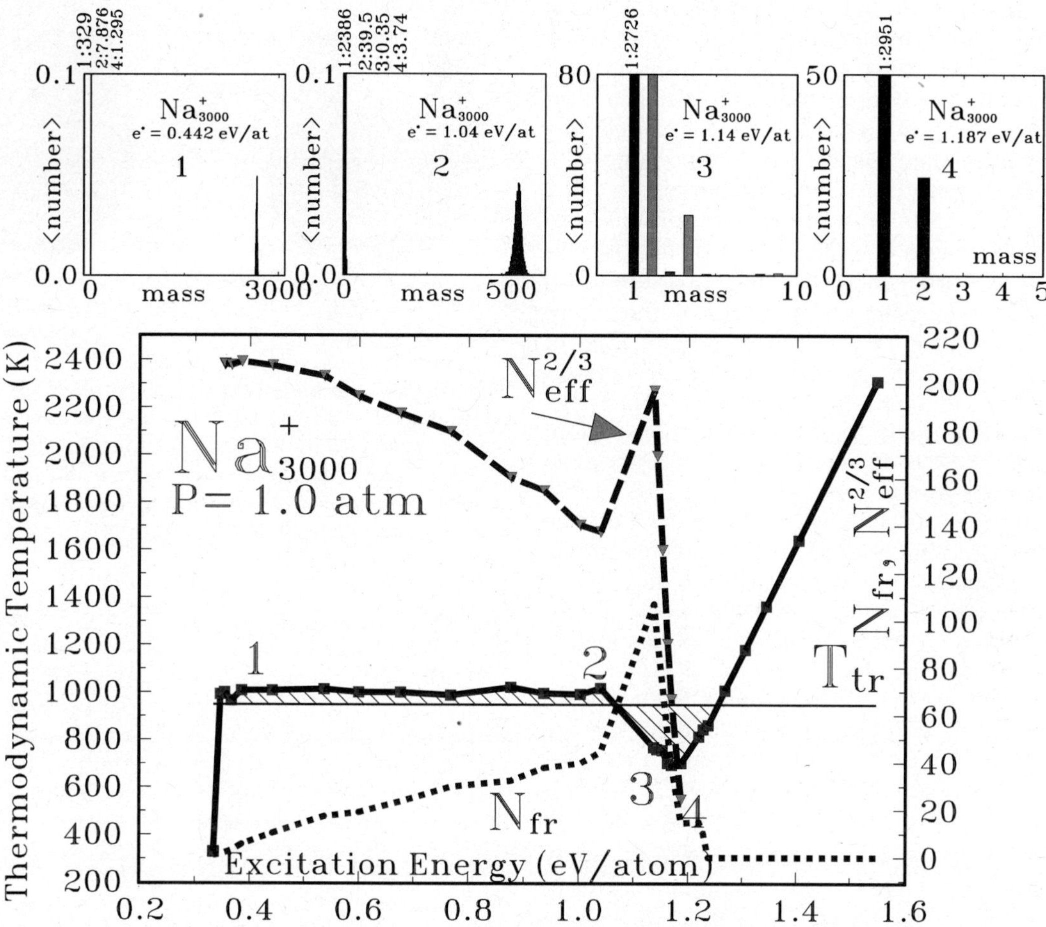

Fig. 2. Atomic cluster fragmentation, see sect. 3.2.

when we are going to extend standard statistical mechanics into the domain of systems far off the thermodynamic limit. There is no infinity of information needed. Of course the box potential must be included in the Hamiltonian which characterizes the system under consideration.

In our MMMC model we have a very specific physical picture in mind. It saves us not to enter dangerous new grounds of "dynamical" statistical mechanics. Moreover it gives us an idea why and how the system may explore statistically the whole accessible phase space. We took care of the fact that the fragments are trapped for $\approx 100\,\text{fm}/c$ behind the Coulomb barrier. This can clearly be seen in BUU dynamical calculations [4]. The Coulomb barrier defines the freeze-out volume. The fragmenting system is assumed to be in statistical equilibrium during this time. So the ensemble imagined is a standard (N, E, V) ensemble and NOT some unbound unconfined system. Of course this is a simplifying approximation to a much more complicated *dynamical* situation. But judged from its great success this is a very reasonable simplification.

Now, certainly neither the phase of the whole multifragmented nucleus nor the individual fragments themselves can be considered as macroscopic homogeneous phases in the sense of chemical thermodynamics (ChTh). Consequently, ChTh cannot and should not be applied to fragmenting nuclei and the microcanonical description is ultimately demanded. This becomes explicitly clear by the fact that the configurations of a multifragmented nucleus have a *negative* heat capacity at constant volume C_V [42,43], and further references therein, and also at constant pressure C_P (if at all a pressure can be associated to nuclear fragmentation [4]).

The existence of well-defined and separated peaks (phases, *if distinguished by conserved control parameters*) in the event distribution of nuclear fragmentation data is demonstrated in [44] from various points of view. This signal is in a small system like a nucleus by far more sophisticated and detailed than the simple jumping from liquid to gas in traditional macroscopic systems in chemistry. A lot more physics about the mechanism of phase transitions can be learned from such studies. This will be the topic of the contributions by P. Chomaz, F. Gulminelli and by O. Lopez and M.F. Rivet, as also by B. Tamain, to this topical issue.

3.2 Atomic clusters

As there are several examples for nuclear multifragmentation in this paper I will show the analogous development of the fragmentation of a single charged cluster of 3000 Na atoms with rising excitation energy from the evaporation of a few Na atoms over multifragmentation into monomers, dimers up to 10-mers towards finally the total vaporisation of the original cluster (see fig. 2). Notice that this occurs all within the range of the backbending (*i.e.* the negative heat capacity) of the caloric curve.

To compare with usual macroscopic conditions, the calculations were done at each energy using a volume $V(E)$ such that the microcanonical pressure $P = \frac{\partial S}{\partial V} / \frac{\partial S}{\partial E} = 1\,\mathrm{atm}$. The inserts on the top of the figure give the mass distribution at the various points. *E.g.*, in insert 1 the label "4:1.295" means 1.295 quadrimers on average. This gives a detailed insight into what happens with rising excitation energy over the transition region: At the beginning ($e^* \sim 0.442\,\mathrm{eV}$) the liquid sodium drop evaporates 329 single atoms and 7.876 dimers and 1.295 quadrimers on average. At energies per atom $e \gtrsim 1\,\mathrm{eV}$ the drop starts to fragment into several small droplets ("intermediate mass fragments") *e.g.* at point 3: 2726 monomers, 80 dimers, ~ 5 trimers, ~ 15 quadrimers and a few heavier ones up to 10-mers. The evaporation residue disappears. This multifragmentation finishes at point 4. It induces the strong backward swing of the caloric curve $T(E)$. Above point 4 one has a gas of free monomers and at the beginning a few dimers. This transition scenario has a lot of similarity with nuclear multifragmentation. The total inter-phase surface area $\propto N_{eff}^{2/3} = \sum_i N_i^{2/3}$ with $N_i \geq 2$ (N_i the number of atoms in the i-th cluster) stays roughly constant up to point 3 even though the number of fragments ($N_{fr} = \sum_i$) rises monotonically. Notice, the caloric curve between point 1 and 2 looks like the "compound nucleus for ever" proposed by [45], though the temperature is higher than T_{tr} and the decay is not evaporation for ever. In contrast to claims in [45] the phase transiton finishes with considerable multifragmentation and a deep back-bend of the caloric curve $T(E)$.

3.3 Fragmentation of astrophysical systems

Self-gravitation leads to a non-extensive potential energy $\propto N^2$. No thermodynamic limit exists for E/N and no canonical treatment makes sense. At negative total energies these systems have a negative heat capacity. This was for a long time considered as an absurd situation within canonical statistical mechanics with its thermodynamic "limit". However, within our geometric theory this is just a simple example of the pseudo-Riemannian topology of the microcanonical entropy $S(E, N)$ provided that we restrict to densities $\leq$ the density of normal hydrogen burning stars, *i.e.* to ordinary visible stars. We treated the various phases of a self-gravitating cloud of particles as a function of the total energy and angular momentum as shown in fig. 3. Clearly, these are the most important constraints in stellar physics. The necessity of using "extensive" instead of "intensive" control parameter is explicit in astrophysical problems. *E.g.*, for the description of rotating stars one conventionally works at a given temperature

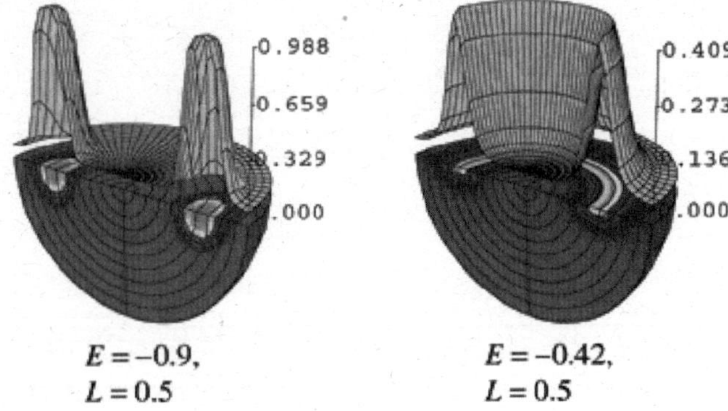

Fig. 3. Contour plots and density profiles of a rotating, self-gravitating N-body system showing the formation of a stable double cluster (left) and an unstable ring (right) at different energies. The double-cluster structure illustrates the spontaneous breaking of rotational symmetry at intermediate energy and high angular momentum (from [47]).

and fixed angular velocity Ω, cf. [46]. Of course in reality there is neither a heat bath nor a rotating disk. Moreover, the latter scenario is fundamentally wrong as at the periphery of the disk the rotational velocity may even become larger than velocity of light. Non-extensive systems like astro-physical ones do not allow a "field-theoretical" description controlled by intensive fields!

E.g., configurations with a maximum of random energy

$$E_{random} = E - \frac{\Theta \Omega^2}{2} - E_{pot} \qquad (20)$$

and consequently with the largest entropy are the ones with smallest moment of inertia Θ, compact single stars. Just the opposite happens when the angular momentum L and not the angular velocity Ω are fixed:

$$E_{random} = E - \frac{L^2}{2\Theta} - E_{pot}. \qquad (21)$$

Then configurations with large moment of inertia are maximizing the phase space and the entropy. *I.e.* eventually double or multistars are produced, as observed in reality.

In fig. 4 one clearly sees the rich and realistic microcanonical phase diagram of a rotating gravitating system controlled by the "extensive" parameters energy and angular momentum [47].

3.4 Outlook

It is a deep and fascinating aspect of *nuclear* fragmentation: First, in nuclear fragmentation we can measure the *whole statistical distribution* of the ensemble event by event including eventual inter-phase fluctuations. This is interesting as the character of the distribution, deeply bimodal *vs.* energy or more equal, tells about the constraint in the experiment, *e.g.* by temperature (unlikely) or by energy. Not only their mean values are of physical interest. Statistical mechanics can be explored from its first

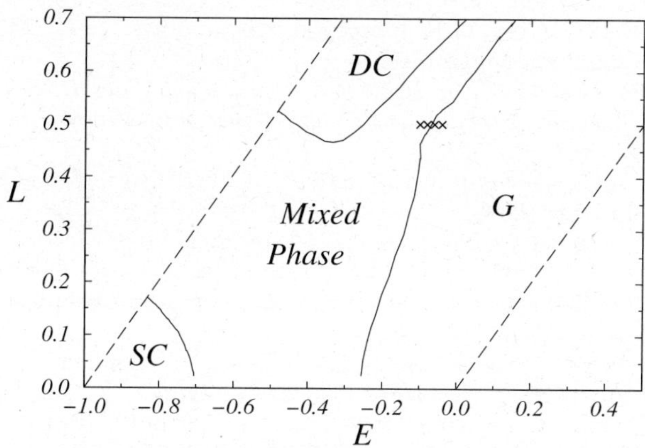

Fig. 4. Phase diagram of rotating self-gravitating systems in the energy angular-momentum (E, L)-plane [47]. DC: region of double stars, G: gas phase, SC: single stars. In the mixed region one finds various exotic configurations like ring systems in coexistence with gas, double stars or single stars. In this region of phase separation the heat capacity is negative and the entropy $S(E, L)$ is convex. The dashed lines $E - L = -1$ (left) and $E = L$ (right) delimit the region where systematic calculations were carried out.

microscopic principles in any detail well away from the thermodynamic limit. By our studies of nuclear fragmentation we found [30,31] the very general appearance of a negative heat capacity and the necessary convexity of the entropy $S(E)$ at any phase separation which seems to be little known in thermodynamics. *Clausius' version of the second law "heat always flows from hot to cold" is in general violated at any phase separation even in macroscopic systems.*

In nuclear fragmentation there may be other conserved control parameters besides the energy: *e.g.* in the recent paper by Lopez *et al.* [48] a bimodality in the mass asymmetry of the fragments is demonstrated to be controlled by the transferred spin and not by excitation energy. This is an interesting, though still theoretical, example of the rich facets of the fragmentation phase transition in *finite* systems which goes beyond the liquid-gas transition and *does not exist in chemistry.* Angular momentum is a very crucial control parameter in stellar systems.

Second, and this may be more important: For the first time phase transitions to *non-homogeneous phases* can be studied where these phases are within themselves composed of several nuclei. This situation is very much analogous to multistar systems like rotating double stars during intermediate times, when nuclear burning prevents their final implosion. The occurrence of negative heat capacities is an old well-known peculiarity of the statistics of self-gravitating systems [26,49]. Also these cannot be described by a canonical ensemble. It was shown in [21,27] how the *microcanonical* phase space of these self-gravitating systems has many of the realistic configurations which are observed. Of course, the question whether these systems really fill uniformly this phase space, *i.e.* whether they are interim equilibrated or not is not proven by this observation though it is rather likely.

Microcanonical thermostatistics is proven to give a realistic, objective picture of a broad scenario of real physical phenomena, much broader than conventional canonical thermodynamics. Moreover, one of the original objects of thermodynamics, the description of the liquid-gas phase *separation* in steam engines, can now be understood within statistical mechanics.

I am very grateful to Francesca Gulminelli and to Wolfgang Trautmann for many helpful comments that improved this manuscript.

References

1. J. Randrup, S.E. Koonin, Nucl. Phys. A **356**, 223 (1981).
2. D.H.E. Gross, Meng Ta-chung, *Production mechanism of large fragments in high energy nuclear reactions*, in *Proceedings of the 4th Nordic Meeting on Intermediate and High Energy Nuclear Physics, Geilo, Norway, Jan 5-9, 1981* (University of Lund, 1981) p. 29.
3. D.H.E. Gross, Phys. Scr. T **5**, 213 (1983).
4. D.H.E. Gross, *Microcanonical thermodynamics: Phase transitions in "Small" systems*, Lect. Notes Phys., Vol. **66** (World Scientific, Singapore, 2001).
5. J.P. Bondorf, R. Donangelo, I.N. Mishustin, H. Schulz, K. Sneppen, Nucl. Phys. A **444**, 460 (1985).
6. Ph. Chomaz, F. Gulminelli, Nucl. Phys. A **647**, 153 (1999).
7. Al.H. Raduta, Ad.R. Raduta, Phys. Rev. C **56**, 2059 (1997).
8. A. Le Fevre, M. Ploszajczak, V.D. Toneev, Phys. Rev. C **60**, 051602 (1999) http://xxx.lanl.gov/abs/nucl-th/9901099.
9. D.H.E. Gross, *Thermo-Statistics or Topology of the Microcanonical Entropy Surface*, Lect. Notes Phys., Vol. **602** (Springer, 2002) pp. 21–45; http://xxx.lanl.gov/abs/cond-mat/0206341.
10. C.N. Yang, T.D. Lee, Phys. Rev. **87**, 404 (1952).
11. L. Boltzmann, Sitzungsber. Akad. Wiss. Wien **II**, 67 (1877).
12. L. Boltzmann, Wien. Ber. **76**, 373 (1877).
13. E.A. Guggenheim, *Thermodynamics, An Advanced Treatment for Chemists and Physicists* (North-Holland Personal Library, Amsterdam, 1967).
14. F. Gulminelli, Ph. Chomaz, V. Duflot, *Abnormal kinetic energy fluctuations and critical behaviors in the microcanonical lattice gas model*, Caen preprint, LPCC 99-17, 1999.
15. R. Clausius, Ann. Phys. (Poggendorff) Chem. **79**, 368; 500 (1850).
16. R. Clausius, Ann. Phys. (Poggendorff) Chem. **93**, 481 (1854).
17. R. Clausius, Ann. Phys. (Poggendorff) Chem. **125**, 353 (1865).
18. I. Prigogine, *Thermodynamics of Irreversible Processes* (John Wiley & Sons, New York, 1961).
19. J.E. Kilpatrick, *Classical thermostatistics*, in *Statistical Mechanics*, edited by H. Eyring, No. II (Academic Press, New York, 1967) Chapt. 1, pp. 1–52.
20. D.H.E. Gross, *Ensemble probabilistic equilibrium and non-equilibrium thermodynamics without the thermodynamic*

limit, in *PQ-QP: Quantum Probability, White Noise Analysis*, edited by Andrei Khrennikov (ACM, World Scientific, Boston, 2001) pp. 131–146.

21. D.H.E. Gross, Entropy **6**, 158 (2004).

22. R.D. Carlitz, Phys. Rev. D **5**, 3231 (1972).

23. C. Bréchignac, Ph. Cahuzac, F. Carlier, J. Leygnier, J.Ph. Roux, J. Chem. Phys. **102**, 1 (1995).

24. P. Hertz, Ann. Phys. (Leipzig) **33**, 537 (1910).

25. D.H.E. Gross, *Nuclear statistics, microcanonical or canonical? the physicists vs. the chemists approach*, in *Proceedings of the XLIII International Winter Meeting on Nuclear Physics, Bormio, Italy, 2005*, edited by I. Iori, A. Bortolotti, Ricerca Scientifica ed Educazione Permanente, Suppl. no. **124** (Milano, 2005) p. 148; `http://arXiv.org/abs/nucl-th/0503065` (2005).

26. D. Lynden-Bell, R. Wood, Mon. Not. R. Astron. Soc. **138**, 495 (1968).

27. D.H.E. Gross, Physica A **340**, 76 (2004) cond-mat/0311418.

28. L.G. Moretto, J.B. Elliot, L. Phair, G. J. Wozniak, Phys. Rev. C **66**, 041601(R) `http://arXiv.org/abs/nucl-th/0208024`.

29. O. Schapiro, D.H.E. Gross, A. Ecker, *Microcanonical Monte Carlo*, in *Proceedings of the First International Conference on Monte Carlo and Quasi-Monte Carlo Methods in Scientific Computing, Las Vegas, Nevada, 1994*, Lect. Notes Statistics, Vol. **106** (Springer, 2005) pp. 346–353.

30. D.H.E. Gross, M.E. Madjet, *Microcanonical vs. canonical thermodynamics*, `http://xxx.lanl.gov/abs/cond-mat/9611192` (1996).

31. D.H.E. Gross, M.E. Madjet, *Cluster fragmentation, a laboratory for thermodynamics and phase-transitions in particular*, in *Proceedings of Similarities and Differences between Atomic Nuclei and Clusters, Tsukuba, Japan 97*, edited by Abe, Arai, Lee, Yabana (The American Institute of Physics, 1997) pp. 203–214.

32. D.H.E. Gross, Phys. Chem. Chem. Phys. **4**, 863 (2002) `http://arXiv.org/abs/cond-mat/0201235`.

33. D.H.E. Gross, J.F. Kenney, J. Chem. Phys. **122**, 224111 (2005) cond-mat/0503604.

34. P. Glansdorff, I. Prigogine, *Thermodynamic Theory of Structure, Stability and Fluctuations* (John Wiley & Sons, London, 1971).

35. C. Bustamante, J. Liphardt, F. Ritort, Phys. Today **58**, 43, July (2005).

36. C. Jarzynski, Phys. Rev. Lett. **78**, 2690 (1997).

37. D.H.E. Gross, *Flaw of Jarzynski's equality when applied to systems with several degrees of freedom*, cond-mat/0508721 (2005).

38. D.H.E. Gross, Reply to Jarzynski's comment cond-mat/0509344, cond-mat/0509648 (2005).

39. D.H.E. Gross, Phys. World **17**, 23, October (2004).

40. J.E. Finn *et al.*, Phys. Rev. Lett. **49**, 1321 (1982).

41. F. Gulminelli, Ph. Chomaz, O. Julliet, M.J. Ison, C.O. Dorso, *Information theory of open fragmenting systems*, `lanl.arXiv.org/nucl-th/0511012` (2005).

42. M. D'Agostino *et al.*, Phys. Lett. B **473**, 219 (2000).

43. M. D'Agostino, M. Bruno, F. Gulminelli, R. Bougault, F. Cannata, Ph. Chomaz, R. Bougault, F. Gramegna, N. le Neindre, A. Moroni, G. Vannini, Nucl. Phys. A **734**, 512 (2004).

44. M. Pichon, the INDRA and ALADIN Collaborations, *Bimodality in binary* Au + Au *collisions from* 60 *to* 100 MeV/u, in *Proceedings of the XLI Winter Meeting on Nuclear Physics, Bormio, Italy, 2003*, edited by I. Iori, A. Moroni, Ricerca Scientifica ed Educazione Permanente, Suppl. no. **120** (Milano, 2003) p. 149.

45. D.R. Bowman *et al.*, Phys. Lett. B **189**, 282 (1987).

46. P.H. Chavanis, M. Rieutord, Astron. Astrophys. **412**, 1 (2003) `http://arXiv.org/abs/astro-ph/0302594`.

47. E.V. Votyakov, H.I. Hidmi, A. De Martino, D.H.E. Gross, Phys. Rev. Lett. **89**, 031101 (2002) `http://arXiv.org/abs/cond-mat/0202140`.

48. O. Lopez, D. Lacroix, E. Vient, *Bimodality as signal of liquid-gas phase transition in nuclei?*, `http://arXiv.org/abs/nucl-th/0504027` (2005).

49. W. Thirring, Z. Phys. **235**, 339 (1970).

Eur. Phys. J. A **30**, 303–310 (2006)
DOI 10.1140/epja/i2006-10124-7

Links between heavy ion and astrophysics

C.J. Horowitz[a]

Nuclear Theory Center and Department of Physics, Indiana University, Bloomington, IN 47405, USA

Received: 13 February 2006 /
Published online: 31 October 2006 – © Società Italiana di Fisica / Springer-Verlag 2006

Abstract. Heavy-ion experiments provide important data to test astrophysical models. The high-density equation of state can be probed in HI collisions and applied to the hot protoneutron star formed in core collapse supernovae. The parity radius experiment (PREX) aims to accurately measure the neutron radius of ^{208}Pb with parity-violating electron scattering. This determines the pressure of neutron-rich matter and the density dependence of the symmetry energy. Competition between nuclear attraction and Coulomb repulsion can form exotic shapes called nuclear pasta in neutron star crusts and supernovae. This competition can be probed with multifragmentation HI reactions. We use large-scale semiclassical simulations to study nonuniform neutron-rich matter in supernovae. We find that the Coulomb interactions in astrophysical systems suppress density fluctuations. As a result, there is no first-order liquid-vapor phase transition. Finally, the virial expansion for low-density matter shows that the nuclear vapor phase is complex with significant concentrations of alpha particles and other light nuclei in addition to free nucleons.

PACS. 26.50.+x Nuclear physics aspects of novae, supernovae, and other explosive environments – 26.60.+c Nuclear matter aspects of neutron stars

1 Introduction

Most of the visible mass and energy of the Universe is in atomic nuclei. This suggests some common goals for heavy-ion (HI) research. We can study nuclear matter under extreme conditions of density (both high and low), temperature, size, and isospin. The insight gained from this study can then be applied to: 1) the fundamental behavior of many-particle quantum systems such as cold atoms in laboratory traps, 2) quantum chromodynamics at high densities, and 3) compact objects in astrophysics such as neutron stars, supernovae, gamma-ray bursts, accretion disks, and the origin of the chemical elements.

In this article we discuss links between HI and astrophysics. We need to extrapolate HI data to astrophysical conditions. First, one must extrapolate to longer times. Core collapse supernovae (SN) are giant stellar explosions that produce neutron stars and chemical elements and accelerate cosmic rays. In SN the core of a massive star collapses in milliseconds. This is a remarkably short time scale for a planet-sized object that is more massive then the Sun. However a ms is 10^{20} fm/c! and very long compared to the time scale of a few hundred fm/c for a HI collision. Therefore, SN involve matter that has had plenty of time to reach thermodynamic equilibrium, while this is not always the case in HI collisions.

Second, one must extrapolate to larger systems. A neutron star is a giant atom with a mass number of 10^{57} and an atomic number of 10^{56}. It is about 10 km in radius, or 18 orders of magnitude larger then a conventional atomic nucleus. For this nearly infinite system, Coulomb interactions play a crucial role and require charge neutrality between positively charged nuclear matter and a background electron gas. Thus, one must consider the differences in Coulomb interactions of finite HI collisions compared to those of an infinite system.

Third, one must extrapolate to larger isospin. Astrophysical systems are often more neutron rich than the heavy ions that are available in the laboratory. This extrapolation depends on the symmetry energy. The symmetry energy $S(\rho)$ describes how the energy of nuclear matter rises when one moves away from equal numbers of neutrons and protons. The density dependence of $S(\rho)$ is very important for many astrophysical systems, and can be determined from HI experiments [1]. Furthermore, future experiments with more neutron-rich radioactive beams may provide additional information.

There are errors associated with these extrapolations. Nevertheless, laboratory HI experiments provide real data that can be used to place important constraints on many astrophysical models. Without the HI data, one may be forced to use untested theoretical assumptions that have large errors.

In this paper, we discuss links between HI and astrophysics. Section 2 discusses the high-density equation of

---

[a] e-mail: `horowit@indiana.edu`

state (EOS) and its implications for neutron star structure and supernovae. Next, we consider the EOS at subnuclear densities. Section 3 discusses the parity radius experiment (PREX) to measure the neutron skin thickness in ^{208}Pb. This determines the density dependence of the symmetry energy and the neutron matter EOS at low densities. Section 4 presents molecular-dynamics simulations of the nonuniform neutron-rich matter in the inner crusts of neutron stars. These nuclear-pasta phases may be closely related to multifragmentation in HI collisions. Finally, Section 5 discusses the nuclear-matter liquid-vapor phase transition in supernovae.

2 The high-density equation of state

The equation of state (EOS) describes the pressure P of nuclear matter as a function of density ρ, temperature T, and proton fraction Y_p. Heavy-ion experiments can probe the EOS at high T and ρ and for proton fractions near $Y_p \approx 1/2$. For example, flow observables can be used to constrain the EOS with the help of semiclassical simulations [2]. In addition, yields of other particles such as kaons can provide additional probes of the EOS [3].

Unfortunately, it does not appear possible to directly produce cold dense matter in the laboratory. The energy needed to produce high compression always seems to produce high temperatures because there is no way to get the entropy out. Therefore the authors of ref. [2] assumed the temperature dependence of the EOS was that predicted by some simple mean-field models.

Neutron stars (NS), on the other hand, provide unique probes of the EOS of cold dense matter. Although they are formed hot in SN explosions, they have plenty of time to cool via neutrino emission. Thus NS can probe new forms of cold dense matter such as color superconductors that may not be accessible in the laboratory.

It is an exciting time to study neutron stars [4]. Powerful X-ray telescopes such as Chandra and XMM-Newton and other instruments are slowly turning NS from theoretical curiosities to detailed, well-observed, worlds. Some NS in binary systems have well-measured masses near $1.4M_\odot$. However there are now indications of more massive stars [4,5]. The structure of a neutron star depends only on the EOS of cold neutron-rich matter. The stiffer the EOS (higher pressure for given density), the larger the radius $R(M)$ of a NS, of given mass M. A typical neutron matter EOS may give $R(M) \approx 11$–12 km for $M = 1.4M_\odot$, while a stiff EOS could give $R(M) \approx 13$–14 km.

There is great interest in possible exotic phases for high-density matter. The central density of a NS can be several times the nuclear density. An exotic phase such as strange matter or a color superconductor could lead to a soft high-density EOS. (If the exotic phase has a higher pressure than conventional matter, it may not be thermodynamically favored.) This could lead to a NS radius of 10 km or less.

Astronomers are working hard to measure the radii of NS, see, for example, [6]. One approach follows from thermodynamics and the properties of a blackbody radiator.

The luminosity L (total energy radiated per unit time) of an isolated star is related to the surface temperature T and apparent radius R as follows:

$$L = 4\pi R^2 \sigma T^4, \tag{1}$$

where σ is the Stephen Boltzmann constant. The surface temperature can be deduced from X-ray spectra, while L follows from the apparent magnitude of the star and an accurate measurement of its distance. Unfortunately, there are a number of complications with this simple formula. Neutron stars are not perfect blackbodies, so corrections from realistic stellar atmosphere models may need to be included. Interstellar absorption can influence estimates of both L and T. The temperature may not be uniform over the stars surface. For example T can be larger at the magnetic poles compared to the equator because the thermal conductivity is larger along the magnetic-field direction. The distance to the star may depend on a very delicate measurement of parallax. Finally, gravity is so strong that the curvature of space is important. Some light emitted from the far side of the star can be detected and contributes to L because of this curvature. This increases the apparent radius by about 30%. Nevertheless, astronomers hope to have a number of increasingly accurate measurements of NS radii. Comparing results from several different NS measurements may provide a good check of these corrections.

In addition to cold NS, one is also interested in the structure of very young neutron stars as they are being formed in supernova explosions. These hot, lepton-rich, protoneutron stars can have maximum temperatures as high as 50 MeV. The EOS of protoneutron stars may be directly related to the EOS deduced from energetic HI collisions because the temperature, density, and proton fraction can be similar. Furthermore, this protoneutron star EOS is important for SN simulations [7].

3 The parity radius experiment and the low-density EOS

We now discuss the EOS at subnuclear densities. This has many implications for the structure of NS crusts. One can obtain information on the low-density EOS from both HI collisions and from precision measurements on stable nuclei. The parity radius experiment (PREX) aims to measure the neutron radius of ^{208}Pb, accurately and model independently, via parity-violating electron scattering. As we discuss below, the neutron radius in Pb determines the density dependence of the symmetry energy and the EOS of low-density neutron matter. This information, from a precision experiment on a stable nucleus, nicely complements the information from HI or radioactive beam experiments.

Parity violation probes neutrons because the weak charge of a neutron is much larger than the weak charge of a proton [8]. In the standard model the proton weak charge is proportional to the small factor $1 - 4\sin^2\theta_W$, where θ_W

is the weak mixing angle. One can isolate weak contributions by measuring the parity-violating asymmetry A for elastic electron nucleus scattering. This is the cross-section difference for the scattering of positive $d\sigma/d\Omega_+$ and negative $d\sigma/d\Omega_-$ helicity electrons,

$$A = \frac{d\sigma/d\Omega_+ - d\sigma/d\Omega_-}{d\sigma/d\Omega_+ + d\sigma/d\Omega_-} . \tag{2}$$

In Born approximation A is [8]

$$A = \left(\frac{G_F Q^2}{4\pi\alpha 2^{1/2}} \right) \frac{F_W(Q)}{F_{ch}(Q)} , \tag{3}$$

where G_F is the Fermi constant, α the fine structure constant and Q the momentum transfer. The charge form factor $F_{ch}(Q)$ is the Fourier transform of the charge density, that is known from electron scattering. The weak form factor $F_W(Q)$ is the Fourier transform of the weak charge density. This is dominated by the neutron density and thus the neutron density can be deduced from measurements of A. Note, Coulomb distortions make $\approx 30\%$ corrections to A for scattering from a heavy nucleus [9]. However these can be accurately calculated.

The Jefferson laboratory PREX [10] aims to measure elastic scattering of $850\,\mathrm{MeV}$ electrons from $^{208}\mathrm{Pb}$ at six degrees in the laboratory. The goal is to measure $A \approx 0.6\,\mathrm{ppm}$ with an accuracy of 3%. This allows the neutron r.m.s. radius of $^{208}\mathrm{Pb}$ to be deduced to 1%. A full discussion of the experiment and many possible corrections is contained in [11].

We now discuss the implications of the radius measurement. Heavy nuclei are expected to have a neutron-rich skin. The thickness of this skin depends on the pressure of neutron-rich matter. The larger the pressure, the larger the neutron radius as neutrons are forced out against surface tension. Alex Brown showed that there is a strong correlation between the neutron radius in Pb and the EOS of pure neutron matter, as predicted by many different mean-field interactions [12]. Therefore, the neutron radius in Pb determines P for neutron matter at $\rho \approx 0.1\,\mathrm{fm}^{-3}$. (This is about $2/3\rho_0$ and represents some average of the surface and interior density of Pb.) The pressure depends on the derivative of the energy with respect to density. The energy of pure neutron matter $E_{neutron}$ is the energy of symmetric nuclear matter $E_{nuclear}$ plus the symmetry energy $S(\rho)$,

$$E_{neutron} \approx E_{nuclear} + S(\rho). \tag{4}$$

The pressure depends on $dE_{nuclear}/d\rho$ (which is small and largely known near nuclear density ρ_0) and $dS(\rho)/d\rho$. Therefore, *the neutron radius in* Pb *determines the density dependence of the symmetry energy* $dS(\rho)/d\rho$ *for densities near* ρ_0.

Neutron stars are expected to have a solid neutron-rich crust over a liquid interior, while heavy nuclei have a neutron-rich skin. Both the skin of a nucleus, and the NS crust are made of neutron-rich matter at similar densities. The common unknown is the EOS of low-density neutron

matter. As a result, we find a strong correlation between the neutron radius of $^{208}\mathrm{Pb}$ and the transition density of NS crusts [13]. The thicker the skin in Pb, the faster the energy of neutron matter rises with density, and the more quickly the uniform liquid phase is favored. Therefore, a thick neutron skin in Pb implies a low transition density (maximum density) for the NS crust.

The composition of a neutron star depends on the symmetry energy. In beta equilibrium the neutron chemical potential μ_n is equal to that for protons μ_p plus electrons μ_e, $\mu_n = \mu_p + \mu_e$. Neutron stars are about 90% neutrons and 10% protons plus electrons. However, a large symmetry energy will favor more equal numbers of neutrons and protons and increase the proton fraction. Thus, the composition of matter in the center of a neutron star depends on the symmetry energy at high density.

Neutron stars cool by neutrino emission from the interior. If the proton fraction is large, above about 0.13, then neutrons near the Fermi surface can beta decay to protons and electrons near their Fermi surfaces and conserve both momentum and energy. This leads to the direct URCA process $n \to p + e + \bar{\nu}_e$ followed by $e + p \to n + \nu_e$ that will efficiently cool a NS by rapidly radiating $\nu\bar{\nu}$ pairs. The neutron radius of Pb constrains the density dependence of the symmetry energy near ρ_0. This is the crucial piece of information for extrapolating to find the symmetry energy at large densities. We find that if the neutron minus proton r.m.s. radii in $^{208}\mathrm{Pb}$ is larger then $0.25\,\mathrm{fm}$, all of the mean-field EOS models considered allow direct URCA for a $1.4 M_\odot$ NS [14]. Alternatively, if this skin thickness is less then $0.2\,\mathrm{fm}$, none of the mean-field models allow direct URCA.

Note, the direct URCA process takes place in the high-density interior of a NS at a few or more ρ_0. Therefore, the above relation with the skin thickness in Pb involves an extrapolation to higher density. Alternatively, energetic HI collisions can directly produce high densities. Therefore it would be extremely useful if one could infer the high-density symmetry energy from HI observables. Although potentially difficult and model dependent, *measuring the symmetry energy at high density is perhaps the single most important HI experiment for the structure of NS.*

We close this section with a short discussion of other ways to determine the density dependence of the symmetry energy. If one assumes the symmetry energy depends on a power of the density,

$$S(\rho) \approx S_0 \rho^\gamma, \tag{5}$$

then the power γ can be approximately related to the skin thickness in $^{208}\mathrm{Pb}$ as follows,

$$\langle r_n^2 \rangle^{1/2} - \langle r_p^2 \rangle^{1/2} \approx 0.22\gamma + 0.06\,\mathrm{fm}. \tag{6}$$

This relation is a simple fit to several mean-field calculations, see also [15]. As discussed by Li *et al.* [16] and by Colonna and Tsang [17] in the section on isospin properties of this topical issue, the power γ can be deduced from HI data involving observables such as isoscaling and isospin diffusion. Finally we mention a recent review article which discusses the symmetry energy in astrophysics [18].

4 Nuclear pasta and multifragmentation

Nuclei involve an important interplay between Coulomb and nuclear interactions. Indeed, all baryonic matter is *frustrated*. Nucleons tend to be correlated at short distance, because of short-range nuclear attraction, and anticorrelated at long distances because of Coulomb repulsion. Normally, the nuclear and atomic (or Coulomb) length scales are well separated so nucleons bind into nuclei that are segregated on a crystal lattice.

However, at densities just below ρ_0, in the inner crust of neutron stars and in supernovae, Coulomb and nuclear scales become comparable. Under these conditions, the surface energy, from nuclear attraction that favors spherical shapes, and the Coulomb energy, that can favor nonspherical shapes, compete. This results in exotic nuclear-pasta phases [19] that can involve spherical (meat ball), rod (spaghetti), plate (lasagna), or other shapes.

The Coulomb frustration in nuclear pasta is similar to the frustration found in many condensed-matter systems. Frustrated systems cannot satisfy all of their elementary interactions [20]. Examples range from magnetism [21] to protein folding [22]. Because frustration raises the energy of the ground state, these systems are characterized by a very large number of low-energy excitations that lead to unusual dynamics.

Nuclear pasta may be important for a number of neutron star observables. For example, r-modes are collective oscillations of NS that can radiate gravitational waves and may control pulsar spin periods [23]. The sheer viscosity of the nuclear pasta at the interface between the solid crust and liquid interior of a NS may determine the damping of r-modes. This viscosity in turn may depend crucially on the exotic shapes of the pasta. Some other relevant pasta properties include thermal conductivity, sheer modules, and neutrino emissivity.

Core collapse supernovae radiate of order 10^{58} neutrinos. The very large gravitational binding energy of the newly formed neutron star (100 to $200\,\mathrm{MeV}/A$) is released, almost entirely, in neutrinos. No other known particles can transport the energy out of the very dense core during the few second duration of the explosion. These 10 to $20\,\mathrm{MeV}$ neutrinos can scatter coherently from the nuclear pasta because their wavelengths are comparable to the sizes of the pasta shapes. Thus, neutrino-pasta scattering [24] may be important for supernova dynamics.

Nuclear pasta in astrophysics may be closely related to multifragmentation in laboratory heavy-ion collisions. Heavy ions, at moderate excitation energy, are observed to break apart into several large fragments [25]. This process may occur at the same, slightly subnuclear, densities where nuclear pasta forms. Furthermore, both pasta formation and multifragmentation are driven by the same nuclear and Coulomb energies. One may be able to tune

the interactions used in semiclassical simulations of multifragmentation, in order to reproduce laboratory data. Then, the same simulations and interactions can be used to describe nuclear pasta. This allows laboratory data to be used to constrain astrophysical models.

It is important to go beyond mean-field models in describing nuclear pasta. Mean-field interactions, fit to conventional nuclei, may not be appropriate for complex nonuniform pasta. Furthermore, pasta may not be described well by a Maxwell construction, such as in ref. [26] involving uniform liquid and uniform gas phases. In addition, the Coulomb interaction plays a crucial role in astrophysics. The system must be electrically neutral. Therefore, the positive charge density of the pasta is constrained to be equal and opposite to the electron density. Finally, one should consider a wide variety of possible shapes for the nuclear pasta. Variational calculations involving a few simple shapes, such as rods or plates, may miss more complicated configurations.

In ref. [24] we consider a simple semiclassical model where neutrons and protons interact via short-ranged nuclear and screened Coulomb forces. The electrons form a very degenerate Fermi gas and are not included explicitly. Instead, the very slight polarization of the electrons lead to a Thomas Fermi screening length λ for the Coulomb interactions between protons. Our model Hamiltonian is

$$H = \sum_i \frac{p_i^2}{2m} + \sum_{i<j} V(i,j), \qquad (7)$$

where the two-body potential is

$$V(i,j) = ae^{-r_{ij}^2/\Lambda} + [b+c\tau_z(i)\tau_z(j)]e^{-r_{ij}^2/2\Lambda} + V_c(i,j). \quad (8)$$

Here the distance between the particles is $r_{ij} = |\mathbf{r}_i - \mathbf{r}_j|$ and the isospin of the j-th particle is $\tau_z(j) = 1$ for a proton and $\tau_z(j) = -1$ for a neutron. The model parameters a, b, c, and Λ have been fit to reproduce the binding energy and saturation density of nuclear matter along with a reasonable symmetry energy [24]. The screened Coulomb interaction is

$$V_c(i,j) = \frac{e^2}{r_{ij}}e^{-r_{ij}/\lambda}\tau_p(i)\tau_p(j), \qquad (9)$$

where $\tau_p(j) = (1 + \tau_z(j))/2$ is the nucleon charge and λ is the screening length from the slight polarization of the electrons.

This model yields large nuclei or pieces of pasta that are heavy and have thermal Compton wavelengths much shorter than their inter-particle spacing. This motivates our semiclassical approximation. More elaborate interactions can be employed, such as the QMD calculations of Watanabe *et al.* [27]. However, our simple interaction reproduces nuclear saturation and includes Coulomb interactions. We believe these are the most important features that determine the long-range structure of the nuclear-pasta phases.

The wavelength of a $10\,\mathrm{MeV}$ supernova neutrino is $120\,\mathrm{fm}$. To determine the pasta structure at this long

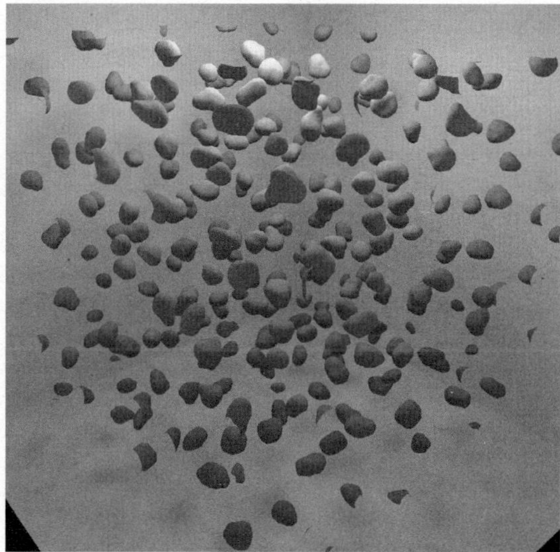

Fig. 1. Proton density iso-surface for a sample configuration of 40000 nucleons at $\rho = 0.01\,\mathrm{fm}^{-3}$, $T = 1\,\mathrm{MeV}$ and proton fraction 0.2. The simulation volume is about 160 fm on a side.

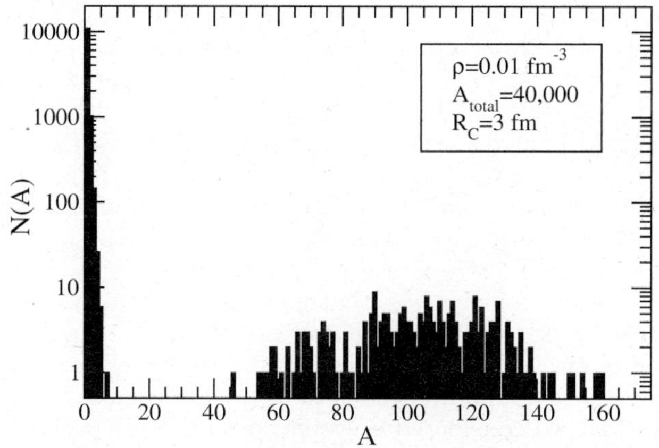

Fig. 2. Fragment size distribution for the sample configuration of fig. 1, see text.

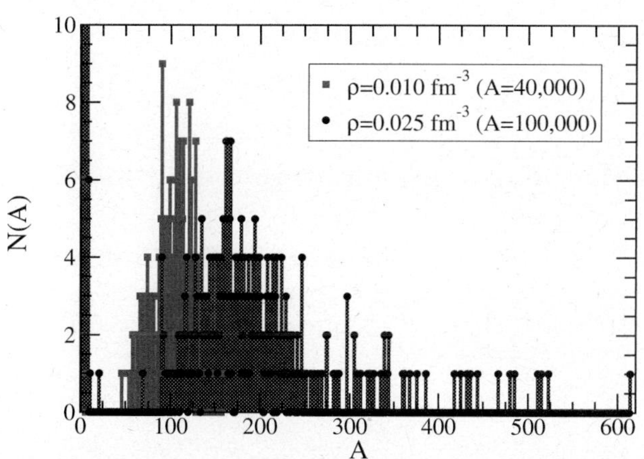

Fig. 3. Fragment size distributions at $\rho = 0.01$ and $0.025\,\mathrm{fm}^{-3}$.

length scale may require simulations involving many particles. For example, at $1/3\rho_0$ there are 100000 nucleons in a cube 120 fm on a side. We have used special purpose MDGRAPE computer hardware to perform molecular-dynamics simulations with 40000 to 200000 nucleons [28].

We are interested in the neutron-rich matter during a supernova. The proton fraction starts near $1/2$ and drops to low values as electron capture proceeds and electron neutrinos diffuse out of the core. Figure 1 shows a sample configuration of 40000 nucleons at a density of $0.01\,\mathrm{fm}^{-3}$, a proton fraction of 0.2, and a temperature of $T = 1\,\mathrm{MeV}$. An iso-surface of the proton density is shown. At this density, most of the protons cluster into neutron-rich nuclei. Between these nuclei, there is a low-density neutron gas that is not shown in fig. 1.

To characterize the heavy nuclei in fig. 1 we have used a clustering algorithm. A nucleon is said to belong to a cluster if it is within a cutoff radius $R_C \approx 3\,\mathrm{fm}$ of at least one other nucleon in the cluster. This divides the 40000 nucleons into about 12000 free neutrons, a collection of light nuclei, and about 250 heavy nuclei as shown in fig. 2. The heavy nuclei have an average mass near $\langle A \rangle \approx 100$ and a $Z/A \approx 0.3$. Note, this Z/A is somewhat greater then the total proton fraction of 0.2 because of isospin distillation. The rest of the neutrons go into the low-density neutron gas. Our simulation results are qualitatively similar to many statistical models such as those of Botvina [29]. The distribution of clusters reflects a balance between binding energy, favoring large clusters, and entropy, that favors light clusters. However in detail, the distribution can be sensitive to the nuclear masses predicted by a given model.

As the density increases, the background electron gas cancels more of the Coulomb interaction. This allows the formation of larger clusters. In fig. 3 we compare the cluster distribution at $\rho = 0.01\,\mathrm{fm}^{-3}$ to that at $\rho = 0.025\,\mathrm{fm}^{-3}$ (for the same $T = 1\,\mathrm{MeV}$ and proton fraction 0.2). At $\rho = 0.025\,\mathrm{fm}^{-3}$ the average mass is now $\langle A \rangle \approx 200$ and there is a tail in the distribution to very heavy nuclei.

Finally, as the density is increased further the nuclei start to strongly interact. Figure 4 shows an iso-surface of the proton density at $\rho = 0.05\,\mathrm{fm}^{-3}$ ($\approx 1/3\rho_0$). Now spherical nuclei are no longer favored. Instead, long spaghetti-like strands are seen that have complex shapes. The fragment distribution now includes very large clusters whose size scales with the simulation volume. Thus, heavy nuclei have percolated together to form a complex pasta phase. Note that increasing the density still further to $\rho = 0.075\,\mathrm{fm}^{-3}$ ($1/2\rho_0$) results in a transition to uniform nuclear matter, not shown.

The clusters seen in figs. 1 and 4 can be characterized by the static structure factor S_q [24,28]. This describes the degree of coherence for neutrino scattering from the nonuniform system. This is directly analogous to S_q for many complex condensed-matter systems that can be deduced from neutron or X-ray scattering. The static struc-

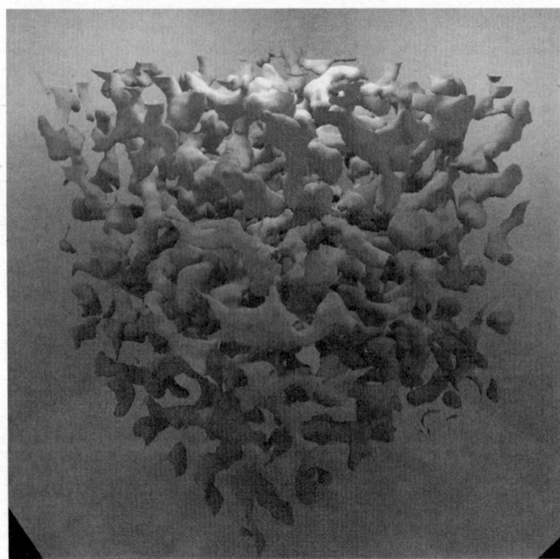

Fig. 4. Proton density iso-surface for a sample configuration of 100000 nucleons at $\rho = 0.05\,\mathrm{fm}^{-3}$, $T = 1\,\mathrm{MeV}$ and proton fraction 0.2. The simulation volume is about 120 fm on a side.

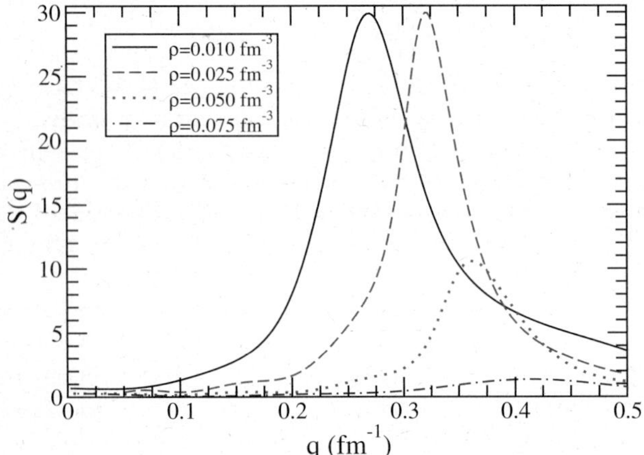

Fig. 5. The static structure factor S_q for $T = 1\,\mathrm{MeV}$ and $Y_p = 0.2$ for the indicated densities.

ture factor coherently sums the reflected waves for neutrino scattering from each neutron in the system,

$$S_q = \sum_{i,j} \exp[i\mathbf{q} \cdot (\mathbf{r}_i - \mathbf{r}_j)], \qquad (10)$$

where $\mathbf{q}$ is the momentum transferred from the neutrino to the system. In fig. 5 we show S_q for densities of 0.01, 0.025, 0.05, and 0.075 fm^{-3}. This scans the density range from largely isolated nuclei (in fig. 1) through the complex pasta phases (fig. 4) to uniform nuclear matter. A large peak is seen in S_q for $q \approx 0.3\,\mathrm{fm}^{-1}$. This corresponds to neutrino nucleus elastic scattering at $\rho = 0.01\,\mathrm{fm}^{-3}$ or coherent neutrino-pasta scattering at $\rho = 0.05\,\mathrm{fm}^{-3}$. Here the neutrino scatters coherently from all of the neutrons in a cluster. This peak largely vanishes for the uniform system at $\rho = 0.075\,\mathrm{fm}^{-3}$.

At low q, S_q is small in fig. 5 because of ion screening. If one places an impurity heavy nucleus or piece of pasta into the system, the other clusters will rearrange because of Coulomb interactions until they act to screen the charge of the impurity. This leads to a reduction of S_q. In the next section, we will use these results for S_q to discuss the liquid-vapor transition.

One can use the time dependence of the molecular-dynamics simulations to calculate the dynamical response function $S(q, w)$ that measures how likely it is for a neutrino to transfer momentum q and energy w to the system. At $\rho = 0.05\,\mathrm{fm}^{-3}$, we find a high-energy peak in $S(q, w)$ that represents plasma oscillations of the charged pasta and a peak at low w that may correspond to nucleons diffusing between the pasta and the vapor [30].

5 Liquid-vapor transition

There is great interest in the transition between a nucleon vapor at low densities and liquid nuclear matter at high density, see, for example, [31]. Often this is described as a first-order phase transition. However, here we would like to discuss two complications to this simple first-order picture that arise in the thermodynamic limit. First, we believe the low-density vapor must necessarily be complex and involve heavier nuclei such as alpha particles in addition to free nucleons. Second, Coulomb interactions replace a first-order liquid-vapor phase transition with complex mixed phases such as nuclear pasta.

The vapor phase, in the limit of very low densities, can be described exactly with the virial expansion [32,33]. Here, the pressure P is expanded in powers of the fugacity $z = \exp(\mu/T)$ where μ is the chemical potential. The second virial coefficient b_2, that gives the z^2 contribution to the pressure, can be calculated exactly in terms of the two-body elastic scattering phase shifts. However, nuclear matter is self-bound and tends to form clusters, see fig. 1. In ref. [33] we considered a system of neutrons, protons, and alpha particles. Because of their large binding energy, alphas tend to be more important than mass-3 nuclei. Furthermore at very low densities, heavy nuclei are disfavored because of their low entropy. We calculated the relevant second virial coefficients from NN, N-α, and α-α elastic scattering phase shifts. This allows one to make model-independent predictions for the alpha-particle fraction in the low-density vapor, see fig. 6 [33]. Errors in this fraction can be estimated from neglected third virial coefficients.

The alpha fraction can be large. Therefore, even at very low densities say $0.001\rho_0$, the vapor, in the thermodynamic limit, must contain more than just free nucleons. Note that the virial expansion is exact in the limit of very low density. It shows that the alpha fraction is nonzero and grows with increasing density, without having to pass through a phase transition.

It is interesting to compare this complex nuclear vapor to steam in the H_2O system. This may be the model for a liquid-vapor phase transition. Clusters of multiple H_2O molecules do indeed form, see, for example, [34]. However, their abundance is very low. In contrast, the large alpha

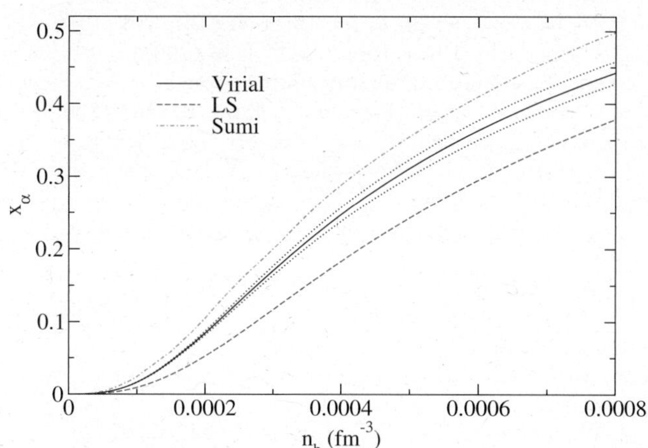

Fig. 6. The alpha-particle mass fraction X_α in symmetric nuclear matter *versus* density at a temperature of 4 MeV as calculated in the virial expansion (solid curve). The error bars are from estimates of the neglected third virial coefficients. The curves labeld LS and Sumi are from phenomenological models, see [33].

binding energy leads to much larger alpha concentrations. Therefore, nuclear vapor may be much more complex than water vapor.

We now discuss a possible first-order liquid-vapor phase transition in astrophysics. A two-phase coexistence region has large density fluctuations as low-density vapor is converted to or from a high-density liquid. Scattering from these fluctuations could greatly reduce the neutrino mean free path in a supernova [35].

The static structure factor, in the long-wavelength limit, $S_{q=0}$ describes fluctuations in the number of neutrons N or density fluctuations,

$$S_{q=0} = \frac{1}{N}(\langle N^2 \rangle - \langle N \rangle^2). \qquad (11)$$

If we assume fluctuations in the neutron density are proportional to fluctuations in the baryon density, this can be written

$$S_{q=0} \approx \left(\frac{N}{N+Z} \right) \frac{T}{\mathrm{d}P/\mathrm{d}n}. \qquad (12)$$

When two phases coexist, the pressure is constant at the vapor pressure, and the derivative of the pressure with respect to density vanishes $\mathrm{d}P/\mathrm{d}n = 0$. Therefore, $S_{q=0}$ diverges in a two-phase coexistence region of a first-order liquid-vapor phase transition.

However, we find in fig. 5 that $S_{q=0}$ is small, from ion screening, instead of diverging from density fluctuations. *Therefore, the system does not undergo a first-order liquid-vapor phase transition.* The complex structures seen in fig. 4 can be viewed as a mixed phase with the positively charged nuclear-pasta liquid in equilibrium with a low-density nucleon vapor that occupies the space between the pasta, and is not shown in fig. 4. However, the average charge density of the pasta must be equal and opposite to the background electron charge density. Therefore,

Coulomb interactions suppress density fluctuations and eliminate a first-order liquid-vapor phase transition.

Note, Coulomb interactions for the relatively small system of a heavy-ion collision, may be smaller and still allow features of a liquid-vapor phase transition. However, Coulomb interactions, in the nearly infinite astrophysical system, may play a larger role suppressing density fluctuations and modifying the liquid-vapor phase transition.

We end this section with some alternative interpretations of our results. One can view the complex density shown in fig. 4 as many microscopic regions of a high-density liquid phase interspersed with a low-density gas phase. Furthermore, this microscopic picture may be useful to describe heavy-ion collisions, with only one (or a few) liquid region(s). However, this picture may have limitations describing large systems in astrophysics. Coulomb interactions strictly limit the size of any single liquid region. Thus, there is no uniform thermodynamic limit. Surface effects will always be important. Furthermore, we do not find the density fluctuations expected of a classical two-phase coexistence region. If by phase, one means a macroscopic region, then fig. 4 cannot represent a macroscopic liquid phase in equilibrium with a macroscopic vapor phase.

This microscopic picture can also be applied to our alpha-particle results in fig. 6. In principle, one can view alpha particles as very tiny drops of liquid. Then the alpha concentration in fig. 6 could represent many tiny regions of a liquid phase in equilibrium with a simple vapor phase composed of only free nucleons. However, given the very small size of alpha particles, we think that this two-phase interpretation is strained.

It may be useful to compare our nonuniform system to a uniform one. As the density is decreased from ρ_0, we find, at some density, the uniform system becomes unstable and a nonuniform system is favored. If one changes the density and temperature very rapidly during a heavy-ion collision, a uniform phase may persist as a metastable state until it reaches a spinodal. At the spinodal, the compressibility is negative and the system may rapidly evolve into a nonuniform state. However there is no way to enforce that the density stays uniform. During a core collapse supernova, the density and temperature change very slowly over a very long time scale of milliseconds. This should allow plenty of time for the system to reach thermodynamic equilibrium. We believe the system will promptly become nonuniform and not pass through a metastable uniform state. As a result the system may never reach the spinodal, and any rapid dynamics associated with the spinodal may not be relevant for astrophysical systems.

Finally, we mention a possible limitation of our results. We have only run simulations for a few densities $\rho = 0.01, 0.025, 0.05,$ and $0.075\,\mathrm{fm}^{-3}$, and a single temperature 1 MeV. Therefore, we cannot rule out a possible critical point, and associated critical fluctuations, for other conditions.

6 Summary

Heavy-ion experiments provide important data to test astrophysical models. In general, one must extrapolate HI data to longer times, larger sizes, and more neutron-rich systems. The high-density equation of state can be probed in HI collisions and applied to the hot protoneutron star formed in core collapse supernovae. The parity radius experiment (PREX) aims to accurately measure the neutron radius of ^{208}Pb with parity-violating electron scattering. This determines the pressure of neutron-rich matter and the density dependence of the symmetry energy. Competition between nuclear attraction and Coulomb repulsion can form exotic shapes called nuclear pasta in neutron star crusts and supernovae. This competition can be probed with multifragmentation HI reactions. A first-order liquid-vapor phase transition has density fluctuations that could impact neutrino interactions in supernovae. We use large-scale semiclassical simulations to study nonuniform neutron-rich matter. We find that the Coulomb interactions in astrophysical systems suppress density fluctuations. As a result, the system does not undergo a first-order liquid-vapor phase transition. Finally, the virial expansion for low-density matter shows that the nuclear vapor phase is complex with significant concentrations of alpha particles and other light nuclei in addition to free nucleons.

Collaborators for this work include Achim Schwenk, Jorge Piekarewicz, Angeles Perez-Garcia, and Don Berry. We thank Marcello Baldo for suggestions. We acknowledge financial support form the U. S. Department of Energy contract DE-FG02-87ER40365.

References

1. M.B. Tsang *et al.*, Phys. Rev. Lett. **92**, 062701 (2004).
2. P. Danielewicz, R. Lacey, W.G. Lynch, Science **298**, 1592 (2002).
3. C. Sturm *et al.*, Phys. Rev. Lett. **86**, 39 (2001).
4. J.M. Lattimer, M. Prakash, Science **304**, 536 (2004).
5. D.J. Niece *et al.*, astro-ph/0508050.
6. R.E. Rutledge *et al.*, Astrophys. J. **559**, 1054 (2001).
7. S. Woosley, H.T. Janka, Nature Phys. **1**, 147 (2005).
8. T.W. Donnelly *et al.*, Nucl. Phys. A **503**, 589 (1989).
9. C.J. Horowitz, Phys. Rev. C **57**, 3430 (1998).
10. http://hallaweb.jlab.org/parity/prex.
11. C.J. Horowitz *et al.*, Phys. Rev. C **63**, 025501 (2001).
12. B. Alex Brown, Phys. Rev. Lett. **85**, 5296 (2000).
13. C.J. Horowitz, J. Piekarewicz, Phys. Rev. Lett. **86**, 5647 (2001).
14. C.J. Horowitz, J. Piekarewicz, Phys. Rev. C **66**, 055803 (2002).
15. Bao-An Li, A.W. Steiner, nucl-th/0511064.
16. M. Di Toro, S.J. Yennello, Bao-An Li, *Isospin flows*, this topical issue.
17. M. Colonna, M.B. Tsang, *Isotopic compositions and scalings*, this topical issue.
18. A.W. Steiner *et al.*, Phys. Rep. **411**, 325 (2005).
19. D.G. Ravenhall, C.J. Pethick, J.R. Wilson, Phys. Rev. Lett. **50**, 2066 (1983).
20. G. Toulouse, Commun. Phys. **2**, 115 (1977); J. Villain, J. Phys. C **10**, 1717 (1977).
21. R. Liebmann, *Statistical Mechanics of Periodic Frustrated Ising Systems*, Vol. **251** (Springer Verlag, Berlin, 1986).
22. C.J. Camacho, Phys. Rev. Lett. **77**, 2324 (1996).
23. J.A. Pons *et al.*, Mon. Not. R. Astron. Soc. **363**, 121 (2005).
24. C.J. Horowitz, M.A. Perez-Garcia, J. Piekarewicz, Phys. Rev. C **69**, 045804 (2004).
25. A.S. Botvina, I.N. Mishustin, Phys. Rev. C **72**, 048801 (2005).
26. Horst Mueller, Brian D. Serot, Nucl. Phys. A **606**, 508 (1996).
27. Gentaro Watanabe, Hidetaka Sonoda, cond-mat/0502515.
28. C.J. Horowitz, M.A. Perez-Garcia, J. Carriere, D.K. Berry, J. Piekarewicz, Phys. Rev. C **70**, 065806 (2004).
29. A.S. Botvina, I.N. Mishustin, Phys. Lett. B **584**, 233 (2004).
30. C.J. Horowitz, M.A. Perez-Garcia, D.K. Berry, J. Piekarewicz, Phys. Rev. C **72**, 035801 (2005).
31. J.B. Natowitz *et al.*, Phys. Rev. Lett. **89**, 212701 (2002); M. Baldo *et al.*, Phys. Rev. C **69**, 034321 (2004).
32. C.J. Horowitz, A. Schwenk, Phys. Lett. B **638**, 153 (2006) nucl-th/0507064.
33. C.J. Horowitz, A. Schwenk, Nucl. Phys. A **776**, 55 (2006) nucl-th/0507033.
34. S. Iyengar *et al.*, J. Chem. Phys. **123**, 084309 (2005).
35. J. Margueron, Phys. Rev. C **70**, 028801 (2004).

Eur. Phys. J. A **30**, 311–316 (2006)
DOI 10.1140/epja/i2006-10125-6

THE EUROPEAN
PHYSICAL JOURNAL A

Possible links between the liquid-gas and deconfinement-hadronization phase transitions

I.N. Mishustin[a]

Frankfurt Institute for Advanced Studies, J.W. Goethe Universität, Max-von-Laue Str. 1, D-60438 Frankfurt am Main, Germany
and Kurchatov Institute, Russian Research Center, Kurchatov Sq. 1, 123182 Moscow, Russia

Received: 14 March 2006 /
Published online: 30 October 2006 – © Società Italiana di Fisica / Springer-Verlag 2006

Abstract. It is commonly accepted that strongly interacting matter has several phase transitions in different domains of temperature and baryon density. In this contribution I discuss two most popular phase transitions which, in principle, can be accessed in nuclear collisions. One of them, the liquid-gas phase transition, is well established theoretically and studied experimentally in nuclear multifragmentation reactions at intermediate energies. The other one, the deconfinement-hadronization phase transition, is at the focus of present and future experimental studies with relativistic heavy-ion beams at SPS, RHIC and LHC. Possible links between these two phase transitions are identified from the viewpoint of their manifestation in violent nuclear collisions.

PACS. 12.38.Mh Quark-gluon plasma – 12.39.-x Phenomenological quark models – 25.70.Pq Multifragment emission and correlations – 25.75.-q Relativistic heavy-ion collisions

1 General remarks

A primary goal of present and future experiments on heavy-ion collisions is to study properties of strongly interacting matter away from the nuclear ground state. Main efforts are focused on searching for possible phase transitions in such collisions. Several phase transitions are predicted in different domains of temperature (T) —baryon density (ρ_B) plane. As is well known, strongly interacting matter has at least one multi-baryon bound state at $\rho_B = \rho_0 \approx 0.16\,\mathrm{fm}^{-3}$ and a binding energy of about $10\,\mathrm{MeV}$, corresponding to atomic nuclei, which can be considered as droplets of nuclear matter. This means that the equation of state of symmetric nuclear matter has a zero-pressure point at $\rho_B = \rho_0$. Since the pressure should also vanish at $\rho_B \to 0$, it must be a non-monotonic function of ρ_B, *i.e.* $\partial P/\partial \rho_B < 0$ in a certain temperature-density domain. This condition signals instability of matter with respect to growing density fluctuations, a characteristic feature of the liquid-gas phase transition. Therefore, it follows from the very existence of the nuclear bound state that there should be a first-order phase transition of the liquid-gas type in normal nuclear matter at subsaturation densities, $\rho_B < \rho_0$, and low temperatures, $T \leq 10\,\mathrm{MeV}$.

The nuclear liquid-gas phase transition manifests itself most clearly in a nuclear multifragmentation phenomenon, observed in intermediate-energy nuclear reactions. Here,

we mention only a few guiding ideas which helped to identify this phase transition. The first one is the anomaly (plateau) in the caloric curve, which was first predicted theoretically [1] and later found experimentally [2]. More recently, an interesting proposal was made [3,4] to look for anomalous energy fluctuations in the multifragmentation events, which might be a good signal of a first-order phase transition in finite systems. Another productive idea proposed in ref. [5] was to search for residual signals of the spinodal decomposition expected in connection with the liquid-gas phase transition. Such a signal, although small, was indeed found experimentally as an enhanced emission of equal-size fragments [6]. Other evidences for the liquid-gas phase transition include large fluctuations in the partition space, bimodality [7,8], or critical behavior near the critical point [8–10].

The situation at high T and non-zero baryon chemical potential μ_B is not so clear, although it is expected that the deconfinement and chiral transitions occur at high enough T and ρ_B. As the result, a new state of matter, the Quark-Gluon Plasma (QGP), should be formed. A rigorous theoretical background for these studies is provided by the QCD-based numerical simulations on a lattice. However, at present reliable lattice calculations exist only for $\mu_B = 0$, *i.e.* $\rho_B = 0$, where they predict a smooth deconfinement transition (crossover) at $T \approx 170\,\mathrm{MeV}$ [11]. As model calculations show, the QCD phase diagram in the (T, μ_B)-plane may contain a first-order transition line (below called the critical line) which ends at a (tri)critical

---

[a] e-mail: `mishustin@fias.uni-frankfurt.de`

point [12–14]. Unfortunately, at finite μ_B the lattice calculations suffer from the so-called "sign problem" and cannot be done easily. Different approximation schemes lead to differing predictions concerning the existence of a critical point (see, *e.g.*, refs. [15–17]). Possible signatures of this point in heavy-ion collisions were discussed in ref. [18]. However, it is unclear at present whether critical fluctuations associated with the second-order phase transition can develop in a rapidly expanding system produced in a relativistic heavy-ion collision, because of the critical slowing-down effect [19]. A more promising strategy would be to search for a first-order phase transition, which may have more spectacular manifestations, as we discuss below.

Relative to the liquid-gas transition, the exploration of the QCD phase diagram is considerably more challenging. On the theoretical side, we have no tractable models to predict how the phase diagram looks in the (T, μ_B)-plane, nor where the dynamical trajectories of expanding matter go. Since, by the nature of a phase transition, the effective degrees of freedom are different in the two phases, often two different models are applied below and above the critical line. Moreover, lattice QCD can only be applied to systems in statistical equilibrium, *i.e.* it cannot be used for dynamical simulations in real time. With regard to dynamical models, the best candidate is perhaps fluid dynamics which needs no specific information about the structure of the matter but merely macroscopic quantities such as the equation of state and kinetic coefficients. However, in its standard form this model is unsuitable for studies of unstable regimes associated with a first-order phase transition. Thus, it is very difficult to provide experimentalists with quantitative guidance to ensure that the parameters of the experiments are those where the phase transition signals are best seen.

On the experimental side, the exploration of the QCD phase structure is made extra complicated by the fact that only the hadronic phase survives in the final state[1], in contrast to the nuclear liquid-gas transition where both phases can occur in the final state. Therefore, the experience accumulated in the liquid-gas phase transition studies may be very useful for designing the analysis techniques for the exploration of the deconfinement-hadronization phase transition.

A similarity between the liquid-gas phase transition and the deconfinement-hadronization transition is the presence of more than one conserved charge: at low energy we have electric charge (Z) and mass number (A), while at high energy, in addition to baryon number B (which is identical to A) and electric charge Q (which corresponds to Z), we have also strangeness (S). Therefore, the lessons learned at low energy regarding multicomponent systems, in particular the isospin degree of freedom, may be helpful for the QGP studies, too.

Finally, notwithstanding the large uncertainty with regard to the value of the critical baryon density (above

which the deconfinement transition is first order), it appears likely that the first-order transition can best be studied experimentally in the region of moderate bombarding energies where compressed matter is characterized by a considerable net baryon density. As we know now, a strongly interacting matter produced at RHIC, presumably a hot quark-gluon plasma, has practically vanishing net baryon density [21]. While more suitable conditions may well have been achieved already at SPS, those data have not been analyzed in a way which would unambiguously demonstrate the QGP formation. To study the first-order transition of the QCD phase diagram, the most promising facility for the future is the planned FAIR at GSI, where compressed baryonic matter is one of the prime areas of intended research.

A striking feature of central heavy-ion collisions at high energies, confirmed in many experiments (see, *e.g.*, [22, 23]), is a very strong collective expansion of matter at later stages of the reaction. This process looks like an explosion with the matter flow velocities comparable with the speed of light. The applicability of equilibrium concepts for describing phase transitions under such conditions becomes questionable and one should expect strong non-equilibrium effects. Below we demonstrate that non-equilibrium phase transitions in rapidly expanding matter can lead to interesting phenomena which, in a certain sense, are even easier to observe.

2 Dynamical fragmentation of a metastable phase

2.1 Nuclear liquid-gas transition

Let us consider a simple model showing how the collective flow can modify the conventional picture of a first-order phase transition [24]. Let us consider first the liquid-gas transition in nuclear matter. We assume that a system expands uniformly with the collective velocity field of a Hubble type, $\mathbf{v}_f(\mathbf{r}) = H\mathbf{r}$, where H is an appropriate Hubble constant. The expansion acts against the attractive forces which keep the nucleons together at normal density. Therefore, instead of uniformly expanding the whole system, it is energetically more favorable to split it into droplets which preserve a sufficiently high density inside, to keep attractive forces acting, and recede from each other according to the Hubble law. The space between the droplets is almost empty so that the energy cost for producing such an inhomogeneous state may be estimated as an extra interface area times a surface tension coefficient σ. One should expect that in violent reactions where thermal excitation is high, σ might be significantly reduced compared to the value of about $1\,\mathrm{MeV/fm}^2$ known for cold nuclei. The shape of the droplets, which is determined by the local density fluctuations, might be also quite complicated. But for our order-of-magnitude estimates we assume that the system splits into more or less spherical droplets of a similar size.

---

[1] Some information about the deconfined phase can be obtained from the electromagnetic probes and from quenching of hard partonic jets, see, *e.g.*, ref. [20].

Now let us imagine that at the stage of the break-up the expanding system is represented by the collection of droplets with density $\rho_B \approx \rho_0$ (nuclear fragments) separated by fully developed surfaces. Within the leptodermous approximation the total energy of an individual spherical droplet of radius $R = (3A/4\pi\rho_B)^{1/3}$ can be decomposed as

$$E = E_{\text{bulk}} + E_{\text{kin}} + E_{\text{sur}}. \tag{1}$$

Here the bulk term at $\rho_B \neq \rho_0$ can be written as

$$E_{\text{bulk}} = \left[a_V + \frac{K}{18}\left(1 - \frac{\rho_B}{\rho_0}\right)^2 \right] \cdot A, \tag{2}$$

where a_V is the bulk coefficient in the Weizsäcker formula and K is the incompressibility modulus. The kinetic energy of an individual droplet, associated with its collective expansion with respect to the center of mass, is easily calculated,

$$E_{\text{kin}} = \int_0^R \frac{1}{2} m_N v_f^2(r)\rho(r)4\pi r^2 \mathrm{d}r = \frac{2\pi}{5} m_N H^2 \rho_B R^5, \tag{3}$$

where m_N is the nucleon mass. The surface energy of a droplet is $4\pi R^2 \sigma$. It is worth noting that the collective kinetic energy acts here as an effective long-range potential similar to the Coulomb potential in nuclei.

To find the optimal droplet size one can apply Grady's argument [25] that the redistribution of matter is a local process that minimizes the energy per droplet volume, $\Delta E/V$. Then, since the bulk contribution does not depend on R, the minimization condition constitutes the balance between the collective kinetic energy and interface energy. This gives for the optimal droplet mass

$$\overline{A} = \frac{4\pi}{3}\rho R^3 = \frac{20\pi}{3}\frac{\sigma}{m_N H^2}. \tag{4}$$

It is determined by only two parameters: the surface tension σ and the Hubble constant H. The latter one can be estimated from flow observables. For instance, in central Au + Au collisions at 150, 250 and 400 MeV/nucleon the measured flow velocities v_f are $0.20c$, $0.26c$ and $0.34c$, respectively [22]. Now one can estimate the Hubble constant as $H^{-1} = R_{\text{Au}}/v_f$, which gives 35, 26 and 20 fm/c, respectively. To get the mean fragment mass $\overline{A} \approx 3$, as seen in experiment, one should take in eq. (4) $\sigma \approx 0.2$ MeV/fm^2, which is about a factor 5 smaller than in cold nuclei! Maybe this is not surprising because at a "temperature" 17 MeV, obtained for this reaction, σ would already vanish in a thermodynamically equilibrated system. One should bear in mind, however, that the observed cold fragments are produced from hot primary fragments after their de-excitation. Therefore, primary fragments produced at the break-up stage should be bigger.

One can use the minimum information principle [26, 27] to find the inclusive fragment mass distribution, $P(A)$. In principle, the information entropy should be defined in terms of microstate probabilities, p_i, as $\sum_i p_i \ln p_i$. Since we are interested only in the inclusive mass distribution, we can sum up all microstates containing the fragment of mass A. Then the information function can be defined simply as $\sum_A P(A)\ln P(A)$. Minimizing this function under constraint that the average fragment mass is fixed, $\overline{A} = \sum_A AP(A)$, we get the normalized mass distribution of the form

$$P(A) = \frac{1}{\overline{A}}\exp\left(-\frac{A}{\overline{A}}\right). \tag{5}$$

This kind of mass distribution has been seen in numerical simulations [28] as well as in the free-jet fragmentation experiments [27]. It is remarkable that exactly this type of mass (charge) distributions is also observed in nuclear experiments! For instance, exponential fragment charge distributions have been found in central Au + Au collisions at 150, 250 and 400 MeV/nucleon [22], discussed above. By applying naively the statistical approach to these reactions one obtains charge distributions which are much too steep (smaller $\overline{A}$).

2.2 Deconfinement-hadronization transition

A similar scenario can also be considered for the deconfinement-hadronization phase transition in relativistic nuclear collisions [29,30]. The difference will be mainly in the parameters characterizing this phase transition. Of course, this consideration is justified only for the first-order phase transition, which is expected at moderate T and high enough ρ_B (see discussion in the introduction). Most likely, this picture does not apply for the RHIC energies, where produced matter is characterized by very high T and very low μ_B [21], corresponding to the crossover transition.

For simplicity, below we use capital letters Q and H (not to be confused with the Hubble constant H) for the deconfined (quark-gluon) phase and the hadronic phase, respectively. Let us assume that the dynamical fragmentation of the deconfined phase has resulted in a collection of Q droplets embedded in a dilute H phase, as illustrated in fig. 1. The optimal droplet size can be determined by applying the same energy balance prescription as discussed above. The only difference is that the droplet mass with respect to the hadronic background is now calculated as $M = \Delta\mathcal{E}V$, where $\Delta\mathcal{E} = \mathcal{E}_Q - \mathcal{E}_H$ is the energy density difference of Q and H bulk phases, and V is the volume of the droplet. Applying Grady's minimization rule we get the optimum droplet radius

$$R^* = \left(\frac{5\sigma}{\Delta\mathcal{E}H^2}\right)^{1/3}. \tag{6}$$

As eq. (6) indicates, the droplet size depends strongly on H. When expansion is slow (small H) the droplets are big. In the adiabatic limit the process may look like a fission of a cloud of plasma. But fast expansion should lead to very small droplets. This state of matter is very far from a thermodynamically equilibrated mixed phase, particularly because the H phase is very dilute. One can say

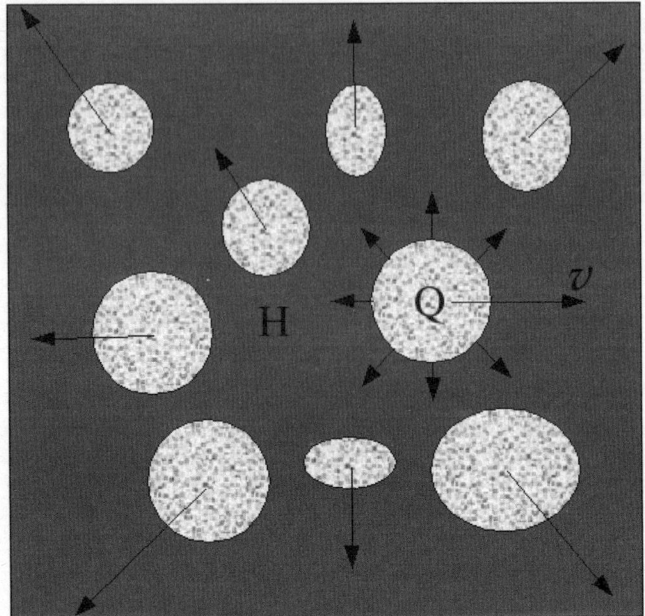

Fig. 1. Schematic view of the multi-droplet state produced after the dynamical fragmentation of a metastable high energy density phase (in this example, the Q phase). The droplets are embedded in the low energy density phase (in this example, the H phase). Each droplet expands individually as well as participates in the overall Hubble-like expansion.

that the metastable Q matter is torn apart by a mechanical strain associated with the collective expansion. This has a direct analogy with the dynamical multifragmentation process, described in the previous section, or with the fragmentation of pressurized fluids leaving nozzles [27].

The driving force for expansion is the pressure gradient, $\nabla P \equiv c_s^2 \nabla \mathcal{E}$, which depends crucially on the sound velocity in matter, c_s. Here we are interested in the expansion rate of the partonic phase, which is not directly observable but predicted by the hydrodynamical simulations. In the vicinity of the phase transition, one may expect a "soft point" [31,32] where the sound velocity is smallest and the ability of matter to generate the collective expansion is minimal. If the initial state of the Q phase is close to this point, its subsequent expansion will be slow. Accordingly, the droplets produced in this case will be big. When moving away from the soft point, one would see smaller and smaller droplets. For numerical estimates we choose two values of the Hubble constant: $H^{-1} = 20\,\mathrm{fm}/c$ to represent the slow expansion from the soft point and $H^{-1} = 6\,\mathrm{fm}/c$ for the fast expansion.

One should also specify two other parameters, σ and $\Delta \mathcal{E}$. The surface tension σ is a subject of debate at present. Lattice simulations indicate that it could be as low as a few $\mathrm{MeV}/\mathrm{fm}^2$ in the vicinity of the critical line. However, for our non-equilibrium scenario, more appropriate values are closer to 10–$20\,\mathrm{MeV}/\mathrm{fm}^2$, which follow from effective chiral models. As a compromise, the value $\sigma = 10\,\mathrm{MeV}/\mathrm{fm}^2$ is used below for rough estimates. Bearing in mind that nucleons and heavy mesons are the smallest droplets of the Q phase, one can take $\Delta \mathcal{E} = 0.5\,\mathrm{GeV}/\mathrm{fm}^3$, *i.e.* the energy

density inside the nucleon. Then one gets $R^* = 3.4\,\mathrm{fm}$ for $H^{-1} = 20\,\mathrm{fm}/c$ and $R^* = 1.5\,\mathrm{fm}$ for $H^{-1} = 6\,\mathrm{fm}/c$. As follows from eq. (6), for a spherical droplet $V \propto 1/\Delta \mathcal{E}$, and in the first approximation its mass,

$$M^* \approx \Delta \mathcal{E} V = \frac{20\pi}{3} \frac{\sigma}{H^2}, \tag{7}$$

is independent of $\Delta \mathcal{E}$ (compare with eq. (4)). For the two values of R^* given above, the optimal droplet mass is $\sim 100\,\mathrm{GeV}$ and $\sim 10\,\mathrm{GeV}$, respectively. As mentioned in the previous section, the distribution of droplet masses should follow an exponential law, $\exp\left(-\frac{M}{M^*}\right)$. Thus, about $2/3$ of droplets have masses smaller than M^*, but with 1% probability one can find droplets as heavy as $5M^*$.

3 Observable manifestations of quark droplets

After separation, the QGP droplets will recede from each other according to the global collective expansion, predominantly in the beam direction. Therefore, their c.m. rapidities y_i will be in one-to-one correspondence with their spatial positions. One may expect that they will be distributed more or less uniformly between the target and the projectile rapidities. Since rescatterings in the dilute H phase are rare, most hadrons produced from individual droplets will go directly into detectors. This may explain why freeze-out parameters extracted from the hadronic yields are close to the phase transition boundary [21]. Indeed, due to the rapid expansion it is unlikely that the thermodynamical equilibrium will be established between the Q and H phases or within the H phase alone. If this were to happen, the final H phase would be more or less uniform, and thus, no traces of the droplet phase would appear in the final state.

The final fate of individual droplets depends on their sizes and on details of the equation of state. Due to the negative Laplace pressure, $2\sigma/R$, the residual expansion of individual droplets will slow down. The smaller droplets may even reverse their expansion and cooling to shrinking and reheating. Then, the conversion of Q matter into H phase may proceed through the formation of the imploding deflagration front [32,33]. Bigger droplets may expand further until they enter the region of spinodal instability At this stage the difference between 1st- and 2nd-order phase transitions or a crossover is insignificant. Since the characteristic "rolling down" time is rather short, $\sim 1\,\mathrm{fm}/c$ [34], the Q droplets will be rapidly converted into the non-equilibrium H phase. In refs. [35–37] the evolution of individual droplets was studied numerically within a hydrodynamical approach including dynamical chiral fields. It has been demonstrated that the energy released at the spinodal decomposition can be transferred directly into the collective oscillations of the (σ, π) fields which give rise to the soft pion radiation. One can also expect the formation of Disoriented Chiral Condensates (DCC) [38] in the voids between the droplets.

It is interesting to note that the surface tension has a stabilizing effect on the droplet evolution. Since the

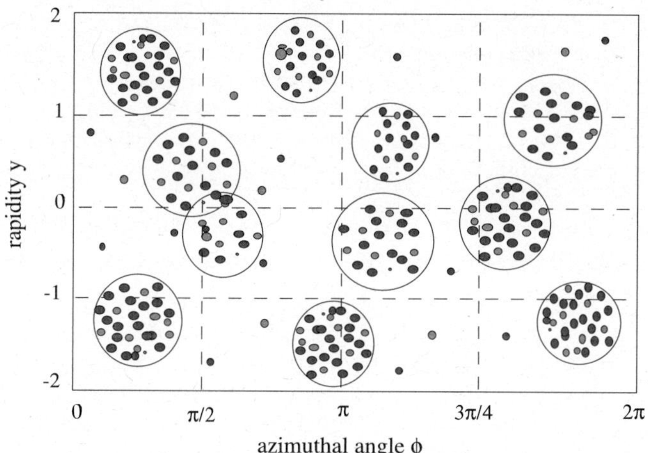

Fig. 2. Schematic view of the momentum space distribution of secondary hadrons produced from an ensemble of droplets. Each droplet emits hadrons (mostly pions) within a rapidity interval $\delta y \sim 1$ and azimuthal angle spreading of $\delta\phi \sim 1$.

droplets are hot, their lifetime will be mainly determined by the rate of hadron evaporation from the surface (see also the discussion in ref. [39]). This will lead to their cooling and shrinking. One can speculate about all kinds of exotic objects, like *e.g.* strangelets, glueballs, formed in this way. The possibility of forming "vacuum bubbles", *i.e.* regions with depleted quark and gluon condensates, was discussed in ref. [35]. All these interesting possibilities deserve further study and numerical simulations.

In the droplet phase the mean number of produced hadrons in a given rapidity interval is

$$\langle N \rangle = \sum_i^{N_D} \overline{n_i} = \langle n \rangle \langle N_D \rangle \,, \tag{8}$$

where $\overline{n_i}$ is the mean multiplicity of hadrons emitted from a droplet i, $\langle n \rangle$ is the average multiplicity per droplet and $\langle N_D \rangle$ is the mean number of droplets produced in this interval. If droplets do not overlap in rapidity space, each of them will give a bump in the hadron rapidity distribution around its center-of-mass rapidity y_i [29,34]. In case of a Boltzmann spectrum the width of the bump will be $\delta y \approx \sqrt{T/m}$, where T is the droplet temperature and m is the particle mass. At $T \sim 100\,\mathrm{MeV}$ this gives $\delta y \approx 0.8$ for pions and $\delta y \approx 0.3$ for nucleons. These spectra might be slightly modified by the residual expansion of droplets. Due to the radial expansion of the fireball the droplets should also be well separated in the azimuthal angle. The characteristic angular spreading of pions produced by an individual droplet is determined by the ratio of the thermal momentum of emitted pions to their mean transverse momentum, $\delta\phi \approx 3T/\langle p_\perp \rangle \sim 1$. The resulting phase-space distribution of hadrons in a single event will be a superposition of contributions from different Q droplets superimposed on a more or less uniform background from the H phase. Such a distribution is shown schematically in fig. 2. It is obvious that such inhomogeneities (clusterization) in the momentum space will be reflected in strong

non-statistical fluctuations of hadron multiplicities measured in a given rapidity and angular window. The fluctuations will be more pronounced if primordial droplets are big, as expected in the vicinity of the soft point. If droplets as heavy as $100\,\mathrm{GeV}$ are formed, each of them will emit up to ~ 200 pions within a narrow rapidity and angular interval, $\delta y \sim 1$, $\delta\phi \sim 1$. If only a few droplets are produced on average per unit rapidity, $N_D \gtrsim 1$, they will be easily resolved and analyzed. On the other hand, the fluctuations will be suppressed by a factor $\sqrt{N_D}$ if many small droplets fall in the same rapidity interval.

It is convenient to characterize the multiplicity fluctuations in a given rapidity window by the scaled variance

$$\omega_N \equiv \frac{\langle N^2 \rangle - \langle N \rangle^2}{\langle N \rangle} \,. \tag{9}$$

Its important property is that $\omega_N = 1$ for the Poisson distribution, and therefore any deviation from unity will signal a non-statistical emission mechanism. As shown in ref. [40], for an ensemble of emitting sources (droplets) ω_N can be expressed in a simple form, $\omega_N = \omega_n + \langle n \rangle \omega_D$, where ω_n is an average multiplicity fluctuation in a single droplet, ω_D is the fluctuation in the droplet mass distribution and $\langle n \rangle$ is the mean multiplicity from a single droplet. Since ω_n and ω_D are typically of order of unity, the fluctuations from the multi-droplet emission are enhanced by the factor $\langle n \rangle$. According to the picture of a first-order phase transition advocated above, this enhancement factor can be as large as 10–100. A more detailed consideration of the multiplicity distributions associated with the hadron emission from an ensemble of droplets is given in ref. [30]. Until now no strong anomalies in hadron multiplicity distributions have been observed in relativistic heavy-ion collisions (see, *e.g.*, ref. [41]).

4 Conclusions

- It is most likely that strongly interacting matter has at least two first-order phase transitions, *i.e.* the nuclear liquid-gas transition and the deconfinement-hadronization transition. Their unambiguous experimental identification is the main goal of heavy-ion collision experiments at present and future facilities. Studying phase transitions in such a dynamical environment should take into account strong non-equilibrium effects.
- A first-order phase transition in rapidly expanding matter should proceed through the nonequilibrium stage when a metastable phase splits into droplets whose size is inversely proportional to the expansion rate. The primordial droplets should be biggest in the vicinity of a soft point when the expansion is slowest.
- Hadron emission from droplets of the quark-gluon plasma should lead to large non-statistical fluctuations in their rapidity and azimuthal spectra, as well as in multiplicity distributions in a given rapidity window. The hadron abundances may reflect directly the chemical composition in the plasma phase.

- To identify the phase transition threshold, the measurements should be done at different collision energies. The predicted dependence on the expansion rate and the reaction geometry can be checked in collisions with different ion masses and impact parameters.
- If the first-order deconfinement/chiral phase transition is only possible at finite baryon densities, one should try to identify it by searching for the anomalous fluctuations in the regions of phase space characterized by a large baryon chemical potential. These could be the nuclear fragmentation regions in collisions with very high energies (high-energy SPS, RHIC, LHC) or the central rapidity region in less energetic collisions (AGS, low-energy SPS, future GSI facility FAIR).
- A rich experience has been accumulated in theoretical and experimental studies of nuclear multifragmentation as a signal of the liquid-gas phase transition in normal nuclear matter. These lessons may be useful in present and future studies of the deconfinement-hadronization and chiral phase transitions in relativistic heavy-ion collisions.

I thank J. Randrup and F. Gulminelli for fruitful discussions and useful advises. This work was supported in part by the grants RFFR 05-02-04013 and NS-8756.2006.2 (Russia).

References

1. J.P. Bondorf, R. Donangelo, I.N. Mishustin, H. Schulz, Nucl. Phys. A **444**, 460 (1985).
2. J. Pochodzalla, ALADIN Collaboration, Phys. Rev. Lett. **75**, 1040 (1995).
3. Ph. Chomaz, F. Gulminelli, V. Duflot, Phys. Rev. E **64**, 046114 (2001).
4. M. D'Agostino *et al.*, Phys. Lett. B **473**, 219 (2000).
5. Ph. Chomaz, M. Colonna, J. Randrup, Phys. Rep. **389**, 263 (2004).
6. B. Borderie *et al.*, Phys. Rev. Lett. **86**, 3252 (2001).
7. A.S. Botvina *et al.*, Nucl. Phys. A **584**, 737 (1995).
8. M. D'Agostino *et al.*, Nucl. Phys. A **650**, 329 (1999).
9. B.K. Srivastava *et al.*, Phys. Rev. C **65**, 054617 (2002).
10. J.B. Elliott *et al.*, Phys. Rev. Lett. **88**, 042701 (2002).
11. F. Karsch, E. Laermann, A. Peikert, Nucl. Phys. B **605**, 579 (2001).
12. M.A. Halasz, A.D. Jackson, R.E. Shrock, M.A. Stephanov, J.J.M. Verbarshot, Phys. Rev. D **58**, 096007 (1998).
13. J. Berges, K. Rajagopal, Nucl. Phys. B **538**, 215 (1999).
14. O. Scavenius, A. Mocsy, I.N. Mishustin, D. Rischke, Phys. Rev. C **64**, 045202 (2001).
15. Z. Fodor, S.D. Katz, JHEP **0203**, 014 (2002).
16. C.R. Allton *et al.*, Phys. Rev. D **66**, 074507 (2002).
17. R.V. Gavai, S. Gupta, Phys. Rev. D **68**, 034506 (2003).
18. M. Stephanov, K. Rajagopal, E. Shuryak, Phys. Rev. Lett. **81**, 4816 (1998).
19. Boris Berdnikov, Krishna Rajagopal, Phys. Rev. D **61**, 105017 (2000).
20. J. Adams, STAR Collaboration, Nucl. Phys. A **757**, 102 (2005).
21. P. Braun-Munzinger, D. Magestro, K. Redlich, J. Stachel, Phys. Lett. B **518**, 41 (2001); P. Braun-Munzinger, J. Stachel, J. Phys. G: Nucl. Part. Phys. **28**, 1971 (2002); A. Andronic, P. Braun-Munzinger, J. Stachel, Nucl. Phys. A **772**, 167 (2006).
22. W. Reisdorf, FOPI Collaboration, Nucl. Phys. A **612**, 493 (1997).
23. Nu Xu, Prog. Part. Nucl. Phys. **53**, 165 (2004).
24. I.N. Mishustin, Nucl. Phys. A **630**, 111c (1998).
25. D.E. Grady, J. Appl. Phys. **53**, 322 (1981).
26. J. Aichelin, J. Hüfner, Phys. Lett. B **136**, 5 (1984).
27. E.L. Knuth, U. Henne, J. Chem. Phys. **110**, 2664 (1999).
28. B.L. Holian, D.E. Grady, Phys. Rev. Lett. **60**, 1355 (1988).
29. I.N. Mishustin, Phys. Rev. Lett. **82**, 4779 (1999).
30. I.N. Mishustin, in *Proceedings of the International Conference New Trends in High-Energy Physics, Crimea, Ukraine, September 10-17, 2005*, hep-ph/0512366; T. Chechak *et al.* (Editors), *Nuclear Science and Safety in Europe* (Springer, 2006) pp. 99-111.
31. C.M. Huang, E.V. Shuryak, Phys. Rev. Lett. **75**, 4003 (1995); Phys. Rev. C **57**, 1891 (1998).
32. D. Rischke, M. Gyulassy, Nucl. Phys. A **597**, 701 (1996); **608**, 479 (1996).
33. S. Digal, A.M. Srivastava, Phys. Rev. Lett. **80**, 1841 (1998).
34. L.P. Csernai, I.N. Mishustin, Phys. Rev. Lett. **74**, 5005 (1995).
35. I.N. Mishustin, O. Scavenius, Phys. Rev. Lett. **83**, 3134 (1999).
36. O. Scavenius, A. Dumitru, E.S. Fraga, J.T. Lenaghan, A.D. Jackson, Phys. Rev. D **63**, 116003 (2001).
37. K. Paech, H. Stoecker, A. Dumitru, Phys. Rev. C **68**, 044907 (2003).
38. J.D. Bjorken, K.L. Kowalski, C.C. Taylor, SLAC-PUB-6413, Sep. 1993, hep-ph/9309235.
39. M. Alford, K. Rajagopal, F. Wilczek, Phys. Lett. B **422**, 247 (1998).
40. G. Baym, H. Heiselberg, Phys. Lett. B **469**, 7 (1999).
41. M. Rybczynski, NA49 Collaboration, J. Phys. Conf. Ser. **5**, 74 (2005).

Eur. Phys. J. A **30**, 317–331 (2006)
DOI 10.1140/epja/i2006-10126-5

THE EUROPEAN
PHYSICAL JOURNAL A

The challenges of finite-system statistical mechanics

P. Chomaz[1] and F. Gulminelli[2,a]

[1] GANIL (DSM-CEA/IN2P3-CNRS), Blvd. H. Becquerel, F-14076 Caen cédex, France
[2] LPC (IN2P3-CNRS/Ensicaen et Université), F-14076 Caen cédex, France

Received: 22 March 2006 /
Published online: 27 October 2006 – © Società Italiana di Fisica / Springer-Verlag 2006

Abstract. In this paper, we review the main challenges associated with the statistical mechanics of finite systems, with a particular emphasis on the present understanding of phase transitions in the framework of information theory. We show that this is a very powerful formalism allowing to treat in a thermodynamically consistent way many difficult problems in the statistical treatment of finite, open, transient and expanding systems. The first point we analyze is the problem of boundary conditions, which in the framework of information theory must also be treated statistically. We recall that the different ensembles do not lead to the same equation of states, in particular in the region of a first-order phase transition, and we stress the fact that different statistical ensembles may be relevant to heavy-ion physics depending upon the actual experimental conditions. Finally, we present a coherent description of first-order phase transitions demonstrating the equivalence between the Yang-Lee theorem, the occurrence of bimodalities in the intensive ensemble and the presence of inverted curvatures of the thermodynamic potential of the extensive ensemble. We stress that this discussion is not restricted to the possible occurrence of negative specific heat, but can also include negative compressibilities and negative susceptibilities, and in fact any curvature anomaly of the thermodynamic potential. Since the relevant entropy surface explored in nuclear multifragmentation is not yet well understood and largely debated in the community, the experimental evidence of new thermodynamic anomalies is one of the important challenges of future heavy-ion experiments.

PACS. 05.20.Gg Classical ensemble theory – 25.70.Pq Multifragment emission and correlations – 64.60.-i General studies of phase transitions – 65.40.Gr Entropy and other thermodynamical quantities

1 Introduction

Finite-systems properties, non-extensive thermodynamics, and phase transitions out of the thermodynamic limit are strongly debated issues in many different fields of physics (see for example [1]). This may be the case of non-saturating forces such as the gravitational [2–5] or the Coulombic forces. The system may be too small, as in the case of clusters and nuclei [6–9]. The physics of finite systems is even more complicated since often they are not only small but also open and transient. This implies that the various concepts of thermodynamics and statistical mechanics [10–13] have to be completed and revisited [1,14–18]. Another contribution to this topical issue deals with some aspects of this question [19] and we address the reader to that paper to have a more complete view of the different formalisms that can be applied. In the present paper, we focus on the information theory approach to statistical mechanics [14,16] and we will show that this is a very powerful formalism that allows to ad-

dress in a consistent way the statistical mechanics of open systems evolving in time, independent of their interaction range and number of constituents.

After a short summary of the statistical-physics concepts, we will summarize the discussion about ensemble inequivalence. Statistical ensembles are presented in many elementary textbooks as qualitatively equivalent and quantitatively almost identical, because they differ only at the fluctuation level. However, for finite systems it is now well documented in the literature [18,20–23] that two ensembles which put different constraints on the fluctuations of the order parameter lead to qualitatively different equations of state close to a first-order phase transition.

This will lead us to the discussion of phase transitions in finite systems. As an example, when energy is the order parameter, the microcanonical (at fixed energy) heat capacity diverges to become negative while the canonical one (at fixed temperature) remains always positive and finite [24–36]. If the number of particle is the order parameter, it is the chemical susceptibility which is expected to present a negative branch in between two divergences in

a Member of the Institut Universitaire de France;
e-mail: gulminelli@lpccaen.in2p3.fr

the fixed number of particle ensemble (microcanonical or canonical) while in the grand canonical it should remain positive. This difference between ensembles can be of primordial importance for mesoscopic systems undergoing a phase transition. Such systems are now studied in many fields of physics, from Bose condensates [37,38] to the quark-gluon plasma [39,40], from cluster melting [6,41] to nuclear fragmentation [7]. Moreover, such inequivalences may survive at the infinite size limit for systems involving long-range forces such as self-gravitating objects [3–5].

We will then present some basic characteristics of first-order phase transitions in finite systems. In particular, we will summarize the mathematical connections between the Yang-Lee approach [42] through the zeroes of the partition sum, the bimodality of the order parameter distribution in the same ensemble, and the anomalous (inverted) curvature of the thermodynamic potential of the ensemble where the order parameter is fixed [43–45]. The best documented example in the literature is the bimodality of the canonical energy distribution being equivalent to negative microcanonical heat capacity [46–48].

Finally, we will skim over the time evolution problem, stressing the need to take into account time odd constraint in the statistical picture, and then conclude presenting three challenges for the field of the thermodynamic properties of finite systems.

2 Finite systems and statistical mechanics

A largely debated issue in the nuclear-physics community is the applicability of thermodynamical concepts like equilibrium, temperature, pressure etc. to objects as tiny as nuclei. How large must a system be for a temperature to be defined? It is well known that the different statistical ensembles only converge in the thermodynamic limit: out of this limit, what is the physical meaning of thermodynamic quantities —say, temperature— evaluated through different ensembles? is there a "correct" ensemble to be used? We all know finite systems can change state or shape, a typical example being the case of isomerization; how many degrees of freedom do we need in order to call this change of state a phase transition?

Let us consider a system that can exist in two single microstates of different energy (a single spin in a magnetic field, a two-level atom in a bath of radiation...) The system being much smaller than its environment, let us consider the case for which the interaction between system and environment can be neglected and we have no reason to believe that the environment will be in any specific state. Then the distribution of the system microstates is simply given by the number of states of the environment

$$p^{(n)} = W(E_t - e_n)/(W(E_t - e_1) + W(E_t - e_2))$$
$$\propto \exp\left(S(E_t - e_n)\right), \tag{1}$$

where E_t is the total energy (system + environment) and $S = \log W$ is the (microcanonical) entropy associated with the environment. Since $e_n \ll E_t$, a Taylor expansion of the

entropy gives

$$S(E_t - e_n) \approx S(E_t) - e_n \frac{\partial S}{\partial E}(E_t); \quad p^{(n)} \propto \exp\left(-\beta e_n\right),$$
$$\tag{2}$$

where we have introduced $\beta = \partial S \partial E$, the temperature of the environment.

This very simple textbook exercise gives us a number of interesting informations:

- thermodynamic concepts like temperature can be defined for systems having an arbitrary number of degrees of freedom (the minimum being 2 levels);
- Boltzmann-Gibbs statistics naturally emerges as soon as we observe a limited information constructed from a reduced number of degrees of freedom.

If we now take into account a slightly more complicated system with energy states associated with a degeneracy $w(e)$, the energy distribution will be modified to

$$p(e) = \frac{w(e)W(E_t - e)}{\sum_n w(e_n)W(E_t - e_n)} \approx \frac{w(e)\exp\left(-\beta e\right)}{Z_\beta}, \tag{3}$$

where the canonical approximation is still correct if the system is associated with a much smaller number of degrees of freedom than its environment. Equation (3) gives for instance the energy distribution of a thermometer loosely coupled to an otherwise isolated system. Temperature is defined as the response of the thermometer in the most probable energy state $\bar{e}$; if we maximize the distribution (3) we get, assuming that energy can be treated as a continuous variable,

$$\frac{\partial \log W}{\partial E}\Big|_{E_t - \bar{e}} = \frac{\partial \log w}{\partial E}\Big|_{\bar{e}}. \tag{4}$$

We then learn that the quantity shared at the most probable energy partition is the microcanonical temperature. This shows that there is no ambiguity in the definition of temperature (and any other thermodynamic quantity) when dealing with small systems. It is important to note that eq. (3) is not limited to the observation of energy, but can apply to the distribution of any generic observable $A = \langle \hat{A} \rangle$. We can then expect that canonical-like ensembles (*i.e.* ensembles where distributions are given by Boltzmann factors) will arise each time that we are isolating a small number of degrees of freedom from a more complex system.

More generally, we will recall in the next section that a statistical description is in order each time that the system is complex enough to have a large number of microstates associated with a given set of relevant observables. The proper statistical ensemble will then depend on the way the system is prepared. If the relevant observables are recognized, equilibrium is therefore a very generic concept that certainly applies to the output of a heavy-ion collision independent of the reaction time.

3 Statistical physics and information theory

Information-theory–based statistical mechanics provides a very powerful framework to a consistent treatment of the

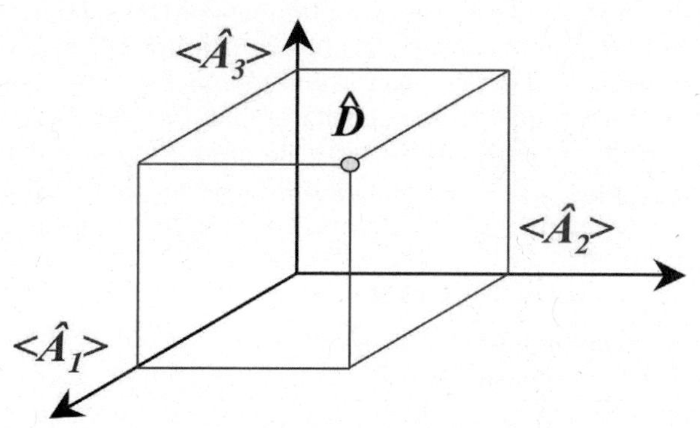

Fig. 1. Illustration of the Liouville space of density matrices: an observation $\langle \hat{A}_i \rangle$ is a projection of $\hat{D}$ on the axis associated with the corresponding observable $\hat{A}_i$.

thermodynamics of finite systems both in the classical and in the quantal world [16,14]. Let us summarize here the essential ingredients. Statistical physics treats statistical ensembles of possible solutions for the considered physical system since the assumption is made that we have only a limited knowledge on it. Such a "macrostate" can be formally be represented by its density matrix

$$\hat{D} = \sum_{(n)} \left| \Psi^{(n)} \right\rangle \, p^{(n)} \, \left\langle \Psi^{(n)} \right|, \tag{5}$$

where the states ("microstates", or "partitions", or "replicas", or simply "events") $|\Psi^{(n)}\rangle$ pertain to the considered Fock or Hilbert space. $p^{(n)}$ is the occurrence probability of the event $|\Psi^{(n)}\rangle$. The result of the measurement of an observable $\hat{A}$ is

$$\langle \hat{A} \rangle_{\hat{D}} = \mathrm{Tr}\,\hat{A}\hat{D}, \tag{6}$$

where Tr means the trace over the quantum Fock or Hilbert space of states $\{|\Psi\rangle\}$.

In the space of Hermitian matrices, the trace provides a scalar product [49,50]

$$\langle\langle \hat{A} || \hat{D} \rangle\rangle = \mathrm{Tr}\,\hat{A}\hat{D}. \tag{7}$$

It is then possible to define an orthonormal basis of Hermitian operators $\{\hat{O}_l\}$ in the observable space, and to interpret the measurement $\langle \hat{O}_l \rangle_{\hat{D}}$ as a coordinate of the density matrix $\hat{D}$ (see fig. 1). The size of the observable space is the square of the dimension of the Hilbert or Fock space, which are in general infinite; therefore in order to describe the system, one is forced to consider a reduced set of (collective) observables $\{\hat{A}_\ell\}$ which are supposed to contain the relevant information. The Gibbs formulation of statistical mechanics can then be derived if the least biased "macrostate" is assumed to be given by the maximization of the entropy[1]

$$S[\hat{D}] = -\,\mathrm{Tr}\,\hat{D}\log\hat{D}, \tag{8}$$

which is nothing but the opposite of the Shannon information [16,14]. It is important to stress that eq. (8) is a microscopic definition of entropy which coincides with the standard thermodynamic entropy only after maximization, see eq. (13) below.

If the system is characterized by L observables (or "extensive" variables[2]), $\hat{\mathbf{A}} = \{\hat{A}_\ell\}$, known in average $\langle \hat{A}_\ell \rangle = \mathrm{Tr}\,\hat{D}\hat{A}_\ell$, the variation is not free and one should maximize the constrained entropy

$$S' = S - \sum_\ell \lambda_\ell \langle \hat{A}_\ell \rangle, \tag{9}$$

where the $\boldsymbol{\lambda} = \{\lambda_\ell\}$ are L Lagrange multipliers associated with the L constraints $\langle \hat{A}_\ell \rangle$.

A maximization of the entropy under constraints gives a prediction for the minimum-biased density matrix (or "event distribution") which can be viewed as a generalization of Gibbs equilibrium:

$$\hat{D}_{\boldsymbol{\lambda}} = \frac{1}{Z_{\boldsymbol{\lambda}}} \exp -\boldsymbol{\lambda} \cdot \hat{\mathbf{A}}, \tag{10}$$

where $\boldsymbol{\lambda} \cdot \hat{\mathbf{A}} = \sum_{\ell=1}^{L} \lambda_\ell \hat{A}_\ell$ and where $Z_{\boldsymbol{\lambda}}$ is the associated partition sum insuring the normalization of $\hat{D}_{\boldsymbol{\lambda}}$:

$$Z_{\boldsymbol{\lambda}} = \mathrm{Tr}\,\exp -\boldsymbol{\lambda} \cdot \hat{\mathbf{A}}. \tag{11}$$

Using this definition, we can compute the associated equations of state (EoS):

$$\langle \hat{A}_\ell \rangle = \partial_{\lambda_\ell} \log Z_{\boldsymbol{\lambda}}. \tag{12}$$

The entropy associated with $\hat{D}_{\boldsymbol{\lambda}}$ is

$$S[\hat{D}_{\boldsymbol{\lambda}}] = \log Z_{\boldsymbol{\lambda}} + \sum_\ell \lambda_\ell \langle \hat{A}_\ell \rangle, \tag{13}$$

which has the structure of a Legendre transform between the entropy and the thermodynamic potential. To interpret the Gibbs ensemble as resulting from the contact with a reservoir or to guarantee the stationarity of eq. (10), it is often assumed that the observables $\hat{A}_\ell$ are conserved quantities such as the energy $\hat{H}$, the particle (or charge) numbers $\hat{N}_i$ or the angular momentum $\hat{L}$ [19]. However, there is no formal reason to limit the state variables to constants of motion. Even more, the introduction of not conserved quantities might be a way to take into account some non-ergodic aspects. Indeed, an additional constraint reduces the entropy, limiting the populated phase space or modifying the event distribution. This point will be developed at length in the next sections.

---

[1] In this article we implicitly use units such that the Boltzmann constant $k = 1$.

[2] In this paper the word "extensive" is used in the general sense of resulting from an observation, *i.e.* the $\langle \hat{A}_\ell \rangle$, and not in the restricted sense of additive variable. Intensive variables are conjugate to extensive variables *i.e.* Lagrange multipliers λ_ℓ imposing the average value of the associated extensive variable.

It should be noticed that microcanonical thermodynamics also corresponds to a maximization of the entropy (8) in a fixed energy subspace. In this case the maximum of the Shannon entropy can be identified with the Boltzmann entropy

$$\max\left(S\right) = \log W\left(E\right), \tag{14}$$

where W is the total state density with the energy E. The microcanonical case can also be seen as a particular Gibbs equilibrium (10) for which both the energy and its fluctuation are constrained. This so-called Gaussian ensemble in fact interpolates between the canonical and microcanonical ensembles depending upon the constraint on the energy fluctuation [24,51], and the same procedure can be applied to any conservation law. In this sense the Gibbs formulation (10) can be considered as the most general.

The ensemble of extensive variables constrained exactly or in average completely defines the statistical ensemble. This means that many different ensembles can be defined, and the most appropriate description of a finite system may be different from the standard microcanonical, canonical or grand-canonical.

4 Finite size and boundary conditions

An important problem when considering finite-size systems is the need to define boundary conditions to define the finite size. This is only a mathematical detail for "condensed" systems, i.e. finite-size self-bound systems in a much larger container, or particles trapped in an external confining potential [52]. In the other cases, finite-size systems can only be defined when proper boundary conditions are specified. Conversely to the thermodynamic limit which, when it exists, clearly isolates bulk properties independent of the actual shape of the container, finite-size systems explicitly depend on boundary conditions.

From a mathematical point of view the system Hamiltonian $\hat{H}$ is not defined until boundary conditions are specified. For example, for a particle problem a boundary can be the definition of a surface given by the implicit equation $\sigma(x, y, z) = 0$. Since the Hamiltonian $\hat{H}_\sigma$ explicitly contains the boundary, the entropy S_σ also directly depends upon the definition of this boundary, according to

$$S_\sigma(E) = \log \operatorname{tr} \delta(E - \hat{H}_\sigma). \tag{15}$$

This brings a severe conceptual problem; the knowledge of the boundary requires an infinite information: the values of the function σ defining the actual surface in each space point. This is easily seen introducing the projector $\hat{P}_\sigma$ over the surface and its exterior. Indeed the boundary conditions applied to each microstate $\hat{P}_\sigma|\Psi^{(n)}\rangle = 0$ is exactly equivalent to the extra constraint $\langle\hat{P}_\sigma\rangle = \operatorname{Tr}\hat{D}\hat{P}_\sigma = 0$. If we note again $\mathbf{A}$, the observables (including the Hamiltonian $\hat{H}_\sigma$) characterizing a given equilibrium, the density matrix including the boundary condition reads

$$\hat{D}_{\boldsymbol{\lambda}\sigma} = \frac{1}{Z_{\boldsymbol{\lambda}\sigma}} \exp -\boldsymbol{\lambda} \cdot \mathbf{A} - b\hat{P}_\sigma, \tag{16}$$

which shows that the thermodynamics of the system does not only depend on the Lagrange multiplier b, but on the whole surface. For the very same global features such as the same average particle density or energy, we will have as many different thermodynamics as boundary conditions. More important, to specify the density matrix, the projector $\hat{P}_\sigma$ has to be exactly known and this is in fact impossible. The nature of $\hat{P}_\sigma$ is intrinsically different from the usual global observables $\hat{A}_\ell$. At variance with $\hat{A}_\ell$, $\hat{P}_\sigma$ is a many-body operator which does not correspond to any physical measurable observable. The knowledge of $\hat{P}_\sigma$ requires the exact knowledge of each point of the boundary surface while no or few parameters are sufficient to define $\hat{A}_\ell$. If we consider statistical physics as founded by the concept of minimum information [14,16], it is difficult to justify such an exact knowledge of the boundary. One should rather apply the minimum information concept also to the boundaries, introducing a hierarchy of collective observables which define the size and shape of the considered system. This amounts to introduce statistical ensembles treating the boundaries as additional extensive variables fixed by conjugated Lagrange parameters [53]. If, for instance, we consider that the relevant size information for an unbound system is its global square radius $\langle\hat{R}^2\rangle$, the adequate partition sum is

$$Z_\lambda = \sum_R W\left(R\right) e^{-\lambda R^2}, \tag{17}$$

where $W(R)$ is the state density associated with each possible value of the system radius.

5 Concept of equilibrium

As we have discussed in the previous sections, a statistical treatment is justified whenever a very large number of microstates exists for a given set of observables. This is always the case for the output of a heavy-ion collision, meaning that at least in principle a statistical approach should always be successful. An ensemble of events coming from similar prepared initial systems and/or selected by sorting always constitutes a statistical ensemble. Indeed, using the entropy concept, different observations are associated with a different information content. If we are able to recognize all the relevant degrees of freedom (i.e. the observations with a strong information content) the ensemble of replicas is by construction a statistical ensemble, i.e. a Gibbs equilibrium in the extended sense of sect. 3.

This generic statement hides the fundamental problem of recognizing the relevant observables. In the statistical models used to describe nuclear multifragmentation [54, 19] the hypothesis is made that all the information content is exhausted by the total energy, number of protons, neutrons, volume occupied at the time when fragments are decoupled, and in some cases angular momentum. This simple set of observables is certainly not sufficient to describe the whole phenomenology of heavy-ion reactions at all impact parameters. Then the remark is often made in the literature that "dynamical effects" dominate [55], meaning that extra constraints have to be put in order to have a statistical description of the final state of the reaction. However, the minimal information of multifragmentation statistical models [54,19] may be enough to describe limited portions of the collision phase space ("sources") properly selected by sorting [56]. The theoretical justification of this minimal statistical picture comes from the fact that complex classical systems subject to a non-linear dynamics are generally mixing [57]. In such a case the statistical ensemble is created by the propagation in time of initial fluctuations. The averages are averages over the initial conditions and the mixing character of the dynamics (if it can be proved) insures that the initial fluctuations are amplified in such a way that the ensemble of events covers the whole phase space uniformly[3]. For a classical dynamics which conserves the phase space volume of the ensemble of events, this means that the initial distribution is elongated and folded in such away that it gets close to any point of the phase space (the so-called baker transformation). This classical picture can be replaced in the quantum case by the idea of projection of irrelevant correlations [50]. The phase space can be described as a subspace of all possible observations. The regular quantum dynamics in the full space is transformed into a complex dynamics by the projection in the relevant observation subspace. Then two different realizations corresponding to the same projection, *i.e.* the same point in the relevant space, may differ in the full space (and consequently in their successive evolution) because of the unobserved correlations. This ensemble of correlations may lead to a statistical ensemble of realizations after a finite time. This phenomenon is often described introducing stochastic dynamics, *i.e.* assuming that the unobserved part of the dynamics which is averaged over is a random process [58,59].

5.1 How far is the system from equilibrium

An important point to be discussed is the justification of the statistical description. As we have just mentioned, the applicability of a statistical picture is in most cases an hypothesis (or a principle like in the thermodynamics second law). Therefore, the equilibrium hypothesis should be *a posteriori* controlled. Different properties can provide tests of equilibration such as

- the comparison with statistical models,
- the consistency of thermodynamical quantities, namely the compatibility of the different intensive variables measurements (*e.g.*, of the different thermometers) or the fulfillment of thermodynamic relations between averages and fluctuations (*e.g.*, $\sigma^2_{A_\lambda} = \partial^2 \log Z_{\boldsymbol{\lambda}}/\partial\lambda^2_\ell = -\partial\langle A_\ell\rangle/\partial\lambda_\ell$),
- the memory loss or the independence of the results on the preparation method of the considered ensemble.

However, it should be stressed that the real question is not whether the system is *at* equilibrium, but rather how far it is from a given equilibrium. Indeed, equilibrium is not unique in a finite system, and moreover exact equilibrium is a theoretical abstraction which cannot be achieved in the real world. To answer this question we should define a distance. The first idea could be to use the Liouville metric

$$d^2_{eq} = \text{tr}\left(\hat{D} - \hat{D}_{\boldsymbol{\lambda}}\right)^2 \tag{18}$$

between the actual ensemble characterized by the density matrix $\hat{D}$, and the equilibrium one $\hat{D}_{\boldsymbol{\lambda}}$ computed for the same collective variables $\langle A_\ell\rangle$. This is a nice theoretical tool, but a rather difficult definition as far as experiments are concerned. Another possibility is to introduce entropy as a metric [58]

$$d_{eq} = \left|S[\hat{D}] - S[\hat{D}_{\boldsymbol{\lambda}}]\right|/S[\hat{D}_{\boldsymbol{\lambda}}]. \tag{19}$$

This is a way to measure how far the system is from the maximum entropy state or in other words to measure how much information on the actual system is included in the collective variables $\{\langle A_\ell\rangle\}$ and how much is out of the considered equilibrium. This is a more physical distance but again it is difficult to implement in real experimental situations. A more practical measurement of the distance to equilibrium is to focus on the information used to deduce physical properties. Since the information about the actual system is contained in the observations $\langle \hat{O}_i\rangle$, the natural space to introduce this distance is the observation space. This is a formally well-defined problem since considering $\text{Tr}\,\hat{O}_i\hat{O}_j$ as the scalar product between observables, the observation space has a well-defined topology. Then, when orthogonal observables are considered[4], the distance to equilibrium is simply

$$d_i = \left|\langle\hat{O}_i\rangle - \langle\hat{O}_i\rangle_{eq}\right|. \tag{20}$$

A typical example is given by the difference between the measured fluctuations $\sigma^2_{A_\ell} = \langle A^2_\ell\rangle - \langle A_\ell\rangle^2$ and the expected ones $\sigma^2_{A_\ell} = -\partial\langle A_\ell\rangle/\partial\lambda_\ell$ in the ensemble controlled by the λ_ℓ.

6 Finite systems and ensemble inequivalence

We have discussed that many different statistical ensembles can be defined when one considers finite systems. A fundamental theorem in statistical mechanics, the Van Hove theorem [60] (see appendix), guarantees the equivalence between different statistical ensembles at the thermodynamic limit. However the theorem does not apply in finite systems. In fact, it is strongly violated in first-order phase transitions if the system is finite, and this violation can persist up to the thermodynamic limit in the case of

---

[3] Meaning that any phase space point gets close to at least one event.

[4] If observables are not orthogonal it is alway possible to use a Schmitt procedure to define a set of orthogonal observables [16].

long-range forces. A consequence of that is that it is possible to give a rigorous definition of phase transitions even in finite systems, with the prediction of fancy phenomena like negative heat capacities, negative compressibilities and negative susceptibilities. The non-equivalence of statistical ensembles has also important conceptual consequences. It implies that the value of thermodynamic variables for the very same system depends on the type of experiment which is performed (*i.e.* on the ensemble of constraints which are put on the system), contrary to the standard thermodynamic viewpoint that water heated in a kettle is the same as water put in an oven at the same temperature. Ensemble inequivalence is the subject of an abundant literature (see, for example, refs. [22, 23, 25–30] for a discussion in a general context, and refs. [31–36] concerning phase transitions).

Generally speaking, for a given value of the control parameters (or *intensive variables*) λ_ℓ, the properties of a substance are univocally defined, *i.e.* the conjugated *extensive variables* $\langle \hat{A}_\ell \rangle$ have a unique value unambiguously defined by the corresponding equation of state ($\langle A_\ell \rangle = -\partial_{\lambda_\ell} \log Z(\{\lambda_\ell\})$). In reality, this fixes only the average value and the event-by-event value of the observation of $\hat{A}_\ell$ produces a probability distribution. The intuitive expectation that extensive variables at equilibrium have a unique value therefore means that the probability distribution is narrow and normal, such that a good approximation can be obtained by replacing the distribution with its most probable value. The normality of probability distributions is usually assumed on the basis of the central limit theorem. However, in finite systems the probability distributions has a finite width and moreover it can depart from a normal distribution. We will discuss in particular the case of a bimodal distribution [43]: in this case two different properties (phases) coexist for the same value of the intensive control variable.

The topological anomalies of probability distributions and the failure of the central limit theorem in phase coexistence imply that in a first-order phase transition the different statistical ensembles are in general not equivalent and different phenomena can be observed depending on the fact that the controlled variable is extensive or intensive. In the following, we will often take as a paradigm of intensive ensembles the canonical ensemble for which the inverse of the temperature β^{-1} is controlled, while the archetype of the extensive ensemble will be the microcanonical one for which energy is strictly controlled.

6.1 The difference between Laplace and Legendre

The relation between the canonical entropy and the logarithm of the partition sum is given by a Legendre transform eq. (13). It is important to distinguish between transformations within the same ensemble, as the Legendre transform, and transformations between different ensembles, which are given by non-linear integral transforms [35]. Let us consider energy as the extensive observable and inverse temperature β as the conjugated intensive

one. The definition of the canonical partition sum is

$$Z_\beta = \sum_n \exp(-\beta E^{(n)}), \qquad (21)$$

where the sum runs over the available eigenstates n of the Hamiltonian. Here, we assume that the partition sum converges; this is not always the case as discussed in ref. [61]. The possible divergence of the thermodynamic potential of the intensive ensemble is already a known case of ensemble inequivalence [19, 61]. Computing the canonical (Shannon) entropy, we get

$$S_{can}(\langle E \rangle) = \log Z_\beta + \beta \langle E \rangle, \qquad (22)$$

which is an exact Legendre transform since the EoS reads $\langle E \rangle = -\partial_\beta \log Z_\beta$. If energy can be treated as a continuous variable, eq. (21) can be written as

$$Z_\beta = \int_0^\infty dE\, W(E) \exp(-\beta E), \qquad (23)$$

where energies are evaluated from the ground state. Equation (23) is a Laplace transform between the canonical partition sum and the microcanonical density of states linked to the entropy by $S_E = \ln W(E)$. If the integrand $f(E) = \exp(E_E - \beta E)$ is a strongly peaked function, it can be approximated by a Gaussian (saddle point approximation) so that the integral can be replaced by the maximum $f(\bar{E})$ times a Gaussian integral. Neglecting this factor, we get

$$Z_\beta \approx W(\bar{E}) \exp(-\beta \bar{E}), \qquad (24)$$

which can be rewritten as

$$\ln Z_\beta \approx S_{\bar{E}} - \beta \bar{E}; \qquad (25)$$

or introducing the free energy $F_T = -\beta^{-1} \ln Z_\beta$,

$$F_T \approx \bar{E} - T S_{\bar{E}}. \qquad (26)$$

Equation (25) has the structure of an approximate Legendre transform similar to the exact expression (22). This shows that in the lowest-order saddle point approximation eq. (24), the ensembles differing at the level of constraints acting on a specific observable (here energy) lead to the same entropy, *i.e.* they are equivalent. We will see in the next section that, however, the saddle point approximation eq. (24) can be highly incorrect close to a phase transition for the simple reason that the integrand is bimodal making a unique saddle point approximation inadequate. In this case eq. (25) cannot be applied, eq. (23) is the only correct transformation between the different ensembles, and ensemble inequivalence naturally arises.

6.2 Ensemble inequivalence and phase transitions

Let us consider the case of a first-order phase transition where the canonical energy distribution

$$P_{\beta_0}(E) = W(E) \exp(-\beta_0 E)/Z_{\beta_0} \qquad (27)$$

has a characteristic bimodal shape [43, 46, 47] at the temperature β_0 with two maxima $\overline{E}_\beta^{(1)}$, $\overline{E}_\beta^{(2)}$ that can be associated with the two phases. It is easy to see that eq. (23) can also be seen as a Laplace transform of the canonical probability $P_{\beta_0}(E)$

$$Z_\beta = Z_{\beta_0} \int_0^\infty \mathrm{d}E \, P_{\beta_0}(E) \exp(-(\beta - \beta_0)E). \qquad (28)$$

A single saddle point approximation is not valid when $P_{\beta_0}(E)$ is bimodal; however it is always possible to write

$$P_\beta = m_\beta^{(1)} P_\beta^{(1)} + m_\beta^{(2)} P_\beta^{(2)}, \qquad (29)$$

with $P_\beta^{(i)}$ mono-modal normalized probability distribution peaked at $\overline{E}_\beta^{(i)}$. The canonical mean energy is then the weighted average of the two energies

$$\langle E \rangle_\beta = \tilde{m}_\beta^{(1)} \overline{E}_\beta^{(1)} + \tilde{m}_\beta^{(2)} \overline{E}_\beta^{(2)}, \qquad (30)$$

with

$$\tilde{m}_\beta^{(i)} = m_\beta^{(i)} \int \mathrm{d}E P_\beta^{(i)}(E) E / \overline{E}_\beta^{(i)} \simeq m_\beta^{(i)}. \qquad (31)$$

Since only one mean energy is associated with a given temperature β^{-1}, the canonical caloric curve is monotonic, and the microcanonical one is not. Indeed it is immediate to see from eq. (27) that the bimodality of P_β implies then a back bending of the microcanonical caloric curve $T^{-1} = \partial_E S$, meaning that in the first-order phase transition region the two ensembles are not equivalent. If instead of looking at the average $\langle E \rangle_\beta$ we look at the most probable energy $\overline{E}_\beta$, this (unusual) canonical caloric curve is identical to the microcanonical one, up to the transition temperature β_t^{-1} for which the two components of $P_\beta(E)$ have the same height. At this point the most probable energy jumps from the low- to the high-energy branch of the microcanonical caloric curve.

The question arises whether this violation of ensemble equivalence survives towards the thermodynamic limit. This limit can be expressed as the fact that the thermodynamic potentials per particle converge when the number of particles N goes to infinity:

$$f_{N,\beta} = \beta^{-1} \frac{\log Z_\beta}{N} \to \bar{f}_\beta; \quad s_N(e) = \frac{S(E)}{N} \to \bar{s}(e), \qquad (32)$$

where $e = E/N$. Let us also introduce the reduced probability $p_{N,\beta}(e) = (P_\beta(N, E))^{1/N}$ which then converges towards an asymptotic distribution

$$p_{N,\beta}(e) \to \bar{p}_\beta(e); \quad \bar{p}_\beta(e) = \exp\left(\bar{s}(e) - \beta e + \bar{f}_\beta\right). \qquad (33)$$

Since $P_\beta(N, E) \approx (\bar{p}_\beta(e))^N$, one can see that when $\bar{p}_\beta(e)$ is normal, the relative energy fluctuation in $P_\beta(N, E)$ is suppressed by a factor $1/\sqrt{N}$. At the thermodynamic limit P_β reduces to a δ-function and ensemble equivalence is

recovered. To analyze the thermodynamic limit of a bimodal $p_{N,\beta}(e)$, let us introduce as before $\beta_{N,t}^{-1}$ the temperature for which the two maxima of $p_{N,\beta}(e)$ have the same height. For a first-order phase transition $\beta_{N,t}^{-1}$ converges to a fixed point $\bar{\beta}_t^{-1}$ as well as the two maximum energies $e_{N,\beta}^{(i)} \to \bar{e}_\beta^{(i)}$. For all temperatures lower (higher) than $\bar{\beta}_t^{-1}$ only the low- (high-) energy peak will survive at the thermodynamic limit, since the difference of the two maximum probabilities will be raised to the power N. Therefore, below $\bar{e}_\beta^{(1)}$ and above $\bar{e}_\beta^{(2)}$ the canonical caloric curve coincides with the microcanonical one in the thermodynamic limit. In the canonical ensemble the temperature $\bar{\beta}_t^{-1}$ corresponds to a discontinuity in the state energy irrespectively of the behavior of the entropy between $\bar{e}_\beta^{(1)}$ and $\bar{e}_\beta^{(2)}$.

The microcanonical caloric curve in the phase transition region may either converge towards the Maxwell construction, or keep a backbending behavior [21], since a negative heat capacity system can be thermodynamically stable even in the thermodynamic limit if it is isolated [25]. Examples of a backbending behavior at the thermodynamic limit have been reported for a model many-body interaction taken as a functional of the hypergeometric radius in the analytical work of ref. [3], and for the long-range Ising model [4]. This can be understood as a general effect of long-range interactions for which the topological anomaly leading to the convex intruder in the entropy is not cured by increasing the number of particles [4, 62]. Conversely, for short-range interactions [15] the backbending is a surface effect which should disappear at the thermodynamic limit. This is the case for the Potts model [32], the microcanonical model of fragmentation of atomic clusters [63] and for the lattice gas model with fluctuating volume [48]. The interphase surface entropy goes to zero as $N \to \infty$ in these models, leading to a linear increase of the entropy in agreement with the canonical predictions.

Within the approach based on the topology of the probability distribution of observables [43] it was shown that ensemble inequivalence arises from fluctuations of the order parameter [22]. Ensembles putting different constraints on the fluctuations of the order parameter lead to a different thermodynamics. In the case of phase transitions with a finite latent heat, the total energy usually plays the role of an order parameter except in the microcanonical ensemble which, therefore, is expected to present a different thermodynamics than the other ensembles [19]. This inequivalence may remain at the thermodynamic limit if the involved phenomena are not reduced to short-range effects.

6.3 Temperature jump at constant energy

In particular, it may happen that the energy of a subsystem becomes an order parameter when the total energy is constrained by a conservation law or a microcanonical sorting. This frequently occurs for Hamiltonians containing a kinetic energy contribution [3, 4, 64]: if the kinetic heat capacity is large enough, it becomes an order parameter in the microcanonical ensemble. Then, the microcanonical caloric curve presents at the thermodynamical limit a temperature jump in complete disagreement with the canonical ensemble.

To understand this phenomenon, let us consider a finite system for which the Hamiltonian can be separated into two components $E = E_1 + E_2$, that are statistically independent ($W(E_1, E_2) = W_1(E_1)W_2(E_2)$) and such that the associated degrees of freedom scale in the same way with the number of particles; we will also consider the case where $S_1 = \log W_1$ has no anomaly while $S_2 = \log W_2$ presents a convex intruder [15] which is preserved at the thermodynamic limit. Typical examples of E_1 are given by the kinetic energy for a classical system with velocity-independent interactions, or other similar one-body operators [4]. The probability to get a partial energy E_1 when the total energy is E is given by

$$P_E\left(E_1\right) = \exp\left(S_1\left(E_1\right) + S_2\left(E - E_1\right) - S\left(E\right)\right). \quad (34)$$

The extremum of $P_E(E_1)$ is obtained for the partitioning of the total energy E between the kinetic and potential components that equalizes the two partial temperatures

$$\overline{T_1}^{-1} = \partial_{E_1} S_1(\overline{E}_1) = \partial_{E_2} S_2(E - \overline{E}_1) = \overline{T_2}^{-1}. \quad (35)$$

If $\overline{E}_1$ is unique, $P_E(E_1)$ is mono-modal and we can use a saddle point approximation around this solution to compute the entropy

$$S\left(E\right) = \log \int_{-\infty}^{E} \mathrm{d}E_1 \exp\left(S_1\left(E_1\right) + S_2\left(E - E_1\right)\right). \quad (36)$$

At the lowest order, the entropy is simply additive so that the microcanonical temperature of the global system $\partial_E S(E) = \overline{T}^{-1}$ is the one of the most probable energy partition. Therefore, the most probable partial energy $\overline{E}_1$ acts as a microcanonical thermometer. If $\overline{E}_1$ is always unique, the kinetic thermometer in the backbending region will follow the whole decrease of temperature as the total energy increases. Therefore, the total caloric curve will present the same anomaly as the potential one. If, conversely, the partial energy distribution is double humped [65], then the equality of the partial temperatures admits three solutions, one of them $\overline{E}_1^{(0)}$ being a minimum. At this point, the partial heat capacities

$$C_1^{-1} = -\overline{T}^2\,\partial_{E_1}^2 S_1\big(\overline{E}_1^{(0)}\big); \quad C_2^{-1} = -\overline{T}^2\,\partial_{E_2}^2 S_2\big(E - \overline{E}_1^{(0)}\big) \quad (37)$$

fulfill the relation

$$C_1^{-1} + C_2^{-1} < 0. \quad (38)$$

This happens when the potential heat capacity is negative and the kinetic energy is large enough ($C_1 > -C_2$) to act as an approximate heat bath: the partial energy distribution $P_E(E_1)$ in the microcanonical ensemble is then bimodal as the total energy distribution $P_\beta(E)$ in the canonical ensemble, implying that the kinetic energy is the order

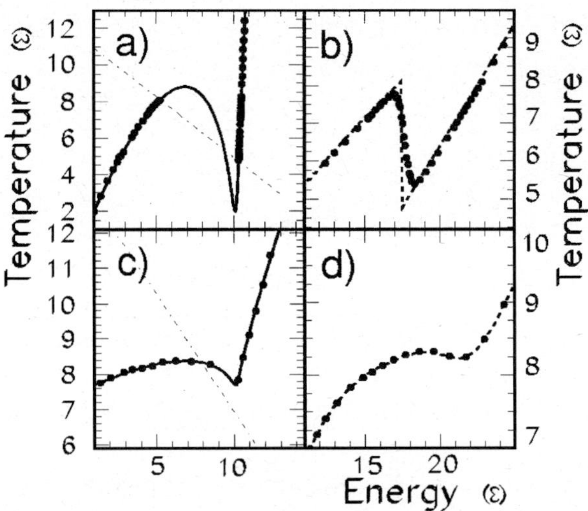

Fig. 2. Left panels: temperature as a function of the potential energy E_2 (full lines) and of the kinetic energy $E - E_2$ (dot-dashed lines) for two model equation of states of classical systems showing a first-order phase transition. Symbols: temperatures extracted from the most probable kinetic energy thermometer from eq. (35). Right panels: total caloric curves (symbols) corresponding to the left panels and thermodynamic limit of eq. (39) (dashed lines).

parameter of the transition in the microcanonical ensemble. In this case the microcanonical temperature is given by a weighted average of the two estimations from the two maxima of the kinetic energy distribution

$$T = \partial_E S(E) = \frac{\overline{P}^{(1)}\sigma^{(1)}/\overline{T}^{(1)} + \overline{P}^{(2)}\sigma^{(2)}/\overline{T}^{(2)}}{\overline{P}^{(1)}\sigma^{(1)} + \overline{P}^{(2)}\sigma^{(2)}}, \quad (39)$$

where $\overline{T}^{(i)} = T_1(\overline{E}_1^{(i)})$ are the kinetic temperatures calculated at the two maxima, $\overline{P}^{(i)} = P_E(\overline{E}_1^{(i)})$ are the probabilities of the two peaks and $\sigma^{(i)}$ their widths. At the thermodynamic limit eq. (38) reads $c_1^{-1} + c_2^{-1} < 0$, with $c = \lim_{N\to\infty} C/N$. If this condition is fulfilled, the probability distribution $P_\beta(E)$ presents two maxima for all finite sizes and only the highest peak survives at $N = \infty$. Let E_t be the energy at which $P_{E_t}(\overline{E}^{(1)}) = P_{E_t}(\overline{E}^{(2)})$. Because of eq. (39), at the thermodynamic limit the caloric curve will follow the high- (low-) energy maximum of $P_E(E_1)$ for all energies below (above) E_t; there will be a temperature jump at the transition energy E_t.

Let us illustrate the above results with two examples for a classical gas of interacting particles. For the kinetic energy contribution we have $S_1(E) = c_1 \ln(E/N)^N$ with a constant kinetic heat capacity per particle $c_1 = 3/2$. For the potential part we will take two polynomial parametrizations of the interaction caloric curve presenting a back bending which are displayed in the left part of fig. 2 in units of an arbitrary scale ϵ. If the decrease of the partial temperature $T_2(E_2)$ is steeper than $-2/3$ (fig. 2a) then eq. (38) is verified [3] and the kinetic caloric curve $T_1(E - E_1)$ (dot-dashed line) crosses the potential one $T_2(E_2)$ (full line) in three different points for all values

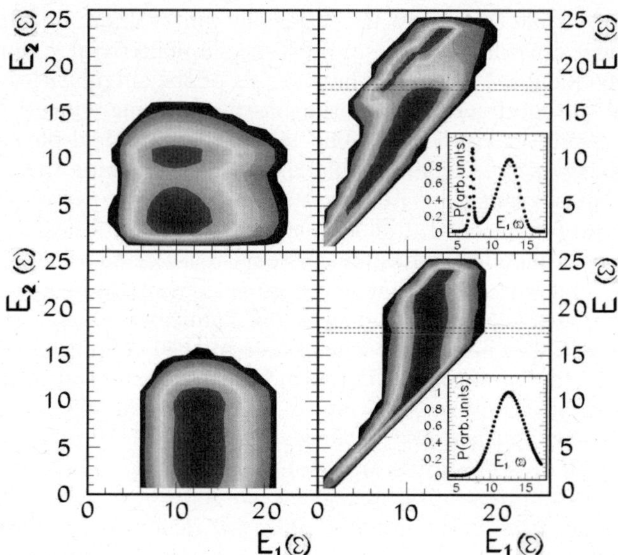

Fig. 3. Canonical event distributions in the potential *versus* kinetic energy plane (left panels) and total *versus* kinetic energy plane (right panels) at the transition temperature for the two model equations of state of fig. 2. The inserts show two constant total energy cuts of the distributions.

of the total energy lying inside the region of coexistence of two kinetic energy maxima. The resulting caloric curve for the whole system is shown in fig. 2b (symbols) together with the thermodynamic limit (line) evaluated from the double saddle point approximation eq. (39). In this case one observes a temperature jump at the transition energy. If the temperature decrease is smoother (fig. 2c) the shape of the interaction caloric curve is preserved at the thermodynamic limit (fig. 2d).

The occurrence of a temperature jump in the thermodynamic limit is easily spotted by looking at the bidimensional canonical event distribution $P_\beta(E_1, E_2)$ shown at the transition temperature $\beta = \beta_t$ in the left part of fig. 3 for the two model equation of states of fig. 2. In the canonical ensemble the kinetic energy distribution is normal. These same distributions are shown as a function of E and E_1, $P_\beta(E, E_1) \propto \exp S_1(E_1) \exp S_2(E - E_1) \exp(-\beta E)$ in the right part of fig. 3. The microcanonical ensemble is a constant energy cut of $P_\beta(E, E_1)$, which leads to the microcanonical distribution $P_E(E_1)$ within a normalization constant. If the anomaly in the potential equation of state is sufficiently important, the distortion of events due to the coordinate change is such that one can still see the two phases coexist even after a sorting in energy.

7 Definitions of phase transitions

Phase transitions are universal properties of matter in interaction. In macroscopic physics, they are singularities (*i.e.* non-analytical behaviors) in the system equation of state (EoS) and hence classified according to the degree of non-analyticity of the EoS at the transition point. Then, a phase transition is an intrinsic property of the system

and not of the statistical ensemble used to describe the equilibrium. Indeed, at the thermodynamic limit all the possible statistical ensembles converge towards the same EoS (see appendix), and the various thermodynamic potentials are related by simple Legendre transformations leading to a unique thermodynamics. On the other side for finite systems, as discussed above, two ensembles which put different constraints on the fluctuations of the order parameter lead to qualitatively different EoS close to a first-order phase transition [15,24]. Thermodynamic observables like heat capacities can, therefore, be completely different depending on the experimental conditions of the measurement. Moreover, such inequivalences may survive at the thermodynamic limit if forces are long ranged as for self-gravitating objects [3,4]. In fact, the characteristic of phase transitions in finite systems, and in particular the occurrence of a negative heat capacity, have first been discussed in the astrophysical context [2,25,30,66–70]. Since these pioneering works in astrophysics, an abundant literature is focused on the understanding of phase transitions in small systems from a general point of view [15,29,35, 71–76] or in the mean-field context [4,77] or for some specific systems such as metallic clusters [47,65] or nuclei [78] and even DNA [79].

7.1 Phase transitions in infinite systems

Let us first recall the definition of phase transitions in infinite systems. At the thermodynamic limit for short-range interactions the statistical ensembles are equivalent and it is enough to reduce the discussion to the ensemble where only one extensive variable A_L is kept fixed, all the others being constrained through the associated Lagrange parameters. The typical example is the grand-canonical ensemble where only the volume $A_L = V$ is kept as an extensive variable. Then all the thermodynamics is contained in the associated potential $\log Z_{\lambda_1,\ldots,\lambda_{L-1}}(A_L)$. Since it is extensive, the potential is proportional to the remaining extensive variable

$$\log Z_{\lambda_1,\ldots,\lambda_{L-1}}(A_L) = A_L \lambda_L(\lambda_1,\ldots,\lambda_{L-1}) \qquad (40)$$

so that all the non-trivial thermodynamic properties are included in the reduced potential, *i.e.* the intensive variable

$$\lambda_L = \partial_{A_L} \log Z_{\lambda_1,\ldots,\lambda_{L-1}}(A_L) = \frac{\log Z_{\lambda_1,\ldots,\lambda_{L-1}}}{A_L} \qquad (41)$$

associated with A_L. In the grand-canonical case $A_L = V$, the reduced potential is the pressure, $\lambda_L \propto P$, which is then a function of the temperature and the chemical potential(s). In this limit all the thermodynamics is included in the single function $\lambda_L(\lambda_1,\ldots,\lambda_{L-1})$, and this is why in the literature $p(V)$ is often loosely referred to as "the" EoS, and the existence of many EoS is ignored. If this EoS is analytical, all the thermodynamic quantities which are all derivatives of the thermodynamic potential, present smooth behaviors, and no phase transition appears. A phase transition is a major modification of the macrostate

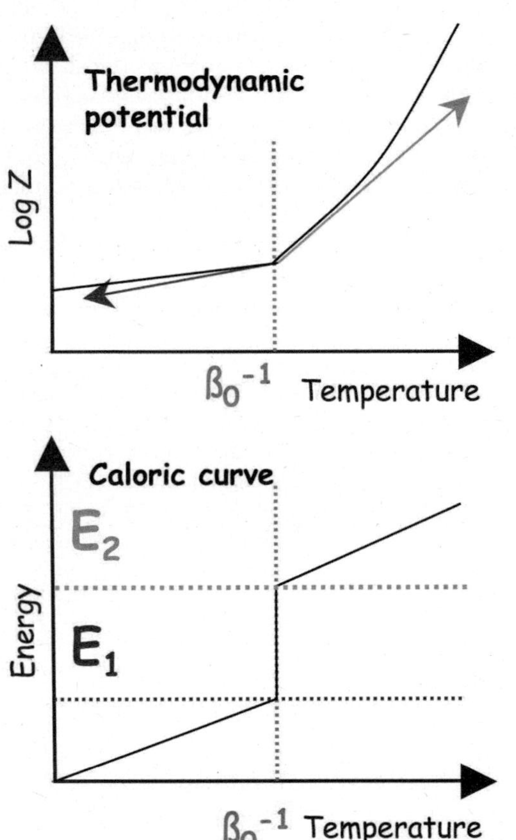

Fig. 4. Schematic representation of a first-order phase transition in the canonical case. Top: the log of the canonical partition sum (*i.e.* the free energy) presents an angular point. Bottom: the first derivative as a function of the temperature (*i.e.* the energy) presents a jump.

properties for a small modification of the control parameters $(\lambda_1, \dots, \lambda_{L-1})$. Such an anomalous behavior can only happen if the thermodynamic potential presents a singularity. This singularity can be classified according to the order of the derivative which presents a discontinuity or a divergence. According to Ehrenfest this is the order of the phase transition. In modern statistical mechanics, all the higher-order transitions are called under the generic name of continuous transitions. Figure 4 schematically illustrates a first-order phase transition in the canonical ensemble.

7.2 Phase transitions in finite systems

As soon as one considers a finite physical system, all the above discussion does not apply. First the thermodynamic potential and observables are not additive, therefore we cannot introduce a reduced potential. Indeed, the inequality

$$\lambda_L(\lambda_1, \dots, \lambda_{L-1}, A_L) \equiv \frac{\partial \log Z_{\lambda_1, \dots, \lambda_{L-1}}(A_L)}{\partial A_L}$$

$$\neq \frac{\log Z_{\lambda_1, \dots, \lambda_{L-1}}(A_L)}{A_L} \qquad (42)$$

shows that the grand potential per unit volume does no longer give the pressure and presents a non-trivial volume dependence. Moreover, the analysis of the singularities of the thermodynamic potential has no meaning, since it is an analytical function. The standard statistical-physics textbooks thus conclude that rigourously speaking there is no phase transitions in finite systems. However, as we have already mentioned, first for self-gravitating objects [2, 25, 30, 66–70] and then in small systems [3, 15, 29, 35, 47, 71–73, 76, 78] it was shown that phase transitions might be associated with the occurrence of negative microcanonical heat capacities. This can be generalized to the occurence of an inverted curvature of the thermodynamic potential of any ensemble keeping at least one extensive variable A_L not orthogonal to the order parameter[5] [35, 80]. In the following we call this ensemble an extensive ensemble. Then, negative compressibility or negative susceptibility should be, like negative heat capacity, observed in first-order phase transitions of finite systems. In the microcanonical ensemble of classical particles, it was proposed that anomalously large fluctuations of the kinetic energy, *i.e.* larger than the expected canonical value, highlight a negative heat capacity [81]. It was then demonstrated that those two signals of a phase transition, negative curvatures and anomalous fluctuations, observed in extensive ensembles where the order parameter is fixed, are directly related to the appearance of bimodalities in the distribution of this order parameter in the intensive ensemble where the order parameter is only fixed in average through its conjugated Lagrange multiplier [6, 43].

The occurrence of bimodalities is discussed in the literature since a long time and is often used as a practical way to look for phase transitions in numerical simulations [17, 46]; however, the general equivalence between negative curvatures and bimodalities was presented in ref. [43]. For intensive ensembles, since the pioneering work of Yang and Lee [42] another definition was proposed considering the zeroes of the partition sum in the complex intensive parameter plane [42, 82]. The idea is simple: the zeroes of Z are the singularities of $\log Z$ and so phase transitions, which are singularities, must come from the zeroes of the partition sum. In a finite system the zeroes of the partition sum cannot be on the real axis since the partition sum Z is the sum of exponential factors which cannot produce a singularity of $\log Z$. However, the thermodynamic limit of an infinite volume may bring the singularity on the real axis. This is schematically illustrated in fig. 5. Only regions where zeroes converge towards the real axis may present phase transitions, while the other regions present no anomalies. The order of the transition can be associated with the asymptotic behavior of the zeroes [82].

The distribution of zeroes has been analyzed in ref. [44] where the transition was studied with a parabolic entropy. In ref. [45] the equivalence of the expected behavior of the zeroes in a first-order phase transition case and the occurrence of bimodalities in the distribution of the associated extensive parameter was demonstrated. To be precise, in

---

⁵ Orthogonality is here defined using the trace as a scalar product between observables following sect. 3.

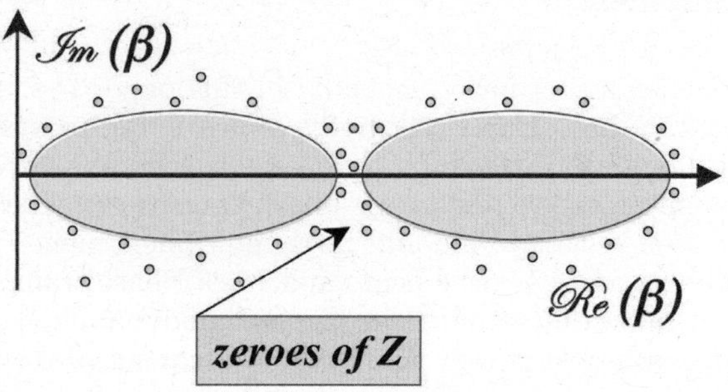

Fig. 5. Schematic representation of the zeroes of the partition sum Z in the complex temperature plane. The regions where no zeroes are coming close to the real axis when the thermodynamic limit is taken will not present singularities of $\log Z$.

this demonstration bimodality means that the extensive variable distribution can be split at the transition point into two distributions of equal height, with the distance between the two maxima scaling like the system size [83].

This global picture of phase transitions in finite systems is summarized in fig. 6 in the case where energy is the order parameter of the transition. The occurrence of a bimodal distribution of the extensive parameter (*e.g.*, energy) in the associated intensive (*e.g.*, canonical) ensemble is a necessary and sufficient condition to asymptotically get the distribution of the Yang-Lee zeroes in the complex Lagrange multiplier (*e.g.*, temperature) plane, which is expected in a first-order transition. The direction of bimodality is the direction of the order parameter. This bimodality is also equivalent to the presence of an anomalous curvature in the thermodynamic potential of the extensive (microcanonical) ensemble obtained constraining the bimodal observable to a fixed value. In the extensive ensemble, the inverted curvature can be spotted looking for anomalously large fluctuations (*e.g.*, larger than the canonical ones) of the partition of the extensive variable (*e.g.*, energy) between two independent subsystems.

8 Statistical description of evolving systems

A major issue in the statistical treatment of finite systems is that most of the time open and transient systems are studied. Therefore, they are not only finite in size but also finite in time and, in fact, they are evolving. The number of degrees of freedom of a quantum many-body problem being infinite, it is impossible to have all the information needed to solve exactly the dynamical problem. Since only a small part of the observation space is relevant, this time evolution may also be treated with statistical tools. This is the purpose of many models: from Langevin approaches to Fokker-Planck equations, from hydrodynamics to stochastic Time-Dependent Hartree-Fock theory. The purpose of this paper is not to review those theoretical approaches, therefore we will not enter here into details about the different recent progresses, and we will rather focus this discussion on general arguments of time-dependent statistical ensembles [50,53].

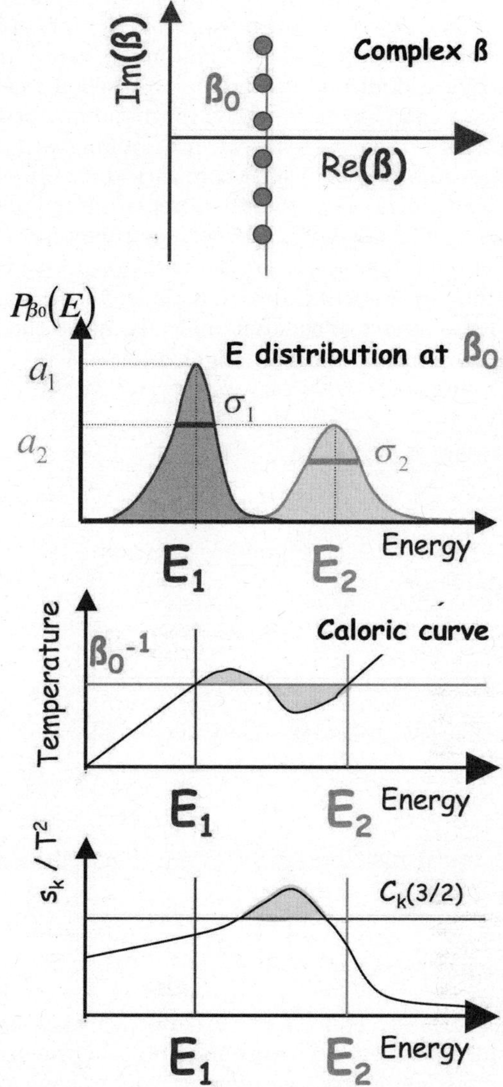

Fig. 6. Schematic representation of the different equivalent definitions of first-order phase transitions in finite systems. From top to bottom: the partition sum's zeroes aligning perpendicular to the real temperature axis with a density scaling like the number of particles; the bimodality of the energy distribution with a distance between the two maxima scaling like the number of particles times the latent heat; the appearence of a back-bending in the microcanonical caloric curve, *i.e.* a negative heat capacity region; and the observation of anomalously large fluctuations of the energy splitting between the kinetic part and the interaction part.

A statistical treatment of a dynamical process is based on the idea that at any time one can consider only the relevant variables A_ℓ, disregarding all the other ones a_m as irrelevant. If only the maximum entropy state is followed in time assuming that all the irrelevant degrees of freedom have relaxed instantaneously, one gets a generalized mean-field approach [58]. If the fluctuations of the irrelevant degrees of freedom are included, this leads to a Langevin dynamics [59]. With those considerations one can see that statistical approaches can be always improved including more and more degrees of freedom to asymptotically become exact. However, before including a huge number of degrees of freedom one should ask himself if only a few observables can take care of the most important dynamical aspects of the systems we are looking at. In a recent paper [53] it was proposed to introduce observations at different times (*e.g.*, different freeze-out/equilibration times) as well as time-odd extensive parameters. The idea is simple: maximizing the Shannon entropy with different observables $\hat{A}_\ell$ known at different times $t_\ell = t_0 + \Delta t_\ell$ is a way to treat a part of the dynamics. Going to the Heisenberg representation, if we propagate all the $\hat{A}_\ell$ to the same time t_0 we get

$$\hat{A}_\ell(t_0) = e^{-i\Delta t_\ell \hat{H}} \hat{A}_\ell e^{i\Delta t_\ell \hat{H}} \tag{43}$$

$$= \hat{A}_\ell - i\Delta t_\ell [\hat{H}, \hat{A}_\ell] + \ldots. \tag{44}$$

This shows that the time propagation introduces new constraining operators

$$\hat{B}_\ell = -i[\hat{H}, \hat{A}_\ell]. \tag{45}$$

If $\hat{A}_\ell$ is a time-even observable, $\hat{B}_\ell$ is a time-odd operator. Let us take the example of an unconfined finite system characterized at a given time by a typical size $\langle \hat{R}^2 \rangle = \langle \hat{S} \rangle$, where $\hat{R}^2$ is the one body operator $\sum_i \hat{r}_i^2$. If the whole information is assumed to be known at the same time, then the statistical distribution of events reads in a classical canonical picture

$$p^{(n)} = \frac{1}{Z} e^{-\beta E^{(n)} - \lambda S^{(n)}}, \tag{46}$$

which is formally equivalent to a particle in a harmonic potential. However, if now we assume that the size information is coming from a different time, then according to eq. (45) we must introduce a new time-odd operator $\hat{v}_r = -i[\hat{H}, \hat{r}^2]$. For a local interaction, this reduces to

$$\hat{v}_r = (\hat{r}\hat{p} + \hat{p}\hat{r})/m, \tag{47}$$

which represents a radial flow. Then the classical canonical probability reads

$$p^{(n)} = \frac{1}{Z} e^{-\beta \left(\mathbf{p}^{(n)} - h(t)\mathbf{r}^{(n)}\right)^2 - \lambda S^{(n)}} \tag{48}$$

which is a statistical ensemble of particles under a Hubblian flow. In the ideal-gas model eq. (48) provides the exact solution at any time of the dynamics. This simple example shows that information theory allows to treat in a statistical picture dynamical processes where observables are defined at different times, by taking into account time-odd components such as flows. This might be a tool to extract thermodynamical quantities from complex dynamics. In particular, the above example shows that in an open system an initial extension in space is always transformed into an expansion, meaning that flow is an essential ingredient even in statistical approaches.

9 Conclusion

In conclusion, we have presented in this paper the actual understanding of the thermodynamics of finite systems from the point of view of information theory. We have put some emphasis on first-order phase transitions which are associated with specific and intriguing phenomena as bimodalities and negative heat capacities. Phase transitions have been widely studied in the thermodynamic limit of infinite systems. However, in the physical situations considered here, this limit cannot be taken and phase transitions should be reconsidered from a more general point of view. This is for example the case of matter under long-range forces like gravitation. Even if these self-gravitating systems are very large they cannot be considered as infinite because of the non-saturating nature of the force. Other cases are provided by microscopic or mesoscopic systems built out of matter which is known to present phase transitions. Metallic clusters can melt before being vaporized. Quantum fluids may undergo Bose condensation or a super-fluid phase transition. Dense hadronic matter should merge in a quark and gluon plasma phase while nuclei are expected to exhibit a liquid-gas phase transition and a superfluid phase. For all these systems the theoretical and experimental issue is how to define and sign a possible phase transition in a finite system. In this review we have presented the synthesis of different works which tend to show that phase transitions can be defined as clearly as in the thermodynamic limit. Depending upon the statistical ensemble, *i.e.* on the experimental situation, one should look for different signals. In the ensemble where the order parameter is free to fluctuate (intensive ensemble), the topology of the event distribution should be studied. A bimodal distribution signals a first-order phase transition. The direction in the observable space in which the distribution is bimodal defines the best order parameter. To survive the thermodynamic limit, the distance between the two distributions, the two "phases", should scale like the number of particles. This occurrence of a bimodal distribution is equivalent to the alignment of the partition sum zeroes as described by the Yang and Lee theorem. In the associated extensive ensemble, the bimodality condition is also equivalent to the requirement of a convexity anomaly in the thermodynamic potential. The first experimental evidences of such a phenomenon have been reported recently in different fields: the melting of sodium clusters [6], the fragmentation of hydrogen clusters [8], the pairing in nuclei [9] and nuclear multifragmentation [7, 84,85]. However, much more experimental and theoretical studies are now expected to progress in this new field of phase transitions in finite systems. Three challenges can thus be assigned to the physics community:

- The statistical description of non-extensive systems and in particular of open transient finite systems.
- The experimental and theoretical study of phase transitions in those systems and of the expected abnormal thermodynamics.
- The confirmation of the observation of the nuclear phase transition and the analysis of the associated

equation-of-state properties and the associated phase diagram.

Appendix A. The Van Hove theorem

Let us consider a system in a volume V for which only the average value of energy and number of particles is defined (grand-canonical ensemble). Let us calculate the grand potential $\Omega = -T \ln Z$:

$$Z_{\beta\mu}(V) = \sum_n \exp\left(-\beta\left(H^{(n)} - \mu N^{(n)}\right)\right), \qquad (A.1)$$

where the sum extends over all the possible configurations of the system, $H^{(n)} = K^{(n)} + U^{(n)}\left(N^{(n)}\right)$ represents the energy (number of particles) of the system in the configuration (n), and β, μ are the associated Lagrange multipliers, the inverse temperature and the chemical potential, respectively. The partition sum results

$$Z_{\beta\mu}(V) = \sum_{N=0}^{\infty} z_k^N Z_\beta(N, V) \qquad (A.2)$$

with $z_k = \exp(\beta\mu)(\frac{2m\pi}{h^2\beta})^{3/2}$ the ideal-gas part and

$$Z_\beta(N, V) = \frac{1}{N!} \int_V \mathrm{d}^{3N} r \exp\left(-\beta U\right) \qquad (A.3)$$

the partition sum associated with the interaction part. Let us divide $V = mV_0 + V_1$ in m equal boxes of volume V_0 separated by "corridors" of width b larger than the range of the force such that the interactions among particles in different boxes can be neglected (see fig. 7). The volume excluded by the corridors is V_1. To calculate $Z_\beta(N, V)$ let us consider the number of particles in the corridor N_1:

$$Z_\beta(N, V) = \sum_{N_1=0}^{N} \frac{1}{N_1!} \frac{1}{(N-N_1)!} \int_{V_1} \mathrm{d}^{3N_1} r, \qquad (A.4)$$

$$\int_{V-V_1} \mathrm{d}^{3(N-N_1)} r \exp\left(-\beta U\right). \qquad (A.5)$$

Let us note ϵ the minimum of the two-body interaction (see fig. 7); the potential energy in the corridor satisfies then the inequality $U_{V_1} \geq \varepsilon\xi N_1$, where $\xi = (b/a)^3$ represents the maximum number of particles interacting with a given particle. For the total potential energy, we can write

$$U \geq \varepsilon\xi N_1 + \frac{1}{2} \sum_{i=N_1+1}^{N} \qquad (A.6)$$

leading to

$$Z_\beta(N, V) \leq \sum_{N_1=0}^{N} \frac{1}{N_1!} \frac{1}{(N-N_1)!} V_1^{N_1} \exp\left(-N_1\beta\varepsilon\xi\right), \qquad (A.7)$$

$$\int_{mV_0} \mathrm{d}^{3(N-N_1)} r \exp\left(-\beta U\right), \qquad (A.8)$$

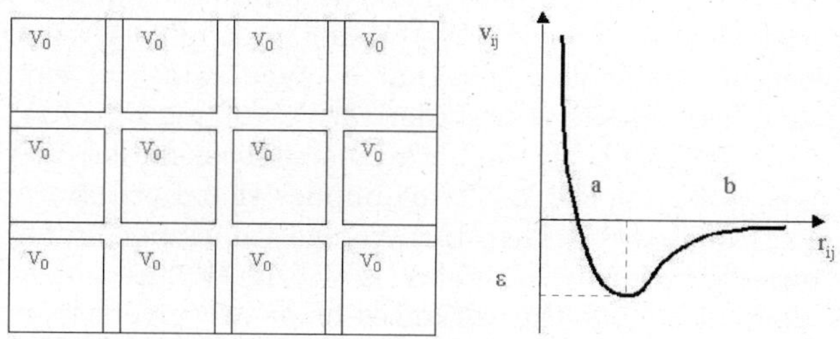

Fig. 7. Schematic representation of the Van Hove theorem demonstration (left) and the corresponding inter-particle interaction (right).

where the last integrals run over the n-independent volumes V_0. Introducing this expression in (A.4) with $N_2 = N - N_1$, the partition sum $Z_{\beta\mu}(V)$ reads

$$Z_{\beta\mu}(V) \leq \sum_{N_1=0}^{\infty} \frac{1}{N_1!} V_1^{N_1} z_k^{N_1} \exp\left(-N_1\beta\varepsilon\xi\right)$$

$$\cdot \sum_{N_2=0}^{\infty} \frac{1}{N_2!} z_k^{N_2} \int_{mV_0} \mathrm{d}^{3N_2} r \exp\left(-\beta U\right) \qquad (A.9)$$

$$= \exp\left(z_k V_1 e^{-\beta\varepsilon\xi}\right) Z_{\beta\mu}^m(V_0), \qquad (A.10)$$

where the last equality stems from the fact that particles interact only within the same box again because of the short range of the force. Finally, we get using $V_1 \propto mV_0^{2/3}$:

$$\log Z_{\beta\mu}(V) \leq kmV_0^{2/3} + m \log Z_{\beta\mu}(V_0), \qquad (A.11)$$

$$\frac{\log Z_{\beta\mu}(V)}{V} \leq kV_0^{-1/3} + \frac{\log Z_{\beta\mu}(V_0)}{V_0}, \qquad (A.12)$$

which gives in the thermodynamic limit (keeping m constant) $V \to \infty$, $V_0 \to \infty$, $V \to mV_0$,

$$\frac{\log Z_{\beta\mu}(V)}{V} \leq \frac{\log Z_{\beta\mu}(V_0)}{V_0}. \qquad (A.13)$$

On the other side, the opposite inequality is trivially true:

$$Z_{\beta\mu}(V) \geq Z_{\beta\mu}^m(V_0) \qquad (A.14)$$

since by neglecting the corridor in the integral (A.5) a positive term in the partition sum is neglected. In conclusion we have demonstrated that

$$\frac{\log Z_{\beta\mu}(V)}{V} \xrightarrow[V \text{ and } V_0 \to \infty]{} \frac{\log Z_{\beta\mu}(V_0)}{V_0}. \qquad (A.15)$$

It is very important to stress that this result is true only for short-range interactions. For these specific systems the implications of eq. (A.15) can be summarized as follows:

- A thermodynamic limit exists for these systems if the thermodynamic potential per unit volume tends to a constant for large volumes $\log Z_{\beta\mu}(V)/V \to \omega$;

– In the thermodynamic limit ensembles are equivalent. Indeed if $\omega = \log Z_{\beta\mu}(V_i)/V_i$ for an arbitrary subsystem V_i, using the fact that average values of extensive variables are first derivatives of $\log Z$ ($\langle A_\ell \rangle = -\partial_{\lambda_\ell} \log Z(\{\lambda_\ell\})$) and variances second derivatives ($\sigma_\ell^2 = \partial_{\lambda_\ell}^2 \log Z(\{\lambda_\ell\})$), this implies that both are proportional to V_i. Then the average values per unit volume of extensive variables ($\rho_\ell = \langle A_\ell \rangle/V$) are independent of V and the variances of ρ_ℓ are inversely proportional to V, approaching zero as V goes to infinity. Since ensembles differ at the level of fluctuations, this demonstrates the equivalence between ensembles. For the explicit demonstration of the equality of the canonical and grand-canonical EoS we refer the reader to refs. [11,35].

References

1. T. Dauxois *et al.*, *Dynamics and Thermodynamics of Systems with Long Range Interactions*, Lect. Notes Phys., Vol. **602** (Springer, 2002).
2. D. Lynden-Bell, R. Wood, Mon. Not. R. Astron. Soc. **138**, 495 (1968); D. Lynden-Bell, Physica A **263**, 293 (1999).
3. R.M. Lynden-Bell, Mol. Phys. **86**, 1353 (1995).
4. J. Barré, D. Mukamel, S. Ruffo, Phys. Rev. Lett. **87**, 030601 (2001); J. Barré, F. Bouchet, T. Dauxois, S. Ruffo, J. Stat. Phys. **119**, 677 (2005); D. Mukamel, S. Ruffo, N. Schreiber, Phys. Rev. Lett. **95**, 240604 (2005).
5. T. Tatekawa, F. Bouchet, T. Dauxois, S. Ruffo, Phys. Rev. E **71**, 056111 (2005).
6. M. Schmidt *et al.*, Phys. Rev. Lett. **86**, 1191 (2001).
7. M. D'Agostino *et al.*, Phys. Lett. B **473**, 219 (2000).
8. F. Gobet *et al.*, Phys. Rev. Lett. **89**, 183403 (2002).
9. E. Melby *et al.*, Phys. Rev. Lett. **83**, 3150 (1999); S. Siem *et al.*, Phys. Rev. C **65**, 044318 (2002).
10. L.D. Landau, E.M. Lifshitz, *Statistical Physics* (Pergamon Press, 1980) Chapt. 3.
11. K. Huang, *Statistical Mechanics* (John Wiley and Sons Inc., 1963) Chapt. 5.
12. R.C. Tolman, *Principles of Statistical Mechanics* (Oxford University Press, London, 1962).
13. M. Rasetti, *Modern Methods in Statistical Mechanics* (World Scientific, Singapore, 1986).
14. E.T. Jaynes, *Information theory and statistical mechanics*, Stat. Phys., Brandeis Lect. **3**, 160 (1963).
15. D.H.E. Gross, *Microcanonical Thermodynamics: Phase transitions in Finite Systems*, Lect. Notes Phys., Vol. **66** (Springer, 2001).
16. R. Balian, *From Microphysics to Macrophysics* (Springer Verlag, 1982).
17. T.L. Hill, *Thermodynamics of Small Systems* (Dover, New York, 1994).
18. S. Abe, Y. Okamoto, *Nonextensive Statistical Mechanics and its Applications*, Lect. Notes Phys., Vol. **560** (Springer, 2001).
19. D.H.E. Gross, this topical issue and references therein.
20. F. Bouchet, J. Barré, J. Stat. Phys. **118**, 1073 (2005).
21. F. Leyvraz, S. Ruffo, Physica A **305**, 58 (2002).
22. F. Gulminelli, Ph. Chomaz, Phys. Rev. E **66**, 046108 (2002).
23. M. Costeniuc, R.S. Ellis, H. Touchette, J. Math. Phys. **46**, 063301 (2005).
24. M.S.S. Challa, J.H. Hetherington, Phys. Rev. Lett. **60**, 77 (1988); R.S. Johal, A. Planes, E. Vives, cond-mat/0307646.
25. W. Thirring, Z. Phys. **235**, 339 (1970).
26. A. Huller, Z. Phys. B **93**, 401 (1994).
27. R.S. Ellis, K. Haven, B. Turkington, J. Stat. Phys. **101**, 999 (2000).
28. T. Dauxois, P. Holdsworth, S. Ruffo, Eur. Phys. J. B **16**, 659 (2000).
29. J. Barré, D. Mukamel, S. Ruffo, Phys. Rev. Lett. **87**, 030601 (2001) cond-mat/0209357.
30. I. Ispolatov, E.G.D. Cohen, Physica A **295**, 475I (2001).
31. M. Kastner, M. Promberger, A. Huller, J. Stat. Phys. **99**, 1251 (2000); M. Kastner, J. Stat. Phys. **109**, 133 (2002); Physica A **359**, 447 (2006).
32. D.H.E. Gross, E.V. Votyakov, Eur. Phys. J. B **15**, 115 (2000).
33. A. Huller, M. Pleimling, Int. J. Mod. Phys. C **13**, 947 (2002).
34. M. Pleimling, H. Behringer, A. Huller, Phys. Lett. A **328**, 432 (2004); H. Behringer, J. Stat. Mech. P06014 (2005).
35. Ph. Chomaz, F. Gulminelli, in Lect. Notes Phys., Vol. **602** (Springer, 2002); F. Gulminelli, Ann. Phys. (Paris) **29**, 6 (2004).
36. P.H. Chavanis, I. Ispolatov, Phys. Rev. E **66**, 036109 (2002).
37. C.J. Pethick, H. Smith, *Bose Einstein Condensation in Dilute Gases* (Cambridge University Press, Cambridge, 2002).
38. A. Minguzzi *et al.*, Phys. Rep. **395**, 223 (2004).
39. E.V. Shuryak, Phys. Rep. **391**, 381 (2004).
40. P. Braun-Munzinger *et al.*, Phys. Lett. B **596**, 61 (2004); F. Becattini *et al.*, Phys. Rev. C **72**, 064904 (2005).
41. C. Brechignac *et al.*, Phys. Rev. Lett. **92**, 083401 (2004).
42. T.D. Lee, C.N. Yang, Phys. Rev. **87**, 404 (1952).
43. Ph. Chomaz, F. Gulminelli, V. Duflot, Phys. Rev. E **64**, 046114 (2001).
44. K.C. Lee, Phys. Rev. E **53**, 6558 (1996).
45. Ph. Chomaz, F. Gulminelli, Physica A **330**, 451 (2003).
46. K. Binder, D.P. Landau, Phys. Rev. B **30**, 1477 (1984).
47. P. Labastie, R.L. Whetten, Phys. Rev. Lett. **65**, 1567 (1990).
48. F. Gulminelli, Ph. Chomaz, V. Duflot, Europhys. Lett. **50**, 434 (2000).
49. H. Reinhardt, Nucl. Phys. A **413**, 475 (1984).
50. R. Balian, Y. Alhassid, H. Reinhardt, Phys. Rep. **131**, 1 (1986).
51. M. Costeniuc, R.S. Ellis, H. Touchette, B. Turkington, Phys. Rev. E **73**, 026105 (2006).
52. C. Menotti, P. Pedri, S. Stringari, Phys. Rev. Lett. **89**, 252402 (2002).
53. F. Gulminelli, Ph. Chomaz, Nucl. Phys. A **734**, 581 (2004); Ph. Chomaz, F. Gulminelli, O. Juillet, Ann. Phys. **320**, 135 (2005).
54. A.S. Botvina, I.N. Mishustin, this topical issue and references therein.
55. M. Di Toro, A. Olmi, R. Roy, this topical issue and references therein.
56. B. Tamain, this topical issue and references therein.
57. J.L. Mc Cauley, *Chaos Dynamics and Fractals*, Cambridge Nonlinear Science Series 2 (Cambridge University Press, 1993).

58. R. Balian, M. Vénéroni, Phys. Rev. Lett. **47**, 1353; 1765 (1981)(E); Ann. Phys. (N.Y.) **164**, 334 (1985).
59. Ph. Chomaz, Ann. Phys. (Paris) **21**, 669 (1996).
60. L. Van Hove, Physica **15**, 951 (1949); C.N. Yang, T.D. Lee, Phys. Rev. **87**, 404 (1952); K. Huang, *Statistical Mechanics* (John Wiley and Sons Inc., 1963) Chapt. 15.2 and appendix C.
61. P.H. Chavanis, M. Rieutord, Astron. Astrophys. **412**, 1 (2003); D.H.E. Gross, Entropy **6**, 158 (2004); A. De Martino, E.V. Votyakov, D.H.E. Gross, Nucl. Phys. B **654**, 427 (2003).
62. L. Casetti, M. Pettini, E.G.D. Cohen, Phys. Rep. **337**, 237 (2000).
63. I. Hidmi, D.H.E. Gross, H.R. Jaqaman, Eur. Phys. J. D **20**, 87 (2002).
64. T. Dauxois, V. Latora, A. Rapisarda, S. Ruffo, A. Torcini, in Lect. Notes Phys., Vol. **602** (Springer, 2002).
65. T.L. Beck, R.S. Berry, J. Chem. Phys. **88**, 3910 (1988); D.J. Wales, R.S. Berry, Phys. Rev. Lett. **73**, 2875 (1994); R.E. Kunz, R.S. Berry, J. Chem. Phys. **103**, 1904 (1995); D.J. Wales *et al.*, Adv. Chem. Phys. **115**, 1 (2001); R.S. Berry, Israel J. Chem. **44**, 211 (2004).
66. V.A. Antonov, Len. Univ. **7**, 135 (1962); IAU Symp. **113**, 525 (1995).
67. P. Hertel, W. Thirring, Ann. Phys. (N.Y.) **63**, 520 (1971).
68. P.H. Chavanis, in Lect. Notes Phys. Vol. **602** (Springer, 2002); Astron. Astrophys. **432**, 117 (2005).
69. T. Padhmanaban, in Lect. Notes Phys., Vol. **602** (Springer, 2002).
70. J. Katz, Not. R. Astron. Soc. **183**, 765 (1978).
71. M. Promberger, A. Huller, Z. Phys. B **97**, 341 (1995); **93**, 401 (1994).
72. H. Behringer, M. Pleimling, A. Hüller, J. Phys. A **38**, 973 (2005); H. Behringer, J. Phys. A **37**, 1443 (2004).
73. M. Kastner, M. Promberger, J. Stat. Phys. **53**, 795 (1988).
74. R. Franzosi, M. Pettini, L. Spinelli, Phys. Rev. E **60**, 5009 (1999); Phys. Rev. Lett. **84**, 2774 (2000).
75. R. Franzosi, M. Pettini, Phys. Rev. Lett. **92**, 60601 (2004); math-ph/0505057, math-ph/0505058.
76. J. Naudts, Europhys. Lett. **69**, 719 (2005) cond-mat/0412683.
77. M. Antoni, S. Ruffo, A. Torcini, Europhys. Lett. **66**, 645 (2004).
78. F. Gulminelli, Ph. Chomaz, A.H. Raduta, A.R. Raduta, Phys. Rev. Lett. **91**, 202701 (2003).
79. T.E. Strezelecka, M.W. Davidson, R.L. Rill, Nature **331**, 457 (1988); Y. Kafri, D. Mukamel, L. Peliti, Phys. Rev. Lett. **85**, 4988 (2000); A. Wynveen, D.J. Lee, A.A. Kornyshev, Eur. Phys. J. E **16**, 303 (2005).
80. F. Gulminelli, Ph. Chomaz, Phys. Rev. Lett. **82**, 1402 (1999).
81. Ph. Chomaz, F. Gulminelli, Nucl. Phys. A **647**, 153 (1999).
82. S. Grossmann, W. Rosenhauer, Z. Phys. **207**, 138 (1967); P. Borrmann *et al.*, Phys. Rev. Lett. **84**, 3511 (2000); H. Stamerjohanns *et al.*, Phys. Rev. Lett. **88**, 053401 (2002).
83. H. Touchette, Physica A **359**, 375 (2005).
84. M. Pichon, B. Tamain, R. Bougault, O. Lopez, Nucl. Phys. A **749**, 93 (2005).
85. O. Lopez *et al.*, Nucl. Phys. A **685**, 246 (2001).

Eur. Phys. J. A **30**, 333–342 (2006)

DOI 10.1140/epja/i2006-10127-4

THE EUROPEAN
PHYSICAL JOURNAL A

Small fermionic systems: The common methods and challenges

J. Navarro[1], P.-G. Reinhard[2], and E. Suraud[3,a]

[1] IFIC (CSIC-Universitad de Valencia), Apdo. 22085, E-46071 Valencia, Spain
[2] Institut für Theoretische Physik, Universität Erlangen, Staudstrasse 7, D-91058 Erlangen, Germany
[3] Laboratoire Physique Théorique, Université Paul Sabatier, 118 Route de Narbonne, F-31062 Toulouse cedex, France

Received: 18 May 2006 /
Published online: 3 November 2006 – © Società Italiana di Fisica / Springer-Verlag 2006

Abstract. We discuss three finite fermion systems in comparison: nuclei, metal clusters, and droplets of liquid ^{3}He. A principle sorting in "natural units" of energy and length scales is given. We address the theoretical description in terms of self-consistent mean-field theories and their effective energy-density functionals. We look at the interplay of the different time scales from the various constituents of either system. Finally, we discuss the prospects of more detailed experimental analyses for the case of metal clusters, in particular in the non-linear domain where truly dynamical behaviors are expected.

PACS. 25.70.-z Low and intermediate energy heavy-ion reactions – 36.40.-c Atomic and molecular clusters – 67.55.-s Normal phase of liquid ^{3}He

1 Introduction

Finite fermion systems are droplets of Fermi liquid. Typical examples are atomic nuclei, ^{3}He droplets, and metal clusters. Fermi liquids constitute one of the basic states of matter [1]. They are highly correlated systems which, however, never freeze out to a crystalline state. They have a well-defined saturation density with low (nuclei and metal clusters) or moderate (liquid ^{3}He) compressibility. The finite drops thus share several key features: scaling of radius with $N^{1/3}$, shell effects (magic numbers, Jahn-Teller deformation, for clusters see [2]), pronounced resonance excitations (giant resonances, plasmons [3]), and fusion/fission [4]. The strong correlations can hardly be dealt with in detail. Effective energy-density functionals are employed for self-consistent calculations of ground state and dynamics, see *e.g.* [5] for the electrons in clusters, [6] for nuclei, and [7] for ^{3}He droplets. This short list of basic properties shows that finite fermion systems have much in common. On the other side, there are several noteworthy differences, *e.g.*, concerning composition or relation of time scales. It is thus most interesting to discuss these systems in comparison. It is the aim of this contribution to provide such a discussion in due brevity. Thereby, we will concentrate on the energetic dynamical aspects and refer to [8] for structure and the low-energy domain. We will first compare the three systems concerning typical scales (length, time, energy), construction of effective energy-density functionals, and available data. In the last section, we will discuss

briefly observables from non-linear dynamics for the particular example of metal clusters. A much more extensive discussion of practically all aspects of cluster dynamics can be found in [9].

2 Nuclei, ^{3}He droplets and metal clusters

Nuclei, helium droplets and the electron cloud of metal clusters are dense fermion systems with strong Pauli correlations. For a more quantitative discussion, let us briefly recall a few key characteristics of nuclei, metal clusters and helium droplets, concerning, in particular, dominant interactions, sizes, structure and dynamics. These characteristics are briefly sketched in table 1. In metal clusters, the Coulomb interaction plays an important role. The repulsive interactions between electrons are compensated by the attraction to ions. In a neutral cluster, it is finally the electronic exchange and correlation part of the interaction which provides most of the binding. In nuclei, the binding is dominated by the short-range nuclear interaction providing more than enough binding to overrule the repulsive Coulomb interaction which grows with the number of protons. In helium droplets, the interaction, originally Coulombic, reduces to a mere (extremely faint) interaction of van der Waals form between atoms (structureless at this energy scale), leading to extremely fragile structures. All three systems furthermore exhibit a "saturating" behavior. Their radii scale with the power 1/3 of the size of the system. The proportionality factor is the Wigner-Seitz radius r_s, often denoted as r_0 in case of

---

[a] e-mail: `suraud@irsamc.ups-tlse.fr`

Table 1. Gross characteristics of nuclei, metal clusters and helium droplets. One successively considers the constituents (all fermions but for the ions in clusters), the interactions at play, the radii of systems of sizes A (nuclei) and N (clusters, helium droplets), the typical distance between constituents, the typical mean free path and de Broglie wavelength as estimated from a Fermi-gas picture of the ground state. Distances are expressed in terms of r_0 for nuclei and r_s for clusters and helium droplets. The parameter r_0 is the parameter entering the systematics of nuclear radii; the parameter r_s is the Wigner-Seitz radius of the material constituting the metal clusters or the helium droplets.

	Nuclei	Clusters	Helium
Constituents	N Neutrons Z Protons	N Electrons N_{ions} Ions	N ^{3}He atoms $(2p, 1n, 2e)$
Interaction	Short-range (nuclear) + Long-range (Coulomb)	Long-range (Coulomb)	Short-range van der Waals
Size	$N+Z = A \leq 300$	$3 \leq N \leq 10^{5-7}$	$30 \leq N$
Radius	$R \sim r_0 A^{1/3}$ $r_s \equiv r_0 \sim 1.2$ fm	$R \sim r_s N^{1/3}$ $r_s \sim 0.1$–0.3 nm	$R \simeq r_s N^{1/3}$ $r_0 \sim 0.25$ nm
Distance constituents	$d \sim 1.5$–$2r_{0,s}$		
Mean free path	$\lambda \sim R$		
de Broglie wavelength	$\lambda_{\text{B}} \sim \pi r_{0,s}$		
Fermi energy ϵ_{F}	40 MeV	1.4–12 eV	5 K

nuclei. This means that each fermion occupies the same volume given by $(4/3)\pi r_s^3$. This also implies that the average density of these systems is $\rho \sim 3/(4\pi r_s^3)$ independent of the system size. The parameters r_0, r_s thus play a key role in fixing the characteristic scales in these systems. One can for example estimate the typical distance between constituents, which amounts to about 1.5–2 r_s. One can also evaluate the mean free path. In all cases it turns out to be of the order of magnitude of the actual size of the system, so that one can adopt the view that the fermions evolve nearly independent of each other, which motivates a mean-field approach. Taking for the sake of simplicity a Fermi-gas picture (which serves reasonably well as a first approximation) one can also introduce an energy scale in terms of r_s through the Fermi momentum $k_{\text{F}} = (3\pi^2\rho)^{1/3} = (9\pi/4)^{1/3}r_s^{-1}$ from which one can deduce a typical Fermi energy $\epsilon_{\text{F}} = (\hbar^2/2m)k_{\text{F}}^2$ and Fermi velocity $v_{\text{F}} = \hbar k_{\text{F}}/m$ (with $m = m_n, m_e, m_{\text{He}}$). This, by the way, also provides a simple estimate of the de Broglie wavelength in the ground state $\lambda_{\text{B}} \sim 2\pi/k_{\text{F}} \sim \pi r_s$, which confirms the essentially quantal nature of these systems.

A comparison of mean fields is presented in fig. 1 for the cluster Na$_{40}$, the nucleus ^{78}Sr (with 40 neutrons) and a helium droplet with 40 ^{3}He atoms. The results are plotted in natural units (see figure caption) for making the systems comparable. The comparison is quite enlightening. First one notes that all three systems fit into one figure, *i.e.* have about the same scales when expressed in natural units. Moreover, they exhibit the same spatial extension, directly connected to the "saturation scale" introduced by r_s. At second glance, one also spots differences, in the depth of the potential wells and in the asymptotic behaviors. The cluster and nucleus share a comparably deep potential while the helium droplet exhibits a much more shallow potential well reflecting the faintness of the interaction between two He atoms. On the other hand, the helium droplet and the nucleus share the same asymp-

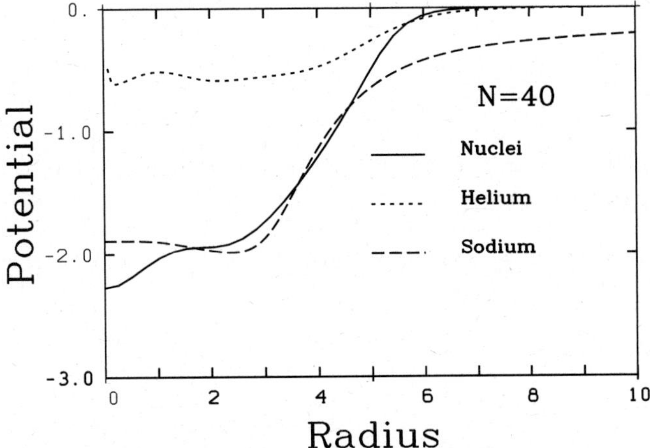

Fig. 1. The mean-field potentials for a Na cluster, a nucleus and a helium droplet, for 40 particles (the results are only shown for the neutron part in the nuclear case). Natural units are used: lengths in units of r_s and potentials in units of ϵ_{F}.

totic behavior characteristic of a system dominated by a relatively short-range interaction, while the cluster case exhibits a typical long-range Coulomb behavior. All in all the comparison nevertheless shows the overall similarity between the various systems, up to details.

3 Mean field: a possible common theory

The free nucleon-nucleon interaction is known to be strongly repulsive at short range [10]. This makes a mean-field theory *a priori* questionable. But the strong Pauli correlations in nuclei significantly suppress low-energy scattering, which renormalizes, in a nuclear medium, the interaction to an effective one. The short-distance repulsion is thus highly suppressed and the effective nucleon-nucleon interaction smooth enough to justify a mean-field

Table 2. One-body Hamiltonian used in mean-field calculations in nuclei (Skyrme-like interaction), metal clusters (DFT LDA) and helium droplets (DFT). In nuclei the density ρ represents the neutron and/or proton density, the latter entering the Coulomb interaction alone (ρ_p). In clusters the density is the electronic one while in helium it is the density of helium atoms. In the case of cluster electrons an important contribution comes from the exchange correlation potential $U_{\mathrm{xc}}[\rho]$ which can be treated in the simplest LDA approximation. The external potential $U_{\mathrm{ext}}(\mathbf{r}, t)$ typically refers to the fields as generated by ions or a possible external field (laser, bypassing ion, ...). In the case of helium droplets one has to introduce an effective mass m^* as indicated, which again depends on the density. The same is true in principle for the nuclear case but the effective mass is much closer to the bare mass (about 20%) than in the helium case (factor 3 typically).

Nuclei	$h[\varrho] = -\nabla \frac{\hbar}{m^*(r)} \nabla + t_0\varrho + t_3\varrho^{1+\sigma} + t_{12}(\nabla\varrho)^2 + \int \frac{\varrho_p(\mathbf{r}')}{\|\mathbf{r}-\mathbf{r}'\|} \mathrm{d}\mathbf{r}' + \cdots$
	$\frac{\hbar^2}{2m^*} = \frac{\hbar^2}{2m} + \alpha\varrho$
Clusters	$h[\varrho] = -\frac{\hbar^2}{2m}\Delta + \int \frac{\varrho(\mathbf{r}')}{\|\mathbf{r}-\mathbf{r}'\|}\mathrm{d}\mathbf{r}' + U_{\mathrm{xc}}[\varrho] + U_{\mathrm{ext}}(\mathbf{r}, t)$
Helium	$h[\varrho] = -\nabla \frac{\hbar^2}{m^*(r)} \nabla + \int V_{\mathrm{eff}}(\mathbf{r} - \mathbf{r}')\varrho(\mathbf{r}')\mathrm{d}\mathbf{r}' + A\varrho^{1+\gamma} + \cdots$
	$\frac{\hbar^2}{2m^*} = \frac{\hbar^2}{2m} + \alpha\varrho + \beta\varrho^2$

picture [10]. The so-called Skyrme interactions, which integrate in an effective way these Pauli correlation effects on a basically zero-range bare interaction, have been a standard microscopic tool for decades, because of their simplicity and because of the many successes they have allowed, at least for stability valley nuclei, for a recent review, see [6]. As can be seen from table 2 the Skyrme interaction appears as a density functional, mostly local, non-locality being usually assumed in terms of a gradient expansion.

A somewhat similar reasoning applies to metal clusters. The general atomic problem is indeed singular (due to the point charge of the atomic nucleus), but a limited number of valence electrons actually take part in the binding of molecular systems or clusters. This is especially true in the case of simple metals as the valence shell is well separated from core levels and usually little bound. Valence electrons can thus easily be delocalized to form the rather "soft" metal bonds. This, by the way, also allows for the packaging of the effect of the core electrons into a pseudo potential: this reduces the many-electron problem to the treatment of the valence electrons only in a reasonably smooth ionic background [9]. This again provides a favorable situation for a mean-field treatment. The success of the many calculations based on Density Functional Theory (DFT) even in its simplest Local Density Approximation (LDA) version indeed proves the reliability of the mean-field approach.

The case of helium droplets is a bit more involved. Indeed the applicability of a DFT approach was long debated for such systems in view of the much stronger correlations. However, it seems today that density functionals can indeed be used in that case provided one introduces a finite range in the construction [7,11]. Once taking that step, the situation is even more favorable than in the case of nuclei. The *ab initio* calculations for ^{3}He matter have achieved a high degree of reliability [12] and there exists experimental access to bulk ^{3}He over a wide range of pressures [13]. Both facts provide well-tested data as input for a proper calibration of density functionals. There exist even *ab initio* calculations for finite droplets which

serve as additional benchmark [14]. For a recent review see also [15].

In the majority of practical cases the mean-field calculations based on the one-body Hamiltonians presented in table 2 are done at quantum level. Each particle (nucleon, electron, helium atom) is attributed a one-particle wavefunction $\phi_i(\mathbf{r})$, from which one deduces the single-particle density matrix $\hat{\rho}(\mathbf{r}, \mathbf{r}')$ and the local one-body density $\varrho(\mathbf{r}) = \hat{\rho}(\mathbf{r}, \mathbf{r}) = \sum_i |\phi_i(\mathbf{r})|^2$ (where the summation runs over all particles). The one-body wave functions then follow an effective Schrödinger equation

$$i\hbar \frac{\partial |\varphi_i\rangle}{\partial t} = h[\varrho(\mathbf{r})]|\varphi_i\rangle \tag{1}$$

with an effective single-particle Hamiltonian h expressed as a functional of the density $\rho(\mathbf{r})$, as given in table 2 for the various Fermi systems. This mean-field equation can be recast in the equivalent matrix form

$$i\hbar\hat{\rho} = [h, \hat{\rho}]. \tag{2}$$

There also exist semi-classical approximations to this quantum scheme. They can be "formally" obtained by transforming the density operator $\hat{\rho}$ into a one-body phase space distribution $f(\mathbf{r}, \mathbf{p}, t)$, which becomes the basic ingredient, and the commutator into Poisson brackets:

$$\begin{aligned} \hat{\rho}(\mathbf{r}, \mathbf{r}') &\longrightarrow f(\mathbf{r}, \mathbf{p}, t), \\ [\cdot, \cdot] &\longrightarrow \{\cdot, \cdot\}. \end{aligned} \tag{3}$$

This leads to the Vlasov equation

$$\frac{\partial f}{\partial t} = \{h, f\}. \tag{4}$$

The one-body Hamiltonian has the same expression in terms of the density $\varrho(\mathbf{r})$ as in the quantal form, but the density is now computed from the phase space density as

$$\varrho(\mathbf{r}, t) = \int \mathrm{d}^3 p\, f(\mathbf{r}, \mathbf{p}, t). \tag{5}$$

This equation can then be extended to account for dynamical correlations. Particle-particle scattering effects can indeed easily be included as a Markovian collision term for the phase space distribution f. This has been worked out in great detail in nuclear-physics applications [16] and it was also extended to the cluster case. In both cases (nuclei, metal clusters) one ends up with the VUU (Vlasov-Uehling-Uhlenbeck) equation

$$\frac{\partial f}{\partial t} + \frac{\mathbf{p}}{m}\frac{\partial f}{\partial \mathbf{r}} - \frac{\partial V}{\partial \mathbf{r}}\frac{\partial f}{\partial \mathbf{p}} = I_{\mathrm{UU}}(\mathbf{r}, \mathbf{p}, t) \qquad (6)$$

with the collision term

$$I_{\mathrm{UU}} = \int \frac{\mathrm{d}^3 p_2 \mathrm{d}\Omega}{(2\pi\hbar)^3}\frac{\mathrm{d}\sigma}{\mathrm{d}\Omega}|v_{12}|$$

$$\cdot\left\{f_1 f_2\left(1 - \frac{f_3}{2}\right)\left(1 - \frac{f_4}{2}\right) - f_3 f_4\left(1 - \frac{f_1}{2}\right)\left(1 - \frac{f_2}{2}\right)\right\}, \qquad (7)$$

where v_{12} is the relative velocity of the colliding particles 1 and 2. The differential cross-section $\mathrm{d}\sigma/\mathrm{d}\Omega$ (depending on the scattering angle Ω) is evaluated in the center-of-mass frame of the two colliding particles. Indices 3 and 4 label the momenta of the two particles after an elementary collision and we use the standard abbreviation $f_i = f(\mathbf{r}, \mathbf{p}_i, t)$. The collision is supposed elastic (conservation of energy, of total momentum). Pauli-blocking factors $(1 - f_i/2)(1 - f_j/2)$ play an important role here, as they provide the necessary preservation of the Pauli principle in the course of fermion collisions. In the ground state, they block correctly all kinematically possible (and thus classically possible) collisions. At high excitation energy the phase space opens up and two-body collisions start to populate it in the course of thermalization. The VUU scheme was very much used in the case of heavy-ion collisions in the Fermi energy domain. As we shall see below, it should also be taken into account in the case of metal clusters, for energetic processes.

4 From one field to the next

4.1 Status of knowledge

The three fields (nuclei, clusters, helium droplets) are by no means at the same stage of developments. Nuclei have been studied for almost a century while studies on free metal clusters have only been started a few decades ago, and even later for helium droplets, although the homogeneous phases of electrons (in bulk metal) or helium had been studied much earlier. As a result the available expertise varies from one field to the next. We have tried to summarize roughly the stages of achievement for each field in table 3, grouping into theory *versus* experiment and structure *versus* dynamics. Studies in nuclei cover each of the four topics widely. The case of clusters is more mixed. While structure properties start to be well known both at theoretical and experimental levels, dynamics is still in its infancy, especially when far from equilibrium. Theoretical studies are here probably a bit more advanced than experimental ones. The latter require the development of very elaborate detectors with which the cluster

Table 3. Schematic status of studies for the three finite fermion systems. The entry "no $\rightarrow$ yes" indicates that research is underway and results may show up soon. The entry "not yet" means that calculations have not been done but are feasible in principle. See text for details.

System	Experiment		Theory	
	Structure	Dynamics	Structure	Dynamics
Nuclei	yes	yes	yes	yes
Clusters	yes	no $\rightarrow$ yes	yes	no $\rightarrow$ yes
^{3}He droplets	no $\rightarrow$ yes	no	yes	not yet

community is not fully familiar and which would become unusually expensive in view of the elsewise rather economical cluster experiments. Data for helium droplets are even more sparse. In fact, there is very little known experimentally. Even the minimum size of such droplets is not fully ascertained experimentally, and it is admittedly very hard to deal with these volatile and neutral objects. The theory side is a bit better developed at structure level. Dynamical studies are conceivable from the theory side, but there is not much effort in that direction because experimental data will not appear all too soon. In the context of this topical issue focusing on nuclear dynamics, there are thus, presently, much stronger possible connexions between metal clusters and nuclei than with helium droplets. In the following, we shall thus focus the discussion on dynamical examples as taken from metal clusters and not further analyze the case of helium droplets.

4.2 Multiscale dynamics

It is interesting, as a starter, to compare nuclear and cluster time scales. In order to make the actual comparison more telling we use reduced units in terms of the Fermi-gas characteristics of both systems, following the values introduced above. Indeed, we define a basic time $r_{s,0}/v_{\mathrm{F}}$ and energy scale ϵ_{F}, as built from the Wigner-Seitz radius r_s for clusters and from the parameter r_0 of nuclear-radius systematics ($R \sim r_0 A^{1/3}$, with $r_0 \sim 1.12$ fm). For the sake of simplicity, we restrict the analysis of the cluster case to Na, thus taking for the Wigner-Seitz radius $r_s = 4a_0$. This leads to $r_s/v_{\mathrm{F}} = 0.2$ fs and $\epsilon_{\mathrm{F}} = 3.2$ eV. Indeed, electronic time scales for other alkalines perfectly match the values obtained in the case of Na. The ionic motion times scale with the square root of the atom mass. In nuclei the basic time and energy scales read $r_0/v_{\mathrm{F}} = 3.3$ fm/c and $\epsilon_{\mathrm{F}} = 40$ MeV. We plot times as a function of temperature. It should be noted that this is rather a measure for the average excitation and does not necessarily imply a full thermalization. The choice of temperature is here practical and allows to overlook, to a large extent, size-dependent effects.

With this system of reduced units we compare nuclear and sodium time scales in fig. 2. The comparison concerns various relevant times: the cluster plasmon period (equivalent to the giant dipole resonance in nuclei), ionic

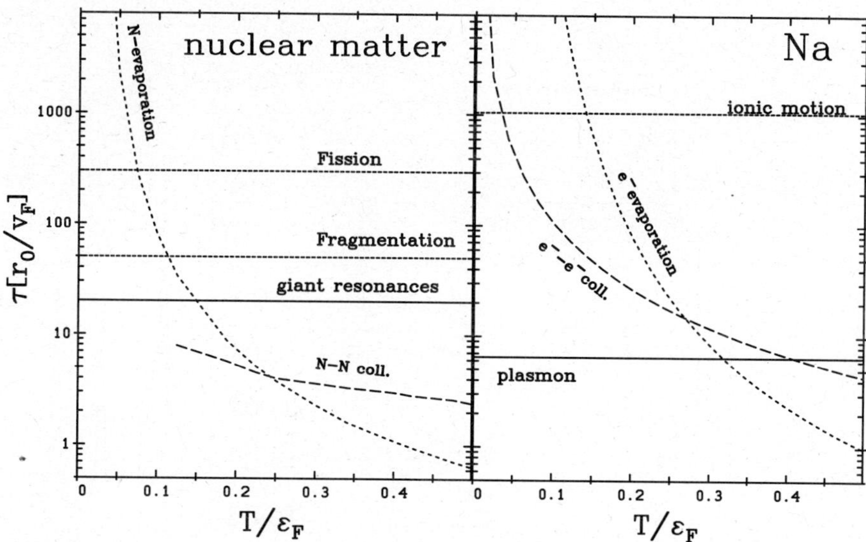

Fig. 2. Comparison of relevant times scales in nuclei and sodium clusters. Reduced units are used in both cases to allow a relevant comparison (see text for details). Various times plotted are: plasmon period in clusters (equivalent to giant dipole resonance in nuclei), ionic time scale (comparable to nuclear fission/fragmentation), electron evaporation time (equivalent to neutron evaporation time), electron-electron (or nucleon-nucleon) collision time scale.

time scale (comparable to nuclear fission/fragmentation), electron (or neutron) evaporation time, electron-electron (or nucleon-nucleon) time scale. This comparison calls for several comments. At first glance one can note a relative similarity between electronic and nuclear time scales, in particular comparable dependences (or independence) of times on temperature. But details differ. Indeed the hierarchy of time scales is pretty different between Na and nuclei. Grossly speaking, nuclear time scales look more similar to each other than cluster ones. This means that there exists a natural hierarchy of well-separated time scales in clusters, while nuclear times tend to be much more mixed up. This has important implications in particular from the theoretical point of view. The lack of a clear time hierarchy in nuclear dynamics makes a clean adiabatic decoupling of slow degrees of freedom difficult. The simple Born-Oppenheimer treatment of slow degrees of freedom has to be replaced by the much more involved generator-coordinate-method [17]. In cluster physics the huge mass difference between electron and ionic masses makes electron time scales an order of magnitude smaller than ionic ones. Electrons are thus more responsive than ions and need to be accounted for in priority in cluster dynamics. One should nevertheless note that the separation of electronic and ionic time scales tends to shrink in strongly non-linear situations where huge electromagnetic fields can be generated. Differences between nuclear and cluster hierarchies of time scales do not only reduce to the hierarchies by themselves but also to the times with respect to each other. One should in particular note the relative importance of electron-electron interactions. They become dominant for much higher temperatures in clusters than in nuclei, which means that mean-field methods can probably be used at much higher excitation energies in clusters than in nuclei. This is a welcome feature in

view of the theoretical difficulties the inclusion of dynamical correlations raises. In a similar way, thermal emission comes into play much earlier in nuclei than in clusters. This again reflects the stronger interference amongst nuclear time scales as compared to cluster ones.

5 Electron dynamics in metal clusters

5.1 Electron emission from irradiated clusters

Experimental observation requires that some objects reach a counter, preferably charged particles. A major tool is here to keep a protocol of emitted electrons. Figure 3 illustrates the various observables which can be drawn from electron emission. The left upper panel symbolizes the photo-ionization cross-section $\sigma(\omega)$ (where ω is the laser frequency) which is a good approximation to the total photo-absorption cross-section and which can be measured easily by tracking the net electron yield as a function of frequency. This quantity is obviously of inclusive nature and thus does not provide very detailed information, in particular at the side of individual electrons. More information can be extracted when measuring the distribution of the kinetic energies from the emitted electrons. This is called photo-electron spectroscopy (PES) and gives access to differential cross-sections $d\sigma/dE$ (where E is the electron kinetic energy). The right panels characterize PES briefly, in terms of mechanism (lower panel) and of typical observable (upper panel). The energy is drawn horizontally in both cases. The zero point is clearly indicated. The lower part shows three vertical lines. The two solid lines indicate the (negative) energies of occupied bound electron states (in the simple case of a small cluster here) and the dashed line stands for the continuum threshold. The horizontal arrow indicates a photon. It transfers a

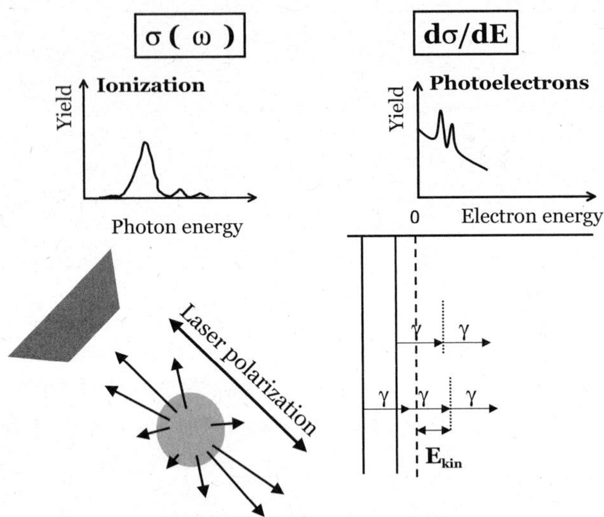

Fig. 3. Schematic view of observables related to electron emission from metal clusters. Left lower panel: irradiated cluster with emission preferentially (to some extent) along the laser polarization; a simple detector is also represented, its width symbolising its capability to measure electron kinetic energies; left upper panel: total ionization cross-section $\sigma(\omega)$ as a function of laser frequency ω; right panels: principle of photoelectron spectroscopy (upper panel: typical data, differential cross-section $d\sigma/dE$; lower panel: mechanism).

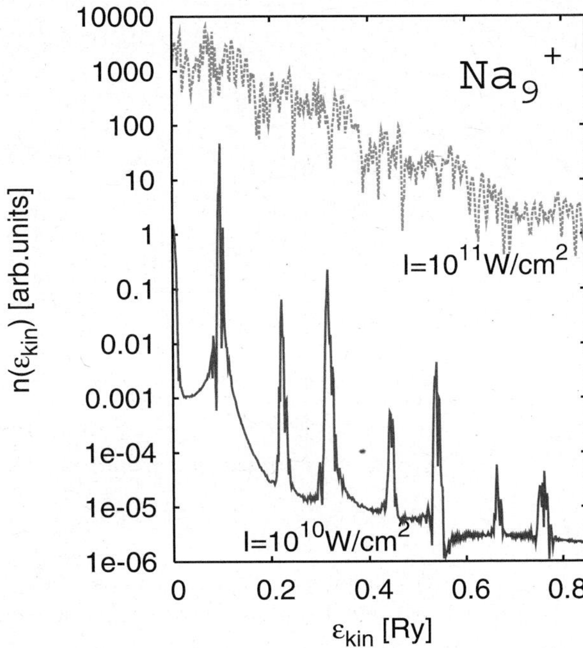

Fig. 4. Photoelectron differential cross-section (arbitrary units) of Na_9^+ at two intensities as indicated. A short laser pulse with FWHM = 25 fs was used.

well-defined amount of energy. Bound-state energy plus photon energy(ies) sum up to the kinetic energy finally observed. The photo-electron spectrum (upper panel) thus shows distinct peaks at those energies. Having these peaks and knowing the photon frequency allows to conclude on the underlying single-electron states. Thus far the simple story in the low-intensity domain. The situation becomes more involved in a more energetic domain. The case of PES in this energy domain will be discussed in sect. 5.2.

5.2 Photoelectron spectroscopy

Figure 4 shows the PES of Na_9^+ for two different laser intensities around the transition to the high field regime. The lower intensity still resolves the detailed single-electron states in repeated sequences (see also the discussion around fig. 6). A moderate enhancement of the laser intensity by an order of magnitude suffices to wipe out the structures. A more or less smooth curve then emerges which fits nicely to an exponential decrease. The smooth pattern persists, of course, for even larger intensities. The slope decreases with increasing intensity. It is interesting, then, to analyze the origin of these smooth patterns, a question which is still a matter of debate. Indeed one could interpret this exponential decrease as a signal of thermalization of the electron cloud. However, this is not applicable to short laser pulses during which thermalization can hardly play a dominant role. Without arguing in terms of thermalization one can also note that, together with the dramatic changes in the pattern of the PES, we see an

equally dramatic increase in the ionization. This large increase in net charge at the cluster site has a side-effect on binding. Due to the growing Coulomb force, the mean field acquires extra binding, which globally down-shifts electronic single-particle energies by the same amount. This happens as a dynamic process. Thus all levels are smeared which eventually generates the smooth pattern seen for the higher intensities. It can be shown that this process does also deliver an exponential decrease of the PES [18]. A better indicator of thermalization is provided by the (more detailed) analysis of electron emission in terms of the angular distribution of the emitted electrons. This will be discussed in sect. 5.3.

The PES change pattern when going to larger systems. Indeed, the larger the system, the denser the density of electronic states, which inhibits a detailed resolution of separate single-electron states, whatever the excitation regime. At best, one can expect step-like structures indicating bands of occupied states, as was observed in the case of C_{60} for short, moderate pulses [19] and for large Ag clusters on a substrate [20]. More recently were also published measurements on Na_{93}^+ from [21], which show smooth trends throughout and are interpreted as thermal emission. Let us thus consider here Na_{93}^+ as an example of a larger cluster. We have computed PES for Na_{93}^+ for a variety of laser intensities (but fixed photon frequency and pulse width). At low intensity we observe step-like patterns related to a dense block of occupied states. And at larger laser intensity PES exhibit smooth patterns with nearly exponential decrease. A simple characteristics of the PES is thus provided by the slope of the exponential decrease. The criterion is unambiguous at large intensity

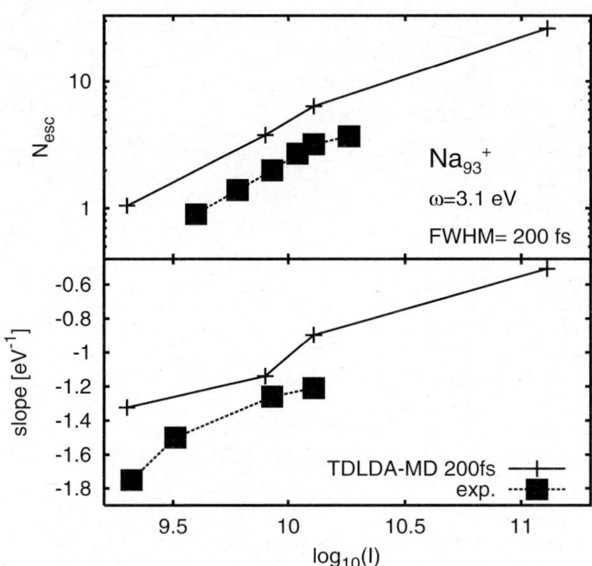

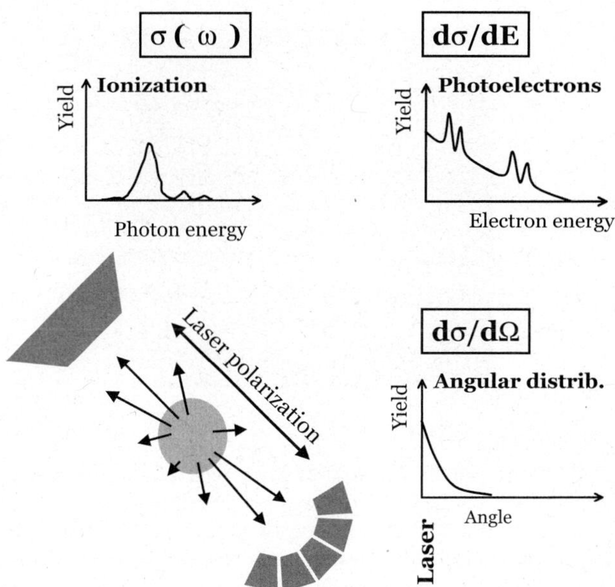

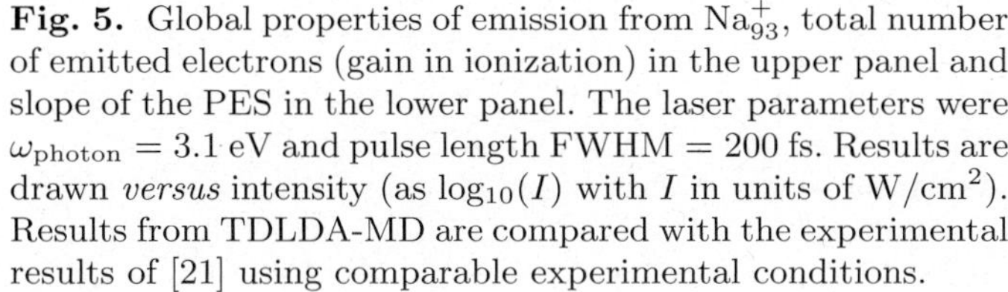

Fig. 5. Global properties of emission from Na_{93}^+, total number of emitted electrons (gain in ionization) in the upper panel and slope of the PES in the lower panel. The laser parameters were $\omega_{\mathrm{photon}} = 3.1$ eV and pulse length FWHM = 200 fs. Results are drawn *versus* intensity (as $\log_{10}(I)$ with I in units of W/cm^2). Results from TDLDA-MD are compared with the experimental results of [21] using comparable experimental conditions.

Fig. 6. Schematic view of detailed characteristics of emitted electrons, in particular in the non-linear domain. The left panels of the figure are similar to the ones of fig. 3 but for the detectors (left lower panel) where we have schematized a series of small detectors to access angular distributions of emitted electrons (and possibly even kinetic energies at the same time). Right upper panel: PES in the multiphoton regime with copies of the series of single-particle peaks separated by the laser frequency, yielding the differential cross-section $\mathrm{d}\sigma/\mathrm{d}E$; Right lower panel: angular distribution of emitted electrons with a more intense yield along the laser polarization axis, yielding the differential cross-section $\mathrm{d}\sigma/\mathrm{d}\Omega$ and possibly the double differential cross-section $\mathrm{d}^2\sigma/\mathrm{d}\Omega\mathrm{d}E$.

but requires some caution at lower intensities because of the step-like pattern. The "staircases" have in fact all the same step height (on logarithmic scale) such that their envelope is a straight line to which it is easy to associate an exponential decrease. We can thus extend the simple slope characterization to any laser intensity. This allows a direct comparison to experimental data.

The experiments [21] were done for rather long pulses with a FWHM of about 200 fs. One can expect to see patterns related to electron-electron collisions, beyond mean field. Nonetheless, it is interesting to compare TDLDA (mean-field) results with those measurements concerning global properties such as net ionization and the slope of the PES. The results are compared with data in fig. 5. Note that for such long pulses one has also to account for ionic motion which sizably alters the sequence of electronic levels and thus the PES. This effect was of course taken into account in the calculations presented in fig. 5. The comparison shows that the calculations reproduce the net ionization (number of emitted electrons N_{esc} within a factor of two, as well as the growth with laser intensity. This has to be considered as a good agreement in view of the fact that ionization also sensitively depends on the pulse shape. Calculations use here a cosine^2 pulse profile while the experimental profile is not so well known, probably having longer tails. The results for the slopes (lower part of fig. 5) are also quite encouraging in size and in trend. Similar results were reported in [22] in the framework of a Vlasov-LDA approach. The remaining differences between the calculations and the experimental results are presum-

ably to be attributed to the lack of account of electron-electron collisions (overlooking details of laser pulse shape, as mentioned above or even cluster temperature control).

5.3 Angular distributions

Besides the kinetic energy, one can also measure the angular distribution of emitted electrons, a quantity which also carries a lot of interesting information. This gives access to the differential cross-section $\mathrm{d}\sigma/\mathrm{d}\Omega$ (where Ω is the solid angle). The principle of such a measurement is presented schematically in fig. 6. In the case of laser irradiation one expects electrons to be emitted preferentially, at least to some extent, along the laser polarization axis. As we shall see below, the amount of anisotropy somewhat depends on the experimental conditions (characteristics of laser pulse in particular). One may even measure simultaneously both angular distributions and kinetic energies of emitted electrons (see the left panel of fig. 7 for an example). This thus gives access to the double differential cross-section $\mathrm{d}^2\sigma/\mathrm{d}\Omega\mathrm{d}E$. Note also that in fig. 6 the PES panel (upper right part) has been modified, with respect to fig. 3, in order to describe the multiphoton regime, with successive copies of the single-electron level sequence.

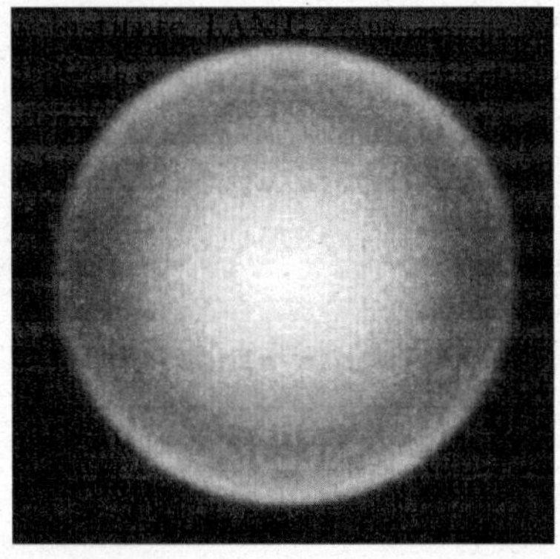
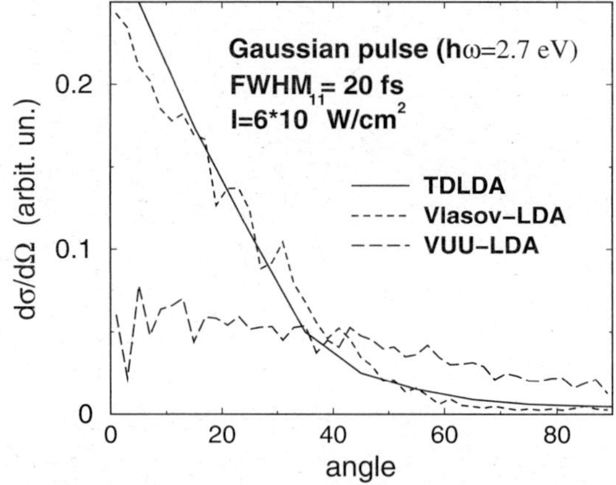

Fig. 7. Left panel: 2D equi-density plot of kinetic energy spectra and angular distribution from a W_4^- cluster anion after irradiation with a laser of frequency 4 eV. The emission angle is mapped in terms of polar coordinates while the kinetic energies grow with radial distance from the center of the plot. The gray scale indicates flux: high emission shines white. The laser polarization is along the vertical axis [23]. Right panel: angular distribution of emitted electrons computed in quantum TDLDA (full line), semi-classical Vlasov-LDA (short-dashed line), and VUU with collision term (long-dashed line). The test case is Na_{41}^+ with ionic structure. Laser parameters are indicated in the figures. The angle is defined relative to the laser polarization axis.

There are only few available experimental data on angular distributions. An example is shown in fig. 7. The left panel exhibits a combined kinetic energy and angular-distribution measurement, after irradiation of a W_4^- cluster anion by a low-intensity ns laser at a frequency of 4 eV. The anion (negatively charged cluster) has a low ionization threshold around 1.6 eV, much lower than the monomer evaporation threshold (larger than 7 eV). As a result thermal ionization is favored over monomer evaporation in this case. The competition remains, though, between direct (in particular, one-photon processes in such an anion) and thermal electron emission. The extremely long laser pulse (as compared to typical electronic or even ionic times) gives thermalization through electron-electron collisions good chances to be activated and efficient. One thus expects a significant contribution from thermal emission. This is indeed what can be seen from the figure, where light grey indicates large emission and dark grey low emission. The broad central spot can be associated with thermal (isotropic) emission and the kinetic energy spectra (not directly visible in the figure) indeed confirm the correct trend $\propto \sqrt{\varepsilon_{\mathrm{kin}}} \exp\left(-\varepsilon_{\mathrm{kin}}/T\right)$ [23]. But at larger kinetic energies (which correspond in this representation to larger radial distances) one can also spot a non-isotropic component in the emission, directed along the laser polarization axis. This is clearly a signal from a direct-emission process which competes with thermal (isotropic) emission.

A full description of such processes should thus account for electron-electron collisions in order to properly access the isotropic component of the electronic emission. This would allow to cover both regimes, direct emission as well as thermal evaporation. A way to include such effects in the TDLDA approach is to rely on the semi-classical version of the theory, properly extended by a collision term to account for electron-electron collisions [24,25], in the spirit of similar extensions worked out several years ago in nuclear physics [16]. The semi-classical approach of course

requires sufficiently high excitation. But, as in the nuclear context, that is the typical situation for thermalization to play a role at all. The right panel of fig. 7 shows an example of angular distributions obtained from such a VUU calculation, and compared to pure mean-field results (a quantum TDLDA one and a semi-classical Vlasov-LDA one). Note that at variance with the left panel of fig. 7 the distribution has been integrated over final kinetic energy. First, we see that TDLDA and Vlasov nicely agree in that excitation regime. Both show an emission clearly peaked along the laser polarization, a behavior characteristic of direct emission. But the electron-electron collisions in VUU also leads to a sizeable isotropic component. Not surprisingly, the delayed emission of the thermalized electrons has lost memory of the original polarization axis and subsequently one obtains a much smoother angular distribution, as can be seen on the VUU curve of the right panel of fig. 7. The distribution is nevertheless not perfectly isotropic: there remains a sizeable fraction of directly emitted electrons for the chosen conditions. But the branching between direct and thermal emission in fact sensitively depends on the details of the excitation. Systematic studies of these influences could thus deliver valuable information on the underlying dynamics. But these studies have yet to be worked out, both theoretically and experimentally, in particular in the combined analysis of kinetic energies and angular distribution as in the case presented in the left panel of fig. 7.

6 Some conclusions and perspectives

Fermi liquids are a generic state of matter denoted by a more or less well-defined saturation point and a long mean free path for particles with low momenta. Wigner-Seitz radius (related to saturation density) and Fermi energy set natural units for length and energy scales. Finite drops of Fermi liquid show several interesting features as, *e.g.*, pronounced shell effects and resonance excitations. Various different materials look very similar when expressed in the natural units of the material. We have exemplified that here for the three systems: nuclei, metal clusters and ^{3}He droplets. There are, of course, also many differences between these systems. These concern mostly availability and experimental access. For example, nuclei are limited in size, but metal clusters can be grown arbitrarily large which allows to study the approach to bulk matter and the evolution of shell effects for large systems. Furthermore, clusters are responsive to laser light which opens up an enormously rich field of dynamical studies yet to come. As another example, ^{3}He droplets have extremely soft surfaces which requires an extension of DFT to incorporate effective interactions with finite range. This shows that the differences are highly welcome as they deliver complementing information for a common understanding of finite fermion systems. It is to be remarked that nuclei are a particularly demanding species in several respects. The nuclear many-body problem (*ab initio* models) has not yet fully converged and similarly the nuclear DFT, although extremely successful in sorting the basic nuclear properties, is still under development. On the other hand, nuclei are probably the best-studied objects from the experimental side. There exist plenty of data in any dynamical regime. Altogether, a combined analysis of the various systems (nuclei, metal clusters, He droplets) will mutually boost the understanding of all systems.

Pursuing the comparative analysis a bit deeper it seems clear, in particular in relation with the examples shown above, that there obviously exist rather clear directions of fruitful enrichment, especially in the domain of dynamics. Indeed, nuclear physics, and heavy-ion collisions, provide remarkable examples of detailed studies of complex dynamical processes requiring sophisticated multi parameter detectors and elaborate many-body theories beyond the mere effective mean field. It seems to us that the field of metal clusters should easily benefit from this knowhow, as it has already benefited in the case of low-energy dynamics or in some structure properties. We have outlined, in particular, the importance of accessing as detailed as possible information on emitted electrons from irradiated clusters. And it seems that all the experience gathered around multidetectors in heavy-ion collisions could here be very useful. The essential step to pass from inclusive to exclusive measurements has been extensively explored in these nuclear studies and the experience gained here appears quite valuable for other domains of physics.

This direction towards exclusive measurements is however not the single direction to be explored. Fundamental cross disciplinary questions also arise in other areas at the interface between nuclear and cluster physics. Let us in particular cite the question of "phase transitions" in finite systems, a problem extensively studied in relation to nuclear fragmentation and to melting in metal clusters. This particular question is addressed elsewhere in this topical issue. We have thus deliberately avoided this

point. From a more formal point of view, one should also mention the basic questions raised by density functional theory in various systems (electronic, nuclear as well as helium), especially in relation, again, to dynamical questions. The development of the so-called Time-Dependent Density Functional Theory (TDDFT) still remains a matter of intense activity and debates. The nuclear-physics approach, for example in terms of truncations of hierarchies of density matrices, brings here an interesting viewpoint, to be merged with the more "bottom-up" methods inherited from standard DFT methods. Finally, we would like again to mention, in continuity to these questions on DFT and TDDFT, the growing importance of dynamical correlations in more and more energetic processes. Again the nuclear-physics experience, for example in terms of kinetic theory, provides valuable assets for other fields of physics and this should be valorized.

The authors thank the French German exchange program PROCOPE, the CNRS program "Matériaux", the Institut Universitaire de France, the Humboldt Foundation and the French Ministry of Research (Gay Lussac) for financial support during the realization of this work.

Appendix A.

We list in the following a few textbooks used in the preparation of this manuscript, as well as the proceedings of a few major conferences on cluster and helium physics. This list is by no means exhaustive.

a) *Cluster physics*
- S. Sugano, *Microclusters* (Springer, Berlin, 1987);
- H. Haberland, *Clusters of Atoms and Molecules 1 —Theory, Experiment, and Clusters of Atoms*, Springer Series in Chemical Physics, Vol. **52**, (Springer, Berlin, 1994);
- H. Haberland, *Clusters of Atoms and Molecules 2 —Solvation and Chemistry of Free Clusters, and Embedded, Supported and Compressed Clusters*, Springer Series in Chemical Physics, Vol. **56**, (Springer, Berlin, 1994);
- W. Ekardt, *Metal Clusters* (Wiley, New York, 1999);
- J. Jellinek, *Theory of Atomic and Molecular Clusters* (Springer, Berlin, 1999);
- P.-G. Reinhard, E. Suraud, *Introduction to Cluster Dynamics* (Wiley-VCH, Berlin, 2003);
- *Proceedings of the ISSPIC Conferences* of the past decade:
  ISSPIC 7: Surf. Rev. Lett. **3** (1996);
  ISSPIC 8: Z. Phys. D **40** (1997);
  ISSPIC 9: Eur. Phys. J. D **9** (1999);
  ISSPIC 10: Eur. Phys. J. D **16** (2001)
  ISSPIC 11: Eur. Phys. J. D **24** (2003);
  ISSPIC 12: Eur. Phys. J. D **34** (2005).

b) *Helium clusters*
- E.R. Dobbs, *Helium Three* (Oxford University Press, New York, 2000);